CURRENT TOPICS IN
HUMAN
GENETICS
Studies in Complex Diseases

CURRENT TOPICS IN
HUMAN
GENETICS
Studies in Complex Diseases

Co-Editors-in-Chief

Hong-Wen Deng
Hui Shen

University of Missouri–Kansas City, USA

Associate Editors

Yong-Jun Liu

University of Missouri–Kansas City, USA

Hai Hu

Windber Research Institute, USA

NEW JERSEY · LONDON · SINGAPORE · BEIJING · SHANGHAI · HONG KONG · TAIPEI · CHENNAI

Published by

World Scientific Publishing Co. Pte. Ltd.

5 Toh Tuck Link, Singapore 596224

USA office: 27 Warren Street, Suite 401-402, Hackensack, NJ 07601

UK office: 57 Shelton Street, Covent Garden, London WC2H 9HE

Library of Congress Cataloging-in-Publication Data
Current topics in human genetics : studies in complex diseases / co-editors-in-chief,
    Hong-Wen Deng and Hui Shen; associate editors, Yong-Jun Liu, Hai Hu.
        p. cm.
    Includes bibliographical references.
    ISBN-13 978-981-270-472-6 (hardcover)
    ISBN-10 981-270-472-8 (hardcover)
        1. Genetic disorders. 2. Medical genetics. I. Deng, Hong-wen.
    [DNLM: 1. Genetic Diseases, Inborn--genetics. 2. Genetic Predisposition to Disease--genetics.
    3. Genetics, Medical--methods. QZ 50 C976 2007]
    RB155.5.C87 2007
    616'.042--dc22

                                                                    2007019047

British Library Cataloguing-in-Publication Data
A catalogue record for this book is available from the British Library.

Typeset by Stallion Press
Email: enquiries@stallionpress.com

Printed in Singapore by Mainland Press Pte Ltd

Preface

The sequencing of the human genome has brought human genetics into a new era of study with an explosive amount of information. The application of genomics, functional genomics, proteomics, and bioinformatics technologies to the study of human genetics has made it possible for human genetic diseases to be studied on a large scale, which is unprecedented, both *in silico* and in the wet lab.

The primary goal of this book is to provide up-to-date coverage of the broad range of research topics in the genetic studies of human diseases. The book contains two major parts. In the first part, a whole spectrum of approaches to human genetics research is reviewed for both background and progress. In the second part, important topics related to the genetic research of various complex human diseases are elaborated. The robust content and diverse array of subjects are meant to allow the book to serve as a concise "encyclopedia" that introduces basic and essential concepts of human genetics, as well as an in-depth review of the current understanding of genetic research in human diseases.

Another important goal of this book is to foster scientific exchange among Chinese-speaking and non–Chinese-speaking scholars. Currently, there are many Chinese-speaking scientists who are doing excellent research in human genetics and have made significant contributions to the field. However, the lack of active communication among scholars and scientists impedes scientific understanding and efficient interaction and collaboration with one another. With this in mind, when we initiated this publication, we invited mainly Chinese-speaking scholars to contribute. As such, the majority of the contributors are scholars of Chinese origin who are active in human genetic research in institutions in the US, Europe, and China, or who are senior scientists from major US pharmaceutical companies. At the same time, we also invited an elite group of internationally renowned human geneticists of non-Chinese origin to contribute some of the important chapters.

This book consists of two parts. Part 1, which includes Chapters 1 To 20, covers the commonly used approaches in the gene identification of complex diseases. Chapters 1 to 10 describe the genetic epidemiology approaches, which include traditional and contemporary linkage and association mapping methods. Chapters 11 and 12 discuss the animal models that are helpful for human disease research. A succinct summary of the statistical methods for QTL mapping in experimental crosses and a detailed overview of the animal models used in the genetic study of human complex diseases are included. Chapters 13 to 17 discuss functional genomics, one of the most promising fields that is poised to boost our understanding of systems-level cellular behavior and the fundamental etiology of human diseases. This segment includes a detailed description on the technology and application of the microarray and proteomics approaches in human genetic studies. Chapters 18 to 20 present a comprehensive discussion of bioinformatics and its potential in human disease research, given that a wealth of molecular information is rapidly accumulating and that the convergence of biology, information science, and computation is becoming critically important. With these tools in hand, Chapters 21 to 33 in Part 2 review its applications in the genetic study of human complex diseases, including osteoporosis, diabetes, metabolic syndrome, obesity, psychiatric disorders, autoimmune diseases, and cancers.

Since the book covers both the basic concepts of human genetics and in-depth reviews updating the current understanding of the genetic dissection of human diseases, the primary audience of this book would be professors, research scientists, predoctoral and postdoctoral scientists, and clinicians who are interested or involved in the genetic research of human diseases, in both academia and industry. We believe the book will benefit scholars in China as well as around the world.

Hong-Wen Deng
Professor
Departments of Orthopaedic Surgery & Basic Medical Sciences
University of Missouri–Kansas City
USA

Acknowledgments

Obviously, this book would not have been possible without the efforts of all the contributing authors, our colleagues and friends. We thank all of them for their overwhelming interest and enthusiasm for this book. We thank Feng Zhang and Fang Yang for their assistance in formatting the book. Last but not least, it was a pleasure working with Ms Sook Cheng Lim and her colleagues at World Scientific Publishing. It is their diligent effort that has resulted in the timely publication of this well-designed book.

Contents

Chapter 1

Regression-based Linkage Analysis Methods

Tao Wang and Robert C. Elston*

Department of Epidemiology and Biostatistics
Case Western Reserve University, Wolstein Research Building
2103 Cornell Road, Cleveland, OH 44106-7281, USA

Regression-based methods of model-free linkage analysis offer a valuable framework for mapping both quantitative and qualitative traits. Beginning with the method proposed by Haseman and Elston,[1] these methods have been widely used in practice because of their simplicity and robustness. Furthermore, the newer methods can utilize full information from trait values, and they are applicable to any type of pedigree data. With the availability of the denser markers and appropriate sampling, these methods give hope that they may play an important role in mapping complex genetic traits. The information yielded by such an analysis can guide and facilitate the design and result inference of further association studies.

1. INTRODUCTION

Model-free linkage methods are commonly used for mapping complex diseases because they do not require the mode of inheritance of the trait under study to be correctly specified in any detail. Among the various model-free linkage methods, regression-based methods offer a valuable framework for both quantitative and qualitative traits. The regression-based linkage analysis method was first proposed by Haseman and Elston.[1] The Haseman and Elston (HE) regression is simple, robust, and therefore widely used in practice. However, there are some limitations to the original HE regression

---

*Correspondence author.

method. One limitation is that the original HE regression may be less powerful compared with other full likelihood–based variance component methods because it does not make full use of all the information available in the trait values. Another major limitation is that the original HE regression is not applicable to any type of pedigree data. To overcome these limitations, in recent years there have been great efforts to extend and enhance the original HE method. The newer regression methods can fully utilize the trait values on multiple types of relative pairs, and hence can play a crucial role in the linkage analysis of complex traits.

Our discussion of regression-based methods will mostly focus on quantitative trait locus (QTL) mapping. The typical regression-based method is to regress a measure of trait similarity between the members of a relative pair on a measure of their genetic similarity, which is usually described by the number of alleles shared identical by descent (IBD) at a genomic location. The rationale behind this method is that, when a marker is not linked to a disease locus, the number of marker alleles shared IBD by a pair of relatives does not depend on the relatives' trait values; on the other hand, when the marker is linked to a disease locus, we expect a correlation between the number of alleles shared IBD at the marker locus and the measure of the relatives' trait similarity. The regression methods evaluate this correlation by examining the regression coefficient. From this point of view, critical for the power and the validity of a regression-based linkage method is how we calculate the trait similarity measure and the number of marker alleles shared IBD.

In Sec. 2, we shall discuss the measure of genetic similarity and the number of alleles shared IBD. We focus on possible pitfalls in the IBD calculation, which may result in deleterious effects for all model-free–based linkage methods. In Sec. 3, we initially focus on independent sib pairs and quantitative traits, and discuss the various regression-based methods based on different definitions of the trait similarity measure in this setup. In Sec. 4, we further introduce regression-based methods recently developed to deal with any family structure. Lastly, we briefly cover regression-based methods for affected sib pairs (ASPs) that incorporate covariates.

2. THE NUMBER (PROPORTION) OF ALLELES SHARED IBD

The various regression-based linkage methods nowadays depend on the crucial concept of the number of alleles shared identical by descent (IBD), although the original model-free linkage method was based on marker identity in state, the marker similarity between two relatives being based solely on their similarity with respect to marker phenotype. The number of alleles shared IBD is defined as the number of alleles at a given locus on a particular pair of chromosomes that two relatives inherit from a common ancestor. Because methods based on IBD are more powerful, we shall not consider methods based on marker identity in state, although the two measures of marker similarity approach each other as a marker becomes more polymorphic. Because the estimation of the IBD probability is often separable from the actual regression, the usual regression-based linkage method can be looked upon as a two-stage procedure: (1) the IBD probabilities are calculated; and (2) the regression model is constructed. Obviously, how well we estimate the IBD probabilities at the first stage is directly related to the validity and power of a regression model at the second stage.

The estimation of IBD sharing between two members of a pedigree is based on all of the observed marker genotypic information. Early regression-based methods only used samples of full sibs. If parents are also typed for a codominant marker, or if the parental marker genotypes can be deduced, it is possible to count the number of alleles a sib pair shares IBD. The early methods eliminated those pairs for which it was not possible to actually count the number of alleles shared IBD, and this approach could lead to bias when the marker is not highly informative. Haseman and Elston[1] proposed estimating IBD probabilities at a marker locus by utilizing the marker information available for the sibs and their available parents. The IBD probabilities for other relative pairs can also be accommodated.[2] Kruglyak and colleagues[3] as well as Kruglyak and Lander[4] proposed using an algorithm (the Lander–Green algorithm) to calculate the IBD allele-sharing probabilities in a multipoint fashion. The amount of computation in their algorithm increases linearly with the number of markers, and exponentially with the size of the family. For large pedigrees, Sobel and Lange[5] implemented a Markov chain Monte Carlo (MCMC) method and, with the assumption of no interference, Fulker and colleagues[6] proposed a fast regression method to obtain approximate multipoint estimates of the proportion of alleles shared IBD by full sibs at any location, based on the estimates at the locations for which marker information is available. This regression method was used by Almasy and Blangero[7] when they implemented a full pedigree variance component likelihood model based on the assumption of trait multivariate normality across pedigree members.

Generally, the informativity of a marker for estimating IBD sharing depends on its degree of polymorphism. For a highly informative marker, the founders of a pedigree have unique alleles and therefore the number of alleles shared IBD can be unambiguously determined. However, the number of alleles for a marker used in practice is limited, with the result that the number of alleles shared IBD often cannot be specified unambiguously. To describe the degree of polymorphism, two measures are commonly used in practice. One is the marker heterozygosity, which is defined by

$$H = 1 - \sum p_i^2,$$

where p_i is the population frequency of the i allele. From this definition, we can see that heterozygosity is simply the probability that a random individual is heterozygous at a locus in a population with Hardy–Weinberg equilibrium. The other measure is the polymorphism information content (PIC) value.[8] The PIC value was first derived for a rare dominant disease and it is defined as

$$\text{PIC} = 1 - \sum p_i^2 - 2\sum_i \sum_{j>i} p_i^2 p_j^2.$$

To measure a marker's informativity for a model-free linkage analysis via pairs of relatives, Guo and Elston[9] developed a third measure, the linkage information content (LIC). The LIC values measure the probability of being able to determine IBD proportions for each particular type of relative pair.

When the number of alleles shared IBD between members of a relative pair in a pedigree cannot be specified unambiguously, conventionally in a regression-based method we describe the genetic similarity by the estimated proportion of alleles shared IBD in place of the exact proportion of alleles shared. For a sib pair, the proportion of alleles shared IBD — π (which in reality can take on only the values 0, $\frac{1}{2}$, or 1) — is taken to be $\hat{\pi} = \hat{f}_2 + \frac{1}{2}\hat{f}_1$, where $\hat{f}_2$ and $\hat{f}_1$ are the estimated probabilities of sharing two and one alleles IBD, respectively, given all of the marker data available. Let f_i be the prior probability that a relative pair shares, by virtue of degree of relationship alone, i alleles IBD. By Bayes' theorem, the estimated probability given the available marker information I_m is simply

$$\hat{f}_i = \frac{f_i P(I_m|i)}{P(I_m)}.$$

When the marker is completely uninformative, we can see that the estimated IBD probabilities are actually the IBD probabilities under the null hypothesis and $\hat{\pi}$ is π_0, by which we denote the proportion of alleles shared IBD when there is no linkage. This approach has been criticized in the literature because it can lead to loss of power. For example, Schork and

Greenwood[10] pointed out that the estimate of the number of alleles IBD after imputation is biased toward the null and therefore the power can potentially be reduced by less informative markers. To avoid/alleviate this loss of power, some authors have proposed to weight relative pairs according to the informativity of the number of alleles shared IBD.[10–12] Although weighting approaches to alleviate the loss of power are possible, care should be taken when any such approach is used in practice because the missing mechanism of IBD sharing is not always at random, e.g. when parents are not observed. In this case, the weighting or deleting approach can lead to a worse outcome as it will result in an invalid or conservative statistic. We do not recommend blindly weighting each pair according to the estimated IBD sharing in practice, because the risk is too large compared with the limited possible gain.

Typically, regression-based linkage methods use the estimated IBD proportion, i.e. $\hat{\pi}$, to summarize the marker similarity between relatives in a pair. Kruglyak and Lander[4] extended the HE method to use the full estimated distribution of 0, 1, or 2 alleles being shared IBD. They argued that the proportion of IBD sharing does not fully utilize the information provided by the whole IBD distribution. Their simulation showed that their modified HE test has better behavior in the presence of uninformative families. The approximation involved in the use of the mean proportion of alleles shared IBD instead of the full distribution has also been tested by Gessler and Xu.[13] They explicitly make the distinction between the distribution approach and the expectation approach, but they found that there is little difference between them in terms of power. Cordell[14] performed simulations to investigate the test statistics of HE and variance component methods. She found that the expectation approach suffers in both precision and power when the squared difference is the dependent variable in HE regression. However, the simulation parameters in her study were rather extreme because there was no sibling resemblance other than that due to a single QTL, and this QTL explained more than 90% of the trait variance.

As pointed out by Cordell,[14] the best way to minimize the ambiguity of allele transmission when it is difficult to recruit a full sample is to use as densely spaced markers as possible. With the recent availability of a dense map of single nucleotide polymorphism (SNP) markers that can be genotyped automatically, economically, and more accurately, there is great hope that these can be used to improve the current linkage approach to finding genes. However, currently most multipoint linkage programs assume linkage equilibrium among the markers. Whereas this assumption may be appropriate for sparsely spaced markers with intermarker distances exceeding a few centimorgans, for densely spaced SNP markers linkage disequilibrium (LD) may exist and, if not appropriately allowed for, may lead to bias in the calculation of the probabilities of IBD sharing when pedigrees have missing founder information. It has been shown that when some or all of the parental genotypes are missing, assuming linkage equilibrium among markers when strong LD exists can cause false-positive evidence from ASP data. Because this bias would not affect the correlation between the proportion of IBD sharing and trait similarity measures, the validity of regression-based linkage analysis for a quantitative trait could be robust to this bias, but nevertheless its power may be affected. A simple way to correct this bias at the stage of IBD calculation is to organize the SNPs into nonoverlapping clusters in such a way that one can assume no LD between markers in different clusters and no recombination within each cluster.[15]

Another bias in the calculation of the probabilities of IBD sharing, which may be more important in the case of regression-based methods for quantitative trait analysis, is related to population stratification. Although the confounding effect of population stratification on a genetic association study has long been recognized, regression-based linkage methods, like all other model-free linkage analysis methods, are usually thought to be robust to population stratification because the estimation of the probabilities of alleles shared IBD does not depend on the population frequencies of the marker alleles when founder marker genotypes are observed or are deducible. Unfortunately, it is not unusual for founders to be missing in the case of late-onset complex diseases. In this case, the distribution of the estimated IBD sharing may no longer be independent of the marker allele frequencies and, therefore, all linkage studies can be potentially biased by population stratification.[16] For an affected sib-pair design, it has been shown that, when some or all of the founder genotypes are missing, heterogeneity of allele frequencies among the subpopulations can cause excess false-positive discoveries — even when the trait distribution is homogeneous among the subpopulations. After incorporating a control group of discordant sib pairs, or in the case of a quantitative trait, two conditions must be met for population stratification to be a confounder in linkage analysis: the distributions of both the marker and the trait must be heterogeneous among the subpopulations. When this occurs, the bias can result in a test that is either liberal (and hence invalid) or conservative. An obvious way to avoid such deleterious effects from population stratification is to include as many founders or other family members as possible to reduce the uncertainty about founder marker information.

3. VARIOUS REGRESSION-BASED METHODS WITH DIFFERENT TRAIT SIMILARITY MEASURES

The regression-based linkage analysis method and the maximum likelihood–based variance component methods are two commonly used approaches for the analysis of quantitative traits. The initial regression-based linkage method is the original HE regression[1] for independent full-sib pairs, which has been shown to be robust to selected samples and the distribution of trait values, although it was derived under the assumption of randomly sampled sib pairs and on a squared sib-pair trait difference that is normally distributed. On the other hand, maximum likelihood variance component analysis can be less robust to selected sampling and to nonnormality of the trait values, although it can be more powerful than the original HE regression. Recently, various regression-based methods have been developed with the aim of extending them to any type of family data and improving their power while retaining robustness.

3.1. *The Original Haseman–Elston Regression*

The Haseman–Elston method, as originally proposed by Haseman and Elston,[1] makes use of the squared trait difference between the two sibs in a pair as the measure of trait similarity (in this case, a measure of trait dissimilarity). Let X, the value of a trait, be composed of an overall mean μ, the major genetic effect of a quantitative trait locus (QTL) g, and an independent random effect e. Let $Y_j = (X_{1j} - X_{2j})^2$, where X_{1j} and X_{2j} are the two trait values for the jth full-sib pair. Let π_{tj} denote the proportion of alleles that the jth full-sib pair shares IBD at the trait locus. With the assumption of no dominant effect, it can be shown that

$$E(Y_j|\pi_{tj}) = (\sigma_e^2 + 2\sigma_g^2) - 2\sigma_g^2 \pi_{tj},$$

where σ_g^2 is the variance of the QTL and σ_e^2 is the random effect variance.

Let π_{mj} denote the proportion of alleles the jth full-sib pair shares IBD at a marker locus, and $\hat{f}_{mij}$ be the estimated probability that the jth full-sib pair shares i alleles IBD ($i = 0, 1,$ or 2) at the marker locus, conditional on the marker information available on the sib pair and their relatives. The estimate of the proportion of alleles shared IBD at the marker locus is

$\hat{\pi}_{mj} = \hat{f}_{m2j} + 0.5\hat{f}_{m1j}$. Assuming linkage equilibrium between the marker and trait loci, it follows that

$$E(Y_j|\hat{\pi}_{mj}) = \sum_{\pi_{tj}} \sum_{\pi_{mj}} E(Y|\pi_{tj}) P(\pi_{tj}|\pi_{mj}) P(\pi_{mj}|\hat{\pi}_{mj}).$$

Then, it can be shown, letting θ be the recombination fraction between the trait and marker loci, that

$$E(Y_j|\hat{\pi}_{mj}) = [\sigma_e^2 + 2(1 - 2\theta + 2\theta^2)\sigma_g^2] - 2(1 - 2\theta)^2 \sigma_g^2 \hat{\pi}_{mj}.$$

This can be simply written in the form

$$E(Y_j|\hat{\pi}_{mj}) = \alpha + \beta \hat{\pi}_{mj}.$$

Therefore, the hypothesis of no linkage can be tested by a one-sided t-test. Because the regression t-test is derived assuming the residuals of the regression equation are normally distributed, there have been concerns about violation of this statistical assumption. However, it has been shown that the type I error rate is quite robust to deviations from normality for reasonably sized samples. In other words, the regression coefficient, being a weighted average, tends to be normally distributed by reason of the central limit theorem. Wan and colleagues[17] proposed a permutation procedure to evaluate the P-value of the original HE method. The permutation procedure keeps the $\hat{\pi}_{mj}$ in the original order, and randomly permutes the values of Y among sib pairs. For a large number of sib pairs, they showed that the permutation variance of the regression slope and the variance estimated by least squares are equal under the null hypothesis. Simulations showed that the conventional t-test approximates the permutation test quite well. These results indirectly addressed concerns about the assumption violation of the conventional t-test.

Due to its robustness and simplicity, the original Haseman–Elston regression has been widely extended to various situations. Amos and colleagues[2] proposed a multivariate extension for the original Haseman–Elston method. Tiwari and Elston[18] extended the original HE procedure to two unlinked quantitative trait loci (QTLs) that might interact epistatically. Hanson and colleagues[19] indicated how the original HE test can be readily extended to accommodate parent-of-origin effects, by estimating separate β coefficients according to the parental source of the allele sharing.

3.2. *New Regression-based Methods*

Compared to the full likelihood–based variance component methods, the original Haseman–Elston regression is less powerful when normality of the trait is approximately correct. This problem raised the question of how to improve the power of regression-based linkage analysis methods, and invoked a series of papers to make use of a more informative measure of trait similarity. Different trait similarity measures are summarized in Table 1 and are implemented in the software package Statistical Analysis for Genetic Epidemiology (S.A.G.E.).[20]

Wright[21] initially indicated that using only the squared difference of a pair of trait values as the trait similarity measure may result in loss of some information for linkage, although this had also been noted by Gaines and Elston[22] in another context. He pointed out that the full likelihood function for a sib pair can be written in terms of both a sum and a difference. Under the bivariate normal assumption, he demonstrated that a nontrivial amount of power can be gained when the sum of the pair of trait values is also included. Responding to Wright,[21] Drigalenko[23] first proposed an extension of the HE method that uses both the sib-pair trait sum and difference as dependent variables. Because the squared pair sum of the trait values (taken with opposite sign) results in a regression line that is parallel to that for the squared pair difference, he suggested estimating the slope by simply averaging the estimates from the two regressions of the squared sum and difference, which is the best estimate under the assumption that the residuals have the same variance for both the squared sum and squared difference. He also showed that such a combination is equivalent to performing a single regression using the pair-trait product. Based on the same

Table 1 **Definitions of the Dependent Variable for Various Forms of Haseman–Elston Regression**

Keyword	Acronym	Dependent Variable	Option in S.A.G.E.
Original	oHE	$-\frac{1}{2}(X_{1j} - X_{2j})^2$	diff
Revisited	rHE	$(X_{1j} - \overline{X})(X_{2j} - \overline{X})$	prod
Weighted[a]	wHE	$\frac{1}{2}[(1-w)(X_{1j} + X_{2j} - 2\overline{X})^2 \\ -w(X_{1j} - X_{2j})^2]$	W2–W4
Sibship sample mean	smHE	$(X_{1j} - \overline{X}_j)(X_{2j} - \overline{X}_j)$	sibship_mean = yes
Shrinkage mean	pmHE	$(X_{1j} - \tilde{\mu}_j)(X_{2j} - \tilde{\mu}_j)$	—

$\overline{X}$: overall mean; $\overline{X}_j$: sibship mean; $\tilde{\mu}_j$: shrinkage mean; w: weight.

[a]Various slightly different weighting options are available in S.A.G.E.; W4, which is asymptotically optimal and adjusts for all the nonindependence of full-sib-pair squared sums and differences in larger sibships, is the one used for simulation here.

idea, the overall mean–centered cross-product of sib-pair traits was adopted as the trait similarity measure in the revisited HE method.[24]

Although the revisited method was presumed to be more powerful than the original HE method, Palmer and colleagues[25] showed with an interesting simulation result that the empirical power of the revisited method could be substantially lower than that of the original method when there are strong familial polygenic or environmental correlations. Their simulation result motivated further work by several groups. It is easy to show that the assumption of the same residual variance for the squared sum and difference cannot be attained, and therefore they are not equally informative when there are familial polygenic or environmental correlations. So, the overall mean–corrected cross-product — which weights equally (but with opposite sign) the two slope estimates, i.e. that from the squared difference and that from sum — is not optimal.

Several groups proposed weighting the two slope estimates inversely according to the variances of the squared sum and the squared difference. Let the estimator of the slope from the squared difference be $-\hat{\beta}_1$, and that of the slope from the squared sum be β_2. Xu and colleagues[26] considered a class of estimators for β of the form $w\hat{\beta}_1 + (1 - w)\hat{\beta}_2$, where w is a given weight. Let $\hat{\sigma}_{12}$ be the estimated covariance of $-\hat{\beta}_1$ and β_2, and $\hat{\sigma}_{11}$ and $\hat{\sigma}_{22}$ be the estimated variances of $-\hat{\beta}_1$ and β_2. Then, for a large sample size, the estimator β with weight $w = (\hat{\sigma}_{22} - \hat{\sigma}_{11})/(\hat{\sigma}_{22} + \hat{\sigma}_{11} - \hat{\sigma}_{12})$ has the smallest variance among all linear combinations of $-\hat{\beta}_1$ and β_2. Visscher and Hopper[27] proposed a similar method, which also weights the slope estimates based on the variance estimates separately from the regression models. The method of Forrest[28] simultaneously calculates the estimates of the two intercepts and the single slope as well as the two variances, using least squares iteratively. Shete and colleagues[29] further proposed a weighting method for larger sibships, allowing for the correlation between pairs within a sibship. Instead of weighting the slope estimates based on the empirical variances from the regression model, the approach of Sham and Purcell[30] used variance estimates that they derived as a function of the population trait-pair covariance. They also showed that this method can be used for the selection of maximally informative sib pairs.

Recognizing the loss of power that can result from using the overall sample mean–corrected cross-product as the trait similarity measure, due to the existence of the correlation arising from sibship-specific effects such as common environmental and polygenic effects, investigators proposed using a sibship-specific mean correction to absorb this correlation. Wang and colleagues,[31] in particular, proposed the use of a trait product centered by the family-specific sample mean rather than by the whole sample mean. Their simulations showed that centering the trait by the family-specific sample mean results in more power than centering the trait by the whole sample mean when the size of sibship is large and there are family-specific phenotype effects (common environmental and additional QTL effects).

A shrinkage sibship mean–corrected cross-product was also proposed as the trait similarity measure to further improve power.[32,33] It was shown that the shrinkage sibship mean is actually a weighted average of the overall sample mean and the sibship sample mean, where the weights depend on the variance within and among sibships. When neither the variance within nor among sibships dominates, the shrinkage mean is a combination of both estimators; when the variance among sibships dominates, this trait similarity measure is equivalent to the squared difference used in the original HE regression; and when the variance within sibships dominates, i.e. there is no common environmental or polygenic effect and the QTL effect is not too strong, this trait similarity measure is equivalent to the overall sample mean–corrected cross-product. Simulation results showed that the empirical power of the shrinkage mean–corrected cross-product and the weighted squared sum and difference are similar in most situations.

Another extension of HE regression is the "reversed" method. Sham and colleagues[34] noticed that the regression coefficients would be biased when sampling is through the dependent variable. Because it is common to sample subjects according to their trait values, they proposed to avoid biased estimators of the regression coefficient by a regression method that uses the proportion of IBD sharing as the dependent variable and the squared sum and squared difference as independent (predictor) variables. However, in the absence of bivariate normality, there may exist collinearity between the squared sums and differences for pairs within a sibship. To remove this collinearity, they arbitrarily trim the squared differences in such a way that each individual is represented at least once. An alternative is to use both the squares and the cross-products as independent (predictor) variables. This is equivalent to the use of the squared sum and difference, but is more conveniently extended to multivariate phenotypes. When the aim is to test for linkage, Schaid and colleagues[35] pointed out that, for the calculation of an LOD score, how the Y value (the transformed trait values in a regression model) is computed from the traits is critical, but whether it is on the left-hand side or right-hand side of the regression equation is not important if there are no other covariates in the regression model. In another context, Gray-McGuire and colleagues[36] noted that, in the presence of other covariates, reversal of the regression equation will, in general, affect the test of significance. More extensive simulations are still needed to examine this "reversed" method.

3.3. Regression-based Linkage Methods for Any Type of Pedigree Structure

The maximum likelihood–based variance component approach, which is based on an assumption about the distribution of trait values across pedigree members, utilizes the whole information from a pedigree in a relatively simple manner. However, the original HE method is based on the assumption of independent full-sib pairs only. In linkage analysis, large sibships and other types of family structures are also often included. It is desirable to extend the original HE to those cases.

For sibships of size larger than two, Blackwelder[37] showed that the correlation between pairs of squared sib-pair differences with no sib in common is 0, and that it usually lies between $\frac{1}{4}$ and $\frac{1}{3}$ when there is one sib in common. Assuming multivariate normality of the sibs' trait values, this correlation is $\frac{1}{4}$.[38] Single and Finch[39] showed that, when sibships of size larger than two are analyzed, a generalized least squares regression can improve the power of the original HE method by allowing for the correlations between the pairs of squared sib-pair trait differences. In the revisited HE method,[24] the same approach was adopted. The correlation matrix W of the dependent variable is not an identity matrix, but rather is block diagonal in form with each diagonal block being a matrix W_ρ, where ρ is the correlation between two sib pairs that have one sib in common. Furthermore, it has also been shown that, for a given correlation ρ and sibship size s, the inverse of the correlation matrix can be obtained algebraically, so a generalized least squares estimate can be computed quickly. Shete and colleagues[29] adopted this idea to obtain an optimally weighted HE method.

Amos and Elston[40] extended the original HE method to any type of noninbred relative pair. Let X_j, the value of a trait measured on the jth noninbred individual, be composed of an overall mean μ, a major genetic effect g, and an independently distributed random effect e, with zero mean and variance σ_e^2. Let π_{tj} denote the proportion of IBD sharing at the trait locus by the jth relative pair, and π_{mj} denote the proportion of IBD sharing at the marker locus. For unilineal relative pairs, we estimate π_{mj} by $\hat{\pi}_{mj} = 0.5\hat{f}_{m1j}$, where $\hat{f}_{m1j}$ is the estimated probability that the jth relative pair shares one allele IBD. Let $Y_j = (X_{1j} - X_{2j})^2$. Following the same reasoning as for full-sib pairs, it can be shown that for unilineal relative pairs,

$$E(Y_j|I) = \alpha_1 + \beta_1 \hat{\pi}_{mj}.$$

The coefficients of these regressions for each type of unilineal relative pair are given in Table 2. Note that the coefficient β_1 is negative in each case if $\theta < 0.5$ and $\sigma_a^2 > 0$.

Although regression coefficients of various types of relative pairs were derived as early as 1989, an optimal test that combines the information obtained from the various types of relative pairs was not well developed for some time. One of the obvious problems is the dependence among the test statistics derived from the various types of relative pairs. Schaid and colleagues[41] extended the original HE to combine the information from full sibs and half sibs into a single test for linkage. The method estimates the common regression coefficient for full-sib and half-sib pairs, allowing the intercepts and residual variances to differ for full-sib and half-sib pairs. Assuming a multivariate normal distribution, the nondiagonal elements of the variance–covariance matrix, V, was classified into one of six categories, each one being approximated by a function of the variance of the squared difference of full-sib pairs, $V(Y_F)$. A simulation study demonstrated that this approach worked well in many situations, but there are some conditions where the statistic can have a slightly inflated type I error rate.

In linkage analysis, correlated traits of relative pairs in a pedigree form a cluster of correlated observations. In theory, it should be appropriate to use the generalized estimating equation (GEE)[42] methodology for this situation. Olson and Wijsman[43] first considered a HE method to combine various types of relative pairs using GEE. In their paper, the methodology

Table 2   Coefficients of the Regression of Squared Pair Difference on the Proportion of Alleles Shared IBD for Various Relative Pairs

Relative Type	α_1	β_1
Grandchild	$\sigma_e^2 + 2\sigma_g^2 + \theta\sigma_a^2$	$-2\theta(1-2\theta)\sigma_a^2$
Half sib	$\sigma_e^2 + 2\sigma_g^2 - 2\theta(1-\theta)\sigma_a^2$	$-2\theta(1-2\theta)\sigma_a^2$
Avuncular	$\sigma_e^2 + 2\sigma_g^2 - \left(\frac{5}{2}\theta - 4\theta^2 + 2\theta^3\right)\sigma_a^2$	$-2\theta(1-2\theta)(1-\theta)\sigma_a^2$
First cousin	$\sigma_e^2 + 2\sigma_g^2 - \left(\frac{4}{3}\theta - \frac{5}{2}\theta^2 + 2\theta^3 - \frac{2}{3}\theta^4\right)\sigma_a^2$	$-2\theta(1-2\theta)\left(1 - \frac{4}{3}\theta + 2\theta^2\right)\sigma_a^2$

of GEE was used to provide an estimate of the robust covariance matrix of the set of estimated relative-pair-type–specific regression coefficients. Using this covariance matrix, the asymptotically most powerful test of linkage that optimally combined the information contained in the different types of relative pairs was constructed.

Let W be the set $\{s, g, h, a, c\}$ of subscripts denoting the relative-pair types: (full) sibling, grandparent–grandchild, half sib, avuncular, and first cousin, respectively. Letting Z_{ik} be the squared difference for relative pair i in family k, then

$$E(Z_{ik}) = \mu_{ik} = \alpha_w + \beta_w \hat{\pi}_{ik},$$

where α_w and β_w are parameters that need to be estimated, and $\hat{\pi}_{ik}$ is the estimated proportion of IBD sharing. Let $\lambda = (\alpha_s, \beta_s, \ldots, \alpha_c, \beta_c)^T$. An estimate of λ, $\hat{\lambda}$, may be obtained by solving the GEE. Note that $\hat{\lambda}$, conditional on $\hat{\pi}_{ik}$, is consistent and asymptotically normal with a covariance matrix that may be consistently estimated by the robust sandwich estimator. These asymptotic properties hold even if the form of the working covariance matrix is misspecified.

Olson and Wijsman[43] discussed possible working covariance matrices, including the independent covariance matrix and a partially specified matrix, and pointed out that some efficiency may be gained by modeling the correlations between relative pairs more precisely. Chen and colleagues[44] described a more general framework for general pedigrees using GEE. In particular, they showed that the Haseman–Elston methods, the variance component model, and some score tests are all closely related in that the different choices of the working covariance matrix lead to the different methods. Wang and Elston[45] developed a two-level HE for quantitative trait linkage analysis and general pedigrees under the framework of multiple-level regression. They adopted an iterative generalized least squares (IGLS) algorithm to tractably handle variance–covariance structures varying across families. They showed that the two-level HE can compete favorably with any current version of HE in that it can naturally make use of all the information available in any general pedigree, simultaneously incorporating individual-level and pedigree-level effects and feasibly modeling various complex genetic effects.

4. DISCUSSION

In the literature on QTL mapping, three approaches — regression-based methods, variance component methods, and the newly developed score statistics — are often used. Although each approach has its own advantages and disadvantages and is usually viewed separately, it is worth noting that they are closely related in large samples because of the similarity of the underlying trait model.

Based on a general trait model, Putter and colleagues[46] derived a score test for the proportion of the total phenotypic variance due to the quantitative trait locus in a variance component model and showed that, for sib pairs, it is mathematically equivalent to the HE method that optimally combines the squared sum and the squared difference of the centered phenotype values of the sib pairs. Because score tests and likelihood-ratio tests are equivalent for large sample sizes, the variance component likelihood-ratio test is also asymptotically equivalent to this optimal HE test. Their results gave a theoretical explanation of the empirical observations found in simulation studies reporting similar power of the variance component likelihood-ratio test and the optimal HE method.

Based on a trait model in which the trait value is generated by a family-specific effect, a QTL, and a random effect, Tritchler and colleagues[47] proposed a score test for genetic linkage in nuclear families that applies to any trait having a distribution belonging to the exponential family. They also showed that the score test is closely related to HE methods. For sib pairs, their score test is proportional to the regression estimate of β in the model $\pi_i - 1 = \beta(X_{i1} - \hat{\mu}_i)(X_{i2} - \hat{\mu}_i) + \varepsilon_i$. The HE regression tests are obtained from the above regression by interchanging the response variable and the predictor variable and adding an intercept: $(X_{i1} - \hat{\mu}_i)(X_{i2} - \hat{\mu}_i) = \alpha + \beta(\pi_i - 1) + \varepsilon_i$. Different estimates of $\hat{\mu}_i$ yield different HE test statistics. For example, this test is the original HE regression if $\hat{\mu}_i$ is estimated by the sample mean of a sib pair; while it is the revisited HE regression if $\hat{\mu}_i$ is estimated by the overall mean.

Chen and colleagues[44] viewed various methods for QTL mapping from the framework of GEE, and different choices of the working matrix lead to the different methods. Although there is a close relationship among the various linkage test statistics for a large sample, the question of which method should be used for a given data set is difficult to answer. The recent computer simulation results from Cuenco and colleagues,[48] Szatkiewicz and colleagues,[49] and Chen and colleagues[50] provide some comparison of the performance of the various new methods in terms of power and the type I error rates.

Here, we have only focused on using the regression framework to detect linkage for a quantitative trait. In the linkage literature, a regression method is also used in affected sib-pair (ASP) designs to incorporate covariates. Common complex diseases are likely to be genetically heterogeneous, with different genetic and environmental factors contributing to the disease.[51] It is critical to take account of heterogeneity in a linkage analysis. The regression-based methods can be used to deal with heterogeneity in ASP linkage analysis by allowing the IBD sharing probabilities for an ASP to depend on covariate information. For example, Olson[52] showed that the ASP LOD score can be reparameterized in terms of the natural logarithms of relationship-specific relative recurrence risks. The method incorporates locus heterogeneity by allowing the genetic relative risk to be conditional on indicators of heterogeneity, so that the allele sharing at the marker locus differs for different values of the indicators. The original model of Olson[52] required two additional parameters for each covariate, and therefore may not be optimal in terms of power. To reduce the number of regression parameters, different approaches have been proposed. Currently, the LOD-PAL software in S.A.G.E. constrains the relative risks in a manner that reduces both the number of parameters in the basic model and the number of additional parameters for each heterogeneity indicator.[53]

We have discussed many features related to regression-based linkage methods. The power of a regression-based method depends on how we measure the marker similarity and the trait similarity between two relatives in a pair. The regression-based statistics, score test statistics, and variance component statistics are related because they all test for a nonzero QTL variance component. In a general sense, they are all in fact variance component methods, although the usual variance component method is to use a maximum likelihood–ratio test to examine the QTL variance, which is based on the critical assumption of multivariate normality for pedigree data. Apart from being robust to nonnormality, regression-based methods also include the advantage of rapid computation, which makes it feasible to evaluate p-values empirically. The regression framework also offers flexibility to model environmental effects as well as gene–gene and gene–environment interactions.

ACKNOWLEDGMENTS

This work was supported in part by a U.S. Public Health Service Resource Grant from the National Center for Research Resources (RR03655) and a Research Grant from the National Institute of General Medical Sciences (GM28356).

REFERENCES

1. Haseman JK, Elston RC. (1972) The investigation of linkage between a quantitative trait and a marker locus. *Behav Genet* **2**:3–19.
2. Amos CI, Dawson DV, Elston RC. (1990) The probabilistic determination of identity-by-descent sharing. *Am J Hum Genet* **47**:842–853.
3. Kruglyak L, Daly MJ, Lander ES. (1995) Rapid multipoint linkage analysis of recessive traits in nuclear families, including homozygosity mapping. *Am J Hum Genet* **56**:519–527.
4. Lander E, Kruglyak L. (1995) Complete multipoint sib-pair analysis of qualitative and quantitative traits. *Am J Hum Genet* **57**:439–454.
5. Sobel E, Lange K. (1996) Descent graphs in pedigree analysis: applications to haplotyping, location scores, and marker-sharing statistics. *Am J Hum Genet* **58**:1323–1337.
6. Fulker DW, Cherny SS, Cardon LR. (1995) Multipoint interval mapping of quantitative trait loci, using sib pairs. *Am J Hum Genet* **56**:1224–1233.
7. Almasy L, Blangero J. (1998) Multipoint quantitative-trait linkage analysis in general pedigrees. *Am J Hum Genet* **62**:1198–1211.
8. Botstein D, White RL, Skolnick M, Davis RW. (1980) Construction of a genetic linkage map in man using restriction fragment length polymorphisms. *Am J Hum Genet* **32**:314–331.
9. Guo X, Elston RC. (1999) Linkage information content of polymorphic genetic markers. *Hum Hered* **49**:112–118.
10. Schork NJ, Greenwood TA. (2004) Inherent bias toward the null hypothesis in conventional multipoint nonparametric linkage analysis. *Am J Hum Genet* **74**:306–316.
11. Jacobs KB, Gray-McGuire C, Cartier KC, Elston RC. (2003) Genome-wide linkage scan for genes affecting longitudinal trends in systolic blood pressure. *BMC Genet* **4**(Suppl 1):S82.
12. Franke D, Ziegler A. (2005) Weighting affected sib pairs by marker informativity. *Am J Hum Genet* **77**:230–241.
13. Gessler DD, Xu S. (1996) Using the expectation or the distribution of the identity by descent for mapping quantitative trait loci under the random model. *Am J Hum Genet* **59**:1382–1390.
14. Cordell HJ. (2004) Bias toward the null hypothesis in model-free linkage analysis is highly dependent on the test statistic used. *Am J Hum Genet* **74**:1294–1302.
15. Abecasis GR, Wigginton JE. (2005) Handling marker–marker linkage disequilibrium: pedigree analysis with clustered markers. *Am J Hum Genet* **77**:754–767.
16. Wang T, Elston RC. (2005) The bias introduced by population stratification in IBD-based linkage analysis. *Hum Hered* **60**:134–142.
17. Wan Y, Cohen JC, Guerra R. (1997) A permutation test for the robust sib-pair method. *Ann Hum Genet* **61**:7987.
18. Tiwari HK, Elston RC. (1997) Linkage of multi-locus components of variance of polymorphic markers. *Am J Hum Genet* **61**:253–261.

19. Hanson RL, Kobes S, Lindsay RS, Knowler WC. (2001) Assessment of parent-of-origin effects in linkage analysis of quantitative traits. *Am J Hum Genet* **68**:951–962.

20. S.A.G.E. Statistical Analysis for Genetic Epidemiology. Available at http://darwin.cwru.edu/sage/

21. Wright FA. (1997) The phenotypic difference discards sib-pair QTL linkage information. *Am J Hum Genet* **60**:740–742.

22. Gaines RE, Elston RC. (1969) On the probability that a twin pair is monozygotic. *Am J Hum Genet* **21**:457–465.

23. Drigalenko E. (1998) How sib-pairs reveal linkage. *Am J Hum Genet* **63**: 1242–1245.

24. Elston RC, Buxbaum S, Jacobs KB, Olson JM. (2000) Haseman and Elston revisited. *Genet Epidemiol* **19**:1–17.

25. Palmer LJ, Jacobs KB, Elston RC. (2000) Haseman and Elston revisited: the effects of ascertainment and residual familial correlations on power to detect linkage. *Genet Epidemiol* **19**:456–460.

26. Xu X, Weiss S, Wei LJ. (2000) A unified Haseman–Elston method for testing linkage with quantitative traits. *Am J Hum Genet* **67**:1025–1028.

27. Visscher P, Hopper J. (2001) Power of regression and maximum likelihood methods to map QTL from sib pair and DZ twin data. *Ann Hum Genet* **65**:583–601.

28. Forrest WF. (2001) Weighting improves the "new Haseman–Elston" method. *Hum Hered* **52**:47–54.

29. Shete S, Jacobs KB, Elston RC. (2003) Adding further power to the Haseman and Elston method for detecting linkage in larger sibships: weighting sums and differences. *Hum Hered* **55**:79–85.

30. Sham PC, Purcell S. (2001) Equivalence between Haseman–Elston and variance components linkage analyses for sib pairs. *Am J Hum Genet* **68**: 1527–1532.

31. Wang D, Lin S, Cheng R, *et al.* (2001) Transformation of sib-pair values for the Haseman–Elston method. *Am J Hum Genet* **68**:1238–1249.

32. Wright FA. (2003) Information perspectives of the Haseman–Elston method. *Hum Hered* **55**:132–142.

33. Wang T, Elston RC. (2004) A modified revisited Haseman–Elston method to further improve power. *Hum Hered* **57**:109–116.

34. Sham PC, Purcell S, Cherny SS, Abecasis GR. (2002) Powerful regression-based quantitative-trait linkage analysis of general pedigrees. *Am J Hum Genet* **71**:238–253.

35. Schaid DJ, Olson JM, Gauderman WJ, Elston RC. (2003) Regression models for linkage: issues of traits, covariates, heterogeneity, and interaction. *Hum Hered* **55**:86–96.

36. Gray-McGuire C, Bochud M, Elston RC. (2007) Genetic association tests based on family data. In: *Proceedings of the Third Seattle Symposium in Biostatistics: Statistical Genetics and Genomics*, Schadt E, Storey JD (eds.), Springer, New York, NY (in press).

37. Blackwelder WC. (1977) Statistical methods for detecting genetic linkage from sibship data. Institute of Statistics Mimeo Series No. 1114, The University of North Carolina at Chapel Hill, Chapel Hill, NC.

38. Wilson AF, Elston RC. (1993) Statistical validity of the Haseman–Elston sib-pair test in small samples. *Genet Epidemiol* **10**:593–598.

39. Single RM, Finch SJ. (1995) Gain in efficiency from using generalized least squares in the Haseman–Elston test. *Genet Epidemiol* **12**:889–894.

40. Amos CI, Elston RC. (1989) Robust methods for the detection of genetic linkage for quantitative data from pedigrees. *Genet Epidemiol* **6**:349–360.

41. Schaid DJ, Elston RC, Tran L, Wilson AF. (2000) Model-free sib-pair linkage analysis: combining full-sib and half-sib pairs. *Genet Epidemiol* **19**:30–51.

42. Zeger SL, Liang KY. (1986) Longitudinal data analysis for discrete and continuous outcomes. *Biometrics* **42**:121–130.

43. Olson JM, Wijsman EM. (1993) Linkage between quantitative trait and marker loci: methods using all relative pairs. *Genet Epidemiol* **10**:87–102.

44. Chen WM, Broman KW, Liang KY. (2004) Quantitative trait linkage analysis by generalized estimating equations: unification of variance components and Haseman–Elston regression. *Genet Epidemiol* **26**:265–272.

45. Wang T, Elston RC. (2005) Two-level Haseman–Elston regression for general pedigree data analysis. *Genet Epidemiol* **29**:12–22.

46. Putter H, Sandkuijl LA, Houwelingen JC. (2002) Score test for detecting linkage to quantitative traits. *Genet Epidemiol* **22**:345–355.

47. Tritchler D, Liu Y, Fallah S. (2003) A test of linkage for complex discrete and continuous traits in nuclear families. *Biometrics* **59**:382–392.

48. T Cuenco K, Szatkiewicz JP, Feingold E. (2003) Recent advances in human quantitative-trait-locus mapping: comparison of methods for selected sibling pairs. *Am J Hum Genet* **73**:863–873.

49. Szatkiewicz JP, T Cuenco K, Feingold E. (2003) Recent advances in human quantitative-trait-locus mapping: comparison of methods for discordant sibling pairs. *Am J Hum Genet* **73**:874–885.

50. Chen WM, Broman KW, Liang KY. (2005) Power and robustness of linkage tests for quantitative traits in general pedigrees. *Genet Epidemiol* **28**:11–23.

51. Risch N. (1990) Linkage strategies for genetically complex traits. II. The power of affected relative pairs. *Am J Hum Genet* **46**:229–241.

52. Olson JM. (1999) A general conditional-logistic model for affected-relative-pair linkage studies. *Am J Hum Genet* **65**:1760–1769.

53. Goddard KA, Witte JS, Suavez BK, *et al.* (2001) Model-free linkage analysis with covariates confirms linkage of prostate cancer to chromosomes 1 and 4. *Am J Hum Genet* **68**:1197–1206.

Chapter 2

Admixture Mapping: Methods and Applications

Xiaofeng Zhu

*Loyola University Medical Center, 2160 S. First Ave
Maywood, IL 60153, USA*

Linkage disequilibrium arising from the recent admixture of genetically distinct populations can be used to map genes influencing complex diseases. Mapping genes based on locus ancestral information is called admixture mapping. Admixture mapping requires much less markers than association studies, and is less affected by allelic heterogeneity. Theoretical works demonstrate that admixture mapping can be more powerful, with a higher mapping resolution than the traditional linkage studies. In this chapter, I will review the recent theoretical developments, the applications to localize genes influencing complex traits, and the technological challenges faced in using this method.

1. INTRODUCTION

Complex diseases are caused by a combination of many genes and environmental factors as well as gene–gene and gene–environment interactions, which result in a weak genotype–phenotype correlation. Currently, there are two main approaches to identify the genes underlying complex diseases, namely linkage and association studies. While linkage studies identify a locus linked to the trait through studying the cosegregation of the locus and the trait within families, association studies identify a particular variant associated with the trait at the population level. The linkage analysis approach has proved to be successful in Mendelian traits. However, the study of complex traits has been frustratingly difficult.[1] Association analysis has been promoted because of its greater power compared to linkage studies when a disease variant is in strong linkage disequilibrium (LD) with a testing marker.[1,2] The efficiency of association studies depends critically on the strength of LD between the trait locus and the test marker, which varies unpredictably across the genome in various populations. The number of SNPs required to perform a genomewide association study depends on the studied population.[3,4] With the complement of the International HapMap Project, tagging SNPs can be selected to perform genomewide association studies at a reduced cost,[5] although such cost may still be prohibitive for many laboratories.

Besides linkage and association studies, an alternative approach is association study using the information generated by recent population admixture for localizing susceptibility genes.[6–10] This strategy was first proposed by Rife[8] to detect linkage in an admixed population using correlation analysis. Since then, various theoretical methods of admixture mapping have been developed.[10–17] However, applications are limited because this method requires a large number of ancestry-informative markers (AIMs), which were not available until recently when abundant SNPs across the genome became available.

More recently, both the classical likelihood-based methods[18,19] and the Bayesian approaches[20–22] have been proposed specifically for genomewide or regionwide admixture mapping. According to these theoretical studies, admixture mapping could be more powerful than traditional linkage studies when the population risk ratio between parental populations is high and at a much lower genotyping cost than genomewide association studies. As the first practical exercise in admixture mapping, Zhu *et al.*[23] used a panel of microsatellite markers designed for traditional linkage analysis and identified two regions on chromosomes 6 and 21 that may potentially harbor genes influencing hypertension. More recently, Reich *et al.*[24] conducted an admixture mapping for multiple sclerosis using Smith's SNP panel[25] and identified a region on chromosome 1 that may potentially harbor genes affecting multiple sclerosis. These successes demonstrate that admixture mapping could be an important alternative to the traditional linkage and association methods used in mapping complex traits.

2. ADMIXTURE MAPPING

The idea behind admixture mapping is very straightforward. Figure 1 shows a simplified admixture process, where mating between two parental populations occurs only at the beginning. The red and blue bars represent the chromosomes from two ancestral populations. At each generation,

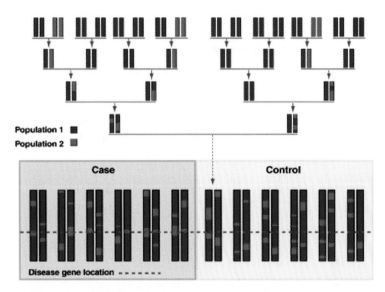

Fig. 1.   The crossovers of ancestral chromosomes occur during the admixture process. Adapted from Davasi and Shifman.[26]

chromosome segment exchanges occur because of chromosome crossovers during meiosis. At the current generation, each chromosome consists of many segments originating from two ancestral populations. For the purpose of illustrating an admixture mapping, we further assume a disease trait that only occurs in one of the two ancestral populations (for example, population 2, Fig. 1). When the cases and the controls are collected in a study, we expect to observe chromosome segments from population 2 more frequently at the disease gene location in the cases than in the controls. This is the basis of admixture mapping using a case-control design. Admixture mapping can also be performed within cases. Because linkage disequilibrium can decay quickly if two markers are unlinked, we expect to observe chromosome segments more frequently from population 2 at the disease location than outside it, which is the basis of case-only admixture mapping.

Zhu *et al.*[18] demonstrated the above idea mathematically. Suppose we have an admixture population C resulting from two parental populations, X and Y. We further assume the admixture process follows a continuous gene flow (CGF) model as illustrated in Fig. 2, where admixture occurs at a steady rate in each generation. Let $\Pi_d(\theta)$ be the proportion of marker alleles X by descent (the "ancestral probability" from population X) at a position of genetic distance θ from the disease location among affected individuals

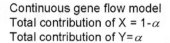

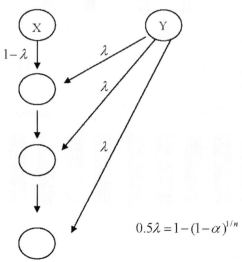

Fig. 2.   A continuous gene flow model.

at the current generation where admixture occurs. Zhu *et al.*[18] showed that $\Pi_d(\theta)$ is a strict monotonic function of θ. Whether it is an increasing or a decreasing function depends on which ancestral population has a higher disease allele frequency. Thus, a test of linkage $\theta = 0.5$ between a marker locus and a disease locus is equivalent to a test of the null hypothesis $\Pi_d(\theta) = \Pi_d(0.5)$. Without confusion, let $\hat{\Pi}_d(t)$ be the estimated proportion of ancestry from population X at chromosome location t, conditional on the observed marker genotypes. To test linkage, Zhu *et al.*[18] and Montana and Pritchard[21] used the statistic

$$ Z_C(t) = \frac{\hat{\Pi}_d(t) - \hat{\Pi}_d(\theta = 0.5)}{\sigma[\hat{\Pi}_d(t)]}. $$

Under the null hypothesis of no association, $Z_C(t)$ follows approximately a standard normal distribution, where $\hat{\Pi}_d(\theta = 0.5)$ is the estimate of ancestral proportion from population X at a marker locus unlinked to the disease locus, and $\sigma[\hat{\Pi}_d(t)]$ is the standard deviation of $\hat{\Pi}_d(t)$. To estimate $\hat{\Pi}_d(\theta = 0.5)$ and $\sigma[\hat{\Pi}_d(t)]$, Zhu *et al.*[18] suggested using the mean and standard error of the proportions of population X by descent at a set of unlinked marker loci, while Montana and Pritchard[21] suggested a parametric bootstrap method which required intensive computation time.

Similarly for the case-control design, let $\Pi_c(\theta)$ be the proportion of population X by descent at a marker locus in the controls with a recombinant fraction θ from a disease locus. $\Pi_c(\theta)$ is also a strict monotonic function of θ with an opposite increasing or decreasing direction of $\Pi_d(\theta)$.[27] Again, testing linkage $\theta = 0.5$ between a marker locus and a disease locus is equivalent to testing the null hypothesis $\Pi_d(\theta) - \Pi_d(0.5) = \Pi_c(\theta) - \Pi_c(0.5)$. A test statistic of the case-control design at marker locus t is

$$Z_{CC}(t) = \frac{[\hat{\Pi}_d(t) - \hat{\Pi}_d(\theta = 0.5)] - [\hat{\Pi}_c(t) - \hat{\Pi}_c(\theta = 0.5)]}{\sigma[\hat{\Pi}_d(t) - \hat{\Pi}_c(t)]}.$$

Again, the null distribution of $Z_{CC}(t)$ follows approximately a standard normal distribution, where $\hat{\Pi}_c(t)$ and $\hat{\Pi}_c(\theta = 0.5)$ are the estimates of the proportion of population X by descent at marker locus t and at an unlinked marker locus in the controls, respectively. The denominator, $\sigma[\hat{\Pi}_d(t) - \hat{\Pi}_c(t)]$, is the standard error of $\hat{\Pi}_d(t) - \hat{\Pi}_c(t)$. The standard error can be similarly estimated as in a case-only analysis.

McKeigue[13] and Hoggart *et al.*[20] proposed a likelihood-based approach for case-only analysis. The likelihood of the observed ancestral allele A (A = 1 if the allele is X by descent and 0 otherwise) can be calculated by

$$L(r; A) = \frac{(\lambda\sqrt{r})^A (1 - \lambda)^{1-A}}{\lambda\sqrt{r} + 1 - \lambda},$$

where r is the ancestry risk ratio at the locus under study, defined by the ratio of carrying two versus zero alleles X by descent under the assumption of multiplicative mode, and λ is the admixture proportion from the high-risk parental population (i.e. population X). A standard likelihood-ratio test or a score test can be carried out to test the null hypothesis $r = 0$.[20] For a case-control test, a logistic regression can be applied (Hoggart *et al.*[20]):

$$\log\frac{P(y \mid x)}{1 - P(y \mid x)} = \log\frac{\mu}{1 - \mu} + \beta(x - \lambda),$$

where y is the disease status, x is the proportion of alleles X by descent, μ is the disease prevalence in study sample, and β is the log odds ratio of disease for individuals with two vs. zero alleles X by descent. An alternative test for admixture mapping was introduced by Patterson *et al.*[22]

3. HIDDEN MARKOV MODEL

We can estimate $\Pi_d(t)$ and $\Pi_c(t)$ using a set of ancestral informative markers (AIMs) through a hidden Markov model (HMM). Consider M marker loci with known locations on the genome. Let A and a denote the two alleles at marker locus t, where $t = 1, \ldots, M$ and the M markers are located on one chromosome. Let $g_t \in \{0, 1, 2\}$ denote the genotype at the marker locus t for an individual with

$$g_t = \begin{cases} 0 & \text{for genotype } aa \\ 1 & \text{for genotype } Aa \\ 2 & \text{for genotype } AA. \end{cases}$$

Let $v_t \in \{0, 1, 2\}$ denote the number of alleles being X by descent at marker locus t. $v_1, v_2, \ldots, v_M$ then arises from a hidden Markov chain:

$$\begin{array}{cccc} \text{Observed genotype} \quad g_1 & g_2 & \cdots & g_M \\ \uparrow & \uparrow & & \uparrow \\ \text{Hidden states} \quad v_1 \rightarrow & v_2 \rightarrow & \cdots \rightarrow & v_M. \end{array}$$

The likelihood of the observed genotypes depends on the following quantities:

(1) Distribution of the initial hidden state $\delta_l = \Pr(v_1 = l); l = 0, 1, 2$.
(2) Emission probabilities $q_{jl}^t = \Pr(g_t = j \mid v_t = l); j, l = 0, 1, 2$ and $t = 1, \ldots, M$.
(3) Transition probabilities $t_{jl}^t = \Pr(v_{t+1} = j \mid v_t = l); j, l = 0, 1, 2$ and $t = 1, \ldots, M-1$.

The emission probabilities are the functions of allele frequencies in ancestral populations. Zhu *et al.*[18] formulated the transition probabilities for admixture processes according to the intermixing or continuous gene flow model, while Falush *et al.*[28] and Patterson *et al.*[22] assumed that the intermixing model and the crossover of ancestral chromosomes followed a Poisson distribution.[29] The transition probabilities used in Zhu *et al.*[18,27] are more accurate for modeling an African-American population than other populations such as a Hispanic population. The model by Falush *et al.*[28] and Patterson *et al.*[22] is much more flexible; it can be extended to an admixed population with more than two ancestral populations.

4. EM ALGORITHM FOR ESTIMATING ANCESTRY

Several methods have been proposed to calculate the likelihood of the hidden Markov model. There are basically two typical approaches: the Bayesian resampling approach[13,20–22] and the classical likelihood approach.[18,19,27,30]

Here, we introduce a simple expectation maximization (EM) algorithm to estimate the model parameters. Define $p = (p_{X1}, \ldots, p_{XM}; p_{Y1}, \ldots, p_{YM})$, where p_{Xt} and p_{Yt} are the allele frequencies of A at the tth marker locus in the parental populations, X and Y, respectively. Assume there are N individuals in the sample. Let v_{it} and g_{it} be the hidden ancestral state and the observed genotype at marker locus t of the ith individual. To estimate the parameters, we can apply the following iteration[27]:

Step 1: Given $\lambda = \lambda^{(n)}$ and an initial p, we estimate the conditional probabilities of the hidden state given "the genotypes" by the standard forward–backward algorithm for each individual. We then estimate the allele frequencies in the parental populations by

$$p_t^{(n+1)} = \frac{\sum_{i=1}^{N} \sum_{l=1}^{2} [\Pr(g_{it} = 1, v_{it} = l) + l\Pr(g_{it} = 2, v_{it} = l)]}{\sum_{i=1}^{N} [\Pr(v_{it} = 1) + 2\Pr(v_{it} = 2)]}.$$

Step 2: Given $p = p^{(n+1)}$, we maximize the likelihood to estimate λ and denote it $\lambda^{(n+1)}$.

Step 3: Repeat steps 1 and 2 until the difference in log likelihood is less than a predefined value between successive iterations.

5. MARKOV CHAIN MONTE CARLO (MCMC) APPROACH

The MCMC approach is an alternative way to estimate an individual's ancestry.[20–22] Here, we introduce the approach by Patterson *et al.*,[22] who conducted simulation studies and suggested that the method is robust and efficient. In general, the first step of the MCMC approach is to pick starting values of the unknown parameters, including

(1) the ancestral allele frequencies $p = (p_{X1}, \ldots, p_{XM}; p_{Y1}, \ldots, p_{YM})$, which are set to be the values estimated from the parental populations;
(2) the proportion of ancestry for each individual, which is set to be the value estimated by maximum likelihood estimates when treating all markers as unlinked; and
(3) the number of generations, which is set to be six.

MCMC then repeats the following steps:

Step A: Conditional on the current set of parameters, $p = (p_{X1}, \ldots, p_{XM}, p_{Y1}, \ldots, p_{YM})$, the proportion of ancestry for each individual, and the number of generations, we use HMM to generate a sequence of ancestry states across the genome.

Step B: Loop over all of the unknown parameters, updating each in turn by (1) fixing the values for all other unknown parameters; (2) calculating the likelihood distribution for the unknown, conditional on the fixed values of the other; and (3) randomly sampling a value for the unknown parameter from the conditional distribution.

6. INFORMATIVE MARKERS FOR ANCESTRY

The power of admixture mapping depends critically on the type of markers used. Microsatellite markers can be more informative for ancestry than SNPs in general. However, far more SNPs are available than microsatellite markers across the genome, leading to easy selection of highly informative SNPs for ancestry. There are several measurements for evaluating the marker information content of two ancestral populations: the absolute allele frequency difference δ, F_{ST} distance, Fisher information content, pairwise Kullback–Leible divergence, and informativeness for assignment (I_n).[31] Rosenberg *et al.*[31] compared the performance of these measurements and concluded that all of the measurements are highly correlated, although I_n performs slightly better. By screening 450 000 SNPs across the genome, Smith *et al.*[25] assembled a panel of 3075 ancestry-informative SNPs from populations of mixed West African and European descent. The average information content of this panel for distinguishing a European from a West African population was 28% (measured by I_n), corresponding to $\delta = 0.56$. Zhu *et al.*[27] studied the multiple-marker information content of this panel by simulations. They found that this panel could extract 89% of ancestral information, although there were some gaps that needed to be filled.

Phase 1 of the current HapMap Project[5] released over one million SNPs for four populations: Yoruba in Ibadan, Nigeria (YRI); Centre d'Etude du Polymorphisme Humain (CEU) in Utah, USA; Han Chinese in Beijing, China (CHB); and Japanese in Tokyo, Japan (JPT). Thus, a panel of highly dense ancestral informative SNPs can be selected from HapMap. We compared the allele frequency difference among YRI, CEU, CHB, and JPT on chromosomes 6 and 21, and concluded that there are abundant ancestral informative SNPs available across the genome (Table 1). Interestingly, the

Table 1   Distributions of AIMs in HapMap Data (δ: Allele Frequency Difference between Two Populations)

	CEU − YRI		YRI − (CHB + JPN)		CEU − (CHB + JPN)	
	Chr 6	Chr 21	Chr 6	Chr 21	Chr 6	Chr 21
$\delta > 0.4$	15246	3390	21157	4270	7690	1248
No. of AIMs						
Maximum distance between adjacent SNPs (kb)	3728	522.3	3248	457	7067	1007
Average distance (kb)	11.2	9.9	8.1	7.8	22.2	26.5
Average δ	0.51 ± 0.09	0.51 ± 0.09	0.52 ± 0.10	0.52 ± 0.10	0.48 ± 0.07	0.49 ± 0.08
$\delta > 0.5$	6404	1526	10262	2000	2537	466
No. of AIMs						
Maximum distance between adjacent SNPs (kb)	3980	1365	3299	616	7067	1961
Average distance (kb)	26.6	22.1	16.6	16.6	67.3	70.7
Average δ	0.59 ± 0.08	0.59 ± 0.08	0.61 ± 0.09	0.60 ± 0.08	0.57 ± 0.06	0.57 ± 0.07

average distances between AIMs are almost the same on chromosomes 6 and 21 for any two compared populations, suggesting that selection only occurs in a small portion of the genome.

7. POWER OF ADMIXTURE MAPPING

The power to detect linkage for admixture mapping is dependent on various model parameters, including a parental population risk ratio that is attributed to the disease locus, model of inheritance, admixture rate, and distance between the testing marker and disease loci. Admixture mapping will have no power if the difference in the parental population risk is the resulted of distinct environmental or social influences instead of an underlying genetic difference. Conversely, if the disease gene allele frequencies are different in the parental populations, admixture mapping would have power even if the disease prevalence is similar in parental populations.[32] An important issue in the study of complex diseases is the phenomenon of genetic heterogeneity, including allelic heterogeneity or locus heterogeneity that produces the same, indistinguishable phenotype.[1] Similar to traditional linkage analysis, allelic heterogeneity is less likely to cause a problem for admixture mapping because the different mutations in a gene will likely fall in the same block created by the admixture process. However, locus heterogeneity may cause a serious problem in admixture mapping as in linkage analysis. For example, a disease is caused by many genes, with a small effect from each gene; even when the population risk ratio is large, a large sample size is still required to detect each gene.

In Table 2, we present the power of 800 cases in an admixture mapping on the population risk ratio based on an African-American population, in which we assume a mixing of 18% European and 82% West African ancestries. We assumed that the admixture process occurred 12 generations ago under the continuous gene flow model. This model fits the African-American population well.[27] For comparison, we also presented the power of 800 sib pairs in linkage studies for the best case, that is, a closely linked marker locus and a fully informative marker. We calculated the power of linkage studies according to the method described by Risch and colleagues.[1,33,34] In order to make a reasonable comparison, we used a set of parameters such as genotype relative risks and disease allele frequencies compatible with the population risk ratio. Since the power of admixture mapping is dependent on the population risk ratio, and different disease allele frequencies and genotype relative risks can produce the same population risk ratio, we then presented the power of linkage analysis for different disease risk allele frequencies. When the difference of risk allele frequencies in the parental populations is large (e.g. 0.5 vs. 0.1), admixture mapping is more powerful than linkage analysis. Conversely, linkage analysis overpowers admixture mapping when the risk allele frequency difference is less than 0.2 between parental populations. The power comparison suggests that admixture mapping and linkage mapping can be complementary, and they map different kinds of genes in dissecting complex diseases.

8. APPLICATION

The ideal populations with a high chance of success for admixture mapping are those where an admixture process occurred recently, such as the African-American and Hispanic populations. So far, only hypertension and multiple sclerosis have been reported with the use of admixture mapping.[23,24] The risk of hypertension in African-Americans has been reported to be 2.6 times higher than that of European-Americans.[35] Conversely, European-Americans have a risk of multiple sclerosis two times higher than that of African-Americans.[36] The different risks across the parental populations are critical to the success of admixture mapping. Smith and O'Brien[32] listed 25 diseases that have different risks in African-Americans and European-Americans, including various cancers, immunological pathologies, diabetes, hypertension, and multiple sclerosis. This information can be helpful for the design of an admixture mapping study. Admixture mapping in a Hispanic population faces a great challenge because of the lack of ancestral

Table 2   Comparison of Power between Admixture Mapping and Linkage Analysis

Risk ratio	Admixture Mapping				Linkage Analysis							
					0.5 vs. 0.1				0.3 vs. 0.1			
	Mult	Add	Rec	Dom	Mult	Add	Rec	Dom	Mult	Add	Rec	Dom
1.1	1	1	2	0	0	0	0	0	0	0	0	0
1.2	3	3	13	1	0	0	0	0	0	0	0	0
1.3	10	9	41	1	0	0	0	0	0	0	13	0
1.4	22	18	72	2	0	0	0	0	0	0	83	0
1.5	38	31	92	2	0	0	0	0	0	0	100	0
1.6	55	46	98	3	0	0	1	0	2	0	100	0
1.7	70	59	100	4	0	0	6	0	12	1	100	1
1.8	82	71	100	5	0	0	18	0	39	4	100	4
1.9	90	80	100	6	0	0	39	0	71	10	100	10
2	95	87	100	8	0	0	62	0	92	20	100	21
2.1	98	91	100	9	0	0	80	0	98	34	100	35
2.2	99	95	100	10	0	0	91	0	100	49	100	51
2.3	100	97	100	11	0	0	96	0	100	63	100	64
2.4	100	98	100	13	1	0	99	0	100	74	100	76
2.5	100	99	100	14	3	0	100	0	100	83	100	84

population information. In general, the Hispanic population has a genetic mixing of three ancestral populations: African, European, and Native American ancestries.

There are substantial fluctuations in ancestry depending on the geographic region. African ancestry ranged from less than 10% to over 50%, European ancestry from less than 20% to over 80%, and Native American ancestry fell between 5% and 20%.[37] The case-control design can be more robust than the case-only design for a Hispanic population because the case-only design is heavily dependent on the estimation of allele frequency estimates in parental populations.[27]

9. LIMITATIONS

Several authors have pointed out the limitations of admixture mapping.[32,38] For example, admixture mapping will likely miss the disease variants with similar allele frequencies in the parental populations. In this case, linkage studies will be more powerful in identifying these variants than admixture mapping. Admixture mapping is also dependent on the accuracy of allele frequency estimates in the parental populations because the information is unknown and has to be estimated. These estimates are made through the HMM with the assumed population admixture model. The continuous gene flow model is known to fit the African-American population well; hence, the corresponding allele frequencies in the parental populations can be accurately estimated.[27] However, it may be more difficult for a Hispanic population because of the lack of information in ancestral populations such as the Native American. The current HapMap Project provides abundant information for identifying AIMs that are useful for African-American populations. How useful this is for the Hispanic population and other admixed groups remains unknown. Montana and Pritchard[21] also suggested that a large set of random SNPs could be equally informative as a small set of AIMs for admixture mapping studies. However, biased estimates of allele frequencies in the parental populations can arise as a result of background linkage disequilibrium among the markers in the parental populations. In addition, computation becomes very intensive. Tang et al.[39] developed a Markov hidden Markov model by allowing for background LD, which is very easy to apply.

Because of the possible biases of admixture mapping studies, guidelines for declaring the significance in admixture mapping studies have been proposed.[32,40] These guidelines are helpful in reducing false-positive findings in practice; however, power may also be sacrificed. For example, a problem of mapping complex diseases is the accurate definition of the phenotype, especially those phenotypes that depend on age and other environmental factors. Therefore, it may be difficult to identify true controls. As a result, the deviation of certain ancestries may be seen in both cases and controls.[24] The study by Zhu et al.[23] has been criticized for not meeting these guidelines.[40] However, the significant result of this study is the cumulative evidence from three separate network studies as well as a combined analysis. As linkage and association studies, we believe replication studies will provide more convincing information than a single study with a great statistical significance.

10. CONCLUSIONS

Defining the genetic architecture of complex diseases remains a daunting challenge. A common feature of the underlying architecture for complex diseases is that it involves multiple genetic variants, each with a modest contribution. Nonetheless, resolving the genetics of these disorders will be a crucial step forward for biomedical research. In addition to the basic biological insights that could be gained, this knowledge could help to speed up the development of new drugs and treatments.

The rapid advance of computational and molecular technological developments opens great opportunities for the study of complex diseases, and also brings new challenges including the development of more complex statistical methods. Traditional linkage-based studies are still optimal in the search for some genetic susceptibility variants. Although genomewide association studies are promising in the search for all susceptibility variants, the cost is still prohibitive for most labs. Admixture mapping has the advantages of both linkage and association analyses, and can be powerful, especially in the search for ethnic-specific variants.

ACKNOWLEDGMENT

This work was supported by a grant from the National Human Genome Research Institute (HG003054).

REFERENCES

1. Risch NJ. (2000) Searching for genetic determinants in the new millennium. *Nature* **405**:847–856.
2. Risch N, Merikangas K. (1996) The future of genetic studies of complex human diseases. *Science* **273**:1516–1517.
3. Carlson CS, Eberle MA, Rieder MJ, *et al.* (2004) Selecting a maximally informative set of single-nucleotide polymorphisms for association analyses using linkage disequilibrium. *Am J Hum Genet* **74**:106–120.
4. Gabriel SB, Schaffner SF, Nguyen H, *et al.* (2002) The structure of haplotype blocks in the human genome. *Science* **296**:2225–2229.
5. International HapMap Consortium. (2005) A haplotype map of the human genome. *Nature* **437**:1299–1320.
6. Lautenberger JA, Stephens JC, O'Brien SJ, Smith MW. (2000) Significant admixture linkage disequilibrium across 30 cM around the FY locus in African Americans. *Am J Hum Genet* **66**:969–978.
7. Pfaff CL, Parra EJ, Bonilla C, *et al.* (2001) Population structure in admixed populations: effect of admixture dynamics on the pattern of linkage disequilibrium. *Am J Hum Genet* **68**:198–207.
8. Rife DC. (1954) Populations of hybrid origin as source material for the detection of linkage. *Am J Hum Genet* **6**:26–33.
9. Risch N. (1992) Mapping genes for complex disease using association studies with recently admixed populations. *Am J Hum Genet* **51**(Suppl):13.
10. Stephens JC, Briscoe D, O'Brien SJ. (1994) Mapping by admixture linkage disequilibrium in human populations: limits and guidelines. *Am J Hum Genet* **55**:809–824.
11. Chakraborty R, Weiss KM. (1988) Admixture as a tool for finding linked genes and detecting that difference from allelic association between loci. *Proc Natl Acad Sci USA* **85**:9119–9123.

12. McKeigue PM. (1997) Mapping genes underlying ethnic differences in disease risk by linkage disequilibrium in recently admixed populations. *Am J Hum Genet* **60**:188–196.

13. McKeigue PM. (1998) Mapping genes that underlie ethnic differences in disease risk: methods for detecting linkage in admixed populations, by conditioning on parental admixture. *Am J Hum Genet* **63**:241–251.

14. McKeigue PM, Carpenter JR, Parra EJ, Shriver MD. (2000) Estimation of admixture and detection of linkage in admixed populations by a Bayesian approach: application to African-American populations. *Ann Hum Genet* **64**:171–186.

15. Shriver MD, Parra EJ, Dios S, *et al.* (2003) Skin pigmentation, biogeographical ancestry and admixture mapping. *Hum Genet* **112**:387–399.

16. Thomson G. (1995) Mapping disease genes: family-based association studies. *Am J Hum Genet* **57**:487–498.

17. Zheng C, Elston RC. (1999) Multipoint linkage disequilibrium mapping with particular reference to the African-American population. *Genet Epidemiol* **17**:79–101.

18. Zhu X, Cooper RS, Elston RC. (2004) Linkage analysis of a complex disease through use of admixed populations. *Am J Hum Genet* **74**:1136–1153.

19. Zhang C, Chen K, Seldin MF, Li H. (2004) A hidden Markov modeling approach for admixture mapping based on case-control data. *Genet Epidemiol* **27**:225–239.

20. Hoggart CJ, Shriver MD, Kittles RA, *et al.* (2004) Design and analysis of admixture mapping studies. *Am J Hum Genet* **74**:965–978.

21. Montana G, Pritchard JK. (2004) Statistical tests for admixture mapping with case-control and cases-only data. *Am J Hum Genet* **75**:771–789.

22. Patterson N, Hattangadi N, Lane B, *et al.* (2004) Methods for high-density admixture mapping of disease genes. *Am J Hum Genet* **74**:979–1000.

23. Zhu X, Luke A, Cooper RS, *et al.* (2005) Admixture mapping for hypertension loci with genome-scan markers. *Nat Genet* **37**:177–181.

24. Reich D, Patterson N, De Jager PL, *et al.* (2005) A whole-genome admixture scan finds a candidate locus for multiple sclerosis susceptibility. *Nat Genet* **37**:1113–1118.

25. Smith MW, Patterson N, Lautenberger JA, *et al.* (2004) A high-density admixture map for disease gene discovery in African Americans. *Am J Hum Genet* **74**:1001–1013.

26. Darvasi A, Shifman S. (2005) The beauty of admixture. *Nat Genet* **37**:118–119.

27. Zhu X, Zhang S, Tang H, Cooper RS. (2006) A classical likelihood based approach for admixture mapping using EM algorithm. *Hum Genet* **120**:431–445.

28. Falush D, Stephens M, Pritchard JK. (2003) Inference of population structure using multilocus genotype data: linked loci and correlated allele frequencies. *Genetics* **164**:1567–1587.

29. Long JC. (1991) The genetic structure of admixed populations. *Genetics* **127**:417–428.

30. Tang H, Peng J, Wang P, Risch NJ. (2005) Estimation of individual admixture: analytical and study design considerations. *Genet Epidemiol* **28**:289–301.
31. Rosenberg NA, Li LM, Ward R, Pritchard JK. (2003) Informativeness of genetic markers for inference of ancestry. *Am J Hum Genet* **73**:1402–1422.
32. Smith MW, O'Brien SJ. (2005) Mapping by admixture linkage disequilibrium: advances, limitations and guidelines. *Nat Rev Genet* **6**:623–632.
33. Risch N, Zhang H. (1995) Extreme discordant sib pairs for mapping quantitative trait loci in humans. *Science* **268**:1584–1589.
34. Risch N, Merikangas K. (1996) The future of genetic studies of complex human diseases. *Science* **273**:1516–1517.
35. Gupta V, Nanda NC, Yesilbursa D, *et al.* (2003) Racial differences in thoracic aorta atherosclerosis among ischemic stroke patients. *Stroke* **34**:408–412.
36. Hogancamp WE, Rodriguez M, Weinshenker BG. (1997) Identification of multiple sclerosis–associated genes. *Mayo Clin Proc* **72**:965–976.
37. Burchard GE, Borrell LN, Choudhry S, *et al.* (2005) Latino populations: a unique opportunity for the study of race, genetics, and social environment in epidemiological research. *Am J Public Health* **95**:2161–2168.
38. McKeigue PM. (2005) Prospects for admixture mapping of complex traits. *Am J Hum Genet* **76**:1–7.
39. Tang H, Coram M, Wang P, *et al.* (2006) Reconstructing genetic ancestry blocks in admixed individuals. *Am J Hum Genet* **79**:1–12.
40. Reich D, Patterson N. (2005) Will admixture mapping work to find disease genes? *Philos Trans R Soc Lond B Biol Sci* **360**:1605–1607.

Chapter 3

Survival Analysis Methods in Genetic Epidemiology

Hongzhe Li

Department of Biostatistics and Epidemiology
University of Pennsylvania School of Medicine
Philadelphia, PA 19104-6021, USA

Mapping genes for complex human diseases is a challenging problem because many such diseases are the result of both genetic and environmental risk factors. Many also exhibit phenotypic heterogeneity, such as variable age of onset. Information on the variable age of disease onset is often a good indicator for disease heterogeneity. The incorporation of such information together with environmental risk factors in genetic analysis should lead to more powerful tests. Because of the problem of censoring, survival analysis methods have proved to be very useful for genetic analysis. In this paper, I review some recent methodological developments on integrating modern survival analysis methods and human genetics in order to rigorously incorporate both the age of onset and the environmental covariate data into aggregation analysis, segregation analysis, linkage analysis, association analysis, and gene risk characterization. I also briefly discuss the issue of ascertainment correction and survival analysis methods for high-dimensional genomic data. Finally, I outline several areas that need further methodological developments.

1. INTRODUCTION

The major burden of ill health in Western society, and to a growing extent in developing societies, is attributed to complex chronic diseases such as coronary heart disease, stroke, breast cancer, prostate cancer, and diabetes. It is believed that both genetic and environmental factors contribute to both the risk of developing many of these common human diseases and the response to treatments. Because multiple genetic and environmental factors may play important roles in the susceptibility of individuals to develop these diseases and in the variation in treatment responses, they are often referred to as complex traits. While the data necessary to study different complex traits are trait-specific, the underlying principles and statistical methods of analysis of the genetic component are applicable to a variety of traits.

One important feature of many complex human diseases is disease heterogeneity, which is attributed to genetic and other etiological factors. For example, many complex diseases exhibit variability in the age of onset, and early age of onset has been the hallmark of genetic predisposition in many diseases that aggregate in families. Therefore, age of onset outcomes such as the age at diagnosis are frequently gathered in genetic and epidemiological studies, including both genetic association and linkage studies. An important feature of age of onset data is the censorship resulting from being too young to develop the disease or death before developing the disease. This makes it possible for some of the unaffected siblings to share the disease gene with the affected siblings, who might be too young to exhibit the trait. In fact, affected relatives with different ages of onset may also be the result of different genetic etiologies. Age of onset data have been used to distinguish between two subforms of breast cancer[1,2] and prostate cancer.[3] For these adult-onset cancers, carriers of high-risk alleles were estimated to have an earlier onset of cancer than noncarriers (sporadic cases). Taking into account age of onset information has been shown to be important in studying disease correlation and aggregation,[4,5] in parametric linkage analysis,[6,7] in segregation analysis,[4,5] and in allele-sharing–based linkage analysis.[8–11] A study by Li and Hsu[12] also indicates that ignoring the age of onset can reduce the power of both the allele-sharing–based linkage test and the transmission/disequilibrium test (TDT).[13,14]

Another important feature of many complex traits is that many of these traits are known or suspected to be influenced by various environmental risk factors and interactions between genetic and environmental risk factors (G × E), e.g. breast cancer[15] and rheumatoid arthritis.[16] From a statistical standpoint, ignoring existing gene–environment interactions can result in underestimation of both genetic and environmental effects,[17] incorrect conclusions with regard to the mode of inheritance and the magnitude of genetic effects in segregation analysis,[18] and lower power in detecting genetic linkage.[19–21]

Information on the variable age of disease onset is often a good indicator for disease heterogeneity. The incorporation of such information together with environmental risk factors in genetic analysis should lead to more powerful tests. Because of the problem of censoring, survival analysis methods, which are developed particularly for handling censoring, have proved to be very useful for genetic analysis. In this paper, I review some recent methodological developments in genetic epidemiology in order to rigorously take into account the age of onset and the environmental risk factors in aggregation analysis, segregation analysis, linkage and family-based association analysis, and gene risk characterization in the population. I also briefly discuss the issue of ascertainment correction and survival analysis methods for high-dimensional genomic data. Although I attempt a full and balanced treatment of most of the available literature, certain parts of the presentation lean naturally towards my own work. At the end of this review, I outline several areas that I think need further methodological developments, in particular the areas where high-throughput genomic data such as the genomewide single nucleotide polymorphism (SNP) data are available.

2. SURVIVAL ANALYSIS METHODS FOR AGGREGATION ANALYSIS

The purpose of aggregation analysis is to test whether disease aggregates within a family after some known environmental risk factors are taken into account. The ideal design is to collect a random sample of N families from the study population and to collect both age of disease onset or age at censoring data and the environmental risk factors of all the individuals within the families sampled. Then, the test of disease aggregation within the family will be equivalent to testing whether the ages of onset of family members are correlated after adjusting for environmental risk factors.

I first define some notations that are used throughout this review. Suppose we have a collection of N families collected randomly or by some ascertainment criteria. Let the subscript ik indicate the ith individual in kth family: $i = 1, \ldots, m_k$; $k = 1, \ldots, N$. T_{ik} is the age at onset, C_{ik} the censoring age, $t_{ik} = \min(T_{ik}, C_{ik})$, and $\delta_{ik} = I(t_{ik} = T_{ik})$ where $I(.)$ is the indicator function. The observed data are $(t_{ik}, \delta_{ik}, X_{ik})$, where X_{ik} is a p-dimensional vector of covariates that are independent of the genotype.

2.1. *Shared Frailty Models Based on Random Sample of Families*

The most commonly used model for assessing disease aggregation is the shared frailty model, which assumes the following conditional hazard function:

$$\lambda_{ik}(t|Z_k) = \lambda_0(t) \exp(X_{ik}\beta) Z_k, \tag{1}$$

where $\lambda_0(t)$ is the baseline hazard function, X_{ik} is the individual-specific covariate vector, β is the corresponding risk ratio parameters, and Z_k is the family-specific random effect or shared frailty. If $Z_1, \ldots, Z_k, \ldots, Z_N$ are assumed to be i.i.d. from a gamma distribution $\Gamma(\nu, \eta)$, where ν is the shape parameter and η is the scale parameter, then the model is also called the gamma frailty, Clayton, or Clayton–Oakes model.[22,23] For the identifiability of $\lambda_0(t)$, it is assumed that $\nu = \eta$ so that $E(Z) = 1$. The estimation of such a model has been the subject of active research since the mid-1980s.[24–30] The most common nonparametric maximum likelihood estimates (NPMLEs) of the parameters can be obtained by the EM algorithm. Other assumptions on the frailty distribution include a positive stable distribution[31] and a log-normal distribution. Different distributions induce different dependency structures of the age of onset within a family. Glidden and Self[32] provide a model-checking procedure for the gamma frailty model.

Under the shared frailty model, the null hypothesis of no disease aggregation can be formulated as testing

$$H_0 : \mathrm{var}(Z) = \nu = 0.$$

For randomly sampled families, the standard inference procedure for the shared frailty model is based on the EM algorithm and the likelihood ration test for H_0.[26,29] The theoretical development was given by Murphy[28] for the shared gamma frailty models.

2.2. *Multivariate Frailty Models*

The limitation of the shared frailty model for investigating disease aggregation within a family is that such a model assumes the same degree of dependency for any pair of individuals within a family, which is likely to be violated when disease aggregation is due to genetic segregation within the families. One way of extending the shared frailty model is to assume an individual-specific additive frailty. For example, Peterson[33] defined the following additive frailty model:

$$\lambda_{ik}(t|\eta_k) = \lambda_0(t)\exp(X_{ik}\beta)Z_{ik}, \tag{2}$$
$$Z_{ik} = Z_{k0} + Z_i, \tag{3}$$

where Z_{ik} is the individual-specific frailty, the frailty Z_{k0} is the shared frailty by family members in the kth family and is assumed to follow a $\Gamma(\nu_0, \eta)$, and Z_i are the individual-specific frailties and are assumed to follow $\Gamma(\nu_1, \eta)$. For the purpose of identifiability of $\lambda_0(t)$, it is often assumed that $\nu_1 + \nu_0 = \eta$ so that $E(Z) = 1$. Under this additive gamma frailty model, the null hypothesis of interest is

$$H_0 : \nu_0 = 0.$$

Peterson[33] presented an EM algorithm to obtain the NPMLEs for the parameters, and Parner[34] developed the asymptotic theorem for the estimators and the likelihood-ratio test for H_0.

If the main goal is to test for disease aggregation due to genetic segregation, one should explicitly model the family dependence. Extensions of the frailty models to account for kinship relationship have been developed in recent years. One approach by Korsgaard and Andersen[35] is in the framework of additive gamma frailty models, where the additive individual-specific frailties are explicitly defined based on gene segregation. Another approach is to assume that the log of the family-specific frailty, vector $\log(Z_k) = \{\log(Z_{1k}), \ldots, \log(Z_{n_k k})\}$, follows a multivariate normal distribution $MVN(0, \Sigma)$, where the variance–covariance matrix is defined by the kinship coefficient matrix. The estimation of such multivariate normal frailty models includes the Monte Carlo or approximate EM algorithm[36] or the penalized partial likelihood approach.[37]

2.3. *Familiar Aggregation Based on Case-Control Family Design*

As an alternative to the family cohort design for assessing disease aggregation, the case-control sampling design is often used for rare diseases because a sufficient number of cases can be ascertained. In family studies, investigators use the case-control design to enroll relatives for more detailed information, obtain medical records to validate reported disease, and obtain biospecimens for studies of genetic markers associated with the disease. Case-control family studies allow a direct examination of the disease outcomes in relatives and the collection of both risk factor and exposure data on each individual; and with measured genetic markers, permit a more complete assessment of genetic and environmental factors through segregation and linkage analyses. Such a case-control study identifies a sample of diseased cases and, for each case, an independent sample of age-matched disease-free controls. For each identified individual (the proband), his environmental covariates, his family structure, the disease status, the age of onset (or age at censoring), and the environmental covariates of his relatives are obtained.

When analyzing such case-control family data, one has to account for both the sampling issue and the dependency of age of onset within the family. Estimating the marginal hazard function from the correlated failure time data arising from case-control family studies is complicated by non-cohort study design and risk heterogeneity due to unmeasured, shared risk factors among the family members. By assuming a Clayton multivariate survival model, Li *et al.*[38] developed a procedure based on Prentice and Breslow's[39] retrospective likelihood formulation assuming a parametric baseline hazard function. The method provides a way to combine the information relating disease incidence to risk factors in relatives with the information contained in the case-control contrasts in order to obtain more precise estimates of the effects of the putative risk factors. Shih and Chatterjee[40] developed a similar estimation procedure, but left the baseline hazard function unspecified. Hsu *et al.*[41] proposed a two-stage estimation procedure. At the first stage, the dependence parameter in the distribution for risk heterogeneity is estimated without obtaining the marginal distribution first or simultaneously. Assuming that the dependence parameter is known, at the second stage, the marginal hazard function is estimated by iterating between the estimation of the risk heterogeneity (frailty) for each family and the maximization of the partial likelihood function with an offset to account for the risk heterogeneity.

3. SURVIVAL ANALYSIS METHODS FOR SEGREGATION ANALYSIS

The goal of genetic segregation analysis is to develop a genetic model that best describes the disease aggregation within a family. Often, it is assumed that a major gene with or without polygenes is involved in disease segregation. Segregation analysis based on the parametric distributional assumptions on age of onset distribution is simple; instead, I review two semiparametric models developed for segregation analysis.

3.1. *The Cox Gene Model for Gene Segregation*

Assuming that a single major Mendelian diallelic locus governs the age-specific disease rate and that the corresponding alleles are a and A, where A is the dominant disease allele with allele frequency $P(A) = q$, let g_{ik} be the genotype of ikth individual, taking one of three values: aa, Aa, or AA. Li and Thompson[4] developed the following Cox gene model. They assume that, conditional on the unobserved major genotype g_{ik}, ages of onset are assumed to be independent with a hazard function for the ikth individual:

$$\lambda_{ik}(t|X_{ik}, g_{ik}) = \lambda_0(t)\exp(\beta'X_{ik} + \mu_{ik}), \tag{4}$$

where

$$\mu_{ik} = \begin{cases} 0 & \text{if } g_{ik} = aa \\ \mu & \text{if } g_{ik} = Aa \text{ or } AA \end{cases}$$

under the assumption of a dominant mode of inheritance. The vector parameter β specifies the log of the risk ratios associated with the covariates, and $\lambda_0(t)$ is an unspecified baseline hazard function. Let $\Lambda_0(t) = \int_0^t \lambda(s)ds$ be the cumulative hazard function. Since A is the disease allele, we assume that $\mu \geq 0$. Li and Thompson[4] further developed a Monte Carlo EM algorithm for estimating the parameters, especially for large pedigrees when the exact computation of the EM algorithm is not feasible.

Similar models were also developed and studied by Gauderman and Thomas [42] and by Siegmund and McKnight. [43] Chang *et al.* [44] established the asymptotic properties of the NPMLEs from the EM algorithm. Chang *et al.* [45] developed a faster algorithm for computing the NPMLEs than the EM algorithm.

3.2. Cox Model with Major Gene and Random Environmental Effects for Age of Onset

The Cox gene model assumes that disease aggregation is due to the segregation of one major gene, which accounts for all of the correlations among family members. To account for possible shared environmental effects, Li *et al.*[5] defined a model to allow for both major gene effects and shared environmental effects by introducing a family-specific gamma random effect. Specifically, conditional on individual-specific major genotype g_{ik} and family-specific random environment ϵ_k, ages of onset are independent with the hazard function for the ikth individual:

$$\lambda_{ik}(t|X_{ik}, g_{ik}, \epsilon_k) = \lambda_0(t)\epsilon_k \exp(\beta'X_{ik} + \mu_{g_{ik}}), \tag{5}$$

where $\mu_{g_{ik}} = 0$ if $g_{ik} = aa$, or $\mu - 1$ if $g_{ik} = Aa$ or AA, is the genetic effect. The vector parameter β specifies risk ratios associated with the covariate X_{ik}, and $\lambda_0(t)$ is an unspecified baseline hazard function; $\Lambda_0(t) = \int_0^t \lambda(s)ds$ is the cumulative hazard function. The family effect, ϵ_k, is assumed to be *i.i.d.* gamma variate with mean $= 1$ and unknown variance ν. This model incorporates the dependencies attributed to gene segregation and to a shared environment. It is appropriate only for data on many families; variance ν is estimable only with a set of at least three families. The full model is specified by $\Theta = [\mu, q, \theta, \beta, \Lambda_0(t)]$. If $\theta = 0$, then $\epsilon_k = 1.0$ with probability $= 1$ for all families and the model in Eq. (5) reduces to the Cox gene model [Eq. (4)]. If $\mu = 0$ or $q = 0$, then the model in Eq. (5) reduces to the gamma frailty model. The parameters associated with the frailties are $\{\mu, q, \theta\}$. The genetic effects are measured by two parameters q and μ, where q measures the frequency of genetic susceptibility and μ measures the extent of genetic effects. The family-specific effects are measured by parameter ν; a larger value of ν corresponds to a stronger familial dependence due to common environmental effects and greater heterogeneity between families. Mendelian dependence of $\mu_{g_{ik}}$ makes it possible in theory to separate the genetic effects from the shared environmental effects, and thus identify and estimate the model parameters.

Li *et al.*[5] developed an MCEM algorithm for estimating the model parameters, and applied this model to the analysis of a breast cancer family data set. The null hypothesis of interest includes $H_0 : \nu = 0$; when rejected, it implies that the major gene segregation cannot explain all of the correlations of disease risks within families and that additional genes or shared environmental factors may exist.

4. SURVIVAL ANALYSIS METHODS FOR LINKAGE ANALYSIS

Linkage analysis examines the cosegregation of disease locus and markers or genomic loci within a family. Model-based linkage analysis often assumes a penetrance function and a specific mode of inheritance, and tests whether the recombination fraction between the candidate disease locus and the marker locus is 0.5. Model-free allele-sharing–based linkage analysis is based on testing whether the probability distribution of identity by descent (IBD) among affected sib pairs deviates from the null probability or whether the distribution of the inheritance vector at a putative disease locus deviates from the null distribution under Mendelian segregation among the affected relatives.[46] Incorporating age of onset or covariate data into parametric model–based linkage analysis is easy; it is done by simply introducing age-dependent and covariate-dependent penetrance functions. In the following, I review only the survival analysis methods for allele-sharing–based linkage analysis based on the inheritance vectors.

4.1. *Construction of Genetic Frailties for Sibship*

In order to adequately model the within-family dependency of the age of onset variable to the segregation of genes, the genetic frailties should be defined according to the law of Mendelian segregation. Li[10] gave the following construction of the genetic frailties based on the concept of inheritance vectors.[46,47] Consider a sibship with n sibs — $1, 2, \ldots, n$ — and denote their parents as F for the father and M for the mother. Assuming that the father and mother are unrelated, there are only four unique alleles that are distinct by descent at a given locus. Consider the setting of Kruglyak *et al.*,[46] where we have a series of markers on a chromosomal region that may harbor the disease-causing locus or loci. Suppose d is a point in this test chromosomal region. We are interested in testing whether there is a disease-susceptible (DS) gene linked to locus d. Arbitrarily label the paternal chromosomes containing the locus of interest by (1,2), and label the maternal chromosomes by (3,4). The inheritance vector[46,47] of a sibship at the d locus is the vector

$$V_d = (v_1, v_2, \ldots, v_{2j-1}, v_{2j}, \ldots, v_{2n-1}, v_{2n}),$$

where $v_{2j-1} = 1$ or 2, and $v_{2j} = 3$ or 4 for $j = 1, 2, \ldots, n$. The inheritance vector indicates the parts of the genome at locus d that are transmitted to the n children from the father and the mother.

Li and Zhong[9] defined the additive genetic frailties due to the gene linked to locus d for the father and mother as

$$Z_{dF} = U_{d1} + U_{d2},$$
$$Z_{dM} = U_{d3} + U_{d4},$$

where U_{d1} and U_{d2} are used to represent the genetic frailties due to part of the genome on the two chromosomes of the father at locus d, and U_{d3} and U_{d4} are analogous for the mother. For a given inheritance vector v_d at the d locus for a sibship, we define the frailty for the jth sib as

$$Z_{dj} = U_{dv_{2j-1}} + U_{dv_{2j}}$$

for $j = 1, 2, \ldots, n$. This definition is based on the fact that it is the parts of the genome of the parents that are transmitted to the sibs, and the inheritance vectors indicate which parts are transmitted. We further assume that U_{d1}, U_{d2}, U_{d3}, and U_{d4} are independently and identically distributed across different families as $\Gamma(\nu_d/2, \eta)$, where the parameter η is the inverse scale parameter and ν_d is the shape parameter. Then, Z_{dj} is distributed as $\Gamma(\nu_d, \eta)$ for $j = 1, 2, \ldots, n$.

Taking into account possible genetic contributions to the disease that are not due to the single disease locus linked to d, e.g. those that are due to loci unlinked to locus d or contributions to shared familial effects, we add another random frailty term U_p to the genetic frailty, and define the genetic frailty for the jth sib as

$$Z_j = Z_{dj} + U_p$$
$$= U_{dv_{2j-1}} + U_{dv_{2j}} + U_p.$$

We assume that U_p is distributed as $\Gamma(\nu_p, \eta)$ over different sibships. Then, Z_j follows a $\Gamma(\nu_d + \nu_p, \eta)$ distribution. It is easy to verify that both the conditional (on V_d) and the marginal means of the frailties are

$$E(Z_1) = E(Z_2) = \cdots = E(Z_n) = \frac{\nu_d + \nu_p}{\eta},$$

and that both the conditional and the marginal variances of the frailties are

$$\mathrm{Var}(Z_1) = \mathrm{Var}(Z_2) = \cdots = \mathrm{Var}(Z_n) = \frac{\nu_d + \nu_p}{\eta^2}.$$

So, the parameter ν_d can be interpreted as the proportion of the variance of the genetic frailty that can be explained by the gene linked to the locus d.

The frailties for a sibship can be written into a matrix form as

$$Z = HU, \tag{6}$$

where

$$Z = \{Z_1, Z_2, \ldots, Z_n\}',$$

$$H = \begin{pmatrix} a_{11} \ a_{12} \ a_{13} \ a_{14} \ 1 \\ \vdots \\ a_{n1} \ a_{n2} \ a_{n3} \ a_{n4} \ 1 \end{pmatrix},$$

$$U = \{U_{d1}, U_{d2}, U_{d3}, U_{d4}, U_p\}',$$

where $a_{j1} = I(v_{2j-1} = 1), a_{j2} = I(v_{2j-1} = 2), a_{j3} = I(v_{2j} = 3)$, and $a_{j4} = I(v_{2j} = 4)$ for $j = 1, 2, \ldots, n$, where $I(.)$ is the indicator function.

4.2. *The Additive Genetic Gamma Frailty Model for Sibship Data*

Consider a sibship with n sibs. Let T_j be the random variable of age at disease onset for the jth sib. Let (t_j, δ_j) be the observed data, where t_j is the observed age at onset if $\delta_j = 1$ and age at censoring if $\delta_j = 0$. We assume that the hazard function of the developing disease for the jth individual at age t_j is modeled by the proportional hazards model with random effect Z_j:

$$\lambda_j(t_j|Z_j) = \lambda_0(t)\exp(X_j'\beta)Z_j, \text{ for } j = 1, 2, \ldots, n, \qquad (7)$$

where $\lambda_0(t)$ is the unspecified baseline hazard function, X_j is a vector of observed covariates for the jth sib, and β is a vector of regression parameters associated with the covariates. Z_j is the unobserved genetic frailty constructed by Eq. (6) in the previous section. Since $Z_1, Z_2, \ldots, Z_n$ are dependent due to gene segregation and shared frailty, $T_1, T_2, \ldots, T_n$ are therefore dependent. Finally, to make the baseline hazard $\lambda_0(t)$ identifiable, we let $\nu_d + \nu_p = \eta$, which sets $E(Z_j) = 1$ for $j = 1, 2, \ldots, n$ and prevents arbitrary scaling in the proportional hazards model [Eq. (7)]. Under this restriction, there are two free parameters, ν_d and ν_p; $Z_{dj} \sim \Gamma(\nu_d, \nu_d + \nu_p)$ and $Z_p \sim \Gamma(\nu_p, \nu_d + \nu_p)$. We may also consider reparameterization in terms of the two frailty variances, $\sigma_d = \text{Var}(U_{dj}) = \nu_d/\eta^2$ and $\sigma_p = \text{Var}(U_p) = \nu_p/\eta^2$. Let $\sigma_{dp} = \sigma_d + \sigma_p$ denote the variance of Z_j. We then have $Z_{dj} \sim \Gamma(\sigma_d\sigma_{dp}^{-2}, \sigma_{dp}^{-1})$ and $Z_p \sim \Gamma(\sigma_p\sigma_{dp}^{-2}, \sigma_{dp}^{-1})$.

Based on this additive genetic gamma frailty model, Li and Zhong[9] derived the joint survival function of age of onset data within a family as a function of the baseline hazard function, the covariate effects, and the parameters related to the frailty. In addition, the null hypothesis that the disease locus is not linked to the candidate locus d can be reformulated as

$$H_0 : \nu_a = 0,$$

which is equivalent to assuming that the additive variance due to the gene linked to locus d is zero. In order to test this hypothesis, an estimate of the baseline hazard function is often required. However, the data collected for linkage analysis, such as affected sib pairs or affected relatives, do not often provide enough information for estimating such population-level baseline hazard functions. Instead, Li and Zhong[9] developed a retrospective likelihood-ratio test for this null hypothesis based on the assumption that the baseline hazard function can be estimated from external data such as the SEER database for various types of cancers.

Zhong and Li[48] further extended the additive genetic gamma frailty model to simultaneously consider the linkage to two unlinked loci, and demonstrated that simultaneously searching for two loci can result in increased power to detect linkage when the disease risk is affected by two unlinked genes. Instead of assuming gamma frailty models, one can also assume a multivariate log-normal frailty, where the variance–covariance matrix is specified by kinship coefficients and pairwise IBD-sharing proportions.[49] Furthermore, Pankratz *et al.*[49] proposed a procedure using Laplace approximation for estimating model parameters and testing linkages, i.e. testing whether the additive variance due to a given locus is zero.

5. SURVIVAL ANALYSIS METHODS FOR FAMILY-BASED GENETIC ASSOCIATION ANALYSIS

Association studies look for specific alleles at a marker locus that are more frequent in affected individuals (cases) than in the unaffected population (controls). Population-based studies compare allele frequencies in cases and controls, but this methodology has been criticized as being prone to false-positives due to population admixture. To eliminate the effect of disequilibrium created by population stratification, thereby eliminating the false-positive mapping results, family-based association methods such as haplotype relative risk,[50] the transmission disequilibrium test (TDT),[13,14] and a likelihood-based method[51,52] using affected and family-based controls are often used. Li and Hsu[12] demonstrated the importance of incorporating age of onset data into family-based genetic association analysis.

5.1. *Survival Analysis Methods for Family-based Association Tests*

There are several approaches that extend the TDT to handle age of onset or age at censoring. Li and Fan[53] proposed a linkage disequilibrium–based Cox (LDCox) model for nuclear family data, and used a robust Wald's test for association. Mokliatchouk *et al.*[54] and Shih and Whittemore[55] developed likelihood-based score statistics to test for association between a disease and a genetic marker. The score statistic can be written as a weighted sum over family members of their observed minus expected genotypes. Age of onset data can be used in the weight, which is the difference between the observed and expected values, $\delta_i - \Lambda_0(t_i)$, for individual i, where $\Lambda_0(t_i)$ is the cumulative hazard function at age t_i, which is assumed to be known from external data sources. Both methods of Li and Fan[53] as well as Shih and Whittemore[55] assume that the genetic effects on the risk of onset are proportional in the framework of the Cox regression model. Jiang *et al.*[56] developed a family-based association test for time-to-onset data, assuming time-dependent differences between the hazard functions among different genotype groups by using the weighted log-rank approach of Flemming and Harrington.[57]

5.2. *Test of Genetic Association in the Presence of Linkage*

It is well known that genetic linkage induces within-family association of phenotypes such as disease onset or age at disease onset. A limitation of most family-based association tests is that, although they remain valid tests of linkage, they are not valid tests of association if related nuclear families or sibships from larger pedigrees are used. The allele-sharing–based linkage analysis only considers allele sharing by descent pattern among the sibs within a sibship; however, it does not differentiate which allele they share as long as they share it by descent. In other words, linkage analysis does not consider which particular allele is shared by the sibs. On the other hand, the association that we are interested in is the association attributable to LD. For association analysis and LD analysis, the particular allele that an individual carries determines his or her risk of developing a disease, since different marker alleles have different coupling frequencies with the disease variant if LD exists. In typical tests of association, it is very rare that the genetic marker itself is the disease-susceptible locus (DSL). When the marker locus is not the DSL but is in LD associated with it, all sibling resemblance or lack of resemblance and within-sibship correlation of age of onset cannot be fully accounted for by the genotypes at the marker locus.

Motivated by this key difference between linkage and LD, Zhong and Li[8] defined a joint model for the risk of disease to account for both the allele-sharing information and the genotype information at the candidate marker locus by including the genetic frailties derived from the inheritance vector. Specifically, consider a candidate marker d in the linked region, and let $g = (g_1, \ldots, g_n)$ denote the vector of genotypes at marker locus d of n sibs of known age at onset or censoring. Zhong and Li[8] assumed that the hazard function of developing disease for the jth individual at age t_j is modeled by the proportional hazards model with random effect Z_j:

$$\lambda_j(t_j | Z_j) = \lambda_0(t_j) \exp(X_{g_j}\beta) Z_j, \text{ for } j = 1, 2, \ldots, n, \tag{8}$$

where $\lambda_0(t)$ is the unspecified baseline hazard function and X_{g_j} denotes some function of the jth offspring's marker genotype in the family. For example, for an additive model, $X_{g_j} = l$, $l = 0, 1, 2$, which counts the number of putative high-risk marker alleles and corresponds to the genotype of the jth member of the family who carries l copies of the putative high-risk marker allele. Z_j is the unobserved genetic frailty, which is defined as in Eq. (6).

When $\beta = 0$, the hazard function and the joint survival function for a sibship do not depend on the genotype at the marker locus d. Therefore, tests of allelic association between locus d and the disease, or the null hypothesis that the genotype at the marker locus is not associated with the risk of disease, can be formulated as testing

$$H_0 : \beta = 0.$$

Zhong and Li[8] developed a score test for H_0 based on a retrospective likelihood function, which is a weighted sum over family members of their observed minus expected genotypes, where weights depend on both age of onset and IBD sharing between the sibs within a family. Different from score tests for linkage and association, the score test for testing association in the presence of linkage is also a function of IBD allele sharing among the sib pairs or the inheritance distribution among the sibships. Zhong and Li[8] demonstrated by simulations that such a score test indeed results in a correct type 1 error rate when testing for association in the linked region.

6. SURVIVAL MODELS FOR HAPLOTYPE EFFECTS BASED ON COHORT, CASE-COHORT, AND NESTED CASE-CONTROL DESIGNS

The most commonly used design for population-based haplotype analysis is the case-control design. Although case-control studies can potentially identify disease-predisposing variants, such studies have certain limitations. These include the tendency for clinically diagnosed cases to represent more severe ends of the whole disease spectrum and the difficulty in selecting unbiased controls. In addition, such designs may suffer recall bias in disease status and other covariates such as family history.[58,59] In contrast, large-scale population-based cohort studies can overcome these limitations. The prospective population cohorts can also enable the study of many complex diseases in the same cohort. Large cohort studies are designed to study the gene and environmental effects for relatively rare diseases. However, since many of the environmental covariates of interest are expensive to obtain, in order to reduce the cost of large cohort studies, several alternative sampling schemes within the framework of cohort studies have been suggested, well studied, and widely applied in the traditional epidemiological investigations. Among these, the most popular ones are the case-cohort design proposed by Prentice[60] and the nested case-control design proposed by Thomas[61] and Liddell *et al.*[62]

I review some recent methods for haplotype association analysis for cohort data, case-cohort data, and nested case-control data. In order to account for variable age of onset, survival analysis methods are required to test the haplotype effects for cohort, case-cohort, and nested case-control designs.

6.1. *Survival Model for Haplotype Inference Based on Cohort Data*

Assume that N individuals are collected from a cohort and are typed over K SNP markers. Consider the proportional hazards model to relate the disease risk to haplotype. Specifically, for the ith individual in the cohort, we assume the Cox proportional hazards model

$$\lambda(t_i|X_i, H_i) = \lambda_0(t_i) \exp[\beta' \mathcal{F}(X_i, H_i)] \tag{9}$$

to relate the hazard function to the covariate vector X_i and the haplotype H_i, where $\mathcal{F}(X_i, H_i)$ is a known function to parameterize the covariates and the haplotype. Here, the haplotype H_i can be over a set of SNPs in a candidate gene or in a sliding window in a whole-genome study. Depending on the model we choose, there are many different ways to parameterize the function $\mathcal{F}(X_i, H_i)$. For example, if h_0 is a particular haplotype of interest, we can assume the following multiplicative model with haplotype and covariate interaction:

$$\mathcal{F}(X_i, H_i) = \beta_1 h(l, m, h_0) + \beta_2 X_i + \beta_3 X_i h(l, m, h_0),$$

where $h(l, m, h_0) = I(h_l = h_0) + I(h_m = h_0)$, and (h_l, h_m) are the haplotype pair of H_i.

Lin [63] proposed a likelihood-based approach and an EM algorithm for estimating the parameter β and for haplotype inference for the Cox proportional hazards model [Eq. (9)] in full cohort studies of unrelated individuals. Chen *et al.*[64] derived a score test based on the partial likelihood function for testing the null hypothesis $H_0 : \beta = 0$, which is much easier to implement than the likelihood-based approach of Lin.[63] However, although the method of Lin provides an estimate of the haplotype risk ratio parameters and the baseline hazard function, the method may suffer computational instability due to many possible rare haplotypes.

6.2. *Test of Haplotype Association for Case-Cohort and Nested Case-Control Designs*

Liddell *et al.*[62] and Thomas[61] suggested an alternative design called a nested case-control design, in which a cohort is followed to identify cases of some disease of interest; controls are then selected for each case from within the cohort (i.e. controls are a random sample of unaffected individuals from the risk set in the cohort at the event time). Cases and controls can be matched on some covariates. In such a design, the covariates of interest are only measured for the cases and controls. For nested case-control data, Chen *et al.*[64] developed a score test for $H_0 : \beta = 0$ in the Cox proportional hazards model [Eq. (9)]. Alternatively, if the disease onset information is available for the full cohort, one can develop an EM algorithm to obtain the NPMLEs for the parameters associated with the model [Eq. (9)]. An alternative design to the nested case-control design is the case-cohort design, as proposed by Prentice[60] for large survey studies such as the Women's Health Study, where the population size makes it unfeasible to collect data on all individuals in the cohort. If there is a concurrent registry that can be used to identify all of the subjects who experience an event, then it is possible to collect covariate data on only a subcohort of the subjects, randomly sampled from the population at large, and (perhaps at a later date) on those subjects who experience an event. The subcohort in a given stratum constitutes the comparison set of cases occurring at a range of failure times.[60]

Since detailed procedures for haplotype analysis for case-cohort data have not been seen in the literature, I provide some details on estimating the parameters under the case-cohort setting for the haplotype/disease risk model [Eq. (9)]. For a case-cohort design, the data for individuals in R^+ (including case set R_1 and controls in the subcohort) are $D_i = (t_i, \delta_i, M_i, Z_i)$. However, the haplotype H_i may not be known for all individuals in R^+. Let $S(M_i)$ be the set of haplotype pairs consistent with genotype M_i. For individuals in R^- (those in the cohort, but not in R^+), we only observe $D_i = (t_i, \delta_i)$; and for these individuals, let $S(M_i)$ be the set of all possible haplotypes. Denote $D = \{D_1, \ldots, D_N\}$ as the observed data. N is the number of individuals in the full cohort.

Let $f(Z)$ be the marginal distribution of the covariate Z in the population, and $G(t) = Pr(T > t)$. The likelihood function of the observed data is given by

$$L(\theta) = \prod_{i \in R^+} \left\{ \sum_{l,m} I[H_i(l,m) \in S(M_i)][\lambda_0(t_i)e^{\beta' \mathcal{F}[Z_i, H_i(l,m)]}]^{\delta_i} \right.$$
$$\left. \times \exp[-\Lambda_0(t_i)e^{\beta' \mathcal{F}[Z_i, H_i(l,m)]}]\pi_l \pi_m \right\} \times f(Z_i) \prod_{i \in R^-} G(t_i), \qquad (10)$$

where π_l and π_m are the haplotype frequencies of the haplotypes h_l and h_m, respectively; $H_i(l,m)$ represents the two haplotypes h_l and h_m for the ith individual; and $\theta = \{\Lambda_0(t), f(Z), \pi_l, \pi_m, \beta\}$. Note that here we assume that the Hardy–Weinberg equilibrium holds for the haplotypes, although this can be relaxed by introducing additional parameters. Instead of assuming a particular distribution of the covariates, we propose to deal with the distribution of Z nonparametrically, as in Wellner and Zhan[65] and in Scheike and Juul.[66]

Since there are two nonparametric terms in this likelihood function, it is difficult to maximize it directly. We can develop an EM algorithm instead. To write down the full-data likelihood, we define W_j as the observed jth distinct combination among the set $\{Z_i : i \in R^+\}$ for $j \in j = 1, \ldots, J$, and the corresponding point mass as $p_1, \ldots, p_J$ such that $\sum_j p_j = 1$. Then, the missing data are the phases of the haplotypes for some individuals in R^+ and both the haplotypes and covariates for individuals in R^-. The corresponding log full-data likelihood is

$$l(\theta) = \sum_{i=1}^{N} \{\delta_i [\log \lambda_0(t_i) + \beta' \mathcal{F}(Z_i, H_i)] - \Lambda_0(t_i) e^{\beta' \mathcal{F}(Z_i, H_i)}\}$$

$$+ \sum_{j=1}^{J} \sum_{i=1}^{N} I(Z_i = W_j) \log p_j + \sum_{l,m} \sum_{i=1}^{N} I[H_i = (h_l, h_m)] \log(\pi_l \pi_m).$$

$$(11)$$

E-Step: To implement the EM algorithm, we need to obtain the expectation of Eq. (11), which requires the following expectations. First, for an individual $i \in R^+$, (M_i, Z_i) are known, but H_i may not be known. We have

$$E\{I[H_i = (h_l, h_m)] | D_i\}$$

$$= I[H_i(l, m) \in S(M_i)] \exp\{\delta_i(\beta' \mathcal{F}[Z_i, H_i(l, m)])$$

$$- \Lambda_0(t_i) e^{\beta' \mathcal{F}[Z_i, H_i(l,m)]}\} \pi_l \pi_m \Big/ \sum_{H_i(l', m') \in S(M_i)}$$

$$\times \exp\{\delta_i(\beta' \mathcal{F}[Z_i, H_i(l', m')]) - \Lambda_0(t_i) e^{\beta' \mathcal{F}[Z_i, H_i(l', m')]}\} \pi_{l'} \pi_{m'},$$

and with this probability, $E[\mathcal{F}(Z_i, H_i)]$ and $E\{\exp[\mathcal{F}(Z_i, H_i)]\}$ can be derived. For an individual $i \in R^-$, we only observe $(t_i, \delta_i = 0)$:

$$E\{I[Z_i = w_j, H_i = (h_l, h_m)] | T_i > t_i\}$$

$$= \frac{p_j \exp\{-\Lambda_0(t_i) e^{\beta' \mathcal{F}[W_j, H_i(l,m)]}\} \pi_l \pi_m}{\sum_{j'=1}^{J} p'_j \sum_{H_{i'}(l,m)} \exp\{-\Lambda_0(t_i) e^{\beta' \mathcal{F}[W_{j'}, H_{i'}(l'm')]}\} \pi_{l'} \pi_{m'}},$$

and with this probability, $E[\mathcal{F}(Z_i, H_i)]$ and $E\{\exp[\mathcal{F}(Z_i, H_i)]\}$ can be derived.

M-Step: It is easy to see that the EM equations in the M-step are

$$\hat{p}_j = \frac{\sum_{i=1}^{N} E[I(Z_i = W_j)|D]}{N}, \text{ for } j = 1, \ldots, J,$$

$$\hat{\pi}_l = \frac{\sum_{i=1}^{N} \sum_{j=1}^{J} E[I(Z_i = W_j)|D] \sum_{H_i(l,m) \in S(M_i)} E\{I[H_i = (h_l, h_m)]|D\}}{2N},$$

$$\hat{\Lambda}_0(t) = \sum_{i=1}^{N} \frac{I(t_i \le t)\delta_i}{\sum_{j \in Y(t_i)} \sum_{j'=1}^{J} E(j, j') \sum_{H_j(l,m)} E(j, l, m) e^{\beta' \mathcal{F}[W_{j'}, H_j(l,m)]}},$$

where

$$E(j, j') = E[I(Z_j = W_{j'})|D]$$
$$E(j, l, m) = E[H_j = (h_l, h_m)|D]$$

and $Y(t_i)$ is the set of individuals who are at risk at time t_i. Finally, the estimator of β is the root of the estimating function

$$U(\beta)$$

$$= \sum_{i=1}^{N} \delta_i \sum_{j=1}^{J} E[I(Z_i = W_j)|D] \sum_{l,m} E\{I[H_i = (h_l, h_m)]|D\}\mathcal{F}(Z_i, H_i)^T$$

$$- \frac{\sum_{j \in Y(t_i)} \sum_{j'=1}^{J} E[I(Z_j = W_{j'})|D] \sum_{H_j(l,m)} E(j, j', l, m)\mathcal{F}(Z_j, H_j)^T}{\sum_{j \in Y(t_i)} \sum_{j'=1}^{J} E[I(Z_j = W_{j'})|D] \sum_{H_j(l,m)} E(j, j', l, m)},$$

where

$$E(j, j', l, m) = E[H_j = (h_l, h_m)|D]e^{\beta' \mathcal{F}[W_{j'}, H_j(l,m)]}.$$

This is the score equation corresponding to a Cox model with an individual-specific offset term, which can be easily solved by using the Newton–Ralphson iteration.

Based on different ways of parameterizing the haplotype effects, the test of haplotype effects can generally be formulated as testing $H_0 : \beta_1 = 0$, where β_1 is a subvector of $\beta = \{\beta_1, \beta_2\}$. Similar to Lin[63] and to Scheike and Juul,[66] the likelihood-ratio test can be applied for this null hypothesis.

7. SURVIVAL ANALYSIS METHODS FOR GENE CHARACTERIZATION

After the genetic variants related to the risk of disease are identified, it is important to estimate the penetrance of the variants and other population-based parameters such as the allele frequencies. Cohort or case-control family designs can be used for gene characterization and for estimation of population parameters such as genotype relative risk and age-dependent penetrance functions. For rare diseases, a large cohort is often required to estimate such population parameters. For case-control family designs, if the genotypes of the disease variants are available for all of the family members, the methods by Li *et al.*[38] and Shih and Chatterjee[40] can be used to estimate the age-dependent penetrance functions.

When genotypes of the family members are not available, the kin-cohort design[67] is a promising alternative to traditional cohort or case-control family designs for estimating the penetrance of an identified rare autosomal mutation. In such a design, a suitably selected sample of participants provides genotype and detailed family history information on the disease of interest; however, the genotypes of the family members are not known. Gail *et al.*[68] used the term "genotyped probands" to emphasize that the probands are genotyped in the kin-cohort design. To estimate penetrance of the mutation, Chatterjee and Wacholder[69] considered a marginal likelihood approach that is computationally simple to implement, more flexible than the original analytic approach proposed by Wacholder *et al,*[67] and more robust than the likelihood approach considered by Gail *et al.*[68] to the presence of residual familial correlation. Chatterjee *et al.*[70] further extended the approach of Shih and Chatterjee[40] to data from the kin-cohort design with both case and control probands as well as the kin-cohort design with only cases in order to account for residual correlations. In order to allow for residual familial aggregation–given genotypes, Chatterjee *et al.*[70] considered a copula model[71] to specify joint risks of the disease among the proband and family members. The key to these various approaches is to make an inference based on the likelihood function that corrects for ascertainment.

8. ASCERTAINMENT CORRECTION

Different from traditional multivariate survival analysis, one of the most difficult problems in analyzing family data in genetic studies is that the families for genetic analysis are often not random samples from the population; rather, they are often ascertained because one or more of the family members are affected with the disease of interest. This ascertainment problem makes statistical inferences for the proposed models in this paper difficult. For the ascertained family samples, estimating the population baseline hazard function becomes even more difficult.

One way to go around this problem is by using a retrospective likelihood, which is defined as the probability of marker data given the observed age of onset data. In order to maximize such a likelihood function, the baseline hazard function is often assumed to be known or to follow some parametric form. Due to conditioning, one may expect some loss of efficiency in parameter estimates. Sun and Li[72] recently proposed and evaluated two approaches based on conditional prospective likelihood and conditional ascertainment-corrected likelihood for the additive genetic gamma frailty model in order to estimate the baseline hazard function based on the family data collected for linkage analysis. However, such an ascertainment correction procedure requires knowledge of the population distribution of the family structures and family sizes, which can be difficult to obtain.

9. SURVIVAL ANALYSIS METHODS IN THE GENOMICS ERA

The recent development of new high-throughput technologies for generating very high-dimensional genomic data such as microarray gene expression data raises other important and interesting problems that require the development of new survival analysis methods. One such area is to link the microarray gene expression data to censored survival outcomes such as cancer recurrence. Due to the high dimensionality of the data, traditional survival analysis methods either cannot be applied directly to such data sets or are expected to perform poorly.

Currently, there are several classes of approaches for these types of censored data regression problems in the high-dimension and low-sample-size settings. One class of approaches is dimension-reduction–based methods, such as the extension of the partial least squares regression method to censored data regression problems,[73,74] the extension of the slice inverse regression method,[76] and supervised principal component analysis.[75] While these methods may perform well in prediction, they usually do not provide a direct way of selecting genes that are potentially related to time-to-event.

Another class of approaches is based on regularized estimation procedures such as L_2 penalized estimation,[77] the extension of the least absolute shrinkage and selection operator (Lasso) of Tibshirani[78] to censored survival data using the least angle regression (LARS),[79-81] and the threshold gradient descent procedure.[82,83] These methods provide a way of selecting genes whose expression might be related to clinical outcomes such as time-to-event. In addition, these methods can be used to build a model for predicting future patients' time-to-event.

Survival ensembles, based on extensions of the random forests[84] and the gradient descent boosting procedure[85] to censored survival data, have also been recently developed.[86,87] These procedures are more flexible, and usually perform better in predicting future patients' time-to-event.

10. CONCLUSION AND FUTURE DIRECTIONS

Since many complex diseases show a large variation in age at disease onset, consideration of the variable age of disease onset is an important aspect of the genetic analysis of complex diseases. Methods in survival analysis provide a natural framework for incorporating the age of onset and environmental risk factors into genetic analysis. In this paper, I have reviewed some recently developed survival analysis methods for aggregation, segregation, linkage and association analysis, and gene characterization analysis in genetic epidemiology. Most of these methods were developed in the last 10 years and have been shown to be able to offer additional insights into the genetic study of complex diseases. As user-friendly software packages implementing these methods become available, we should expect to see more applications of these methods in mapping genes for complex diseases.

With the completion of the Human Genome Project and the HapMap Project, genomewide association studies of complex traits are now possible and have already been proposed for several complex diseases. Under such studies, hundreds of thousands of SNPs are typed for a large set of patients and controls. In addition, large-scale cohort studies are under discussion or are already underway in the UK (UK Biobank), Iceland (Decode), Germany, Canada, and Japan. The US is also considering to propose its own large-scale population cohort.[59] We therefore expect that large amounts of data will be generated from these large cohort studies in the near future. Besides large cohort data, case-cohort and nested case-control designs offer alternatives to cohort and case-control designs.

An important research question is how to identify SNPs, SNP–SNP interactions, gene–gene interactions, and gene–environment interactions among the hundreds of thousands of SNPs that may affect the disease risk based on case-cohort or nested case-control data in the framework of survival analysis. In addition, many common diseases are known to be affected by certain genotype combinations; therefore, statistical methods to detect the influential genes along with their interaction structures are also required. Finally, new statistical methods are also required in order to fully utilize the genomewide linkage disequilibrium patterns and the haplotype block structures available from the HapMap Project. New ideas from statistical learning[88] hold great promise in addressing these important issues.

Since genes and proteins almost never work alone, they interact with each other and with other molecules in highly structured but incredibly complex ways. Understanding this interplay of the human genome and environmental influences is crucial to developing a systemic understanding of human health and disease. An important venue for future research is to develop methods that can incorporate known biological knowledge such as pathways into statistical modeling in order to limit the search space for gene–gene and gene–environment interactions.[89,90] Wei and Li[90] proposed a nonparametric pathway-based regression model to incorporate pathway information into regression analysis. As biological knowledge accumulates, one should expect to see the development of new methods and more applications of these models in identifying genes and environmental risk factors that are related to the risk of developing disease.

ACKNOWLEDGMENTS

This research was supported by NIH grant ES009911. I thank Mr Edmund Weisberg, MS, at Penn CCEB for his editorial assistance.

REFERENCES

1. Claus EB, Risch NJ, Thompson WD. (1990) Using age of onset to distinguish between subforms of breast cancer. *Ann Hum Genet* **54**:169–177.
2. Hall JM, Lee MK, Newman B, *et al.* (1990) Linkage of early-onset familial breast cancer to chromosome 17q21. *Science* **250**:1684–1689.
3. Carter BS, Beaty HB, Steinberg GD, *et al.* (1992) Mendelian inheritance of familial prostate cancer. *Proc Natl Acad Sci USA* **89**:3367–3371.
4. Li H, Thompson EA. (1997) Semiparametric estimation of major gene and random familial effects for age of onset. *Biometrics* **53**:282–293.
5. Li H, Thompson EA, Wijsman EA. (1998) Semiparametric estimation of major gene effects for age of onset. *Genet Epidemiol* **15**:279–298.
6. Morton LA, Kidd KK. (1980) The effects of variable age-of-onset and diagnostic criteria on the estimates of linkage: an example using manic-depressive illness and color blindness. *Soc Biol* **27**:1–10.
7. Haynes C, Pericak-Vance MA, Dawson D. (1986) Genetic analysis workshop IV, analysis of Huntington's disease linkage and age of onset curves. *Genet Epidemiol* **1**:109–122.

8. Zhong X, Li H. (2004) Score tests of genetic association in the presence of linkage based on the additive genetic gamma frailty model. *Biostatistics* **5**:307–327.

9. Li H, Zhong X. (2002) Multivariate survival models induced by genetic frailties, with application to linkage analysis. *Biostatistics* **3**:57–75.

10. Li H. (2002) An additive genetic gamma frailty model for linkage analysis of diseases with variable age of onset using nuclear families. *Lifetime Data Anal* **8**:315–334.

11. Hsu L, Li H, Houwing J. (2002) A method for incorporating ages at onset in affected sib-pair linkage studies. *Hum Hered* **54**:1–12.

12. Li H, Hsu L. (2000) Effects of ages at onset on the power of the affected sib pair and transmission/disequilibrium tests. *Ann Hum Genet* **64**:239–254.

13. Spielman RS, Ewens WJ. (1996) The TDT and other family-based tests for linkage disequilibrium and association. *Am J Hum Genet* **59**:983–989.

14. Spielman RS, Ewens WJ. (1998) A sibship test for linkage in the presence of association: the sib transmission/disequilibrium test. *Am J Hum Genet* **62**:450–458.

15. Andrieu N, Demenais F. (1997) Interaction between genetic and reproductive factors in breast cancer risk in a French family sample. *Am J Hum Genet* **61**:678–690.

16. Brennan P, Ollier B, Worthington J, *et al.* (1996) Are both genetic and reproductive associations with rheumatoid arthritis linked to prolactin? *Lancet* **348**:106–109.

17. Ottman R. (1990) An epidemiologic approach to gene–environment interaction. *Genet Epidemiol* **11**:75–86.

18. Tiret L, Rigat B, Visvikis S, *et al.* (1992) Evidence, from combined segregation and linkage analysis, that a variant of the angiotensin I–converting enzyme (ACE) gene controls plasma ACE levels. *Am J Hum Genet* **51**:197–205.

19. Towne B, Siervogel RM, Blangero J. (1997) Effects of genotype-by-sex interaction on quantitative trait linkage analysis. *Genet Epidemiol* **14**:1053–1058.

20. Guo SW. (2000) Gene–environment interactions and the affected-sib-pair designs. *Hum Hered* **50**:271–285.

21. Guo SW. (2000) Gene–environment interaction and the mapping of complex traits: some statistical models and their implications. *Hum Hered* **50**:286–303.

22. Clayton DG. (1978) A model for association in bivariate life tables and its application in epidemiological studies of familial tendency in chronic disease incidence. *Biometrika* **65**:141–151.

23. Oakes D. (1982) A model for association in bivariate survival data. *J R Stat Soc Ser B* **44**:414–422.

24. Clayton DG, Cuzick J. (1985) Multivariate generalizations of the proportional hazards model. *J R Stat Assoc Ser A* **148**:82–117.

25. Self SG, Prentice RL. (1986) Incorporating random effects into multivariate relative risk regression models. In: *Modern Statistical Methods in Chronic*

Disease Epidemiology, Moolgavkar SH, Prentice RL (eds.), John Wiley & Sons, New York, NY, pp. 167–178.

26. Klein JP. (1992) Semiparametric estimation of random effects using the Cox model based on the EM algorithm. *Biometrics* **48**:795–806.

27. McGilchrist CA. (1993) REML estimation for survival models with frailty. *Biometrics* **47**:461–466.

28. Murphy SA. (1994) Consistency in a proportional hazards model incorporating random effects. *Ann Stat* **22**:712–731.

29. Nielsen GG, Gill RD, Andersen PK, Sorensen TIA. (1992) A counting process approach to maximum likelihood estimation in frailty models. *Scand J Stat* **19**:25–43.

30. Gliden DV. (1999) Checking the adequacy of the gamma frailty model for multivariate failure times. *Biometrika* **86**:381–393.

31. Houggaard P. (1995) Frailty models for survival data. *Lifetime Data Anal* **1**:255–273.

32. Glidden DV, Self SG. (1999) Semiparametric likelihood estimation in the Clayton–Oakes model. *Scand J Stat* **26**:363–372.

33. Petersen JH. (1998) An additive frailty model for correlated life times. *Biometrics* **54**:646–661.

34. Parner E. (1998) Asymptotic theory for the correlated gamma-frailty model. *Ann Stat* **26**:183–214.

35. Korsgaard IR, Anderson AH. (1998) The additive genetic gamma frailty model. *Scand J Stat* **25**:255–269.

36. Ripatti S, Larsen K, Palmgren J. (2002) Maximum likelihood inference for multivariate frailty models using a Monte Carlo EM algorithm. *Lifetime Data Anal* **8**:349–360.

37. Ripatti S, Palmgren J. (2000) Estimation of multivariate frailty models using penalized partial likelihood. *Biometrics* **56**:1016–1022.

38. Li H, Yang P, Schwartz AG. (1998) Analysis of age of onset data from case-control family studies. *Biometrics* **54**:1030–1039.

39. Prentice R, Breslow N. (1978) Retrospective studies and failure time models. *Biometrika* **65**:153–158.

40. Shih JH, Chatterjee N. (2002) Analysis of survival data from case-control family studies. *Biometrics* **58**:502–509.

41. Hsu L, Chen L, Gorfine M, Malone K. (2004) Semiparametric estimation of marginal hazard function from case-control family studies. *Biometrics* **60**:936–944.

42. Gauderman WJ, Thomas D. (1994) Censored survival models for genetic epidemiology: a Gibbs sampling approach. *Genet Epidemiol* **11**:171–188.

43. Siegmund K, McKnight B. (1998) Modeling hazard functions in families. *Genet Epidemiol* **15**(2):147–171.

44. Chang IS, Hsiung CA, Wang MC, Wen CC. (2005) An asymptotic theory for the nonparametric maximum likelihood estimator in the Cox gene model. *Bernoulli* **11**(5):863–892.

45. Chang IS, Wen CC, Wu YJ, Yang CC. (2007) Fast algorithm for the nonparametric maximum likelihood estimate in the Cox-gene model. *Statistica Sinica* (in press).

46. Kruglyak L, Daly MJ, Reeve-Daly MP, Lander ES. (1996) Parametric and nonparametric linkage analysis: a unified multipoint approach. *Am J Hum Genet* **8**:1347–1363.

47. Lander E, Green P. (1987) Construction of multilocus genetic maps in humans. *Proc Natl Acad Sci USA* **84**:2363–2367.

48. Zhong X, Li H. (2002) An additive genetic gamma frailty model for two-locus linkage analysis using sibship age of onset data. *Stat Appl Genet Mol Biol* **1**:article 2.

49. Pankratz VS, de Andrade M, Therneau TM. (2005) Random-effects Cox proportional hazards model: general variance components methods for time-to-event data. *Genet Epidemiol* **28**(2):97–109.

50. Falk CT, Rubinstein P. (1987) Haplotype relative risks: an easy reliable way to construct a proper control sample for risk calculations. *Ann Hum Genet* **51**:227–233.

51. Schaid DJ. (1996) General score tests for associations of genetic markers with disease using cases and their parents. *Genet Epidemiol* **13**:423–449.

52. Schaid DJ, Li H. (1997) Genotype relative-risks and association tests for nuclear families with missing parental data. *Genet Epidemiol* **14**:1113–1118.

53. Li H, Fan JJ. (2000) A general test of association for complex diseases with variable age of onset. *Genet Epidemiol* **19**:S43–S49.

54. Mokliatchouk O, Blacker D, Rabinowitz D. (2001) Association tests for traits with variable age at onset. *Hum Hered* **51**:46–53.

55. Shih MC, Whittemore AS. (2002) Tests for genetic association using family data. *Genet Epidemiol* **22**:128–145.

56. Jiang H, Harrington D, Raby BA, *et al.* (2006) Family-based association test for time-to-event data with time-dependent difference between hazard functions. *Genet Epidemiol* **30**:124–132.

57. Flemming TR, Harrington DP. (1981) A class of hypothesis tests for one and two sample censored survival data. *Commun Stat Theory Methods* **10**:763–794.

58. Doll R. (1964) Retrospective and prospective studies. In: *Medical Surveys and Clinical Trials*, 2nd ed., Witts LJ (ed.), Oxford University Press, London, England, pp. 71–98.

59. Collins FS. (2004) The case for a US prospective cohort study of genes and environment. *Nature* **429**:475–477.

60. Prentice RL. (1986) A case-cohort design for epidemiologic cohort studies and disease prevention trials. *Biometrika* **73**:1–11.

61. Thomas DC. (1977) Addendum to: methods of cohort analysis: appraisal by application to asbestos mining, by Liddell FDK, McDonald JC, Thomas DC. *J R Stat Soc* **4**:469–491.

62. Liddell FDK, McDonald JC, Thomas DC. (1977) Methods of cohort analysis: appraisal by application to asbestos mining. *J R Stat Soc* **4**:469–491.

63. Lin DY. (2004) Haplotype-based association analysis in cohort studies of unrelated individuals. *Genet Epidemiol* **26**:255–265.

64. Chen J, Peters U, Foster C, Chatterjee N. (2004) Haplotype-based test of genetic association using data from cohort and nested case-control epidemiologic studies. *Hum Hered* **58**:18–29.

65. Wellner JA, Zhan Y. (1997) A hybrid algorithm for computing the nonparametric maximum likelihood estimator from censored data. *J Am Stat Assoc* **92**:945–959.

66. Scheike TH, Juul A. (2004) Maximum likelihood estimation for Cox's regression model under nested case-control sampling. *Biostatistics* **5**:193–206.

67. Wacholder S, Hartge P, Sruewing JP, *et al.* (1998) The kin-cohort study for estimating penetrance. *Am J Epidemiol* **148**:623–630.

68. Gail M, Pee D, Benichou J, Carroll R. (1999) Designing studies to estimate the penetrance of an identified autosomal dominant mutation: cohort, case-control, and genotyped-proband designs. *Genet Epidemiol* **16**:15–39.

69. Chatterjee N, Wacholder S. (2001) A marginal likelihood approach for estimating penetrance from kin-cohort designs. *Biometrics* **57**(1):245–252.

70. Chatterjee N, Kalaylioglu Z, Shih JH, Gail MH. (2006) Case-control and case-only designs with genotype and family history data: estimating relative risk, residual familial aggregation, and cumulative risk. *Biometrics* **62**: 36–48.

71. Genest C, MacKay J. (1986) The joy of copulas: bivariate distributions with given margins. *Am Stat* **40**:280–283.

72. Sun W, Li H. (2004) Ascertainment-adjusted maximum likelihood estimation for the additive genetic gamma frailty models. *Lifetime Data Anal* **10**:229–245.

73. Park PJ, Tian L, Kohane IS. (2002) Linking expression data with patient survival times using partial least squares. *Bioinformatics* **18**:S120–S127.

74. Li H, Gui J. (2004) Partial Cox regression analysis for high-dimensional microarray gene expression data. *Bioinformatics* **20**:i208–i215.

75. Bair E, Tibshirani R. (2004) Semi-supervised methods for predicting patient survival from gene expression papers. *PLoS Biol* **2**:5011–5022.

76. Li L, Li H. (2004) Dimension reduction methods for microarrays with application to censored survival data. *Bioinformatics* **20**:3406–3412.

77. Li H, Luan Y. (2003) Kernel Cox regression models for linking gene expression profiles to censored survival data. *Pac Symp Biocomput* **8**:65–76.

78. Tibshirani R. (1996) Regression shrinkage and selection via the Lasso. *J R Stat Soc B* **58**:267–288.

79. Efron B, Johnston I, Hastie T, Tibshirani R. (2004) Least angle regression. *Ann Stat* **32**:407–499.

80. Gui J, Li H. (2005) Penalized Cox regression analysis in the high-dimensional and low-sample size settings, with applications to microarray gene expression data. *Bioinformatics* **21**:3001–3008.

81. Segal MR. (2006) Microarray gene expression data with linked survival phenotypes: diffuse large-B-cell lymphoma revisited. *Biostatistics* **7**:268–285.

82. Friedman JH, Popescu BE. (2004) Gradient directed regularization for linear regression and classification. Technical Report, Statistics Department, Stanford University, Palo Alto, CA.

83. Gui J, Li H. (2005) Threshold gradient descent method for censored data regression, with applications in pharmacogenomics. *Pac Symp Biocomput* **10**:272–283.

84. Breiman L. (2001) Random forests. *Mach Learn* **45**:5–32.
85. Friedman JH. (2001) Greedy function approximation: a gradient boosting machine. *Ann Stat* **29**:1189–1232.
86. Li H, Luan Y. (2005) Boosting proportional hazards models using smoothing splines, with applications to high-dimensional microarray data. *Bioinformatics* **21**:2403–2409.
87. Hothorn T, Buhlmann P, Dudoit S, *et al.* (2006) Survival ensembles. *Biostatistics* **7**:355–373.
88. Hasite T, Tibshirani R, Friedman J. (2001) *The Elements of Statistical Learning.* Springer, New York, NY.
89. Conti DV, Cortessis V, Molitor J, Thomas DC. (2003) Bayesian modeling of complex metabolic pathways. *Hum Hered* **56**:83–93.
90. Wei Z, Li H. (2007) Nonparametric pathway-based regression models for analysis of genomic data. *Biostatistics* **8**:265–284 .

Chapter 4

Genetic Association Studies: Concepts and Applications

Jun Li

Stanford Human Genome Center, Department of Genetics
Stanford University School of Medicine
975 California Ave, Palo Alto, CA 94304, USA

1. INTRODUCTION

Association studies are a widely used and potentially very powerful strategy for finding genetic and environmental risk factors underlying human diseases and quantitative traits. The underlying principle is a very simple one: if a DNA variant or an environmental factor increases disease susceptibility, it is expected to be observed more frequently among those who are affected than those who are not. For example, a recent study found that at-risk alleles of the transcription factor 7-like 2 gene occurred at a frequency of 27.6% in controls and 36.4% in patients of type 2 diabetes, for an increased risk of about 50%.[1] In most applications, the term "association" is synonymous with "correlation" or "dependency". An association is said to be found between two variables if it can be shown that their values are correlated, i.e. they are dependent on each other. The dependency is initially established in the statistical sense, but it may also lead to (and be explained by) actual mechanistic dependencies; that is, the biological processes described by the two variables may be interacting directly with each other. For example, genes involved in melanin metabolism may be associated with skin pigmentation, both statistically and in the biological sense.[2,3]

It is important to emphasize that an association is defined at the level of the population rather than that of the individual. Correlation describes a relationship between the distributions of two variables involving many observed data points (such as the heights and weights in a population), but it does not automatically explain or predict individual outcomes for any given person. It is from the standpoint of disease likelihood and public health management that we are interested in such populationwide trends. Equally important to keep in mind is that association does not imply causality, just as having a greater height does not cause a person to have a greater weight or vice versa. Until proven otherwise, the dependency between two variables is mutual; both may be the outcome of a third factor that is their common cause. In some situations, it is possible to show that one variable in fact affects the second variable in a unidirectional fashion, as in the case of a reactant and a product in a biochemical reaction (which has no feedback control).

In most situations, however, especially when the mechanism of interaction is not known, conclusions regarding causality require much more stringent conditions to be met than those regarding association.[a] A good example can be found in the comparison of mRNA, protein, or metabolite levels between normal and disease tissues in what is known as gene expression profiling and proteomic/metabalomic analyses. These molecular phenotypes can be either the cause or the outcome of the disease. As a result, an observed association must not be automatically equated with disease etiology or be referred to as the genetic basis of the disease. In this regard, genetic association is actually an exception because, unlike gene expression or protein levels, one's genotype does not usually change with time or circumstance and is always considered as preceding the outcome.

Some authors have argued that such Mendelian randomization is similar to the randomized assignment of risk exposure in controlled experiments or clinical trials.[8] But still, one's genotype often interacts with the

---

[a]Hill[4] suggested a set of nine criteria to distinguish causal from noncausal association: (1) strength (of association), (2) consistency, (3) specificity, (4) temporality (cause must precede effect), (5) biological gradient (a monotonic dose response), (6) plausibility, (7) coherence, (8) experimental evidence, and (9) analogy. While the requirement of (4) — cause must precede effect — is necessary for inferring causality, none of the other criteria is applicable or necessary in all situations. Conversely, sometimes even when all of the criteria are satisfied, the causality is still not established. Unlike logical deduction, the induction of a causal relationship from empirical evidence can never be absolutely certain. This viewpoint about the empirical basis of natural sciences can be traced to the epistemological arguments by Hume[5] and Popper.[6] In studies involving human subjects in particular, ethical considerations preclude experiments that may shed light on the cause–effect relationship among associated variables. In animal models, on the other hand, one can apply techniques such as experimental crosses and mutagenesis to complement the purely observational human studies. The Hill criteria were discussed in the context of epidemiology by Rothman and Greenland.[7]

environment in many subtle ways, such as when radiation injury causes structural changes in genomic DNA in somatic tissues (an example being the loss of heterozygosity in some cancers), when the pattern of epigenetic regulation is perturbed during development, or when a person's lifestyle and behavior gradually alter his or her environment. An added caveat is that it is often difficult to pinpoint the actual disease-causing nucleotide, even after an association is established for a gene or for a broader region of the genome. Many adjacent polymorphisms may be similarly associated and are in linkage disequilibrium to each other, but only one of these associations is causal (see below). In short, much caution is warranted when drawing conclusions regarding genetic causes.

Although this chapter deals specifically with the association between genetic variation and inherited diseases, the analyses of nongenetic factors in the realm of epidemiology has a longer tradition and a broader scope of considered risk factors. In epidemiology, one is interested in the relationship between a disease outcome, such as lung cancer, and any environmental, cultural, or behavioral risk factors, such as the habit of smoking. The principles and statistical tools are very similar between genetics and epidemiology[9,10] such that in many quantitative models the genetic factors are equivalent to (in fact, analyzed together with) the environmental risk factors and demographic characters such as age, gender, and race/ethnicity. However, genetics has certain special features, such as the transmission of discrete genetic material across generations, the finiteness of the physical genome, and the nonrandom allelic association between nearby polymorphisms (linkage disequilibrium or LD), resulting in many special considerations in studying genetic association in humans. This review will introduce the basic concepts, common study designs, the issue of population stratification, and some recent developments in whole-genome association studies that are made possible by the technical and statistical innovations taking place in the genome sciences.

2. BASIC METHODS AND COMMON APPLICATIONS

2.1. *Basic Analysis Methods*

Typically, the factors to be examined in an association study include both genetic and nongenetic factors, the latter including cultural conditions or lifestyle choices such as exercise and diet; behavioral differences such as smoking and alcohol drinking; and environmental hazards including exposure to radiation, pollutant, and infection. Genetic risk factors are usually analyzed in the form of alleles, genotypes, or haplotypes at polymorphic loci. Naturally, the independent genetic variables are discrete or categorical variables. The dependent variables, on the other hand, can be either discrete or continuous phenotypes.

For discrete outcomes, such as the dichotomous disease status or drug responsiveness, the most basic unit of analysis is a 2-by-2 contingency table, tabulating the numbers of observations for the four allele disease combinations (for a biallelic marker) in an adequately sized sample of the population. The evidence of association can be evaluated, in the simplest case, by comparing genotype or allele frequencies between cases and controls with a chi-square test of independence. In contrast, for continuous phenotypes, such as serum levels of triglyceride and bone marrow densities, the simplest comparison is between the trait value distributions in people with different alleles or genotypes. The means of these distributions can be compared by several statistical methods, such as the analysis of variance (ANOVA) or linear regression. Continuous outcomes are also called quantitative traits; they are therefore the subject of quantitative trait locus (QTL) analyses.

In practice, the simplest situations described above almost always need to be extended in order to accommodate the levels of complexity encountered in real studies. When a categorical variable has more than two levels, such as for the three genotypes at any locus in a diploid genome, a 2-by-3 contingency table is used, with the option to examine recessive, dominant, and multiplicative models of gene action. Covariates are included when there are additional independent variables representing other contributing risk factors. For example, if body mass index (BMI) is added as a covariate in an analysis of genetic association of cardiovascular disorders, one is essentially asking whether the action of the gene is through its effect on BMI such that genetic association is no longer present after controlling BMI or, independent of such an intermediate effect, such that the evidence of association is still present among people of similar BMI.

Sometimes, multiple outcomes can be chosen as the primary phenotype under study; these may be imperfectly but significantly correlated with each other, such as in studies of diabetes, obesity, hypertension, or hypercholesterolemia. Sometimes, one wishes to examine interactions between variables, for example, when using a 2-2-2 table as the simplest unit for gene–environment interaction for a disease. There are also situations when other methods are more appropriate, such as logistic regression, survival analysis, and resampling-based methods for determining statistical significance.

2.2. Common Applications

In genetic research, there are three main applications of the association study design: candidate gene analyses, gene mapping via linkage disequilibrium (LD), and disease gene characterization. All three share the basic goal of detecting and quantifying a correlation between DNA variants and disease status in the population.

2.2.1. Candidate Gene Studies

The goal of a candidate gene study is to test the hypothesis that DNA variants in certain genes affect disease susceptibility. Candidate gene studies can be divided into either the direct approach or the indirect approach. In the direct approach, DNA variants that have a good chance of affecting the activity or expression level of the gene products are tested for association. In the indirect approach, anonymous (and presumably neutral) markers that are in LD with the unknown disease variants are tested.

The LD-based, indirect approach applies to both the study of candidate SNPs and the positional mapping efforts described below. The feasibility of this approach derives from the evolution history of human populations: as specific combinations of alleles at neighboring loci (i.e. haplotypes) are transmitted from one generation to the next in a population, the lengths of such cotransmitted segments are reduced by recombination. Many parts of the human genome, however, exhibit a clearly block-like structure,[11-14] which is largely due to the presence of recombination hotspots and coldspots,[15,16] the relative youth of our species, and the specific demographic history of individual populations that include drift and selection.[17-19] Within such blocks, the haplotype diversity is lower than expected because alleles at nearby polymorphic loci are correlated, i.e. in LD. In effect, if the risk of a common disease is affected by common DNA variants, most affected individuals in the population may have inherited the mutation from a common, albeit distant, ancestor, and are likely to share a segment of the ancient haplotype where this mutation first arose. This feature makes it possible to identify association indirectly, via testing any of the markers in strong LD with the as yet unidentified disease variant.

The success of the candidate gene approach depends most crucially on the ability of the investigators to develop good hypotheses involving a reasonable number of candidate genes. This requires valid prior knowledge of not only disease etiology and the function of the genes, but also the molecular mechanism that plausibly links the two. For diseases such as autism and schizophrenia, however, there is no clearly described biological mechanism; consequently, there is a scarcity of particularly promising candidates that are more plausible than all of the other genes. Instead, one usually has to start with hundreds to thousands of genes that cover broad functional categories such as neurodevelopment, metabolism, or synaptic function.

To prioritize such a formidable collection of candidates would require the combined knowledge of all the earlier results from many related fields, including genetic linkage analysis, pharmacological evidence, imaging, and studies of molecular phenotypes such as gene expression patterns or protein and metabolite abnormalities. Still, since our knowledge about gene function and disease origin is still at its infancy, a lack of well-chosen candidates remains one of the major obstacles in studying most complex diseases by the candidate gene approach. In fact, most of the existing findings from association studies were, in retrospect, either not obvious pretest candidates or genes whose functions were still not known.

2.2.2. *LD-based Gene Mapping*

LD mapping belongs to the category of genetic methods commonly known as positional cloning.[b] The goal of positional cloning, as the name implies, is

---

[b]Some authors classify these related research strategies in terms of forward vs. reverse genetics, or linkage analyses and association studies. The association studies would then be further divided into candidate gene-based and whole-genome studies. While these manners of categorization capture the essential contrast between different genetic methods, this review has elected to highlight positional cloning (gene mapping) vs. function cloning as the two main lines of pursuit, depending on whether functional hypotheses have been prespecified. Positional cloning includes linkage analysis and LD-based association studies (also called fine mapping), which can be either regional or whole-genome. Functional cloning indicates candidate gene studies by either marker association or resequencing. Resequencing allows the analysis of both common and rare variants,[20,21] but its cost is currently prohibitively high on a genome scale; thus, it can only be carried out for selected candidate genes until new sequencing technologies can increase throughput and reduce cost by several orders of magnitude.[22] All formats of association studies can be either direct or indirect, i.e. relying on LD. Confusion may arise because in candidate gene studies, one may also test SNPs indirectly, i.e. by LD among SNPs. The distinction between different approaches can thus be blurred in practice. For example, to fine map a broad region of positive linkage, one may carry out positional cloning by examining random or tagging SNPs in an LD-based screening for signals of association; but in doing so, one may also emphasize the most promising candidate genes in this region by adding SNPs that are functionally important. In another example, when a large number of adjacent SNPs are in LD and are similarly associated with the disease, one may revert to linkage analysis to identify the allele that truly provides a linkage signal in the family data.

to identify disease genes through the estimation of their chromosomal locations without any prior knowledge of gene function. Its two main approaches are linkage analysis and association studies.[23]

Before the identification of DNA variations in the human genome in 1980, the main tool for gene discovery in humans was the use of blood group systems and protein markers. With the discovery of restriction fragment length polymorphisms (RFLPs)[24] and microsatellite markers, linkage analyses were made possible and were routinely adopted to examine the cosegregation of low-density DNA markers with disease status in human pedigrees. In one of its designs, the rates of allele sharing between affected sib pairs are compared to the null expectation of 50%. Linkage analysis is most effective for single-gene Mendelian disorders, where the gene effects are so large that the mutation is both necessary and sufficient for the disease. As the marker segregates tightly with the disease, linkage analysis has led to the successful identification of disease genes for thousands of inherited rare diseases.[25]

The drawback of this approach, however, is (1) it has limited resolution (5–10 cM), requiring substantial effort of fine mapping and candidate gene analyses to follow up the positive linkage regions; and (2) for complex diseases, which usually involve many genes of moderate effect (i.e. locus heterogeneity and low penetrance) as well as gene–environment interactions, it has limited power under realistic sample sizes. The power is further reduced by diagnostic uncertainty and phenocopy (i.e. disease phenotype induced by nongenetic factors resembles the genetically influenced phenotype). As a result, with the exception of a few isolated cases,[26–30] linkage analyses have been largely unsuccessful for complex diseases.[31]

In contrast, population-based association studies do not require family samples, often have greater power for common variants,[32] and provide finer resolution.[33] This is because the lengths of LD in the human genome are in the order of 60–200 kb in general populations (those that are not historically isolated or recently admixed), reflecting recombination events accumulated over hundreds of generations; whereas linkage analyses rely on meiotic recombination events in one or two generations in pedigrees. In this sense, individuals who share an ancient haplotype by descent are genetically related at this particular chromosomal region, even though they are not family members. Association studies thus mark the application of population genetic and epidemiological principles to disease studies.

LD mapping can be divided into two categories: regional and whole-genome. In the former, broad regions of the genome that are believed to contain the disease gene are examined for association. Such regions of the genome become LD mapping candidates either because of known chromosomal abnormalities or from some earlier linkage analyses. LD mapping is therefore considered as the natural follow-up after a linkage signal is found in order to eventually identify the disease gene. Historically, many disease genes were identified through traditional linkage analysis followed by LD-based fine mapping, for example, cystic fibrosis transmembrane conductance regulator (CFTR)[34,35] and Huntington's disease.[36,37]

In the second category, which is known as whole-genome association (WGA) studies, the scope of screening is expanded to include most of the known common variants in the entire genome. Ten years ago, Risch and Merikangas[32] pointed out that an unbiased scanning of the whole genome for signals of association in population samples would have greater statistical power, or a smaller required sample size in most situations, than traditional linkage analysis methods such as sib sharing. Because the lengths of haplotype blocks are in the order of 100 kb or less, the first requirement of such a study is the technology to genotype several hundred thousand to several million SNPs efficiently. There are about 10 million SNPs in humans with minor allele frequency > 0.01 (and many more rare ones), but not all of them are independent of each other; nearby markers may be strongly correlated (in LD). As a result, it is possible (in fact, desirable) to genotype only a subset of common SNPs, allowing the typed SNPs to "tag" the untyped SNPs by virtue of the strong LD between them.[11,12,38–40] A second requirement for WGA is, therefore, the knowledge of LD patterns to guide the selection of tagging SNPs. LD is influenced not only by locus-specific factors such as recombination,[15,16] mutation, and gene conversion, but also by population-specific factors such as recent migration and admixture, local forces of selection,[17] mating practices, expansions and bottlenecks, and random drift.[41] As a result, LD patterns vary across different regions of the genome and across different populations.[42]

To facilitate SNP selection, the International HapMap Project was initiated to characterize patterns of common DNA variation in 270 individuals from four populations: African, Caucasian, Chinese, and Japanese.[43,44] In late 2005, the first phase of HapMap was finished and led to the preliminary selection of tagging SNPs for each of these populations. The result showed that about 300 000 well-chosen SNPs can capture $\sim$80% of the common SNPs (minor allele frequency $> 5\%$) in Caucasians. Similar numbers are needed for the Asian populations, while up to 600 000 are needed for the

African samples.[c] Also in 2005–2006, high-throughput genotyping platforms that can assay 300 000–500 000 SNPs became widely available at a cost of about US\$500–US\$1000 per sample, with the current unit cost of ∼0.1 cent per genotype expected to drop further in the near future. Thus, all requirements for WGA are met, and a large number of WGA studies will be carried out in the next several years for a variety of complex diseases including bipolar disorder, heart disease, diabetes, and hypertension. By February 2006, early examples of WGA studies started to emerge,[46,47] with more surely to follow. We will soon find out whether this new strategy of pursuing disease genes will bring significant findings and valuable biological insights.

2.2.3. *Characterizing Established Disease Genes*

The third application of the association strategy is to quantitatively assess risk ratios for known genetic factors and for different alleles of established disease genes. Many disease genes show high levels of allelic heterogeneity, with many different variants, conferring a spectrum of risks. Association studies are needed, even after the disease gene is found, to define the allele-specific risks, their changes with age, genetic background, environmental risk factors, as well as the interactions between alleles and with other genes. Sometimes, the disease gene is initially discovered through linkage analyses or by association studies that use family-based controls (see below); but the population attributable risks, penetrance levels, or modifying factors are still not characterized. Sometimes, a large number of nearby SNPs are similarly associated, and it requires replication studies in different populations to confirm the finding and identify the true causative variant. In practice, such knowledge forms the basis of diagnostic testing, drug response prediction, and risk profiling in genetic counseling for specific populations and specific age groups.

---

[c]While many commercial platforms and currently planned studies can be described as whole-genome, the extent of genome coverage varies across different genotyping systems. Commercially available methods that involve whole-genome amplification and variant reduction immediately lose a significant portion of the genome due to the choice of the enzyme used and variations in amplification efficiency.[45] A panel of 300 000–500 000 SNPs can only cover most of the common variants in a specific population; it takes more to cover African populations. Even when the tagging SNPs are chosen based on the HapMap data, there are "un-HapMappable" portions of the genome that we have not been able to characterize. Truly comprehensive coverage is a goal that can be gradually approached, but theoretically not achieved until, ultimately, every base of the genome in every DNA sample is determined by sequencing.

3. GENERAL DESIGNS

For any given disease, the cases are recruited according to a predefined set of diagnostic criteria, sometimes from a population-based disease registry. The controls, however, can be sampled from the population in many different ways. The choice of controls distinguishes the main study designs: population-based cohort-control or case-control design, or family-based control design.

3.1. *Population-based Controls*

3.1.1. *Cohort-Control Design*

In a cohort study, a large number of at-risk individuals living in a defined geographic area are followed for many years to (1) characterize their exposure to risk factors at regular intervals, (2) record the progression of intermediate phenotypes (such as LDL levels for cardiovascular disorders), and (3) identify subjects who develop the disease. The goal is to discover an association between risk exposure and the disease (or its associated phenotypes). During the years or decades of follow-up, changes in risk profiles are documented for every subject, even though at the end of the study period most of them will not have developed the disease. Essentially, the controls will be the unaffected subjects (or sometimes a randomly chosen subset of the unaffected subjects) from the same cohort as the cases.

Because all covariates are "banked" prior to the assembly of the case and control groups, this design is most effective in reducing the selection bias and recall bias that case-control studies are often prone to (see below). Selection bias comes from inaccurate ascertainment of the disease outcome. Recall bias refers to the tendency, during retrospective assessment of exposure history in case-control studies, that people who have the disease often remember past behavior or risk exposure differently than those who do not have the disease. Although people's genotypes do not change over time, their environmental, cultural, and behavioral factors do. Thus, minimizing recall bias of the nongenetic factors is crucial for studying the role of gene–environment interactions. Studies that explicitly consider environmental variables are generally more useful than those focusing solely on the genetic factors.[8] For this reason, it has been increasingly realized that large-scale, prospective cohort studies are needed for the eventual unraveling of the genetic and environmental causes of common diseases.[48–51]

Sometimes, in order to ensure unbiased estimates of risk ratios, the controls are preselected at random from the cohort even though some of

them may later become cases. This is called the case-cohort design. It has
the attractive features that multiple diseases can be analyzed by comparing
with the same control group and that the information on controls can be
obtained early in the study, particularly for genetic data, which do not
change with time.

The main disadvantage of the cohort design is its cost. Generally, tens
to hundreds of thousands of individuals are enrolled and followed for many
years. This requires significant commitment of funds and personnel to track
the subjects, ascertain disease status, store the specimen, and collect data
on risk factors. Inevitably, some subjects will withdraw from the study due
to death, moving out of the study area, or reluctance to participate further.

3.1.2. *Case-Control Design*

For rare diseases, only a tiny fraction of the population will develop the disease. To collect a sufficient number of cases would require very large cohorts,
involving up to several million subjects. In such situations, the cost would
be prohibitively high, and the investigators may have no choice but to adopt
the case-control design. A case-control study begins with the ascertainment
of a group of affected individuals, followed by the collection of unaffected
individuals from the same source population as the cases who are matched
on known risk factors that are not the direct interest of the study. Unlike in
a cohort study, where these ancillary risk factors have been collected without the knowledge of disease status, past risk exposure in a case-control
study is obtained retrospectively, typically by a structured questionnaire or
analysis of medical records, with the disease status already known. When
the disease is rare, this design is much more efficient in terms of cost and
takes up much less time than enrolling the full cohort, with only a small
loss of precision in most situations.

A special issue with genetic case-control studies is the possibility of
confounding caused by the ancestry or ethnic origin of the enrolled subjects;
this is known as population stratification. Spurious association may arise,
or odds ratio estimates may become biased, when allele frequencies as well
as baseline disease rates vary across different subpopulations in the study.[52]
An often cited example is an association study of a human leukocyte antigen
haplotype and non–insulin-dependent diabetes in Pima Indians.[53] Both the
rate of diabetes and the haplotype frequencies differ between full-heritage
Indians and those with European ancestry. A stratified reanalysis revealed
that an initial positive finding for the HLA haplotype could be explained by

the unequal degree of admixture between cases and controls. In general, if a disease is more common in a particular ethnic group, this group (including people who have partial ancestry from this group) would be more likely to be recruited as the cases than as the controls. Any genetic variants that are more common in this ethnic group, including those unrelated to the disease, would appear to be associated. Some authors, however, have argued that population stratification may not be a major concern, at least for non-Hispanic whites of European descent in the US.[54–56] Its effect becomes a problem only when the disease rates and allele frequencies have significant and correlated differences across ethnic groups, and when the conventional self-reported origin is grossly inaccurate.

Nevertheless, minimizing confounding due to ancestry is an important component of a good study design. This is particularly true for studies that include African-Americans or US Hispanics, who are recently admixed, have large interindividual variations in admixture proportions due to continuing gene flow and social segregation, and are often not able to report their ancestry fully or accurately.[57,58] This is also true for those genetic loci that have large allele frequency gradients across populations,[3,59–61] and for diseases that exhibit population-specific prevalence.[62,d] In fact, the correlation between locus-specific ancestry and the traits of interest constitutes the essential material for admixture mapping.[64]

3.2. Family-based Studies: Family Member Controls and the Transmission/Disequilibrium Test

The most effective way to match cases and controls by ancestry is to use family controls, such as unaffected sibs, cousins, or parents. Members of a nuclear family share the same grandparents' gene pool; hence, the risk of population stratification is essentially avoided. The use of sibs and cousins has the advantages of age matching and ease of recruitment for late-onset diseases (for which many parents of the patients are deceased). Using sibs

---

[d]The extent to which different populations differ genetically, as well as the meaning and utility of race in biomedical research, is still the subject of intense debate. The incidence rates of many common diseases, such as asthma, prostate cancer, and diabetes, clearly vary in different parts of the world and among different ethnic groups in the US.[63] Although disparities in socioeconomic status, environmental factors, and access to health-care account for some of the differences, DNA variants that confer higher disease risks may also differ in frequencies among populations. The relative importance of genetic and environmental or cultural factors will certainly depend on the disease in question. This type of research is still at an early stage, and will continue to expand in the coming years.

provides a closer match on environmental factors, but it is overmatched on genotype and is therefore less efficient than using cousins. However, cousins bring some risks of stratification, as they share only one pair of grandparents (for first cousins), while the unshared grandparents might come from a different ethnic background.

In the case-parent design, one can compare the genotypes between cases and their parents; or in the "pseudosibling" design, compare the hypothetical sibs carrying the other three genotypes as three controls for each single case. Often, the parental alleles transmitted to the affected children are compared to those transmitted to unaffected children, and the allele (or haplotype) relative risks are evaluated. Similarly, in the transmission/disequilibrium test (TDT),[65,66] one of the most widely endorsed designs, only the heterozygous parents are considered and the allele transmitted to the affected child is compared to the allele not transmitted, resulting in a 2×2 test (for biallelic markers) of preferential transmission of the susceptibility allele against the other allele. Numerous further extensions have appeared to deal with missing parental data,[67–69] multiple alleles per marker,[70] and continuous trait values.[71,72]

The disadvantage of family-based tests is that the family samples are more difficult and expensive to recruit than unrelated samples from a population. Parental data are often unavailable for late-onset diseases, and the statistical power is reduced due to the genetic relatedness of cases and controls, particularly if only the heterozygous parents are used. This is because family members are genetically related such that the first-degree relatives, for example, share 50% of their genetic material and sometimes result in a 50% higher required sample size. The robust protection against population stratification thus comes with a significant cost in statistical power and logistic efficiency.

4. GENOMIC CONTROL AND STRUCTURED ASSOCIATION

Even in a well-designed case-control study, and even when self-reported ancestry is recorded as carefully as possible, there may still be residual genetic structure left that can drive the association results.[59,73,74] Furthermore, relying on reported ethnic data may not be adequate because many culturally defined ethnic groups, such as the Hispanics in the US, have complex histories of migration and recent admixture[75,76] and contain cryptic population structures due to past socioeconomic segregation that are not well described by the coarse labels of race, ethnicity, or nationality. For example, Puerto Ricans and Mexican-Americans are both described as Hispanics, but the former have a greater European and African ancestry, while the latter have a greater Native American contribution. Even among a random sample of Puerto Ricans, individual ratios of admixture vary significantly, ranging from 20% to 80% European ancestry.[58,62]

Poorly controlled stratification may impair an association study in two ways. First, a subtle imbalance of ancestry between cases and controls may cause many neutral markers to appear associated if these markers have different frequencies across different ancestral groups. Second, unobserved heterogeneity in both cases and controls often dilutes the association signal if the association is present only in one of the subpopulations. Mainly for these concerns, many researchers are skeptical about case-control results and believe that only family-based controls are safe in protecting against this confounder. However, as discussed above, family-based controls are more expensive and time-consuming to collect, and are not as efficient as population-based controls. Furthermore, for late-onset diseases, parents or sibs of the index cases are often no longer available.

In response to the concern of false-positives and false-negatives due to stratification, two main approaches have been developed to detect and correct the effects of population structure: genomic control and structured association. Both aim to make use of multilocus genotype data to enable valid case-control studies, even in the presence of population structure. A shared assumption is that, when many unlinked genetic loci are screened for signals of association, population structure will affect most loci across the genome similarly even when they are unlinked, while the true association to the disease variant will only affect the small set of markers that are in LD with it.

4.1. *Genomic Control*

Population stratification can sometimes be detected from genetic data when the data have been collected for a large number of loci. Devlin and Roeder,[77] Bacanu *et al.*,[78] and Reich and Goldstein[79] proposed a framework called "genomic control", which uses unlinked markers (the majority of which are presumably neutral with regard to the trait or disease under study) to detect overdispersion of the standard test statistics due to stratification; and, if overdispersion does occur, to adjust the observed statistics by the empirical distribution.

In the simple example of a 2×2 test, under the null hypothesis of no association, the observed χ^2 statistics should follow a theoretical χ^2

distribution with one degree of freedom. However, if cases and controls have different compositions from several subpopulations, the distribution of χ^2 of the unassociated markers will, under a range of assumptions, be simply inflated by a scaling factor of λ, which can be estimated empirically if a large number of unlinked markers are genotyped. Subsequently, the observed χ^2 statistics for every marker can be adjusted by this multiplicative factor and compared to the usual χ^2 distribution for making inference and for declaring degrees of significance. The scaling factor λ increases with the case-control differences in population makeup and with the sample size. For studies involving several thousand samples, this inflation factor can be fairly large, e.g. more than 50%.

A power analysis indicated that the genomic control method can be more powerful than the TDT (described earlier) in most situations.[78] However, the estimation of λ may have large uncertainties, particularly if the number of loci is small (several hundred). Since it averages information across not only subpopulations but also all of the studied loci, its use can be problematic if the effects of the disease loci vary across subpopulations, especially for large sample sizes and small P values.[80,81]

4.2. *Structured Association*

In the second approach, the genotype data are used to infer the detailed population structure. Specifically, the proportions of every individual's genome derived from each of K possible ancestral populations are estimated. When the ancestral populations are not measured, there are two sets of unknowns to be simultaneously estimated from the observed genotype data: the allele frequencies of each locus in each of the K founding populations; and each individual's ancestral proportions, from 1 to K. Pritchard *et al.*[82] used a Markov chain Monte Carlo (MCMC) algorithm to estimate individual admixture, and proposed a method[83] to test for genetic association while incorporating the inferred ancestral proportions of individual samples. Other MCMC- or likelihood-based methods have also been proposed for inferring ancestry.[84–86]

In all of these methods, each individual is probabilistically assigned to a number of latent subpopulations, which are assumed to be homogeneous. Subsequently, one can carry out tests of association for each marker conditional on the admixture ratios, essentially using such ratios as covariates and testing for association within subpopulations. The inferred ancestry can also be "plugged in" in *post hoc* sample matching and regression analyses (for continuous outcomes), in which the inferred ancestry acts as a covariate.[83,87,88]

Genomic control and structured association can be applied interactively in a single study. For example, one may use the inferred ancestral ratios for retrospective and more fine-tuned sample matching, then use genomic controls to detect any residual stratification. In practice, DNA markers vary in their allele frequencies in different populations and have different informativeness for inferring ancestry. It is now feasible to identify the population-informative markers for any set of population that may be involved in a study, and to routinely include such markers in genotyping panels.

In essence, population stratification presents the problem of sample classification and matching. In genetic association studies, as in any epidemiology study, cases and controls should be matched on all established confounders. Genomic control monitors the impact of poor matching on observed association, whereas structure association empirically estimates the details of the confounder and provides the option to use the data-derived individual admixture as a better surrogate for the underlying genetic diversity. This option is especially helpful when the initial classification is based on crudely categorized and occasionally inaccurate self-reporting. The development of genomic control and structured association methods has offered researchers much greater freedom in study design by allowing the recruitment of large population-based samples at low cost in both genetic association studies and clinical drug trials.

5. PRACTICAL CONSIDERATIONS IN WHOLE-GENOME STUDIES

5.1. *Power and Sample Size*

The past decade has witnessed an explosion of candidate gene–based association studies. The overall success, however, has been plagued by a general lack of replication.[89–91] For example, in 166 putative associations that have been studied three or more times, it was found that only six have been consistently replicated.[89] The likely causes of these apparent failures include inadequate sample size, suboptimal selection of DNA markers, uncertainties in diagnosis, lack of power in replication attempts, uncontrolled confounding due to population structure, population-specific LD patterns, publication bias, overinterpretation of marginal results, and sheer biological complexity.

The power to detect association depends crucially on four parameters: the sample size, the effect size in terms of odds ratios of individual alleles or genotypes, the marker allele and disease allele frequencies, and the degree of LD between them. For any predetermined set of parameter values, the statistical power also depends on the type of controls (unrelated or family controls), the type 1 error allowed after correcting for multiple testing, and the extent of allelic heterogeneity (i.e. different at-risk alleles, occurring on different haplotype backgrounds, may independently account for the disease in different cases). Because of the complexities involved in such calculations, most power analyses would examine individual scenarios that cover a range of study designs and the most likely parameter values such as sample sizes of 500, 1000, or 2000; odds ratios of 1.2, 1.5, or 2; allele frequencies of 0.05, 0.1, 0.2, or 0.4; use of unrelated, parental, or sib controls; and assumption of either perfect LD between the typed marker and the untyped disease variant or strong LD with r^2 of 0.5, 0.6, 0.7, or 0.8.[92,93] Often, to simplify the presentation of results, it is also assumed that the typed and untyped markers are matched in allele frequencies, and that there is no allelic heterogeneity. Typically, unadjusted significance levels of 10^{-6} to 10^{-8} are used to account for testing hundreds of thousands of markers.

Taken as a whole, these calculations reveal that large samples are needed to detect association with alleles which are not too rare in the population and are of moderate effects. For example, for a statistical power of 80%, at a significance level of 10^{-6}, assuming perfect LD between the marker and disease variant and that their minor allele frequencies are 20%, it would take 5000–6000 cases and an equal number of controls to detect an odds ratio of 1.2.[94] The required sample size would decrease for larger odds ratios. However, it would increase with imperfect LD between the typed marker and the untyped disease variant, and with their allele frequency differences.[93,95]

To increase power (and the efficiency of genotyping), some authors have explored the sample pooling method[96] and the staged genotyping design.[97–99] With sample pooling, the allele frequency differences between the case and control groups can be directly observed at a fraction of the cost when compared to individual genotyping. The drawback of this approach, however, is at least twofold. First, the ability to detect small allele frequency differences (such as 2%) is limited due to the inherent noise of the technology and sample pooling errors. Second, having an individual genotype is necessary not only for a variety of stratified or subset analyses that incorporate other variables (such as each subject's environmental and comorbidity factors), but also for haplotype-based analyses.

In the multistaged design, only a subset of available samples are genotyped on a large number of markers in stage 1, and only those markers that appear to be associated at a moderate significance level are followed up by genotyping them on the remaining samples. The cost saving in typing only a small number of markers in stage 2 can be substantial. However, if the call rate for stage 2 is not very high (e.g. not above 99%), there will be a small number of markers that need to be done a third time, diminishing the cost advantage.

5.2. *Allelic Heterogeneity*

Allelic heterogeneity refers to the scenario in which multiple variants in the same gene are independently sufficient for increasing disease risk in different individuals, yet each variant has a different history and haplotype background. As a result, the study suffers a severe drop in power because the signal of genotype–phenotype correlation is diluted over multiple marker alleles. This is not a problem for linkage analysis because the genetic marker that cosegregates with the disease is the same in different families, even though the actual high-risk allele of the disease gene might be linked to different alleles of this marker in different families. In other words, a linkage analysis essentially attempts to detect an association in the form of a random effect within families, and is able to tolerate allelic heterogeneity across families. Association studies, on the other hand, rely on the same allele being associated with higher disease risks in most of the affected individuals in the population. This issue is intimately tied to the question of common vs. rare variants, which we will discuss next.

5.3. *Common versus Rare Variants*

A central debate regarding the genetic basis of heritable traits and common diseases is the relative importance of common and rare DNA variants. Although most geneticists agree that this distinction is artificial, the frequency spectrum of the causative variants clearly has a major impact on the study design and the likelihood of success. While linkage analysis is suited for identifying variants of high penetrance even if they are rare, association studies are effective only for identifying common variants of moderate effects in the absence of significant allelic heterogeneity.

The common-disease, common-variant (CDCM) hypothesis is based on the notion that diseases which are of high prevalence in multiple populations are likely to be due to common and evolutionarily old genetic variants, and that many patients in the population develop the disease by carrying the same at-risk allele. As these variants tend to have moderate effects and sometimes manifest as late-onset disorders, they are not under strong negative selection and tend to exhibit high frequencies in the population. The alternative hypothesis is that a heterogeneous collection of rare and high-penetrance alleles, sometimes in a group of functionally related genes, independently account for a considerable proportion of cases even if these patients develop the disease by carrying different alleles and often from different genes.[100,101] These variants tend to cause severe and early-onset phenotypes, and are maintained at low frequencies due to mutation–selection balance.[e]

The CDCM hypothesis is easier to test empirically. This is because common SNPs have been more thoroughly ascertained and have become not only the focus of molecular biological studies and candidate gene–based association studies, but also the natural content of high-density SNP genotyping sets.

Currently, the exact genetic mechanisms underlying most complex diseases are not known, and it is entirely possible that a constellation of rare, high-penetrance variants (including structural variations such as deletions and insertions) play a major role in a considerable proportion of the cases. To examine the role of rare variants requires either appropriately powered linkage analysis involving large pedigrees and high-density markers or complete resequencing of candidate genes; in the most favorable scenario, finding a significantly greater number of mutations among cases than among controls.[102] This method, however, relies on the prescient selection of the correct genes and is far more expensive than genotyping known common variants in the same genes.

In the foreseeable future, as the sequencing throughput continues to improve[103,104] and the cost continues to drop, more and more candidate genes will be completely sequenced in all cases and controls, effectively combining variant discovery, scoring, and analysis of association. Eventually, the scope of resequencing will expand from candidate genes to candidate regions and to the entire genome, replacing genotyping as the main experimental approach to study association. Such resequencing data will inevitably contain a large number of rare or even unique variants, all in the presence of background levels of genotyping error. This approach will raise new

---

[e]Occasionally, deleterious variants may still be observed at high frequencies in specific populations, mostly due to a heterozygote advantage (selection in favor of the heterozygous individuals) in past epidemics in these populations or in late onset of the disease.

methodological challenges, including testing multiple disease mechanisms involving a broad spectrum of mutations on a complex haplotype background. Clearly, new statistical and bioinformatic tools will be needed to meet these challenges.

Rare alleles are more likely to be population-specific, and are therefore more difficult to be replicated in a different population. However, different variants in the same gene might be associated in other populations. In fact, the demonstration of allelic heterogeneity in independent subjects and populations, although a major inconvenience in terms of statistical power, stands as strong proof of a functional link between the gene and the disease. This is true for both family-based linkage analyses of single-gene disorders and population-based studies of complex diseases.

5.4. *Multiple Testing Corrections*

The availability of high-density SNP maps and high-throughput genotyping technologies has rekindled the hope that WGA studies may be the last push needed to unravel the genetic basis of inherited non-Mendelian disorders. However, the simultaneous testing of hundreds of thousands of SNP markers also means that many thousands of "hits" would occur by chance alone, and the challenge of multiple testing correction becomes far more acute than in conventional epidemiological studies. Standard procedures such as the Bonferroni correction are considered too conservative, as it ignores the dependency between markers such as those in LD with each other. Because WGA studies tend to be exploratory in nature, it is usually not necessary to ensure that not a single false-positive result is obtained; rather, what is important is to assess the proportion of false-positives so as to inform and prioritize downstream validation experiments. In this context, a false discovery rate approach has gained much popularity.[105,106] To deal with the correlation structure among SNPs, new statistical techniques that rely on sample permutation are being developed to provide genomewide empirical *p* values.[107] Still, successful replication in independent samples and additional populations remains the ultimate standard of validation.

In candidate gene or candidate region studies that examine several hundred genes or thousands of SNPs, it may become difficult to decide whether the number of tests to be corrected for should take into account all SNPs in the positive genes, all SNPs in all candidate genes analyzed in the study, or all candidate genes that could have been considered in the future. Often, the prior likelihood of a DNA variant to bring about functional changes can be considered in a Bayesian framework, as it is often the case that a nonconservative amino acid change in the protein is more likely to have a phenotypical consequence than an intronic SNP that may or may not affect the expression levels of the gene product.[108] The main limitation of this approach is that the prior probability of functional change of individual variants is not known, particularly for the large numbers of regulatory variants. Even for those that result in nonsynonymous coding changes in gene products, the biological impact is difficult to quantify without experimental data. Because different genes serve vastly different molecular functions, the experiments that examine the functional consequences of DNA variation tend to be one of a kind and difficult to scale up. The lack of functional knowledge of DNA variants remains one of the greatest challenges in interpreting association results.

As genotyping data accumulate, increasing attention will be placed on analyzing gene–gene interaction in light of plausible biological networks. To examine all pairwise epistatic interactions for 300 000 SNPs would yield 90 billion tests, certainly too many for naive adjustment such as the Bonferroni correction. However, sophisticated methods are constantly being developed to address such needs, as in, for example, Marchini *et al.*[109]

5.5. *Uncertainties in Phenotyping*

For many complex disorders, the uncertainty in diagnosis and the selection of a good phenotype to study remain important obstacles. Behavior traits and psychiatric disorders — in particular, the lack of reliable biochemical, pharmacological, or imaging assays — have led to a complete reliance on subjective and often ambiguous criteria that are based on observation, interview, and occasionally drug response. Issues such as interrater reliability, personal bias, and the constantly evolving diagnostic guidelines hamper combined analysis across studies that are carried out at different laboratories and at different times. Disorders of the heart, lung, and blood, on the other hand, do have biochemical and physiological markers. These include, for example, serum levels of resting glucose and BMI for diabetes; and C-reactive protein, carotid wall thickness, and lipid levels for atherosclerosis (as well as obesity and a host of related conditions). Most of these covariates or comorbidity factors are not only highly correlated among themselves, but also have overlapping contributions to multiple disorders. From the genetic standpoint, they also have different heritability estimates, depending on the samples or populations under study.

There are therefore numerous choices and compromises in study design and variable selection, presenting both opportunities and significant challenges. For instance, if three disease outcomes (such as hypertension, diabetes, and obesity) were considered along with 10–20 physiological covariates (HDL, TG, BMI, etc.), several demographic and socioeconomic factors (age, gender, race, income, education, and occupation), and hundreds to thousands of SNPs in many equally plausible candidate genes, one is faced with a series of difficult decisions in variable selection as well as remarkable complexity in multiple testing correction, even before the second- and third-order interactions are considered. Despite the difficulties, the relevant alternative phenotypes must be explored because some of the physiological measures for cardiovascular diseases, for example, represent intermediate phenotypes that are more proximal to the genetic effects. The impact of genetic variation on the more distal, clinically manifest outcomes is often much less consistent and much harder to dissect.

5.6. *Future Outlook*

The successful discovery of genetic association often requires great effort in identifying the truly causal variant out of a larger number of polymorphic loci that are in LD and are similarly associated. As discussed above, the priority of validation can be ordered by the estimated functional consequences of the variants, such as the resulting amino acid changes, splicing differences, and variabilities in promoter activity. The knowledge of functional consequences at the molecular level is usually obtained by a plethora of *in vitro* assays. Alternatively, a different population might have a different LD pattern in the gene; in the case of African populations or those with significant African admixture, the associated haplotype blocks tend to be narrower, effectively shortening the list of candidate SNPs to be validated. After a WGA study, much follow-up work is expected to take place in these areas.

Understanding the genetic contribution to complex diseases at the population level is only the first step toward improving diagnosis, prevention, and treatment. Often, the single most effective action any individual can take to improve his or her health is not to amend a mild genetic deficiency, but to adjust health-related behavior and improve the risk profile of the environment. This is because if the marginal effect of a causal allele is only a 20% higher risk, the medical relevance of this knowledge, particularly with regard to each individual carrier, is not immediately obvious.

The disease-predisposing variant must act in the context of development, adaptation, and medical intervention. It is therefore vitally important to integrate genetic association studies with the large-scale characterization of gene–environment interactions. For example, a recent study found that people carrying a functional polymorphism in the promoter region of the serotonin transporter are more likely to develop depression, but only if they also experienced stressful life events.[110] To this end, many longitudinal population-based studies have been initiated to implement this ambitious strategy. There is considerable expectation that these very large, well-planned prospective studies will be far more powerful than the current salvo of case-control studies, and will be far more predictive in the clinical setting.

At this point, several well-powered WGA studies are underway and the technical, statistical, and logistic infrastructures are constantly improving. We fully expect that, in the next several years, the genotyping and sequencing cost will continue to drop, while the throughput will continue to increase. The optimal panel of tagging SNPs as well as the best ancestry-informative markers will continue to be refined for the major populations and consolidated into standard genotyping platforms. The genetic structure of most populations, as well as the subtle differences between them, will be characterized in light of the evolutionary processes that gave rise to the modern-day genetic diversity. Functional annotation will continue to mature as efforts such as the ENCODE project[111] will gradually expand to cover the entire human genome. New technologies for phenotyping, novel diagnostic instruments, and the discovery of more informative or more heritable endophenotypes will become increasingly standardized, allowing more efficient collection of phenotype data. Well-characterized population controls that can be contrasted to multiple disease groups will be available, along with carefully and often prospectively recorded environmental factors. Animal models, particularly those amenable for QTL mapping of complex phenotypes,[112] will continue to provide new insight into the genetic underpinnings of mammalian biology and human health.

The merger of high-throughput technologies and the compelling strength of association study designs have improved our ability to find the genetic predictors of complex diseases, new targets of treatment, and new preventive strategies. These will help us to better understand the basic mechanisms in biology and provide the knowledge foundation for personalized medicine.

ACKNOWLEDGMENTS

I would like to thank Dr Richard Myers for his long-standing encouragement and mentoring, as well as members of the Myers lab and my colleagues at the Stanford Human Genome Center and Department of Genetics of the Stanford University School of Medicine for their generous support.

REFERENCES

1. Grant SF, Thorleifsson G, Reynisdottir I, *et al.* (2006) Variant of transcription factor 7-like 2 (*TCF7L2*) gene confers risk of type 2 diabetes. *Nat Genet* **38**:320–323.
2. Lamason RL, Mohideen MA, Mest JR, *et al.* (2005) SLC24A5, a putative cation exchanger, affects pigmentation in zebrafish and humans. *Science* **310**:1782–1786.
3. Parra EJ, Kittles RA, Shriver MD. (2004) Implications of correlations between skin color and genetic ancestry for biomedical research. *Nat Genet* **36**:S54–S60.
4. Hill AB. (1965) The environment and disease: association or causation? *Proc R Soc Med* **58**:295–300.
5. Hume D. (1888) *A Treatise of Human Nature.* Oxford University Press, Oxford, England.
6. Popper KR. (1959) *The Logic of Scientific Discovery*, 2nd ed. Routledge, London, England.
7. Rothman KJ, Greenland S. (1998) *Modern Epidemiology*, 2nd ed. Lippincott Williams & Wilkins, Philadelphia, PA.
8. Clayton D, McKeigue PM. (2001) Epidemiological methods for studying genes and environmental factors in complex diseases. *Lancet* **358**:1356–1360.
9. Khoury MJ, Beaty TH, Cohen BH. (1993) *Fundamentals of Genetic Epidemiology.* Oxford University Press, New York, NY.
10. Thomas DC. (2004) *Statistical Methods in Genetic Epidemiology.* Oxford University Press, New York, NY.
11. Gabriel SB, Schaffner SF, Nguyen H, *et al.* (2002) The structure of haplotype blocks in the human genome. *Science* **296**:2225–2229.
12. Patil N, Berno AJ, Hinds DA, *et al.* (2001) Blocks of limited haplotype diversity revealed by high-resolution scanning of human chromosome 21. *Science* **294**:1719–1723.
13. Reich DE, Cargill M, Bolk S, *et al.* (2001) Linkage disequilibrium in the human genome. *Nature* **411**:199–204.
14. Phillips MS, Lawrence R, Sachidanandam R, *et al.* (2003) Chromosome-wide distribution of haplotype blocks and the role of recombination hot spots. *Nat Genet* **33**:382–387.
15. Crawford DC, Bhangale T, Li N, *et al.* (2004) Evidence for substantial fine-scale variation in recombination rates across the human genome. *Nat Genet* **36**:700–706.

16. McVean GA, Myers SR, Hunt S, *et al.* (2004) The fine-scale structure of recombination rate variation in the human genome. *Science* **304**:581–584.
17. Sabeti PC, Reich DE, Higgins JM, *et al.* (2002) Detecting recent positive selection in the human genome from haplotype structure. *Nature* **419**: 832–837.
18. Voight BF, Kudaravalli S, Wen X, Pritchard JK. (2006) A map of recent positive selection in the human genome. *PLoS Biol* **4**:e72.
19. Wang ET, Kodama G, Baldi P, Moyzis RK. (2006) Global landscape of recent inferred Darwinian selection for *Homo sapiens. Proc Natl Acad Sci USA* **103**:135–140.
20. Botstein D, Risch N. (2003) Discovering genotypes underlying human phenotypes: past successes for Mendelian disease, future approaches for complex disease. *Nat Genet* **33**(Suppl):228–237.
21. Hirschhorn JN, Altshuler D. (2002) Once and again — issues surrounding replication in genetic association studies. *J Clin Endocrinol Metab* **87**: 4438–4441.
22. Shendure J, Mitra RD, Varma C, Church GM. (2004) Advanced sequencing technologies: methods and goals. *Nat Rev Genet* **5**:335–344.
23. Lander ES, Schork NJ. (1994) Genetic dissection of complex traits. *Science* **265**:2037–2048.
24. Botstein D, White RL, Skolnick M, Davis RW. (1980) Construction of a genetic linkage map in man using restriction fragment length polymorphisms. *Am J Hum Genet* **32**:314–331.
25. OMIM. Online Mendelian Inheritance in Man. Available at http://www.ncbi.nlm.nih.gov/entrez/query.fcgi?db=OMIM/
26. Hugot JP, Chamaillard M, Zouali H, *et al.* (2001) Association of *NOD2* leucine-rich repeat variants with susceptibility to Crohn's disease. *Nature* **411**:599–603.
27. Ogura Y, Bonen DK, Inohara N, *et al.* (2001) A frameshift mutation in *NOD2* associated with susceptibility to Crohn's disease. *Nature* **411**: 603–606.
28. Rioux JD, Daly MJ, Silverberg MS, *et al.* (2001) Genetic variation in the 5q31 cytokine gene cluster confers susceptibility to Crohn disease. *Nat Genet* **29**:223–228.
29. Stefansson H, Sigurdsson E, Steinthorsdottir V, *et al.* (2002) Neuregulin 1 and susceptibility to schizophrenia. *Am J Hum Genet* **71**:877–892.
30. Stoll M, Corneliussen B, Costello CM, *et al.* (2004) Genetic variation in *DLG5* is associated with inflammatory bowel disease. *Nat Genet* **36**: 476–480.
31. Altmuller J, Palmer LJ, Fischer G, *et al.* (2001) Genomewide scans of complex human diseases: true linkage is hard to find. *Am J Hum Genet* **69**: 936–950.
32. Risch N, Merikangas K. (1996) The future of genetic studies of complex human diseases. *Science* **273**:1516–1517.
33. Jorde LB. (1995) Linkage disequilibrium as a gene-mapping tool. *Am J Hum Genet* **56**:11–14.

34. Riordan JR, Rommens JM, Kerem B, *et al.* (1989) Identification of the cystic fibrosis gene: cloning and characterization of complementary DNA. *Science* **245**:1066–1073.

35. Kerem B, Rommens JM, Buchanan JA, *et al.* (1989) Identification of the cystic fibrosis gene: genetic analysis. *Science* **245**:1073–1080.

36. Gusella JF, Wexler NS, Conneally PM, *et al.* (1983) A polymorphic DNA marker genetically linked to Huntington's disease. *Nature* **306**:234–238.

37. Group HsDCR. (1993) A novel gene containing a trinucleotide repeat that is expanded and unstable on Huntington's disease chromosomes. The Huntington's Disease Collaborative Research Group. *Cell* **72**:971–983.

38. Carlson CS, Eberle MA, Rieder MJ, *et al.* (2004) Selecting a maximally informative set of single-nucleotide polymorphisms for association analyses using linkage disequilibrium. *Am J Hum Genet* **74**:106–120.

39. Johnson GC, Esposito L, Barratt BJ, *et al.* (2001) Haplotype tagging for the identification of common disease genes. *Nat Genet* **29**:233–237.

40. Ke X, Cardon LR. (2003) Efficient selective screening of haplotype tag SNPs. *Bioinformatics* **19**:287–288.

41. Cavalli-Sforza LL, Menozzi P, Piazza A. (1994) *History and Geography of Human Genes*. Princeton University Press, Princeton, NJ.

42. Wall JD, Pritchard JK. (2003) Haplotype blocks and linkage disequilibrium in the human genome. *Nat Rev Genet* **4**:587–597.

43. HapMap. (2003) The International HapMap Project. *Nature* **426**:789–796.

44. Tanaka T. (2005) [International HapMap project]. *Nippon Rinsho* **63**(Suppl 12):29–34.

45. Kennedy GC, Matsuzaki H, Dong S, *et al.* (2003) Large-scale genotyping of complex DNA. *Nat Biotechnol* **21**:1233–1237.

46. Klein RJ, Zeiss C, Chew EY, *et al.* (2005) Complement factor H polymorphism in age-related macular degeneration. *Science* **308**:385–389.

47. Maraganore DM, de Andrade M, Lesnick TG, *et al.* (2005) High-resolution whole-genome association study of Parkinson disease. *Am J Hum Genet* **77**:685–693.

48. Collins FS. (2004) The case for a US prospective cohort study of genes and environment. *Nature* **429**:475–477.

49. Study TMW. (1999) The Million Women Study: design and characteristics of the study population. The Million Women Study Collaborative Group. *Breast Cancer Res* **1**:73–80.

50. Slimani N, Kaaks R, Ferrari P, *et al.* (2002) European Prospective Investigation into Cancer and Nutrition (EPIC) calibration study: rationale, design and population characteristics. *Public Health Nutr* **5**:1125–1145.

51. Wright AF, Carothers AD, Campbell H. (2002) Gene–environment interactions — the BioBank UK study. *Pharmacogenomics J* **2**:75–82.

52. Thomas DC, Witte JS. (2002) Point: population stratification: a problem for case-control studies of candidate-gene associations? *Cancer Epidemiol Biomarkers Prev* **11**:505–512.

53. Knowler WC, Williams RC, Pettitt DJ, Steinberg AG. (1988) Gm3;5,13,14 and type 2 diabetes mellitus: an association in American Indians with genetic admixture. *Am J Hum Genet* **43**:520–526.

54. Wacholder S, Rothman N, Caporaso N. (2000) Population stratification in epidemiologic studies of common genetic variants and cancer: quantification of bias. *J Natl Cancer Inst* **92**:1151–1158.

55. Wacholder S, Rothman N, Caporaso N. (2002) Counterpoint: bias from population stratification is not a major threat to the validity of conclusions from epidemiological studies of common polymorphisms and cancer. *Cancer Epidemiol Biomarkers Prev* **11**:513–520.

56. Cardon LR, Palmer LJ. (2003) Population stratification and spurious allelic association. *Lancet* **361**:598–604.

57. Reiner AP, Ziv E, Lind DL, *et al.* (2005) Population structure, admixture, and aging-related phenotypes in African American adults: the Cardiovascular Health Study. *Am J Hum Genet* **76**:463–477.

58. Salari K, Choudhry S, Tang H, *et al.* (2005) Genetic admixture and asthma-related phenotypes in Mexican American and Puerto Rican asthmatics. *Genet Epidemiol* **29**:76–86.

59. Campbell CD, Ogburn EL, Lunesta KL, *et al.* (2005) Demonstrating stratification in a European American population. *Nat Genet* **37**:868–872.

60. Lao O, van Duijn K, Kersbergen P, *et al.* (2006) Proportioning whole-genome single-nucleotide-polymorphism diversity for the identification of geographic population structure and genetic ancestry. *Am J Hum Genet* **78**:680.

61. Tishkoff SA, Varkonyi R, Cahinhinan N, *et al.* (2001) Haplotype diversity and linkage disequilibrium at human G6PD: recent origin of alleles that confer malarial resistance. *Science* **293**:455-462.

62. Choudhry S, Coyle NE, Tang H, *et al.* (2006) Population stratification confounds genetic association studies among Latinos. *Hum Genet* **118**:652–664.

63. Risch N, Burchard E, Ziv E, Tang H. (2002) Categorization of humans in biomedical research: genes, race and disease. *Genome Biol* **3**:comment2007.

64. McKeigue PM. (2005) Prospects for admixture mapping of complex traits. *Am J Hum Genet* **76**:1–7.

65. Spielman RS, McGinnis RE, Ewens WJ. (1993) Transmission test for linkage disequilibrium: the insulin gene region and insulin-dependent diabetes mellitus (IDDM). *Am J Hum Genet* **52**:506–516.

66. Ewens WJ, Spielman RS. (1995) The transmission/disequilibrium test: history, subdivision, and admixture. *Am J Hum Genet* **57**:455–464.

67. Spielman RS, Ewens WJ. (1998) A sibship test for linkage in the presence of association: the sib transmission/disequilibrium test. *Am J Hum Genet* **62**:450–458.

68. Curtis D. (1997) Use of siblings as controls in case-control association studies. *Ann Hum Genet* **61**(Pt 4):319–333.

69. Boehnke M, Langefeld CD. (1998) Genetic association mapping based on discordant sib pairs: the discordant-alleles test. *Am J Hum Genet* **62**:950–961.

70. Sham PC, Curtis D. (1995) An extended transmission/disequilibrium test (TDT) for multi-allele marker loci. *Ann Hum Genet* **59**(Pt 3):323–336.

71. Allison DB. (1997) Transmission-disequilibrium tests for quantitative traits. *Am J Hum Genet* **60**:676–690.

72. Rabinowitz D. (1997) A transmission disequilibrium test for quantitative trait loci. *Hum Hered* **47**:342–350.

73. Freedman ML, Reich D, Penney KL, *et al.* (2004) Assessing the impact of population stratification on genetic association studies. *Nat Genet* **36**: 388–393.

74. Marchini J, Cardon LR, Phillips MS, Donnelly P. (2004) The effects of human population structure on large genetic association studies. *Nat Genet* **36**:512–517.

75. Burchard EG, Borrell LN, Choudhry S, *et al.* (2005) Latino populations: a unique opportunity for the study of race, genetics, and social environment in epidemiological research. *Am J Public Health* **95**:2161–2168.

76. Gonzalez Burchard E, Borrell LN, Choudhry S, *et al.* (2005) Latino populations: a unique opportunity for the study of race, genetics, and social environment in epidemiological research. *Am J Public Health* **95**: 2161–2168.

77. Devlin B, Roeder K. (1999) Genomic control for association studies. *Biometrics* **55**:997–1004.

78. Bacanu SA, Devlin B, Roeder K. (2000) The power of genomic control. *Am J Hum Genet* **66**:1933–1944.

79. Reich DE, Goldstein DB. (2001) Detecting association in a case-control study while correcting for population stratification. *Genet Epidemiol* **20**: 4–16.

80. Devlin B, Bacanu SA, Roeder K. (2004) Genomic control to the extreme. *Nat Genet* **36**:1129–1130; author reply 1131.

81. Pritchard JK, Donnelly P. (2001) Case-control studies of association in structured or admixed populations. *Theor Popul Biol* **60**:227–237.

82. Pritchard JK, Stephens M, Donnelly P. (2000) Inference of population structure using multilocus genotype data. *Genetics* **155**:945–959.

83. Pritchard JK, Stephens M, Rosenberg NA, Donnelly P. (2000) Association mapping in structured populations. *Am J Hum Genet* **67**:170–181.

84. Satten GA, Flanders WD, Yang Q. (2001) Accounting for unmeasured population substructure in case-control studies of genetic association using a novel latent-class model. *Am J Hum Genet* **68**:466–477.

85. Tang H, Peng J, Wang P, Risch NJ. (2005) Estimation of individual admixture: analytical and study design considerations. *Genet Epidemiol* **28**: 289–301.

86. Purcell S, Sham P. (2004) Properties of structured association approaches to detecting population stratification. *Hum Hered* **58**:93–107.

87. Hinds DA, Stokowski RP, Patil N, *et al.* (2004) Matching strategies for genetic association studies in structured populations. *Am J Hum Genet* **74**:317–325.

88. Hoggart CJ, Parra EJ, Shriver MD, *et al.* (2003) Control of confounding of genetic associations in stratified populations. *Am J Hum Genet* **72**:1492–1504.

89. Hirschhorn JN, Lohmueller K, Byrne E, Hirschhorn K. (2002) A comprehensive review of genetic association studies. *Genet Med* **4**:45–61.

90. Lohmueller KE, Pearce CL, Pike M, *et al.* (2003) Meta-analysis of genetic association studies supports a contribution of common variants to susceptibility to common disease. *Nat Genet* **33**:177–182.
91. Ioannidis JP, Ntzani EE, Trikalinos TA, Contopoulos-Ioannidis DG. (2001) Replication validity of genetic association studies. *Nat Genet* **29**:306–309.
92. Teng J, Risch N. (1999) The relative power of family-based and case-control designs for linkage disequilibrium studies of complex human diseases. II. Individual genotyping. *Genome Res* **9**:234–241.
93. Zondervan KT, Cardon LR. (2004) The complex interplay among factors that influence allelic association. *Nat Rev Genet* **5**:89–100.
94. Wang WY, Barratt BJ, Clayton DG, Todd JA. (2005) Genome-wide association studies: theoretical and practical concerns. *Nat Rev Genet* **6**:109–118.
95. Hirschhorn JN, Daly MJ. (2005) Genome-wide association studies for common diseases and complex traits. *Nat Rev Genet* **6**:95–108.
96. Sham P, Bader JS, Craig IW, *et al.* (2002) DNA pooling: a tool for large-scale association studies. *Nat Rev Genet* **3**:862–871.
97. Satagopan JM, Venkatraman ES, Begg CB. (2004) Two-stage designs for gene–disease association studies with sample size constraints. *Biometrics* **60**:589–597.
98. Skol AD, Scott LJ, Abecasis GR, Boehnke M. (2006) Joint analysis is more efficient than replication-based analysis for two-stage genome-wide association studies. *Nat Genet* **38**:209–213.
99. Thomas D, Xie R, Gebregziabher M. (2004) Two-stage sampling designs for gene association studies. *Genet Epidemiol* **27**:401–414.
100. Blangero J. (2004) Localization and identification of human quantitative trait loci: king harvest has surely come. *Curr Opin Genet Dev* **14**:233–240.
101. Pritchard JK, Cox NJ. (2002) The allelic architecture of human disease genes: common disease–common variant ... or not? *Hum Mol Genet* **11**: 2417–2423.
102. Cohen JC, Kiss RS, Pertsemlidis A, *et al.* (2004) Multiple rare alleles contribute to low plasma levels of HDL cholesterol. *Science* **305**:869–872.
103. Margulies M, Egholm M, Altman WE, *et al.* (2005) Genome sequencing in microfabricated high-density picolitre reactors. *Nature* **437**:376–380.
104. Shendure J, Porreca GJ, Reppas NB, *et al.* (2005) Accurate multiplex polony sequencing of an evolved bacterial genome. *Science* **309**:1728–1732.
105. Benjamini Y, Hochberg Y. (1995) Controlling the false discovery rate: a practical and powerful approach to multiple testing. *J R Stat Soc* **57**: 289–300.
106. Storey JD, Tibshirani R. (2003) Statistical significance for genomewide studies. *Proc Natl Acad Sci USA* **100**:9440–9445.
107. Doerge RW, Churchill GA. (1996) Permutation tests for multiple loci affecting a quantitative character. *Genetics* **142**:285–294.
108. Wacholder S, Chanock S, Garcia-Closas M, *et al.* (2004) Assessing the probability that a positive report is false: an approach for molecular epidemiology studies. *J Natl Cancer Inst* **96**:434–442.

109. Marchini J, Donnelly P, Cardon LR. (2005) Genome-wide strategies for detecting multiple loci that influence complex diseases. *Nat Genet* **37**: 413–417.
110. Caspi A, Sugden K, Moffitt TE, *et al.* (2003) Influence of life stress on depression: moderation by a polymorphism in the *5-HTT* gene. *Science* **301**:386–389.
111. The ENCODE Project Consortium. (2004) The ENCODE (ENCyclopedia Of DNA Elements) Project. *Science* **306**:636–640.
112. Churchill GA, Airey DC, Allayee H, *et al.* (2004) The Collaborative Cross, a community resource for the genetic analysis of complex traits. *Nat Genet* **36**:1133–1137.

Chapter 5

Tag SNP Selection and Its Applications in Association Studies

Kui Zhang[*] and Fengzhu Sun[†]

Section on Statistical Genetics, Department of Biostatistics
School of Public Health, University of Alabama at Birmingham
Birmingham, AL 35294, USA

†*Molecular and Computational Biology Program*
Department of Biological Sciences, University of Southern California
Los Angeles, CA 90089, USA

Linkage disequilibrium (LD) plays a central role in association studies for identifying the genetic variation responsible for complex human diseases. Recent studies on LD patterns of the human genome using several large-scale genotype data have suggested that the human genome has a haplotype block structure: it can be decomposed into long regions with strong LD and relatively few haplotypes, separated by short regions of low LD. This observation has practical and important implications in association studies because a small fraction of single nucleotide polymorphisms (SNPs), referred to as "tag SNPs", can be chosen in each block to map the genetic variation responsible for complex human diseases. The use of tag SNPs can significantly reduce genotyping effort without much loss of power. Therefore, it has recently drawn considerable attention and become a very active research field. Many methods have been developed, and new methods for tag SNP selection are continuously being developed. In this chapter, we review and discuss recent developments for haplotype block partitioning and tag SNP selection, and their applications in association studies. The aim of this chapter is not to enumerate and detail all available methods for haplotype block partitioning and tag SNP selection, but rather to focus on how to use the

---

[*] Correspondence author.

available methods, tools, and resources to facilitate tag SNP selection in association studies.

1. INTRODUCTION

Over the past decade, many complex human diseases such as hypertension, diabetes, and obesity[1-3] have increased in incidence in both the United States and other developed countries. They pose a striking threat to human health. Considerable efforts have been expended to dissect the genetic etiology of such diseases in order to help us better understand their pathogenesis, with the intent of yielding improved strategies for prevention, diagnosis, and treatment.[4] The first step toward this ultimate goal is to determine the genetic variants responsible for these diseases. Linkage analysis and association analysis are the two main strategies used by researchers in this context. Both have been successfully applied to dissect the genes responsible for simple Mendelian diseases in which only one or a few genes have a major impact. These include Huntington's disease,[5] cystic fibrosis,[6] Fanconi anemia,[7] and breast cancer.[8,9] However, both strategies have thus far been less successful in identifying the genes responsible for complex diseases such as hypertension, diabetes, obesity, schizophrenia, alcoholism, etc., which likely originate from the small effects of many genes as well as gene–gene and gene–environment interactions.

The underlying principle for association analysis is based on linkage disequilibrium (LD), which refers to the nonrandom association between alleles at different, tightly linked loci in the population.[10] Due to their inherent nature, association studies for both genomewide mapping and fine mapping based on LD using single nucleotide polymorphisms (SNPs) have become increasingly popular, as they offer a potentially more cost-effective and powerful approach for gene mapping than linkage analysis.[11-15] However, the optimal design of efficient and powerful association studies requires detailed characterization of LD patterns for the regions of interest, which in turn requires the survey of a large number of SNP markers. Therefore, a primary question frequently asked by researchers in designing genomewide or candidate gene association studies concerns how many and which SNPs should be used for analysis. This certainly depends on the extent of LD across the human genome. Currently, it is prohibitively expensive to genotype all SNPs for a large number of samples. Given the number of SNPs that we can genotype, the judicial selection of SNPs that can achieve the largest power is therefore of paramount importance. Recent studies showed that the human genome can be partitioned into blocks with limited haplotype diversity such that only a small number of tag SNPs can capture most haplotypes and predict the other SNPs.[16–21] This observation suggests one way of selecting SNPs in association studies. One of the objectives of the Human HapMap Project is to describe the set of haplotype blocks and the SNPs that tag them (http://www.hapmap.org/). This is a very active research field, and many methods have been proposed[20,22–28] in the past several years.

The procedures for haplotype block partitioning and tag SNP selection in genomewide association studies, as well as in candidate gene studies, can be divided into four steps. In the first step, the samples with SNP genotype data at a relatively dense map in the regions of interest are collected. Hereafter, these samples are referred to as reference samples, which are used to obtain the haplotype block structure and tag SNPs. Reference samples with their genotype can be drawn from publicly available data, such as those from the HapMap Project[21,29] and other large-scale genotyping projects,[20,30,31] or they can be generated from researchers' personal labs. In the second step, algorithms for haplotype block partitioning and tag SNP selection are employed to identify the haplotype block structure and a set of well-spaced tag SNPs. In the third step, a larger number of samples are genotyped only at these tag SNP markers. In the fourth step, association studies are conducted using all of the genotyped samples, with knowledge of the haplotype block structure.

The aforementioned procedures raise many fundamental questions. How to select reference samples? If researchers decide to generate the preliminary data from their personal labs, how many samples and what density of SNPs should be used? If researchers decide to use the publicly available data, which data set should be used? How well do the selected tag SNPs from a population perform in mapping samples from another different population? Given so many available methods for tag SNP selection, what is the best method? These questions must be answered before the experiments can be conducted. In this chapter, we review and discuss recent developments in designing optimal strategies for haplotype block partitioning and tag SNP selection in association studies. The aim of this chapter is not to enumerate and detail all of the available methods for haplotype block partitioning and tag SNP selection, but rather to focus on how to use available methods, tools, and resources for tag SNP selection in order to facilitate association studies.

2. METHODS FOR HAPLOTYPE BLOCK PARTITIONING

Recent studies have shown that the human genome can be partitioned into discrete blocks of high-LD regions separated by shorter regions of low LD.[16–20] Within those high-LD regions, which are referred to as "blocks" in the literature, the LD between markers is relatively high and only a few common haplotypes can account for the majority of haplotype diversity. Between haplotype blocks, however, LD is observed to decay rapidly with distance between markers. This observation of haplotype block-like structure across the human genome has important implications in association studies because only a small fraction of tag SNPs can be used to efficiently represent the other SNPs and capture most of the haplotype diversity in each block. Accordingly, a variety of methods have been developed to identify haplotype blocks using SNP data.[16,18,20,22,26,32] These methods can be classified into two major categories: those that define blocks mainly on the basis of LD measures within blocks,[18,20,26] and those that identify blocks according to the decay of LD between blocks.[16,22]

Several measures have been proposed to quantify the strong LD within blocks and used to define blocks.[18,20,26,33] Patil and colleagues[20] as well as Zhang and colleagues[26] defined haplotype blocks based on the coverage of common haplotypes, in which at least a certain percentage of observed or inferred haplotypes must be common haplotypes. Gabriel and colleagues[18] as well as Phillips and colleagues[34] proposed to partition the human genome into blocks based on the pairwise LD measure D'.[10] In each block, they required that at least a certain proportion of SNP pairs have strong LD (e.g. the pairwise $|D'|$ greater than the threshold). Wang and colleagues[33] defined a set of consecutive SNPs as a block if there are no historical recombination events, which is determined by using the four-gamete test.[35] Nothnagel and colleagues[32] proposed the normalized entropy difference as a new multilocus measure for LD; then, they calculated the values of this measure with respect to different sets of consecutive SNPs, and regarded the set of SNPs with maximal normalized entropy difference as a haplotype block. Although these methods can ensure strong LD and low haplotype diversity within each block, they do not consider the LD decay between blocks. Therefore, the haplotype blocks identified by these methods may not have definite boundaries and generally contain overlaps.[20,26] It is also difficult to connect biological meanings to the position of these block boundaries.

Another group of methods for haplotype block partitioning considers not only the strong LD within each block, but also the LD decay between adjacent blocks. Daly and colleagues[16] developed a hidden Markov model (HMM) to estimate the historical recombination frequency between each pair of markers using the identified ancestral haplotypes. Then, the blocks were separated by a pair of SNPs that have significant historical recombinations. Within each block, the strong LD and low haplotype diversity are maintained because there is no significant recombination between adjacent SNPs. Anderson and Novembre[22] also developed a family of Markov models to identify haplotype blocks that can simultaneously utilize information about LD decay between blocks and about the haplotype diversity within a block. In their method, the occurrence of haplotypes within blocks and their relationship between blocks are first characterized by a set of HMMs, then the block boundaries are selected using the minimum description length (MDL) criterion.[36] Compared with the method developed by Daly and colleagues,[16] the method by Anderson and Novembre[22] does not require the information of ancestral haplotypes; it models LD between whole haplotypes carried at adjacent blocks rather than LD between two adjacent SNP markers. The MDL criterion has also been used in other methods to identify haplotype blocks.[37]

All of these methods have been applied to real data sets with two major applications: aiding tag SNP selection for association studies, and characterizing underlying biological processes such as recombination. Haplotype block identification can be performed either separate from or coupled with tag SNP selection. The first situation is quite simple: haplotype blocks are preidentified, and then tag SNPs are selected within each block.[18] In this situation, definitions for tag SNPs do not affect the identification of haplotype blocks. Patil and colleagues[20] and Zhang and colleagues[26-28] took another way of using haplotype blocks. Patil and colleagues[20] developed a greedy algorithm to minimize the number of tag SNPs. They considered all potential blocks of consecutive SNPs, and selected the block that maximizes the ratio of the number of SNPs to the number of tag SNPs in the block. The process was repeated until the entire region was covered. Subsequently, Zhang and colleagues[26,27] developed a set of dynamic programming algorithms that guarantee finding a minimum set of tag SNPs. In this situation, the identification of haplotype blocks and tag SNPs are coupled together, and definitions of tag SNPs do affect the identification of blocks.

Although the main interest in haplotype block analysis is its potential utility in tag SNP selection, it is of great interest to understand the relationship between the haplotype block structure and its underlying biological processes such as recombination. To this end, studies based on simulations and real data sets were conducted.[33,34,38–42] It has been clearly shown that recombination hot and cold spots can give rise to haplotype block-like patterns.[39–41,43] Through single-sperm genotyping on a 216-kb segment of the class II region of the major histocompatibility complex (MHC), Jeffreys and colleagues[40] conducted a study of the influences of recombination rates on LD patterns. They found six recombination hot spots that separate the high-LD regions and low-LD regions. Kauppi and colleagues[41] studied the same MHC region using samples from three populations. While they found that population history has some influence on LD patterns, they confirmed that the recombination rate plays a primary role in shaping LD patterns. Studies on the other genes, such as low-density lipoprotein-receptor–related protein gene ($LRP5$)[44] and $PGM1$ on 1p31,[45] provided additional evidence that haplotype blocks are separated by regions of higher recombination rates. In a larger-scale study, Greenwood and colleagues[43] studied the relationship between block size and recombination rate using the haplotype block data collected from the genome by Gabriel and colleagues[18] and the genomic recombination rates calculated by Kong and colleagues.[46] They found a negative correlation between the average recombination rate and the average block length, and suggested that the recombination rate plays a central role in forming haplotype block structures across the human genome. By detailed analysis of recombination rates and LD patterns for a 2.5-Mb region on chromosome 21, Greenawalt and colleagues[47] also found a strong statistically positive association between recombination rates and haplotype block boundaries.

On the other hand, empirical studies and simulations showed that many other factors, such as genetic drift, variable mutation rates, and population history, could contribute to haplotype block structures.[33,34,38,48] By analyzing chromosome 21 data[20] and using coalescent simulations,[49] Wang and colleagues[33] found that the population demographic history, recombination, and mutation together contribute to haplotype block structures and that haplotype blocks can arise in the absence of recombination hot spots. Zhang and colleagues[48] studied the effect of other factors that can contribute to haplotype block structures using simulations. They observed similar haplotype block structures even when recombination rates were uniformly distributed across the human genome, indicating that the genetic drift and other factors can generate a haplotype block-like structure. Phillips and colleagues[34] constructed a haplotype map on chromosome 19, and found that recombination hot spots are not required to explain most of the observed blocks. The analysis results from the published 5q21 data[16] by

Anderson and Slatkin[38] also showed that the block patterns found by Daly and colleagues[16] could be consistent with the absence of recombination hot spots if there was a period of rapid population growth. Jeffreys and colleagues[50] studied a 206-kb region on chromosome 1 and detected several recombination hot spots. Surprisingly, they also found that some hot spots are in regions of higher LD, suggesting that there may be a rapid population growth.

The relationship between haplotype block structures and their underlying biological mechanisms has important implications in tag SNP sections for association studies. In regions where recombination is the primary factor that creates the haplotype block structure, the haplotype structure obtained from one population is likely applicable to another population. This is evidenced by studies on the MHC region[41] and on *PGM1* on chromosome 1q31.[45] Both studies have shown that recombination hot spots play a central role in creating haplotype block structures and that haplotype blocks are similar, even identical, across the different populations studied. In this situation, only a few populations need to be surveyed to obtain haplotype structures. In regions where other factors that are unique to different populations, such as genetic drift and population history, contribute to generate haplotype block structures, it may be necessary to obtain the block structure for each population of interest separately.

Other than tag SNP selection, haplotype block structures can be useful for haplotype inference and may improve the power of association studies. Niu and colleagues[51] observed that the utilization of haplotype block boundaries can improve the accuracy of haplotype inference. Kimmel and Shamir[52] proposed a likelihood method to simultaneously model the haplotype inference and haplotype block partition using genotype data. Their simulation results indicated that their method was more efficient and accurate than PHASE,[53] a leading software for haplotype inference. In association studies, haplotype-based methods are generally regarded as being more powerful than methods based on single markers,[54,55] since the former fully exploit LD information from multiple markers. In these methods, each haplotype is treated as an allele at a multiallelic marker. Thus, the traditional x^2 test or logistic regression can be used.[56] However, the method to use in choosing the number of SNPs to form haplotypes, given a set of consecutive SNPs, is still not clear. Thus, the sliding window approach, which uses haplotypes formed by several adjacent SNPs, is generally employed. The main drawback of this approach is that it increases the number of tests and thus reduces the power.

Haplotype blocks provide a possible way to conduct such haplotype analysis: each block can be treated as a unit and the haplotype within it can be used. Daly and colleagues[16] provided a compelling example by studying a 500-kb region of the human chromosome that may contain a genetic variant responsible for Crohn's disease. They found that the association pattern based on single markers often yielded irregular, nonmonotonic pictures; while when haplotypes within each block were considered as a unit and the association was calculated based on haplotypes in other blocks using the multiallelic D', a clear, monotonic LD pattern appeared. Unfortunately, only a few methods have been proposed to tackle this problem[57] and most of them are not tailored to do this. Therefore, it is of great interest to develop new methods for this purpose in the future.

A remaining question is which method is most appropriate for block partitioning. The answer certainly depends on how inferred blocks will be used. Although a gold standard for the systematic comparison of haplotype blocks identified using these various methods does not exist, several empirical studies have been conducted to assess their performance based on various criteria. Schwartz and colleagues[58] studied the robustness of haplotype blocks identified from three methods: haplotype diversity–based method from Patil and colleagues[20] as well as Zhang and colleagues,[26] LD-based method from Gabriel and colleagues,[18] and four-gamete method from Wang and colleagues.[33] They developed a statistical method to assess the concordance of two block partitions and found that the absolute magnitude of the concordance was low, although there was non-random concordance. Based on this observation, they suggested that it might be better to select tag SNPs using methods which are not strongly dependent on block boundaries. Indap and colleagues[59] compared three methods for haplotype block partitioning,[18,19,22,26] using 1000 haplotypes generated from the coalescent model under different allele frequencies and marker densities from three populations. Although they found that the haplotype blocks identified from each method differed in terms of number, size, and coverage, and had few matching boundaries, these blocks still had overlaps and a high portion of chromosome region was common to all methods. Zeggini and colleagues[60] compared several methods for block partitioning[18,33] using 137 SNPs spanning 6.1 Mb of chromosome 17q in 189 unrelated healthy individuals. Although they found that the blocks varied in terms of number and size, the number of tag SNPs that needed to be typed were the same. Therefore, they concluded that the methods for block partitioning perform almost equally well when they are used for tag SNP selection.

However, we should be cautious when generalizing their conclusions because most of these studies were based on specific sample sizes, SNP densities, and allele frequencies. In actuality, identified haplotype blocks are affected by a number of factors, including sample size, density or number of SNPs studied, allele frequencies, fraction of missing data, and genotyping error rate. So far, only a few studies have been conducted to directly study the effects of these factors on block partitioning. Wang and colleagues[33] studied the effect of sample size on haplotype blocks identified from the four-gamete test. They found that block sizes increased with smaller sample sizes, and that the change in block size became much less when more than 100 haplotypes were used. This observation is consistent with the observation of Zhang and colleagues,[28] who studied the effect of sample size based on blocks identified using Patil's method and the dynamic programming algorithm.[26,28] Both Schulze and colleagues[61] as well as Indap and colleagues[59] studied the effect of allele frequencies on block identification. Schulze and colleagues[61] observed that the inclusion of rarer SNPs increased the number of blocks and decreased the size of the blocks. However, the inclusion of rarer SNPs did not necessarily increase the number of blocks if the density or number of SNPs was fixed, as was observed by Indap and colleagues.[59] The SNP density has been shown to be another critical factor that can affect block identification. Ke and colleagues[62] genotyped ~5000 SNPs in a 10-Mb region of chromosome 20 for samples from African-American, Asian, and UK Caucasian populations. Block boundaries using the different algorithms were identified with marker densities from 1 SNP per 2 kb to 1 SNP per 10 kb. Their results showed that longer blocks at sparser densities were broken into smaller blocks as denser SNPs were used, suggesting that dense SNPs are needed to dissect fine-scale LD patterns across the human genome.

In summary, various methods for haplotype block partitioning have been proposed that yield different numbers, block sizes, and block boundaries. It is clear that there is no single gold standard to select optimal methods for block partitioning. In addition, identified blocks are affected by a number of factors, including population history, sample size, and density or number of SNPs. Therefore, it is advisable to collect a large number of samples with a high density of SNPs and employ various methods for haplotype block partitioning in order to identify regions with common block structures.

3. METHODS FOR TAG SNP SELECTION

Many methods have recently been developed for tag SNP selection. Available methods can be classified into two groups, block-dependent methods and block-free methods, although all of them are based on LD patterns of the human genome. The first group relies on methods for haplotype block partitioning to first partition the haplotypes into blocks and then select tag SNPs within each resulting block.[18,19,26-28] The other methods select tag SNPs directly in accordance with LD patterns[24,63-67] or through comprehensive power computations[23,68] across the human genome. Following the discussion of Halldorsson and colleagues,[69] tag SNP selection can be divided into three largely independent steps: (1) identifying genomic segments where the tag SNP selection will be performed; (2) defining a measure to quantify how well a set of tag SNPs can predict all observed and/or unobserved SNPs; and (3) searching a minimum set of tag SNPs that meets a desired threshold.

It has been argued whether haplotype blocks are in fact necessary for tag SNP selection for several reasons. First, there are several available methods for haplotype block partitioning, but there is no agreement on which is the best method. Second, block boundaries are different using various methods, and there is no universal criterion to construct a set of consensus block boundaries. Third, substantial LD still remains between adjacent blocks, thus there are redundant SNPs for the tag SNPs selected over several adjacent blocks. On the other hand, the use of blocks still has several compelling advantages over block-free methods. First, the search for a minimum set of tag SNPs over a region of interest is generally NP-hard,[26,69] but such computation is tractable within blocks. For example, Zhang and colleagues[26] employed an enumeration method to detect a minimum set of SNPs that can distinguish all common haplotypes within a block. Second, block-free methods will generally keep only one of two distant SNPs as the tag SNP when they are in strong LD. This may result in imprecise estimates of disease gene location because a disease gene can be close to either one of these two SNPs. Therefore, it is desirable to constrain the tag SNP section within a certain region, even for block-free methods.

To overcome the inconsistent block boundaries while still taking the averages of block-like structure, Halldorsson and colleagues[70] assumed that any pair of SNPs within a certain distance can potentially belong to the same block. This is similar to the tag SNP selection method implemented in HaploView,[71] in which the values of r^2 between tag SNPs and all SNPs are calculated based on pairs of SNPs within a certain distance. Meng and colleagues[64] proposed to use a sliding window of a fixed number of SNPs to select tag SNPs. In summary, the tag SNP selection should be constrained within blocks, although these blocks can be identified from the methods for block partitioning or just be relatively small candidate regions that do not exceed a certain length.

A variety of measures have been developed to quantify how well a subset of SNPs represents other observed and/or unobserved SNPs in a data set. These measures can be based on either the correlation between a pair of SNPs or the correlation of a set of SNPs. Carlson and colleagues[24] suggested using the LD measure r^2 for tag SNP selection[24,72] because the statistical power of association studies is proportional to the value of r^2 under certain simplified assumptions.[73] Here, for any given subset of SNPs, all pairwise r^2 values between the SNPs in this subset and the SNPs absent in this subset are calculated. For a given SNP absent in the subset, we take the maximum value of r^2 as its individual prediction power. The minimum individual prediction power over all SNPs absent in this subset is taken as the overall prediction power. The set of SNPs with the overall prediction power greater than the threshold is defined as the set of tag SNPs. In the literature, the threshold of 0.80 has been suggested for tag SNP selection.[24,65,74] The covariance between all pairs of SNPs has been used by several researchers[63,64,75] to construct a positive definite matrix. Then, principal component analysis is performed to obtain the eigenvalues and corresponding eigenvectors. The set of SNPs that contributes the most to the eigenvectors is selected as the set of tag SNPs.

Statistical power can serve as another measure to quantify the performance of tag SNPs. The basic assumption is that every observed SNP in the initial data set has an equal probability of being the potential disease locus. For each pair of SNPs, the power of one SNP being the disease locus and one SNP being the mapping marker can be calculated. Byng and colleagues[23] proposed to rank SNPs based on such power. Hu and colleagues[76] proposed two measures for tag SNP selection: the average power and minimum power between all SNPs and the subset of SNPs. Here, for an SNP and a given subset of SNPs, all pairwise power values between this SNP being the disease locus and the SNP in this subset being the mapping marker are calculated, and the maximum power is taken as the individual power of this SNP. Then, the average power is defined as the average of the individual power over all of the SNPs, while the minimum power is defined as the minimum individual power over all of the SNPs. A subset of SNPs with the average power or the minimum power greater than a prespecified threshold is considered as a set of tag SNPs.

Instead of using the correlation between pairs of SNPs, another group of methods for tag SNP selection considers the correlation of a set of SNPs. Examples of some commonly used measures for tag SNP selection are as follows:

(1) Haplotype diversity. Haplotype diversity has been widely used as a measure for tag SNP selection.[19,26–28,66] This includes several distinct, but tightly related, measures. Patil and colleagues[20] defined the minimum set of SNPs that can uniquely distinguish a certain percentage of all the common haplotypes as a set of tag SNPs. This definition was subsequently used by Zhang and colleagues[26–28] and other researchers.[66,77] Johnson and colleagues[19] first defined haplotype diversity as the total number of differences recorded in all pairs of haplotypes, and considered the minimum set of SNPs that can account for a certain percentage of overall haplotype diversity as the set of tag SNPs. Halldorsson and colleagues[70] suggested using the informativeness measure for tag SNP selection.

(2) Haplotype entropy. Haplotype entropy has recently been used for haplotype block partitioning[32] and tag SNP selection.[78] If there are n haplotypes and the frequency of haplotype i is denoted by p_i, then entropy of these haplotypes is defined as $S = -\sum_{i=-p_i}^{n} \log p_i$. So, the set of tag SNPs is the minimum set of SNPs that can account for a certain percentage of overall haplotype entropy.

(3) Haplotype determination coefficient R_h^2. Stram and colleagues[25] proposed a formal measure, R_h^2, to characterize the uncertainty in the prediction of haplotypes from genotype data, and used it for tag SNP selection. In this method, R_h^2 for each common haplotype is calculated based on a subset of SNPs over all SNPs within a block. The minimum R_h^2 among all common haplotypes is taken as the overall haplotype prediction strength. The minimum set of SNPs with the overall haplotype prediction strength exceeding a prespecified threshold is taken as a set of tag SNPs.

(4) Haplotype r^2. Instead of using pairwise r^2, Weale and colleagues[74] proposed a method to calculate the value of r^2 between an SNP and the haplotypes composed of a subset of SNPs. A set of SNPs is considered as the set of tag SNPs if the minimum value of haplotype r^2 over all of the SNPs is greater than a prespecified threshold (e.g. ≥ 0.80).

Given so many available measures that can be used to quantify how well a set of tag SNPs represents all of the SNPs in the data set, it is important to be aware of their strengths and weaknesses. First, the pairwise LD measures and the power between pairs of SNPs are generally easy to calculate. Methods based on the correlation of multiple SNPs, on the other hand, often involve the use of haplotypes. Due to the prohibitive cost of molecular haplotyping, most large-scale projects only provide the genotype data and statistical and computational methods that are needed for haplotyping. Despite recent advancements in this area, the computation can be intensive and there is uncertainty for inferred haplotypes. Therefore, care must be exercised when haplotype-based methods are used. Second, methods for tag SNP selection that are based on the pairwise measures only consider the relationship between pairs of observed SNPs, so the unobserved SNPs in the flanking regions may not be well captured for data sets with relative sparse SNP density. Third, appropriate statistical tests should be used in combination with methods for tag SNP selection. Single marker–based association methods have been shown to be inefficient.[79,80] Fourth, measures based on power calculation can be considered as the most relevant ones. In addition, the power calculation in a simple situation outlined by Hu and colleagues[76] as well as Byng and colleagues[23] can be readily extended to adopt more complex realities, including the use of more complex disease models and study designs as well as complicated statistical tests. However, many factors, including the disease inheritance models that can affect the power calculation, are difficult to obtain, especially for complex diseases. This reduces its applicability in general situations. Fifth, the use of a single measure may not be the optimal way for tag SNP selection. There are generally multiple sets of tag SNPs of the same size given a measure. All of these sets of tag SNPs are equal under this measure; other measures can be used to obtain a more optimal set of tag SNPs.[65,81]

Given a threshold and a measure, there are many possible sets of tag SNPs. To minimize cost, a minimum set of tag SNPs is desired. Unfortunately, the search for a minimum set of tag SNPs over a region of interest is generally NP-hard,[26,70] implying that there is no algorithm which is guaranteed to be fast and able to find the optimal solution in all situations. However, when the number of SNPs over a region is not large, the enumeration method or the branch-bound method can efficiently explore all subsets of SNPs to find a minimum set of tag SNPs. Zhang and colleagues[26] developed a set of dynamic programming algorithms for tag SNP selection. In their methods, the regions of interest are first partitioned into blocks, then the tag SNPs are selected within each block using an enumeration method. The total number of tag SNPs is minimized by the dynamic programming algorithm over all possible partitions of blocks. Halldorsson and colleagues[70] proposed a similar method, but considered the haplotype blocks as sets of consecutive SNPs within a certain range. Several papers have used the branch-bound method, and have shown that it can be quite efficient when only a small region and/or a moderate number of SNPs are considered.[66,82]

Although the enumeration method and the branch-bound method are guaranteed to find the minimum set of tag SNPs, they are inefficient when a large region and/or a large number of SNPs are considered simultaneously. Therefore, many greedy algorithms have been proposed to speed up this search process.[24,65,76,83–85] Carlson and colleagues[24] proposed to sequentially add SNPs into the set of tag SNPs. Specifically, the pairwise r^2 values for a given SNP and other SNPs are calculated. The SNP with the maximum number of r^2 values greater than the specified threshold is taken as the tag SNP. Then, this tag SNP and those SNPs having r^2 values, like this tag SNP, greater than the threshold are removed and not considered in the next selection step. The process is repeated until no SNPs are left. Similar algorithms have also been proposed by Ao and colleagues[83] as well as Hu and colleagues.[76] Qin and colleagues[65] developed an exhaustive search algorithm where all SNP combinations are evaluated. They found that their method gives fewer tag SNPs than the greedy method developed by Carlson and colleagues[24] under the same criteria. Hao and colleagues[84] developed a sparse tree method based on haplotype heterogeneity or haplotype entropy. Their simulation results demonstrated that their method is more efficient than the enumeration method. Pardi and colleagues[85] formulated tag SNP selection as an optimization problem, where the objective is to maximize the statistical power given a fixed number of SNPs. They adopted a genetic algorithm to achieve this goal. As pointed out by themselves and Hubley and colleagues,[86] one of the advantages of the genetic or evolution algorithms is that they are flexible to incorporate multiple criteria into one objective function.

Finally, it is worth noting that the step for defining measures and the step for searching a minimum set of tag SNPs are largely independent. The proposed algorithms for searching tag SNPs based on one measure can be easily applied to another measure without, or only with minor, modifications. For example, the greedy algorithm based on the set cover problem and the clustering algorithm have been proposed to select tag SNPs based on the pairwise r^2 [83] or on the statistical power of tag SNPs.[76,83]

4. POWER OF ASSOCIATION STUDIES USING TAG SNPs

The use of tag SNPs in association studies can significantly reduce the genotyping effort, but it also raises important questions. The primary concern is how much power is preserved using tag SNPs instead of all SNPs. In general, there are two ways to assess the power of tag SNPs: direct and indirect ways. In direct assessment, the statistical power is calculated based on simulated samples using the tag SNPs and all of the SNPs. In most cases, the simulation procedures are designed to mimic how the tag SNPs are selected and used in practice.[87,88] Using this strategy, we can explore the power of tag SNPs under different disease models and complicated test methods. In indirect assessment, the correlation between the tag SNPs and the disease locus, which can be observed or not observed in the data, is calculated and used to measure the power of tag SNPs. Among them, the LD measure r^2 is most commonly used because the required sample size for a tag SNP to detect an indirect association with a disease is inversely proportional to the r^2 value under some simplified assumptions.[73] Such an assessment is easy to conduct and extend to large-scale studies; but it may not account for the complexities and heterogeneities in the LD mapping of disease genes, and it also cannot be readily extended to haplotype-based methods.[24,68,74]

Both direct and indirect assessment methods have been extensively used to evaluate the power loss of using tag SNPs instead of all the SNPs in the literature. Zhang and colleagues[28,88] studied this problem using simulations and found that the loss of power was moderate — certainly much smaller than if an equivalent number of markers had been randomly chosen. Similar results have been observed by other researchers[87] and in studies of quantitative trait loci.[89] Zhai and colleagues[80] argued that the method to randomly select markers by Zhang and colleagues[28,88] is not practical. They did the same simulation, but used evenly distributed SNPs with minor allele frequency (MAF) greater than the threshold for comparison. They found that the power using evenly distributed SNPs can be similar to the power using tag SNPs, sometimes even better. However, Zhai and colleagues[80] did not use haplotype-based methods, which can be more powerful than single marker–based methods.

Carlson and colleagues[24] genotyped 3178 common SNPs (with MAF $\geq$ 10%) in 24 African-American individuals and 2193 in European-American individuals by resequencing 100 genes, respectively. They then identified 1366 and 689 corresponding tag SNPs using the threshold of 0.80 for r^2 to reduce the genotyping effort of 57% and 69% while preserving 80% of the power. This observation is consistent with observations from several large-scale studies using the same method for tag SNP selection.[21,90,91] Using a stringent threshold of 1.00 for r^2 to select tag SNPs in 10 ENCODE regions from the HapMap Project, de Bakker and colleagues[91] and the International HapMap Consortium[21] found that the genotypes can be reduced by at least 46%. In addition, with a relaxed threshold of 0.80 for r^2, the genotyping effort can be further reduced by about another 30%.

Other than tag SNP selection methods themselves, many other factors — such as sample size,[28,33,87] density or number of SNPs used,[62,90,92] MAF at the disease locus,[61] fraction of missing data,[28] fraction of genotyping error,[28,93] and methods for testing the association between SNPs and the disease locus[91] — can affect the power of association studies using tag SNPs. The required sample size depends on the MAF of SNPs. Ahmadi and colleagues[90] observed that the performance of tag SNPs is good and is affected only slightly by the sample size for SNPs with MAF $\geq$ 20%, even for a sample size as low as 16. For SNPs with MAF $\geq$ 5%, several simulation studies have found that the number of tag SNPs identified from more than 50 samples is almost identical and that the power of association tests is almost the same.[28,84,87] Zeggini and colleagues[94] simulated SNP data from 500 unrelated individuals under the neutral coalescent model. They then randomly selected 200 individuals as the training data set for tag SNP selection and 300 individuals as the test data. They found that with a sample size of 45 in the training set, more than 80% of common SNPs in the test set were captured at $r^2 \geq 0.8$. But, if we would like to capture 80% of SNPs with $1\% \leq$ MAF $\leq 5\%$, the required sample size would reach 100 individuals.

SNP density is another important factor that can affect tag SNP selection. Using either a pairwise r^2 or haplotype r^2 with a threshold of 0.80 for tag SNP selection, several studies have observed that the tag SNPs selected in different SNP densities have similar power to capture SNPs observed in the data.[90–92,95] With the decrease in SNP density, a smaller set of tag SNPs is selected, but a subset of unobserved SNPs cannot be captured by tag SNPs.[91,92] Zhang and colleagues[28] also showed that the power of using tag SNPs to detect the association between the tag SNPs and an unobserved disease locus drops quickly when the SNP density decreases, as fewer and fewer SNPs are included. These observations have important implications for the HapMap Project. The identified tag SNPs only explain the variation in other already typed SNPs, and may not explain the other (common) SNPs in the population. Therefore, it has been suggested that a set of high-density SNPs is required to yield reliable sets of tag SNPs.[62,92]

Although there is no doubt that tag SNPs can be used in association studies to reduce the genotyping effort and still preserve the relatively high power, it is still not clear which method is best for tag SNP selection. Here, we emphasize that the gold standard in such comparison is the power in association studies. One method for tag SNP selection is better than the other if, and only if, the tag SNPs selected from this method have more power than the same number of tag SNPs selected from the other method.

Several studies have been conducted to compare methods for tag SNP selection using this criterion. Zhang and Sun[89] assessed the power of tag SNPs identified using the haplotype diversity and the pairwise r^2 in quantitative trait locus (QTL) mapping.[24,26] For the common disease allele and individual marker analysis, the performance of methods for tag SNP selection depends on the LD pattern around the disease locus; there is no clear pattern to indicate which approach for tag SNP selection gives the best performance. For the common disease allele and haplotype-based analysis, the tag SNPs identified based on haplotype diversity outperform the tag SNPs selected based on r^2. For the rare disease allele, the tag SNPs based on r^2 generally perform better. Ke and colleagues[95] compared three methods for tag SNP selection: the haplotype r^2,[74] the pairwise r^2,[24] and the haplotype diversity.[82] They found that there is no one method that always outperforms the other two, and a method's performance is strongly affected by the LD pattern in the regions of interest. However, some studies have shown that the method based on the haplotype r^2 can be more efficient and effective than the method based on the pairwise r^2 [74,91,96] by comparing the power to detect the single disease susceptibility locus using tag SNPs from four different methods for tag SNP selection: two based on haplotype information[19,25] and two based on pairwise r^2.[24,68] This is consistent with the previous observation that no one method consistently outperforms the others.

So far, only a few studies have been conducted and a few methods for tag SNP selection have been compared. Due to the high heterogeneity of LD patterns across the human genome and the numerous factors that can affect the power of tag SNPs, it is unlikely that there exists a single best approach. However, it is still of great interest to find out in what situation a particular method for tag SNP selection performs better than the others; hence, new studies based on simulated and/or empirical data sets are warranted.

5. COLLECTION OF REFERENCE SAMPLES

In the first step of tag SNP selection, we must decide how a subset of samples can be selected and how many individuals should be genotyped at a dense set of SNPs. There are two different designs for selecting the samples for tag SNP identification: population-specific design and case-by-case design. For the population-specific design, as proposed in the International HapMap Project (http://www.hapmap.org/), a set of samples from a population is used to determine the set of tag SNPs; any association study based on this population can use this set of tag SNPs. In the case-by-case design, a large number of samples are available, but only a subset is chosen to be genotyped at a very dense SNP map in the regions of interest.

Both designs have their advantages and disadvantages. The use of the population-based design has no additional cost, but tag SNP selection is limited to available populations, number of samples, SNP density, and genomic regions of interest. The tag SNPs selected from such data may not represent SNP variation well in studied samples. While the case-by-case design may avoid the inefficiency of tag SNPs selected from the population-based design, it has the additional cost of generating its own genotype data. Therefore, such a design is limited only to studies of small genomic regions. In addition, the cost may limit the number of samples and the density of SNPs that will be genotyped, thus also affecting the efficiency and effectiveness of tag SNP identification.

For the population-based design, genomewide-scale data[21,29,31] as well as the more detailed SNP map on some specific regions are available for haplotype block partitioning and tag SNP selection.[30] The International HapMap Project launched in 2002 is a typical example. In the HapMap project, 269 individuals are genotyped, including 90 individuals (30 parent–offspring trios) from the Yoruba in Ibadan, Nigeria; 90 individuals (30 parent–offspring trios) from the Centre d'Etude du Polymorphisme Humain (CEPH) collection in Utah, USA; 45 Han Chinese in Beijing, China; and 44 Japanese in Tokyo, Japan. In phase I of the HapMap project, 1 007 329 SNPs were sucessfully genotyped. Among these SNPs, only 3.3% of inter-SNP distances are longer than 10 kb, spanning 11.9% of the genome. In addition, 10 ENCODE regions are genotyped at 17 944 SNPs across the five megabases (Mb) (1 per 279 kb), in which the density of SNPs is approximately 10 times higher than the genomewide density of SNP. In addition to the HapMap Project, other genomewide SNP data sets are also available. Hinds and colleagues[31] have genotyped 1 586 383 SNPs in 71 Americans of European, African, and Asian ancestry (24 European-Americans, 23 African-Americans, and 24 Han Chinese from the Los Angeles area). The average distance between adjacent SNPs is about 1.9 kb. More than 95% of the genome is covered by SNPs with inter-SNP distances less than 50 kb, and about 66% of the sequenced genome is covered by SNPs with inter-SNP distances of 10 kb or less.

There are also many publicly available genotype data at specific genes and/or genomic regions for tag SNP selection.[30] Since these data contain only SNPs in one or several regions, they usually have a much higher SNP density than that in genomewide SNP data. For example, Crawford and colleagues[30] have completely resequenced 100 genes for inflammation, lipid metabolism, and blood pressure regulation in two populations (23 individuals selected from the CEPH and 24 individuals from the African-American Human Variation Panel). The average density of SNPs identified in the African decedent and European decedent populations across all 100 genes are 4.7 and 2.8 SNPs per kb, respectively. They have integrated their data and analysis results, including the genotype data, the allele frequency, the haplotypes inferred from PHASE,[53] the tag SNP selected using the pairwise r^2,[24] etc., at http://pga.gs.washington.edu/finished_genes.html. As another example, Ahmadi and colleagues[90] have genotyped 55 genes at 904 SNPs in 64 unrelated CEPH individuals of European ancestry and 64 unrelated individuals of Japanese ancestry. The average densities for SNPs with MAF $\geq$ 5% are 1 SNP per 3.7 kb and 4.27 kb for the European decedent and Japanese decedent populations, respectively. Other similar data sets have also been generated for the purpose of tag SNP selection,[44,97,98] and we expect more data sets to be available for tag SNP selection in the near future.

In the population-specific design, there is no additional cost for genotyping, but tag SNP selection is limited by available populations, number of samples, SNP densities, and genomic regions of interest. Researchers may use samples for mapping from a population, but select tag SNPs based on available samples from a different population. Several studies based on the real data sets clearly showed that the haplotype block structure varies among different populations.[99] Consequently, the set of tag SNPs varies among different populations, and a set of tag SNPs identified from a population may not perform well in another population.[74,99] Even when the samples used for tag SNP selection and for mapping are from the same population, the selected tag SNPs may not perform well in the independent mapping samples. Therefore, the success of the population-specific design depends heavily on the transferability of the selected SNPs — how well the selected SNPs perform in another set of samples from the same or a different population.

Recently, many empirical studies that have been conducted to assess the transferability of tag SNPs have resulted in controversial conclusions. Some studies have shown that the tag SNPs selected from one population can be used in samples from the same population or other different populations.[90,100–104] Ahmadi and colleagues[90] genotyped 11 genes in 64 CEPH samples, and identified a set of SNPs using a multiple-marker criterion (haplotype r^2) within each gene.[74] They then genotyped these tag SNPs and an additional 16 candidate functional SNPs in 238 females from another north European population, and found that haplotype r^2 between the tag SNPs and 16 candidate SNPs is at least 0.68 with an average of 0.89. Therefore, they concluded that selected tag SNPs perform well in samples from a population that is different from the one in which the tag SNPs were selected, and expected those tag SNPs to perform well in other northern European samples. Gonzale-Neira and colleagues[100] analyzed 144 SNPs in a 1-Mb region of chromosome 22 in 1055 individuals from 38 worldwide populations, classified into seven continental groups. They found that the selected tag SNPs are highly informative in other populations within each continental group. They suggested that the tag SNPs defined from the HapMap project would be useful for other populations in the world. Huang and colleagues[101] genotyped over 20 000 SNPs in chromosome 21 for samples from seven populations. They showed that the tag SNPs selected from one population can be applied to the other populations. Ribas and colleagues[102] studied SNPs in 175 cancer-associated genes, and found that the tag SNPs chosen from the HapMap Project performed generally well in the Spanish population. Tenesa and Dunlop[103] simulated the LD pattern on a 10-Mb region on chromosome 20 for two European and two Asian populations. Their results showed that the power is not reduced using tag SNPs selected from other populations. Willer and colleagues[104] also demonstrated that the HapMap CEU samples provide an adequate basis for tag SNP selection in Finnish individuals.

In contrast, other studies suggested that tag SNPs selected from one population are inadequate for use in another population.[24,99,105–107] Beaty and colleagues[105] genotyped 11 candidate genes for 283 individuals from four different populations (European-Americans, Chinese-Singaporeans, Malay-Singaporeans, and Indian-Singaporeans). Although allele frequencies and overall patterns of LD were similar across these four populations, haplotype frequencies varied significantly across populations. Therefore, they concluded that the selected tag SNPs from one population may not be applicable to another population. Based on a threshold of $r^2 = 0.5$, Carlson and colleagues[24] identified 435 European-American tag SNPs. Only 1028 (32%) of 3178 SNPs in the African-American population had $r^2 \geq 0.5$ with the tag SNPs from the European-American population, implying that the tag SNPs selected from the European-American population cannot capture the common SNPs in the African-American population well. Mueller and colleagues[107] genotyped four genomic regions using samples from eight European populations. They observed the conservation of the LD across eight populations; and found that the tag SNPs selected from 30 CEPH trios of the HapMap Project performed well in only two populations, but not in the others.

In summary, the transferability of selected SNPs is highly subject to the populations that are going to be studied. In addition, tag SNP selection in the population-specific design is also limited by the availability and number of samples, SNP densities, and genomic regions of interest. Therefore, for any tag SNP selection using the population-specific design, a careful evaluation of available population samples should be conducted first.

6. DISCUSSION

In this chapter, we have reviewed and discussed recent developments in designing optimal strategies for haplotype block partitioning and tag SNP selection in association studies, especially in reference sample collection and analytical methods for block partitioning and tag SNP selection. In the analytical methods for tag SNP selection, we implicitly assume that all SNPs are equally important from biological and technical perspectives, so tag SNP selection is only subject to the LD among them. However, from a biological point of view, some SNPs such as nonsynonymous SNPs in coding regions and SNPs in regulatory regions are more important than the other SNPs. In this situation, these tag SNPs can be forced to be tag SNPs.[62,65] From a technical point of view, some SNPs may be difficult to genotype using a particular technology. In this situation, these SNPs can be simply excluded from tag SNP selection. More complex methods, such as the evolution algorithms,[86] have been proposed to take the "quality" of SNPs into the selection procedure.

Most of the available methods for haplotype block partitioning and tag SNP selection have been implemented into efficient software that is publicly available.[71,81–83,108,109] Some of them have integrated the HapMap data download, haplotype block partitioning and tag SNP selection, and the biological information of SNPs into a tool suit or a comprehensive web server[71] (http://gvs.gs.washington.edu/GVS/). However, they have limited options for different data resources and methods for tag SNP selection. Although it is impractical to integrate all available data resources and methods for tag

SNP selection, tools that can provide multiple data sources and tag SNP selection methods are still warranted.

In summary, tag SNP selection is highly subjective and complicated. Many open questions remain, and some of them can only be answered when more genotyping data become available. Currently, there are multiple choices available, but none of them is appropriate in all situations in each step of tag SNP selection. Thus, the optimal procedure for tag SNP selection in practice can only be achieved by the careful consideration of each step and the extensive collaborative work of experts from many different fields.

ACKNOWLEDGMENTS

The work was partially supported by the NIH grant R01GM074913 (K. Zhang) and the NIH grant P50 HG 002790 (F. Sun).

REFERENCES

1. Ogden CL, Flegal KM, Carroll MD, Johnson CL. (2002) Prevalence and trends in overweight among US children and adolescents, 1999–2000. *JAMA* **288**:1728–1732.
2. Flegal KM, Carroll MD, Ogden CL, Johnson CL. (2002) Prevalence and trends in obesity among US adults, 1999–2000. *JAMA* **288**:1723–1727.
3. Harris MI, Flegal KM, Cowie CC, *et al.* (1998) Prevalence of diabetes, impaired fasting glucose, and impaired glucose tolerance in US adults — the Third National Health and Nutrition Examination Survey, 1988–1994. *Diabetes Care* **21**:518–524.
4. Risch N. (2000) Searching for genetic determinants in the new millennium. *Nature* **405**:847–856.
5. Gusella JF, Wexler NS, Conneally PM, *et al.* (1983) A polymorphic DNA marker genetically linked to Huntington's disease. *Nature* **306**:234–238.
6. Kerem B, Rommens JM, Buchanan JA, *et al.* (1989) Identification of the cystic fibrosis gene: genetic analysis. *Science* **245**:1073–1080.
7. Strathdee CA, Gavish H, Shannon WR, Buchwald M. (1992) Cloning of cDNAs for Faconi's anaemia by functional complementation. *Nature* **356**:763–767.
8. Miki Y, Swensen J, Shattuck-Eidens D, *et al.* (1994) A strong candidate for the breast and ovarian cancer susceptibility gene *BRCA1*. *Science* **253**: 66–71.
9. Wooster R, Bignell G, Lancaster J, *et al.* (1995) Identification of the breast cancer susceptibility gene *BRCA2*. *Nature* **378**:789–792.
10. Lewontin RC. (1964) The interaction of selection and linkage. I. General considerations. *Genetics* **49**:49–67.
11. Ardlie K, Kruglyak L, Seielstad M. (2002) Patterns of linkage disequilibrium in the human genome. *Nat Rev Genet* **3**:299–309.

12. Botstein D, Risch N. (2003) Discovering genotypes underlying human phenotypes: past successes for Mendelian disease, future approaches for complex disease. *Nat Genet* **33**:228–237.

13. Clark AG. (2003) Finding genes underlying risk of complex disease by linkage disequilibrium mapping. *Curr Opin Genet Dev* **13**:296–302.

14. Nordborg M, Tavaré S. (2002) Linkage disequilibrium: what history has to tell us. *Trends Genet* **18**:83–90.

15. Weiss KM, Clark AG. (2002) Linkage disequilibrium and the mapping of complex human traits. *Trends Genet* **18**:19–24.

16. Daly MJ, Rioux JD, Schaffner SE, *et al.* (2001) High-resolution haplotype structure in the human genome. *Nat Genet* **29**:229–232.

17. Dawson E, Abecasis GR, Bumpstead S, *et al.* (2001) A first-generation linkage disequilibrium map of chromosome 22. *Nature* **418**:544–548.

18. Gabriel SB, Schaffner SF, Nguyen H, *et al.* (2002) The structure of haplotype blocks in the human genome. *Science* **296**:2225–2229.

19. Johnson GCL, Esposito L, Barratt BJ, *et al.* (2001) Haplotype tagging for the identification of common disease genes. *Nat Genet* **29**: 233–237.

20. Patil N, Berno AJ, Hinds DA, *et al.* (2001) Blocks of limited haplotype diversity revealed by high-resolution scanning of human chromosome 21. *Science* **294**:1719–1723.

21. The International HapMap Consortium. (2005) A haplotype map of the human genome. *Nature* **437**:1299–1320.

22. Anderson EC, Novembre J. (2003) Finding haplotype block boundaries by using the minimum-description-length principle. *Am J Hum Genet* **73**: 336–354.

23. Byng MC, Whittaker JC, Cuthbert AR, *et al.* (2003) SNP subset selection for genetic association studies. *Ann Hum Genet* **67**:543–556.

24. Carlson CS, Eberle MA, Rieder MJ, *et al.* (2004) Selecting a maximally informative set of single-nucleotide polymorphisms for association analyses using linkage disequilibrium. *Am J Hum Genet* **74**:106–120.

25. Stram DO, Haiman CA, Hirschhorn JN, *et al.* (2003) Choosing haplotype-tagging SNPs based on unphased genotype data using a preliminary sample of unrelated subjects with an example from the Multiethnic Cohort Study. *Hum Hered* **55**:27–36.

26. Zhang K, Deng MH, Chen T, *et al.* (2002) A dynamic programming algorithm for haplotype block partitioning. *Proc Natl Acad Sci USA* **99**: 7335–7339.

27. Zhang K, Sun FZ, Waterman MS, Chen T. (2003) Haplotype block partition with limited resources and applications to human chromosome 21 haplotype data. *Am J Hum Genet* **73**:63–73.

28. Zhang K, Qin Z, Ting C, *et al.* (2004) Haplotype block partitioning and tag SNP selection using genotype data and their applications to association studies. *Genome Res* **14**:908–916.

29. The International HapMap Consortium. (2003) The International HapMap Project. *Nature* **426**:789–796.

30. Crawford DC, Carlson CS, Rieder MJ, *et al.* (2004) Haplotype diversity across 100 candidate genes for inflammation, lipid metabolism, and blood pressure regulation in two populations. *Am J Hum Genet* **74**:610–622.

31. Hinds DA, Stuve LL, Nilsen GB, *et al.* (2005) Whole-genome patterns of common DNA variation in three human populations. *Science* **307**: 1072–1079.

32. Nothnagel M, Furst R, Rohde K. (2002) Entropy as a measure for linkage disequilibrium over multilocus haplotype blocks. *Hum Hered* **54**:186–198.

33. Wang N, Akey JM, Zhang K, *et al.* (2002) Distribution of recombination crossovers and the origin of haplotype blocks: the interplay of population history, recombination, and mutation. *Am J Hum Genet* **71**: 1227–1234.

34. Phillips MS, Lawrence R, Sachidanandam R, *et al.* (2003) Chromosome-wide distribution of haplotype blocks and the role of recombination hot spots. *Nat Genet* **33**:382–387.

35. Hudson RR, Kaplan NL. (1985) Statistical properties of the number of recombination events in the history of a sample of DNA sequences. *Genetics* **111**:147–164.

36. Rissanen J. (1978) Modeling by shortest data description. *Automatica* **14**:465–471.

37. Greenspan G, Geiger D. (2004) Model-based inference of haplotype block variation. *J Comput Biol* **11**:495–506.

38. Anderson EC, Slatkin M. (2004) Population-genetic basis of haplotype blocks in the 5q31 region. *Am J Hum Genet* **74**:40–49.

39. Crawford DC, Bhangale T, Li N, *et al.* (2004) Evidence for substantial fine-scale variation in recombination rates across the human genome. *Nat Genet* **36**:700–706.

40. Jeffreys AJ, Kauppi L, Neumann R. (2001) Intensely punctate meiotic recombination in the class II region of the major histocompatibility complex. *Nat Genet* **29**:217–222.

41. Kauppi L, Sajantila A, Jeffreys AJ. (2003) Recombination hotspots rather than population history dominate linkage disequilibrium in the MHC class II region. *Hum Mol Genet* **12**:33–40.

42. Wall JD, Pritchard JK. (2003) Assessing the performance of the haplotype block model of linkage disequilibrium. *Am J Hum Genet* **73**:502–515.

43. Greenwood TA, Rana BK, Schork NJ. (2004) Human haplotype block sizes are negatively correlated with recombination rates. *Genome Res* **14**: 1358–1361.

44. Twells RCJ, Mein CA, Philips MS, *et al.* (2003) Haplotype structure, LD blocks, and uneven recombination within the *LRPS* gene. *Genome Res* **13**:845–855.

45. Rana NA, Ebenezer ND, Webster AR, *et al.* (2004) Recombination hotspots and block structure of linkage disequilibrium in the human genome exemplified by detailed analysis of *PGM1* on 1p31. *Hum Mol Genet* **13**:3089–3102.

46. Kong A, Gudbjartsson DF, Sainz J, *et al.* (2002) A high-resolution recombination map of the human genome. *Nat Genet* **31**:241–247.

47. Greenawalt DM, Cui XF, Wu YJ, *et al.* (2006) Strong correlation between meiotic crossovers and haplotype structure in a 2.5-Mb region on the long arm of chromosome 21. *Genome Res* **16**:208–214.

48. Zhang K, Akey JM, Wang N, *et al.* (2003) Randomly distributed crossovers may generate block-like patterns of linkage disequilibrium: an act of genetic drift. *Hum Genet* **113**:51–59.

49. Hudson RR. (1983) Properties of a neutral-allele model with intragenic recombination. *Theor Popul Biol* **23**:183–201.

50. Jeffreys AJ, Neumann R, Panayi M, *et al.* (2005) Human recombination hot spots hidden in regions of strong marker association. *Nat Genet* **37**: 601–606.

51. Niu T, Qin Z, Xu X, Liu JS. (2002) Bayesian haplotype inference for multiple linked single-nucleotide polymorphisms. *Am J Hum Genet* **70**:157–159.

52. Kimmel G, Shamir R. (2005) GERBIL: genotype resolution and block identification using likelihood. *Proc Natl Acad Sci USA* **102**:158–162.

53. Stephens M, Smith NJ, Donnelly P. (2001) A new statistical method for haplotype reconstruction from population data. *Am J Hum Genet* **68**: 978–989.

54. Akey J, Jin L, Xiong M. (2001) Haplotypes vs. single marker linkage disequilibrium tests: what do we gain? *Eur J Hum Genet* **9**:291–300.

55. Morris RW, Kaplan NL. (2002) On the advantage of haplotype analysis in the presence of multiple disease susceptibility alleles. *Genet Epidemiol* **23**: 221–233.

56. Wallenstein S, Hodge SE, Weston A. (1998) Logistic regression model for analyzing extended haplotype data. *Genet Epidemiol* **15**:173–181.

57. Kimmel G, Shamir R. (2005) A block-free hidden Markov model for genotypes and its application to disease association. *J Comput Biol* **12**: 1243–1260.

58. Schwartz R, Halldorsson BV, Bafna V, *et al.* (2003) Robustness of inference of haplotype block structure. *J Comput Biol* **10**:13–19.

59. Indap AR, Marth GT, Struble CA, *et al.* (2005) Analysis of concordance of different haplotype block partitioning algorithms. *BMC Bioinformatics* **6**:303.

60. Zeggini E, Barton A, Eyre S, *et al.* (2005) Characterisation of the genomic architecture of human chromosome 17q and evaluation of different methods for haplotype block definition. *BMC Genet* **6**:21.

61. Schulze TG, Zhang K, Chen YS, *et al.* (2004) Defining haplotype blocks and tag single-nucleotide polymorphisms in the human genome. *Hum Mol Genet* **13**:335–342.

62. Ke X, Hunt S, Tapper W, *et al.* (2004) The impact of SNP density on fine-scale patterns of linkage disequilibrium. *Hum Mol Genet* **13**:577–588.

63. Lin Z, Altman RB. (2004) Finding haplotype tagging SNPs by use of principal components analysis. *Am J Hum Genet* **75**:850–861.

64. Meng ZL, Zaykin DV, Xu CF, *et al.* (2003) Selection of genetic markers for association analyses using linkage disequilibrium and haplotypes. *Am J Hum Genet* **73**:115–130.

65. Qin ZS, Gopalakrishnan S, Abecasis GR. (2006) An efficient comprehensive search algorithm for tagSNP selection using linkage disequilibrium criteria. *Bioinformatics* **22**:220–225.

66. Sebastiani P, Lazarus R, Weiss ST, *et al.* (2003) Minimal haplotype tagging. *Proc Natl Acad Sci USA* **100**:9900–9905.

67. Nicolas P, Sun F, Li LM. (2006) A model-based approach to selection of tag SNPs. *BMC Bioinformatics* **7**:303.

68. Cousin E, Genin E, Mace S, *et al.* (2003) Association studies in candidate genes: strategies to select SNPs to be tested. *Hum Hered* **56**:151–159.

69. Halldorsson BV, Istrail S, De la Vega FM. (2004) Optimal selection of SNP markers for disease association studies. *Hum Hered* **58**:190–202.

70. Halldorsson BV, Bafna V, Lippert R, *et al.* (2004) Optimal haplotype block-free selection of tagging SNPs for genome-wide association studies. *Genome Res* **14**:1633–1640.

71. Barrett JC, Fry B, Maller J, Daly MJ. (2005) Haploview: analysis and visualization of LD and haplotype maps. *Bioinformatics* **21**:263–265.

72. Zhang K, Jin L. (2003) HaploBlockFinder: haplotype block analyses. *Bioinformatics* **19**:1300–1301.

73. Pritchard JK, Przeworski M. (2001) Linkage disequilibrium in humans: models and data. *Am J Hum Genet* **69**:1-14.

74. Weale ME, Depondt C, Macdonald SJ, *et al.* (2003) Selection and evaluation of tagging SNPs in the neuronal-sodium-channel gene *SCN1A*: implications for linkage-disequilibrium gene mapping. *Am J Hum Genet* **73**:551–565.

75. Horne BD, Camp NJ. (2004) Principal component analysis for selection of optimal SNP-sets that capture intragenic genetic variation. *Genet Epidemiol* **26**:11–21.

76. Hu XL, Schrodi SJ, Ross DA, Cargill M. (2004) Selecting tagging SNPs for association studies using power calculations from genotype data. *Hum Hered* **57**:156–170.

77. Huang YT, Zhang K, Chen T, Chao KM. (2005) Selecting additional tag SNPs for tolerating missing data in genotyping. *BMC Bioinformatics* **6**:263.

78. Liu ZQ, Lin SL. (2005) Multilocus LD measure and tagging SNP selection with generalized mutual information. *Genet Epidemiol* **29**:353–364.

79. Zhang WH, Collins A, Morton NE. (2004) Does haplotype diversity predict power for association mapping of disease susceptibility? *Hum Genet* **115**: 157–164.

80. Zhai WW, Todd MJ, Nielsen R. (2004) Is haplotype block identification useful for association mapping studies? *Genet Epidemiol* **27**:80–83.

81. Zhang K, Qin Z, Chen T, *et al.* (2005) HapBlock: haplotype block partitioning and tag SNP selection software using a set of dynamic programming algorithms. *Bioinformatics* **21**:131–134.

82. Ke XY, Cardon LR. (2003) Efficient selective screening of haplotype tag SNPs. *Bioinformatics* **19**:287–288.

83. Ao SI, Yip K, Ng M, *et al.* (2005) CLUSTAG: hierarchical clustering and graph methods for selecting tag SNPs. *Bioinformatics* **21**:1735–1736.

84. Hao K, Liu SM, Niu TH. (2005) A sparse marker extension tree algorithm for selecting the best set of haplotype tagging single nucleotide polymorphisms. *Genet Epidemiol* **29**:336–352.

85. Pardi F, Lewis CM, Whittaker JC. (2005) SNP selection for association studies: maximizing power across SNP choice and study size. *Ann Hum Genet* **69**:733–746.

86. Hubley RM, Zitzler E, Roach JC. (2003) Evolutionary algorithms for the selection of single nucleotide polymorphisms. *BMC Bioinformatics* **4**:30.

87. Thompson D, Stram D, Goldgar D, Witte JS. (2003) Haplotype tagging single nucleotide polymorphisms and association studies. *Hum Hered* **56**: 48–55.

88. Zhang K, Calabrese P, Nordborg M, Sun FZ. (2002) Haplotype block structure and its applications to association studies: power and study designs. *Am J Hum Genet* **71**:1386–1394.

89. Zhang K, Sun FZ. (2005) Assessing the power of tag SNPs in the mapping of quantitative trait loci (QTL) with extremal and random samples. *BMC Genet* **6**:51.

90. Ahmadi KR, Weale ME, Xue ZYY, *et al.* (2005) A single-nucleotide polymorphism tagging set for human drug metabolism and transport. *Nat Genet* **37**:84–89.

91. de Bakker PIW, Yelensky R, Pe'er I, *et al.* (2005) Efficiency and power in genetic association studies. *Nat Genet* **37**:1217–1223.

92. Iles MM. (2005) The effect of SNP marker density on the efficacy of haplotype tagging SNPs — a warning. *Ann Hum Genet* **69**:209–215.

93. Liu WL, Zhao W, Chase GA. (2006) The impact of missing and erroneous genotypes on tagging SNP selection and power of subsequent association tests. *Hum Hered* **61**:31–44.

94. Zeggini E, Rayner W, Morris AP, *et al.* (2005) An evaluation of HapMap sample size and tagging SNP performance in large-scale empirical and simulated data sets. *Nat Genet* **37**:1320–1322.

95. Ke XY, Miretti MM, Broxholme J, *et al.* (2005) A comparison of tagging methods and their tagging space. *Hum Mol Genet* **14**:2757–2767.

96. Cousin E, Deleuze JF, Genin E. (2006) Selection of SNP subsets for association studies in candidate genes: comparison of the power of different strategies to detect single disease susceptibility locus effects. *BMC Genet* **7**:20.

97. Kamatani N, Sekine A, Kitamoto T, *et al.* (2004) Large-scale single-nucleotide polymorphism (SNP) and haplotype analyses, using dense SNP maps, of 199 drug-related genes in 752 subjects: the analysis of the association between uncommon SNPs within haplotype blocks and the haplotypes constructed with haplotype-tagging SNPs. *Am J Hum Genet* **75**:190–203.

98. Lee HJ, Kim KJ, Park MH, *et al.* (2005) Single-nucleotide polymorphisms and haplotype LD analysis of the 29-kb *IGF2* region on chromosome 11p15.5 in the Korean population. *Hum Hered* **60**:73–80.

99. Liu NJ, Sawyer SL, Mukherjee N, *et al.* (2004) Haplotype block structures show significant variation among populations. *Genet Epidemiol* **27**:385–400.

100. Gonzalez-Neira A, Ke XY, Lao O, *et al.* (2006) The portability of tagSNPs across populations: a worldwide survey. *Genome Res* **16**:323–330.
101. Huang W, He YG, Wang HF, *et al.* (2006) Linkage disequilibrium sharing and haplotype-tagged SNP portability between populations. *Proc Natl Acad Sci USA* **103**:1418–1421.
102. Ribas G, Gonzalez-Neira A, Salas A, *et al.* (2006) Evaluating HapMap SNP data transferability in a large-scale genotyping project involving 175 cancer-associated genes. *Hum Genet* **118**:669–679.
103. Tenesa A, Dunlop MG. (2006) Validity of tagging SNPs across populations for association studies. *Eur J Hum Genet* **14**:357–363.
104. Willer CJ, Scott LJ, Bonnycastle LL, *et al.* (2006) Tag SNP selection for Finnish individuals based on the CEPH Utah HapMap database. *Genet Epidemiol* **30**:180–190.
105. Beaty TH, Fallin MD, Hetmanski JB, *et al.* (2005) Haplotype diversity in 11 candidate genes across four populations. *Genetics* **171**:259–267.
106. Lohmueller KE, Wong LJC, Mauney MM, *et al.* (2006) Patterns of genetic variation in the hypertension candidate gene *GRK4*: ethnic variation and haplotype structure. *Ann Hum Genet* **70**:27–41.
107. Mueller JC, Lohmussaar E, Magi R, *et al.* (2005) Linkage disequilibrium patterns and tagSNP transferability among European populations. *Am J Hum Genet* **76**:387–398.
108. Ding KY, Zhang J, Zhou KX, *et al.* (2005) HtSNPer1.0: software for haplotype block partition and htSNPs selection. *BMC Bioinformatics* **6**:38.
109. Xu H, Gregory SG, Hauser ER, *et al.* (2005) SNPselector: a web tool for selecting SNPs for genetic association studies. *Bioinformatics* **21**:4181–4186.

Chapter 6

Haplotype Inference and Association Analysis in Unrelated Samples

Zeny Feng*, Nianjun Liu† and Hongyu Zhao‡

*Department of Mathematics and Statistics, University of Guelph
Guelph, Ontario N1G 2W1, Canada

†Department of Biostatistics, School of Public Health
University of Alabama at Birmingham, Birmingham, AL 35294, USA

‡Department of Epidemiology & Public Health and Department of Genetics
Yale University School of Medicine, New Haven, CT 06520, USA

Haplotype, the combination of marker alleles on the same chromosome that were inherited as a unit from one parent, plays a very important role in the study of the genetic basis of diseases and population genetics. Because haplotypes may carry more information than single markers, a variety of statistical methods, such as those for assessing haplotype–disease associations, have been proposed for studying haplotypes. With unambiguous haplotype information, haplotypes may be treated as alleles of a single locus, and many existing methods for single markers can be used in analysis. Because obtaining haplotype information directly from experiments can be costly, haplotype analyses present many unique challenges and opportunities for human geneticists. In this chapter, we focus on two issues: haplotype inference and haplotype association analysis. We first review several methods for haplotype inference, including Clark's algorithm, methods based on the expectation-maximization algorithm, Bayesian methods, and others. In the context of haplotype association analysis, we cover a number of statistical methods. In addition, we discuss the advantages and limitations of the different methods.

---

‡Correspondence author.

1. INTRODUCTION

A haplotype is the combination of marker alleles on a single chromosome. With the availability of high-density markers, haplotype-based analysis plays an ever-increasing important role in various genetics studies. For markers located close to each other in a chromosomal region, e.g. markers within the same gene, there is often allelic association or linkage disequilibrium (LD) among them, i.e. certain alleles of multiple markers on the same chromosome tend to be associated with each other. When there is LD among markers, analyzing haplotypes constructed from individual markers, e.g. single nucleotide polymorphisms (SNPs), may provide more power for disease association studies. In the case of single marker–based analysis, Sham and colleagues[1] showed that the statistical power to identify disease-susceptible variants increases as the number of alleles increases, and this effect is particularly strong when more than one pathogenic mutation event has occurred in the gene during evolution.

As SNPs in general are biallelic, it is relatively uninformative to test them individually. When there is LD among SNPs, haplotypes constructed from SNPs may provide more information for association studies. For these reasons, haplotype-based methods are preferable to single SNP–based analysis. For example, Fallin and colleagues[2] demonstrated that within the *APOE* gene, two SNPs, C19M3 and C19M4, showed the strongest association with Alzheimer's disease. Even though these two SNPs were not genotyped for association study, haplotype-based association analysis could recapture this association signal in the region. In some cases, when multiple closely linked markers are tested individually, none of them shows a significant association with disease, but analysis based on haplotypes formed by these markers can show significant results.[3] Moreover, single marker–based analysis provides no phase information on whether there is a *cis*-acting (mutated alleles on the same chromosome, i.e. haplotype) or *trans*-acting (mutated alleles on the opposite homologous chromosome) effect on the trait of interest. It is often the case that *cis*-acting mutations have a stronger effect on the trait.

Biologically, several mutations on a haplotype may cause a series of changes in amino acid coding, leading to a larger joint effect on the trait of interest than the single amino acid change caused by a single mutation.[4] Examples include the lipoprotein lipase gene in human,[5] the *HPC2/ELAC2* gene which influences prostate cancer,[6] and a gene influencing initial lactase activity in human.[7] In this case, haplotypes should be more informative than individual genotypes in revealing disease-causing mechanisms for a candidate gene.

It is well known that SNPs are the most common type of DNA variant in the human genome. Compared to other types of polymorphisms, SNPs have a relatively low mutation rate, high genotyping accuracy, and are most likely to be functionally relevant. In addition, recent technological developments have enabled researchers to genotype dense SNPs at an affordable cost. The International HapMap Project (http://www.hapmap.org/) has characterized the allele frequencies and dependence patterns of millions of SNPs in the human genome in four populations in three geographical areas. As discussed above, genetic association analysis based on individual SNPs in the presence of high-density SNP data may not be efficient; many genetics studies now collect and jointly analyze information from dense SNPs.

When haplotype information (phase information) is collected unambiguously, haplotypes across a set of markers can be treated in the same way as alleles at a single marker. As such, traditional association methods for a single marker can be applied to analyze haplotypes, although haplotypes comprising information from many markers do carry more information. However, in general, unambiguous haplotype information is not easily obtainable because it is both laborious and expensive to obtain haplotypes through molecular methods. In practice, analytical methods are commonly used to infer haplotypes based on individual marker genotype data. In Sec. 2, we will review several commonly used techniques for haplotype reconstruction: parsimony-based methods and likelihood-based methods.

In general, even with the more sophisticated methods, there are often uncertainties associated with haplotype inference. If an association analysis based on the inferred haplotypes does not account for haplotype inference uncertainty, there can be biases in association analysis.[8] Thus, several methods have been proposed in the literature that incorporate haplotype inference into haplotype association analyses, where unphased genotype data are used directly. In Sec. 3, we will review the existing methods, including both regression approaches and haplotype-sharing approaches, for testing and estimating specific haplotype effects on disease risk. Finally, in Sec. 4, we will discuss and compare the various approaches.

2. HAPLOTYPE INFERENCE

In the past decade, different analytical methods have been developed to resolve haplotypes based on the observed genotype data. Suppose genotype data at m SNPs are collected from N individuals. Without loss of generality, we denote the two alleles for each SNP as 0 and 1. For example, an individual with heterozygous genotypes at two SNPs can be represented as $\{(0, 1), (0, 1)\}$.

One issue in haplotype inference is to assign a pair of binary vectors of length m that best explain the observed genotypes of each individual. If a genotype vector contains zero or only one site that is heterozygous, then this genotype vector gives a pair of unambiguous haplotypes; we call this genotype "resolved" or "unambiguous". If a genotype vector contains more than one heterozygous site, then this genotype vector gives more than one possible pair of haplotypes that are compatible with it; we call this genotype "unresolved" or "ambiguous". For example, if the observed genotype vector of an individual at four SNPs is $\{(0,0),(0,1),(1,1),(0,1)\}$, there are two possible pairs of haplotypes that are compatible with the observed genotypes: $\{(0010,0111),(0110,0011)\}$. For a genotype vector with $k(k > 0)$ heterozygous sites, there are 2^{k-1} possible haplotype pairs that are compatible with the observed genotypes. The goal of haplotype inference includes both the estimate of haplotype frequencies in the sampled population and the assignment of the most likely haplotype pair(s) to each individual.

In general, the existing methods for haplotype inference can be categorized into two major approaches: parsimony approaches and statistical approaches. The parsimony approaches attempt to use the least number of distinct haplotypes to describe the observed genotypes. The statistical methods, on the other hand, are based on statistical models that describe the distributions of haplotypes in the population. Several commonly used algorithms belonging to these approaches are discussed in Secs. 2.1 and 2.2, respectively. Some extensions and other methods are discussed in Sec. 2.3.

2.1. *Parsimony Approaches*

2.1.1. *Clark's Algorithm*

Clark's algorithm[9] is probably the first one in the literature for haplotype inference. This algorithm starts by identifying all of the individuals who are homozygotes and single-site heterozygotes, and enumerating all of the resulting unambiguous haplotypes from these genotype vectors to form an initial set of resolved haplotypes denoted by R. Then, an inference rule is applied to infer the remaining unresolved genotypes.

Clark's inference rule: Suppose an ambiguous genotype vector V has k ambiguous sites (i.e. k heterozygous sites). Among 2^{k-1} pairs of haplotypes that are potential solutions of V, let r be the one that is a member in the resolved set R. We infer that V is the combination of r and another uniquely determined complement of r, denoted by r'. Then, vector

r' is added to the set of known resolved haplotypes R if it is not in the set, and V is removed from the ambiguous genotype vector set.

In practice, we can apply Clark's inference rule repeatedly and stop the algorithm when either all of the genotypes have been resolved or no further genotype can be resolved. Unresolved genotypes are called "orphans". In the application of Clark's inference rule, for a given ambiguous vector V, there may be multiple choices of resolved haplotypes such that each choice gives a different new resolved haplotype adding to the resolved set R. Thus, there may be multiple solutions to resolve all of the ambiguous genotype vectors. In some cases, a series of resolved haplotypes picked from R to resolve a series of ambiguous genotype vectors may lead to some unresolved ambiguous genotype vectors (orphans), while some other series of resolved haplotypes picked for resolving may not lead to any orphan. Clark's algorithm can also produce different solutions, depending on the order of ambiguous genotype vectors to be resolved. Then, a natural question is which solution is the most accurate one. Clark showed, through simulation studies, that the solution with the fewest orphans is the most accurate one, and suggested that a solution resolving the maximum number of ambiguous genotypes is likely to be unique.

However, there are often multiple solutions to a genotype data set from real studies. To handle this problem, Gusfield and Orzach[10] suggested a two-step approach to obtaining an accurate solution. In step 1, run Clark's algorithm many times (say 10 000), each time with a different (permuted) order of the input genotype data. This step gives a set of different solutions. In step 2, select the solutions that use the smallest or close to the smallest number of different haplotypes to resolve all or the most genotype vectors. Among these selected solutions, a set of haplotype pairs called "consensus", which was most commonly used to resolve each genotype, is considered as the final solution. Simulation studies[10] showed that the consensus solution using the smallest or next-to-smallest number of distinct haplotypes, and with each haplotype having a high frequency of showing up in all solutions ($> 85\%$), was almost always correct. Thus, this strategy provides one solution to improve accuracy.

Another limitation of Clark's algorithm is that when there is no homozygote or single-site heterozygote in the sample, the algorithm cannot start. Clark[9] expressed the probability of failing to get the algorithm started as a function of the sample size N and the expected number of heterozygous sites K (ambiguous sites). When the sample size is large enough, say greater than 24, even for K as large as 10, the probability of failing to get the algorithm started is less than 1%. One may question that, although increasing the sample size may reduce the chance for algorithm failure to start, it may increase the chance of having orphans. However, simulation results by Clark[9] showed that as the sample size increases, orphans are less likely to remain. This is because large samples are more likely to contain paths connecting all alleles to homozygotes, and hence generate a large set of resolved haplotypes R to start the algorithm.

The inference rule in Clark's algorithm implies two major assumptions in population genetics studies.[10] First, Clark's algorithm resolves identical genotype vectors identically. This implies the "infinite site" assumption that only one mutation can occur at any given site in the history of the sample sequences. Secondly, the sampled individuals are considered as a random sample representing the whole population. The inference rule is consistent with the "random mating" assumption of a population such that the sequentially increased set R is very likely to represent the order of the commonness of haplotypes in the population.

2.1.2. *The Pure Parsimony Approach*

Clark's rule-based inference method can be considered as a parsimony approach. The original paper empirically showed that the solution with the fewest orphans and the smallest number of distinct haplotypes is usually the most accurate one. However, unless the algorithm is repeated many times to obtain a consensus solution, a solution from a single execution of the algorithm is generally poor in terms of the proportion of correctly inferred haplotypes. In practice, it is also computationally intensive to repeat the algorithm many times for obtaining a "consensus" solution. A different approach to handling this problem, called the pure parsimony approach, has been proposed for haplotype inference by Gusfield.[11] The pure parsimony approach in haplotype inference problems refers to finding a solution in order to minimize the total number of distinct haplotypes used to resolve all genotypes. Gusfield[11] proposed to use integer linear programming to solve this pure parsimony problem. We briefly outline integer linear programming to the pure parsimony solution as follows.

Assume we have N individuals. For each individual's genotype, say g_i for the ith individual, suppose there are k_i ambiguous sites. So, there are 2^{k_i-1} haplotype pairs compatible with g_i. Let y_{ij} be a binary variable denoting each of these potential haplotype pairs, $j = 1, \ldots, 2^{k_i-1}$. For example, suppose we have $g_i = \{(0,0), (0,1), (1,1), (0,1)\}$. As illustrated above, there are two compatible haplotype pairs: $\{(0010,0111),(0110,0011)\}$. We can enumerate these haplotypes as $y_{i1} = 1$ if $(0010,0111)$, and 0 otherwise; and $y_{i2} = 1$ if $(0110,0011)$, and 0 otherwise. Suppose some of these haplotypes are not listed in the resolved list R. Then, a binary variable, denoted by h, is generated for each unseen haplotype. For example, if (0010) is not in R, then $h = 1$ and 0 otherwise. The haplotype (0010) is added to R. All of the h variables are collected to a set denoted by H. Note that each h in the set H represents only one distinct haplotype. With these defined notations, the integer linear programming is set up as follows:

Objective function: $$\min \sum_{h \in H} h$$

Subject to: $$\sum_{j=1}^{x_1} y_{1j} = 1, \quad x_1 = \max(1, 2^{k_1 - 1}),$$

$$\vdots$$

$$\sum_{j=1}^{x_N} y_{N,j} = 1, \quad x_N = \max(1, 2^{k_N - 1}),$$

$$y_{11} - h_1 \leq 0,$$
$$y_{11} - h_2 \leq 0,$$

$$\vdots$$

$$y_{N,x_N} - h_l \leq 0.$$

For the first N constraints, each constraint explains that only one of the haplotype pairs y_{ij}'s would be used to explain the observed genotype g_i. The expression of $x_i = \max(1, 2^{k-1})$ ensures that at least one haplotype pair would be used to explain g_i. If g_i contains zero or one heterozygous site, then $x_i = 1$, indicating that one haplotype pair needs to be used to explain g_i unambiguously. Constraints written in the form of inequality describe the situation as follows. For example, for g_1, if a haplotype pair denoted by variable y_{11} is used to explain g_1, then $y_{11} = 1$. The two haplotypes denoted by variables h_1 and h_2 making up the haplotype pair y_{11} must also be set at 1. So, the constraints $y_{11} - h_1 \leq 0$ and $y_{11} - h_2 \leq 0$ are set to describe this situation. Note that if the haplotype pair denoted by y_{11} is homozygous, then we have only one inequality: $y_{11} - h_1 \leq 0$. As the h variable is valued at 0 or 1, the objective function is to minimize the total number of h's. In other words, integer linear programming is formulated in this way in order to find a solution that uses a set of the minimum number of distinct haplotypes to explain the entire set of genotypes. This formulation solves the pure parsimony haplotype problem. Gusfield[11] called this formulation the TIP formulation, as this integer programming formulation truly solves the pure parsimony problem.

For data sets with more than 50 individuals and more than 30 SNPs, which require a large number of constraints to be generated for formulation, more practical formulations are given by Gusfield.[11] First, if both haplotypes in a haploype pair denoted by y_{ij} are not involved in explaining any other genotype g_s for $s \neq i$, then the variable y_{ij} and the corresponding h variable are removed from the constraint set. For a genotype g_i, if all compatible haplotype pairs have both haplotypes that are not involved in the explanation of any other genotype, then an arbitrarily chosen y variable and its corresponding h variable will remain in the constraint set. This reduced formulation generates fewer variables and fewer inequalities, allowing us to reduce the computational time effectively. This reduced formulation is called the RTIP formulation by Gusfield.[11]

Based on experimental results, Gusfield also showed that for genotype data of less than 50 individuals and less than 30 SNPs, this pure parsimony approach gives about 80%–95% of correctly inferred haplotype pairs. In general, the integer program takes several seconds to minutes to obtain the solution. Brown and Harrower[12] proposed a polynomial-sized integer linear programming (ILP) formulation that takes additional inequalities to reduce the running time. Results from extensive experiments by Brown and Harrower showed that the performances of RTIP and ILP formulations are comparable.

2.2. *Statistical Approaches*

Likelihood, coupled with the Hardy–Weinberg equilibrium (HWE) assumption, is the core of the statistical approaches for haplotype reconstruction. Under the assumption of random mating, a likelihood function can be derived; the objective is to obtain haplotype frequency estimates that maximize the likelihood of the observed genotypes. One popular likelihood-based approach is based on the expectation-maximization (EM) algorithm.[13–17] Because of computational limitations, earlier EM methods could not handle a large number of markers (e.g. no more than 20 SNPs). A method that combines the idea of partition ligation and EM algorithm (PL-EM)[17] was developed to overcome this limitation. Another approach is based on Bayesian inference, which incorporates biologically informed prior and other modeling assumptions into the inference procedure. PHASE[18] and Haplotyper[19] are two programs developed within the Bayesian framework with different prior and modeling assumptions. In this section, we provide an overview of these statistical approaches. EM methods are discussed in Sec. 2.2.1, and Bayesian methods are discussed in Sec. 2.2.2.

2.2.1. *EM Methods*

The EM algorithm formalized by Dempster and colleagues[20] fits naturally with haplotype inference.[13–16] In these EM methods, genotypes at multiple markers, denoted by $G = (g_1, \ldots, g_N)$, are observed for N unrelated individuals. Suppose that in the population pool, there are M possible haplotypes (unknown) with population frequencies $p = (p_1, \ldots, p_M)$. Here, M depends on the number of loci and the number of heterozygous sites for each individual's genotype. Then, the likelihood of observing N genotypes conditional on the population haplotype frequencies is

$$L(\boldsymbol{p}; G) = \prod_{i=1}^{N} \Pr(g_i \mid \boldsymbol{p}) = \prod_{i=1}^{N} \sum_{(hh') \in H_i} \Pr(hh'),$$

where H_i is the set of all haplotype pairs that are compatible with g_i, and h and h' are general notations for the two haplotypes that make up the haplotype pairs. Under the assumption of random mating,

$$\Pr(hh') = \begin{cases} p_h^2, & \text{if } h = h', \\ 2p_h p_{h'}, & \text{if } h \neq h'. \end{cases}$$

The haplotype frequencies can be estimated by an EM procedure, which is outlined as follows:

Given: Initiate a set of values of $p_1^{(0)}, p_2^{(0)}, \ldots, p_M^{(0)}$.

E-step: Calculate the posterior probability of each possible haplotype pair that is compatible with the observed genotype by

$$\Pr(hh') = \begin{cases} [p_h^{(0)}]^2, & \text{if } h = h', \\ 2p_h^{(0)} p_{h'}^{(0)}, & \text{if } h \neq h'. \end{cases}$$

M-step: Update $p_j^{(1)}$ for each j by aggregating the posterior probabilities that involve the jth haplotype:

$$p_j^{(1)} = \frac{1}{2N} \sum_{i=1}^{N} \frac{\sum_{hh' \in H_i} \{\Pr(hh') I_{(h=j)} + \Pr(hh') I_{(h'=j)}\}}{\sum_{hh' \in H_i} \Pr(hh')} \quad \text{for } j = 1, \ldots, M,$$

where $I_{(h_1=j)}$ is an indicator variable that is equal to 1
if the jth haplotype is involved in the haplotype pair
which is compatible with g_i, and 0 otherwise.
    Repeat the E-step and M-step until convergence.

    The EM algorithm may not converge to the global maximum. To avoid
this problem, it is always good practice to try several sets of initial haplotype frequencies before choosing the one that has the largest maximum
likelihood value. The EM method often gives satisfactorily accurate estimates of the haplotype frequencies, even when the HWE is not satisfied.[19]
In practice, this EM algorithm is limited by the number of possible haplotypes, which grows exponentially with the number of loci. It can generally
handle no more than 20 loci.

2.2.2. Bayesian Methods

The Bayesian approaches calculate the posterior distribution of the unobserved haplotypes conditional on the observed genotypes by incorporating
biologically relevant information into the priors. Stephens and colleagues[18]
used Gibbs sampler, a type of Markov chain Monte Carlo (MCMC) algorithm, to assign a haplotype pair to each individual genotype based on its
posterior distribution. This algorithm, which updates the posterior distribution of the haplotype pairs for each individual by incorporating population
genetics models and coalescent theory, may lead to better estimation of
haplotype frequencies. Similarly, Niu and colleagues[19] used Gibbs sampler
for haplotype inference, but assumed a Dirichlet prior for updating the posterior distribution of haplotypes. Niu and colleagues[19] also introduced the
idea of partition ligation (PL) to overcome the limitation of computational
complexity for inference on a large number of loci. This idea was also used to
develop a partition-ligation–expectation-maximization (PL-EM) algorithm
for haplotype inference.[17] Various methods adopting the PL idea have subsequently been developed. These include those by Stephens and Donnelly[21]
and by Lin and colleagues.[22] In this subsection, our discussion will focus
mainly on the Bayesian methods of Stephens and colleagues[18] as well as
Niu and colleagues.[19]

    Stephens and colleagues[18] proposed two procedures for haplotype infer-
ence, where each is based on a Gibbs sampling procedure with a different
prior. The Gibbs sampling procedure starts with an initial haplotype reconstruction $H^{(0)}$ for all individuals. Then, it repeatedly selects an individual
and assigns a pair of haplotypes to this individual based on the posterior

distribution of all the possible haplotypes. In the following, we outline a pseudoprocedure to illustrate the Gibbs sampling algorithm.

Given an initial haplotype reconstruction $H^{(0)} = (H_1^{(0)}, H_2^{(0)}, \ldots, H_N^{(0)})$ for each of the N individuals, for $t = 0, 1, 2, \ldots$, use the following procedure to update $H^{(t)}$:

(1) Uniformly and randomly choose an individual, say i, from all ambiguous individuals.

(2) Assign a pair of haplotypes $H_i^{(t+1)}$ to individual i based on $\Pr(H_i \backslash G, H_{-i}^{(t)})$, where $H_{-i}^{(t)}$ is the set of haplotype pairs without individuals i's haplotype pair.

(3) Set $H_j^{(t+1)} = H_j^{(t)}$ for $j \neq i$.

Stephens and colleagues[18] used a Dirichlet prior distribution — which is based on the assumption of parent-independent mutation, i.e. the type of mutant haplotype is independent of its parental haplotypes — to compute $\Pr(H_i \mid G, H_{-i}^{(t)})$. They pointed out that the performance of the algorithm based on this Dirichlet prior is roughly comparable to the EM algorithm and that the results from this algorithm provide a good starting point for their second improved algorithm, which uses a more practical prior distribution, the approximate coalescent prior distribution. The approximate coalescent prior is based on the idea that a future sampled haplotype is more likely to be the same as or be similar to one of those haplotypes that have been observed. Simulation results in Stephens and colleagues[18] showed that their algorithm for haplotype reconstruction outperformed Clark's algorithm[9] and the EM algorithm.[13–15] For a large number of markers, Stephens and colleagues[18] updated only a subset of loci for each randomly selected individual to reduce the computational complexity.

Niu and colleagues[19] proposed another Bayesian procedure based on an alternative Gibbs sampling algorithm and a Dirichlet prior that "imposes no assumptions on the population evolutionary history." Their procedure starts with an initial set of pseudocounts, denoted by $\boldsymbol{\beta}$, to obtain an initial guess on the haplotype frequencies, denoted by $\boldsymbol{p}$, based on the *a priori* assumption that $\boldsymbol{p} \sim \text{Dirichlet}(\boldsymbol{\beta})$. Conditional on $\boldsymbol{p}$, each individual is assigned a pair of compatible haplotypes according to

$$\Pr[H_i^{(t+1)} = (h_g h_k) \mid \boldsymbol{p}, G] = \frac{p_{h_g} p_{h_k}}{\sum_{(h_g' h_k') \in H_i} p_{h_g'} p_{h_k'}}.$$

A Gibbs sampler is applied to update the assignment of a haplotype pair to each individual with posterior distribution, computed by

$$\Pr[H_i^{(t+1)} = (h_g h_k) \mid H_{-i}^{(t)}, G] \propto (N_{h_g}^{(t)} + \beta_g^{(t)})(N_{h_k}^{(t)} + \beta_k^{(t)}),$$

where N_{h_g} and N_{h_k} are the counts of haplotypes h_g and h_k in H_{-i}, respectively. Niu and colleagues[19] applied the prior annealing technique to shrink the pseudocount $\beta^{(t)}$ at a fixed rate. In this way, the resulting haplotype assignment for each individual depends less on the pseudocount β.

To handle a large number of loci, Niu and colleagues[19] first introduced PL in haplotype inference. Suppose a sequence has m loci. Partition m loci into B consecutive blocks, and apply the Gibbs sampling procedure to each block to reconstruct the partial haplotypes formed by markers within the block. In the ligation step, two strategies — progressive ligation and hierarchical ligation — can be employed. In progressive ligation, Gibbs sampler is applied to ligate the partial haplotypes in the first two blocks, where each block is treated as a multiallelic marker. Then, Gibbs sampler is applied recursively by ligating the next partial haplotype block to the current ligated partial haplotypes until all of the blocks are ligated to form the entire haplotypes spanning all loci. In hierarchical ligation, L loci are first partitioned into $B = 2^\alpha$ blocks, where α is a nonnegative integer. Then, in the ligation step, apply the Gibbs sampler to ligate the $2j - 1$th and $2j$th blocks for $j = 1, \ldots, 2^{\alpha-1}$, to form $2^{\alpha-1}$ larger blocks. Continue joining the $2j - 1$th and $2j$th of the newly formed blocks for $j = 1, \ldots, 2^{\alpha-2}$, to form $2^{\alpha-2}$ larger blocks. Repeat the ligation procedure until the whole haplotype spanning L loci is formed.

Niu and colleagues[19] stated that incorporating the PL strategy enables the algorithm to not only handle a larger number of loci, but also reduce the computational time. They also demonstrated that, in the case of departure from HWE, their method and the EM method performed more robustly than those of Clark[9] and of Stephens and colleagues.[18] They also claimed that, in all real data applications and in a simulation study where the population conformed to a bottleneck demographic history, their method outperformed the methods of Clark,[9] EM,[13–15] and Stephens and colleagues.[18]

2.3. Some Extensions and Other Methods

Even though the Gibbs sampling–based algorithms are commonly used for haplotype reconstruction, the EM algorithm is still attractive. If multiple sets of initial haplotype frequencies are used to feed the EM algorithm, the chance of finding maximum likelihood estimates (MLEs) of haplotype frequencies increases. An EM algorithm is also computationally less intensive than the Gibbs sampling algorithm. However, as we mentioned earlier, most of the EM methods can handle only a moderate number of loci (< 20). Qin and colleagues[17] used the PL strategy in an EM-based algorithm to handle this problem. However, during the ligation step, the EM method ligates only on reconstructed partial haplotypes, whereby those partial haplotypes with low EM-estimated frequencies would most likely be dropped for ligation. In this case, the EM algorithm in the ligation step would not reach the MLEs of all haplotype frequencies. To overcome this difficulty, Qin and colleagues[17] reserved a "backup buffer" to keep not only those partial haplotypes whose EM algorithm–estimated frequencies are high, but also some of the partial haplotypes whose estimated frequencies are low. The buffer size and the total number of partial haplotypes in the buffer for each block is kept constant in the ligation step. Qin and colleagues[17] also presented a fast and robust method for computing the standard error of the estimated haplotype frequencies.

Since the LD structures in the human genome appear in block structures, where markers within the blocks are in high LD and markers between the blocks are in relatively low LD, the locations between the blocks may correspond to recombination hotspots and are likely to be more suitable for partition sites. Thus, it is desirable that the block structure information be incorporated to customize the partition sites. One of the features of the PL-EM program is that it allows the user to specify partition sites accordingly.

However, the block description of LD structure in the human genome is somewhat subjective. In addition, none of the abovementioned methods takes into account the effects of LD decay within a block and the spacing of markers. Stephens and Scheet[23] extended the algorithm of Stephens and colleagues[18] to incorporate LD decay information. In their algorithm, the underlying recombination rates between each pair of consecutive markers are treated as the "nuisance parameters". The posterior distribution of haplotypes is evaluated conditional on the recombination rates. The posterior distribution of haplotypes and the recombination rates are simultaneously updated by the algorithm. This algorithm is implemented in PHASE v2.1.1.

At this point, we have described several commonly used computational and statistical algorithms for haplotype inference. In addition to these algorithms, many others have been developed. These include Arlequin (Bayesian inference for each locus on data in two windows of nearby loci),[24] *hap* (an *ad hoc* modification of Dirichlet prior),[22,25] and HAP (based on imperfect phylogeny: partition loci into blocks and assume that haplotypes within blocks approximate to a perfect phylogeny, i.e. no recombination or infinite-site model).[26-28] In the case of missing alleles or genotypes, a number of algorithms accommodate the feature of imputing the missing genotypes under the missing-at-random assumption. In general, statistical methods are able to handle multiallelic markers, while pure parsimony approaches are able to handle biallelic markers only.

Stephens and Scheet[23] compared the performances among PHASE v2, HAP, *hap*, PL-EM, SNPHAP,[29] Haplotyper, and Arlequin. They demonstrated that PHASE v2 outperformed all of the other methods, and that the improvement was consistently achieved across many of the genes studied. However, when the coalescent model was violated, Niu and colleagues[19] showed that Haplotyper and EM-DECODER are more accurate than PHASE v1.0. Excoffier and colleagues[24] also reported that Arlequin v3.0 outperformed PHASE v1.0 when the coalescent model was violated. Overall, algorithms based on the Bayesian approach, EM algorithm, and imperfect phylogeny performed similarly. Salem and colleagues[30] conducted a comprehensive literature review on the methods and software for haplotype inference using unrelated individuals. They indicated that there is no clear evidence to suggest a distinguished algorithm. Therefore, we feel that it is always a good practice to compare the results from different methods and then use the consensus result from different algorithms. For readers' convenience, we list some of the most commonly used algorithms for haplotype inference in unrelated individuals and their implemented programs or software (if they are available) in Table 1.

An alternative approach for haplotype inference is through related individuals. Incorporating pedigree information in haplotype inference may reduce haplotype ambiguity, and improve the accuracy and efficiency for haplotype inference.[31-34] Many computational and statistical methods have been developed for haplotype inference using related individuals.[32,35-53] Zhang and Zhao[34] compared the performance of four methods based on related individuals: HAPLORE,[53] GENEHUNTER,[54] PedPhase,[46] and MERLIN.[55,56] These methods were also compared with PHASE. Simulation results showed that HAPLORE and MERLIN performed similarly and outperformed the other methods. Several studies[31,32,34,57] showed that incorporating parental information (if available) could improve the accuracy of haplotype inference.

Table 1 Common Softwares for Haplotype Inference

Software	Method	Web Link	Platform	Reference
Arlequin 3.0	Bayesian	anthro.unige.ch/arlequin	MAC	Excoffier et al.,[24] Schneider et al.[58]
BPPH	Imperfect phylogeny	http://wwwcsif.cs.ucdavis.edu/~gusfield/bpph.html	MAC	Chung and Gusfield[59]
CHAPLIN	EM	http://server2k.genetics.emory.edu/chaplin/chaplin.php	PC	Epstein and Satten[60]
DPPH	Perfect phylogeny	http://wwwcsif.cs.ucdavis.edu/~gusfield/dpph.html	LINUX/MAC	Gusfield,[61] Bafna et al.[62]
EH	EM	http://linkage.rockefeller.edu/ott/	PC	Xie and Ott,[63] Terwilliger and Ott[64]
EH+	EM	http://statgen.iop.kcl.ac.uk/	PC/UNIX	Zhao et al.[65]
EM-DeCODER	EM	http://www.people.fas.harvard.edu/~junliu/em/	UNIX	Niu et al.[19]
FASTEHPLUS	EM	http://www.ucl.ac.uk/~rmjdjhz/software.htm	PC/UNIX	Zhao and Sham[66]
GCHAP	EM	http://bioinformatics.med.utah.edu/~alun/gchap/docs/alun/gchap/GCHap.html	JRE on PC/UNIX	Thomas[67,68]
Genecounting/Hap	EM	http://www.mrc-epid.cam.ac.uk/Personal/jinghua.zhao/software.htm	PC/UNIX	Zhao et al.,[69] Zhao[70]
GPPH/PPH	Perfect phylogeny	http://wwwcsif.cs.ucdavis.edu/~gusfield/pph.html	LINUX/MAC/PC/UNIX	Gusfield,[61] Chung and Gusfield[59]
GS-EM	EM	http://www.people.fas.harvard.edu/~junliu/genotype/	Web-based	Kang et al.[71]
HAP	Imperfect phylogeny	http://research.calit2.net/hap/	Web-based	Eskin et al.,[26,27] Halperin and Eskin[28]
HAPAR	Parsimony	http://theory.stanford.edu/~xuying/hapar/	PC/UNIX	Wang and Xu[72]
HAPINFERX	Clark's	www.nslij-genetics.org/soft/hapinferx.f	UNIX	Clark[9]

(Continued)

Table 1   *(Continued)*

Software	Method	Web Link	Platform	Reference
HAPLO/ PERMUTE	EM	krunch.med.yale.edu/haplo/	UNIX	Hawley and Kidd[14]
HAPLOTYPER	Bayesian	http://www.people.fas.harvard.edu/~junliu/Haplo/docMain.htm	UNIX	Niu et al.[19]
HAPLOREC	Bayesian	http://www.cs.helsinki.fi/group/genetics/haplotyping.html	Java Virtual Machine, v1.4 or newer	Eronen et al.[73]
HAPLOVIEW	EM + PL	http://www.broad.mit.edu/mpg/haploview/	JRE on MAC/PC/UNIX	Barrett et al.[74]
HAPMAX	MLE	http://www.uni-kiel.de/medinfo/mitarbeiter/krawczak/download/index.html	PC	Krawczak[75]
HAPSCOPE	EM/Bayesian	http://lpg.nci.nih.gov/lpg_small/protocols/HapScope/	UNIX/Windows	Zhang et al.[76]
Helix Tree Genetics Analysis	EM	http://www.goldenhelix.com/pharmhelixtreesummary.html	MAC	Excoffier and Slatkin,[13] implemented by Gold Helix, Inc.
HPLUS	EM + EE + PL	http://qge.fhcrc.org/hplus/	MATLAB on PC/UNIX	Li et al.,[77] Zhao et al.[78]
LPPH	Perfect phylogeny	http://www.csif.cs.ucdavis.edu/~gusfield/	PC/LINUX	Ding et al.[79]
PHASE v2.1.1	Bayesian	www.stat.washington.edu/stephens/software.html	PC/MAC/UNIX	Stephens et al.,[18] Stephens and Sheet[23]
PL-EM	PL-EM	www.people.fas.harvard.edu/~junliu/plem	PC/UNIX	Qin et al.[17]
SASGenetics	EM		SAS on PC/UNIX	Czika et al.[80]
SNPEM	EM	http://polymorphism.ucsd.edu/snpem/	UNIX	Fallin et al.[2]
SNPHAP	EM	www.gene.cimr.cam.ac.uk/clayton/software/	UNIX	Clayton[29]
THESIAS	Stochastic EM	http://genecanvas.ecgene.net/downloads.php	PC/UNIX	Tregouet et al.[81]
WHAP	EM	http://pngu.mgh.harvard.edu/~purcell/whap/	PC/UNIX	Purcell and Sham[82]

3. HAPLOTYPE ASSOCIATION ANALYSIS

Studies have shown that haplotype-based analyses using the inferred haplotypes may be biased if haplotype inference uncertainties are not considered in the analysis.[78,83] Thus, methods that incorporate haplotype inference into haplotype association analysis have been proposed in the literature, where unphased genotype data are used. In this section, we will review haplotype analysis methods under a regression framework, and methods based on haplotype sharing and clustering.

3.1. *Haplotype Association Analysis Using Regression Models*

When haplotype information is known, i.e. phase is known, it is intuitive to perform haplotype association analysis under a regression framework, where haplotypes are treated as categorical covariates. This framework is very flexible and has several advantages[8]: with the link function in generalized linear models (GLMs), many traits (such as binary traits, quantitative traits, and survival traits) can be handled in a similar way; other covariates and interactions can be evaluated; many established statistical regression techniques (such as model selection and diagnosis) can be readily used; and well-developed software exists. Using a simple example, Schaid[8] also showed that the haplotype model has a close relationship with the locus model. However, as stated in previous sections, haplotype phase information is usually unknown.

3.1.1. *Regression Models for Unphased Haplotypes*

Even the sophisticated haplotype inference methods discussed in Sec. 2 cannot unambiguously resolve the haplotypes most of the time. If only the most likely haplotype pair is used for each individual without accounting for the discarded haplotype pairs that are possible, there may be substantial loss of information with this two-stage approach.[8,33] In addition, when only the most likely haplotype pair is used, it may introduce measurement error and induce bias into the estimates of haplotype effects.[78,83]

One way to deal with this problem is to use all possible haplotype pairs that are consistent with the observed genotype data.[4,8,16,78,84,85] Statistically, more powerful methods have also been proposed to simultaneously estimate haplotypes and their effects.[60,86,87] Some of these methods are based on prospective likelihood.[16,78,84,88–91] To illustrate the methods, we use G_i to denote genotype information for the ith individual who is directly observed; $H_i = \{h_i, h_i'\}$ to denote haplotype pair for the ith individual; $S(g)$ to denote the set of haplotype pairs that is consistent with genotype $G = g$;

X to represent environmental covariates (here, we use the term "environmental covariates/factors" in a general way, including risk factors such as age, gender, and body mass index — which are not environmental factors in a strict way — to follow the conventional usage) that may affect disease risk; $P(H)$ to denote the prior probability of haplotype pair H; and Y to denote the dependent trait. Then, the prospective likelihood of the data can be expressed as[8]

$$L = \prod_{i=1}^{N} \sum_{H_i \in S(G_i)} P\{Y_i \mid X_i, G_i(H_i), \beta\} P(H_i),$$

where N is the number of individuals in the sample and β is the vector of regression coefficients (including the effects of X, H, and possibly their interactions).

To fit the above regression model, maximum likelihood methods are usually used, such as that described by Lake and colleagues,[84] where haplotype frequencies and regression parameters are jointly estimated. An alternative way is to use estimating equations,[78] where the logistic regression parameters (ξ, β) ($\xi = \alpha + \log\{[(1 - \theta)\eta]/[\theta(1 - \eta)]\}$) represent a shifted intercept on the logit scale, θ is a fraction of cases in the study, and η is the probability of disease in the general population that can be estimated based on the following score equations derived from the prospective likelihood:

$$U(\xi, \beta) = \sum_{i=1}^{N} \begin{pmatrix} U_i(\xi) \\ U_i(\beta) \end{pmatrix}$$

$$= \sum_{i=1}^{N} E_{\Omega_i} \left[\left(\begin{matrix} 1 \\ I'(h_i, h'_i, x_i, \beta) \end{matrix} \right) (d_i - \mu_i) \mid g_i, d_i, x_i \right]$$

$$= \sum_{i=1}^{N} \sum_{\Omega_i} \begin{pmatrix} 1 \\ X'_i \end{pmatrix} (d_i - \mu_i) \mathrm{Pr}_\pi(\Omega_i \mid g_i, d_i, x_i) = 0,$$

where $X'_i = I'(h_i, h'_i, x_i, \beta)$ is the partial derivative of $I(h_i, h'_i, x_i, \beta)$, which is a function chosen according to the hypothesis of interest, with respect to β; $\mathrm{Pr}_\pi(\Omega_i \mid g_i, d_i, x_i)$ is the posterior probability of the latent variables indexed by haplotype frequencies (π), Ω_i (a vector of phase indicators), given the ith individual's observed data (g_i, d_i, x_i); and π is a vector of the population frequencies of haplotypes. However, Zhao and colleagues[78] used an EM algorithm similar to that proposed by Excoffier and Slatkin[13] to estimate haplotype frequencies required for evaluation of the prospective score equations using only controls.

It is straightforward to incorporate environmental factors in these methods under the assumption of independence between genes and environmental factors.[78,84] This assumption is likely to be satisfied when an exposure to external environmental factors is not directly controlled by an individual's own behavior; however, when an exposure depends on an individual's behavior, this assumption could be violated.[87] Lin and Zeng[86] indicated that this assumption may not hold in practice and is not statistically efficient.

For case-control studies, the cases are oversampled because of sample ascertainments; as a consequence, the estimated haplotype frequencies will be biased under the alternative hypothesis.[8] The lack of consideration of the ascertainment is a limitation of the prospective likelihood methods. Although Zhao and colleagues[78] proposed to use only controls to estimate haplotype frequencies, this may work only for rare diseases, where the control population can be viewed as representative of the general population. In fact, Spinka and colleagues[87] showed through simulations that the method of Zhao and colleagues[78] could produce substantial biases for model parameter estimates when the underlying rare disease assumption is violated. This bias induced by ascertainment does not occur when phase is known, i.e. for phased haplotype data.[8] The magnitude of the bias depends on the accuracy of the estimated haplotypes. Studies have shown that the stronger the LD among the markers, the more accurate the inferences of haplotypes,[92,93] resulting in little bias.[89]

Stram and colleagues[89] proposed to use sampling weights based on the population disease prevalence to correct biased estimates. Another way is to use retrospective likelihood.[60,87] Epstein and Satten[60] proposed an approach based on retrospective likelihood, where regression parameters and haplotype frequencies can be jointly estimated. Their retrospective likelihood can be expressed as a product of multinomial observed genotype data as follows:

$$
L = \prod_g [\Pr(G = g \mid D = 0)]^{c_g} [\Pr(G = g \mid D = 1)]^{d_g}
$$

$$
= \prod_g \left\{ \sum_{(h,h') \in S(g)} \Pr[H = (h, h') \mid D = 0] \right\}^{c_g}
$$

$$
\times \left\{ \sum_{(h,h') \in S(g)} \Pr[H = (h, h') \mid D = 1] \right\}^{d_g},
$$

where c_g and d_g are the number of control subjects and case subjects with genotype g in the sample, respectively; and D is the disease status ($D = 0$ indicates free of disease; $D = 1$ indicates presence of disease). Another advantage of this approach is that the assumption of HWE is required only for the controls, yet both cases and controls are still used to estimate haplotype frequency and haplotype odds ratio. Satten and Epstein[94] demonstrated that this retrospective approach has equal or better power than some of the previously proposed prospective approaches. This may be due to the fact that the retrospective approach fully exploits the HWE assumption for the underlying population.[87]

However, the incorporation of environmental factors (and haplotype–environment interactions) is complicated in this approach, as pointed out by Spinka and colleagues.[87] This is because retrospective likelihood involves nuisance parameters that specify the distribution of the environmental factors. Another disadvantage of the current retrospective likelihood methods is that they are not robust to departures from HWE among the controls when haplotype effects are dominant or recessive,[94] in contrast to some prospective methods that are robust to departures from HWE.[4,78] However, Satten and Epstein[94] proposed to introduce an additional parameter (i.e. fixation index) to account for the average departure from HWE in order to reduce bias.

Spinka and colleagues[87] extended the retrospective maximum likelihood approach of Chatterjee and Carroll[95] to incorporate both genetic and environmental factors, which can account for the presence of missing data in the genetic factors, with a special emphasis on haplotype analysis where missing data arise due to unknown phase information. They used the following likelihood function[87]:

$$L = \prod_{i=1}^{N} P(G_i, X_i \mid D_i) = \prod_{i=1}^{N} P(D_i \mid G_i, X_i)P(G_i)P(X_i)/P(D_i),$$

where

$$P(D) = \int_x \sum_{h \in H} P(D_i \mid X = x, H = h)P(H = h)dF(x)$$

$$P(D \mid X, G) = \sum_{h \in S(G)} P(D \mid X, H = h, G)P(H = h \mid X, G)$$

$$= \sum_{h \in S(G)} P(D \mid X, H = h)P(H = h \mid G)$$

$$= \sum_{h \in S(G)} \frac{P(D \mid X, H = h)q(h; \theta)}{\sum_{h \in S(G)} q(h'; \theta)},$$

where $q(.)$ is a known function, θ is a vector of parameters, and H is the set of all possible values of genetic covariate of interest H (in this context, haplotype). They derived a relatively simple parameter estimation procedure using a profile likelihood technique and an appropriate EM algorithm. They also described two robust approaches that are less sensitive to the genetic–environmental independence and HWE assumptions. In addition, they showed that their retrospective maximum likelihood procedure is equivalent to an extension of the ascertainment-corrected joint likelihood method of Stram and colleagues.[89]

Recently, Lin and Zeng[86] proposed a more general framework for haplotype association analysis based on likelihood methods. They considered all commonly used study designs (including cross-sectional, case-control, and cohort studies) and phenotypes (including binary, quantitative, and survival traits). Their regression models can evaluate the effects of haplotypes as well as gene–environment interactions, while accommodating a variety of genetic mechanisms (recessive, dominant, additive, and codominant models). Like Satten and Epstein,[94] they included a fixation index to allow for departure from HWE. They provided a comprehensive and mathematically rigorous way for haplotype association analysis. Under certain conditions, they proved the identifiability of the model parameters as well as the consistency, asymptotic normality, and efficiency of the maximum likelihood estimators. However, they still assumed that environmental factors are independent of haplotypes conditional on genotypes, although this conditional independence assumption is much weaker than the gene–environment independence assumption and is more likely to hold in practice. In addition, they only focused on SNP data.[86]

3.1.2. Score Statistics for Haplotype Analysis

As indicated by Schaid,[8] an advantage of the GLM framework is that it provides a way to construct score statistics to test the null hypothesis of no haplotype effects. Let $\tilde{y}_i$ be a fitted value from a GLM with only environmental covariates, and $E[]$ be the conditional expectation over the posterior probability of haplotype pairs under the null hypothesis given the observed genotype data. The score statistics can be represented as[8]

$$ U = \sum_{i=1}^{N} \frac{(y_i - \tilde{y}_i)}{a(\phi)} E_i[X], $$

where $a(\phi)$ is used to scale the distribution, with a value of σ^2_{mse} for normal and 1 for binomial and Poisson distributions.[4,8] So, the score statistics measure the covariance of the residuals of a GLM model that fits only the environmental covariates with the expected haplotypes. The weights for the expected haplotypes are the posterior probabilities of the haplotype pairs given the observed genotypes.[8]

As stated by Schaid,[8] the main advantage of the score statistics is that they are rapid to compute. This computation efficiency makes it feasible to estimate significance level by simulations. The P values from simulations are usually more robust for sparse haplotypes than those from asymptotic theory. Zaykin and colleagues[85] proposed the use of expected haplotype scoring in standard regression packages. Xie and Stram[96] proved that the two score tests of the null hypothesis are asymptotically equivalent under standard regularity conditions. They also showed by simulations that the two methods are also nearly equivalent away from the null when the sample size is small.

In this section, we have mainly introduced some regression-based methods in haplotype association analysis. Table 2 summarizes some of the methods that fall under this general scheme; it is an expansion of Table II in Schaid.[8] These regression approaches offer several advantages[8] and constitute a very important part of haplotype analysis.

3.2. *Haplotype Sharing and Clustering*

We know that mutations underlying simple genetic disorders are often found to be in tight LD with their surrounding markers.[105] This information can be used for disease gene fine mapping or even detection.[106,107] For complex disorders, we still expect to find unusual haplotype sharing among affected individuals in the neighborhood of some disease mutations more than that shared among randomly selected individuals, although the sharing patterns become more complicated.[101,108]

Many statistical methods that are based on searching for excess similarity among haplotypes from affected individuals compared to that from unaffected individuals have been proposed.[101,102,109–113] These methods are based on pairwise comparisons of all haplotypes in a sample of individuals, in contrast to the usual comparisons of allele or haplotype frequencies that are usually goodness-of-fit (GOF) tests.[100] The initial idea was put forth by Van der Meulen and Te Meerman,[110] who proposed a haplotype-sharing statistic based on the variance of the shared lengths of haplotypes surrounding a possible location across all possible pairs of case haplotypes. Thomas[114] gave a brief introduction of the development

Table 2 Regression-based Haplotype Association Methods

Trait	Method	HWE	Covariates	Software	Platform	Reference
Binary	Prospective likelihood	Pool	No			Chiano and Clayton[16]
Binary	Prospective likelihood	Pool	Yes			Mander[97]
Binary/ Quantitative	Prospective likelihood	Pool	Yes	THESIAS	Windows, Linux	Tregouet et al.[98]
Quantitative	Prospective likelihood	Pool	Yes	qhapipf	Stata	Mander[99]
Quantitative	Prospective likelihood	Pool	Yes	Weighted penalized log-likelihood maximization routine	MatLab	Tanck et al.[83]
Binary	Prospective likelihood	Controls	Yes	HPlus	MatLab	Zhao et al.[78]
GLM	Prospective likelihood	Pool	Yes	Haplo Stats	Splus/R	Lake et al.,[84] Schaid et al.[4]
GLM	Prospective likelihood	Pool	Yes	eHap	Windows	Seltman et al.[43]
Binary	Prospective likelihood	Pool	Yes			Stram et al.[89]
Binary	Retrospective likelihood	Controls	No	CHAPLIN	Windows	Epstein and Satten[60]
Censored	Prospective likelihood	Pool	Yes	HAPSTAT	Windows	Lin[88]
Binary	Prospective likelihood	Pool	Yes	CLADHC	Unix	Durrant et al.[100]

(Continued)

Table 2 *(Continued)*

Trait	Method	HWE	Covariates	Software	Platform	Reference
Binary	Restrospective likelihood, modified prospective score equations		Yes	HapLogistic	R	Spinka et al.[87]
Binary, quantitative, censored	Likelihood	Pool	Yes	HAPSTAT	Windows	Lin and Zeng[86]
Binary	Haplotype similarity	Pool	No	QSHS	R	Tzeng et al.[101]
Binary	Haplotype similarity	Pool	No	Evolutionary-based haplotype clustering	R	Tzeng[102]
Binary	Likelihood ratio test	Pool	No	FAMHAP	Unix	Becker and Knapp,[48] Becker et al.,[103] Becker et al.[104]

of the haplotype-sharing methods. Many of these haplotype-sharing statistics fall under the general framework proposed by Tzeng and colleagues[101] using the following quadratic form:

$$D = \hat{\Pi}_a^t A_a \hat{\Pi}_a - \hat{\Pi}_u^t A_u \hat{\Pi}_u, \quad T = \frac{D}{\sigma(\hat{\Pi})},$$

where $\sigma^2(\Pi) = \text{Var}(D)$; $\Pi_a = (\pi_{a1}, \ldots, \pi_{aR})$ and $\Pi_u = (\pi_{u1}, \ldots, \pi_{uR})$ are haplotype distributions (assume that there are R distinct haplotypes in the population) for cases and controls, respectively; and A is a symmetric matrix containing the entries defined by a symmetric kernel function of some feature in the comparison of the ith and jth haplotypes.

Tzeng and colleagues[101] showed by evolutionary simulation that haplotype-sharing tests can be more powerful than GOF tests for common disease haplotypes, but less powerful for rare haplotypes. In addition, they showed that the power of both approaches can be enhanced by clustering rare haplotypes appropriately from the distributions prior to testing.

Because haplotype-sharing methods take the LD information between multiple markers into account and usually have fewer degrees of freedom, they may have good power to detect predisposing genes.[101,109] Another advantage of the similarity measures is that they are easy to implement in standard software.[8]

When there are many haplotypes, we often face two difficulties: how to account for the rare haplotypes, and how to deal with the large number of degrees of freedom.[8,101,102,115–117] The estimates for the rare haplotypes can have large variances, which may result in model instability[8]; while too many degrees of freedom can weaken the power to detect association.[8,101,102,116,117] Haplotype similarity provides a natural way to group (cluster) haplotypes, which offers one promising solution to these difficulties. Haplotype grouping may enhance the efficiency of haplotype analysis by using a small number of haplotype clusters that can reduce the degrees of freedom and get rid of rare haplotypes. In addition, the haplotype similarity measure may provide information on the covariances of haplotype effects.[8]

Many statistical methods have been proposed for haplotype association analysis using haplotype clustering information.[43,100,102,116–128] In a series of papers, Templeton and colleagues[123–128] proposed cladistic analyses that use a cladogram (an unrooted tree) to search for alleles associated to traits. This approach is potentially more powerful than an omnibus test of all haplotypes. Seltman and colleagues[43] extended this cladistic approach to the GLM framework, allowing for unphased haplotypes. Durrant and colleagues[100] proposed the use of standard hierarchical clustering to create a hierarchical tree of haplotype for use together with a logistic regression method for case-control data. A final tree is obtained by trimming the tree back towards its root, until a poor fit of the model is encountered.

Tzeng and colleagues[117] generalized the probabilistic clustering methods of Tzeng[102] to the GLM framework proposed by Schaid and colleagues.[4] The clustering methods of Tzeng[102] are based on an evolutionary tree of haplotypes. The tree is sequentially trimmed by combining rare haplotypes with their one-step neighboring haplotypes, from the leaves of the tree back to the major nodes. An information criterion is used to find the final tree.[102] The score tests of Tzeng and colleagues[117] are based on clustered haplotypes that can incorporate both phase uncertainty (if phase is unknown) and clustering uncertainty, and that can evaluate the overall haplotype association and the individual haplotype effect. In addition, the method can include covariates and model many traits (such as binary, quantitative, and survival traits). The method seems to be powerful; however, its performance still needs further evaluation in practice.

The main advantages of the haplotype-sharing or haplotype-clustering methods are twofold: they reduce the degrees of freedom and get rid of rare haplotypes. As a consequence, they are incapable of detecting rare variants with large effects since rare haplotypes are not retained in the clustered haplotype space, as pointed out by Tzeng and colleagues.[117] Recently, a method based on haplotype clustering, which claims to work well when the disease-predisposing allele is moderately rare, has been proposed.[129] Schaid[8] indicated that currently most of such methods do not account for ascertainment, and it is not clear how well they would work for case-control studies of complex disorders. In addition, these methods depend highly on the clustering scheme — hence, the similarity measures. More research is needed to define the best type of similarity measure.

4. DISCUSSION

Haplotypes have played a key role in the study of simple Mendelian diseases, and will likely continue to play an important role in the study of complex traits.[8] A haplotype map of the human genome[130] has begun to provide a wealth of data that can motivate the development and testing of new haplotype methods.

There are many methods for haplotype analysis in genetic studies. In this chapter, we have only focused on haplotype association analysis using unrelated individuals. LD mapping is another approach where many methods using haplotype information have been proposed. According to Schaid,[8]

> Both haplotype association methods and LD fine mapping methods take advantage of the fact that in the vicinity of a causative locus, haplotypes of diseased subjects tend to share more ancestry than haplotypes of unaffected subjects, and this excess sharing decreases with distance from the causative locus. Both approaches share the common goal of localizing the disease susceptibility locus. However, the means to accomplish this goal differs between the two approaches.

Schaid[8] also distinguished between the two approaches: "Although the strategies to include this historical information border between purely haplotype association methods and LD fine mapping methods that attempt to estimate the trait locus position, many are based conceptually on the coalescent process." Good introductions about LD mapping can be found in the papers of Schaid[8] and Thomas.[114]

Many studies suggest that haplotypes tend to have block-like structures throughout the human genome. Within such haplotype blocks, there is low haplotype diversity, with a few common haplotypes capturing most of the genetic variation. These haplotypes can be represented by a small number of haplotype-tagging SNPs (htSNPs). Many methods have been proposed for haplotype block partitioning and tagging SNP identification; these methods and concepts have been used in haplotype association analysis. This issue is covered in Chapter 5.

One very important issue in haplotype analysis is to get the right-sized haplotypes. If the haplotypes are too long and include too many markers, they may be composed of too many alleles, resulting in many haplotypes and thereby diluting the association signals with diseases.[8] Some researchers use sliding windows of markers[8,100]; some use haplotype block information.[131] An ideal way is to scan the haplotype for the subhaplotype that has the strongest association with disease. However, this may induce intensive computation and multiple comparison issues.[47,103,104] It may be more meaningful to incorporate biological information into this search. This is a very important issue; however, it has not attracted enough attention.

Most of the methods in haplotype analysis (including haplotype inference and haplotype association analysis) are developed under the assumption of HWE on either the whole sample or part of the sample. Departure of the haplotype pairs from HWE can result from many sources, such as genotyping errors, gene flow, gene selection, and population stratification. Studies have shown that only departure from HWE in the direction of excessive heterozygosity can increase the errors.[19,92,93] Satten and Epstein[94] showed that prospective methods are more robust to departures from HWE than retrospective methods. Some methods[86,94] include an extra parameter, the fixation index, to relax the assumption and thus reduce bias. Since the human population is by no means the result of random mating, the HWE assumption needs to be carefully evaluated in haplotype analysis.

Missing genotypes and genotyping errors are common in genetic studies,[57,93,132–137] where variation in DNA quality or molecular effects, experimental techniques used, unknown DNA variations, and conduction of studies can cause some individuals (e.g. cases vs. controls in the case-control studies) and some sites to have more or less than their fair share of missing data and genotyping errors. Although several studies have shown that ignoring missing genotypes and genotyping errors could significantly decrease the accuracy of estimates in haplotype analysis,[57,93,135] most current methods do not take these into consideration (or simply assume that what is missing is at random). This issue needs to be taken into consideration more seriously in order to have valid inference.

Many methods for haplotype analysis assume that people are interested in a small number of haplotypes which are relatively frequent, especially the methods based on haplotype clustering.[86,100–102,117] When there are many haplotypes, we are confronted not only with the problem of multiple comparisons, but also with the problem of stability of estimates for the rare haplotypes. This poses a challenge in haplotype analysis.

In this chapter, we have focused on haplotype association analysis for unrelated individuals. Analogous to the introduced methods, many methods for haplotype association analysis have been developed for related individuals, i.e. pedigree data.[40,42,43,45,47–49,51,52] We want to emphasize here that this is a very important research area that deserves equal attention.

We would like to indicate that the area of haplotype inference and analysis is still very active. Some new research work has just appeared.[138–142] Apart from the work of Souverein and colleagues,[142] which is on haplotype analysis, the rest are on haplotype inference. We will not introduce these methods in detail here, but have provided the citations to the latest work to show that the field is vigorous and still evolving.

Although haplotypes are promising in the study of the genetic basis of complex traits, the argument about the efficiency of using haplotypes vs. single markers always exists. Schaid[8] gave an excellent summary of this issue in situations where haplotype-based methods are more powerful or less powerful than the single-locus methods. We believe that no one method is universally preferable. Users should choose the methods based on their problems, and may need to try different methods.

ACKNOWLEDGMENTS

This work was supported in part by GM59507 and GM57672 from the NIH.

REFERENCES

1. Sham PC, Zhao JH, Curtis D. (2000) The effect of marker characteristics on the power to detect linkage disequilibrium due to single or multiple ancestral mutations. *Ann Hum Genet* **64**:161–169.
2. Fallin D, Cohen A, Essioux L, *et al.* (2001) Genetic analysis of case/control data using estimated haplotype frequencies: application to *APOE* locus variation and Alzheimer's disease. *Genome Res* **11**:143–151.
3. Kaňková K, Stejskalová A, Hertlová M, Znojil V. (2005) Haplotype analysis of *RAGE* gene: identification of a haplotype marker for diabetic nephropathy in type 2 diabetes mellitus. *Nephrol Dial Transplant* **20**(6):1093–1102.
4. Schaid DJ, Rowland CM, Tines DE, *et al.* (2002) Score tests for association between traits and haplotypes when linkage phase is ambiguous. *Am J Hum Genet* **70**(2):425–434.
5. Clark AG, Weiss KM, Nickerson DA, *et al.* (1998) Haplotype structure and population genetic inferences from nucleotide-sequence variation in human lipoprotein lipase. *Am J Hum Genet* **63**:595–612.
6. Tavtigian SV, Simard J, Teng DH, *et al.* (2001) A candidate prostate cancer susceptibility gene at chromosome 17p. *Nat Genet* **27**(172):180.
7. Hollox EJ, Poulter M, Zvarik M, *et al.* (2001) Lactase haplotype diversity in the Old World. *Am J Hum Genet* **68**:160–172.
8. Schaid DJ. (2004) Evaluating associations of haplotypes with traits. *Genet Epidemiol* **27**(4):348–364.
9. Clark AG. (1990) Inference of haplotypes from PCR-amplified samples of diploid population. *Mol Biol Evol* **7**:111–122.
10. Gusfield D, Orzack SH. (2005) Haplotype inference. In: *Handbook on Bioinformatics*, Aluru S (ed.), CRC Press, Boca Raton, FL, pp. 1–25.
11. Gusfield D. (2003) Haplotype inference by pure parsimony. In: *14th Annual Symposium on Combinatorial Pattern Matching (CPM'03)*, Springer, Berlin, Germany, pp. 144–155.
12. Brown D, Harrower I. (2004) A new integer programming formulation for the pure parsimony problem in haplotype analysis. In: *Proceedings of the 4th Annual Workshop on Algorithms in Bioinformatics (WABI'04)*, Springer, Berlin, Germany, pp. 254–265.
13. Excoffier L, Slatkin M. (1995) Maximum-likelihood-estimation of molecular haplotype frequencies in a diploid population. *Mol Biol Evol* **12**(5):921–927.
14. Hawley M, Kidd K. (1995) HAPLO: a program using the EM algorithm to estimate the frequencies of multi-site haplotypes. *J Hered* **86**:409–411.

15. Long JC, Williams RC, Urbanek M. (1995) An E-M algorithm and testing strategy for multiple-locus haplotypes. *Am J Hum Genet* **56**:799–810.

16. Chiano MN, Clayton DG. (1998) Fine genetic mapping using haplotype analysis and the missing data problem. *Ann Hum Genet* **62**:55–60.

17. Qin Z, Niu T, Liu JS. (2002) Partition-ligation–expectation-maximization algorithm for haplotype inference with single-nucleotide polymorphism. *Am J Hum Genet* **71**:1242–1247.

18. Stephens M, Smith NJ, Donnelly P. (2001) A new statistical method for haplotype reconstruction from population data. *Am J Hum Genet* **68**:978–989.

19. Niu TH, Qin ZS, Xu X, Liu JS. (2002) Bayesian haplotype inference for multiple linked single-nucleotide polymorphisms. *Am J Hum Genet* **70**(1): 157–169.

20. Dempster AP, Laird NM, Rubin DB. (1977) Maximum likelihood from incomplete data via EM algorithm. *J R Stat Soc Ser B* **39**:1–38.

21. Stephens M, Donnelly P. (2003) A comparison of Bayesian methods for haplotype reconstruction from population genotype data. *Am J Hum Genet* **73**:1162–1169.

22. Lin S, Chakravarti A, Cutler DJ. (2004) Haplotype and missing data inference in nuclear families. *Genome Res* **14**:1624–1632.

23. Stephens M, Scheet P. (2005) Accounting for decay of linkage disequilibrium in haplotype inference and missing-data imputation. *Am J Hum Genet* **76**:449–462.

24. Excoffier L, Laval G, Balding DN. (2003) Gametic phase estimation over large genomic regions using an adaptive window approach. *Hum Genomics* **1**:7–19.

25. Lin S, Cutlet DJ, Zwick ME, Chakravarti A. (2002) Haplotype inference in random population samples. *Am J Hum Genet* **71**:1129–1137.

26. Eskin E, Halperin E, Karp R. (2003) Efficient reconstruction of haplotype structure via perfect phylogeny. *J Bioinform Comput Biol* **1**:1–20.

27. Eskin E, Halperin E, Karp R. (2003) Large scale reconstruction of haplotypes from genotype data. In: *Proceedings of the 7th Annual International Conference on Research in Computational Molecular Biology (RECOMB 2003)*, ACM Press, New York, NY, pp. 104–113.

28. Halperin E, Eskin E. (2004) Haplotype reconstruction from genotype data using imperfect phylogeny. *Bioinformatics* **20**:1842–1949.

29. Clayton D. (2001) SNPHAP: a program for estimating frequencies of haplotypes of large numbers of diallelic markers from unphased genotype data from unrelated subjects. Available at http://www-gene.cimr.cam.ac.uk/ clayton/software/snphap.txt/

30. Salem R, Wessel J, Schork N. (2005) A comprehensive literature review of haplotyping software and methods for use with unrelated individuals. *Hum Genomics* **2**(1):39–66.

31. Rohde K, Fuerst R. (2001) Haplotyping and estimation of haplotype frequencies for closely linked biallelic multilocus genetic phenotypes including nuclear family information. *Hum Mutat* **17**:289–295.

32. Becker T, Knapp M. (2002) Efficiency of haplotype frequency estimation when nuclear family information is included. *Hum Hered* **54**(1): 45–53.

33. Schaid DJ. (2002) Relative efficiency of ambiguous vs. directly measured haplotype frequencies. *Genet Epidemiol* **23**(4):426–443.

34. Zhang K, Zhao H. (2006) A comparison of several methods for haplotype frequency estimation and haplotype reconstruction for tightly linked markers from general pedigrees.*Genet Epidemiol* **30**(5):423–437.

35. Lander ES, Green P. (1987) Construction of multilocus genetic-linkage maps in humans. *Proc Natl Acad Sci USA* **84**:2363–2367.

36. Wijsman EM. (1987) A deductive method of haplotype analysis in pedigrees. *Am J Hum Genet* **41**:356–373.

37. Sobel E, Lange K, O'Connell JR, Weeks DE. (1995) Haplotype algorithm. In: *Genetic Mapping and DNA Sequencing (IMA Volumes in Mathematics and Its Applications)*, Speed TP, Waterman MS (eds.), Springer, New York, NY, pp. 89–110.

38. Sobel E, Lange K. (1996) Descent graphs in pedigree analysis: application to haplotyping, location scores and marker-sharing statistics. *Am J Hum Genet* **58**:1323–1337.

39. Kruglyak L, Daly MJ, Reeve-Daly MP, Lander ES. (1996) Parametric and nonparametric linkage analysis: a unified multipoint approach. *Am J Hum Genet* **58**:1347–1363.

40. Dudbridge F, Koeleman BP, Todd JA, Clayton DG. (2000) Unbiased application of the transmission/disequilibrium test to multilocus haplotypes. *Am J Hum Genet* **66**(6):2009–2012.

41. O'Connell JR. (2000) Zero-recombinant haplotyping: applications to fine mapping using SNPs. *Genet Epidemiol* **19**:S64–S70.

42. Seltman H, Roeder K, Devlin B. (2001) Transmission/disequilibrium test meets measured haplotype analysis: family-based association analysis guided by evolution of haplotypes. *Am J Hum Genet* **68**(5):1250–1263.

43. Seltman H, Roeder K, Devlin B. (2003) Evolutionary-based association analysis using haplotype data. *Genet Epidemiol* **25**(1):48–58.

44. Qian D, Beckmann L. (2002) Minimum-recombinant haplotyping in pedigrees. *Am J Hum Genet* **70**:1434–1445.

45. Chapman JM, Cooper JD, Todd JA, Clayton DG. (2003) Detecting disease associations due to linkage disequilibrium using haplotype tags: a class of tests and the determinants of statistical power. *Hum Hered* **56**(1–3):18–31.

46. Li J, Jiang T. (2003) Efficient rule-based haplotyping algorithm for pedigree data. In: *Proceedings of the 7th Annual International Conference on Research in Computational Molecular Biology (RECOMB 2003)*, ACM Press, New York, NY, pp. 197–206.

47. Becker T, Knapp M. (2004) A powerful strategy to account for multiple testing in the context of haplotype analysis. *Am J Hum Genet* **75**(4):561–570.

48. Becker T, Knapp M. (2004) Maximum-likelihood estimation of haplotype frequencies in nuclear families. *Genet Epidemiol* **27**(1):21–32.

49. Horvath S, Xu X, Lake SL, *et al.* (2004) Family-based tests for associating haplotypes with general phenotype data: application to asthma genetics. *Genet Epidemiol* **26**(1):61–69.

50. Gao G, Hoeschele I, Sorensen P, Du FX. (2004) Conditional probability methods for haplotyping in pedigrees. *Genetics* **167**:2055–2065.

51. Allen AS, Satten GA, Tsiatis AA. (2005) Locally-efficient robust estimation of haplotype-disease association in family-based studies. *Biometrika* **92**(3):559–571.

52. Yu K, Zhang S, Borecki I, *et al.* (2005) A haplotype similarity based transmission/disequilibrium test under founder heterogeneity. *Ann Hum Genet* **69**:455–467.

53. Zhang K, Sun F, Zhao H. (2005) HAPLORE: a program for haplotype reconstruction in general pedigrees without recombination. *Bioinformatics* **21**:90–103.

54. Kruglyak L, Lander ES. (1995) High-resolution of genetic mapping of complex traits. *Am J Hum Genet* **56**:1212–1223.

55. Abecasis GR, Cherny SS, Cookson WO, Cardon LR. (2002) Merlin — rapid analysis of dense genetic maps using sparse gene flow trees. *Nat Genet* **30**:97–101.

56. Abecasis GR, Wigginton JE. (2005) Handling marker–marker linkage disequilibrium: pedigree analysis. *Am J Hum Genet* **77**:754–767.

57. Kirk KM, Cardon LR. (2002) The impact of genotyping error on haplotype reconstruction and frequency estimation. *Eur J Hum Genet* **10**(10):616–622.

58. Schneidler S, Roessli D, Excoffier L. (2002) Arlequin version 2.001: a software for population genetics data analysis. Genetics and Biometry Laboratory, University of Geneva, Geneva, Switzerland.

59. Chung RH, Gusfield D. (2003) Empirical exploration of perfect phylogeny haplotyping and haplotypers. In: *Lecture Notes in Computer Science*, Warnow T, Zhu B (eds.), Springer, Big Sky, MT, pp. 5–19.

60. Epstein MP, Satten GA. (2003) Inference on haplotype effects in case-control studies using unphased genotype data. *Am J Hum Genet* **73**(6):1316–1329.

61. Gusfield D. (2002) Haplotyping as perfect phylogeny: conceptual framework and efficient solutions. In: *Proceedings of the Sixth Annual Conference on Research in Computational Molecular Biology*, ACM Press, New York, NY, pp. 166–175.

62. Bafna V, Gusfield D, Lancia G, Yooseph S. (2003) Haplotyping as perfect phylogeny: a direct approach. *J Comput Biol* **10**:323–340.

63. Xie X, Ott J. (1993) Testing linkage disequilibrium between a disease and marker locus. *Am J Hum Genet* **53**:1107.

64. Terwilliger J, Ott J. (1994) *Handbook for Human Genetics Linkage.* Johns Hopkins University Press, Baltimore, MD.

65. Zhao JH, Curtis D, Sham PC. (2000) Model-free analysis and permutation tests for allelic associations. *Hum Hered* **50**:133–139.

66. Zhao JH, Sham PC. (2002) Faster haplotype frequency estimation using unrelated subjects. *Hum Hered* **53**(1):36–41.

67. Thomas A. (2003) Accelerated gene counting for haplotype frequency estimation. *Ann Hum Genet* **67**:608–612.
68. Thomas A. (2003) GCHap: fast MLEs for haplotype frequencies by gene counting. *Bioinformatics* **19**:2002–2003.
69. Zhao JH, Lissarrague S, Essioux L, Sham PC. (2002) GENECOUNTING: haplotype analysis with missing genotypes. *Bioinformatics* **18**:1694–1695.
70. Zhao JH. (2004) 2LD, GENECOUNTING and HAP: computer programs for linkage disequilibrium analysis. *Bioinformatics* **20**:1325–1326.
71. Kang H, Qin ZS, Niu T, Liu JS. (2004) Incorporating genotyping uncertainty in haplotype inference for single-nucleotide polymorphisms. *Am J Hum Genet* **74**:495–510.
72. Wang L, Xu Y. (2003) Haplotype inference by maximum parsimony. *Bioinformatics* **19**:1773–1780.
73. Eronen L, Geerts E, Toivonen H. (2004) A Markov chain approach to reconstruction of long haplotype. *Pac Symp Biocomput* **9**:104–115.
74. Barrett JC, Fry B, Maller J, Daly MJ. (2005) Haploview: analysis and visualization of LD and haplotype maps. *Bioinformatics* **21**:263–265.
75. Krawczak M. (1994) HAPMAX documentation. Available at http://www.uni-kiel.de/medinfo/mitarbeiter/krawczak/download/
76. Zhang J, Rowe WL, Struewing JP, Buetow KH. (2002) HapScope: a software system for automated and visual analysis of functionally annotated haplotypes. *Nucleic Acids Res* **30**:5213–5221.
77. Li SS, Khalid N, Carlson C, Zhao LP. (2003) Estimating haplotype frequencies and standard errors for multiple single nucleotide polymorphisms. *Biostatistics* **4**:513–522.
78. Zhao LP, Li SS, Khalid N. (2003) A method for the assessment of disease associations with single-nucleotide polymorphism haplotypes and environmental variables in case-control studies. *Am J Hum Genet* **72**(5): 1231–1250.
79. Ding Z, Filkov V, Gusfield D. (2005) A linear-time algorithm for the perfect phylogeny haplotyping problem. In: *Proceedings of the 9th Annual International Conference on Research in Computational Molecular Biology (RECOMB 2005)*, ACM Press, New York, NY, pp. 585–600.
80. Czika W, Yu X, Wolfinger RD. (2003) An introduction to genetic data analysis using SAS/genetics. Available at http://support.sas.com/rnd/papers/sugi27/genetics.pdf/
81. Tregouet DA, Escolano S, Tiret L, *et al.* (2004) A new algorithm for haplotype-based association analysis: the stochastic-EM algorithm. *Ann Hum Genet* **68**:165–177.
82. Purcell S, Sham P. (2003) WHAP: SNP haplotype analysis package. Available at http://www.genome.wi.mit.edu/~shaun/whap/
83. Tanck MW, Klerkx AH, Jukema JW, *et al.* (2003) Estimation of multilocus haplotype effects using weighted penalised log-likelihood: analysis of five sequence variations at the cholesteryl ester transfer protein gene locus. *Ann Hum Genet* **67**(Pt 2):175–184.

84. Lake SL, Lyon H, Tantisira K, *et al.* (2003) Estimation and tests of haplotype–environment interaction when linkage phase is ambiguous. *Hum Hered* **55**(1):56–65.

85. Zaykin DV, Westfall PH, Young SS, *et al.* (2002) Testing association of statistically inferred haplotypes with discrete and continuous traits in samples of unrelated individuals. *Hum Hered* **53**(2):79–91.

86. Lin DY, Zeng D. (2006) Likelihood-based inference on haplotype effects in genetic association studies. *J Am Stat Assoc* **101**(473):89–104.

87. Spinka C, Carroll RJ, Chatterjee N. (2005) Analysis of case-control studies of genetic and environmental factors with missing genetic information and haplotype-phase ambiguity. *Genet Epidemiol* **29**(2):108–127.

88. Lin DY. (2004) Haplotype-based association analysis in cohort studies of unrelated individuals. *Genet Epidemiol* **26**(4):255–264.

89. Stram DO, Pearce CL, Bretsky P, *et al.* (2003) Modeling and E-M estimation of haplotype-specific relative risks from genotype data for a case-control study of unrelated individuals. *Hum Hered* **55**(4):179–190.

90. Mander AP. (2001) Haplotype analysis in population-based association studies. *Stata J* **1**(1):58–75.

91. Mander AP. (2002) Analysis of quantitative traits using regression and log-linear modeling when phase is unknown. *Stata J* **2**(1):65–70.

92. Fallin D, Schork NJ. (2000) Accuracy of haplotype frequency estimation for biallelic loci, via the expectation-maximization algorithm for unphased diploid genotype data. *Am J Hum Genet* **67**(4):947–959.

93. Liu N, Beerman I, Lifton R, Zhao H. (2006) Haplotype analysis in the presence of informatively missing genotype data. *Genet Epidemiol* **30**:290–300.

94. Satten GA, Epstein MP. (2004) Comparison of prospective and retrospective methods for haplotype inference in case-control studies. *Genet Epidemiol* **27**(3):192–201.

95. Chatterjee N, Carroll RJ. (2005) Semiparametric maximum likelihood estimation exploiting gene–environment independence in case-control studies. *Biometrika* **92**(2):399–418.

96. Xie RR, Stram DO. (2005) Asymptotic equivalence between two score tests for haplotype-specific risk in general linear models. *Genet Epidemiol* **29**(2):166–170.

97. Mander PA. (2001) Haplotype analysis in population-based association studies. *Stata J* **1**(1):58–75.

98. Tregouet DA, Barbaux S, Escolano S, *et al.* (2002) Specific haplotypes of the P-selectin gene are associated with myocardial infarction. *Hum Mol Genet* **11**(17):2015–2023.

99. Mander PA. (2002) Analysis of quantitative traits using regression and log-linear modeling when phase is unknown. *Stata J* **2**(1):65–70.

100. Durrant C, Zondervan KT, Cardon LR, *et al.* (2004) Linkage disequilibrium mapping via cladistic analysis of single-nucleotide polymorphism haplotypes. *Am J Hum Genet* **75**(1):35–43.

101. Tzeng JY, Devlin B, Wasserman L, Roeder K. (2003) On the identification of disease mutations by the analysis of haplotype similarity and goodness of fit. *Am J Hum Genet* **72**(4):891–902.

102. Tzeng JY. (2005) Evolutionary-based grouping of haplotypes in association analysis. *Genet Epidemiol* **28**(3):220–231.

103. Becker T, Cichon S, Jönson E, Knapp M. (2005) Multiple testing in the context of haplotype analysis revisited: application to case-control data. *Ann Hum Genet* **69**(Pt 6):747–756.

104. Becker T, Schumacher J, Cichon S, *et al.* (2005) Haplotype interaction analysis of unlinked regions. *Genet Epidemiol* **29**(4):313–322.

105. Jorde LB. (2000) Linkage disequilibrium and the search for complex disease genes. *Genome Res* **10**(10):1435–1444.

106. Mcpeek MS, Strahs A. (1999) Assessment of linkage disequilibrium by the decay of haplotype sharing, with application to fine-scale genetic mapping. *Am J Hum Genet* **65**(3):858–875.

107. Houwen RH, Baharloo S, Blankenship K, *et al.* (1994) Genome screening by searching for shared segments: mapping a gene for benign recurrent intrahepatic cholestasis. *Nat Genet* **8**(4):380–386.

108. Fan R, Lange K. (1998) Models for haplotype evolution in a nonstationary population. *Theor Popul Biol* **53**(3):184–198.

109. Beckmann L, Thomas D, Fischer C, Chang-Claude J. (2005) Haplotype sharing analysis using Mantel statistics. *Hum Hered* **59**(2):67–78.

110. Van der Meulen MA, Te Meerman GJ. (1997) Haplotype sharing analysis in affected individuals from nuclear families with at least one affected offspring. *Genet Epidemiol* **14**(6):915–920.

111. Bourgain C, Génin E, Holopainen P, *et al.* (2001) Use of closely related affected individuals for the genetic study of complex diseases in founder populations. *Am J Hum Genet* **68**(1):154–159.

112. Bourgain C, Genin E, Quesneville H, Clerget-Darpoux F. (2000) Search for multifactorial disease susceptibility genes in founder populations. *Ann Hum Genet* **64**:255–265.

113. Devlin B, Roeder K, Wasserman L. (2000) Genomic control for association studies: a semiparametric test to detect excess-haplotype sharing. *Biostatistics* **1**(4):369–387.

114. Thomas DC. (2004) *Statistical Methods in Genetic Epidemiology.* Oxford University Press, New York, NY.

115. Clayton D, Chapman J, Cooper J. (2004) Use of unphased multilocus genotype data in indirect association studies. *Genet Epidemiol* **27**(4):415–428.

116. Yu K, Xu J, Rao DC, Province M. (2005) Using tree-based recursive partitioning methods to group haplotypes for increased power in association studies. *Ann Hum Genet* **69**:577–589.

117. Tzeng JY, Wang CH, Kao JT, Hsiao CK. (2006) Regression-based association analysis with clustered haplotypes through use of genotypes. *Am J Hum Genet* **78**(2):231–242.

118. Molitor J, Marjoram P, Thomas D. (2003) Fine-scale mapping of disease genes with multiple mutations via spatial clustering techniques. *Am J Hum Genet* **73**(6):1368–1384.

119. Thomas DC, Stram DO, Conti D, *et al.* (2003) Bayesian spatial modeling of haplotype associations. *Hum Hered* **56**(1–3):32–40.

120. Morris AP. (2005) Direct analysis of unphased SNP genotype data in population-based association studies via Bayesian partition modelling of haplotypes. *Genet Epidemiol* **29**(2):91–107.

121. Morris AP, Whittaker JC, Balding DJ. (2002) Fine-scale mapping of disease loci via shattered coalescent modeling of genealogies. *Am J Hum Genet* **70**(3):686–707.

122. Yu K, Gu CC, Province M, *et al.* (2004) Genetic association mapping under founder heterogeneity via weighted haplotype similarity analysis in candidate genes. *Genet Epidemiol* **27**(3):182–191.

123. Templeton AR, Boerwinkle E, Sing CF. (1987) A cladistic analysis of phenotypic associations with haplotypes inferred from restriction endonuclease mapping. I. Basic theory and an analysis of alcohol dehydrogenase activity in *Drosophila*. *Genetics* **117**(2):343–351.

124. Templeton AR, Sing CF, Kessling A, Humphries S. (1988) A cladistic analysis of phenotype associations with haplotypes inferred from restriction endonuclease mapping. II. The analysis of natural populations. *Genetics* **120**(4):1145–1154.

125. Templeton AR, Crandall KA, Sing CF. (1992) A cladistic analysis of phenotypic associations with haplotypes inferred from restriction endonuclease mapping and DNA sequence data. III. Cladogram estimation. *Genetics* **132**(2):619–633.

126. Templeton AR, Sing CF. (1993) A cladistic analysis of phenotypic associations with haplotypes inferred from restriction endonuclease mapping. IV. Nested analyses with cladogram uncertainty and recombination. *Genetics* **134**(2):659–669.

127. Templeton AR. (1995) A cladistic analysis of phenotypic associations with haplotypes inferred from restriction endonuclease mapping or DNA sequencing. V. Analysis of case/control sampling designs: Alzheimer's disease and the apoprotein E locus. *Genetics* **140**(1):403–409.

128. Templeton AR, Maxwell T, Posada D, *et al.* (2005) Tree scanning: a method for using haplotype trees in phenotype/genotype association studies. *Genetics* **169**(1):441–453.

129. Waldron ERB, Whittaker JC, Balding DJ. (2006) Fine mapping of disease genes via haplotype clustering. *Genet Epidemiol* **30**(2):170–179.

130. Altshuler D, Brooks LD, Chakravarti A, *et al.* (2005) A haplotype make of the human genome. *Nature* **437**:1299–1320.

131. Greenspan G, Geiger D. (2004) High density linkage disequilibrium mapping using models of haplotype block variation. *Bioinformatics* **20**(Suppl 1): I137–I144.

132. Becker T, Valentonyte R, Croucher PJ, *et al.* (2006) Identification of probable genotyping errors by consideration of haplotypes. *Eur J Hum Genet* **14**(4):450–458.

133. Becker T, Knapp M. (2005) Impact of missing genotype data on Monte-Carlo simulation based haplotype analysis. *Hum Hered* **59**(4):185–189.
134. Cardon LR, Abecasis GR, Cherny SS. (2000) The effect of genotype error on the power to detect linkage and association with quantitative traits. *Am J Hum Genet* **67**(4):310.
135. Abecasis GR, Cherny SS, Cardon LR. (2001) The impact of genotyping error on family-based analysis of quantitative traits. *Eur J Hum Genet* **9**(2):130–134.
136. Knapp M, Becker T. (2004) Impact of genotyping errors on type I error rate of the haplotype-sharing transmission/disequilibrium test (HS-TDT). *Am J Hum Genet* **74**(3):589–591.
137. Niu TH. (2004) Algorithms for inferring haplotypes. *Genet Epidemiol* **27**(4):334–347.
138. Brinza D, Zelikovsky A. (2006) Phasing of 2-SNP genotypes based on non-random mating model. In: *Proceedings of International Conference on Computational Science 2006,* pp. 767–774.
139. Zhang Y, Niu TH, Liu JS. (2006) A coalescence-guided hierarchical Bayesian method for haplotype inference. *Am J Hum Genet* **79**(2):313–322.
140. Halperin E, Hazan E. (2006) HAPLOFREQ — estimating haplotype frequencies efficiently. *J Comput Biol* **13**(2):481–500.
141. Scheet P, Stephens M. (2006) A fast and flexible statistical model for large-scale population genotype data: applications to inferring missing genotypes and haplotypic phase. *Am J Hum Genet* **78**(4):629–644.
142. Souverein OW, Zwinderman AH, Tanck MWT. (2006) Estimating haplotype effects on dichotomous outcome for unphased genotype data using a weighted penalized log-likelihood approach. *Hum Hered* **61**(2):104–110.

Chapter 7

A Survey on Haplotyping Algorithms for Tightly Linked Markers

Jing Li

Electrical Engineering and Computer Science Department
Case Western Reserve University
Cleveland, OH 44106, USA

Two grand challenges in the postgenomic era are the development of a detailed understanding of heritable variation in the human genome, and robust strategies for identifying genetic contribution to diseases and drug responses.[1] Haplotypes of single nucleotide polymorphisms (SNPs) have been suggested as an effective representation of human variation, and various haplotype-based association mapping methods for complex traits have been proposed in the literature. However, humans are diploid and, in practice, genotype data instead of haplotype data are collected directly. Therefore, efficient and accurate computational methods for haplotype reconstruction are needed and have recently been investigated intensively, especially for tightly linked markers such as SNPs. This paper reviews statistical and combinatorial haplotyping algorithms using pedigree data, unrelated individuals, or pooled samples.

1. INTRODUCTION

With the completion of the Human Genome Project,[2,3] an almost complete human genomic DNA sequence has become available, which is essential to understanding the functions and characteristics of human genetic material. An important next step in human genomics is to determine genetic variation among humans as well as the correlation between genetic variation and phenotypic variation such as disease status, quantitative traits, etc. To achieve this goal, an international collaboration, namely the International

HapMap Project,[4] was launched in October 2002. The main objective of the HapMap Project is to identify the haplotype structure of humans and common haplotypes among populations. However, the human genome is a diploid and, in practice, haplotype data are not collected directly, especially in large-scale sequencing projects, mainly due to cost considerations. Instead, genotype data are collected routinely in large sequencing projects. Hence, efficient and accurate computational methods and computer programs for the inference of haplotypes from genotypes are greatly needed.

The existing computational methods for haplotyping fit into two broad categories: statistical methods and combinatorial (or rule-based) methods. Both methodologies can be applied to pedigree data, population data, or pooled samples. An earlier review paper[5] discussed haplotype inference on pedigree data, but the methods mentioned did not directly address the problem for tightly linked markers. Many developments have been made since then. There have been a few review papers on haplotype inference in recent years,[6–8] but all of them focus mainly on combinatorial formulations and solutions. Two of them[7,8] deal only with unrelated population data.

This paper will review both statistical and combinatorial algorithms for three different types of data: pedigree data, population data, and pooled samples. It is organized as follows. Biological backgrounds of the problem and notations used later will be first introduced in Sec. 1. Haplotype inference algorithms for three different types of input data will be discussed in three separate sections. Some commonly used haplotyping programs on the Internet will be listed at the end.

1.1. *Genetic Backgrounds*

The genome of an organism consists of chromosomes, which are double-stranded DNA. Locations on a chromosome can be identified using markers, which are small segments of DNA with some specific features (or a single nucleotide for single nucleotide polymorphisms [SNPs]). The position of a marker on the chromosome is called a marker locus, and a marker state is called an allele. A set of markers and their positions define a genetic map of chromosomes. There are many types of markers. The two most commonly used markers are microsatellite markers and SNP markers. Different sets of markers have different properties, such as the total number of different allelic states at one locus, frequency of each allele, distance between two adjacent loci, etc. A microsatellite marker usually has several

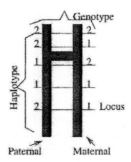

Fig. 1.   The structure of a pair of chromosomes from a mathematical point of view.

different alleles at a locus (called multiallele); while an SNP marker can be treated as a biallele, which has two alternative states. There are millions of SNPs but only hundreds of microsatellite markers, so the average distance between two SNPs is much smaller than the average distance between two microsatellite marker loci. With the advances in genotyping techniques, SNP markers are increasingly common in gene fine mapping and whole-genome association studies.

In diploid organisms, chromosomes come in pairs. The status of two alleles at a particular marker locus of a pair of chromosomes is called a marker genotype. The genotype information of a locus is denoted using a set (a, b), where a and b are integers representing allele IDs. For example, for biallelic markers like SNPs, $a, b \in \{1, 2\}$. If the two alleles are the same, the genotype is homozygous; otherwise, it is heterozygous. A haplotype consists of all alleles, one from each locus, that are on the same chromosome. Figure 1 illustrates the above concepts.

The Mendelian law of inheritance states that the genotype of a child must come from the genotypes of his or her parents at each marker locus. In other words, the two alleles at each locus of a child have different origins: one is from the father (the paternal allele), and the other one is from the mother (the maternal allele). Such information is also called the phase of the two alleles, which cannot be obtained directly from genotypes. Usually, for a tightly linked region, a child inherits a complete haplotype from each parent. However, recombination may occur, whereby the two haplotypes of a parent get shuffled due to a crossover of chromosomes and one of the shuffled copies is passed on to the child. Such an event is called a recombination event and its result is called a recombinant. Figure 2 illustrates an example where the paternal haplotype of member 3 is the result of a recombinant.

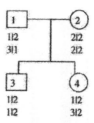

Fig. 2. An example recombination event. A family with two parents (members 1 and 2) and two children (members 3 and 4) is represented by a pedigree graph. The data consist of two loci with genotype/haplotype information listed below each member. The notation $a \mid b$ means that the phase information at the locus has been resolved, and we know that allele a is from the father and allele b is from the mother.

1.2. Notations

For a given region with m loci, the genotypes of an individual can be represented by its genotype vector $g = \langle g[1], g[2], \ldots, g[m] \rangle$, where $g[i]$ is a pair of alleles with possible missing alleles — i.e. $g[i] = (a_s, a_t)$, $a_s, a_t \in A_i \cup \{0\}$, where A_i is the set of all alleles at locus i and 0 stands for missing alleles. In the case of SNPs, $A_i = \{1, 2\}$ for all i. A haplotype is a vector of alleles $h = \langle h[1], h[2], \ldots, h[m] \rangle$, where $h[i] \in A_i$. The genotype vector is actually composed of a maternal haplotype (h_m) and a paternal haplotype (h_p), i.e. $g = (h_m, h_p)$. But, due to the limitations of the current genotyping techniques, such information is lost when obtaining an individual's genotype.

The goal of haplotype inference is to reconstruct the haplotype pair based on constraints imposed by the genotypes of family members or some mathematical models. It is easy to see that without further constraints, an individual with k heterozygous loci will have $2^{(k-1)}$ different haplotype pairs that are consistent with his or her genotype. For a pair of consistent haplotypes, we denote $g = h_1 \oplus h_2$.

2. HAPLOTYPE INFERENCE FROM PEDIGREE DATA

It is generally assumed that pedigree data consist of no mutations; thus, the genotypes are consistent with the Mendelian law of inheritance. This is realistic in practice, given the moderate sizes of most available human pedigrees. With genotype information from parents, the phase of a child at a particular locus may be determined in many cases. But, there are other cases where the phase of a child cannot be determined (e.g. when both parents and a child have the same heterozygous genotype). Missing data further complicate the situation and increase the total freedom in a pedigree.

2.1. *Maximum Likelihood Approach*

The maximum likelihood (ML) principle can be naturally applied here. Given a pedigree with genotype information for each member, with possibly missing data, the goal of the ML approach is to identify the most likely haplotype pair for each individual.

For each haplotype assignment of a pedigree, the calculation of the probability involves two terms: the founder probability and the transmission probability. More specifically, let H denote a consistent haplotype assignment of the pedigree. For individual i in the pedigree, let $(_i h_m, {}_i h_p)$ denote its haplotype pair. Then,

$$\Pr(H) = \prod_i \Pr(_i h_m, {}_i h_p) \prod_{t,j,k} \Pr(_t h_m|_j h_m, {}_j h_p)\Pr(_t h_p|_k h_m, {}_k h_p), \quad (1)$$

where the product on i ranges over all founders and the product on $\{t, j, k\}$ ranges over all offspring-mother-father trios. Under Hardy–Weinberg equilibrium and linkage equilibrium assumptions, the founder probability can be calculated from population frequencies of alleles, i.e. $\Pr(_i h_m, {}_i h_p) = p(_i h_m)p(_i h_p)$, where $p(_i h_m)$ and $p(_i h_p)$ are haplotype frequencies that are products of their allele frequencies. The gamete transmission probabilities $\Pr(_t h_m|_j h_m, {}_j h_p)$ and $\Pr(_t h_m|_j h_m, {}_j h_p)$ can be calculated based on recombination fractions between marker intervals. Notice that for tightly linked markers, the probability of recombination events between two adjacent markers is small, so the ML approach favors haplotype configurations with few recombinations.

In traditional linkage analysis, the likelihood calculation involves the summation over all possible haplotype assignments. So, theoretically, one can simultaneously output the haplotype assignment with maximum likelihood when performing linkage analysis. Two exact algorithms have been proposed to calculate the probability of a pedigree for linkage analysis. The Elston–Stewart algorithm[9] took advantage of the Markov property based on pedigree structure — i.e. given parents' genotype information, the genotype of a child is independent of the genotype of his or her ancestors. The algorithm is linear in pedigree size, but exponential in the number of genetic loci. The Lander–Green algorithm[10] took advantage of the Markov property based on marker loci — i.e. under the assumption of no interference, the phase of a marker depends only on the phase of its previous locus. This algorithm is linear in the number of genetic loci, but exponential in pedigree size. Although much improvement[11–13] has been made for both algorithms, the exact calculation of likelihood and haplotype inference for complex pedigrees with substantial missing data are still computationally unfeasible. Even approximation algorithms employing important sampling techniques, such as simulated annealing,[5] are not efficient enough for complex pedigrees with large sizes.

The calculation of the likelihood of a pedigree, as well as the calculation of haplotype configurations, in most available tools for linkage analysis (for example, GeneHunter, Simwalk2, and S.A.G.E.; the links to these programs are provided in Sec. 5) assumes linkage equilibrium between markers. The assumption is unrealistic for tightly linked markers such as SNPs because it is well known that most SNP markers are in linkage disequilibrium (LD). The effect of violation of such an assumption has been investigated only very recently. Abecasis and Wigginton[14] have proposed a new approach that can directly model LD between markers during multipoint analysis of human pedigrees. The algorithm first partitions all of the SNPs into clusters based on their LD. For each cluster, the authors assume there is no recombination and use haplotypes to incorporate LD within a cluster. LD between clusters is ignored and the likelihood can be calculated using the Lander–Green algorithm, while taking each cluster essentially as a marker. Their simulation results show that the approach resolves previously described biases in multipoint linkage analysis with SNPs that are in LD.

2.2. *Recombination/Crossover Minimization*

Given the fact that maximum likelihood–based approaches are usually time-consuming and the assumptions they require do not always hold for tightly linked markers, rule-based approaches to minimize the recombination/crossover in a pedigree have recently received much attention. The minimum recombination principle basically states that genetic recombination is rare, thus haplotypes with fewer recombinants should be preferred in a haplotype reconstruction.[15–17] For tightly linked markers, the principle is well supported by experimental data. For example, recently published results[23–25] demonstrate that, in humans, the number of distinct haplotypes is very limited relative to the number of all possible haplotype combinations. Moreover, the genomic DNA can probably be partitioned into long blocks such that recombination within each block is rare or even nonexistent during the history. For pedigree data, one can safely assume that recombination is rare for a much larger region. In the literature, the crossover minimization formulation is also called the minimum recombinant haplotype configuration (MRHC) problem.[16,18,19]

Sobel *et al.*[5] proposed a simulated annealing algorithm while taking a pseudo-likelihood function regarding the number of recombinants as the energy function. O'Connell[15] worked on a special case of the MRHC problem: he assumed that the input data contain haplotype solutions with zero recombinants, and proposed a genotype elimination algorithm to eliminate all impossible genotypes. Tapadar *et al.*[17] utilized the genetic algorithm to tackle the same problem. Qian and Beckmann[16] proposed a rule-based algorithm to reconstruct haplotype configurations for pedigree data, based on local minimization of each nuclear family. Although their program MRH performs well for small pedigrees and achieves better results than some previous algorithms,[15,17] its effectiveness scales poorly, especially for data with biallelic markers.

In a series of papers,[18−22] we have proposed several algorithms based on different assumptions about the real data. We have developed an iterative heuristic algorithm, called block extension, for MRHC that is much more efficient than MRH. Our experiments have shown that the block-extension algorithm can compute an optimal solution or nearly optimal solution when the minimum number of recombinants required is small.[18,19] However, its performance deteriorates significantly when the input data require more (e.g. four or more) recombinants. We have also devised an efficient, exact algorithm for solving MRHC on pedigree data that requires no recombinants. The algorithm first identifies all of the necessary constraints based on Mendelian law and the zero recombinant assumption, and represents them using a system of linear equations over binary variables indicating the phase information at each locus. All possible feasible haplotype configurations can be obtained using a variant of the Gaussian elimination algorithm. For pedigrees with small sizes or a small number of markers, we have developed two dynamic programming (DP) algorithms[20]: the running time for the first DP algorithm is linear in the size of a pedigree, and the time for the second one is linear in the number of markers.

For the most general case of the problem, we have designed an effective integer linear programming (ILP) formulation of MRHC. It integrates missing data imputation and haplotype inference together, and employs a branch-and-bound strategy that utilizes a partial order relationship and some other special relationships among variables to decide the branching order. The partial order relationship is discovered in the preprocessing of constraints by taking advantage of some special properties in the ILP formulation. A directed graph is built based on the variables and their partial order relationship. By identifying and collapsing strongly connected components in the graph, we may greatly reduce the size of an ILP instance. Nontrivial lower and upper bounds on the optimal number of recombinants are estimated at each branching node to prune the branch-and-bound search tree. When multiple solutions exist, the best haplotype configuration is selected based on a maximum likelihood approach. The algorithm also incorporates the marker interval distance into the formulation whenever it is known, which overcomes the inadequacy of many rule-based algorithms that ignore this important information. The test results on simulated data demonstrate that the algorithm is very efficient in practice. A comparison of the algorithm with a well-known statistical approach — SimWalk2[5] — on simulated data with evenly and unevenly spaced markers also demonstrates the effectiveness and soundness of the ILP algorithm.[22]

3. HAPLOTYPE INFERENCE FROM POPULATION DATA

In this section, we consider the haplotype inference (HI) problem based on population data, i.e. unrelated individuals. In addition to the haplotype assignment of each individual, we are also interested in the estimation of population haplotype frequencies. These two problems are related and are usually solved simultaneously. Without constraints from family members, an individual with m heterozygous loci has 2^{m-1} consistent haplotype pairs. One is not able to directly infer which pair is more likely to be the true haplotypes for a given individual. Instead, genetic models based on the evolution of human population history have to be adopted, directly or indirectly, to define the extent of optimality of a particular consistent solution for given samples.

The HI problem from unrelated individuals was first addressed in 1990 by Clark,[26] who proposed a rule-based algorithm in his paper. We will start our discussion from this simple but widely used algorithm. Gusfield[27] investigated the mathematical properties of Clark's algorithm and later proposed a discrete model called the perfect phylogeny haplotype (PPH) problem,[28] which implicitly adopts the coalescent model with no recombination. Another commonly used discrete model in the literature is called the pure parsimony approach,[29-31] which intends to find a solution with the smallest number of distinct haplotypes. In addition to discrete models, statistical approaches have also been applied to the HI problem. Two different groups[32,33] have proposed a maximum likelihood approach and employed the expectation-maximization (EM) algorithm to find a haplotype solution. Bayesian approaches[34,35] have also been applied to the haplotype inference problem by incorporating an informative prior based on population genetics models.

3.1. *Clark's Algorithm*

Observe that for a set of m markers, if an individual is homozygous at all loci or has only one heterozygous locus, the haplotype pair of the individual is trivially determined since there is only one consistent pair of haplotypes. If there is at least one such individual, one can obtain an initial set of haplotypes H from the input data. For each genotype vector g with more than one heterozygous locus, Clark's idea was to use a haplotype h_1 in H that is consistent with g to obtain the haplotype pair for g, i.e. $g = h_1 \oplus h_2$, and include h_2 in H. The algorithm will then iterate until no more genotypes can be resolved. It is possible that the algorithm will not even start if there are no individuals with zero or one heterozygous locus. The algorithm also does not guarantee that every genotype will eventually be resolved. Different sequences in the application of Clark's rule may give different results. Gusfield,[27] who investigated the mathematical properties of Clark's algorithm, has proved that finding the sequence(s) in the application of Clark's rule with a minimum number of unresolved genotypes is NP-hard.

3.2. *Perfect Phylogeny Model*

Later, Gusfield[28] introduced a perfect phylogeny model for the HI problem, based on two assumptions. First, the model assumes that for a set of tightly linked SNPs, historical recombination events do not exist. Experimental results and population genetics models generally support this assumption. Researchers have discovered block-like haplotype structures in humans, where no recombination has been observed within such long blocks. Second, the model adopts the standard assumption of infinite sites in population genetics; this basically means that at each SNP site, mutation can only occur at most once. Under these two assumptions, the $2n$ haplotypes from the n individuals can be organized into a rooted tree called perfect phylogeny. Each leaf of the tree represents a haplotype. Each interior edge is labeled by at least one SNP, and each SNP labels exactly one edge. A path from the root to a leaf spells all of the mutant sites of the haplotype at the leaf from the ancestry haplotype at the root (usually not given). The PPH problem is, given a set of genotypes, to find a set of haplotypes that admits a perfect phylogeny.

In his paper, Gusfield[28] presented an algorithm by reducing the problem to the graph realization with an almost linear running time in theory. But, the implementation of an algorithm is too complex to be practical. Since then, a couple of algorithms have been proposed. Two groups[38,39] have independently proposed two algorithms with the same running time $O(nm^2)$, where n is the number of individuals and m is the number of SNPs. More recently, Ding *et al.*[40] presented a linear algorithm for the problem; the algorithm has been implemented in a program called linear PPH (LPPH). For future directions, it is desirable to extend the perfect phylogeny model in order to allow recombination and missing data.

3.3. *Pure Parsimony*

The pure parsimony approach has also been investigated by researchers[29-31] in the computational biology community. Under this criterion, the goal is to find a minimum set of distinct haplotypes that can resolve all of the given genotypes. The rationale of the parsimony principle for the HI problem is based on the observation that in human populations, the number of observed distinct haplotypes is far smaller than the total number of all possible haplotypes. Theories in population genetics and experimental data also support the hypothesis of limited haplotype diversity. Unlike the perfect phylogeny model, which has an optimal linear time algorithm, the computation of the diversity minimization problem turns out to be hard. It has been shown[31] that in theory, the problem not only has no practical exact algorithms, but it also has no practical approximation algorithms. Gusfield[29] formulated the problem using the ILP approach, which is able to find optimal solutions for instances with small sizes. Wang and Xu[30] proposed a branch-and-bound algorithm, and experimental results have shown it to be effective for practical problems.

All three combinatorial formulations for HI using population data have been reviewed in detail by Gusfield.[7] In addition to the discrete approaches, statistical models have been studied in the literature. We will introduce the maximum likelihood model and Bayesian approaches in the next two subsections.

3.4. *Maximum Likelihood*

The maximum likelihood approach[32,33] takes haplotype population frequencies as unknown parameters that need to be inferred. The goal is to estimate the values of haplotype frequencies that maximize the probability of observing the given genotype data. Assume that all of the individuals are independent; the likelihood of the data is then just the multiplication of the probability of each individual. Under the assumption of random mating and Hardy–Weinberg equilibrium, the probability of observing a particular genotype from an individual is the summation of the product of two haplotype frequencies for all haplotype pairs that are consistent with the genotype:

$$L(G) = \prod_i^n \sum_{h_s \oplus h_t = g_i} p(h_s)p(h_t). \qquad (2)$$

When the maximum likelihood estimates (MLEs) cannot be readily obtained from an analytical derivation as in this case, numerical methods are commonly used. The expectation-maximization (EM) algorithm is a widely accepted approach for obtaining the MLEs. The EM algorithm is an iterative method that consists of two steps (E-step and M-step) in each iteration. In the context of HI, it takes the haplotype frequencies as the parameters and the phase of each individual as missing data. If the phase of each individual is known, the MLE of the frequency of a particular haplotype is just the fraction of that haplotype occurring in the samples. On the other hand, if haplotype frequencies are known, the probability of observing an individual with a phased haplotype pair is just the product of the frequencies of the two haplotypes under the assumption that haplotypes are in Hardy–Weinberg equilibrium.

The EM algorithm starts with an initial (probably arbitrary) assignment of haplotype frequencies, $p^0(h_1), p^0(h_2), \ldots, p^0(h_k)$. In the E-step of the ith iteration, it calculates the expected counts $(n_{h_s}^i)$ of a haplotype h_s from samples, assuming that the haplotype frequencies are true values:

$$n_{h_s}^i = \sum_{g:g=h_s \oplus h_t} \frac{p^i(h_s)p^i(h_t)}{\sum_{h_u,h_v:h_u \oplus h_v=g} p^i(h_u)p^i(h_v)}, \qquad (3)$$

where the first summation on the right-hand side is over all individuals whose genotypes are consistent with h_s, and the second summation is over all consistent haplotype pairs for a particular individual with genotype g. In the M-step, the haplotype frequencies are updated based on the expected counts $n_{h_s}^i$:

$$p^{i+1}(h_s) = n_{h_s}^i/2n. \qquad (4)$$

The algorithm iterates until it converges or reaches the maximum number of iterations allowed. To estimate the haplotype pair of each individual, one can pick the pair with the largest probability based on the estimated haplotype frequencies.

Theoretically, the EM algorithm guarantees to converge to a (local) maximum in a linear time, but the number of variables in this case (i.e. haplotype frequencies) can be exponentially large with respect to the number of loci in the region. So, a direct implementation of the EM algorithm for the haplotype inference problem usually cannot deal with data of more than 25 loci. It is also a known fact that the EM algorithm might converge to a local optimal point instead of a global one. Users are recommended to start with different initial values, and then pick the solution with the maximum likelihood probability. Furthermore, the EM algorithm cannot provide the estimates of variances of the MLEs in general, unless the number of loci is small.

3.5. Bayesian Approaches

Unlike the maximum likelihood method, where parameters are unknown points in a parameter space, Bayesianists treat parameters as random variables. The goal of Bayesian inference is to estimate the posterior distribution of parameters given data observed, assuming some known prior knowledge of parameters before seeing the data. Point estimations can be obtained by taking expectations of the posterior distribution. Let $\Pr[p(H)]$ denote the prior distribution of haplotype frequencies, and $\Pr[p(H)|G]$ denote the posterior distribution of haplotype frequencies given genotype data G. The posterior distribution can be calculated via Bayes' theorem:

$$\Pr[p(H)|G] = \frac{\Pr[G|p(H)]\Pr[p(H)]}{\Pr(G)}. \tag{5}$$

The prior probability $\Pr[p(H)]$ is assumed known, and the probability of data given a particular set of parameters $\Pr[G|p(H)]$ is easy to calculate; while the calculation of the overall probability $\Pr(G)$ involves multidimensional integrations or a summation over an exponentially large number of terms, thus making it unfeasible in many cases. Important sampling techniques such as Markov chain Monte Carlo (McMC) are commonly used in such situations.

Two Bayesian approaches[34,35] that have been proposed for HI from population data use Gibbs sampling techniques to obtain an estimation of the posterior distribution of haplotype frequencies. The algorithm of Stephens *et al.*[34] starts from an arbitrary haplotype solution of the given genotypes, then iteratively updates a randomly selected individual while assuming all the other individuals have their correct haplotype assignments. The algorithm of Niu *et al.*[35] starts from an initial assignment of haplotype frequencies. In each iteration step, it first samples a pair of compatible haplotypes for each individual and then updates the haplotype frequencies based on the haplotype solution of each individual. The two methods differ mainly in the prior distributions they assumed[36]: Stephens *et al.*[34] used a prior approximating the coalescent model, while Niu *et al.*[35] used the Dirichlet prior. Under the coalescent model, haplotypes to be sampled tend to be more similar to previously sampled haplotypes, a property that has been used in Clark's algorithm. Experiments[36] have shown that the estimations based on the coalescent model are more accurate than those based on the Dirichlet prior. Both algorithms have been implemented into computer programs (i.e. Phase and Haplotyper) that are widely used.

An important contribution by Niu *et al.*[35] the introduction of the partition–ligation technique, which is an application of the divide–conquer technique that can reduce the computational burden for large data sets. Subsequently, this idea was incorporated into other algorithms such as Phase V2.0 and the EM algorithm.[37]

4. HAPLOTYPE INFERENCE FROM POOLED SAMPLES

As a strategy for reducing genotyping cost, pooling individual samples has been shown to be efficient in estimating population allele frequencies and LD coefficients.[41] For HI, it is obvious that pooling samples adds more ambiguities. Nevertheless, several groups[42,44] have investigated the efficiency and cost-effectiveness of estimating haplotype frequencies from pooled DNA data. In general, suppose $K \geq 1$ independent individuals are pooled together, where $K = 1$ corresponds to the strategy with no pooling. The genotype at each locus can be represented by the number of allele 1, i.e. an integer g such that $0 \leq g \leq 2K$. The primary goal is to estimate haplotype frequencies for the region of interest. Although the most likely haplotype configurations (the $2K$ haplotypes in each pool that are consistent with the input genotypes) might be inferred, haplotypes for each individual cannot be constructed in general. Because the accuracy of haplotype frequency estimates decreases with an increase of K, it is not worthwhile to pool a large number of samples if the accuracy deteriorates too much. To compare the cost-effectiveness of different strategies, Yang *et al.*[44] defined a simple measure named relative efficiency $R(K) = K \times v_1/v_K$, where v_1 and

v_K are the mean squared errors for samples without pooling and with pooling of a size K, respectively. Pooling is only meaningful when $R(K) \geq 1$.

To obtain haplotype frequency estimates from pooled samples, in theory, both the maximum likelihood approach and Bayesian-based approaches can be applied, with the haplotype pair of an individual replaced by the haplotype configuration of a pool. As a matter of fact, Quade *et al.*[43] have proposed an algorithm that views pedigree data and population data as special cases of pooled samples. Yang*et al.*[44] also applied the maximum likelihood method on pooled data and adopted the EM algorithm for the estimation of haplotype frequencies. The algorithm iteratively updates the population haplotype frequencies and haplotype configurations of each pool. Because the number of distinct haplotypes increases exponentially with the number of loci m, and the number of distinct haplotype configurations in each pool increases exponentially with the size of each pool K, the algorithm is only practical for problems with small sizes ($m \leq 15$ and $K \leq 6$). In terms of cost-effectiveness, simulation results by Yang *et al.*[44] have shown that the relative efficiency $R(K)$ increases the most when $K = 2$ or 3, and the pooling strategy is more effective for SNPs with high LD and moderate or large minor allele frequencies. In addition to the loss of genotype and haplotype information for each individual, one other limitation of pooling strategies in genome association studies is the loss of phenotype information, especially when multiple measures of different quantitative traits have been recorded for each individual.

5. SOFTWARE AVAILABLE ON THE WEB

Simwalk2: http://watson.hgen.pitt.edu/docs/simwalk2.html/

Merlin: http://www.sph.umich.edu/csg/abecasis/Merlin/

S.A.G.E.: http://darwin.case.edu/

GeneHunter: http://www.fhcrc.org/science/labs/kruglyak/Downloads/index.html/

PedPhase: http://www.eecs.case.edu/jxl175/haplotyping.html/

Phase: http://www.stat.washington.edu/stephens/software.html/

Haplotyper: http://www.people.fas.harvard.edu/junliu/Haplo/docMain.htm/

PPH and LPPH: http://wwwcsif.cs.ucdavis.edu/gusfield/

ACKNOWLEDGMENTS

This work was supported in part by the NIH R01 LM008991 and a start-up fund from the Case Western Reserve University.

REFERENCES

1. Collins FS, Green ED, Guttmacher AE, Guyer MS. (2003) A vision for the future of genomics research. *Nature* **422**:835–847.
2. International Human Genome Sequencing Consortium. (2001) Initial sequencing and analysis of the human genome. *Nature* **409**(6822):860–921.
3. Venter JC, Adams MD, Myers EW, *et al.* (2001) The sequence of the human genome. *Science* **291**(5507):1304–1351.
4. The International HapMap Consortium. (2003) The International HapMap Project. *Nature* **426**:789–796.
5. Sobel E, Lange K, O'Connell J, Weeks D. (1996) Haplotyping algorithms. In: *Genetic Mapping and DNA Sequencing, IMA Volumes in Mathematics and Its Applications*, Speed T, Waterman M (eds.), Springer-Verlag, New York, NY, pp. 89–110.
6. Bonizzoni P, Vedova GD, Dondi R, Li J. (2003) The haplotyping problem: an overview of computational models and solutions. *J Comput Sci Technol* **18**(6):675–688.
7. Gusfield D. (2004) An overview of combinatorial methods for haplotype inference. In: *Computational Methods for SNPs and Haplotype Inference*, Istrail S, Waterman MS, Clark AG (eds.), Springer, New York, NY, pp. 9–25.
8. Halldórsson BV, Bafna V, Edwards N, *et al.* (2004) A survey of computational methods for determining haplotypes. In: *Computational Methods for SNPs and Haplotype Inference*, Istrail S, Waterman MS, Clark AG (eds.), Springer, New York, NY, pp. 26–47.
9. Elston RC, Stewart J. (1971) A general model for the genetic analysis of pedigree data. *Hum Hered* **21**:523–542.
10. Lander ES, Green P. (1987) Construction of multilocus genetic linkage maps in humans. *Proc Natl Acad Sci USA* **84**:2363–2367.
11. Kruglyak L, Daly MJ, Reeve-Daly MP, Lander ES. (1996) Parametric and nonparametric linkage analysis: a unified multipoint approach. *Am J Hum Genet* **58**:1347–1363.
12. Gudbjartsson DF, Jonasson K, Frigge ML, Kong A. (2000) Allegro, a new computer program for multipoint linkage analysis. *Nat Genet* **25**(1):12–13.
13. Abecasis GR, Cherny SS, Cookson WO, Cardon LR. (2002) Merlin — rapid analysis of dense genetic maps using sparse gene flow trees. *Nat Genet* **30**(1):97–101.
14. Abecasis GR, Wigginton JE. (2005) Handling marker–marker linkage disequilibrium: pedigree analysis with clustered markers. *Am J Hum Genet* **77**:754–767.
15. O'Connell JR. (2000) Zero-recombinant haplotyping: applications to fine mapping using SNPs. *Genet Epidemiol* **19**(Suppl 1):S64–S70.

16. Qian D, Beckmann L. (2002) Minimum-recombinant haplotyping in pedigrees. *Am J Hum Genet* **70**(6):1434–1445.
17. Tapadar P, Ghosh S, Majumder PP. (2000) Haplotyping in pedigrees via a genetic algorithm. *Hum Hered* **50**(1):43–56.
18. Li J, Jiang T. (2003) Efficient rule-based haplotyping algorithms for pedigree data. In: *Proceedings of the 7th Annual International Conferene on Research in Computational Molecular Biology (RECOMB 2003)*, ACM Press, New York, NY, pp. 197–206.
19. Li J, Jiang T. (2003) Efficient inference of haplotypes from genotypes on a pedigree. *J Bioinform Comput Biol* **1**(1):41–69.
20. Doi K, Li J, Jiang T. (2003) Minimum recombinant haplotype configuration on pedigrees without mating loops. In: *Proceedings of the 3rd Annual Workshop on Algorithms in Bioinformatics (WABI'03)*, Springer, Berlin, Germany, pp. 339–353.
21. Li J, Jiang T. (2004) An exact solution for finding minimum recombinant haplotype configurations on pedigrees with missing data by integer linear programming. In: *Proceedings of the 8th Annual International Conference on Research in Computational Molecular Biology (RECOMB 2004)*, ACM Press, New York, NY, pp. 101–110.
22. Li J, Jiang T. (2005) Computing the minimum recombinant haplotype configuration from incomplete genotype data on a pedigree by integer linear programming. *J Comput Biol* **12**:719–739.
23. Daly MJ, Rioux JD, Schaffner SF, *et al.* (2001) High-resolution haplotype structure in the human genome. *Nat Genet* **29**(2):229–232.
24. Gabriel SB, Schaffner SF, Nguyen H, *et al.* (2002) The structure of haplotype blocks in the human genome. *Science* **296**(5576):2225–2229.
25. Helmuth L. (2001) Genome research: map of the human genome 3.0. *Science* **293**(5530):583–585.
26. Clark AG. (1990) Inference of haplotypes from PCR-amplified samples of diploid populations. *Mol Biol Evol* **7**(2):111–122.
27. Gusfield D. (2001) Inference of haplotypes from samples of diploid populations: complexity and algorithms. *J Comput Biol* **8**:305–323.
28. Gusfield D. (2002) Haplotyping as perfect phylogeny: conceptual framework and efficient solutions. In: *Proceedings of the 6th Annual International Conference on Research in Computational Molecular Biology (RECOMB 2002)*, ACM Press, New York, NY, pp. 166–175.
29. Gusfield D. (2003) Haplotype inference by pure parsimony. In: *14th Annual Symposium on Combinational Pattern Matching (CPM'03)*, Springer, Berlin, Germany, pp. 144–155.
30. Wang L, Xu L. (2003) Haplotype inference by pure parsimony. *Bioinformatics* **19**:1773–1780.
31. Lancia G, Pinotti MC, Rizzi R. (2004) Haplotyping populations by pure parsimony. *INFORMS J Comput* **16**(4):348–359.
32. Excoffier L, Slatkin M. (1995) Maximum-likelihood estimation of molecular haplotype frequencies in a diploid population. *Mol Biol Evol* **12**:921–927.

33. Hawley ME, Kidd KK. (1995) HAPLO: a program using the EM algorithm to estimate the frequencies of multi-site haplotypes. *J Hered* **86**:409–411.

34. Stephens M, Smith NJ, Donnelly P. (2001) A new statistical method for haplotype reconstruction from population data. *Am J Hum Genet* **68**(4):978–989.

35. Niu T, Qin Z, Xu X, Liu JS. (2002) Bayesian haplotype inference for multiple linked single-nucleotide polymorphisms. *Am J Hum Genet* **70**:157–159.

36. Stephens M, Donnelly P. (2003) A comparison of Bayesian methods for haplotype reconstruction from population genotype data. *Am J Hum Genet* **73**:1162–1169.

37. Qin Z, Niu T, Liu J. (2002) Partitioning-ligation–expectation-maximization algorithm for haplotype inference with single-nucleotide polymorphisms. *Am J Hum Genet* **71**:1242–1247.

38. Bafna V, Gusfield D, Lancia G, Yooseph S. (2002) Haplotyping as perfect phylogeny: a direct approach. Technical Report, University of California–Davis, Davis, CA.

39. Eskin E, Halperin E. (2003) Large-scale recovery of haplotypes from genotype data using imperfect phylogeny. In: *Proceedings of the 7th Annual International Conference on Research in Computational Molecular Biology (RECOMB 2003)*, ACM Press, New York, NY, pp. 104–113.

40. Ding Z, Filkov V, Gusfield D. (2005) A linear-time algorithm for perfect phylogeny haplotyping. In: *Proceedings of the 9th Annual International Conference on Research in Computational Molecular Biology (RECOMB 2005)*, ACM Press, New York, NY, pp. 585–600.

41. Pfeiffer RM, Rutter JL, Gail MH, *et al.* (2002) Efficiency of DNA pooling to estimate joint allele frequencies and measure linkage disequilibrium. *Genet Epidemiol* **22**(1):94–102.

42. Wang S, Kidd KK, Zhao H. (2003) On the use of DNA pooling to estimate haplotype frequencies. *Genet Epidemiol* **24**(1):74–82.

43. Quade SR, Elston RC, Goddard KA. (2005) Estimating haplotype frequencies in pooled DNA samples when there is genotyping error. *BMC Genet* **6**:25.

44. Yang Y, Zhang J, Hoh J, *et al.* (2003) Efficiency of single-nucleotide polymorphism haplotype estimation from pooled DNA. *Proc Natl Acad Sci USA* **100**(12):7225–7230.

Chapter 8

DNA Pooling: Methods and Applications in Association Studies

Jiexun Wang*, Guohua Zou* and Hongyu Zhao[†]

*Academy of Mathematics and Systems Science
Chinese Academy of Sciences, Beijing 100080, P. R. China

[†]Department of Epidemiology and Public Health
Yale University School of Medicine
New Haven, CT 06520, USA

DNA pooling is a cost-effective approach for estimating genetic marker allele frequencies. It is often advocated as a screening tool to identify candidate markers, followed by individual genotyping. This paper reviews recent developments in DNA pooling methods for association studies. Various DNA pooling designs and association tests using pooled DNA samples are introduced. Guidelines are provided on using this strategy, and open problems warranting further research are also discussed.

1. INTRODUCTION

Genetic studies using molecular markers to identify genetic variants associated with diseases are an established paradigm in human genetics. Although linkage studies have had great success in cloning disease genes, they have been largely restricted to simple Mendelian cases, which are mostly due to single genes. However, most genetic disorders are non-Mendelian, and linkage analysis has been far less successful in these diseases.[1] As a result, genetic association studies have in recent years enjoyed much popularity as

---

[†] Correspondence author.

an alternative and more powerful strategy to identify disease-susceptibility genes. The most common and simple design is the case-control study, where affected individuals (cases) and unaffected individuals (controls) are collected and genotyped either at candidate gene regions or throughout the whole genome to identify genes/markers that show different distributions between the two groups. The genomic regions harboring these markers are possible candidates for further validation studies. The markers under such studies are not necessarily functional variants; their different distributions may simply result from associations between these markers and functional variants, a phenomenon called linkage disequilibrium (LD). LD, which can be defined as the nonindependence between alleles at nearby loci, is also known as gametic phase disequilibrium or allelic association.

Different from linkage analyses, which capitalize on cosegregation between a genetic marker and disease affection status within individual pedigrees, association studies rely on population association between two loci that are closely associated with each other across pedigrees. It is well recognized that association studies have some appealing advantages over linkage studies in identifying susceptibility genes for complex traits, which are generally believed to be influenced by many genes, environmental factors, and the interactions among them. First, association studies may be more powerful than linkage studies for genetic variants carrying small-to-moderate effects, since the excess sharing across affected individuals in the general population over what is expected by chance is generally greater in such cases than the excess sharing between affected individuals within a family over what is expected by chance. Second, association studies generally provide higher precision in identifying the location of susceptibility loci. Last but not least, it is easier to obtain unrelated samples under the case-control design compared to the need to collect groups of relatives or extended pedigrees for linkage studies, especially for late-onset and/or rare diseases.[2] Therefore, the focus is on association studies in this review.

Despite the relative superiority of association studies, a large sample size (hundreds or even thousands of subjects) may still be needed to have adequate power due to disease heterogeneity, small genetic effects, and gene–environment interactions, among other factors. As a result, the cost of these studies can be substantial. Even with the rapid reduction in genotyping cost in recent years, it is still costly to genotype a large number of subjects at many markers, especially for a large-scale genomewide scan. One important strategy to reduce genotyping cost is DNA pooling, whereby DNA samples from multiple individuals are pooled together before genotyping.

To our knowledge, pooling was originally introduced as an efficient method to screen for a rare disease[3]; it was first applied in a case-control association study on HLA class II DR and DQ alleles in type I (insulin-dependent) diabetes mellitus by Arnheim and colleagues[4] (see also Sham and colleagues[5]). In that study, based on the idea that marker alleles in LD with a disease-causing variant would be more enriched or deficient in a pooled sample of affected cases than those in a pooled sample of controls, the authors compared pooled DNA samples from insulin-dependent diabetes mellitus (IDDM) patients with those from randomly selected controls, and successfully detected enrichments of HLA class II DR and DQ alleles in the IDDM population. Subsequently, the DNA pooling strategy has been used in linkage analysis,[6–8] physical mapping,[9–12] and association analysis.[13–15]

With the completion of the sequencing and identification of millions of genetic variations in the human genome, genomewide association study has become a viable and potentially powerful approach to identify disease genes, although such studies require tens or hundreds of thousands of markers, depending on the highly variable LD patterns across different populations and in different genomic regions.[16–18] The genomewide approach is also largely driven by great developments in molecular methods for genotyping microsatellite markers[13,15,19–21] and single nucleotide polymorphisms (SNPs).[22–30]

Generally, the utilization of SNP markers has several advantages over that of microsatellite markers in both individual and pooled DNA genotyping. For example, because of high density, SNPs provide more adequate genomewide coverage than microsatellite markers for systematic screens in association studies. Also, the estimation of microsatellite allele frequencies from pooled DNA may be complicated by the occurrence of stutter bands.[13,20] Furthermore, in contrast to microsatellite markers, genotyping SNPs is relatively easy, fast, and inexpensive. Therefore, like Sham and colleagues,[5] who summarized the quantitative assays for DNA pooling as well as the design and analysis of pooling data in association studies, we will focus on SNP-based methodologies in this review. We will systematically discuss various approaches for detecting associations between one or multiple marker loci and qualitative or quantitative traits based on pooled data. We will consider different designs (population-based and family-based) and their applications, especially the latest developments since 2002.

This paper is organized as follows. In Sec. 2, we will briefly introduce experimental methods for measuring allele frequencies using pooled DNA samples. SNP-based association analyses using pooled DNA are then discussed in the next three sections, including the development of statistical methods of single marker–based association studies (Sec. 3), haplotype-based association studies (Sec. 4), and association analyses incorporating genotyping errors and the effects of other confounders (Sec. 5). In the last section, we will summarize this review and discuss open problems warranting further developments.

2. DNA POOLING EXPERIMENTAL METHODS

A worldwide effort to collect and validate SNP markers has attained great achievements since the Human Genome Project (HGP) formally began in 1990. Public databases have mushroomed, including dbSNP, EnsEMBL (joint project between the EMBL-EBI and the Sanger Centre, England), GenSNPs (USA), JSNP (University of Tokyo, Japan), HGVbase (European consortium), and others. Note that the frequencies of the genetic polymorphisms are crucially important because they are the foundation of successful association studies and other genetic studies. The estimation of allele frequency can be achieved by many experimental technologies.[23,24,26,27,31–34] In this section, we will first introduce the procedure of forming DNA pools, and then briefly discuss two aspects of SNP genotyping assays: biochemical reaction principles and detection methods. Finally, the approaches to allele frequency estimation will be presented.

2.1. *Formation of DNA Pools*

We first describe a protocol for constructing DNA pools. In the first step, the concentrations of DNA samples are crudely measured; this can be done by several approaches such as the PicoGreen dsDNA Quantitation Reagent (Molecular Probes) in a CytoFluor fluorimeter (Applied Biosystems) or ultraviolet (UV) light spectroscopy.[5,35] In the second step, the DNA samples are diluted to 40–80 ng/uL, and then their concentration can be quantified more accurately by a fluorimetric assay if all DNA samples constructing pools have high purity. In the third step, samples are rediluted, requantified, and readjusted until they are diluted to a final required concentration of 4 ng/uL with a bias of 0.5 ng/uL. Check samples and see whether the amounts of each sample are equally pooled; if not, then repeat steps 1 to 3 until only those at 4 ng/uL (± 0.5 ng/uL) are accepted for pooling. Finally, pools are constructed by combining equal volumes of each sample. More details about the procedure can be found in Sham and colleagues.[5]

2.2. SNP Genotyping Methodologies

A large number of methods are available for SNP genotyping in pooled samples. Although each has its own special characteristics, they all consist of a series of biochemical steps and a product detection step. Here, we briefly introduce allele discrimination in a genotyping assay. We refer to Kwok[36] and Syvanen[37] for details.

In a genotyping assay for allele-specific discrimination, two aspects are most relevant: biochemical reaction principles and detection methods that make use of monitoring light emission, measuring the mass of the products or detecting a change in electrical property. The first reaction principle is primer extension, which is based on the ability of DNA polymerase to incorporate specific deoxyribonucleotides complementary to the sequence of the template DNA. This approach is most widely used for SNP analysis. A DNA fragment containing the SNP of interest is amplified, and then — in the presence of specific dideoxyribonucleotides — an oligonucleotide primer, whose 3′ end is perfectly complementary to the template DNA, can be extended. This process can be obtained using several methods (e.g. chain termination and fluorescent tagging, MALDI-TOF).[27,38] By determining whether allele-specific extension products are produced or not, one can infer the allele found on the target DNA. The products can be distinguished by numerous detection methods including luminous detection,[39] colorimetric enzyme-linked immunosorbent assay (ELISA),[40] gel-based fluorescent assay,[41] homogeneous fluorescent detection,[42] flow cytometry–based assay,[43] high-performance liquid chromatographic (HPLC) analysis, time-of-flight mass spectrometry,[44,45] and microarray.[28,30,46]

The second reaction principle is endonuclease cleavage, in which a DNA fragment is cleaved by a specific enzyme for an SNP. Generally, there are two methods to detect an SNP at a polymorphic site. In the first strategy, a fragment containing the site is first amplified by polymerase chain reaction (PCR), then a restriction enzyme endonuclease is used to digest the PCR fragment from the SNP. In the second strategy, modified nucleotides can be included during PCR, so they can be incorporated into the PCR fragment in allele-specific patterns (note that the sites resistant to chemical cleavage are generated); then, an SNP allele can be routinely discriminated by conventional electrophoresis on gels or in capillary systems utilizing the characteristics of the sites.[31,47]

The third reaction principle is to use allele-specific hybridization to exploit the SNP site. With the hybridization approach, two allele-specific probes are designed to hybridize to the target sequence; only the probes that perfectly match the target DNA containing the SNP are stable. There are several ways to determine whether hybridization has occurred, including monitoring the cleavage event of the probe annealed to the target DNA that is amplified during PCR, or distinguishing fluorescence-tagged primers by their fluorescence signal. Then, the stable probe–target hybrids are used to discriminate between local sequences at the polymorphic site. Two methods are commonly used for this type of analysis: one uses allele-specific hybridization of primers in a PCR reaction; and the other uses allele-specific hybridization of primers attached to a solid support, such as microarray.[28,30]

2.3. *Estimation of Allele Frequency*

Regardless of biochemical reaction principles and detection methods, an SNP allele frequency can be estimated by sequencers or software using its signal strengths based on some detection methods. However, different SNPs are generally not equally amplified or extended during primer extension PCR. Many factors, such as the allele-specific differential efficiency of hybridization and the differential detection efficiency of emission energies of different fluorescent dyes, can lead to one allele being more efficiently amplified than the other, thus resulting in biased allele frequency estimates.

In order to obtain unbiased allele frequency estimates in the presence of preferential amplification, we can correct the strengths of the allele-specific signal by a factor k, which can be obtained by analyzing heterozygous samples since they have the same number of copies of the two alleles. Let k denote the ratio of signal strengths of allele A and allele a in heterozygotes, where A is the allele of interest. Hoogendoorn and colleagues[24] suggested estimating k from a number of independent heterozygotes, denoted by $\hat{k}$, which can be used to calculate the corrected allele frequency for each SNP, i.e.

$$\hat{p} = H_{A,\text{pool}} / (H_{A,\text{pool}} + \hat{k} H_{a,\text{pool}}),$$

where $\hat{p}$ is the estimated frequency of allele A in pools; $H_{A,\text{pool}} = \sum_{i=1}^{n_{\text{heter}}} h^i_{A,\text{pool}}$ and $H_{a,\text{pool}} = \sum_{i=1}^{n_{\text{heter}}} h^i_{a,\text{pool}}$ are the accumulated signal strengths corresponding to the two alleles at one SNP site; n_{heter} is the number of heterozygous individuals; $h^i_{A,\text{pool}}$ and $h^i_{a,\text{pool}}$ are the signal strengths of alleles A and a, respectively, for the ith heterozygous individual; $i = 1, 2, \ldots, n_{\text{heter}}$; and the adjustment factor $\hat{k}$ is the arithmetic mean of the ratio $h^i_{A,\text{pool}}/h^i_{a,\text{pool}}$ observed in all independent heterozygous individuals, i.e.

$$\hat{k} = n^{-1}_{\text{heter}} \sum_{i=1}^{n_{\text{heter}}} (h^i_{A,\text{pool}}/h^i_{a,\text{pool}}).$$

Following this approach, Yang and colleagues[48] proposed to use the geometric mean of ratios as adjustment factor $\hat{k}_{Y1}$. Also, when a bias-reduction technique was used, they suggested that another adjustment factor $\hat{k}_{Y2}$ could be given by

$$\hat{k}_{Y2} = \hat{k} + \frac{n_{\text{heter}}}{n_{\text{heter}} - 1}\left(\frac{\bar{h}_{A,\text{pool}}}{\bar{h}_{a,\text{pool}}} - \hat{k}\right),$$

where $\bar{h}_{A,\text{pool}}$ and $\bar{h}_{a,\text{pool}}$ are the means of the signal strengths for alleles A and a among n_{heter} heterozygous individuals, respectively.

3. SINGLE MARKER–BASED ASSOCIATION STUDIES USING DNA POOLING

In genetic association studies, there are primarily two types of traits: qualitative traits and quantitative traits. For a qualitative trait, the phenotype falls into different categories, e.g. affected or unaffected; in contrast, a quantitative trait shows continuous variation. In this section, we will discuss the association between a single SNP and qualitative or quantitative traits based on population and pooled family samples, respectively.

3.1. *Single Marker–based Association Studies for Qualitative Traits*

In this subsection, we will mainly discuss the statistical methods for detecting the association between qualitative traits and an SNP using pooled DNA samples. Consider a typical qualitative phenotype: affected or unaffected, a two-pool design can be employed in which affected individuals form one pool and unaffected ones form the other pool. Similar to Arnheim and colleagues,[4] allele frequencies at the SNP can be compared between the two pools to study the association between the qualitative trait and the marker.

3.1.1. *The Case of Unrelated Population Data*

When the population data are available, for simplicity, we focus on the traditional case-control design. Suppose there are two alleles A and a at an SNP with respective frequencies p and $1 - p$; and two alleles D and d at a disease locus, where D is the predisposing allele. The penetrances for the genotypes DD, Dd, and dd are f_2, f_1, and f_0, respectively. Note that A and a may be true functional alleles or may be in LD with functional alleles. Let A be the allele of interest. We form two pools with n cases and n controls in each pool. Let X_i and Y_i denote the numbers of allele A carried by the ith individual in the case group and control group, respectively. Assuming Hardy–Weinberg equilibrium in the general population, under the null hypothesis of no disease–marker association, X_i and Y_i have a value of 2, 1, or 0 with respective probabilities p^2, $2p(1 - p)$, and $(1 - p)^2$, where $i = 1, \ldots, n$.

To test the null hypothesis of no association between the marker and the disease, the following one-sided test statistic can be used (note that two-sided tests can also be used here, but we focus only on one-sided tests in the following discussion):

$$t_1 = \frac{\hat{p}_1 - \hat{p}_2}{\sqrt{\hat{p}(1 - \hat{p})/n}},$$

where $\hat{p}_1 = \sum_{i=1}^{n} X_i/(2n)$ is the sample frequency of allele A in the cases, $\hat{p}_2 = \sum_{i=1}^{n} Y_i/(2n)$ is the sample frequency of allele A in the controls, and $\hat{p} = (\hat{p}_1 + \hat{p}_2)/2$. Under the null hypothesis, the test statistic t_1 asymptotically follows the standard normal distribution.

Considering a significance level of α, the power of the test statistic t_1 is given by

$$\Phi\left(\frac{-z_\alpha \sqrt{\tilde{p}(1 - \tilde{p})} + \sqrt{n}\mu}{\sigma}\right),$$

where Φ is the cumulative standard normal distribution function; z_α is its upper 100α percentile; $\tilde{p}$ is the expected frequency of allele A; and μ and σ^2/n are the expectation and variance of $\hat{p}_1 - \hat{p}_2$ under the alternative hypothesis, respectively, whose calculation formulas can be found in Zou and Zhao.[49]

It is clear that for the test statistic t_1 shown above, allele information at the individual level is not useful, so DNA pooling has the same efficiency as individual genotyping in this case; however, for other tests, e.g. genotype-based tests, pooled data are not sufficient to allow these tests. In addition, we have considered an ideal case where no genotyping errors are assumed, when in fact genotyping errors are unavoidable and considerable in current DNA pooling technologies. Related issues will be discussed in detail in Sec. 5.

3.1.2. *The Case of Family Data*

As is well known, the major limitation of the case-control design based on population samples is the potential for confounding caused by unknown population structure (although this design is the most traditional, simplest, and extensively used). In this regard, family-based association analysis has some advantages because it can eliminate the effect of this confounding.[50–54] Risch and Teng[14] originally proposed DNA pooling strategies in family-based association studies. They discussed the two-pool design, where all of the affected offspring comprise one pool while the parents, unaffected sibs, or unrelated controls form the other pool. In the following, we will consider the trio family design, i.e. an affected child and his/her parents.

For the ith family in a sample of n family trios, let $X^{(i)}$, $X_f^{(1)}$, and $X_m^{(i)}$ represent the numbers of allele A in the child, father, and mother, respectively, where $i = 1, \ldots, n$. To test the null hypothesis of no association between the disease and the marker locus, a one-sided test statistic can be constructed as

$$t_2 = \frac{\hat{p}_1 - \hat{p}_2}{\sqrt{\hat{p}_2(1 - \hat{p}_2)/(4n)}},$$

where $\hat{p}_1 = \sum_{i=1}^{n} X^{(i)}/(2n)$ is the sample frequency of allele A in the pool of children, and $\hat{p}_2 = \sum_{i=1}^{n}(X_f^{(i)} + X_m^{(i)})/(4n)$ is the sample frequency of allele A in the pool of parents. Under the null hypothesis of no association, the test statistic t_2 asymptotically follows the standard normal distribution.

Considering the significance level of α, the power of the test statistic t_2 is given by

$$\Phi\left(\frac{-z_\alpha \sqrt{\tilde{p}(1 - \tilde{p})/4} + \sqrt{n}\mu}{\sigma}\right),$$

where $\tilde{p}$ is the expected frequency of allele A in the parents, and μ and σ^2/n are the expectation and variance of $\hat{p}_1 - \hat{p}_2$ under the alternative hypothesis, respectively, whose calculation formulas are given in Risch and Teng.[14]

Sometimes, however, it is difficult to obtain information from parents, especially for late-onset diseases. Therefore, other designs based on affected and unaffected sibs or unrelated controls can be considered. For a detailed discussion of designs with the same family structures using pooled DNA, readers can refer to Risch and Teng.[14] A general conclusion is that the sample sizes required to attain the desired power for various family structures depend on the disease model, and using unrelated controls often leads to higher power. However, practical studies often collect families of different structures; it is more flexible if a study does not constrain the types of family structures to be collected. More importantly, as mentioned above, the efficiency for various family structures depends on the disease model, which is often unknown to researchers. Therefore, Zou and Zhao[55] proposed a more general strategy to incorporate different family types; interested readers can refer to their paper.

3.2. Single Marker–based Association Studies for Quantitative Traits

Different from qualitative traits, variation in phenotypes is continuous for quantitative traits. For association studies on quantitative traits, we can still follow the idea used by Arnheim and colleagues[4]; the key is how to choose appropriate individuals to form two pools. In this regard, selective DNA pooling strategy is often advocated as a more efficient design, where appropriate individuals are selected to limit the pool size in order to achieve enough power to detect associations between quantitative phenotypes and markers. The selective strategy was initially proposed by Darvasi and Soller,[8] whereby individuals with extreme phenotypes were genotyped to form the two pools. In addition, the power of selective DNA pooling for detecting genes with large effects and small effects was obtained in that paper. Since then, many researchers have developed various approaches and applications for association studies using selective DNA pooling.[56–61]

In the following two subsections, we will discuss how to test the associations between quantitative traits and genes based on DNA pools formed from population and family data.

3.2.1. The Case of Unrelated Population Data

Suppose N unrelated individuals are either randomly sampled from a population or available from established sources (e.g. hospital or healthcare records). Consider a two-pool selective design: one is called the upper pool, and the other the lower pool. With this selective DNA pooling strategy, an upper threshold Z_U and a lower threshold Z_L are initially specified; individuals with phenotypic values (Z) greater than Z_U are then selected for the upper pool, and individuals with phenotypic values (Z) smaller than Z_L are selected for the lower pool. For simplicity, consider a symmetric design, where the number of individuals in the upper pool is equal to that in the lower pool. Note that the numbers of individuals selected from N individuals to form the upper and lower pools are still n, respectively.

To test the null hypothesis of no association between the markers of interest and the quantitative trait, a one-sided test statistic can be constructed as

$$t_3 = \frac{\hat{p}_U - \hat{p}_L}{\sqrt{\hat{p}(1 - \hat{p})/n}},$$

where $\hat{p}_U$ and $\hat{p}_L$ are the estimated frequencies of allele A in the upper and lower pools, respectively,

$$\hat{p} = \frac{1}{2}(\hat{p}_U + \hat{p}_L).$$

Under the null hypothesis of no association, t_3 asymptotically follows the standard normal distribution.

The power of the test statistic t_3 with a significance level of α is given by

$$\Phi\left(\frac{-z_\alpha\sqrt{\tilde{p}(1 - \tilde{p})} + \sqrt{n}\mu}{\sigma}\right),$$

where $\tilde{p}$ is the expectation of the sample frequency of allele A, and μ and σ^2/n are the expectation and variance of $\hat{p}_U - \hat{p}_L$ under the alternative hypothesis, respectively, whose calculation formulas can be found in Bader and colleagues.[56]

In addition, unlike the association study for qualitative phenotypes, the efficiency of association studies for quantitative phenotypes using DNA pooling concerns not only the statistical power but also the overall cost, as a total of N individuals (instead of the $2n$ individuals used in DNA pooling) need to be selected to form two pools. The total size N required to attain the desired levels of type I error rate of α and power of $1 - \beta$ can be calculated by the formula

$$N = \frac{(z_\alpha\sqrt{\tilde{p}(1 - \tilde{p})} - z_{1-\beta}\sigma)^2}{\mu^2 f},$$

where f is the pooling fraction.

Apparently, for given samples, the selective DNA pooling strategy can reduce the study costs, since only a proportion of the samples are used. However, the overall objective of a genetic study is to achieve some desired statistical power. Assuming no genotyping errors, Bader and colleagues[56] found that the optimal selective design for most markers was to pool the top and bottom 27% of individuals, respectively.

For rare or recessive alleles, however, the symmetric design for selecting individuals may not be appropriate. In this case, an asymmetric design can be suggested; that is, the pooled fractions are not necessarily the same for the upper pool and lower pool.[58]

3.2.2. *The Case of Family Data*

Consider N families with each sibship of size s, so that there are a total of Ns children. A DNA pooling strategy similar to that described in the previous subsection can be applied in two distinct designs: between-family design and within-family design. In the former case, all sibs from the $n = fN$ families with the highest and lowest mean phenotypic values are selected for the upper and lower pools, respectively, where f is the pooling fraction. In the latter case, s' sibs with the highest and lowest phenotypic values within each family are selected for the upper and lower pools, respectively, where $s' = fs$.[57]

To test the null hypothesis of no association between genes and quantitative traits, a one-sided test statistic for the between-family design can be formed as

$$t_4^{(b)} = \frac{\hat{p}_U - \hat{p}_L}{\sqrt{(s+1)\hat{p}(1-\hat{p})/(2ns)}},$$

and a one-sided test statistic for the within-family design is

$$t_4^{(w)} = \frac{\hat{p}_U - \hat{p}_L}{\sqrt{\hat{p}(1-\hat{p})/(2Ns')}}.$$

Under the null hypothesis, both $t_4^{(b)}$ and $t_4^{(w)}$ have an approximate standard normal distribution.

The powers of the test statistics $t_4^{(b)}$ and $t_4^{(w)}$ with a significance level of α are given by

$$\Phi\left(\frac{-z_\alpha\sqrt{(s+1)\tilde{p}(1-\tilde{p})/(2s)} + \sqrt{n}\mu^{(b)}}{\sigma^{(b)}}\right)$$

for the between-family design and

$$\Phi\left(\frac{-z_\alpha\sqrt{\tilde{p}(1-\tilde{p})} + \sqrt{Ns'}\mu^{(w)}}{\sigma^{(w)}}\right)$$

for the within-family design, respectively, where as before $\tilde{p}$ is the expectation of the sample frequency of allele A, and $\mu^{(b)}$ and $\mu^{(w)}$ as well as $\sigma^{(b)2}n$ and $\sigma^{(w)2}/(Ns')$ are the expectations and variances of $\hat{p}_U - \hat{p}_L$ for the two designs under the alternative hypothesis, respectively. We refer to Bader and Sham[57] for their calculation formulas.

As in the case of population data, the total number of families N required to attain the desired levels of type I error rate of α and power of $1 - \beta$ for the two designs can be calculated by the above formulas. Thus, the optimal pooling fractions can be obtained by minimizing the total size N. Obviously, the optimal fractions can also be derived by maximizing the power for given α and N.

4. HAPLOTYPE-BASED ASSOCIATION STUDIES USING DNA POOLING

With the availability of dense SNPs across the genome, it is natural to simultaneously use multiple nearby SNP markers for detecting the associations between genes and phenotypes. The haplotype, which is a specific combination of alleles at a series of closely linked SNPs on the same chromosome, is more informative than single markers. Many investigators[62-67] have demonstrated that the use of haplotypes may be more powerful than individual SNPs for detecting an association. Haplotype analysis has become an important tool in genetic studies, and can be considered as a powerful extension of single marker–based association studies. The studies of haplotype can also be used to address other biological questions, such as migration and immigration rates, LD strength, and the relationships among populations. A comprehensive review was given by Zhao and colleagues[68] on this aspect.

In association studies, the haplotype frequencies in a population should be evaluated, as they are crucial for the success of association studies and other genetic studies. The estimation of haplotype frequencies in a population is the first step in haplotype-based association analysis. However, phase information from sampled individuals is often unknown because it is technologically demanding and cost-prohibitive to obtain haplotypes directly from experiments, although it can sometimes be established by individually genotyping family members of each subject to infer parental chromosomes or by employing laboratory techniques.[69]

In the case of individual genotyping, many statistical and computational methods have been proposed to infer haplotypes or estimate haplotype frequencies (see Niu[70] for a review). When family data are available, phases of haplotypes can be either estimated or determined by using software such as Linkage,[71] GeneHunter,[72] SimWalk2,[73] Merlin,[74] and HAPLORE.[75] When population data are available and Hardy–Weinberg equilibrium is assumed, a large number of methods have been developed to infer haplotypes.[76–81] For example, Clark[76] proposed a sequential haplotype inference algorithm based on the principle of maximum parsimony. Based on coalescence theory, Stephens and colleagues[78] introduced a pseudo-Gibbs sampler for reconstructing haplotypes from genotype data. Gusfield[82] described a "perfect" phylogeny haplotyping algorithm by assuming no recombination and infinite SNP sites. Following Gusfield,[82] Halperin and Eskin[83] developed an "imperfect" phylogeny method by allowing for both recurrent mutations and recombinations. To phase long-range haplotypes, a partition–ligation algorithm implemented by a Bayesian model was described by Niu and colleagues[80]; this algorithm was further investigated using the EM algorithm by Qin and colleagues.[81]

In the case of DNA pooling, some statistical approaches to haplotype frequency estimation, mainly by the EM algorithm, have been proposed in recent years.[2,84–87] For example, Ito and colleagues[84] developed an EM algorithm, which is implemented in the computer program LDPooled, to infer haplotype frequencies. They used the bootstrap method to estimate the variances of the maximum likelihood estimates (MLEs) of haplotype frequencies in each pool. In the scenario of two SNPs, Wang and colleagues[85] considered the estimation of haplotype frequencies by pooling two or three individuals. They also compared the individual genotyping strategy and the DNA pooling strategy with respect to the overall study cost, and found that using two individuals per pool could be more cost-effective than individual genotyping, especially when the sampling cost is not significantly higher than the genotyping cost.

In the case of multiple SNPs and multiple pools, Yang and colleagues[86] computed the MLEs of haplotype frequencies by using the EM algorithm. For the variance estimates of the haplotype frequency estimates, they suggested that if the number of pools is larger than 30, asymptotic variance approximates true variance quite well, and so there is no need to use computer simulation to estimate variance; if the number of pools is less than 30, asymptotic variances of the haplotype frequency estimates can be estimated by using the bootstrap method. In addition, to gauge the efficiency of the DNA pooling design, they defined the relative efficiency of DNA pooling vs. individual genotyping as $R(K) = Kv_1/v_k$, where v_k and v_1 are the mean squared errors of haplotype frequency estimates for DNA pooling and individual genotyping designs, respectively, and K is the pool size. If $R(K) > 1$, then the DNA pooling design can be treated as an efficient one. They found that pool sizes of three to four seem to be optimal for the estimation of haplotype frequencies.

Note that all of the studies above assumed Hardy–Weinberg equilibrium, which is sometimes unrealistic. Zeng and Lin[87] accommodated Hardy–Weinberg disequilibrium, and considered cohort and case-control studies of unrelated samples with arbitrary pool sizes. They developed numerical algorithms to compute the MLEs of haplotype frequencies and their variances. In addition, Zeng and Lin[88] also considered gene–environment interactions and utilized a general approach to infer the effects of haplotypes on different kinds of phenotypes, which can be discrete or continuous, univariate or multivariate, or a potential censored time-to-disease variable.

After the haplotype frequencies are estimated based on pooled DNA, haplotype-based association studies using pooled DNA can be conducted similar to those using individual genotyping (see, for example, Akey and colleagues[64]) or, more effectively, haplotype inference and association analysis can be conducted simultaneously.[88]

5. ASSOCIATION STUDIES USING DNA POOLING IN THE PRESENCE OF GENOTYPING ERRORS AND CONFOUNDERS

Although the DNA pooling strategy is considered a cost-effective screening tool in association studies, it has several disadvantages. First, a discrepancy between the observed and true allele frequencies always occurs due to genotyping errors. In general, the standard deviations of the estimated allele frequencies by most technologies, including PCR amplification, kinetic PCR, and mass spectrometry, are in the range of 1% and 4%.[5,56,89] Barratt and colleagues[90] investigated the sources of errors at different experimental stages in allele frequency estimation, and quantified three types of genotyping errors arising from DNA pooling. Furthermore, many researchers showed that such errors could have large effects on the precision of allele frequency estimation, power of tests, information retained, and so on.[49,58,90–94] For example, Zou and Zhao[49] reported that the majority of positive results identified from DNA pooling may represent false-positives if the measurement errors (a part of genotyping errors; the other part of genotyping errors is from unequal quantity of DNA contributed by individuals) are not appropriately taken into consideration in the design of the association study.

Second, when unrelated population samples are used, there are difficulties in controlling population stratification for pooled DNA. It is well known that neglecting population stratification can lead to false-negative and false-positive findings in association studies. Besides, studies of complex diseases and quantitative traits are confounded by the effects of genetic heterogeneity, gene–gene interactions, and gene–environment interactions. The current pooling approach is not efficient enough in this regard.

5.1. *Association Studies in the Presence of Genotyping Errors*

Genotyping errors can have a large impact on association analysis when DNA pooling technology is used.[49] Therefore, incorporating such errors in our analysis is important. In the next subsection, we will discuss the corresponding testing approaches for various data types in the presence of genotyping errors.

5.1.1. *Single Marker–based Association Studies for Qualitative Traits*

In general, genotyping errors do not lead to a large estimation bias, but mainly affect the estimation variance for the difference in allele frequencies between the cases and controls. Based on this, we can make some modifications to the methods introduced in previous sections so as to incorporate genotyping errors. For example, in the unrelated population data, the test statistic t_1 can be modified to

$$t_1^* = \frac{\hat{p}_1 - \hat{p}_2}{\sqrt{V_S + V_U + V_M}}, \tag{5.1}$$

in the case of genotyping errors, where $\hat{p}_1$ and $\hat{p}_2$ are still the estimated frequencies of allele A in the case and control pools, respectively; V_S represents the sampling variance arising from the unavoidable error in estimating the allele frequency from a finite sample (which is the part we have previously considered); V_U is the variance arising from the unequal quantity of DNA contributed by individuals; and V_M is the variance arising from measurement errors (which are the two parts added due to genotyping errors). That is,

$$V_S = \frac{\hat{p}(1 - \hat{p})}{n}$$

$$V_U \approx \frac{\tau^2 \hat{p}(1 - \hat{p})}{n}$$

$$V_M = 2\varepsilon^2$$

$$\hat{p} = \frac{1}{2}(\hat{p}_1 + \hat{p}_2),$$

where τ is the coefficient of variation of the number of DNA molecules at the marker locus contributed by each individual, and ε is the standard deviation for the estimate of allele frequency in each pool.[58] Under the null hypothesis of no association, t_1^* has an approximate standard normal distribution.

When related family data are available, we can make similar modifications, even when combining different family structures. See Zou and Zhao[55] for the case of incorporating measurement errors.

5.1.2. *Single Marker–based Association Studies for Quantitative Traits*

We first consider the case of population data. Note that for both scenarios, with or without genotyping errors, the pooling designs for quantitative traits are the same; that is, N subjects are either sampled from the population or available from established sources. Individuals with trait values above the upper threshold are then selected for the upper pool, and those below the lower threshold are selected for the lower pool. The modification of the test statistic t_3 is straightforward. The corresponding test statistic has a form similar to t_1^*.

For family data, we still consider two strategies: between-family and within-family designs. The test statistic $t_4^{(b)}$ for the between-family design can be adjusted to

$$t_4^{(b)*} = \frac{\hat{p}_U - \hat{p}_L}{\sqrt{V_S + V_U + V_M}},$$

where $V_S + V_U + V_M = \dfrac{(s+1)\hat{p}(1-\hat{p})}{2ns} + \dfrac{\tau^2 \hat{p}(1-\hat{p})}{ns} + 2\varepsilon^2$. The test statistic $t_4^{(w)}$ for the within-family design can be adjusted to

$$t_4^{(w)*} = \frac{\hat{p}_U - \hat{p}_L}{\sqrt{V_S + V_U + V_M}},$$

where $V_S + V_U + V_M = \dfrac{\hat{p}(1-\hat{p})}{2Ns'} + \dfrac{\tau^2 \hat{p}(1-\hat{p})}{Ns'} + 2\varepsilon^2$.

Under the null hypothesis of no association, both test statistics have an approximately standard normal distribution.

5.1.3. *Haplotype–based Association Studies in the Presence of Genotyping Errors*

Recognizing the possible effects of genotyping errors on haplotype frequency estimation, Zou and Zhao[95] developed an EM algorithm to estimate haplotype frequencies by incorporating genotyping errors in individual genotyping. For DNA pooling, Quade and colleagues[2] combined the errors to estimate haplotype frequencies using simulated pooled samples. Note that Quade and colleagues[2] considered only two diallelic loci, and binomial distribution was assumed to be the error distribution. Making use of this error model, pooled data were simulated under a large number of genetic models using combinations of different allele frequencies at each marker and a measurement of LD between the markers. Then, the EM algorithm was used to estimate the haplotype frequencies.

As in the case of no genotyping errors, haplotype-based association analysis in the presence of genotyping errors can be done using approaches similar to those for individual genotyping once the haplotype frequencies based on pooled DNA are estimated, or can be simultaneously conducted with haplotype inference.

5.2. *Association Studies in the Presence of Confounders*

As discussed above, human populations often exhibit substructure or stratification, which may lead to spurious associations in genetic association studies. This may cause an association between a phenotype and a marker locus to be described, when in fact the marker is unlinked to any causative loci. Therefore, it is especially important to eliminate the confounders in association studies.

For individual genotyping, if family-based data are available, the population substructure can be obviated by using the transmission of alleles from parents to offspring.[52,53,96] When only population data are available, different approaches to eliminate the effect of population substructure have been proposed, e.g. the genomic control (GC) approach proposed by Devlin and Roeder[97] as well as the structured association (SA) approach by Pritchard and colleagues.[98,99] Consider a case-control study of candidate genes. In the GC approach, chi-squared test statistics for independence of candidate loci and null loci (assumed to have no effect on the disease under study, i.e. polymorphisms unlikely to affect liability) are computed. By virtue of testing all markers at the null and candidate loci, a multiplier is derived and the critical values of the significance tests for the candidate loci are adjusted by observing the values of test statistics at the null loci inflated by the impact of population stratification. In this way, GC permits the analysis of stratified case-control data without an increase in false-positives. If population stratification is not detected from the null loci, GC is identical to a standard test of independence for a case-control design.

In the SA approach,[98] three key assumptions are made: (1) marker loci are unlinked to the candidate genes under study; (2) markers are in LD with one another within each population; and (3) Hardy–Weinberg equilibrium is within each population. Then, the number of subpopulations K can be inferred within a Bayesian framework, and each subject's membership probability in each of the K subpopulations can be estimated. The two tasks of computing K and the membership probability vector for each individual can be performed by the program STRUCTURE.[98] Given an estimated membership vector for each of the subjects in the study, it is easy to compute a likelihood ratio test based on calculating the likelihood of the data under the null and alternative hypotheses. The program STRAT[99] can be used to conduct the association tests using the output from STRUCTURE as input.[100]

For DNA pooling, family-based association studies are still usually robust to population structure.[14] For population pooled DNA data, the SA approach cannot be applied directly to a case-control study, although it is theoretically possible to use this method in a study that compares two sets of multiple distinct pools.[5] On the other hand, the GC approach can be used directly because it involves only allele frequencies among many unlinked markers.[5,100] Of course, this may be affected by genotyping errors. However, in this approach, the impact of population structure and admixture is assumed to be approximately constant over all loci[100]; it is inappropriate if loci are under strong subpopulation-specific selection.[100,101] Therefore, developing more efficient approaches that are robust to the hidden substructure is still a challenging and important task.

In addition, despite much concern over population stratification, false-positive and false-negative findings may still result from other confounders, e.g. gene–environment interactions. Alternative pooling strategies can be used to address this problem. For example, for environmental risk factors such as smoking status or level of alcohol intake, DNA could be repooled based on an individual's environmental exposure (e.g. "exposed" vs. "unexposed"). This will lead to the formation of four pools: "unaffected, unexposed"; "unaffected, exposed"; "affected, unexposed"; and "affected, exposed". Note that these pools should be matched for possible confounding factors such as sex, age, and ethnicity. Association studies can then be carried out to incorporate gene–environment interaction.[102]

6. DISCUSSION

In this review, we have discussed recent developments in association studies based on DNA pooling. The experimental aspects of DNA pooling were briefly introduced, and the uses of DNA pooling in association studies for both qualitative and quantitative phenotypes were presented. We considered both population (unrelated) data and family (related) data, and discussed the scenarios of the genotyping errors and population stratification.

A major limitation of the current DNA pooling technology is the errors associated with measuring allele frequencies in the pooled samples. The impact of genotyping errors on DNA pooling is much greater than that on individual genotyping. Therefore, genotyping error reduction in association studies is very important to make DNA pooling a useful strategy for association studies. Practical methods to reduce genotyping errors for DNA pooling include forming multiple pools and/or using multiple measurements.[5] Another important issue, when a case-control design based on population data is used, is how to eliminate spurious association as a result of population stratification. To reduce the impact of population structure, it has been suggested that individuals be partitioned into pools by certain factors such as disease status, gender, age, and ethnic characteristics; alternatively, the sampled individuals can also be pregenotyped by a set of ancestry-informative markers to establish pools with similar genetic backgrounds. Clearly, these may not be efficient ways. Therefore, to reduce or completely eliminate the effect of hidden stratum structures on association studies, some better methods must be developed.

Despite the limitations mentioned above, DNA pooling has been applied successfully and extensively in SNP-based association studies.[14,15,28,29,31,103–107] For example, this strategy has been shown to be quite effective in identifying disease-causing loci with Mendelian traits,[15] non-Mendelian traits,[4,106,108,109] quantitative trait loci (QTLs) using association studies,[29,30,56–58] and parallel phenotype analysis.[108,110] An example of DNA pooling application in complex traits is the study of mild mental impairment (MMI).[29] MMI is hypothesized to represent the low extreme of the quantitative trait of general intelligence. Butcher and colleagues[29] conducted one study of a low vs. high general intelligence extreme group comparison, and implemented another study of case vs. control comparison by using SNP-MaP (SNP microarrays and pooling strategy). After several loci were screened and individually genotyped, they identified four candidate loci, some of which are close to known genes and may be in LD with them. DNA pooling has also been widely applied in other genetic marker–based association studies.[16,21,22,61,111–113] For instance, Fisher and Spelman[61] screened microsatellite markers on bovine chromosome 14 by utilizing a selective DNA pooling strategy, and successfully identified the gene *DGAT1*, which was previously known to affect bovine fat yield, protein yield, and total milk yield. Furthermore, DNA pooling has been proposed in linkage studies[7,8] and physical mapping studies.[9,10,114]

Note that for some complex diseases such as cancer and heart disease, the studies would require many thousands of individuals for investigation[115]; however, there is a limit of approximately 1000 individuals for forming a pool.[116] This would be a limitation for the large-scale studies of diseases using DNA pooling. One solution to this problem is to create multiple subpools, which would allow the pooling approach to be applied to large studies.

Just as the above example of Butcher and colleagues[29] shows that individual genotyping can provide more accurate estimates of allele frequencies and allows for the study of genetic interactions, DNA pooling is often advocated as a screening tool for identifying candidate markers that is to be followed up by individual genotyping. This is in fact a two-stage design that may offer an attractive strategy to balance power and cost.[5,20,89,90,117] In such a design, the first stage evaluates a very large number (e.g. one million) of markers using DNA pooling. Only the most promising ones are selected and studied in the second stage through individual genotyping. However, when DNA pooling is used as a screening tool in the first stage, the following issues should be addressed:

(1) How many markers should be chosen after the first stage so that there is a high probability that all or some of the disease-associated markers are included in the individual genotyping stage?

(2) What is the statistical power by which a disease-associated marker is identified when the overall false-positive rate is appropriately controlled? What is the benefit of doing so?

(3) When the primary goal is to ensure that some of the disease-associated markers are ranked among the top L markers after the two-stage analysis, what is the probability that at least one of the disease-associated markers is ranked among the top?

Zuo and colleagues[118] provided some answers to the above practical questions under the assumption of independence among markers. They found that for a two-stage design, measurement errors have a large impact only on the DNA pooling stage; once the markers are selected, the effect of measurement errors on the overall power can be very small. In genetic studies, the sample in the first stage can be either reused or not in the second stage, where a set of new samples is studied through individual genotyping. Both of these strategies were considered in their article. However, in a genomewide association study, the marker density tends to be high and adjacent markers tend to be highly correlated. How to correctly model the correlation among adjacent markers warrants further research.

Note that only two types of traits, qualitative and quantitative traits, have been considered in the literature so far. In fact, the traits can come in more complex forms, and different models can be assumed to describe the associations between trait values and the underlying genetic compositions. For example, traits can be measured as a survival outcome, and the studied individuals can be monitored through multiple time points; instead of measuring a single trait value, all practical studies could essentially collect information from multiple (related) traits.

In addition to qualitative and quantitative traits, there is another type of trait in practice: threshold traits (also called polyphenisms). These traits have polygenetic inheritance and occur as one of several distinct alternative forms, each expressed under predictable environmental circumstances.[119] This means that the threshold traits are discrete in phenotype, but have the same polygenic genetic basis as quantitative traits. Traditional linkage analysis for quantitative traits is invalid for threshold traits due to their special characters. To our knowledge, association studies for survival outcomes, longitudinal traits, correlated traits, and threshold traits using DNA pooling have been largely unexplored. Thus, there is a great need to extend the existing methods or develop new approaches for studying the association between genes and these other types of traits using pooled DNA. We envision that the DNA pooling strategy will be considerably more cost-effective, powerful, and widely adopted in future genetics studies.

ACKNOWLEDGMENTS

This work was supported in part by grants from the Scientific Research Foundation for Returned Overseas Chinese Scholars No. 70625004 from the

National Natural Science Foundation of China (to G. Zou), and GM59507 and GM57672 from the National Institutes of Health (to H. Zhao).

REFERENCES

1. Risch N. (2000) Searching for genetic determinants in the new millennium. *Nature* **405**:847–856.
2. Quade SRE, Elston RC, Goddard KAB. (2005) Estimating haplotype frequencies in pooled DNA samples when there is genotyping error. *BMC Genet* **6**:25.
3. Dorfman R. (1943) The detection of defective members of large populations. *Ann Math Stat* **14**:436–440.
4. Arnheim N, Strange C, Erlich H. (1985) Use of pooled DNA samples to detect linkage disequilibrium of polymorphic restriction fragments and human diseases: studies of the HLA class II loci. *Proc Natl Acad Sci USA* **82**:6970–6974.
5. Sham P, Bader J, Craig I, *et al.* (2002) DNA pooling: a tool for large-scale association studies. *Nat Rev Genet* **3**:862–871.
6. Michelmore R, Paran I, Kesseli R. (1991) Identification of markers linked to disease-resistance genes by bulk segregant analysis: a rapid method to detect markers in specific genomic regions using segregating populations. *Proc Natl Acad Sci USA* **88**:9828–9832.
7. Churchill GA, Giovannoni JJ, Tanksley SD. (1993) Pooled-sampling makes high-resolution mapping practical with DNA markers. *Proc Natl Acad Sci USA* **90**:16–20.
8. Darvasi A, Soller M. (1994) Selective DNA pooling for determination of linkage between a molecular marker and a quantitative trait locus. *Genetics* **138**:1365–1373.
9. Barillot E, Lacroix B, Cohen D. (1991) Theoretical analysis of library screening using a N-dimensional pooling strategy. *Nucleic Acids Res* **19**:6241–6247.
10. Bruno WJ, Knill E, Balding DJ, *et al.* (1995) Efficient pooling designs for library screening. *Genomics* **26**:21–30.
11. Carmi R, Rokhlina T, Kwitek-Black AE, *et al.* (1995) Use of DNA pooling strategy to identify a human obesity syndrome locus on chromosome 15. *Hum Mol Genet* **4**:9–13.
12. Scott DA, Carmi R, Elbedour K, *et al.* (1996) An autosomal recessive nonsyndromic-hearing loss locus identified by DNA pooling using two inbred Bedouin kindreds. *Am J Hum Genet* **59**:385–391.
13. Daniels J, Holmans P, Williams N, *et al.* (1998) A simple method for analyzing microsatellite allele image patterns generated from DNA pools and its application to allelic association studies. *Am J Hum Genet* **62**:1189–1197.
14. Risch N, Teng J. (1998) The relative power of family-based and case-control designs for linkage disequilibrium studies of complex human diseases. I. DNA pooling. *Genome Res* **8**:1273–1288.

15. Shaw SH, Carrasquillo MM, Kashuk C, *et al.* (1998) Allele frequency distributions in pooled DNA samples: applications to mapping complex disease genes. *Genome Res* **8**:111–123.

16. Collins HE, Li H, Inda SE, *et al.* (2000) A simple and accurate method for determination of microsatellite total allele content differences between DNA pools. *Hum Genet* **106**:218–226.

17. Kruglyak L. (1999) Prospects for whole-genome linkage disequilibrium mapping of common disease genes. *Nat Genet* **22**:139–144.

18. Altshuler D, Brooks LD, Chakravarti A, *et al.* (2005) A haplotype map of the human genome. *Nature* **437**:1299–1320.

19. Pacek P, Sajantila A, Syvanen AC. (1993) Determination of allele frequencies at loci with length polymorphism by quantitative analysis of DNA amplified from pooled samples. *PCR Methods Appl* **2**:313–317.

20. Barcellos LF, Klitz W, Field LL, *et al.* (1997) Association mapping of disease loci, by use of a pooled DNA genomic screen. *Am J Hum Genet* **61**:737–747.

21. Schnack HG, Bakker SC, van't Slot R, *et al.* (2004) Accurate determination of microsatellite allele frequencies in pooled DNA samples. *Eur J Hum Genet* **12**:925–934.

22. Breen G, Sham P, Li T, *et al.* (1999) Accuracy and sensitivity of DNA pooling with microsatellite repeats using capillary electrophoresis. *Mol Cell Probes* **13**:359–365.

23. Germer S, Holland MJ, Higuchi R. (2000) High-throughput SNP allele frequency determination in pooled DNA samples by kinetic PCR. *Genome Res* **10**:258–266.

24. Hoogendoorn B, Norton N, Kirov G, *et al.* (2000) Cheap, accurate and rapid allele frequency estimation of single nucleotide polymorphisms by primer extension and DHPLC in DNA pools. *Hum Genet* **107**:488–493.

25. Ross P, Hall L, Haff LA. (2000) Quantitative approach to single-nucleotide polymorphism analysis using MALDI-TOF mass spectrometry. *Biotechniques* **29**:620–626, 628–629.

26. Sasaki T, Tahira T, Suzuki A, *et al.* (2001) Precise estimation of allele frequencies of single-nucleotide polymorphisms by a quantitative SSCP analysis of pooled DNA. *Am J Hum Genet* **68**:214–218.

27. Norton N, Williams NM, Williams HJ, *et al.* (2002) Universal, robust, highly quantitative SNP allele frequency measurement in DNA pools. *Hum Genet* **110**:471–478.

28. Ye BC, Zuo P, Yi BC, Li SY. (2004) Estimation of relative allele frequencies of single-nucleotide polymorphisms in different populations by microarray hybridization of pooled DNA. *Anal Biochem* **333**:72–78.

29. Butcher LM, Meaburn E, Knight J, *et al.* (2005) SNPs, microarrays and pooled DNA: identification of four loci associated with mild mental impairment in a sample of 6000 children. *Hum Mol Genet* **14**:1315–1325.

30. Meaburn E, Butcher LM, Liu L, *et al.* (2005) Genotyping DNA pools on microarrays: tackling the QTL problem of large samples and large numbers of SNPs. *BMC Genomics* **6**:52.

31. Breen G, Harold D, Ralston S, *et al.* (2000) Determining SNP allele frequencies in DNA pools. *Biotechniques* **28**:464–470.
32. Buetow KH, Edmonson M, MacDonald R, *et al.* (2001) High-throughput development and characterization of a genomewide collection of gene-based single nucleotide polymorphism markers by chip-based matrix-assisted laser desorption/ionization time-of-flight mass spectrometry. *Proc Natl Acad Sci USA* **98**:581–584.
33. Zhou G, Kamahori M, Okano K, *et al.* (2001) Quantitative detection of single nucleotide polymorphisms for a pooled sample by a bioluminometric assay coupled with modified primer extension reaction (BAMPER). *Nucleic Acids Res* **29**:e93.
34. Zhang S, Van Pelt CK, Huang X, Schultz GA. (2002) Detection of single nucleotide polymorphisms using electrospray ionization mass spectrometry: validation of a one-well assay and quantitative pooling studies. *J Mass Spectrom* **37**:1039–1050.
35. Le Hellard S, Ballereau SJ, Visscher PM, *et al.* (2002) SNP genotyping on pooled DNAs: comparison of genotyping technologies and a semi automated method for data storage and analysis. *Nucleic Acids Res* **30**:e74.
36. Kwok PY. (2001) Methods for genotyping single nucleotide polymorphisms. *Annu Rev Genomics Hum Genet* **2**:235–258.
37. Syvanen AC. (2001) Accessing genetic variation: genotyping single nucleotide polymorphisms. *Nat Rev Genet* **2**:930–942.
38. Werner M, Sych M, Herbon N, *et al.* (2002) Large-scale determination of SNP allele frequencies in DNA pools using MALDI-TOF mass spectroscopy. *Hum Mutat* **20**:57–64.
39. Nyren P, Pettersson B, Uhlen M. (1993) Solid phase DNA minisequencing by an enzymatic luminometric inorganic pyrophosphate detection assay. *Anal Biochem* **208**:171–175.
40. Nikiforov TT, Rendle RB, Goelet P, *et al.* (1994) Genetic bit analysis: a solid phase method for typing single nucleotide polymorphisms. *Nucleic Acids Res* **22**:4167–4175.
41. Pastinen T, Partanen J, Syvanen AC. (1996) Multiplex, fluorescent, solid-phase minisequencing for efficient screening of DNA sequence variation. *Clin Chem* **42**:1391–1397.
42. Chen X, Levine L, Kwok PY. (1999) Fluorescence polarization in homogeneous nucleic acid analysis. *Genome Res* **9**:492–498.
43. Cai H, White PS, Torney D, *et al.* (2000) Flow cytometry–based minisequencing: a new platform for high-throughput single-nucleotide polymorphism scoring. *Genomics* **66**:135–143.
44. Haff LA, Smirnov IP. (1997) Single-nucleotide polymorphism identification assays using a thermostable DNA polymerase and delayed extraction MALDI-TOF mass spectrometry. *Genome Res* **7**:378–388.
45. Griffin TJ, Smith LM. (2000) Single-nucleotide polymorphism analysis by MALDI-TOF mass spectrometry. *Anal Chem* **72**:3298–3302.

46. Fan JB, Chen X, Halushka MK, *et al.* (2000) Parallel genotyping of human SNPs using generic high-density oligonucleotide tag arrays. *Genome Res* **10**:853–860.

47. Vaughan P, McCarthy TV. (1998) A novel process for mutation detection using uracil DNA-glycosylase. *Nucleic Acids Res* **26**:810–815.

48. Yang HC, Pan CC, Lu RC, Fan CS. (2005) New adjustment factors and sample size calculation in a DNA-pooling experiment with preferential amplification. *Genetics* **169**:399–410.

49. Zou G, Zhao H. (2004) The impacts of errors in individual genotyping and DNA pooling on association studies. *Genet Epidemiol* **26**:1–10.

50. Falk CT, Rubinstein P. (1987) Haplotype relative risks: an easy reliable way to construct a proper control sample for risk calculations. *Ann Hum Genet* **51**:227–233.

51. Terwilliger JD, Ott J. (1992) A haplotype-based "haplotype relative risk" approach to detecting allelic associations. *Hum Hered* **42**:337–346.

52. Spielman RS, McGinnis RE, Ewens WJ. (1993) Transmission test for linkage disequilibrium: the insulin gene region and insulin-dependent diabetes mellitus (IDDM). *Am J Hum Genet* **52**:506–516.

53. Curtis D. (1997) Use of siblings as controls in case-control association studies. *Ann Hum Genet* **61**:319–333.

54. Zhao H. (2000) Family-based association studies. *Stat Methods Med Res* **9**:563–587.

55. Zou G, Zhao H. (2005) Family-based association tests for different family structures using pooled DNA. *Ann Hum Genet* **69**:1–14.

56. Bader JS, Bansal A, Sham P. (2001) Efficient SNP-based tests of association for quantitative phenotypes using pooled DNA. *Genescreen* **1**:143–150.

57. Bader JS, Sham P. (2002) Family-based association tests for quantitative traits using pooled DNA. *Eur J Hum Genet* **10**:870–878.

58. Jawaid A, Bader JS, Purcell S, *et al.* (2002) Optimal selection strategies for QTL mapping using pooled DNA samples. *Eur J Hum Genet* **10**:125–132.

59. Carleos C, Baro JA, Canon J, Corral N. (2003) Asymptotic variances of QTL estimators with selective DNA pooling. *J Hered* **94**:175–179.

60. Varga L, Muller G, Szabo G, *et al.* (2003) Mapping modifiers affecting muscularity of the myostatin mutant ($Mstn^{Cmpt-dl1Abc}$) compact mouse. *Genetics* **165**:257–267.

61. Fisher PJ, Spelman RJ. (2004) Verification of selective DNA pooling methodology through identification and estimation of the *DGAT1* effect. *Anim Genet* **35**:201–205.

62. Fallin D, Schork NJ. (2000) Accuracy of haplotype frequency estimation for biallelic loci, via the expectation-maximization algorithm for unphased diploid genotype data. *Am J Hum Genet* **67**:947–959.

63. Zhao H, Zhang S, Merikangas KR, *et al.* (2000) Transmission/disequilibrium tests using multiple tightly linked markers. *Am J Hum Genet* **67**:936–946.

64. Akey J, Jin L, Xiong M. (2001) Haplotypes vs. single marker linkage disequilibrium tests: what do we gain? *Eur J Hum Genet* **9**:291–300.

65. Fallin D, Cohen A, Essioux L, *et al.* (2001) Genetic analysis of case/control data using estimated haplotype frequencies: application to *APOE* locus variation and Alzheimer's disease. *Genome Res* **11**:143–151.

66. Morris RW, Kaplan NL. (2002) On the advantage of haplotype analysis in the presence of multiple disease susceptibility alleles. *Genet Epidemiol* **23**:221–233.

67. Zaykin DV, Westfall PH, Young SS, *et al.* (2002) Testing association of statistically inferred haplotypes with discrete and continuous traits in samples of unrelated individuals. *Hum Hered* **53**:79–91.

68. Zhao H, Pfiffer R, Gail MH. (2003) Haplotype analysis in population genetics and association studies. *Pharmacogenomics* **4**:171–178.

69. Michalatos-Beloin S, Tishkoff SA, Bentley KL, *et al.* (1996) Molecular haplotyping of genetic markers 10 kb apart by allele-specific long-range PCR. *Nucleic Acids Res* **24**:4841–4843.

70. Niu T. (2004) Algorithms for inferring haplotypes. *Genet Epidemiol* **27**:334–347.

71. Lathrop GM, Lalouel JM, Julier C, Ott J. (1985) Multilocus linkage analysis in humans: detection of linkage and estimation of recombination. *Am J Hum Genet* **37**:482–498.

72. Kruglyak L, Daly MJ, Reeve-Daly MP, Lander ES. (1996) Parametric and nonparametric linkage analysis: a unified multipoint approach. *Am J Hum Genet* **58**:1347–1363.

73. Sobel E, Lange K. (1996) Descent graphs in pedigree analysis: applications to haplotyping, location scores, and marker sharing statistics. *Am J Hum Genet* **58**:1323–1337.

74. Abecasis GR, Cherny SS, Cookson WO, Cardon LR. (2002) Merlin — rapid analysis of dense genetic maps using sparse gene flow trees. *Nat Genet* **30**:97–101.

75. Zhang K, Sun F, Zhao H. (2005) HAPLORE: a program for haplotype reconstruction in general pedigrees without recombination. *Bioinformatics* **21**:90–103.

76. Clark AG. (1990) Inference of haplotypes from PCR-amplified samples of diploid populations. *Mol Biol Evol* **7**:111–122.

77. Excoffier L, Slatkin M. (1995) Maximum-likelihood estimation of molecular haplotype frequencies in a diploid population. *Mol Biol Evol* **12**:927–931.

78. Stephens M, Smith NJ, Donnelly P. (2001) A new statistical method for haplotype reconstruction from population data. *Am J Hum Genet* **68**:978–989.

79. Zhang S, Pakstis AJ, Kidd KK, Zhao H. (2001) Comparisons of two methods for haplotype reconstruction and haplotype frequency estimation from population data. *Am J Hum Genet* **69**:906–912.

80. Niu T, Qin ZS, Xu X, Liu JS. (2002) Bayesian haplotype inference for multiple linked single-nucleotide polymorphisms. *Am J Hum Genet* **70**:157–169.

81. Qin ZS, Niu T, Liu JS. (2002) Partition-ligation–expectation-maximization algorithm for haplotype inference with single-nucleotide polymorphisms. *Am J Hum Genet* **71**:1242–1247.

82. Gusfield D. (2002) Haplotyping as perfect phylogeny: conceptual framework and efficient solutions. In: *Proceedings of the 6th Annual International Conference on Research in Computational Molecular Biology (RECOMB 2002)*, ACM Press, New York, NY, pp. 166–175.

83. Halperin E, Eskin E. (2004) Haplotype reconstruction from genotype data using imperfect phylogeny. *Bioinformatics* **20**:1842–1849.

84. Ito T, Chiku S, Inous E, *et al.* (2003) Estimation of haplotype frequencies, linkage disequilibrium measures, and combination of haplotype copies in each pool by use of pooled DNA data. *Am J Hum Genet* **72**:384–398.

85. Wang S, Kidd KK, Zhao H. (2003) On the use of DNA pooling to estimate haplotype frequencies. *Genet Epidemiol* **24**:74–82.

86. Yang Y, Zhang J, Hoh J, *et al.* (2003) Efficiency of single-nucleotide polymorphism haplotype estimation from pooled DNA. *Proc Natl Acad Sci USA* **100**:7225–7230.

87. Zeng D, Lin DY. (2005) Estimating haplotype–disease associations with pooled genotype data. *Genet Epidemiol* **28**:70–82.

88. Zeng D, Lin DY. (2006) Likelihood-based inference on haplotype effects in genetic association studies. *J Am Stat Assoc* **101**:89–104.

89. Bansal A, van den Boom D, Kammerer S, *et al.* (2002) Association testing by DNA pooling: an effective initial screen. *Proc Natl Acad Sci USA* **99**:16871–16874.

90. Barratt BJ, Payne F, Rance HE, *et al.* (2002) Identification of the sources of error in allele frequency estimations from pooled DNA indicated an optimal experimental design. *Ann Hum Genet* **66**:393–405.

91. Gordon D, Matise TC, Heath SC, Ott J. (1999) Power loss for multiallelic transmission/disequilibrium test when errors introduced: GAW11 simulated data. *Genet Epidemiol* **17**:S587–S592.

92. Abecasis GR, Cherny SS, Cardon LR. (2001) The impact of genotyping error on family-based analysis of quantitative traits. *Eur J Hum Genet* **9**:130–134.

93. Kirk KM, Cardon LR. (2002) The impact of genotyping error on haplotype reconstruction and frequency estimation. *Eur J Hum Genet* **10**:616–622.

94. Gordon D, Yang Y, Haynes C, *et al.* (2004) Increasing power for tests of genetic association in the presence of phenotype and/or genotype error by use of double-sampling. *Stat Appl Genet Mol Biol* **3**:a26.

95. Zou G, Zhao H. (2003) Haplotype frequency estimation in the presence of genotyping errors. *Hum Hered* **56**:131–138.

96. Ewens WJ, Spielman RS. (1995) The transmission/disequilibrium test: history, subdivision, and admixture. *Am J Hum Genet* **57**:455–464.

97. Devlin B, Roeder K. (1999) Genomic control for association studies. *Biometrics* **55**:788–808.

98. Pritchard JK, Stephens M, Donnelly P. (2000) Inference of population structure using multilocus genotype data. *Genetics* **155**:945–959.

99. Prichard JK, Stephens M, Rosenberg NA, Donnelly P. (2000) Association mapping in structured populations. *Am J Hum Genet* **67**:170–181.

100. Devlin B, Roeder K, Bacanu SA. (2001) Unbiased methods for population-based association studies. *Genet Epidemiol* **21**:273–284.

101. Robertson A. (1975) Gene frequency distribution as a test of selective neutrality. *Genetics* **81**:775–785.

102. Chen J, Germer S, Higuchi R, *et al.* (2002) Kinetic polymerase chain reaction on pooled DNA: a high-throughput, high-efficiency alternative in genetic epidemiological studies. *Cancer Epidemiol Biomarkers Prev* **11**:131–136.

103. Marnellos G. (2003) High-throughput SNP analysis for genetic association studies. *Curr Opin Drug Discov Devel* **6**:317–321.

104. Rautanen A, Zucchelli M, Makela S, Kere J. (2005) Gene mapping with pooled samples on three genotyping platforms. *Mol Cell Probes* **19**:408–416.

105. Brookes KJ, Knight J, Xu X, Asherson P. (2005) DNA pooling analysis of ADHD and genes regulating vesicle release of neurotransmitters. *Am J Med Genet B Neuropsychiatr Genet* **139**:33–37.

106. Lee WC. (2005) A DNA pooling strategy for family-based association studies. *Cancer Epidemiol Biomarkers Prev* **14**:958–962.

107. Moskvina V, Norton N, Williams N, *et al.* (2005) Streamlined analysis of pooled genotype data in SNP-based association studies. *Genet Epidemiol* **28**:273–282.

108. Norton N, Williams NM, O'Donovan MC, Owen MJ. (2004) DNA pooling as a tool for large-scale association studies in complex traits. *Ann Med* **36**:146–152.

109. Castle PE, Schiffman M, Herrero R, *et al.* (2005) PCR testing of pooled longitudinally collected cervical specimens of women to increase the efficiency of studying human papillomavirus infection. *Cancer Epidemiol Biomarkers Prev* **4**:256–260.

110. Williams NM, Bowen T, Spurlock G, *et al.* (2002) Determination of the genomic structure and mutation screening in schizophrenic individuals for five subunits of the N-methyl-D-aspartate glutamate receptor. *Mol Psychiatry* **7**:508–514.

111. Lipkin E, Mosig MO, Darvasi A, *et al.* (1998) Quantitative trait locus mapping in dairy cattle by means of selective milk DNA pooling using dinucleotide microsatellite markers: analysis of milk protein percentage. *Genetics* **149**:1557–1567.

112. Laaksonen M, Jonasdottir A, Fossdal R, *et al.* (2003) A whole genome association study in Finnish multiple sclerosis patients with 3669 markers. *J Neuroimmunol* **143**:70–73.

113. Tamiya G, Shinya M, Imanishi T, *et al.* (2005) Whole genome association study of rheumatoid arthritis using 27 039 microsatellites. *Hum Mol Genet* **14**:2305–2321.

114. Tait E, Simon MC, King S, *et al.* (1997) A *Candida albicans* genome project: cosmid contigs, physical mapping, and gene isolation. *Fungal Genet Biol* **21**:308–314.

115. Risch N, Merikangas K. (1996) The future of genetic studies of complex human diseases. *Science* **273**:1516–1517.

116. Law GR, Rollinson S, Feltbower R, *et al.* (2004) Application of DNA pooling to large studies of disease. *Stat Med* **23**:3841–3850.

117. Zhao ZZ, Nyholt DR, James MR, *et al.* (2005) A comparison of DNA pools constructed following whole genome amplification for two-stage SNP genotyping designs. *Twin Res Hum Genet* **8**:353–361.
118. Zuo Y, Zou G, Zhao H. (2006) Two-stage designs in case-control association analysis. *Genetics* **173**:1747–1760.
119. Emlen DJ, Nijhout HF. (2001) Hormonal control of male horn length dimorphism in *Onthophagus taurus* (Coleoptera: Scarabaeidae): a second critical period of sensitivity to juvenile hormone. *J Insect Physiol* **47**:1045–1054.

Chapter 9

Linkage Disequilibrium and Test for Interaction Between Two Loci

Momiao Xiong

Laboratory of Theoretic Systems Biology, School of Life Science
Fudan University, Shanghai 200433, China

Human Genetics Center
The University of Texas Health Science Center at Houston
Houston, TX 77225, USA

1. INTRODUCTION

The traditional paradigm for genetic studies of complex diseases is to study one marker at a time. Statistical genetic theories and tools developed in the past several decades primarily use this paradigm to dissect the genetic structure of complex traits. However, living systems are complex systems. Most phenotypic variations, including those involved in complex diseases and differences in drug response, are generated by the integrated actions of multiple genetic and environmental factors through dynamic, epigenetic, and regulatory mechanisms.[1] Although the traditional paradigm has been successfully applied to simple Mendelian diseases, it has failed to identify and replicate significant genetic effects due to the complex mechanisms of diseases. A comprehensive delineation of the complicated interplay between genetic and environmental factors that influences complex traits will require the complete characterization of DNA variation in the population, the collection of individuals' histories of environmental exposure, and the development of mathematical tools to unravel the dynamic interaction between genetic variation and environmental exposure.

Complex diseases are caused by multiple genes, primarily through nonlinear gene–gene interactions and gene–environment interactions.[2]

Gene–gene interaction or gene–environment interaction means that the genetic effect at one locus is modified (enhanced or masked) by the genotypes at other loci or environments. The study of gene–gene interactions and gene–environment interactions is essential to the genetic study of complex diseases for the following reasons.[3] First, joint analysis of multiple genes and environments will increase the power to detect the association of genes or environments with the disease. Second, it will improve the estimation of the population-attributable risks of the genetic and environmental factors. Third, it will help to uncover the mechanisms of the diseases by examining the DNA variations of the genes in the biological pathways causing diseases and how environmental exposure influences the function of the biological pathways. Fourth, assessing the degree of the response of biological pathways to the perturbation of environmental factors will help in the design of new preventive and therapeutic strategies. Therefore, developing statistical methods for the detection of gene–gene and gene–environment interactions, which will lead to the dissection of complex patterns of the genetic structure of complex diseases, is a key task in the genetic study of complex diseases.

2. TEST FOR INTERACTION BETWEEN TWO LOCI

2.1. *Introduction*

Despite growing consensus on the importance of gene interactions in the genetic studies of complex diseases, classical genetic analysis either ignores gene interactions or defines the effect of gene interactions as a deviance from genetic additive effects, which is essentially treated as a residual term in genetic analysis.[4] Fisher[5] mathematically defined the effect of gene interactions as a statistical deviance from the additive effects of alleles, which is referred to as the statistical gene interaction and additive model. This was further developed by Cockerham[6] and Kempthorne[7] into the modern representation, which treats statistical gene interactions as interaction terms in a regression or generalized linear model on allelic effects.[3] Therefore, genetic interactions are represented not as genetic effects and appropriately measured, but as noise terms in the additive models.

In the traditional mathematical model of gene interaction including logistic regression,[8–12] a trait is modeled as the additive combination of its single-locus main effects and a residual term. The residual term in the model is defined as a statistical interaction. As a consequence, the major part of functional (or biological) gene interaction is included into the main effects of the trait; whereas the remaining part of functional gene interaction, which is treated as a residual term in the mathematical model of the trait, is small and hard to detect. Therefore, the widely used classical models of gene interaction, which have dealt mainly with its statistical effect, are inadequate to detect true functional gene interaction. New models of gene interaction and statistical methods for the detection of functional gene interaction are needed.

Recently, efforts to define gene interaction as the influence of a gene or genes on the effects of other genes have been made.[3,13] This functional gene interaction definition treats gene interaction as a property of the genotype–phenotype relationship, and is thus fundamentally different from the definition of statistical gene interaction. The mathematical representation of functional gene interaction was originally designed for quantitative traits. There are two major problems in incorporating such a functional definition of gene interaction into the genomewide association studies of qualitative traits. First, we need to extend the functional definition of gene interaction for a quantitative trait to a qualitative trait. Second, the detection of gene interaction for a qualitative trait under the current mathematical representation of functional gene interaction requires intensive computation. Since genomewide association studies involve the computation of an extremely large number of potential pairwise and high-order gene interactions, to incorporate such a defined functional gene interaction into genomewide association studies requires a prohibitively large computation.

In the past several years, combinatorial partitioning[14] and various data mining methods[2,15–23] have been used to detect gene–gene interaction. These methods are designed mainly to study high-order gene–gene interactions. The limitations of these methods are (1) they usually lack clear biological interpretation of gene–gene interaction, (2) they require intensive computation, and (3) their power to detect gene–gene interaction may depend on the dataset structure.

To overcome these limitations, a multiplicative model for defining gene–gene interaction that defines the interaction between two loci (or genes) as a deviance of the penetrance for a haplotype at two loci from the product of the marginal penetrance of the individual alleles spanning the haplotype has recently been proposed.[24]

2.2. Linkage Disequilibrium Generated by Gene–Gene Interactions

Similar to a single disease gene that can cause linkage disequilibrium (LD) between the disease and marker loci in the disease population, two interacting disease loci will create LD between them. To investigate the LD pattern generated by gene–gene interaction, we assume that two disease susceptibility loci are in Hardy–Weinberg equilibrium. Let D_1 and d_1 be the two alleles at the first disease locus with frequencies P_{D_1} and P_{d_1}, respectively. Let D_2 and d_2 be the two alleles at the second disease locus with frequencies P_{D_2} and P_{d_2}, respectively. Alleles D_1 and d_1 can be indexed by 1 and 2, respectively. At the first disease locus, let $D_1 D_1$ be genotype 11, $D_1 d_1$ be genotype 12, and $d_1 d_1$ be genotype 22. The genotypes at the second disease locus are similarly defined. Two-locus genotypes are simply denoted by *ijkl* for individuals carrying the *ij* genotype at the first disease locus and the *kl* genotype at the second locus. Let f_{ijkl} be the penetrance of the genotype spanned by the haplotypes *ik* and *jl*. Let P_{11}, P_{12}, P_{21}, and P_{22} be the frequencies of the haplotypes $H_{D_1 D_2}$, $H_{D_1 d_2}$, $H_{d_1 D_2}$, and $H_{d_1 d_2}$ in the general population, respectively. Let P_{11}^A, P_{12}^A, P_{21}^A, and P_{22}^A be their corresponding haplotype frequencies in the disease population. Let $P_{D_1}^A$, $P_{d_1}^A$, $P_{D_2}^A$, and $P_{D_2}^A$ be the frequencies of the alleles D_1, d_1, D_2, and d_2 in the disease population, respectively.

For ease of discussion, we introduce the concept of haplotype penetrance. Consider a haplotype with allele *i* at the first disease locus and allele *k* at the second disease locus. Then, the penetrance of haplotype H_{ik} is defined as

$$h_{ik} = P_{11} f_{i1k1} + P_{12} f_{i1k2} + P_{21} f_{i2k1} + P_{22} f_{i2k2}.$$

Let $\delta = P_{11} - P_{D_1} P_{D_2}$ be the measure of LD in the general population. Define

$$r_{ik} = f_{i1k1} - f_{i1k2} - (f_{i2k1} - f_{i2k2}).$$

Then, the penetrance of haplotype h_{ik} can be expressed as

$$h_{ki} = g_k^i + r_{ik}\delta,$$

where $g_k^i = P_{D_1} P_{D_2} f_{i1k1} + P_{D_1} P_{d_2} f_{i1k2} + P_{d_1} P_{D_2} f_{i2k1} + P_{d_1} P_{d_2} f_{i2k2}$.

In Appendix A, we show that the frequencies of the haplotypes in the disease population are given by

$$
\begin{aligned}
P_{11}^A &= \frac{P_{11} h_{11}}{P_A} \\[2mm]
P_{12}^A &= \frac{P_{12} h_{12}}{P_A} \\[2mm]
P_{21}^A &= \frac{P_{21} h_{21}}{P_A} \\[2mm]
P_{22}^A &= \frac{P_{22} h_{22}}{P_A},
\end{aligned}
\tag{1}
$$

where P_A is the prevalence of disease and is given by

$$P_A = P_{11}^2 f_{1111} + 2P_{11}P_{12}f_{1112} + P_{12}^2 f_{1122} + 2P_{11}P_{21}f_{1211}$$
$$+ 2(P_{11}P_{22} + P_{12}P_{21})f_{1212} + 2P_{12}P_{22}f_{1222} + P_{21}^2 f_{2211}$$
$$+ 2P_{21}P_{22}f_{2212} + P_{22}^2 f_{2222}.$$

Now, we calculate the measure of LD in a disease population under the general two-locus disease model. The measure of LD in the disease population is defined as $\delta^A = P_{11}^A - P_{D_1}^A P_{D_2}^A$. We can show that it is given by (see Appendix A)

$$\delta^A = \frac{P_{11}P_{22}h_{11}h_{22} - P_{12}P_{21}h_{12}h_{21}}{P_A^2}. \tag{2}$$

From Eq. (2), we can see that, if $h_{11}h_{22} \neq h_{12}h_{21}$, even if two loci are in linkage equilibrium in the general population, two loci will be in LD in the disease population. This provides the basis for testing the interaction between two unlinked loci, as we will show in the section below. If we assume $h_{11}h_{22} = h_{12}h_{21}$, then we have

$$P_{11}P_{22}\delta^A = P_{11}^A P_{22}^A \delta. \tag{3}$$

Now, we examine a special case: a marker and a disease locus. Suppose that the first locus postulated above is a disease susceptibility locus and the second is a marker locus which does not predispose carriers to a disease phenotype. Then, we have

$$
\begin{aligned}
h_{11} &= P_{D_1}f_{11} + P_{d_1}f_{12} \\
h_{22} &= P_{D_1}f_{21} + P_{d_1}f_{22} \\
h_{12} &= P_{D_1}f_{11} + P_{d_1}f_{12} \\
h_{21} &= P_{D_1}f_{21} + P_{d_1}f_{22} \\
r_{11} &= r_{12} = r_{21} = r_{22} = 0.
\end{aligned}
\tag{4}
$$

Thus, combining Eqs. (2) and (4) yields

$$\delta^A = \frac{(P_{D_1}f_{11} + P_{d_1}f_{12})(P_{D_1}f_{21} + P_{d_1}f_{22})}{P_A^2}\delta,$$

i.e. the measure of LD between a disease locus and a marker locus in the disease population (δ^A) can be expressed in terms of the measure of LD in the general population and a multiplicative factor.

2.3. *Measure of Interaction between Genes*

Equation (2) leads us to propose a multiplicative model for the interaction between two loci. Gene–gene interaction can be defined in terms of penetrance of haplotype. Specifically, we define a measure of interaction between two loci that quantifies the magnitude of interaction as

$$I = h_{11}h_{22} - h_{12}h_{21}. \tag{5}$$

To gain further understanding of the interaction measure between two loci, we study the interactions between two loci for five two-locus disease models. Table 1 summarizes the measure of interaction between two disease loci for five disease models. The measure of interaction between two loci depends not only on the penetrance, but also on the frequencies of the disease alleles.

Table 1   **Interaction between Two Unlinked Disease Loci under Six Two-Locus Disease Models**

Model and First Locus	Second Locus			Interaction Measure
	$D_2 D_2$	$D_2 d_2$	$d_2 d_2$	
Dom $\cup$ Dom				$-P_{d_1} P_{d_2} f^2$
$D_1 D_1$	f	f	f	
$D_1 d_1$	f	f	f	
$d_1 d_1$	f	f	0	
Rec $\cup$ Rec				$-P_{D_1} P_{D_2} f^2$
$D_1 D_1$	f	f	f	
$D_1 d_1$	f	0	0	
$d_1 d_1$	f	0	0	
Threshold				$-P_{D_1}^2 P_{D_2}^2 f^2$
$D_1 D_1$	f	f	0	
$D_1 d_1$	f	0	0	
$d_1 d_1$	0	0	0	
Dom $\cup$ Rec				$-P_{d_1} P_{D_2} f^2$
$D_1 D_1$	f	f	f	
$D_1 d_1$	f	f	f	
$d_1 d_1$	f	0	0	
Epistasis				$-P_{D_1} P_{D_2}^2 f^2$
$D_1 D_1$	f	f	0	
$D_1 d_1$	f	0	0	
$d_1 d_1$	f	0	0	
Modifying				$-P_{D_1}^2 P_{D_2} f^2$
$D_1 D_1$	f	f	f	
$D_1 d_1$	f	0	0	
$d_1 d_1$	0	0	0	

The measure of LD in the disease population can then be expressed in terms of the measure of interaction as follows:

$$\delta^A = \frac{h_{11}h_{22}}{P_A^2}\delta + \frac{P_{12}P_{21}}{P_A^2}I. \tag{6}$$

The absence of interaction between two loci is thus defined as

$$h_{11}h_{22} = h_{12}h_{21}. \tag{7}$$

In the absence of interaction between two loci, Eq. (3) will hold.

In Appendix B, we show that if Eq. (7) holds, then we have

$$g_1^1 g_2^2 = g_2^1 g_1^2$$

$$r_{11}r_{22} = r_{12}r_{21},$$

$$g_1^1 r_{22} + g_2^2 r_{11} = g_2^1 r_{21} + g_1^2 r_{12}.$$

Therefore, in the absence of interaction between two loci, the measure δ^A of LD between two disease loci in the disease population is given by (see Appendix B)

$$\delta^A = \alpha\delta + \beta\delta^2 + \gamma\delta^3, \tag{8}$$

where

$$\alpha = \frac{g_1^1 g_2^2}{P_A^2}, \quad \beta = \frac{g_1^1 r_{22} + g_2^2 r_{11}}{P_A^2}, \quad \text{and} \quad \gamma = \frac{r_{11}r_{22}}{P_A^2}.$$

Equation (8) shows that when two loci are in linkage equilibrium in the general population, and in the absence of interaction between the two loci, they are also in linkage equilibrium in the disease population.

2.4. *Relationships between Interaction Measure and Traditional Quantitative Genetic Effects*

To improve our understanding of the new definition of interaction, it is helpful to reveal the relationships between the proposed interaction measure and the statistical genetic effects. We treat penetrance as a quantitative trait. Let f_{ijkl} be the penetrance of an individual carrying genotype *ijkl*. The penetrance f_{ijkl} can be modeled by

$$
\begin{aligned}
f_{ijkl} = {}& \mu + \alpha_i^A + \alpha_j^A + d_{ij}^A + \alpha_k^B + \alpha_l^B + d_{kl}^B + (\alpha^A\alpha^B)_{ik} + (\alpha^A\alpha^B)_{jk} \\
& + (\alpha^A\alpha^B)_{il} + (\alpha^A\alpha^B)_{jl} + (\alpha^A d^B)_{ikl} + (\alpha^A d^B)_{jkl} + (d^A\alpha^B)_{ijk} \\
& + (d^A\alpha^B)_{ijl} + (d^A d^B)_{ijkl} \\
& i = 1, 2, j = 1, 2, k = 1, 2, l = 1, 2,
\end{aligned} \tag{9}
$$

where the superscript represents the locus (A locus or B locus); μ represents the overall population mean; α_i^A and α_j^A represent the additive effects of the alleles at the first locus A; α_k^B and α_l^B represent the additive effects of the alleles at the second locus B; d_{ij}^A and d_{id}^B represent the dominance deviation for the alleles at the loci A and B, respectively; $(\alpha^A \alpha^B)_{ik}$, $(\alpha^A \alpha^B)_{jk}$, $(\alpha^A \alpha^B)_{il}$, and $(\alpha^A \alpha^B)_{jl}$ represent the additive $\times$ additive effects of the corresponding alleles at the two loci; $(\alpha^A d^B)_{ikl}$ and $(\alpha^A d^B)_{jkl}$ represent the additive $\times$ dominance effects of the corresponding alleles at the two loci; $(d^A \alpha^B)_{ijk}$ and $(d^A \alpha^B)_{ijl}$ represent the dominance $\times$ additive effects of the corresponding alleles at the two loci; and $(d^A d^B)_{ijkl}$ denotes the dominance $\times$ dominance effects.

In Appendix C, we show that the contribution of the traditionally defined genetic effects to the interaction measure is given by

$$e = -\alpha^A \alpha^B + e_{\mathrm{D}} + e_{\mathrm{AA}} + e_{\mathrm{AD}} + e_{\mathrm{DA}} + e_{\mathrm{DD}}, \tag{10}$$

where α^A and α^B are the substitution effects at the first locus A and the second locus B, respectively. The contributions of the genetic dominant, additive $\times$ additive, additive $\times$ dominant, dominant $\times$ additive, and dominant $\times$ dominant effects to the interaction measures e_{D}, e_{AA}, e_{AD}, e_{DA}, and e_{DD} are defined in Appendix C. In general, e_{D}, e_{AA}, e_{AD}, e_{DA}, and e_{DD} are small. The interaction measure e is mainly influenced by $\alpha^A \alpha^B$, which implies that the interaction between two loci requires the joint action of two loci.

Figure 1 plots the measure of interaction as a function of the variance of statistical interaction in which the penetrance is taken as a quantitative trait. It demonstrates that as the variance of statistical interaction increases, the measure of interaction also increases, but the measure of interaction increases more rapidly than the variance of statistical interaction.

3. INTERACTION BETWEEN TWO UNLINKED LOCI

In Sec. 2, we discussed the interaction between two linked loci. The pattern of LD created by the interaction between two linked loci is complicated, thus causing difficulty in developing statistics to test for interaction between two linked loci. To simplify the problem, we now focus on studying the interaction between two unlinked loci.

3.1. *Linkage Disequilibrium Generated by Interaction between Two Unlinked Loci*

Now, we calculate the measure of LD in a disease population under the general two-locus disease model. It follows that the measure of LD between

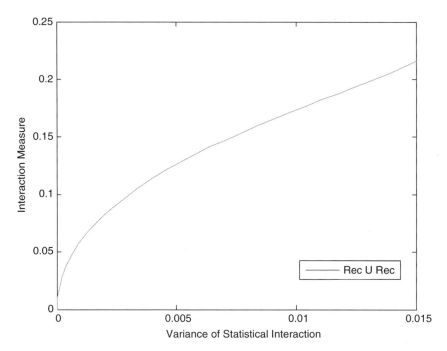

Fig. 1.   The measure of interaction as a function of variance of statistical interaction under a recessive × recessive two-disease locus model.

two unlinked loci in the disease population is given by

$$\delta^A = \frac{P_{D_1} P_{D_2} P_{d_1} P_{d_2}}{P_A^2} I,\tag{11}$$

where $I = h_{11}h_{22} - h_{12}h_{21}$ is a measure of the interaction between two unlinked loci defined in Eq. (5). Under this definition, in the absence of interaction, two unlinked loci in the disease population will be in linkage equilibrium.

From Eq. (11), we can see that, if $h_{11}h_{22} \neq h_{12}h_{21}$, even if two loci are in linkage equilibrium in the general population, two loci will be in LD in the disease population. The LD between two unlinked loci in the disease population is created by the interaction between two unlinked loci. This provides a foundation for testing the interaction between two unlinked loci, as we will show in the section below.

Define

$$h_{D_1} = P(\textit{Affected} \mid D_1) \quad \text{and} \quad h_{D_2} = P(\textit{Affected} \mid D_2).$$

We can show that the absence of interaction between two unlinked loci implies that[24]

$$h_{11} = \frac{1}{P_A} h_{D_1} h_{D_2} \quad \text{or} \quad \frac{h_{11}}{P_A} = \frac{h_{D_1}}{P_A} \frac{h_{D_2}}{P_A}.$$

Similar to linkage equilibrium, where the frequency of a haplotype is equal to the product of the frequencies of the component alleles of the haplotype, absence of interaction between two unlinked loci implies that the proportion of individuals carrying a haplotype in the disease population is equal to the product of the proportions of individuals carrying the component alleles of the haplotype in the disease populations if we assume that the disease is caused by only two investigated disease loci. In other words, interaction between two disease susceptibility loci occurs when the contribution of one locus to the disease depends on the second locus.

Recall that in Sec. 2, we showed that the LD between a disease locus and a marker locus in the disease population is given by

$$\delta^A = \frac{(P_{D_1} f_{11} + P_{d_1} f_{12})(P_{D_1} f_{21} + P_{d_1} f_{22})}{P_A^2} \delta,$$

i.e. the measure of LD between a disease locus and a marker locus in the disease population (δ^A) can be expressed in terms of the measure of LD in the general population and a multiplicative factor. If the disease locus and the marker locus are unlinked, then the disease and marker loci will be in linkage equilibrium. This demonstrates that in the absence of interaction between an unlinked marker and a disease locus, the LD between them in the disease population cannot be created.

To further understand the measure of interaction between two unlinked loci, we studied the interactions between two unlinked loci for six two-locus disease models. Table 1, in which each cell represents the penetrance of the given genotypes,[24] summarizes the measure of interaction between two unlinked disease loci for six disease models. The measure of interaction between two unlinked loci depends not only on the penetrance parameter f, but also on the frequencies of the disease alleles.

3.2. *Indirect Interaction between Two Unlinked Marker Loci*

In the previous subsection, we studied the interaction between two unlinked disease loci. Now, we consider two marker loci, each of which is in LD with either of two interacting loci. Although the two marker loci do not have any physiological interaction between them, if each marker locus is in LD with one of two unlinked interacting loci, we can still observe LD between two unlinked marker loci in the disease population. Assume that the marker M_1 is in LD with the disease locus D_1 and the marker M_2 is in LD with the disease locus D_2. Furthermore, assume that the two disease loci D_1 and D_2 are unlinked. Let δ_M^A be the measure of LD between two marker loci in the disease population. Let δ_i be the measure of LD between the marker M_i and the disease locus $D_i(i = 1, 2)$ in the general population. Then, we can show that[24]

$$\delta_M^A = \frac{\delta_1 \delta_2 (h_{11} h_{22} - h_{21} h_{12})}{P_A^2}$$

$$= \frac{\delta_1 \delta_2}{P_{D_1} P_{D_2} P_{d_1} P_{d_2}} \delta^A, \tag{12}$$

where δ^A is the measure of LD between two unlinked disease loci in the disease population. It is clear that when the marker loci are the disease loci themselves, δ_M^A is reduced to δ^A. Equation (12) can also be written in terms of the measure of interaction between two unlinked loci:

$$\delta_M^A = \frac{\delta_1 \delta_2}{P_A^2} I. \tag{13}$$

Since $\delta_i \leq P_{D_i} P_{d_i}$, the absolute value of the measure of LD between two unlinked marker loci in the disease population $|\delta_M^A|$ will be less than or equal to the absolute value of the measure of LD between two unlinked disease loci in the disease population.

Equation (12) shows that the LD between unlinked marker loci in the disease population is proportional to the product of LD between each marker locus and its linked disease locus $\delta_1 \delta_2$. Since the criteria for tag single nucleotide polymorphism (SNP) selection are based on only one pairwise LD between the marker and disease loci, the LD between tag SNPs and interacting loci may not be large enough to ensure that indirect interaction between two unlinked marker loci will be detected. Thus, if the interacting disease loci are not selected as tag SNPs, many loci with interactions will be missed. This will have profound implications on tag SNP selection.

3.3. Test Statistic

In the previous subsection, we showed that interaction between unlinked loci will create LD between them. Intuitively, we can test the interaction between two unlinked loci by comparing the difference in the levels of LD between two unlinked loci in the cases and controls. If we denote the estimators of the measure of LD in the cases and controls as $\hat{\delta}_A$ and $\hat{\delta}_N$, respectively, then the test statistic can be defined as

$$T_{\mathrm{I}} = \frac{(\hat{\delta}_A - \hat{\delta}_N)^2}{\hat{V}_A + \hat{V}_N}, \tag{14}$$

where

$$\hat{\delta}_A = \hat{P}_{11}^A - \hat{P}_{D_1}^A \hat{P}_{D_2}^A$$

$$\hat{\delta}_N = \hat{P}_{11} - \hat{P}_{D_1} \hat{P}_{D_2}$$

$$\hat{V}_A = \frac{\hat{P}_{D_1}^A (1 - \hat{P}_{D_1}^A) \hat{P}_{D_2}^A (1 - \hat{P}_{D_2}^A) + (1 - 2\hat{P}_{D_1}^A)(1 - 2\hat{P}_{D_2}^A)\hat{\delta}_A - \hat{\delta}_A^2}{2n_A}$$

$$\hat{V}_N = \frac{\hat{P}_{D_1} (1 - \hat{P}_{D_1}) \hat{P}_{D_2} (1 - \hat{P}_{D_2}) + (1 - 2\hat{P}_{D_1})(1 - 2\hat{P}_{D_2})\hat{\delta}_N - \hat{\delta}_N^2}{2n_G},$$

where n_A and n_G are the number of sampled individuals in the cases and controls, respectively; P_{11}^A, $P_{D_1}^A$, $P_{D_2}^A$, P_{11}^N, $P_{D_1}^N$, and $P_{D_2}^N$ are defined as before; $\hat{P}_{11}^A$, $\hat{P}_{D_1}^A$, $\hat{P}_{D_2}^A$, $\hat{P}_{11}$, $\hat{P}_{D_1}$, and $\hat{P}_{D_2}$ are their estimators, respectively; the variance of the measure of LD is the large-sample variance[22]; and $\hat{V}_A$ and $\hat{V}_N$ are the estimators of the variances V_A and V_N, respectively. This statistic will be referred to as the LD-based statistic throughout the paper.

We can show that the test statistic T_{I} is asymptotically distributed as a central $\chi^2_{(1)}$ distribution under the null hypothesis of no interaction between two unlinked loci.

When two loci are unlinked, in theory, when there is no interaction between two loci, the LD between them should be equal to zero. Then, we can use a case-only design to study the interaction between two loci, whereby Eq. (14) is reduced to

$$T_{\mathrm{I}} = \frac{(\hat{\delta}^A)^2}{\hat{V}_A}. \tag{15}$$

However, in practice, background LD between two unlinked loci may exist in the population due to many unknown factors. Therefore, the use of Eq. (15) to test for interaction between two unlinked loci will increase type 1 error rates. The test statistic for testing interaction between two unlinked loci defined in Eq. (14) is more robust than that in Eq. (15). We show that for an admixed population, if the differences in allele frequencies between two subpopulations at each of the two loci in the cases and controls are the same, then the test statistic T_{I} in Eq. (14) is still a valid test for interaction between two loci in the admixed population.[24]

3.4. *Patterns of Pairwise LD under Two-Locus Disease Models*

Knowledge of the differences in LD patterns between the disease and general populations becomes crucial for association studies of complex diseases. We examined the LD between unlinked loci under several two-locus disease models to illustrate how disease models can influence the differences in LD patterns between the disease and general populations.

We first studied the LD between two unlinked loci under three two-locus disease models: Dom ∪ Dom, Rec ∪ Rec, and threshold models (Table 1). Figure 2 shows the LD between two unlinked loci, which are generated by the joint actions of two disease loci, as a function of the frequency of the allele at the first locus, assuming the frequency of the allele at the second locus $P_{D_2} = 0.1$ and penetrance parameter $f = 1$. From Fig. 2, we know that although two unlinked loci in the general population are in linkage equilibrium, LD between two unlinked loci in the disease population does

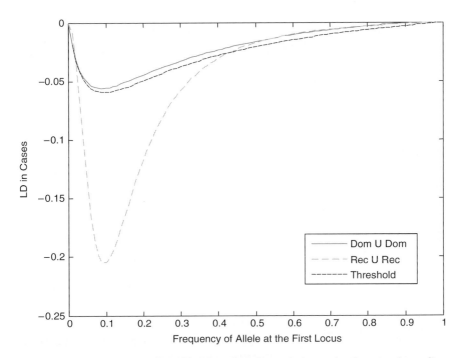

Fig. 2.   LD between two unlinked loci in a disease population under three two-locus disease models as a function of allele frequency at the first locus, assuming allele frequency at the second locus equals to 0.1.

exist. The LD in the disease population depends on the disease models and the frequency of alleles at the two loci.

3.5. *Pairwise Interaction Measure*

The proposed measure of interaction between two unlinked loci quantifies the magnitude of interaction between two unlinked loci. To improve our understanding of the measure of interaction between two unlinked loci, we investigated the impact of two-locus disease models on the measure of interaction between two unlinked loci. Figures 3(a) and 3(b) plot the measure of interaction between two unlinked loci under six two-locus disease models (Table 1) as a function of penetrance parameter f, assuming the frequencies of the alleles at two loci are 0.3 and 0.8 for Fig. 3(a), and 0.2 and 0.4 for Fig. 3(b). The figures show that the measure of interaction is a monotonic function of the penetrance parameter. They demonstrate that the measure of interaction depends on both the disease models and the frequencies of the alleles at two loci.

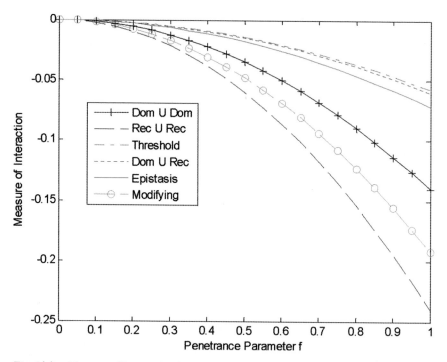

Fig. 3(a).   Measure of interaction between two unlinked loci as a function of penetrance parameter under six two-locus disease models, assuming allele frequencies at the first and second locus equal to 0.3 and 0.8, respectively.

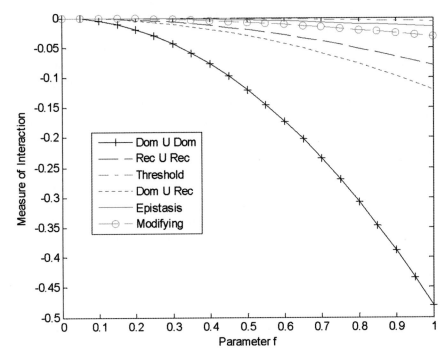

Fig. 3(b).   Measure of interaction between two unlinked loci as a function of penetrance parameter under six two-locus disease models, assuming allele frequencies at the first and second locus equal to 0.2 and 0.4, respectively.

However, the relationship between the measure of interaction and the disease models is complex. For example, when the frequencies of alleles at two loci are 0.2 and 0.4, the measure of interaction for the Dom ∪ Dom model is much larger than that for the Rec ∪ Rec model; but when the frequencies of alleles at two loci are 0.3 and 0.8, the measure of interaction for the Dom ∪ Dom model is smaller than that for the Rec ∪ Rec model. It may explain why replication of the results of interaction in other populations is difficult in reality, due to the difference in allele frequencies between populations.

3.6. *Null Distribution of Test Statistics*

In the previous subsections, we have shown that when the sample size is large enough to apply the large sample theory, the distribution of the statistic T_{I} for testing the interaction between two unlinked loci under the null hypothesis of no interaction between two loci is asymptotically a central

$\chi^2_{(1)}$ distribution. To examine the validity of this statement, we performed a series of simulation studies. The computer program SNaP[25] was used to generate two-locus genotype data of the sample individuals. A total of 10 000 individuals, who were equally divided into cases and controls, were generated in the general population. Of these, 100–500 individuals were randomly sampled from each of the cases and controls. Ten thousand simulations were repeated.

Figures 4(a) and 4(b) plot the histograms of the test statistic T_{I} for testing the interaction between two unlinked loci with sample sizes $n_A = n_G = 150$ and $n_A = n_G = 250$, where n_A and n_G are the numbers of sampled individuals in the cases and the controls, respectively. It can be seen that the distributions of the test statistic T_{I} are similar to the theoretical central $\chi^2_{(1)}$ distribution. Table 2 shows that the estimated type 1 error rates of the statistic T_{I} for testing interaction between unlinked loci were not appreciably different from the nominal levels $\alpha = 0.05, \alpha = 0.01$, and $\alpha = 0.001$.

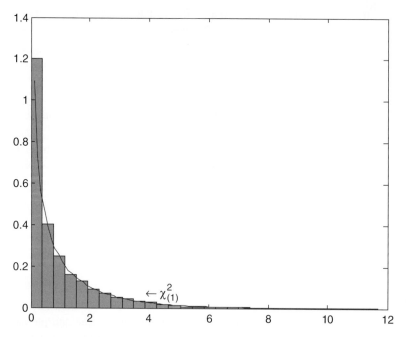

Fig. 4(a). Null distribution of the test statistic T_{I} using 150 individuals in both cases and controls in a homogeneous population.

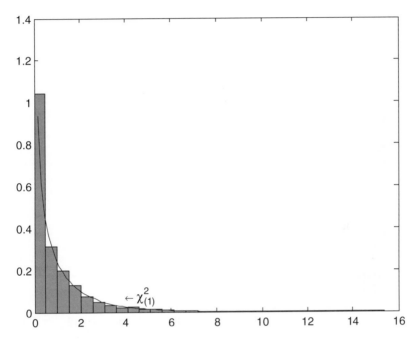

Fig. 4(b). Null distribution of the test statistic T_I using 250 individuals in both cases and controls in a homogeneous population.

Table 2   Type 1 Error Rates of the Test Statistic T_I to Test Interaction between Two Unlinked Loci in a Homogeneous Population

Sample Size	Nominal Levels		
	$\alpha = 0.05$	$\alpha = 0.01$	$\alpha = 0.001$
100	0.0501	0.0108	0.0010
200	0.0502	0.0094	0.0012
300	0.0482	0.0091	0.0010
400	0.0472	0.0091	0.0009
500	0.0466	0.0096	0.0010

To examine the impact of population substructure on the null distribution of the test statistic T_I, we performed simulation to examine the null distribution of the proposed statistic in an admixed population in which we assumed that the frequencies of the alleles at the first locus in population 1 were 0.7 and 0.3, and in population 2 were 0.3 and 0.7; and that the frequencies of the alleles at the second locus in population 1 were 0.2 and

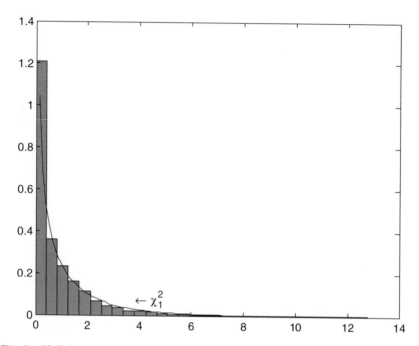

Fig. 5. Null distribution of the test statistic T_{I} using 300 individuals in both cases and controls in an admixed population.

0.8, and in population 2 were 0.8 and 0.2. Ten thousand individuals were sampled from each population and mixed to form an admixed population, which was then equally divided into cases and controls. Three hundred individuals were randomly sampled from each of the cases and controls. Ten thousand simulations were repeated. Figure 5 plots the histograms of the test statistic T_{I}. It can be seen that the distribution of T_{I} is similar to the theoretical central χ^2 distributions. This may show the impact of admixture on the null distribution if the test statistic is mild.

3.7. *Power Evaluation*

To further evaluate the performance of the proposed statistic for testing the interaction between two unlinked loci, we compared the power of the LD-based statistic T_{I} and the logistic model for the detection of interaction. We considered three types of genotype coding (genetic covariate variable). For the recessive model, the homozygous wild-type, heterozygous, and homozygous mutant genotypes were coded as 0, 0, and 1, respectively; for the dominant model, these same genotypes were coded as 0, 1, and 1, respectively; and for the additive model, they were coded as 0, 1, and 2, respectively. We considered two loci: the first locus was denoted by G and the second locus denoted by H. The power for testing interaction between two loci by logistic regression models was calculated by the software QUANTO.[37]

The power for three genetic interaction models — recessive × recessive, dominance × dominance, and additive × additive models — calculated by logistic regression analysis and the LD-based statistic T_I, as a function of interaction odds ratio R_{GH}, is shown in Figs. 6(a), 6(b), and 6(c), respectively. The odds ratio for interaction R_{GH} is a widely used measure to quantify the strength of interaction between two loci. Figs. 6(a)–6(c) show that, similar to logistic regression, the power of the test statistic T_I is also a monotonic function of the odds ratio for interaction, thus implying

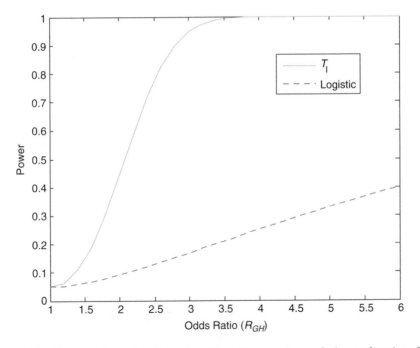

Fig. 6(a).   Power of the test statistic T_I and logistic regression analysis as a function of interaction odds ratio (R_{GH}) under a recessive × recessive model, assuming risk allele frequencies at both loci G and H are 0.2, number of individuals in both cases and controls are 500, population risk is 0.001, significance level is 0.05, and odds ratios $R_G = 5$ and $R_H = 5$.

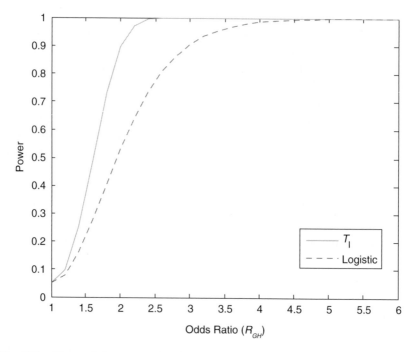

Fig. 6(b).   Power of the test statistic T_I and logistic regression analysis as a function of interaction odds ratio (R_{GH}) under a dominant × dominant model, assuming risk allele frequencies at both loci G and H are 0.1, number of individuals in both cases and controls are 500, population risk is 0.001, significance level is 0.05, and odds ratios $R_G = 2$ and $R_H = 2$.

that the proposed new interaction measure and test statistic are related to the traditional interaction measure. From Figs. 6(a)–6(c), we can see that the power of the test statistic T_I is much higher than that of the logistic regression analysis.

3.8. *Application to a Real Data Example*

To evaluate its performance for detecting interaction between two unlinked loci, the proposed test statistic T_I was applied to two real data sets. The first data set was a breast cancer case-control study. A total of 398 Caucasian breast cancer cases and 372 matched controls were sampled from the Ontario Familial Breast Cancer Registry (OFBCR).[26] A total of 19 SNPs from 18 key genes in DNA repair, cell cycle, carcinogen/estrogen metabolism, and immune system were typed. All SNPs were in Hardy–Weinberg equilibrium. Using multivariate logistic analysis under the

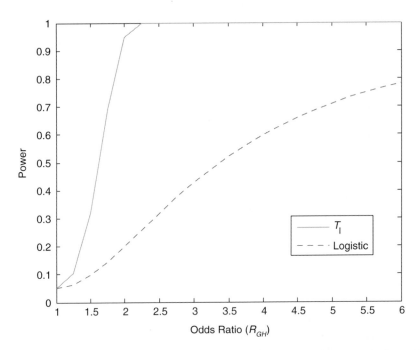

Fig. 6(c). Power of the test statistic T_I and logistic regression analysis as a function of interaction odds ratio (R_{GH}) under an additive × additive model, assuming risk allele frequencies at both loci G and H are 0.1, number of individuals in both cases and controls are 100, population risk is 0.001, significance level is 0.05, and odds ratios $R_G = 2$ and $R_H = 2$.

codominant models, four pairs of genes — *XPD* and *IL10*, *GSTP1* and *COMT*, *COMT* and *CCND1*, and *BARD1* and *XPD* — showed significant interactions.[26] We used the statistic T_I to test interactions between these four pairs of genes.

The test results are summarized in Table 3, where the crude P values taken from Table 4 in Onay *et al.*'s paper.[26] The crude P values were obtained from multivariate logistic regression analysis that included all main effects and only the interaction of interest under the codominant models. It was shown that only *XPD*-[Lys751Gln], out of the 19 SNPs, demonstrated a significant main effect from logistic regression, but the significance of the main effect disappeared after correction for multiple testing using a false discovery rate (FDR).[26] Table 3 shows the use of logistic regression analysis to identify interactions between *XPD*-[Lys751Gln] and *IL10*-[G(-1082)A], *BARD*-[Pro24Ser] and *XPD*-[Lys751Gln], *COMT*-[Met108/158Val] and *CCND1*-[Pro24Pro], and *GSTP1*-[Ile241Val] and *COMT*-[Met108/158Val]. However, after the more conservative Bonferroni

Table 3   Comparison of P values for Testing Gene–Gene Interactions

Pair of Interactions	P values Obtained by	
	Logistic Regression[†]	LD-Based Statistic
XPD-[Lys751Gln] and *IL10*-[G(-1082)A]	0.035	2.7×10^{-4}
BARD1-[Pro24Ser] and *XPD*-[Lys751Gln]	0.024	0.00684
COMT-[Met108/158Val] and *CCND1*-[Pro241Pro]	0.010	0.00395
GSTP1-[Ile105Val] and *COMT*-[Met108/158Val]	0.036	1.15×10^{-5}

P values reported by Onay *et al.*[26]

adjustment, none of these interactions were significant.[26] Table 3 demonstrates that the P values based on the test statistic T_I were much smaller than those based on the traditional logistic regression analysis. Even after the conservative Bonferroni adjustment, *XPD*-[Lys751Gln] and *IL10*-[G(-1082)A] as well as *GSTP1*-[Ile241Val] and *COMT*-[Met108/158Val] still showed significant interactions.

The four identified interactions can be justified by their biological relationships. It was reported that there were protein–protein interactions among *XPDS*, *TP53*, *BRCA1*, and *BARD1*,[38,39] which might account for the interaction between *XPD*-[Lys751Gln] and *IL10*-[G(-1082)A] as well as *BARD*-[Pro24Ser] and *XPD*-[Lys751Gln]. The experiments showed that interactions between *COMT*-[Met108/158Val] and *CCND1*-[Pro24Pro] as well as *GSTP1*-[Ile241Val] and *COMT*-[Met108/158Val] might be implied by estrogen metabolism and cell proliferation. The relation between *COMT* and *GSTP* depends on reduced inactivation of the reactive estrogen intermediates. The expression of estrogen in turn affects cell proliferation via *CCND1* transcription.[26,40,41]

The second data set is a subset of an atherosclerosis case-control data set. It includes 27 SNPs from 22 candidate genes in the inflammatory, antioxidant, and coagulation pathways typed in 916 samples (492 cases and 424 controls) from a Chinese population. Atherosclerosis is the primary cause of coronary heart disease (CHD).[27] Several lines of evidence indicate that the inflammatory, antioxidant, and coagulation pathways play an important role in the pathogenesis of atherosclerosis.[28,29] The P values for testing the association of individual SNPs with atherosclerosis are summarized in Table 4. We statistically tested the interactions between all possible pairs of the 27 SNPs, where all of the SNP pairs were in different chromosomes. The results are summarized in Table 5. In Table 5, we also list the results of a likelihood-ratio test by logistic regression that was implemented by SAS. Specifically, a logistic model is given by[30]

$$\log \frac{P(A = 1)}{1 - P(A = 1)} = b_0 + b_i G_i + b_j G_j + b_{ij} G_i G_j,$$

where A indicates that the individual is affected; and G denotes a genotype coded by 0, 1, and 2, corresponding to the genotypes dd, Dd, and DD, respectively. Association tests were conducted by the standard χ^2 test

Table 4   Test for Association of an Individual SNP with Atherosclerosis

Gene	Chr	SNP	*P* value	Gene	Chr	SNP	*P* value
CD36	7	rs3211913	0.353	EDNRA	8	rs2357924	0.908
CD36	7	rs3211883	0.152	IFNAR2	21	rs2300371	0.0537
CD36	7	rs3211883		PDGFC	4	rs4234903	0.229
CD36	7	rs3211851	0.289	SELL	1	rs4987361	0.936
PON2	7	rs13438526	0.67	TXN	9	rs4135208	0.0632
F5	1	rs9332592	0.266	F7	13	rs3093261	0.759
F5	1	rs2298909	0.167	SERPINE1	7	rs6092	0.770
BLVRA	7	rs2877262	0.125	GCLC	6	rs12524494	0.0551
ITGA2	5	rs3212391	0.603	LAMA4	6	rs9374312	0.291
ITGA2	5	rs27377	0.859	PDGFB	22	rs9611113	0.746
EDNRA	8	rs6842241	0.0236	SCARA3	8	rs505295	0.409
ITGB1	10	rs12256760	0.892	SELP	1	rs3917698	0.109
IL1B	2	rs3917368	0.362	MMP9	20	rs2274755	0.724
ITGB1	10	rs11009157	0.0376	ITGA6	2	rs6743724	0.0809

Table 5   *P* values for Testing Interaction between Two Loci and Association of the SNP Pairs with Atherosclerosis

SNP1	SNP2	*P* value (Interaction)		*P* value (Association)	
		T_{I}	Logistic Reg	χ^2	Logistic Reg
rs3211913	rs2357924	1.71E-09	0.0067	2.79E-02	0.0264
rs3211883	rs2300371	7.18E-09	0.2262	2.77E-08	0.0473
rs3211883	rs4234903	3.94E-08	0.9880	3.81E-07	0.0472
rs3211851	rs4987361	7.59E-08	0.2177	4.88E-03	0.5088
rs13438526	rs4135208	5.21E-08	0.0220	1.33E-07	0.0477
rs9332592	rs3093261	1.31E-07	0.0094	1.59E-06	0.0621
rs2298909	rs6092	1.51E-07	0.0323	2.88E-06	0.1123
rs2877262	rs12524494	6.26E-07	0.1726	6.69E-07	0.3036
rs3212391	rs9374312	5.23E-10	0.8477	2.35E-08	0.9757
rs27377	rs9611113	1.28E-09	0.6807	7.84E-02	0.8518
rs6842241	rs505295	1.49E-09	0.2985	1.17E-08	0.7453
rs12256760	rs3917698	8.13E-08	0.4871	5.82E-07	0.1820
rs3917368	rs2274755	2.98E-07	0.0932	2.93E-06	0.2491
rs11009157	rs6743724	3.20E-07	0.2256	1.80E-07	0.0494

comparing the difference in haplotype frequencies between the cases and controls, and by logistic regression.

Several remarkable features emerged from these results. First, although both genes did not show significant evidence of association with atherosclerosis, they did show significant evidence of interaction. This demonstrates that the effect of one locus can be changed or masked by the effect of another locus. The power to detect association of the individual locus with the disease is likely to be reduced. Therefore, the locus-by-locus paradigm for genomewide association studies is inappropriate. Second, logistic regression only detected the interactions between the SNPs rs3211913 (*CD36*) and rs2357924 (*EDNRA*), rs13438526 (*PON2*) and rs4135208 (*TXN*), rs9332592 (*F5*) and rs3093261 (*F7*), and rs2298909 (*F5*) and rs6092 (*SERPINE1*) with much larger P values than those of the proposed statistic T_I; and failed to detect the interactions between the remaining 10 pairs of genes. Third, the interactions detected by the statistic T_I were confirmed by the association of the standard χ^2 test where, if two SNPs demonstrated interaction, then the two SNPs jointly showed association with atherosclerosis. It is noted again that logistic regression only detected the association of five pairs of SNPs, but failed to detect the association of the remaining nine pairs of SNPs.

The gene interaction results can be indirectly confirmed by experiments. Due to space limitations, we show only part of the interaction. For example, we consider the following three interactions: *EDNRA* and *CD36*, *CD36* and *SELL*, and *IL1B* and *MMP9*. Biological connection between *EDNRA* and *CD36* might be through TGF-beta: TGF-beta downregulates *CD36*,[31] but upregulates *VEGF*[32] which can actively interact with *EDNRA*. The experiments suggest that *CD36* and *SELL* interaction is through IL6. IL6 and TNF-alpha can decrease *SELL* expression in peripheral mononuclear cells,[33] but IL6 can increase *CD36* expression in macrophages.[34] It was reported that keratinocyte MMP-9 production is enhanced by *IL1B*, transforming growth factor beta 1, and tumor necrosis factor-alpha.[35]

4. COMPOSITE MEASURE OF LINKAGE DISEQUILIBRIUM FOR DETECTION OF INTERACTION BETWEEN TWO LOCI

In Secs. 2 and 3, we defined the interaction between two loci in terms of penetrance of haplotype, i.e. the probability of two genes at the same gamete. Although the statistic based on the haplotypes for testing the interaction between two loci is powerful, in practice, only genotype data are available.

Haplotypes estimated from genotype data often have rather large errors, which will increase the false discovery rate. In addition, two genes may be at different gametes; the interaction that exists between two genes at different gametes should also be defined. Composite genotypic disequilibrium, which studies disequilibrium between two genes at either the same gamete or different gametes, can be used to quantify the magnitude of the interaction between two genes at the same gamete or different gametes.

4.1. A General Model for Interaction between Two Loci Including Two Genes at the Same Gamete or Different Gametes

The concept of interaction between two genes at the same gamete can be extended to the interaction between two genes at different gametes. Let D_1 and d_1 be the two alleles at the first disease locus with frequencies P_{D_1} and P_{d_1}, respectively. Let D_2 and d_2 be the two alleles at the second disease locus with frequencies P_{D_2} and P_{d_2}, respectively. Alleles D_1 and d_1 can be indexed by 1 and 2, respectively. At the first disease locus, let $D_1 D_1$ be genotype 11, $D_1 d_1$ be genotype 12, and $d_1 d_1$ be genotype 22. The genotypes at the second disease locus are similarly defined. Two-locus genotypes are simply denoted by *ijkl* for individuals carrying the haplotypes *ik* and *jl* arranged from left to right. Let f_{ijkl} be the penetrance of the individuals with haplotypes *ik* and *jl* arranged from left to right. Let P_{11}, P_{12}, P_{21}, and P_{22} be the frequencies of haplotypes $H_{D_1 D_2}$, $H_{D_1 d_2}$, $H_{d_1 D_2}$, and $H_{d_1 d_2}$ in the general population, respectively. Let P_{11}^A, P_{12}^A, P_{21}^A, and P_{22}^A be their corresponding haplotype frequencies in the disease population. Let $P_{1/1}$, $P_{1/2}$, $P_{2/1}$, and $P_{2/2}$ be the frequencies of H_{D_1/D_2}, H_{D_1/d_2}, H_{d_1/D_2}, and H_{d_1/d_2}, respectively, where the slash denotes that the two alleles are on different chromosomes. Let $P_{1/1}^A$, $P_{1/2}^A$, $P_{2/1}^A$, and $P_{2/2}^A$ be their corresponding haplotype frequencies in the disease population. Let $P_{D_1}^A$, $P_{d_1}^A$, $P_{D_2}^A$, and $P_{d_2}^A$ be the frequencies of the alleles D_1, d_1, D_2, and d_2 in the disease population, respectively.

For ease of discussion, we introduce the concept of haplotype penetrance. Consider a haplotype with allele i at the first disease locus and allele k at the second disease locus. Then, the penetrance of haplotype $H_{D_1 D_2}$ is defined as

$$h_{11} = \left[P_{D_1 D_2}^{D_1 D_2} f_{1111} + \frac{1}{2} \left(P_{D_1 d_2}^{D_1 D_2} f_{1112} + P_{d_1 D_2}^{D_1 D_2} f_{1211} + P_{d_1 d_2}^{D_1 D_2} f_{1212} \right) \right] / P_{D_1 D_2}.$$

The penetrances h_{12}, h_{21}, and h_{22} are similarly defined.

The penetrance of two alleles at different loci on different chromosomes H_{D_1/D_2} can be defined as

$$h_{1/1} = \left[P_{D_1 D_2}^{D_1 D_2} f_{1111} + \frac{1}{2} \left(P_{D_1 d_2}^{D_1 D_2} f_{1112} + P_{d_1 D_2}^{D_1 D_2} f_{1211} + P_{D_1 d_2}^{d_1 D_2} f_{2112} \right) \right] / P_{D_1/D_2}.$$

Similarly, we can define the penetrances $h_{1/2}$, $h_{2/1}$, and $h_{2/2}$. If we assume Hardy–Weinberg equilibrium and genotypic equilibrium in the general population, then we have $h_{11} = h_{1/1}$, $h_{12} = h_{1/2}$, $h_{21} = h_{2/1}$, and $h_{22} = h_{2/2}$.

Let $\delta = P_{11} - P_{D_1} P_{D_2}$ be the LD measure in the general population. In Appendix A, we showed that haplotype frequencies in the disease population can be expressed as

$$P_{11}^A = \frac{P_{D_1 D_2} h_{11}}{P_A}$$

$$P_{12}^A = \frac{P_{D_1 d_2} h_{12}}{P_A}$$

$$P_{21}^A = \frac{P_{d_1 D_2} h_{21}}{P_A} \tag{16}$$

$$P_{22}^A = \frac{P_{d_1 d_2} h_{22}}{P_A}$$

and

$$P_{1/1}^A = \frac{P_{D_1/D_2} h_{11}}{P_A}$$

$$P_{2/1}^A = \frac{P_{d_1/D_2} h_{21}}{P_A} \tag{17}$$

$$P_{2/2}^A = \frac{P_{d_1/d_2} h_{22}}{P_A},$$

where P_A denotes disease prevalence.

Now, we calculate the gametic LD coefficient and the nongametic digenic disequilibrium coefficient in the disease population under a general two-locus disease model. The gametic LD coefficient in the disease population is defined as

$$\delta_{D_1 D_2}^A = P_{11}^A - P_{D_1}^A P_{D_2}^A,$$

and the nongametic digenic disequilibrium coefficient in the disease population is defined as

$$\delta_{D_1/D_2}^A = P_{1/1}^A - P_{D_1}^A P_{D_2}^A.$$

We can show that these can be given by (see Appendix D)

$$\delta^A_{D_1 D_2} = \frac{\delta_{D_1 D_2}}{P_A} h_{11} + \frac{P_{D_1} P_{D_2}}{P_A} \left(h_{11} - \frac{h_{D_1} h_{D_2}}{P_A} \right) \tag{18}$$

and

$$\delta^A_{D_1/D_2} = \frac{\delta_{D_1/D_2}}{P_A} h_{1/1} + \frac{P_{D_1} P_{D_2}}{P_A} \left(h_{1/1} - \frac{h_{D_1} h_{D_2}}{P_A} \right), \tag{19}$$

where $h_{D_1} = P(\textit{Affected} \mid D_1)$ and $h_{D_2} = P(\textit{Affected} \mid D_2)$.

We define the measure of interaction between two loci as

$$I = h_{11} - \frac{h_{D_1} h_{D_2}}{P_A} + h_{1/1} - \frac{h_{D_1} h_{D_2}}{P_A}, \tag{20}$$

which quantifies the magnitude of interaction.

It follows from Eqs. (18)–(20) that the composite measure of LD, $\Delta^A_{D_1 D_2}$,[22] in the disease population is given by

$$\Delta^A_{D_1 D_2} = \delta^A_{D_1 D_2} + \delta^A_{D_1/D_2} = \frac{\delta_{D_1 D_2}}{P_A} h_{11} + \frac{\delta_{D_1/D_2}}{P_A} h_{1/1} + \frac{P_{D_1} P_{D_2}}{P_A} I. \tag{21}$$

The absence of interaction between two loci is then defined as

$$h_{11} = \frac{h_{D_1} h_{D_2}}{P_A} \quad \text{or} \quad \frac{h_{11}}{P_A} = \frac{h_{D_1}}{P_A} \frac{h_{D_2}}{P_A} \tag{22(a)}$$

and

$$h_{1/1} = \frac{h_{D_1} h_{D_2}}{P_A} \quad \text{or} \quad \frac{h_{1/1}}{P_A} = \frac{h_{D_1}}{P_A} \frac{h_{D_2}}{P_A}. \tag{22(b)}$$

Equations [22(a)] and [22(b)] indicate that, similar to linkage equilibrium where the frequency of a haplotype is equal to the product of the frequencies of the component alleles of the haplotype, absence of interaction between two loci implies that the proportion of individuals carrying two alleles (either in the same chromosome or in different chromosomes) in the disease population is equal to the product of the proportion of individuals carrying a single allele in the disease population if we assume that the disease is caused by only two investigated disease loci. In other words, interaction between two disease susceptibility loci occurs when the contribution of one locus to the disease depends on another locus. In contrast to the additive model for interaction introduced by Fisher,[5] the interaction model defined by Eqs. (20), [22(a)], and [22(b)] are referred to as a multiplicative interaction model.

If we assume that two loci are in linkage equilibrium or unlinked in the general population, then — under the multiplicative disease model and in the absence of interaction — two loci in the disease population will be in linkage equilibrium. However, from Eqs. (18) and (19), we can see that when interaction between two loci exists, even if the two loci are in linkage equilibrium in the general population, the two loci will be in LD in the disease population. LD in the disease population is created by the interaction between two unlinked loci. This provides the basis for testing interaction between two unlinked loci, as shown in the section below.

4.2. *Indirect Interaction between Two Unlinked Marker Loci*

In the previous subsection, we studied the interaction between two unlinked disease loci. Now, we consider two marker loci, each of which is in LD with either of two interacting loci. Although there is no physiological interaction between the two marker loci, if each marker locus is in LD with one of the two unlinked interacting loci, we can still observe LD between two unlinked marker loci in the disease population. Assume that marker M_1 is in LD with disease locus D_1 and that marker M_2 is in LD with disease locus D_2. Furthermore, we assume that the two disease loci D_1 and D_2 are unlinked. Let $\delta^A_{M_1 M_2}$ and $\delta^A_{M_1/M_2}$ be the gametic LD coefficient and the nongametic digenic disequilibrium coefficient between two marker loci in the disease population, respectively. We denote the composite measure of LD between two marker loci as $\Delta^A_{M_1 M_2}$. Let δ_i be the LD measure between marker M_i and disease locus $D_i (i = 1, 2)$ in the general population. Then, we can show that (see Appendix D)

$$\delta^A_{M_1 M_2} = \frac{\delta_1 \delta_2}{P_A P_{d_1} P_{d_2}} \left(h_{11} - \frac{h_{D_1} h_{D_2}}{P_A} \right) = \frac{\delta_1 \delta_2}{P_{D_1} P_{D_2} P_{d_1} P_{d_2}} \delta^A_{D_1 D_2} \qquad [23(a)]$$

$$\delta^A_{M_1/M_2} = \frac{\delta_1 \delta_2}{P_A P_{d_1} P_{d_2}} \left(h_{1/1} - \frac{h_{D_1} h_{D_2}}{P_A} \right) = \frac{\delta_1 \delta_2}{P_{D_1} P_{D_2} P_{d_1} P_{d_2}} \delta^A_{D_1/D_2} \qquad [23(b)]$$

or

$$
\begin{aligned}
\Delta^A_{M_1 M_2} &= \delta^A_{M_1 M_2} + \delta^A_{M_1/M_2} \\
&= \frac{\delta_1 \delta_2}{P_A P_{d_1} P_{d_2}} \left(h_{11} - \frac{h_{D_1} h_{D_2}}{P_A} + h_{1/1} - \frac{h_{D_1} h_{D_2}}{P_A} \right) \\
&= \frac{\delta_1 \delta_2}{P_{D_1} P_{D_2} P_{d_1} P_{d_2}} \Delta^A_{D_1 D_2}.
\end{aligned}
\qquad (24)
$$

It is clear that when the marker loci are the disease loci themselves, $\Delta^A_{M_1 M_2}$, $\delta^A_{M_1 M_2}$, and $\delta^A_{M_1/M_1}$ are reduced to $\Delta^A_{D_1 D_2}$, $\delta^A_{D_1 D_2}$, and $\delta^A_{D_1/D_2}$, respectively. Equation (24) can also be written in terms of the measure of interaction between two unlinked loci:

$$\Delta_{M_1 M_2}^A = \frac{\delta_1 \delta_2}{P_A P_{d_1} P_{d_2}} I. \tag{25}$$

Since $\delta_i \leq P_{D_i} P_{d_i}$, the absolute value of the LD measure between two unlinked marker loci in the disease population, e.g. the composite measure of LD between two marker loci $|\Delta_{M_1 M_2}^A|$, will be less than or equal to the absolute value of the composite measure of LD between two unlinked disease loci in the disease population.

Equation (24) shows that the composite measure of LD between unlinked marker loci in the disease population is proportional to the product of LD between each marker locus and its linked disease locus $\delta_1 \delta_2$. Since the criteria for tag SNP selection are based on only one pairwise LD between the marker and disease loci, the LD between tag SNPs and interacting loci may not be large enough to ensure that indirect interaction between two unlinked marker loci will be detected. Thus, if the interacting disease loci are not selected as tag SNPs, many loci with interactions will be missed. This will have profound implications on tag SNP selection.

4.3. *Test Statistic*

In the previous subsection, we showed that under the multiplicative disease model, interaction between unlinked loci will create LD. Intuitively, we can test the interaction by comparing the difference in the composite genotypic disequilibrium between two unlinked loci in the cases and controls. If we denote the estimators of the composite LD measures in the cases and controls as $\hat{\Delta}_A$ and $\hat{\Delta}_N$, respectively, then the test statistic can be defined as

$$T_I = \frac{(\hat{\Delta}_A - \hat{\Delta}_N)^2}{\mathrm{Var}(\hat{\Delta}_A) + \mathrm{Var}(\hat{\Delta}_N)}, \tag{26}$$

where

$$\hat{\Delta}_A = \hat{P}_{11}^A + \hat{P}_{1/1}^A - 2\hat{P}_{D_1}^A \hat{P}_{D_2}^A$$

$$\hat{\Delta}_N = \hat{P}_{11}^N + \hat{P}_{1/1}^N - 2\hat{P}_{D_1}^N \hat{P}_{D_2}^N$$

$$\mathrm{Var}(\hat{\Delta}_A) = \frac{1}{n_A}\left[(\hat{\pi}_{D_1}^A + \hat{\delta}_{D_1}^A)(\hat{\pi}_{D_2}^A + \hat{\delta}_{D_2}^A) + \frac{1}{2}\hat{\tau}_{D_1}^A \hat{\tau}_{D_2}^A \hat{\Delta}_A \right.$$

$$\left. + \hat{\tau}_{D_1}^A \hat{\delta}_{D_1 D_2 D_2}^A + \hat{\tau}_{D_2}^A \hat{\delta}_{D_1 D_1 D_2}^A + \hat{\Delta}_{D_1 D_1 D_2 D_2}^A \right]$$

$$\mathrm{Var}(\hat{\Delta}_N) = \frac{1}{n_G}\left[(\hat{\pi}_{D_1}^N + \hat{\delta}_{D_1}^N)(\hat{\pi}_{D_2}^N + \hat{\delta}_{D_2}^N) + \frac{1}{2}\hat{\tau}_{D_1}^N\hat{\tau}_{D_2}^N\hat{\Delta}_A\right.$$

$$\left. + \hat{\tau}_{D_1}^N\hat{\delta}_{D_1D_2D_2}^N + \hat{\tau}_{D_2}^N\hat{\delta}_{D_1D_1D_2}^N + \hat{\Delta}_{D_1D_1D_2D_2}^N\right]$$

$$\hat{\pi}_{D_1}^A = \hat{P}_{D_1}^A(1 - \hat{P}_{D_1}^A)$$

$$\hat{\pi}_{D_2}^A = \hat{P}_{D_2}^A(1 - \hat{P}_{D_2}^A)$$

$$\hat{\delta}_{D_1}^A = \hat{P}_{D_1D_1}^A - (\hat{P}_{D_1}^A)^2$$

$$\hat{\delta}_{D_2}^A = \hat{P}_{D_2D_2}^A - (\hat{P}_{D_2}^A)^2$$

$$\hat{\tau}_{D_1}^A = (1 - 2\hat{P}_{D_1}^A)$$

$$\hat{\tau}_{D_2}^A = (1 - 2\hat{P}_{D_2}^A)$$

$$\hat{\delta}_{D_1D_1D_2}^A = \hat{P}_{D_1D_1D_2}^A - \hat{P}_{D_1}^A\hat{\Delta}_A - \hat{P}_{D_2}\hat{\delta}_{D_1}^A - (\hat{P}_{D_1}^A)^2\hat{P}_{D_2}^A$$

$$\hat{\delta}_{D_1D_2D_2}^A = \hat{P}_{D_1D_2D_2}^A - \hat{P}_{D_2}^A\hat{\Delta}_A - \hat{P}_{D_1}\hat{\delta}_{D_2}^A - (\hat{P}_{D_2}^A)^2\hat{P}_{D_1}^A$$

$$\hat{\Delta}_{D_1D_1D_2D_2}^A = \hat{P}_{D_1D_1/D_2D_2}^A - 2\hat{P}_{D_1}^A\hat{\delta}_{D_1D_2D_2}^A - 2\hat{P}_{D_2}^A\hat{\delta}_{D_1D_1D_2} - 2\hat{P}_{D_1}^A\hat{P}_{D_2}^A\hat{\Delta}_A$$

$$- (\hat{\Delta}_A)^2 - (\hat{P}_{D_1}^A)^2\hat{\delta}_{D_2}^A - (\hat{P}_{D_2}^A)^2\hat{\delta}_{D_1}^A - \hat{\delta}_{D_1}^A\hat{\delta}_{D_2}^A - (\hat{P}_{D_1}^A\hat{P}_{D_2}^A)^2.$$

$\hat{\pi}_{D_1}^N$, $\hat{\pi}_{D_2}^N$, $\hat{\tau}_{D_1}^N$, $\hat{\tau}_{D_2}^N$, $\hat{\delta}_{D_1}^N$, $\hat{\delta}_{D_2}^N$, $\hat{\delta}_{D_1D_2D_2}^N$, $\hat{\delta}_{D_1D_1D_2}^N$, and $\hat{\Delta}_{D_1D_1D_2D_2}^N$ are similarly defined for controls; P_{11}^A, $P_{1/1}^A$, $P_{D_1}^A$, $P_{D_2}^A$, P_{11}^N, $P_{1/1}^N$, $P_{D_1}^N$, and $P_{D_2}^N$ are defined as before; $\hat{P}_{11}^A$, $\hat{P}_{1/1}^A$, $\hat{P}_{D_1}^A$, $\hat{P}_{D_2}^A$, $\hat{P}_{11}^N$, $\hat{P}_{1/1}^N$, $\hat{P}_{D_1}^N$, and $\hat{P}_{D_2}^N$ are their estimators; the quantities n_A and n_G denote the number of sampled individuals in the cases and the controls, respectively; and the variance of the composite LD measure is the large-sample variance.[25] If we assume genotypic linkage equilibrium between two loci in both cases and controls, then the variances of the composite LD measures between two loci in the cases and controls are reduced to

$$\mathrm{Var}(\hat{\Delta}_A) = \frac{1}{n_A}(\hat{\pi}_{D_1}^A + \hat{\delta}_{D_1}^A)(\hat{\pi}_{D_2}^A + \hat{\delta}_{D_2}^A)$$

$$\mathrm{Var}(\hat{\Delta}_N) = \frac{1}{n_G}(\hat{\pi}_{D_1}^N + \hat{\delta}_{D_1}^N)(\hat{\pi}_{D_2}^N + \hat{\delta}_{D_2}^N).$$

When the sample size is large enough to ensure the application of the large sample theory, the test statistic T_I is asymptotically distributed as a central $\chi_{(1)}^2$ distribution under the null hypothesis of no interaction between two unlinked loci.

In theory, when there is no interaction between two unlinked loci, the LD between them should be zero. Thus, we can use a case-only design to study the interaction between two loci. In this case, Eq. (26) is reduced to

$$T_{\mathrm{I}} = \frac{(\hat{\Delta}^A)^2}{V(\hat{\Delta}_A)}. \tag{27}$$

However, in practice, background LD between two unlinked loci may exist in the population due to many unknown factors. Therefore, using Eq. (27) to test for interaction will increase type 1 error rates. The test statistic defined in Eq. (26) is more robust than that in Eq. (27).

4.4. *Null Distribution of Test Statistics*

In the previous sections, we have shown that when the sample size is large enough to ensure large sample theory, the distribution of the statistic T_{I} for testing the interaction between two loci under the null hypothesis of no interaction between two loci and absence of Hardy–Weinberg disequilibrium in both general and disease populations is asymptotically a central $\chi^2_{(1)}$ distribution. To examine the validity of this statement, we performed a series of simulation studies. The computer program SNaP[25] was used to generate two-locus genotype data of the sample individuals. A total of 10 000 individuals, who were equally divided into cases and controls, were generated from the general population, assuming Hardy–Weinberg equilibrium and genotypic equilibrium between two loci. Ten thousand simulations were repeated.

To investigate the null of distribution of the test statistic, we plot Figs. 7(a) and 7(b) showing the histograms of the test statistic T_{I} for testing the interaction between two loci with sample sizes $n_A = n_G = 200$ and $n_A = n_G = 300$, respectively, where n_A and n_G are the number of sampled individuals in the cases and the controls. It can be seen that the distributions of the test statistic T_{I} are similar to the theoretical central $\chi^2_{(1)}$ distribution.

We randomly sampled 100–400 individuals from each of the cases and controls for calculation of the type I error rates. Table 6 shows that the estimated type I error rates of the statistic T_{I} for testing interaction between two unlinked loci were not appreciably different from the nominal levels $\alpha = 0.05$, $\alpha = 0.01$, and $\alpha = 0.001$.

4.5. *Application to a Real Data Example*

To evaluate its performance for detecting interaction between two unlinked loci, the proposed test statistic T_{I} was applied to two real data sets. The

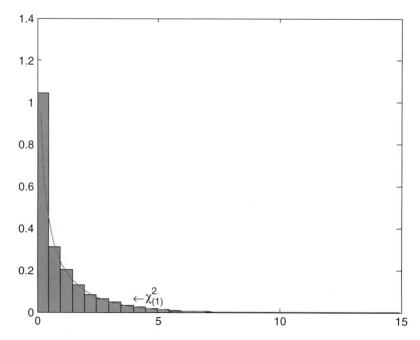

Fig. 7(a).   Null distribution of the test statistic T_{I} using 200 individuals in both cases and controls in a homogeneous population.

Table 6   Type 1 Error Rates of the Test Statistic T_{I} Using Composite Genotypic Disequilibrium Measure for Testing Interaction between Two Unlinked Loci in a Homogeneous Population

Sample Size	Nominal Levels		
	$\alpha = 0.05$	$\alpha = 0.01$	$\alpha = 0.001$
100	0.0545	0.0133	0.0012
150	0.0509	0.0105	0.0011
200	0.0511	0.0112	0.0013
250	0.048	0.0101	0.0008
300	0.053	0.01	0.0011
350	0.051	0.011	0.001
400	0.0472	0.0091	0.0012

first data set is a breast cancer case-control study. A total of 398 Caucasian breast cancer cases and 372 matched controls were sampled from the Ontario Familial Breast Cancer Registry (OFBCR).[26] A total of 19 SNPs from 18 key genes in DNA repair, cell cycle, carcinogen/estrogen

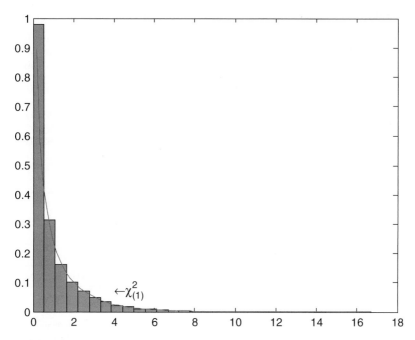

Fig. 7(b). Null distribution of the test statistic T_I using 300 individuals in both cases and controls in a homogeneous population.

metabolism, and immune system were typed. All SNPs were in Hardy–Weinberg equilibrium. Using multivariate logistic analysis under the codominant models, four pairs of genes — *XPD* and *IL10*, *GSTP1* and *COMT*, *COMT* and *CCND1*, and *BARD1* and *XPD* — showed significant interactions.[24] We used the statistic T_I to test interactions between these four pairs of genes. The test results are summarized in Table 7, where the crude P values are taken from Table 4 in Onay *et al.*'s paper.[26] The crude P values were obtained from multivariate logistic regression analysis that included all main effects and only the interaction of interest under the codominant models. Table 7 shows that, in general, the test statistics based on haplotype LD had smaller p values than those of both logistic regressions and test statistics based on genotypic disequilibrium.

However, the statistics based on composite genotypic disequilibrium do not always have smaller P values than logistic regressions. For interactions between *XPD*-[Lys751Gln] and *IL10*-[G(-1082)A], and between *GSTP1*-[Ile105Val] and *COMT*-[Met108/158Val], the P values of the test statistic based on composite genotypic disequilibrium were smaller than those of

Table 7 *P* values for Testing Interactions of Four Pairs of Genes

Interactions between SNPs	Crude *P* Values (Logistic Regression)	*P* Values by T_I (Haplotype LD Analysis)	*P* Values by T_I (Genotype LD Analysis)
XPD-[Lys751Gln] and *IL10*-[G(-1082)A]	0.035	0.00027	0.00531
BARD1-[Pro24Ser] and *XPD*-[Lys751Gln]	0.024	0.00684	0.68944
COMT-[Met108/158Val] and *CCND1*-[Pro241Pro]	0.010	0.00395	0.99468
GSTP1-[Ile105Val] and *COMT*-[Met108/158Val]	0.036	0.0000115	0.00599

Table 8 *P* values for Testing Interaction between Two Unlinked Loci and Association of SNP pairs with Atherosclerosis

SNP1	SNP2	*P* Values		
		T_I (Haplotype)	T_I (Genotype)	Logistic Regression
rs3211913	rs2357924	1.71E-09	0.0006493	0.0067
rs3211883	rs2300371	7.18E-09	0.0002353	0.2262
rs3211883	rs4234903	3.94E-08	0.0005092	0.9880
rs3211851	rs4987361	7.59E-08	0.0309600	0.2177
rs13438526	rs4135208	5.21E-08	0.0038500	0.0220
rs9332592	rs3093261	1.31E-07	0.0003845	0.0094
rs2298909	rs6092	1.51E-07	0.0006936	0.0323
rs2877262	rs12524494	6.26E-07	0.0003369	0.1726
rs3212391	rs9374312	5.23E-10	0.0001381	0.8477
rs27377	rs9611113	1.28E-09	0.0005098	0.6807
rs6842241	rs505295	1.49E-09	0.00002783	0.2985
rs12256760	rs3917698	8.13E-08	0.0001680	0.4871
rs3917368	rs2274755	2.98E-07	0.002323	0.0932
rs11009157	rs6743724	3.20E-07	0.0002365	0.2256

logistic regression; but for other two pairwise interactions, the *P* values of the test statistic based on composite disequilibrium were larger than those of logistic regressions.

The second data set is an atherosclerosis case-control data set, which was discussed in Sec. 3. It includes 27 SNPs from 22 candidate genes in the inflammatory, antioxidant, and coagulation pathways typed in 916 samples (492 cases and 424 controls) from a Chinese population. We statistically tested the interactions between all possible pairs of 27 SNPs using test statistics based on both haplotype LD and composite genotypic disequilibrium, and using logistics regression, where all of the SNP pairs were in different chromosomes. The results are summarized in Table 8. Table 8 shows that, although P values for test statistics based on the composite genotypic disequilibrium measure were larger than those based on haplotype LD, they were still much smaller than those based on logistic regression.

5. CONCLUSIONS

Understanding how genomic information underlies the development of disease is one of the greatest challenges in the 21st century. In the past several decades, genetic studies of human disease have focused on a "locus-by-locus" paradigm.[36] However, biological information is processed in complex networks. The disease emerges as the result of interactions between genes and between genes and environments. Studying one individual gene or polymorphism at a time to explore the cause of the disease while ignoring the interaction between loci (genes) is highly unlikely to deeply unravel the mechanism of disease. Now, the question is how to detect the interaction between genes (loci). Therefore, with the completion of the International HapMap Project, developing statistical methods for the detection of interaction between loci is of great interest. The purpose of this chapter is to present a novel statistical framework for identifying interaction between two loci.

Association studies rely heavily on the LD pattern between pairs of loci. Knowledge of the difference in LD between the disease and general populations is essential for understanding the interaction between two loci and their association with the disease. However, little is known about how the multiple-locus disease models influence the pattern of LD in the disease population, and how the interaction between two functional SNPs generates LD in the disease population. Therefore, before presenting the statistics for the detection of interaction between two loci, we first need to develop the general theory to study LD patterns between two genes in either the same gamete or different gametes in the disease population under two-locus disease models.

We introduced the concept of the penetrance of two genes in the same gamete or in different gametes, which in turn leads to the development of another new concept in the measure of interaction between two loci. It is surprising that the formula for the calculation of the measure of interaction between two loci is very similar to the formula for the calculation of the measure of LD between two loci. The proposed measure of interaction between two loci characterizes the contribution of the interaction between two loci to the cause of disease. We investigated how the two-locus disease models and the population parameters affect the measure of interaction between two loci. Intuitively, the interaction between two loci indicates joint action of two loci in the development of disease. This implies that some haplotypes spanned by the interacting loci will be presented in the disease population more often than expected. In other words, the interaction between two loci will generate LD in the disease population, and the level of haplotype or genotypic LD generated by the interaction depends on the magnitude of the interaction between the two loci. We rigorously proved that the measure of LD between two loci generated by their interaction is proportional to the measure of interaction between them. This motivated us to propose statistics for testing the interaction between two unlinked loci by comparing the difference in haplotype or genotypic LD between the disease and general populations.

To use the proposed LD-based statistic T_I for testing the interaction between two unlinked loci, we first need to study the distribution of the test statistics under the null hypothesis of no interaction. Through extensive simulation (under the assumption of large sample theory), we showed that the distributions of the proposed LD-based statistics are close to a central $\chi^2_{(1)}$ distribution, even for small sample sizes. To validate the test statistics and estimate the false-positive rates, we calculated the type 1 error rates of the LD-based statistics by simulation; the results showed that type 1 error rates were close to the nominal significance levels. We studied the power of the LD-based statistics for the detection of interaction by analytic analysis; it turned out that the power of the LD-based statistics is a function of the measure of interaction. This implies that the proposed LD-based statistics can indeed test the interaction between two unlinked loci. However, how the pattern of the power depends on the measure of interaction is complex. In addition to the measure of interaction, the power of the tests also depends on the allele frequencies. When the measure of interaction is beyond certain ranges, the power is no longer an increasing function of the measure of interaction (data not shown).

To further evaluate its performance for the detection of interaction between two loci, the proposed LD-based statistics were applied to three data sets. The results showed that the LD-based statistics detected more pairwise interactions than the logistic regressions, and that the haplotype LD-based statistic had higher power to detect interaction than the composite genotypic disequilibrium–based statistic. The results also strongly demonstrated that neither of two unlinked individual loci showed any association with diseases, but the test showed significant evidence of the presence of interaction between two loci. These results were confirmed by association analysis and explained by molecular experiments. In the past years, increasingly detailed and comprehensive evidence shows that genetic and molecular interactions govern cell behavior such as cell division, differentiation, and death, and are primary factors for the development of diseases. In many cases, single-locus analysis may fail to unravel the mechanism of the disease. The locus-by-locus paradigm for the genetic study of complex diseases should be shifted to a network paradigm.

The results in this chapter are preliminary. The interaction between two linked loci or the high-order interactions among multiple loci have not been fully studied. Gene–gene interaction is an important but complex concept. There are a number of ways to define gene–gene interaction. How the definition of gene–gene interaction at the population level reflects its biochemical or physiological interaction is still a mystery. Hopefully, this work will further motivate theoretic research in deciphering the genetic and physiological meaning of gene–gene interactions as well as the development of more statistical methods for testing gene–gene interaction. In the coming years, integrating gene–gene interaction into genomewide association analysis will be a major task in the genetic studies of complex diseases.

Appendix A

By definition, we have

$$
\begin{aligned}
P_{11}^A &= P(H_{D_1 D_2} | \textit{Affected}) \\
&= \frac{P(H_{D_1 D_2}, \textit{Affected})}{P_A} \\
&= \frac{P_{11} h_1^1}{P_A}.
\end{aligned}
\tag{A1}
$$

Similarly, we can obtain the remaining formulas in Eq. (1) in the text.

By definition, the measure of LD in the disease population is given by

$$
\begin{aligned}
\delta^A &= P_{11}^A P_{22}^A - P_{12}^A P_{21}^A \\
&= \frac{P_{11} h_{11}}{P_A} \frac{P_{22} h_{22}}{P_A} - \frac{P_{12} h_{12}}{P_A} \frac{P_{21} h_{21}}{P_A} \\
&= \frac{P_{11} P_{22} h_{11} h_{22} - P_{12} P_{21} h_{12} h_{21}}{P_A^2}.
\end{aligned}
\tag{A2}
$$

If we assume $h_{11}h_{22} = h_{12}h_{21}$, then from Eq. (A2), we obtain

$$\delta^A = \frac{h_{11}h_{22}(P_{11}P_{22} - P_{12}P_{21})}{P_A^2}$$

$$= \frac{h_{11}h_{22}\delta}{P_A^2}$$

$$= \frac{P_{11}^A P_{22}^A}{P_{11} P_{22}}\delta,$$

which implies Eq. (3) in the text.

Appendix B

We show that the equation $h_{11}h_{22} = h_{12}h_{21}$ implies

$$r_{11}r_{22} = r_{12}r_{21}$$

$$g_1^1 r_{22} + g_2^2 r_{11} = g_2^1 r_{21} + g_1^2 r_{12}.$$

By definition, we have

$$h_{11}h_{22} = (g_1^1 + r_{11}\delta)(g_2^2 + r_{22}\delta)$$

$$= g_1^1 g_2^2 + (g_1^1 r_{22} + g_2^2 r_{11})\delta + r_{11}r_{22}\delta^2.$$

Similarly, we have

$$h_{12}h_{21} = g_2^1 g_1^2 + (g_2^1 r_{21} + g_1^2 r_{12})\delta + r_{12}r_{21}\delta^2,$$

which implies that

$$h_{11}h_{22} - h_{12}h_{21} = g_1^1 g_2^2 - g_2^1 g_1^2 + (g_1^1 r_{22} + g_2^2 r_{11} - g_2^1 r_{21} - g_1^2 r_{12})\delta$$

$$+ (r_{11}r_{22} - r_{12}r_{21})\delta^2.$$

The equation $h_{11}h_{22} = h_{12}h_{21}$ requires that

$$g_1^1 g_2^2 - g_2^1 g_1^2 + (g_1^1 r_{22} + g_2^2 r_{11} - g_2^1 r_{21} - g_1^2 r_{12})\delta + (r_{11}r_{22} - r_{12}r_{21})\delta^2 = 0$$

holds for any δ. This will lead to

$$g_1^1 g_2^2 = g_2^1 g_1^2$$

$$r_{11}r_{22} = r_{12}r_{21}$$

$$g_1^1 r_{22} + g_2^2 r_{11} = g_2^1 r_{21} + g_1^2 r_{12}.$$

From Eq. (2) in the text, we obtain

$$\delta^A = \frac{h_{11}h_{22}\delta}{P_A^2}$$
$$= \frac{(g_1^1 + r_{11}\delta)(g_2^2 + r_{22}\delta)}{P_A^2}$$
$$= \frac{g_1^1 g_2^2}{P_A^2} + \frac{g_1^1 r_{22} + g_2^2 r_{11}}{P_A^2}\delta + \frac{r_{11}r_{22}}{P_A^2}\delta^2,$$

which proves Eq. (8) in the text.

Appendix C

First, we calculate the contribution of the additive effects in the classical genetic models to the newly defined measure of interaction.

Recall that the penetrance of haplotype H_{11} is given by

$$h_{11} = P_{11}f_{1111} + P_{12}f_{1112} + P_{21}f_{1211} + P_{22}f_{1212}. \qquad (C1)$$

Substituting the additive effects in Eq. (9) into Eq. (C1), we have

$$h_{11} \approx P_{11}(2\alpha_1^A + 2\alpha_1^B) + P_{12}(2\alpha_1^A + \alpha_1^B + \alpha_2^B) + P_{21}(\alpha_1^A + \alpha_2^A + 2\alpha_1^B)$$
$$+ P_{22}(\alpha_1^A + \alpha_2^A + \alpha_1^B + \alpha_2^B)$$
$$= (1 + P_{D_1})\alpha_1^A + P_{d_1}\alpha_2^A + (1 + P_{D_2})\alpha_1^B + P_{d_2}\alpha_2^B.$$

Similarly, we have

$$h_{22} \approx P_{D_1}\alpha_1^A + (1 + P_{d_1})\alpha_2^A + P_{D_2}\alpha_1^B + (1 + P_{d_2})\alpha_2^B$$
$$h_{12} \approx (1 + P_{D_1})\alpha_1^A + P_{d_1}\alpha_2^A + P_{D_2}\alpha_1^B + (1 + P_{d_2})\alpha_2^B$$
$$h_{21} \approx P_{D_1}\alpha_1^A + (1 + P_{d_1})\alpha_2^A + (1 + P_{D_2})\alpha_1^B + P_{d_2}\alpha_2^B.$$

Therefore, the contribution of the additive effect to the interaction measure, denoted by e_A, is given by

$$e_A \approx -(1 + P_{D_1})\alpha_1^A\alpha_1^B + (1 + P_{D_1})\alpha_1^A\alpha_2^B - P_{d_1}\alpha_2^A\alpha_1^B + P_{d_1}\alpha_2^A\alpha_2^B$$
$$+ P_{D_1}\alpha_1^A\alpha_1^B + (1 + P_{d_1})\alpha_2^A\alpha_1^B + 2\alpha_1^B\alpha_2^B - P_{D_1}\alpha_1^A\alpha_2^B$$
$$- (1 + P_{d_1})\alpha_2^A\alpha_2^B - 2\alpha_1^B\alpha_2^B$$
$$\approx -(\alpha_1^A - \alpha_2^A)(\alpha_1^B - \alpha_2^B).$$

Let $\alpha^A = \alpha_1^A - \alpha_2^A$ and $\alpha^B = \alpha_1^B - \alpha_2^B$ be the effect of an allele substitution at the first locus and the second locus, respectively. Then, the contribution of the additive effect to the interaction measure can be expressed as

$$e_A = -\alpha^A\alpha^B.$$

Now, we calculate the contribution of the genetic dominance effects to the interaction measure. Substituting the dominance deviation in Eq. (9) into Eq. (C1) yields

$$h_{11} \approx P_{D_1} d_{11}^A + P_{d_1} d_{12}^A + P_{D_2} d_{11}^B + P_{d_2} d_{12}^B$$
$$h_{22} \approx P_{D_1} d_{12}^A + P_{d_1} d_{22}^A + P_{D_2} d_{12}^B + P_{d_2} d_{22}^B$$
$$h_{12} \approx P_{D_1} d_{11}^A + P_{d_1} d_{12}^A + P_{D_2} d_{12}^B + P_{d_2} d_{22}^B$$
$$h_{21} \approx P_{D_1} d_{12}^A + P_{d_1} d_{22}^A + P_{D_2} d_{11}^B + P_{d_2} d_{12}^B.$$

The contribution of the dominant effects to the interaction measure, denoted by e_D, is given by

$$e_D \approx P_{D_1} P_{D_2} (d_{11}^A - d_{12}^A)(d_{12}^B - d_{11}^B) + P_{d_1} P_{D_2} (d_{12}^A - d_{22}^A)(d_{12}^B - d_{11}^B)$$
$$+ P_{D_1} P_{d_2} (d_{11}^A - d_{12}^A)(d_{22}^B - d_{12}^B) + P_{d_1} P_{d_2} (d_{12}^A - d_{22}^A)(d_{22}^B - d_{12}^B)$$
$$= -[P_{D_1} (d_{11}^A - d_{12}^A) + P_{d_1} (d_{12}^A - d_{22}^A)][P_{D_2} (d_{11}^B - d_{12}^B) + P_{d_2} (d_{12}^B - d_{22}^B)].$$

Next, we calculate the contribution of the genetic additive $\times$ additive effects to the interaction measure. Similarly, substituting the additive $\times$ additive effects in Eq. (9) into Eq. (C1), we obtain

$$h_{11} \approx P_{D_1} P_{D_2} [(\alpha\alpha)_{11} + (\alpha\alpha)_{11} + (\alpha\alpha)_{11} + (\alpha\alpha)_{11}]$$
$$+ 2P_{D_1} P_{d_2} [(\alpha\alpha)_{11} + (\alpha\alpha)_{12}] + 2P_{d_1} P_{D_2} [2(\alpha\alpha)_{11} + 2(\alpha\alpha)_{21}]$$
$$+ P_{d_1} P_{d_2} [(\alpha\alpha)_{11} + (\alpha\alpha)_{12} + (\alpha\alpha)_{21} + (\alpha\alpha)_{22}]$$
$$= (1 + P_{D_1})(1 + P_{D_2})(\alpha\alpha)_{11} + (1 + P_{D_1})P_{d_2}(\alpha\alpha)_{12}$$
$$+ P_{d_1}(1 + P_{D_2})(\alpha\alpha)_{21} + P_{d_1} P_{d_2}(\alpha\alpha)_{22}$$
$$h_{22} \approx P_{D_1} P_{D_2}(\alpha\alpha)_{11} + (P_{D_1} + P_{D_1} P_{d_2})(\alpha\alpha)_{12} + (P_{D_2} + P_{d_1} P_{D_2})(\alpha\alpha)_{21}$$
$$+ (1 + P_{d_1})(1 + P_{d_2})(\alpha\alpha)_{22}$$
$$h_{12} \approx (P_{D_1} P_{D_2} + P_{D_2})(\alpha\alpha)_{11} + (1 + P_{D_1})(1 + P_{d_2})(\alpha\alpha)_{12} + P_{d_1} P_{D_2}(\alpha\alpha)_{21}$$
$$+ (P_{d_1} + P_{d_1} P_{d_2})(\alpha\alpha)_{22}$$
$$h_{21} \approx (P_{D_1} P_{D_2} + P_{D_1})(\alpha\alpha)_{11} + P_{D_1} P_{d_2}(\alpha\alpha)_{12} + (1 + P_{d_1})(1 + P_{D_2})(\alpha\alpha)_{21}$$
$$+ (P_{d_2} + P_{d_1} P_{d_2})(\alpha\alpha)_{22}.$$

The contribution of the additive $\times$ additive effects to the interaction measure, denoted by e_{AA}, is given by

$$
\begin{aligned}
e_{\mathrm{AA}} = {} & 2P_{D_1}(1 + P_{D_1})(\alpha\alpha)_{11}(\alpha\alpha)_{12} + (4 + 2P_{D_1}P_{d_1})(\alpha\alpha)_{11}(\alpha\alpha)_{22} \\
& - 2P_{D_1}(1 + P_{D_1})(\alpha\alpha)_{12}(\alpha\alpha)_{11} - 2(1 + P_{D_1})(1 + P_{d_1})(\alpha\alpha)_{21}(\alpha\alpha)_{21} \\
& + 2P_{D_1}P_{d_1}(\alpha\alpha)_{21}(\alpha\alpha)_{12} + 2P_{d_1}(1 + P_{d_1})(\alpha\alpha)_{21}(\alpha\alpha)_{22} \\
& - 2P_{D_1}P_{d_1}(\alpha\alpha)_{22}(\alpha\alpha)_{11} - 2P_{d_1}(1 + P_{d_1})(\alpha\alpha)_{22}(\alpha\alpha)_{21} \\
= {} & 4[(\alpha\alpha)_{11}(\alpha\alpha)_{22} - (\alpha\alpha)_{12}(\alpha\alpha)_{21}].
\end{aligned}
$$

By the same arguments as above, we can derive the contribution of the additive $\times$ dominant and dominant $\times$ additive effects to the interaction measure, denoted by e_{AD} and e_{DA}, respectively:

$$
\begin{aligned}
e_{\mathrm{AD}} = {} & 2P_{D_2}^2[(\alpha d)_{111}(\alpha d)_{221} - (d\alpha)_{211}(\alpha d)_{121}] \\
& + 2P_{D_1}P_{d_2}[(\alpha d)_{111}(\alpha d)_{222} + (\alpha d)_{112}(dd)_{221} \\
& \quad - (\alpha d)_{211}(\alpha d)_{122} - (\alpha d)_{121}(\alpha d)_{212}] \\
& + 2P_{d_2}^2[(\alpha d)_{112}(\alpha d)_{222} - (\alpha d)_{122}(\alpha d)_{212}] \\
e_{\mathrm{DA}} = {} & 2P_{D_1}^2[(d\alpha)_{111}(d\alpha)_{212} - (d\alpha)_{112}(d\alpha)_{211}] \\
& + 2P_{D_1}P_{d_1}[(d\alpha)_{111}(d\alpha)_{222} + (d\alpha)_{121}(d\alpha)_{212} \\
& \quad - (d\alpha)_{112}(d\alpha)_{221} - (d\alpha)_{112}(d\alpha)_{211}] \\
& + 2P_{d_1}^2[(d\alpha)_{121}(d\alpha)_{222} - (d\alpha)_{122}(d\alpha)_{221}].
\end{aligned}
$$

Let

$$
\begin{aligned}
(dd)_{1.1.} = {} & P_{D_1}P_{D_2}(dd)_{1111} + P_{D_1}P_{d_2}(dd)_{1112} \\
& + P_{d_1}P_{D_2}(dd)_{1211} + P_{d_1}P_{d_2}(dd)_{1212} \\
(dd)_{2.2.} = {} & P_{D_1}P_{D_2}(dd)_{2121} + P_{D_1}P_{d_2}(dd)_{2122} \\
& + P_{d_1}P_{D_2}(dd)_{2221} + P_{d_1}P_{d_2}(dd)_{2222} \\
(dd)_{1.2.} = {} & P_{D_1}P_{D_2}(dd)_{1121} + P_{D_1}P_{d_2}(dd)_{1122} \\
& + P_{d_1}P_{D_2}(dd)_{1221} + P_{d_1}P_{d_2}(dd)_{1222} \\
(dd)_{2.1.} = {} & P_{D_1}P_{D_2}(dd)_{2111} + P_{D_1}P_{d_2}(dd)_{2112} \\
& + P_{d_1}P_{D_2}(dd)_{2211} + P_{d_1}P_{d_2}(dd)_{2212}.
\end{aligned}
$$

Then, the contribution of the dominant $\times$ dominant effect to the interaction measure, denoted by e_{DD}, is given by

$$
e_{\mathrm{DD}} = (dd)_{1.1.}(dd)_{2.2.} - (dd)_{1.2.}(dd)_{2.1.}
$$

Appendix D

Assume that the marker locus M_1 has two alleles M_1 and m_1, and the marker locus M_2 has two alleles M_2 and m_2. Let $q_{M_1}^A$ and $q_{M_2}^A$ be the frequencies of the marker alleles M_1 and M_2 in the disease population, respectively. Let the frequencies of the haplotypes D_1M_1, D_1m_1, d_1M_1, and d_1m_1 be $P_{D_1M_1}$, $P_{D_1m_1}$, $P_{d_1M_1}$, and $P_{d_1m_1}$, respectively. The frequencies of the haplotypes D_2M_2, D_2m_2, d_2M_2, and d_2m_2 can be similarly defined. Let the frequencies of the haplotypes M_1M_2, M_1m_2, m_1M_2, and m_1m_2 in the disease population be q_{11}^A, q_{12}^A, q_{21}^A, and q_{22}^A, respectively. Then, we have

$$
\begin{aligned}
q_{11}^A &= P(M_1M_2 \mid A) \\
&= \frac{P(M_1M_2, A)}{P_A} \\
&= \big(P_{D_1M_1}P_{D_2M_2}h_{11} + P_{D_1M_1}P_{d_2M_2}h_{12} + P_{d_1M_1}P_{D_2M_2}h_{21} \\
&\quad + P_{d_1M_1}P_{d_2M_2}h_{22}\big)/P_A \\
&= P_{M_1}P_{M_2} + \big[P_{M_2}(h_{D_1} - h_{d_1})\delta_1 + P_{M_1}(h_{D_2} - h_{d_2})\delta_2 \\
&\quad + (h_{11} - h_{12} - h_{21} + h_{22})\delta_1\delta_2\big]/P_A.
\end{aligned} \tag{D1}
$$

Similarly, we have

$$
\begin{aligned}
q_{M_1}^A &= P(M_1 \mid A) \\
&= \frac{P_{D_1M_1}h_{D_1} + P_{d_1M_1}h_{d_1}}{P_A} \\
&= P_{M_1} + \frac{h_{D_1} - h_d}{P_A}\delta_1 \\
q_{M_2}^A &= P_{M_2} + \frac{h_{D_2} - h_{D_1}}{P_A}\delta_2.
\end{aligned} \tag{D2}
$$

Note that

$$
\begin{aligned}
h_{D_1} &= \frac{P(D_1D_2, \textit{Affected}) + P(D_1d_2, \textit{Affected})}{P_{D_1}} \\
&= \frac{P_{D_1}P_{D_2}h_{11} + P_{D_1}P_{d_2}h_{12}}{P_{D_1}} \\
&= P_{D_2}h_{11} + P_{d_2}h_{12} \\
h_{D_2} &= P_{D_1}h_{11} + P_{d_1}h_{21} \\
h_{d_1} &= P_{D_2}h_{21} + p_{d_2}h_{22} \\
h_{d_2} &= P_{D_1}h_{12} + P_{d_1}h_{22}.
\end{aligned} \tag{D3}
$$

It follows from Eq. (D3) that

$$
\begin{aligned}
h_{D_1} h_{D_2} &= P_{D_1} P_{D_2} h_{11}^2 + P_{d_1} P_{D_2} h_{11} h_{21} + P_{D_1} P_{d_2} h_{11} h_{12} + P_{d_1} P_{d_2} h_{12} h_{21} \\
&= h_{11}(P_{D_1} P_{D_2} h_{11} + P_{D_2} P_{d_1} h_{21} + P_{D_1} P_{d_2} h_{12} + P_{d_1} P_{d_2} h_{22}) \\
&\quad + P_{d_1} P_{d_2}(h_{12} h_{21} - h_{11} h_{22}) \\
&= h_{11} P_A + P_{d_1} P_{d_2}(h_{12} h_{21} - h_{11} h_{22}).
\end{aligned}
$$

Similarly, we have

$$
\begin{aligned}
h_{D_1} h_{d_2} &= h_{12} P_A + P_{d_1} P_{D_2}(h_{11} h_{22} - h_{12} h_{21}) \\
h_{d_1} h_{D_2} &= h_{21} P_A + P_{D_1} P_{d_2}(h_{11} h_{22} - h_{12} h_{21}) \\
h_{d_1} h_{d_2} &= h_{22} P_A + P_{D_1} P_{D_2}(h_{12} h_{21} - h_{11} h_{22}).
\end{aligned}
\tag{D4}
$$

From Eqs. (D2)–(D4), we obtain

$$
q_{M_1}^A q_{M_2}^A = P_{M_1} P_{M_2} + \frac{P_{M_2}(h_{D_1} - h_{d_1})}{P_A} \delta_1 + \frac{P_{M_1}(h_{D_2} - h_{d_2})}{P_A} \delta_2
$$
$$
+ \left(\frac{h_{11} - h_{12} - h_{21} + h_{22}}{P_A} - \frac{h_{11} h_{22} - h_{12} h_{21}}{P_A^2} \right) \delta_1 \delta_2.
\tag{D5}
$$

Thus,

$$
\begin{aligned}
\delta_{M_1 M_2}^A &= q_{11}^A - q_{M_1}^A q_{M_2}^A \\
&= \frac{1}{P_A P_{d_1} P_{d_2}} \left(h_{11} - \frac{h_{D_1} h_{D_2}}{P_A} \right) \delta_1 \delta_2.
\end{aligned}
\tag{D6}
$$

Similarly, we have

$$
\begin{aligned}
\delta_{M_1/M_2}^A &= q_{1/1}^A - q_{M_1}^A q_{M_2}^A \\
&= \frac{1}{P_A P_{d_1} P_{d_2}} \left(h_{1/1} - \frac{h_{D_1} h_{D_2}}{P_A} \right) \delta_1 \delta_2.
\end{aligned}
\tag{D7}
$$

Combining Eqs. (D6) and (D7) yields

$$
\begin{aligned}
\Delta_{M_1 M_2}^A &= \frac{\delta_1 \delta_2}{P_A P_{d_1} P_{d_2}} I \\
&= \frac{\delta_1 \delta_2}{P_{D_1} P_{D_2} P_{d_1} P_{d_2}} \Delta_{D_1 D_2}^A.
\end{aligned}
$$

REFERENCES

1. Sing CF, Stengard JH, Kardia SL. (2003) Genes, environment, and cardio-vascular disease. *Arterioscler Thromb Vasc Biol* **23**:1190–1196.
2. Cook NR, Zee RY, Ridker PM. (2004) Tree and spline based associa-tion analysis of gene–gene interaction models for ischemic stroke. *Stat Med* **23**:1439–1453.

3. Hunter DJ. (2005) Choosing or losing health? *J Epidemiol Commun Health* **59**:1010–1012.

4. Hansen TF, Wagner GP. (2001) Modeling genetic architecture: a multilinear theory of gene interaction. *Theor Popul Biol* **59**:61–86.

5. Fisher RA. (1918) The correlation between relatives on the supposition of Mendelian inheritance. *Trans R Soc Edinb* **3**:399–433.

6. Cockerham CC. (1954) An extension of the concept of partitioning hereditary variance for analysis of covariances among relatives when epistatis is present. *Genetics* **39**:859–882.

7. Kempthorne O. (1954) The correlation between relatives in a random mating population. *Proc R Soc Lond B Biol Sci* **143**:103–113.

8. Hosmer DW, Lemeshow S. (2000) *Applied Logistic Regression*. John Wiley & Sons, New York, NY.

9. Cheverud JM, Routman EJ. (1995) Epistasis and its contribution to genetic variance components. *Genetics* **139**:1455–1461.

10. Kooperberg C, Ruczinski I, LeBlanc M, Hsu L. (2001) Sequence analysis using logic regression. *Genet Epidemiol* **21**:626–631.

11. Kooperberg C, Ruczinski I. (2005) Identifying interacting SNPs using Monte Carlo logic regression. *Genet Epidemiol* **28**:157–170.

12. Ruczinski I, Kooperberg C, LeBlanc M. (2003) Logic regression. *J Comput Graph Stat* **12**:475–511.

13. Wagner GP, Laubichler MD, Bagheri-Chaichian H. (1998) Genetic measurement theory of epistatic effects. *Genetica* **102/103**:569–580.

14. Nelson MR, Kardia SLR, Ferrell RE, Sing CF. (2001) A combinatorial partitioning method to identify multilocus genotype partitions that predict quantitative trait variation. *Genome Res* **11**:458–470.

15. Ritchie MD, Hahn LW, Roodi N, *et al.* (2001) Multifactor-dimensionality reduction reveals high-order interactions among estrogen-metabolism genes in sporadic breast cancer. *Am J Hum Genet* **69**:138–147.

16. Moore JH, Hahn LW. (2002) A cellular automata approach to detecting interactions among single-nucleotide polymorphisms in complex multifactorial diseases. *Pac Symp Biocomput* **7**:53–64.

17. Bastone L, Reilly M, Rader DJ, Foulkes AS. (2004) MDR and PRP: a comparison of methods for high-order genotype–phenotype associations. *Hum Hered* **58**:82–92.

18. Williams SM, Ritchie MD, Phillips JA 3rd, *et al.* (2004) Multilocus analysis of hypertension: a hierarchical approach. *Hum Hered* **57**:28–38.

19. Soares ML, Coelho T, Sousa A, *et al.* (2005) Susceptibility and modifier genes in Portuguese transthyretin V30M amyloid polyneuropathy: complexity in a single-gene disease. *Hum Mol Genet* **14**:543–553.

20. Foulkes AS, De Gruttola V, Hertogs K. (2004) Combining genotype groups and recursive partitioning: an application to human immunodeficiency virus type 1 genetics data. *Appl Stat* **53**:311–323.

21. Cho YM, Ritchie MD, Moore JH, *et al.* (2004) Multifactor-dimensionality reduction shows a two-locus interaction associated with type 2 diabetes mellitus. *Diabetologia* **47**:549–554.

22. Coffey CS, Hebert PR, Ritchie MD, *et al.* (2004) An application of conditional logistic regression and multifactor dimensionality reduction for detecting gene–gene interactions on risk of myocardial infarction: the importance of model validation. *BMC Bioinformatics* **5**:49.

23. Tsai CT, Lai LP, Lin JL, *et al.* (2004) Renin–angiotensin system gene polymorphisms and atrial fibrillation. *Circulation* **109**:1640–1646.

24. Zhao J, Jin L, Xiong MM. (2006) Test for interaction between two unlinked loci. *Am J Hum Genet* **79**:831–845.

25. Weir BS. (1996) *Genetic Data Analysis.* Sinauer Associates, Sunderland, MA.

26. Onay VU, Briollais L, Knight JA, *et al.* (2006) SNP–SNP interactions in breast cancer susceptibility. *BMC Cancer* **6**:114.

27. Lusis AJ, Mar R, Pajukanta P. (2004) Genetics of atherosclerosis. *Annu Rev Genomics Hum Genet* **5**:189–218.

28. Cordell HJ. (2002) Epistasis: what it means, what it doesn't mean, and statistical methods to detect it in humans. *Hum Mol Genet* **11**:2463–2468.

29. Moore JH, Williams SM. (2005) Traversing the conceptual divide between biological and statistical epistasis: systems biology and a more modern synthesis. *Bioessays* **27**:637–646.

30. Millstein J, Conti DV, Gilliland FD, Gauderman WJ. (2005) A testing framework for identifying susceptibility genes in the presence of epistasis. *Am J Hum Genet* **78**:15–27.

31. Argmann CA, Van Den Diepstraten CH, Sawyez CG, *et al.* (2001) Transforming growth factor-beta1 inhibits macrophage cholesteryl ester accumulation induced by native and oxidized VLDL remnants. *Arterioscler Thromb Vasc Biol* **21**:2011–2018.

32. Cheng D, Lee YC, Rogers JT, *et al.* (2000) Vascular endothelial growth factor level correlates with transforming growth factor-beta isoform levels in pleural effusions. *Chest* **118**:1747–1753.

33. Suwa T, Hogg JC, Quinlan KB, Van Eeden SF. (2002) The effect of interleukin-6 on L-selectin levels on polymorphonuclear leukocytes. *Am J Physiol Heart Circ Physiol* **283**:H879–H884.

34. Keidar S, Heinrich R, Kaplan M, *et al.* (2001) Angiotensin II administration to atherosclerotic mice increases macrophage uptake of oxidized LDL: a possible role for interleukin-6. *Arterioscler Thromb Vasc Biol* **21**: 1464–1469.

35. Salo T, Makela M, Kylmaniemi M, *et al.* (1994) Expression of matrix metalloproteinase-2 and -9 during early human wound healing. *Lab Invest* **70**:176–182.

36. Marchini J, Donnelly P, Cardon LR. (2005) Genome-wide strategies for detecting multiple loci that influence complex diseases. *Nat Genet* **37**: 413–417.

37. Gauderman WJ. (2002) Sample size requirements for association studies of gene–gene interaction. *Am J Epidemiol* **155**:478–484.

38. Wang XW, Vermeulen W, Coursen JD, *et al.* (1996) The XPB and XPD DNA helicases are components of the p53–mediated apoptosis pathway. *Genes Dev* **10**:1219–1232.

39. Fabbro M, Savage K, Hobson K, *et al.* (2004) BRCA1–BARD1 complexes are required for p53^{Ser-15} phosphorylation and a G_1/S arrest following ionizing radiation–induced DNA damage. *J Biol Chem* **279**:31251–31258.
40. Mitrunen K, Hirvonen A. (2003) Molecular epidemiology of sporadic breast cancer. The role of polymorphic genes involved in oestrogen biosynthesis and metabolism. *Mutat Res* **544**:9–41.
41. Lu F, Gladden AB, Diehl JA. (2003) An alternatively spliced cyclin D1 isoform, cyclin D1b, is a nuclear oncogene. *Cancer Res* **63**:7056–7061.

Chapter 10

Association Tests for Complex Disease Genes While Controlling Population Stratification

Shuanglin Zhang[*,†] and Qiuying Sha[*]

*Department of Mathematical Sciences, Michigan Technological University
Houghton, MI 49931, USA

†School of Mathematical Sciences, Heilongjiang University
Harbin 150080, China

Although genetic association studies using unrelated individuals may be subject to bias caused by population stratification, alternative family-based association methods that are robust to population stratification may be less powerful. This chapter discusses methods that use unrelated individuals to identify associations between candidate markers and traits of interest (both qualitative and quantitative), while controlling population stratification through a set of genomic markers of the same individual. These methods can control population stratification and are more powerful than family-based methods. We first introduce association tests based on population samples in a homogeneous population, and discuss why population stratification can cause false-positive results in association studies. Then, we review established methods, which include the genomic control, structure association, and semiparametric approaches, for controlling false-positive results by using a set of unlinked markers of the same individual instead of using family members in family-based association studies. Finally, we discuss some possible extensions of the existing methods and some topics that need further investigation.

---

†Correspondence author.

1. INTRODUCTION

The lack of tangible success in genetic linkage analysis for mapping complex trait loci and the rapid progress in the development of detailed SNP maps of the human genome have led to the suggestion of association studies.[1-3] Association studies can be performed by population- or family-based designs. Population-based designs have many advantages over family-based designs.[4-7] Population-based designs do not require the recruitment of additional family members, which can be expensive and may be impractical for late-onset diseases; and in most cases, population-based designs are more powerful than family-based designs with the same sample size.[5,8-10] Unfortunately, association tests based on population-based designs may be invalid in the presence of population heterogeneity, which can cause false-positive results.

The phenomenon that allele frequencies are different between cases and controls for a qualitative trait, or that trait values are correlated to genotypes for a quantitative trait due to systematic differences in ancestry rather than association of genes with the trait, is known as population stratification.[11-15] There has been much debate,[16-19] but limited data,[20-22] about the impact of population stratification on population-based association studies. The fraction of published associations that is attributed to population stratification is unknown.[23] It has been argued that the effects of stratification can be eliminated simply by carefully matching cases and controls according to self-reported ancestry and geographical origin.[17] However, recent studies[24,25] show that, even in well-designed studies, modest amounts of stratification can exist; and in large-scale studies that are needed to detect typical genetic effects in common diseases, the modest levels of population stratification cannot be safely ignored.

In response to the problem of population stratification, alternative methods utilizing family members as controls to obviate the effect of population stratification have recently become popular.[26-33] A major advantage of family-based designs is their robustness to population stratification; while a major disadvantage is that these designs mitigate against the recruitment of large samples, which in turn limit powers. When comparing population-based and family-based designs, each has its own advantages. Recently, there has been a series of papers that aimed at constructing valid association tests using population-based samples and a set of unlinked genetic markers typed in the same samples to control population stratification.[8-10,15,22,34-42] If the false-positive results caused by population stratification can be controlled, population-based designs will be more promising for mapping complex disease genes.

There are basically three approaches to test the association using population-based samples, while controlling the false-positive results. One is called genomic control (GC), which was proposed by Devlin and Roeder,[15] and extended by Bacanu and colleagues,[34] Devlin and colleagues,[36] Reich and Goldstein,[38] and Zheng and colleagues.[42] The second approach is called structured association (SA), which was developed by Zhang and colleagues,[8] Pritchard and colleagues,[9] Pritchard and Donnelly,[35] Satten and colleagues,[37] and Zhu and colleagues[41] for qualitative traits; by Zhang and Zhao[10] for quantitative traits; and by Hoggart and colleagues[22] for both kinds of traits. The third approach is called semiparametric test (SPT), which was developed by Chen and colleagues[39] for qualitative traits and by Zhang and colleagues[40] for quantitative traits.

All three approaches are based on the idea that demographic factors, such as population structure, are expected to have a similar effect on all loci across the genome. This suggests a strategy of using genotype data from a series of unlinked markers to assess the impact of population structure on the sample. In this article, we first describe the basic methods of association tests and why these methods will lead to false-positive results in the presence of population stratification. Then, we review the existing association tests that control population stratification. Finally, we discuss some possible extensions of the existing approaches and some topics that need further investigation.

2. ASSOCIATION TESTS IN A HOMOGENEOUS POPULATION

Consider two marker loci A and B with alleles $A_1, \ldots, A_L$ and alleles $B_1, \ldots, B_k$, respectively. In total, there are Lk haplotypes $A_1B_1, \ldots, A_1B_k, \ldots, A_LB_1, \ldots, A_LB_k$. If the occurrence of allele A_i at marker locus A and the occurrence of B_j at marker locus B in a haplotype are independent, that is,

$$\Pr(A_iB_j) = \Pr(A_i)\Pr(B_j)$$

for $i = 1, \ldots, L$ and $j = 1, \ldots, k$, then marker loci A and B are said to be in linkage equilibrium; otherwise, loci A and B are said to be in linkage disequilibrium (LD) or associated.

To test the association between a marker locus and a disease (or a trait) is to test the association between the marker locus and a diseased locus (both the location and the genotype at the disease locus are unknown). Consider a biallelic marker A ($L = 2$) with two alleles A_1 and A_2, and a

Table 1 Contingency Table for a Case-Control Study at a Marker Locus with L Alleles

	Number of Alleles				Total Number
	A_1	A_2	$\cdots$	A_L	of Alleles
Cases	n_1	n_2	$\cdots$	n_L	$2n$
Controls	m_L	m_2	$\cdots$	m_L	$2m$

disease locus or a quantitative trait locus (QTL) D with two alleles D_1 and D_2. Let

$$\Delta = \mathrm{Pr}(A_1 D_1) - \mathrm{Pr}(A_1)\mathrm{Pr}(D_1) = \mathrm{Pr}(A_2 D_2) - \mathrm{Pr}(A_2)\mathrm{Pr}(D_2)$$
$$= -[\mathrm{Pr}(A_1 D_2) - \mathrm{Pr}(A_1)\mathrm{Pr}(D_2)] = -[\mathrm{Pr}(A_2 D_1) - \mathrm{Pr}(A_2)\mathrm{Pr}(D_1)]. \quad (1)$$

Then, to test the association between marker locus A and disease locus D is to test $\Delta = 0$.

2.1. Qualitative Traits

For a qualitative trait, consider a case-control study with n cases and m controls. Each sampled individual has been genotyped at locus A with L alleles $A_1, \ldots, A_L$. The genotype data are summarized in Table 1.

Let $p = (p_1, \ldots, p_L)$ and $q = (q_1, \ldots, q_L)$ denote the allele frequencies of alleles $A_1, \ldots, A_L$ in the cases and controls, respectively. Under a case-control study, testing the association between marker A and a disease is equivalent to testing the hypothesis $H_0 : p = q$ vs. $H_a : p \neq q$. To see why testing the association is equivalent to testing $p = q$, consider a biallelic marker A $(L = 2)$. Let f_0, f_1, and f_2 denote penetrances:

$$f_2 = \mathrm{Pr}(Disease \mid D_1 D_1)$$
$$f_1 = \mathrm{Pr}(Disease \mid D_1 D_2)$$
$$f_0 = \mathrm{Pr}(Disease \mid D_2 D_2),$$

where $f_2 \geq f_1 \geq f_0$ and $f_2 \neq f_0$. We have

$$p_1 = \mathrm{Pr}(A_1 \mid Disease) = \mathrm{Pr}(A_1 D_1 \mid Disease) + \mathrm{Pr}(A_1 D_2 \mid Disease)$$
$$= \frac{\mathrm{Pr}(Disease \mid D_1)\mathrm{Pr}(A_1 D_1)}{\mathrm{Pr}(Disease)} + \frac{\mathrm{Pr}(Disease \mid D_2)\mathrm{Pr}(A_1 D_2)}{\mathrm{Pr}(Disease)}$$
$$= \mathrm{Pr}(D_1 \mid Disease)\frac{[\Delta + \mathrm{Pr}(A_1)\mathrm{Pr}(D_1)]}{\mathrm{Pr}(D_1)}$$

$$+ \Pr(D_2|Disease) \frac{[-\Delta + \Pr(A_1)\Pr(D_2)]}{\Pr(D_2)}$$

$$= \Pr(D_1|Disease) \frac{\Delta}{\Pr(D_1)} - \Pr(D_2|Disease) \frac{\Delta}{\Pr(D_2)} + \Pr(A_1). \qquad (2)$$

Similarly,

$$q_1 = \Pr(A_1|Normal) = \Pr(D_1|Normal) \frac{\Delta}{\Pr(D_1)}$$

$$- \Pr(D_2|Normal) \frac{\Delta}{\Pr(D_2)} + \Pr(A_1). \qquad (3)$$

From Eqs. (2) and (3), we have

$$p_1 - q_1 = [\Pr(D_1|Disease) - \Pr(D_1|Normal)] \frac{\Delta}{\Pr(D_1)}$$

$$+ [\Pr(D_2|Normal) - \Pr(D_2|Disease)] \frac{\Delta}{\Pr(D_2)}. \qquad (4)$$

Furthermore, after some algebraic calculations, we have

$$\Pr(D_1|Disease) - \Pr(D_1|Normal)$$

$$= \Pr(D_2|Normal) - \Pr(D_2|Disease)$$

$$= \frac{(f_2 - f_0)\Pr^2(D_1)\Pr^2(D_2) + 2(f_1 - f_0)\Pr(D_1)\Pr^3(D_2)}{\Pr(Disease)[1 - \Pr(Disease)]} > 0. \qquad (5)$$

It is easy to see from Eqs. (4) and (5) that $\Delta = 0$ is equivalent to $p_1 = q_1$ or $p = q$. In other words, testing the association between marker locus A and the disease is equivalent to testing $p = q$.

Let $X = (n_1, \dots, n_L)$ and $Y = (m_1, \dots, m_L)$, and then X and Y will follow multinomial distributions $M_L(2n, p)$ and $M_L(2m, q)$, respectively. The score test to test the hypothesis $H_0 : p = q$ vs. $H_a : p \neq q$ is given by

$$X^2 = \sum_{i=1}^{L} \frac{(\hat{p}_i - \hat{q}_i)^2}{\frac{\hat{p}_i}{2m} + \frac{\hat{q}_i}{2n}}, \qquad (6)$$

which asymptotically follows a χ^2 distribution with degrees of freedom $L-1$, where $\hat{p}_i = \frac{n_i}{2n}$ and $\hat{q}_i = \frac{m_i}{2m}$. When $m = n$, $X^2 = 2n \sum_{i=1}^{L} \frac{(\hat{p}_i - \hat{q}_i)^2}{\hat{p}_i + \hat{q}_i}$. When $m = n$ and $L = 2$,

$$X^2 = 4n \frac{(\hat{p}_1 - \hat{q}_1)^2}{(\hat{p}_1 + \hat{q}_1)[2 - (\hat{p}_1 + \hat{q}_1)]}.$$

2.2. *Quantitative Traits*

Consider a biallelic marker $A(L = 2)$, and let y denote the trait value. Using a similar argument as that for a qualitative trait, we can get that $\Delta = 0$ is equivalent to

$$E(y \mid A_1 A_1) = E(y \mid A_1 A_2) = E(y \mid A_2 A_2).$$

For the ith individual, let y_i and g_i denote the trait value and the genotype at marker A, respectively, and let

$$x_{i1} = \begin{cases} 1, & g_i = A_1 A_1 \\ 0, & g_i = A_1 A_2 \\ -1, & g_i = A_2 A_2 \end{cases} \quad \text{and} \quad x_{i2} = \begin{cases} 0, & g_i = A_1 A_1 \\ 1, & g_i = A_1 A_2 \\ 0, & g_i = A_2 A_2 \end{cases}$$

denote the additive and dominant genotype scores. Under the linear model

$$y_i = \alpha_0 + \alpha_1 x_{i1} + \alpha_2 x_{i2} + e_i, \tag{7}$$

testing the association between marker A and the quantitative trait is equivalent to testing the null hypothesis $H_0 : \alpha_1 = \alpha_2 = 0$. The least-squares (LS) estimators of α_i, denoted by $\hat{\alpha}_i$, are unbiased estimators of $\alpha_i (i = 1, 2)$. The standard F statistic can be used to identify the deviation from the null hypothesis.

3. REASONS FOR FALSE-POSITIVE RESULTS IN THE PRESENCE OF POPULATION STRATIFICATION

For a qualitative trait and a case-control design, consider the simple model of stratified population as described by Pritchard and Rosenberg.[14] Suppose each sampled individual is actually a member of one of two subpopulations, and we select individuals without regard to their origin. Consider a biallelic marker with two alleles A_1 and A_2. Let r_i denote the probability of sampling an individual from subpopulation i $(i = 1, 2)$, and g_i and a_i denote the frequency of disease and the frequency of allele A_1 in subpopulation i $(i = 1, 2)$, respectively. $P = g_1 r_1 + g_2 r_2$ is the population prevalence of the disease. The probability that an affected individual is from subpopulation i is

$$f_{di} = \frac{g_i r_i}{P},$$

and the probability that a healthy individual is from subpopulation i is

$$f_{hi} = \frac{(1 - g_i) r_i}{1 - P}.$$

If allele A_1 is statistically independent of the disease within each subpopulation, it follows that

$\Pr(A_1 \mid Disease) - \Pr(A_1 \mid Normal)$

$$= \sum_{i=1}^{2} \Pr(A_1 \mid \text{subpop } i)[\Pr(subpop \ i \mid Disease) - \Pr(subpop \ i \mid Normal)]$$

$$= (a_1 - a_2)(g_1 - g_2)\frac{r_1 r_2}{P(1 - P)}.$$

Hence, if both a_i and g_i are different in subpopulations, then $E[\hat{\Pr}(A_1 \mid Disease) - \hat{\Pr}(A_1 \mid Normal)] \neq 0$. In this case, when sample size n is large enough, the test statistic given in Eq. (6) will be large enough to lead to false-positive results.

For a quantitative trait, when allele frequencies and mean trait values differ among the subpopulations, the tests based on the linear model in Eq. (7) may also lead to false-positive results [see Zhang and Zhao[10] for details].

4. ASSOCIATION TESTS ROBUST TO POPULATION STRATIFICATION

As discussed in the previous section, the association tests used for a homogeneous population may lead to false-positive results in the presence of population stratification. Recently, several methods have been proposed to test association using population-based samples while controlling population stratification. There are basically three approaches: genomic control (GC), structured association (SA), and semiparametric test (SPT) approaches.

4.1. *GC Approach*

The GC approach to qualitative traits was proposed by Devlin and Roeder[15] and further discussed by Bacanu and colleagues,[34] Devlin and colleagues,[36] Reich and Goldstein,[38] and Zheng and colleagues.[42] Consider a case-control study. The genotype data for a candidate biallelic marker locus are shown in Table 1 ($L = 2$). Additional M unlinked biallelic markers are typed for each of the sampled individuals. In the presence of population stratification, the test statistic X^2 in Eq. (6) may not have a χ^2 distribution with one degree of freedom. The GC approach simply rescales the statistic by a multiplicative factor λ, and considers X^2/λ to have a chi-squared distribution with one degree of freedom. The unlinked marker genotype data can be used to provide an estimator of λ. Let $X_1^2, \ldots, X_M^2$ be the values for the X^2 statistic at M unlinked markers. Devlin and Roeder[15] proposed to use median $\{X_1^2, \ldots, X_M^2\}/0.456$, which is a robust estimator of λ derived from the properties of gamma distribution,[43] as an estimate of λ. Reich and Goldstein[38] proposed to use the mean of $X_1^2, \ldots, X_M^2$ to estimate λ.

The GC approach is computationally simple and allows for a large number of potential subgroups (i.e. it works well with very fine-scale substructures[15]). It can be undertaken with pooled DNA samples, which are substantially less expensive than individual genotyping. An investigation of power indicates that the GC approach can generally be more powerful than the the transmission–disequilibrium test (TDT).[34]

In the presence of population stratification, as discussed in the previous section, $E(\hat{p}_i - \hat{q}_i)$ may not be zero and X^2 — given by Eq. (6) — will asymptotically follow a noncentral chi-squared distribution $\chi^2(\delta)$.[39] If the noncentral parameter δ is small, a central chi-squared distribution adjusted by a suitable constant can be a good approximation of $\chi^2(\delta)$. However, if δ is large, using a central chi-squared distribution adjusted by a constant as an approximation of a noncentral chi-square $\chi^2(\delta)$ will lead to either false-positive results or loss of power, though we do not know yet how large δ will be in practice.

4.2. SA Approach

The SA approach uses the genotypes at a series of unlinked markers to infer details of the population structure; this information is then taken into account by association tests. This approach was first proposed by Pritchard and colleagues,[9] and was further developed by Zhang and colleagues,[8] Satten and colleagues,[37] and Zhu and colleagues[41] for qualitative traits; by Zhang and Zhao[10] for quantitative traits; and by Hoggart and colleagues[22] for both kinds of traits. There are two approaches to infer the population structure: one is the Bayesian approach, which uses Markov chain Monte Carlo (MCMC) to estimate the parameters[9,22]; and the other is the likelihood approach, based on mixture models.[8,10,37,41] We review some typical SA methods below.

4.2.1. STRuctured population Association Test (STRAT) (Method of Pritchard et al.[9])

Suppose we have a sample of cases and controls, each of which is genotyped at M unlinked markers. STRAT has two steps. In the first step, a Bayesian clustering method is used to determine both the number of subpopulations and the fraction of the sampled individual's ancestry in each subpopulation. Specifically, assume that sampled individuals have inherited their genes from a pool of K unstructured subpopulations (where K may be unknown). The allele frequencies at each locus, within each subpopulation, are assumed to be unknown. Let q_{ik} denote the proportion of the ith individual's genome, which originated from subpopulation k. Using the genotypes of n sampled individuals at M unlinked markers, Pritchard and colleagues[9] proposed an MCMC method to estimate the number of subpopulations K, $Q = \{q_{ik} : i = 1, \ldots, n; k = 1, \ldots, K\}$, and the allele frequencies at each locus within each subpopulation. The method can be applied to most of the commonly used genetic markers, including microsatellites and SNPs, and can produce accurate results using modest numbers of markers. The accuracy of the inference depends on the sample size, the number of markers used, and the magnitude of allele frequency differences between the subpopulations.

In the second step, a likelihood ratio test based on the detailed population structure is used to test the null hypothesis H_0 that subpopulation allele frequencies at the candidate locus are independent of phenotype against an alternative hypothesis H_1, where the subpopulation allele frequencies at the candidate locus depend on phenotype. Let G denote the list of genotypes of all sampled individuals at the candidate locus, and P_0 and P_1 denote subpopulation allele frequencies at the candidate locus under H_0 and H_1, respectively. The statistic is the likelihood ratio

$$\Lambda(G) = \frac{\Pr_1(G; \hat{P}_1, \hat{Q})}{\Pr_0(G; \hat{P}_0, \hat{Q})},$$

where $\Pr_0(G; P_0, Q)$ and $\Pr_1(G; P_1, Q)$ are the distributions of G under H_0 and H_1, respectively; and $\hat{P}_0$, $\hat{P}_1$, and $\hat{Q}$ are the estimates of P_0, P_1, and Q, respectively. The values of $\hat{P}_0$, $\hat{P}_1$, and $\hat{Q}$ can be obtained from the MCMC procedure in the first step. See Pritchard and colleagues[9] for the specific models of $\Pr_0(G; P_0, Q)$ and $\Pr_1(G; P_1, Q)$.

The p value of this test is evaluated by the following simulation procedure. Generate new genotypes at the candidate locus under H_0 for each individual as independent random draws from $\Pr_0(\cdot \mid \hat{Q}, \hat{P}_0)$. Repeat this procedure B times and get genotype data sets $G^{(1)}, \ldots, G^{(B)}$. The empirical p value is given by

$$p \text{ value} = \frac{1}{B} \#\{b : \Lambda(G^{(b)}) > \Lambda(G)\},$$

where $\#A$ denotes the number of members in set A. Simulation studies show that this method can control population stratification provided the number of unlinked markers is large enough, and that it is more powerful than the TDT in most cases. One of the difficult problems is the estimation of the number of subpopulations K, especially when there are a large number of potential subgroups.[9]

4.2.2. *Similarity-based Association Test (SAT)*
(Method of Zhang et al.[8])

This method also consists of two steps: inferring population structure, and constructing an association test based on the information of the inferred population structure. In the first step, in order to avoid the difficult problem of estimating the number of subpopulations, this method clusters the similarities between individuals into one or two groups instead of grouping individuals into several subpopulations.

Suppose we have a sample of n individuals, each of which is genotyped at M unlinked biallelic markers with two alleles A_m and a_m at the mth marker. Let z_{im} denote the numerical code (number of copies of allele A_m) of the ith individual at the mth marker, where $i = 1, \ldots, n$ and $m = 1, \ldots, M$. A natural measure of the difference in genotypes between the ith and the jth individuals is $d_{ij} = \sum_{m=1}^{M} |z_{im} - z_{jm}|$. The similarity S_{ij} between the ith and the jth individuals is defined as

$$S_{ij} = \frac{d_{\max} - d_{ij}}{d_{\max}},$$

where $d_{\max}$ is the maximum observed values of d_{ij} across all pairs of individuals. For individuals within the same subpopulation, we expect their similarities to be less than the similarities between individuals from different subpopulations. When the sampled individuals come from a structured population, we may cluster these similarities into two components: a within-subpopulation component and a between-subpopulation component.

To identify possible components among S_{ij}, the following normal mixture model for S_{ij} is proposed:

$$S_{ij} \sim \sum_{k=1}^{K} p_k N(S_{ij}, \mu_k, \sigma_k^2),$$

where K represents the number of components in the mixture model, p_k denotes the proportion of the kth component, and $N(s, \mu_k, \sigma_k^2)$ denotes the Gaussian density function with mean μ_k and variance σ_k^2. The maximum likelihood estimates of the parameters p_k, μ_k, and σ_k, for a given K, can be obtained using the clustering expectation-maximization (CEM) method.[44] Here, we only need to consider $K = 1$ or 2, where $K = 1$ represents the case of no population structure or a homogeneous population and $K = 2$ represents the case that the sampled individuals come from a structured population.

The Bayesian information criterion (BIC)

$$\text{BIC}(K) = -2L(K) + M(K)\log n$$

is used to choose $K = 1$ or $K = 2$, where $L(K) = \sum_{i,j} \log \left[\sum_{k=1}^{K} \hat{p}_k N(S_{ij}, \hat{\mu}_k, \hat{\sigma}_k^2) \right]$ is the maximized log likelihood for a given K, and $M(K)$ is the number of free parameters in the mixture model. When $K = 2$, let $\hat{p}_k$, $\hat{\mu}_k$, and $\hat{\sigma}_k$ denote the maximum likelihood estimates of the parameters p_k, μ_k, and σ_k, respectively. Then,

$$t_{ijk} = \frac{\hat{p}_k N(S_{ij}, \hat{\mu}_k, \hat{\sigma}_k^2)}{\hat{p}_1 N(S_{ij}, \hat{\mu}_1, \hat{\sigma}_1^2) + \hat{p}_2 N(S_{ij}, \hat{\mu}_2, \hat{\sigma}_2^2)}$$

is the conditional probability that S_{ij} arises from the kth mixture component. Assume that $\hat{\mu}_1 > \hat{\mu}_2$. Define the similarity indicator as $W_{ij} = 1$ if $t_{ij1} > 0.5$; and $W_{ij} = 0$, otherwise. If $W_{ij} = 1$, individuals i and j are considered to be in the same subpopulation. If $W_{ij} = 0$, individuals i and j are considered to be in different subpopulations.

To construct a test statistic for a case-control study with n_d cases and n_c controls ($n = n_d + n_c$), let $D_{ii'}$ denote the similarity indicator between affected individuals i and i', B_{ij} denote the similarity indicator between affected individual i and normal individual j, and $N_{jj'}$ denote the similarity indicator between normal individuals j and j'. Denote the two alleles at candidate locus by A and a. Let x_i denote the numerical code (number of copies of allele A) of the genotype of the ith affected individual, and y_j denote the numerical code of genotype of the jth normal individual at the candidate locus. Consider $m_{di} = \sum_{j=1}^{n_c} B_{ij}$, $m_{cj} = \sum_{i=1}^{n_d} B_{ij}$, $k_{di} = \sum_{i'=1}^{n_d} D_{ii'}$, and $k_{cj} = \sum_{j'=1}^{n_c} N_{jj'}$. It is easy to see that $m_{di}(k_{di})$ is the number of normal (affected) individuals in the same subpopulation as the ith affected individual, and that $m_{cj}(k_{cj})$ is the number of affected (normal) individuals in the same subpopulation as the jth normal individual. The statistic of SAT is given by

$$\text{SAT} = \frac{U_s}{\hat{\sigma}},$$

where $U_s = \sum_{i=1}^{n_d} x_i \sqrt{\frac{m_{di}(m_{di}+k_{di})}{k_{di}}} - \sum_{j=1}^{n_c} y_j \sqrt{\frac{m_{cj}(m_{cj}+k_{cj})}{k_{cj}}}$, $\hat{\sigma}^2 = \sum_{i=1}^{n}$ $\sum_{j=1}^{n} z_i (1 - \frac{z_j}{2}) W_{ij}$, and z_i (0, 1, or 2) denotes the numerical code of the genotype at the candidate locus for the ith individual in the sample without regard to the disease status. Under the null hypothesis, the SAT test statistic asymptotically follows the standard normal distribution.

To better understand the meaning of the statistic, we consider a structured population with two subpopulations. If we can correctly identify all pairwise relationships, then it is easy to show that

$$U_s = c_1(\hat{q}_{d_1} - \hat{q}_{c_1}) + c_2(\hat{q}_{d_2} - \hat{q}_{c_2}),$$

where c_1 and c_2 are constants, $\hat{q}_{d_1}$ is the allele A frequency among the affected individuals in the first subpopulation, $\hat{q}_{c_1}$ is the allele A frequency among the normal individuals within the first subpopulation, and $\hat{q}_{d_2}$ and $\hat{q}_{c_2}$ are similarly defined for the second subpopulation. Therefore, U_s is the weighted sum of the allele frequency differences between the affected individuals and the normal individuals within each subpopulation. Thus, $E(U_s) = 0$ under the null hypothesis.

Simulation results[8] show that SAT has a correct type I error rate in the presence of population stratification. The power of SAT is higher than that of family-based TDT, and is also higher than STRAT when the high-risk allele is the same across subpopulations.

4.2.3. *Quantitative Similarity-based Association Test (QSAT)* *(Method of Zhang and Zhao[10])*

The STRAT and SAT discussed above are used for qualitative traits, whereas QSAT was developed for quantitative traits. QSAT is also a two-step procedure. The first step is the same as the first step of SAT.

In the second step, let y_i, A_i, and D_i denote the trait value, additive genotypic score, and dominance genotypic score of the ith individual, respectively, where

$$A_i = \begin{cases} 1, & \text{if genotype is } AA \\ 0, & \text{if genotype is } Aa \\ -1, & \text{if genotype is } aa \end{cases} \text{ and } D_i = \begin{cases} 0, & \text{if genotype is } AA \\ 1, & \text{if genotype is } Aa \\ 0, & \text{if genotype is } aa. \end{cases} \quad (8)$$

Let W_{ij}, obtained from the first step, denote the similarity indicator between individuals i and j, and $n_i = \sum_{j=1}^{n} W_{ij}$. Then, n_i is the number of individuals estimated to be in the same subpopulation as the ith individual.

Using W_{ij}, we can decompose the additive genotypic score A_i into two components, a between-subpopulation component $\overline{A}_i = (\sum_{j=1}^{n} A_j W_{ij})/n_i$ and a within-subpopulation component $A_{wi} = A_i - \overline{A}_i$. Similarly, we can decompose the dominance genotypic score D_i into two components, a between-subpopulation component $\overline{D}_i = (\sum_{j=1}^{n} D_j W_{ij})/n_i$ and a within-subpopulation component $D_{wi} = D_i - \overline{D}_i$. Under the regression model

$$y_i = \mu + \alpha_b \overline{A}_i + \alpha_w A_{wi} + \beta_b \overline{D}_i + \beta_w D_{wi} + e_i, \quad i = 1, \ldots, n, \qquad (9)$$

we test the null hypothesis $H_0 : \alpha_w = \beta_w = 0$ vs. the alternative $H_1 : H_0$ is not true.

The QSAT statistic is the standard F test statistic under a linear model. If $e_1, \ldots, e_n$ are independent normal variables with the same variance, then the QSAT statistic follows an F distribution. In practice, $e_1, \ldots, e_n$ may not follow the normal distributions and may not have the same variance, especially for different genotypes and in different subpopulations. Zhang and Zhao[10] proposed to use a simulation method to evaluate the statistical significance. The basic idea of the simulation method is to permute the trait values of the individuals within the same subpopulation in order to derive an empirical distribution for the test statistic (see Zhang and Zhao[10] for details). Although we describe QSAT in the case of biallelic markers, it is straightforward to extend the method to the case where the markers may have more than two alleles.

4.3. Semiparametric Test (SPT)

Although simulation results have found that the SA methods discussed in previous sections generally perform well under discrete subpopulation models, they may not be effective when the population under study is an admixture of ancestral populations such as the African-American population. In this situation, the inference of population structure (the number of ancestral populations and/or the probability that each individual belongs to a certain subpopulation) is difficult because virtually everyone in the sample is admixed and there is little information about the ancestral populations.[9] Under an admixture population, the sampled individuals may be divided into many subpopulations, reducing the power of the test because of the small sample size within each subpopulation.[39,40]

To avoid the difficulty of clustering, Zhang and colleagues[40] proposed a semiparametric test (SPT) for qualitative traits, and Chen and colleagues[39] proposed a qualitative semiparametric test (QualSPT) for qualitative traits.

Both methods are derived by first estimating the genetic background variable value for each sampled individual using the principal components of many unlinked marker genotypes; and then modeling the relation of trait values, genotypic scores, and the genetic background variable through semiparametric models. We illustrate two SPT methods in the case of allelic markers. However, these two methods are also applicable to other markers with more than two alleles.

4.3.1. *Genetic Background Variable*

The genetic background variable plays an important role in both SPT and QualSPT. The role of the genetic background variable in SPT and QualSPT is similar to the role of the inferred population structure in the SA approach. Suppose n individuals are sampled, each of whom is genotyped at M unlinked markers. The genotypes at typed markers are all of the genetic background information that we know. Chen and colleagues[39] as well as Zhang and colleagues[40] proposed to use principal component analysis on a set of unlinked marker genotypes to summarize the genetic background information. The first principal component (or the first few principal components) is called the genetic background variable. More specifically, suppose all of the M unlinked markers are biallelic markers. Let x_{im} (0, 1, or 2) denote the numerical code of the genotype of individual i at marker m, and $X_i = (x_{i1}, \ldots, x_{iM})^T$. Let $\sum = \sum_{i=1}^{n}(X_i - \overline{X})(X_i - \overline{X})^T$ denote the variance–covariance matrix of $X_i, \ldots, X_n$; and let q, an M-dimensional vector, denote the eigenvector corresponding to the largest eigenvalue of $\sum$. Then, $t_i = q^T X_i$, the first principal component, is used to estimate the genetic background variable value for the ith individual.

4.3.2. *SPT Statistic*

Consider a candidate locus with two alleles. Let y_i, A_i, and D_i denote the quantitative trait value, additive genotypic score, and dominance genotypic score at the candidate locus of the ith individual, respectively. Zhang and colleagues[40] proposed to use the following semiparametric model — also known as partial linear model — to describe the relationship among trait value y_i, genotypic scores A_i and D_i at the candidate locus, and genetic background variable t_i:

$$y_i = \mu(t_i) + \alpha A_i + \beta D_i + e_i \quad (i = 1, \ldots, n), \tag{10}$$

where $\mu(\cdot)$ is an unknown smooth function of the genetic background variable; and $e_1, \ldots, e_n$ are independent of each other and of t_i, A_i, and D_i. The assumption that the function $\mu(\cdot)$ is smooth is based on the consideration that similar genetic backgrounds should lead to similar phenotypic means. This model is more general than the model given in Eq. (9) used by QSAT,[10] which generalizes the idea of decomposing the genotypic scores into two orthogonal components — a between-family (b) component and a within-family (w) component — in the context of family-based association studies.[16,28,31] Under the model in Eq. (10), to test association, we test the hypotheses

$$H_0 : \alpha = \beta = 0 \text{ vs. } H_1 : H_0 \text{ is not true.} \tag{11}$$

The test statistic is based on the estimators of α and β. Various statistical methods have been proposed to estimate α, β, and function $\mu(\cdot)$, including the penalized least squares method,[45–48] the kernel smoothing method,[49] and the local linear method.[50] These methods differ in how to estimate the nonparametric part $\mu(\cdot)$. Zhang and colleagues[40] used the kernel smoothing method proposed by Speckman[49] for computational simplicity and for the well-understood statistical properties of the estimates.

To estimate α, β, and the unknown function $\mu(\cdot)$, let $y_i^* = y_i - \alpha A_i - \beta D_i$. Then, Eq. (10) can be written as

$$y_i^* = \mu(t_i) + e_i. \tag{12}$$

Equation (12) is a standard nonparametric regression model, and the kernel estimator of $\mu(t)$ is given by

$$\hat{\mu}(t) = \sum_{i=1}^{n} w_i(t) y_i^*, \quad w_i(t) = \frac{K(\frac{t_i - t}{h})}{\sum_{j=1}^{n} K(\frac{t_j - t}{h})}, \tag{13}$$

where $K(\cdot)$ is a kernel function and h is the smoothing parameter. Replacing $\mu(t_i)$ in Eq. (10) with $\hat{\mu}(t_i)$, we have, after some simplicities,

$$\hat{y}_i = \alpha \hat{A}_i + \beta \hat{D}_i + e_i, \tag{14}$$

where $\hat{y}_i = y_i - \sum_{j=1}^{n} w_j(t_i) y_j$, $\hat{A}_i = A_i - \sum_{j=1}^{n} w_j(t_i) A_j$, and $\hat{D}_i = D_i - \sum_{j=1}^{n} w_j(t_i) D_j$. Note that $\hat{y}_i (\hat{A}_i$ and $\hat{D}_i)$ is the difference between the trait value (genetic scores) of the ith individual and the weighted local mean of those individuals who have a similar genetic background as the ith individual. This step aims to remove the population stratification effect. If the kernel function $K(\cdot)$ and smoothing parameter h are known, we can calculate $\hat{y}_i$, $\hat{A}_i$, and $\hat{D}_i$. There are many kernel functions available, such as the quadratic kernel[40,49]

$$K(t) = \begin{cases} \frac{15}{16}(1 - t^2)^2, & |t| \le 1 \\ 0, & |t| > 1. \end{cases}$$

We will discuss how to choose h later. Currently, we assume that the smoothing parameter is known. Thus, under the model in Eq. (14), a standard F test statistic can be used to test the hypotheses given by Eq. (11). This test is called SPT.

The statistical significance is evaluated by a simulation procedure. For each simulation, we permute $\hat{y}_1, \ldots, \hat{y}_n$ among the sampled individuals and recalculate the SPT test statistic based on the model in Eq. (14). We repeat this procedure many times to obtain an empirical sample of SPT. The p value of the test can then be estimated from this empirical sample.

4.3.3. *QualSPT Statistic*

For a qualitative trait and a case-control design, let X_i, y_i, and t_i denote the numerical code (may be multidimensional) of genotype at the candidate locus, the trait value (1 for disease status and 0 for normal), and the genetic background value of the ith individual, respectively. Chen and colleagues[39] proposed to model the relation of y_i, X_i, and t_i by the semiparametric logistic model

$$\log \frac{P(y_i = 1 | X_i, t_i)}{1 - P(y_i = 1 | X_i, t_i)} = X_i^T \beta + \mu(t_i),$$

where $\mu(t)$ is an unknown smoothing function of genetic background variable t and is not parameterized. Under this model, the association test is to test the null hypothesis $H_0 : \beta = 0$ vs. the alternative hypothesis $H_1 : \beta \ne 0$. The log-likelihood function is given by

$$L(\beta, \mu) = \sum_{i=1}^{n} l[\beta, \mu(t_i); X_i, y_i]$$

$$= \sum_{i=1}^{n} \{y_i[X_i'\beta + \mu(t_i)] - \log\{1 + \exp[X_i'\beta + \mu(t_i)]\}\}.$$

The QualSPT statistic is a likelihood-ratio test statistic

$$\Lambda = \frac{L[\hat{\beta}, \hat{\mu}_1(t_i)]}{L[0, \hat{\mu}_0(t_i)]},$$

where $\hat{\mu}_0(\cdot)$ and $\hat{\mu}_1(\cdot)$ are the maximum likelihood estimators (MLEs) of $\mu(\cdot)$ under H_0 and H_1, respectively; and $\hat{\beta}$ is the MLE of β under H_1. Under the null hypothesis H_0, the QualSPT statistic follows a chi-squared distribution with degrees of freedom equal to the dimension of β.

The estimation of parameter β and nonparameter function $\mu(t)$ under semiparametric logistic models has been recently developed. Several methods have been proposed in the statistics literature.[51-53] Chen and colleagues[39] proposed to follow the local likelihood approach of Severini and colleagues.[53] For a known smoothing parameter h and a given kernel function $K(\cdot)$, the estimation method is an iterative procedure that follows two steps:

Step 1: For a given β, solve the following equation for η:

$$\sum_{i=1}^{n} K\left(\frac{t_i - t}{h}\right) \frac{\partial}{\partial \eta} l(\beta_m, \eta, X_i, y_i) = 0.$$

Denote $\hat{\mu}_m(t_1), \hat{\mu}_m(t_2), \ldots, \hat{\mu}_m(t_n)$ as the solutions of η for $t = t_1, t = t_2, \ldots, t = t_n$, respectively. Here, β_m is the current estimated value of β.

Step 2: Solve the following equation for β:

$$\sum_{i=1}^{n} \frac{\partial}{\partial \beta} l[\beta, \hat{\mu}_m(t_i), X_i, y_i] = 0.$$

This yields the new parameter estimate β_{m+1}.

Repeat this two-step process until convergence occurs.

4.3.4. *Smoothing Parameter h*

In the above discussion of SPT and QualSPT, we assume that the smooth parameter h is known. We follow Chen and colleagues[39] as well as Zhang and colleagues[40] to choose h that minimizes the Kolmogorov test statistic. Specifically, for a given h, we perform either SPT or QualSPT for all M unlinked markers and get the p values $p_1, \ldots, p_M$. These p values should follow a uniform distribution if population stratification is well controlled. Let F_n be the empirical distribution function of the p values $p_1, \ldots, p_M$ and F be the uniform distribution function. To test the null hypothesis $H_0 : p$ values $p_1, \ldots, p_M$ follow a uniform distribution, the test statistic of the Kolmogorov test is $L(h) = \max_x |F_n(x) - F(x)|$; we reject the null hypothesis when $L(h)$ is large. The Kolmogorov test statistic $L(h)$ is a function of h. Chen and colleagues[39] as well as Zhang and colleagues[40] proposed to choose h^* such that

$$h^* = \arg\min_h L(h). \tag{15}$$

This procedure also provides a method to test if population stratification can be well controlled by the set of unlinked markers. If the p value of the Kolmogorov test ($h = h^*$) is greater than a prespecified significance level, e.g. 0.05, we may consider the population stratification as well controlled; otherwise, we say that the M unlinked markers cannot control the population stratification, and more markers are needed.

In summary, the procedure for testing of association in the presence of population structure with either SPT or QualSPT is as follows. Begin with a sample of n individuals, each of whom is genotyped at a series of unlinked markers and a candidate locus. First, estimate the value of the genetic background variable for each of the individuals by using the principal component analysis of the genotypes at a series of unlinked markers. Then, choose the smoothing parameter $h = h^*$ from Eq. (15) by minimizing $L(h)$. In the minimizing process, for each given value of smoothing parameter h, apply either SPT or QualSPT to all of the unlinked markers and calculate $L(h)$. If the Kolmogorov test ($h = h^*$) shows that the population stratification can be well controlled, apply SPT or QualSPT with $h = h^*$ to the candidate locus. In practice, the unlinked markers might be a series of randomly chosen markers across the genome. The simulation studies by Chen and colleagues[39] as well as Zhang and colleagues[40] showed that both SPT and QualSPT can control the population stratification well in most cases by using 100 or more SNP markers. The power comparisons show that SPT and QualSPT are more powerful than the GC approach, and in most cases are also more powerful than SA approaches. Another advantage of SPT and QualSPT is that other variables, such as environmental factors, are very easily incorporated in the model.

Comparing the three approaches and the methods within each approach, we find that each one has its own advantages. In Table 2, we summarize the comparisons in several aspects. The results of power comparison and the results of the number of unlinked markers needed to control population stratification are based on our simulations in some scenarios; the results may be different in other scenarios. By modeling the association and population structure simultaneously, Hoggart and colleagues[22] as well as Satten and colleagues[37] claimed that their modeling method should be more powerful than the two-step procedure. Without any direct comparison, we are not clear if this claim is true or not.

Table 2 Comparisons of the Three Approaches

Methods		AT* Marker	Power•,°	$M^{\triangle}$	Two-Step Method	ITFC*	AT Traits	Comput. Time
GC		Biallelic	+	+	Yes	No	Qual.	+
SA	Pritchard et al.[9]	General	++	++	Yes	No	Qual.	+++
	Satten et al.[37]	General	?	?	No	No	Qual.	++
	Zhang & Zhao[10]	General	++	+++	Yes	No	Quan.	++
	Zhang et al.[8]	Biallelic	++	+++	Yes	No	Qual.	++
	Zhu et al.[41]	General	++	+	Yes	No	Qual.	++
	Hoggart et al.[22]	General	?	?	No	No	Both	+++
SPT	Zhang et al.[40]	General	+++	+	Yes	Yes	Quan.	++
	Chen et al.[39]	General	+++	+	Yes	Yes	Qual.	++

Note: *, AT means "applicable to"; •, "?" means not clear; °, more "+" means larger or bigger; $\triangle$, M is the number of unlinked markers needed to control the population stratification; *, ITFC means "include a method to test if the false-positive results can be well controlled".

5. EXTENSION AND FUTURE STUDIES

In previous sections, we have discussed several existing association tests that can control false-positive results in the presence of population stratification. In this section, we will discuss some possible extensions of the existing methods and some possible further investigations.

5.1. *Computation Simplicity of SPT Method*

As we have discussed earlier, among the existing methods, the SPT approach proposed by Chen and colleagues[39] and by Zhang and colleagues[40] appears promising under various structured population models. Although the SPT approach is computationally faster than the Bayesian method,[9,22] it is still much more computationally intensive than the GC approach. If we apply the SPT approach to a small number of candidate loci, computation is not a problem. However, when we apply it to a large number of loci or genomewide studies, especially when we consider gene–gene interaction, SPT (particularly QualSPT) will become too computationally intensive. In this case, computationally simple methods are needed to control false-positive results due to population stratification.

Let us take another look at SPT. For a quantitative trait and a candidate locus with $p+1$ alleles, the model in Eq. (10) used by Zhang and colleagues[40] can be written as

$$y_i = x_{i1}\beta_1 + \cdots + x_{ip}\beta_k + \mu(t_i) + \varepsilon_i, \quad i = 1, \ldots, n, \qquad (16)$$

where $X_i = (x_{i1}, \ldots, x_{ip})^T$ is the numerical coding of the genotype of the ith individual, and x_{ij} is the number of copies of jth allele for the ith individual at the candidate locus. Let $\hat{y}_i = \sum_{l=1}^{n} y_i w_l(i)$ and

$$\hat{x}_{ij} = \sum_{l=1}^{n} x_{ij} w_l(i) \quad \text{for } i = 1, \ldots, n \quad \text{and} \quad j = 1, \ldots, p. \qquad (17)$$

Denote the nonparametric regression estimates of y_i and x_{ij}, respectively, where

$$w_l(i) = \frac{K(\frac{t_i - t_l}{h})}{\sum_{j=1}^{n} K(\frac{t_j - t_l}{h})}.$$

The testing procedure proposed by Zhang and colleagues[40] is equivalent to the following two steps:

(1) Calculate the residual y_i^* between y_i and its nonparametric regression estimate $\hat{y}_i$ (i.e. $y_i^* = y_i - \hat{y}_i$). Similarly, calculate the residual $x_{ij}^* = x_{ij} - \hat{x}_{ij}$.

(2) Calculate the standard F test statistic under the linear model

$$y_i^* = x_{i1}^* \beta_1 + \cdots + x_{ip}^* \beta_k + \varepsilon_i,$$

and evaluate the p value by using either the F distribution or a simulation procedure. The smoothing parameter h can be estimated using a series of unlinked marker genotypes, as described in the previous section and in Zhang and colleagues.[40]

This two-step method shows that the effect of population structure can be removed by using residuals y_i^* and x_{ij}^* to replace the original y_i and x_{ij}, respectively. Intuitively, $\hat{y}_i$ is the weighted average of the trait values of those individuals who have a similar genetic background with the ith individual. Thus, under the null hypothesis of no association in each unstructured subpopulation, the expectation of residual $y_i^* = y_i - \hat{y}_i$ will be approximately zero in the presence of population stratification; the same is true for x_{ij}^*. Note that once we obtain the estimate of the smoothing parameter h, using a set of unlinked markers, we can calculate the residual y_i^* and x_{ij}^* for any marker across the genome using the same h. In the second step, we can use various methods for homogeneous populations, including searching for a set of interacting genes,[54-56] to analyze the data as if y_i^* and x_{ij}^* were the original trait values and genotypic scores. In summary, we can use the first step to remove the effect of population structure, while a variety of methods can be used in the second step according to different problems.

For a qualitative trait, let $y = 1$ and $y = 0$ for affected and unaffected individuals, respectively. Then, the two-step method can be applied to qualitative traits in the same way as it is applied to quantitative traits. However, after the first step of adjusting for population stratification, the trait will not be binary. We call this method the quan-adjustment method.

Theoretically, false-positive results due to population stratification occur only when different subpopulations have both different prevalences and different allele frequencies. Therefore, adjusting allele frequencies alone (i.e. replacing x_{ij} with $x_{ij}^* = x_{ij} - \hat{x}_{ij}$ and leaving the disease status unchanged) can also control false-positive results due to population stratification. Based on the above observations, we propose the following two-step method, called the single-adjustment method, for qualitative traits:

(1) Replace the original genotype score by residual $x_{ij}^* = x_{ij} - \hat{x}_{ij}$.

(2) Construct an association test under the logistic linear regression model

$$\log \frac{p_i}{1 - p_i} = x_{i1}^* \beta_1 + \cdots + x_{ip}^* \beta_k,$$

where p_i is the probability that the ith individual is affected.

For a quantitative trait, the two-step method is equivalent to the method proposed by Zhang and colleagues.[40] However, both the quan-adjustment and single-adjustment methods are not equivalent to the QualSPT method proposed by Chen and colleagues[39] for a qualitative trait. We have performed some comparisons of the two-step methods with QualSPT under the admixture population models described in Chen and colleagues.[39] Our simulation studies showed that all three methods can control false-positive results due to population stratification (results are not shown). The power comparisons are summarized in Table 3. The results show that with much more computational efficiency, the quan-adjustment method is only slightly less powerful than QualSPT. The single-adjustment method is a little less powerful than the QualSPT and quan-adjustment methods. However, further investigations are needed for more quantitative comparisons of these methods.

5.2. *Haplotype-based Methods in a Structure Population*

The existing association tests that can control population stratification are all single-marker methods, that is, testing one marker at a time. There is strong evidence that several mutations within a single gene can interact to create a super allele, which has a large effect on the observed traits.[57-61] This emphasizes the importance of haplotype-based methods. Several haplotype-based methods have been proposed, including linear model–based methods[57,62–64] and similarity-based methods,[65] among others. Except for the methods using family-based samples, current haplotype-based methods are valid for homogeneous populations and will lead to false-positive results in the presence of population stratification. Thus, there is a need to develop haplotype-based methods that can control population stratification. In this section, we discuss one possible method based on a linear model for a quantitative trait. The idea can also be similarly applied to a logistic linear model for a qualitative trait.

Consider a sample of n unrelated individuals, where each individual has genotypes at several SNPs on a candidate gene. Let H be the number of possible haplotypes formed by the typed SNPs, and let $h_1, \ldots, h_H$ denote all

Table 3 The Power Comparisons of Three Methods for a Qualitative Trait: QualSPT, Quan-Adjustment, and Single-Adjustment under an Admixture Population

$R_{AA}^{(1)}, R_{AA}^{(2)}$	Disease Model	Significance Level 0.05			Significance Level 0.01		
		QualSPT	Single-Adjustment	Quan-Adjustment	QualSPT	Single-Adjustment	Quan-Adjustment
1, 2	Domi.	0.55	0.49	0.55	0.33	0.28	0.33
	Add.	0.56	0.47	0.50	0.33	0.25	0.29
	Rec.	0.56	0.50	0.53	0.33	0.28	0.31
2, 4	Domi.	0.94	0.89	0.90	0.85	0.78	0.82
	Add.	0.92	0.83	0.87	0.79	0.69	0.74
	Rec.	0.89	0.85	0.88	0.76	0.68	0.75

$R_{AA}^{(1)}$ and $R_{AA}^{(2)}$ denote the relative risks of genotype AA to genotype aa in the first and second ancestral populations, respectively.

haplotypes. Let y_i and g_i denote the trait value and the multimarker genotype of individual i. Following Zaykin and colleagues[62] as well as Sha and colleagues,[63] we code the genotype g_i through the haplotypes that are compatible with g_i and denote the numerical code of g_i by $X_i = (x_{i1}, \ldots, x_{iH})$. If haplotypic phases are available, or in the case of no ambiguity,

$$
x_{ij} = \begin{cases} 2, & \text{if } g_i \text{ is homozygous for haplotype } h_j \\ 1, & \text{if } g_i \text{ is heterozygous and includes haplotype } h_j \\ 0, & \text{otherwise.} \end{cases}
$$

If the phases are unknown and cannot be unambiguously reconstructed, haplotype frequencies are estimated via the EM algorithm.[66-69] In this case, we define the genotype score x_{ij} as the posterior probability that individual i has haplotype h_j given its genotype g_i. x_{ij} can thus be written as

$$
x_{ij} = p(h_j \mid g_i) = \frac{\sum_{k=1}^{j} P(g_i \mid h_k h_j)P(h_k h_j) + \sum_{l=j}^{H} P(g_i \mid h_j h_l)P(h_j h_l)}{\sum \sum_{k \le l} P(g_i \mid h_k h_l)P(h_k h_l)},
$$

where $P(g_i \mid h_k h_j) = 1$ if $h_i h_j$ is compatible with g_i; and $P(g_i \mid h_k h_j) = 0$, otherwise.

Furthermore, suppose each sampled individual has genotypes at a candidate gene as well as at a series of unlinked markers. We estimate the values $t_1, \ldots, t_n$ of the genetic background variable for all sampled individuals by using the principal component analysis discussed in Sec. 4.3.1. Then, we can model trait values y_i, the numerical code of genotypes X_i, and the value of genetic background variable t_i by the semiparametric model

$$
y_i = \beta_0 + \beta_1 x_{i1} + \cdots + \beta_H x_{iH} + \mu(t_i) + \varepsilon_i.
$$

This model is the same as that given in Eq. (16). Thus, we can use SPT to test the association between the candidate gene and the trait.

5.3. *Other Topics*

The number of unlinked markers needed to control population stratification depends on the allele frequency difference between subpopulations, the sample size, and the methods used. Thus, it is very difficult to estimate how many unlinked markers are needed beforehand. According to the simulation results, Chen and colleagues[39] as well as Zhang and colleagues[40] concluded that 100 or more randomly chosen SNPs may well control the population stratification in most cases. If we choose more informative markers (i.e. the markers with larger allele frequency differences among subpopulations), fewer markers will be needed. Smith and colleagues[70] proposed an ancestry-informative marker panel. This marker panel contains the allele frequencies of West African and European populations at more than 2000 markers that are well spaced across the genome. The markers have an average allele frequency difference of 57% between these two populations. The informative markers for other populations are needed. We can search genetic databases available to the public to obtain these informative markers. The number of unlinked markers needed also depends on the methods used. From Table 2, we see that SAT and QSAT need more unlinked markers. If the first few principal components are used to define d_{ij} in the SAT or the QSAT method, SAT and QSAT may need fewer unlinked markers to control population stratification.

Another problem we did not discuss in the previous sections is missing values. In practice, missing values are not unusual. For the GC approach, we estimate λ by median $\{X_1^2, \ldots, X_M^2\}/0.456$, where X_j^2 is the value of the statistic at the jth unlinked marker. Since X_j^2 involves only one marker, we may be able to delete the individuals with missing values at the jth unlinked marker when we calculate X_j^2. For other methods, such as SPT, when we estimate the values of the genetic background variable, we use the genotypes of all unlinked markers simultaneously. In this case, we cannot simply delete the individuals with missing values. How to deal with missing values in these methods is a topic that needs further investigation.

ACKNOWLEDGMENTS

This work was supported by the National Institutes of Health (NIH) grants R01 GM069940, R03 HG 003613, R01 HG003054, and R03 AG024491.

ELECTRONIC DATABASE INFORMATION

The URLs for software:

1. http://www.lshtm.ac.uk/eu/genetics (for AdmixMap program; Hoggart *et al.*[22])
2. http://pritch.bsd.uchicago.edu/software.html (for Structure 2.1 and STRAT programs; Pritchard *et al.*[9])
3. http://bioinformatics.med.yale.edu (for QSAT program; Zhang and Zhao[10])

REFERENCES

1. Gray IC, Campbell DA, Spurr NK. (2000) Single nucleotide polymorphisms as tools in human genetics. *Hum Mol Genet* **9**:2403–2408.
2. Risch N, Merikangas K. (1996) The future of genetic studies of complex human diseases. *Science* **273**:1516–1517.
3. Risch N. (2000) Searching for genetic determinants in the new millennium. *Nature* **405**:847–856.
4. Morton NE, Collins A. (1998) Tests and estimates of allelic association in complex inheritance. *Proc Natl Acad Sci USA* **95**:11389–11393.
5. Risch N, Teng J. (1998) The relative power of family-based and case-control designs for linkage disequilibrium studies of complex human diseases. I. DNA pooling. *Genome Res* **8**:1273–1288.
6. Teng J, Risch N. (1999) The relative power of family-based and case-control designs for linkage disequilibrium studies of complex human diseases. II. Individual genotyping. *Genome Res* **9**:234–241.
7. van den Oord EJCG. (1999) A comparison between different designs and tests to detect QTLs in association studies. *Behav Genet* **29**:245–256.
8. Zhang S, Kidd KK, Zhao HY. (2002) Detecting genetic association in case-control studies using similarity-based association tests. *Stat Sin* **12**(1):337–359.
9. Pritchard JK, Stephens M, Rosenberg NA, Donnelly P. (2000) Association mapping in structured populations. *Am J Hum Genet* **67**:170–181.
10. Zhang S, Zhao H. (2001) Quantitative similarity-based association tests using population samples. *Am J Hum Genet* **69**:601–614.
11. Knowler WC, Williams RC, Pettitt DJ, Steinberg AG. (1988) Gm3:5,13,14 and type 2 diabetes mellitus: an association in American Indians with genetic admixture. *Am J Hum Genet* **43**:520–526.
12. Lander ES, Schork N. (1994) Genetic dissection of complex traits. *Science* **265**:2037–2048.
13. Ewens WJ, Spielman RS. (1995) The transmission/disequilibrium test: history, subdivision, and admixture. *Am J Hum Genet* **57**:455–464.
14. Pritchard JK, Rosenberg NA. (1999) Use of unlinked genetic markers to detect population stratification in association studies. *Am J Hum Genet* **65**:220–228.
15. Devlin B, Roeder K. (1999) Genomic control for association studies. *Biometrics* **55**:997–1004.
16. Thomas DC, Witte JS. (2002) Point: population stratification: a problem for case-control studies of candidate–gene association? *Cancer Epidemiol Biomarkers Prev* **11**:505–512.
17. Wacholder S, Rothman N, Caporaso N. (2002) Counterpoint: bias from population stratification is not a major threat to the validity of conclusions from epidemiological studies of common polymorphisms and cancer. *Cancer Epidemiol Biomarkers Prev* **11**:513–520.
18. Ziv E, Burchard EG. (2003) Human population structure and genetic association studies. *Pharmacogenomics* **4**:431–441.

19. Cardon LR, Palmer LJ. (2003) Population stratification and spurious allelic association. *Lancet* **361**:598–604.

20. Ardlie KG, Lunetta KL, Seielstad M. (2002) Testing for population subdivision and association in four case-control studies. *Am J Hum Genet* **71**:304–311.

21. Schork NJ, Fallin D, Thiel B, *et al.* (2001) The future of genetic case-control studies. *Adv Genet* **42**:191–212.

22. Hoggart CJ, Parra EJ, Shriver MD, *et al.* (2003) Control of confounding of genetic associations in stratified populations. *Am J Hum Genet* **72**:1492–1504.

23. Lohmueller KE, Pearce, CL, Pike M, *et al.* (2003) Meta-analysis of genetic association studies supports a contribution of common variants to susceptibility to common disease. *Nat Genet* **33**:177–182.

24. Freedman ML, Reich D, Penney KL, *et al.* (2004) Assessing the impact of population stratification on genetic association studies. *Nat Genet* **36**:388–393.

25. Marchini J, Cardon LR, Philips MS, Donnelly P. (2004) The effects of human population structure on large genetic association studies. *Nat Genet* **36**:512–517.

26. Abecasis GR, Cardon LR, Cookson OC. (2000) A general test of association for quantitative traits in nuclear families. *Am J Hum Genet* **66**:279–292.

27. Falk CT, Rubinstein P. (1987) Haplotype relative risks: an easy reliable way to construct a proper control sample for risk calculations. *Ann Hum Genet* **51**:227–233.

28. Fulker DW, Cherny SS, Sham PC, Hewitt JK. (1999) Combined linkage and association sib-pair analysis for quantitative traits. *Am J Hum Genet* **64**:259–267.

29. Spielman RS, McGinnis RE, Ewens WJ. (1993) Transmission test for linkage disequilibrium: the insulin gene region and insulin-dependent diabetes mellitus (IDDM). *Am J Hum Genet* **52**:506–513.

30. Curtis D. (1997) Use of siblings as controls in case-control studies. *Ann Hum Genet* **61**:319–333.

31. Sham PC, Cherny SS, Purcell S, Hewitt JK. (2000) Power of linkage versus association analysis of quantitative traits, by use of variance-components models, for sib-ship data. *Am J Hum Genet* **66**:1616–1630.

32. Sun F, Flanders WD, Yang Q, Zhao HY. (2000) Transmission/disequilibrium tests for quantitative traits. *Ann Hum Genet* **64**:555–565.

33. Zhao H, Zhang S, Merikangas KR, *et al.* (2000) Transmission/disequilibrium tests using multiple tightly linked markers. *Am J Hum Genet* **67**(4):936–946.

34. Bacanu SA, Devlin B, Roeder K. (2000) The power of genomic control. *Am J Hum Genet* **66**:1933–1944.

35. Pritchard JK, Donnelly P. (2001) Case-control studies of association in structure using multi-locus genotype data. *Genetics* **155**:945–959.

36. Devlin B, Roeder K, Wasserman L. (2001) Genomic control, a new approach to genetic-based association studies. *Theor Popul Biol* **60**:155–166.

37. Satten GA, Flanders WD, Yang Q. (2001) Accounting for unmeasured population substructure in case-control studies of genetic association using a novel latent-class model. *Am J Hum Genet* **68**:466–477.

38. Reich DE, Goldstein DB. (2001) Detecting association in a case-control study while correcting for population stratification. *Genet Epidemiol* **20**:4–16.

39. Chen HS, Zhu X, Zhao H, Zhang S. (2003) Qualitative semi-parametric test for genetic associations in case-control designs under structured populations. *Ann Hum Genet* **67**:250–264.

40. Zhang S, Zhu X, Zhao H. (2003) On a semi-parametric test to detect associations between quantitative traits and candidate genes using unrelated individuals. *Genet Epidemiol* **24**:45–56.

41. Zhu X, Zhang S, Zhao H, Cooper RS. (2002) Inferring population structure and association mapping for complex traits. *Genet Epidemiol* **23**:181–196.

42. Zheng G, Freidlin B, Gastwirth JL. (2006) Robust genomic control for association studies. *Am J Hum Genet* **78**:350–356.

43. Rice JA. (1988) *Mathematical Statistics and Data Analysis*. Wadsworth & Brooks/Cole, Pacific Grove, CA.

44. Celeux G, Govaert G. (1995) Gaussian parsimonious clustering model. *Pattern Recognit* **28**(5):781–793.

45. Wahba G. (1984) Partial spline models for the semi-parametric estimation of functions of several variables. In: *Statistical Analysis of Time Series*, Institute of Statistical Mathematics, Tokyo, Japan, pp. 319–329.

46. Green P, Jennison C, Seheult A. (1985) Analysis of field experiments by least squares smoothing. *J R Stat Soc Ser B* **47**:299–315.

47. Engle R, Granger C, Rice J, Weiss A. (1986) Nonparametric estimates of the relation between weather and electricity sales. *J Am Stat Assoc* **81**:310–320.

48. Shiau J, Wahba G, Johnson DR. (1986) Partial spline models for the inclusion of tropopause and frontal boundary information in otherwise smooth two- and three-dimensional objective analysis. *J Atmos Ocean Technol* **3**:714–725.

49. Speckman P. (1988) Kernel smoothing in partial linear models. *J R Stat Soc Ser B* **50**:413–436.

50. Hamilton SA, Truong YK. (1997) Local linear estimation in partly linear models. *J Multivar Anal* **60**:1–19.

51. Simonoff JS. (1996) *Smoothing Methods in Statistics*. Springer, New York, NY.

52. Severini TA, Wong W. (1992) Profile likelihood and conditionally parametric models. *Ann Stat* **20**:1768–1802.

53. Severini TA, Severini TA, Staniswalis JG. (1994) Quasi-likelihood estimation in semiparametric models. *J Am Stat Assoc* **89**(426):501–511.

54. Nelson MR, Kardia SLR, Ferrell RE, Sing CF. (2001) A combinatorial partitioning method to identify multilocus genotypic partitions that predict quantitative trait variation. *Genome Res* **11**:458–470.

55. Ritchie MD, Hahn LW, Roodi N, *et al.* (2001) Multifactor-dimensionality reduction reveals high-order interactions among estrogen-metabolism genes in sporadic breast cancer. *Am J Hum Genet* **69**:138–147.

56. Sha Q, Zhu X, Zuo Y, *et al.* (2006) A combinatorial searching method for detecting a set of interacting loci associated with complex traits. *Ann Hum Genet* **70**:677–692.

57. Schaid DJ, Rowland CM, Tines DE, *et al.* (2002) Score test for association between traits and haplotypes when linkage phase is ambiguous. *Am J Hum Genet* **70**:425–434.

58. Hollox EJ, Poulter M, Zvarik M, *et al.* (2001) Lactase haplotype diversity in the Old World. *Am J Hum Genet* **68**(1):160–172.

59. Clark AG, Weiss KM, Nickerson DA, *et al.* (1998) Haplotype structure and population genetic inferences from nucleotide-sequence variation in human lipoprotein lipase. *Am J Hum Genet* **63**:595–612.

60. Tavtigian SV, Simard J, Teng DH, *et al.* (2001) A candidate prostate cancer susceptibility gene at chromosome 17p. *Nat Genet* **27**(2):172–180.

61. Drysdale CM, McGraw DW, Stack CB, *et al.* (2000) Complex promoter and coding region beta 2-adrenergic receptor haplotypes alter receptor expression and predict *in vivo* responsiveness. *Proc Natl Acad Sci USA* **97**(19):10483–10488.

62. Zaykin DV, Westfall PH, Young SS, *et al.* (2002) Testing association of statistically inferred haplotype with discrete and continuous traits in samples of unrelated individuals. *Hum Hered* **53**:79–91.

63. Sha Q, Dong J, Jiang R, Zhang S. (2005) Test of association between quantitative traits and haplotypes in a reduced-dimensional space. *Ann Hum Genet* **69**(6):715–732.

64. Tzeng JY, Wang CH, Kao JT, Hsiao CK. (2006) Regression-based association analysis with clustered haplotypes through use of genotypes. *Am J Hum Genet* **78**:231–242.

65. Tzeng JY, Devlin B, Wasserman L, Roeder K. (2003) On the identification of disease mutations by the analysis of haplotype similarity and goodness of fit. *Am J Hum Genet* **72**:891–902.

66. Excoffier L, Slatkin M. (1995) Maximum-likelihood estimation of molecular haplotype frequencies in diploid population. *Mol Biol Evol* **12**:921–927.

67. Hawley M, Kidd K. (1995) HAPLO: a program using the EM algorithm to estimate the frequencies of multi-site haplotypes. *J Hered* **86**:409–411.

68. Long JC, Williams RC, Urbanek M. (1995) An E-M algorithm and testing strategy for multiple locus haplotypes. *Am J Hum Genet* **56**:799–810.

69. Zhang S, Pakstis AJ, Kidd KK, Zhao H. (2001) Comparisons of two methods for haplotype reconstruction and haplotype frequency estimates from population data. *Am J Hum Genet* **69**:906–912.

70. Smith MW, Patterson N, Lautenberger JA, *et al.* (2004) A high-density admixture map for disease gene discovery in African Americans. *Am J Hum Genet* **74**:1001–1013.

Chapter 11

Statistical Methods for Multiple QTL Mapping in Experimental Crosses

Daniel Shriner[*,†], Solomon Musani[†] and Nengjun Yi[†,‡]

[*] *Department of Nutrition Sciences, University of Alabama at Birmingham Birmingham, AL 35294, USA*

[†] *Department of Biostatistics, Section on Statistical Genetics University of Alabama at Birmingham, Birmingham, AL 35294, USA*

[‡] *Clinical Nutrition Research Center, University of Alabama at Birmingham Birmingham, AL 35294, USA*

Many complex diseases, such as obesity, diabetes, hypertension, and cancer, are determined by multiple genetic and environmental factors. The identification of quantitative trait loci (QTLs) is critical for understanding the biochemical bases of complex diseases, and thus for the identification of drug targets. Animal models have proved to be powerful in elucidating the genetic architectures and etiologies of common human diseases. The ability to control both genotype and environment in inbred populations of animals greatly simplifies the analysis of a complex genetic architecture. This chapter reviews the Bayesian statistical methods and computer software for mapping multiple QTLs in experimental crosses, and comments on several of the statistical issues to consider in the application of these methods.

1. INTRODUCTION

We begin this chapter by discussing the animal models and designs of experimental crosses. We then discuss the genetic architecture of complex traits.

[‡] Correspondence author.

After outlining the basic principles of QTL mapping, we discuss several methods for mapping single QTLs. We then discuss the methods for mapping multiple QTLs, emphasizing our Bayesian method. We conclude this chapter with future directions and a list of freely available software.

2. ANIMAL MODELS

Since the establishment of the Wistar inbred rat line by King in 1909,[1] researchers have increasingly used animal models to study the genetic basis of complex human phenotypes, particularly disease and behavior. Breeding of the first inbred mouse began soon after King's discovery, culminating in the widespread use by mammalian geneticists of what is today a universally proven animal model. Through selective breeding, hundreds of strains (both mouse and rat) have been developed for disorders ranging from hypertension to urological defects. Over 2000 QTLs have been associated with diseases such as obesity, atherosclerosis, diabetes, asthma, and hypertension.[2-5] Other animal models — such as dogs, swine, baboons, zebrafish, and *C. elegans* — have been used to study behavior, atherosclerosis, Parkinson's disease, and developmental biology.

The widespread use of animal models, particularly rodents, in mapping and detecting loci influencing quantitative traits and in dissecting complex polygenic traits is attributable to several reasons. Although the main goal is to map QTLs in humans, ethical considerations cannot permit a direct study thereof. Moreover, mapping QTLs in humans is difficult, expensive, time-consuming, and often compromised by small populations that are genetically diverse and subject to uncontrollable environments. Conversely, mapping QTLs in rodents overcomes these setbacks because of the following features:

(1) Great genetic variation among inbred lines of mice. The existence of several inbred lines, wild-derived strains, and lines from long-term selection for many quantitative traits common to humans is critical to QTL detection.[6] Since an inbred strain is homozygous for all loci, it provides a unique tool to repeatedly access a genetically identical population, thereby enabling a reliable comparison of data across laboratories. Ideally, experimental designs based on inbred lines provide a recommended setting for QTL detection.

(2) Synteny between mouse and human. Although tremendous genetic variation exists among different lines of mice, the order of genes on chromosomes is largely similar between the mouse and human genomes.[7]

This synteny confers utility upon the mouse model in comparative genome studies among mammals. By studying the mouse genome, candidate chromosomal regions in humans are identified at which homologous QTLs may reside.[8] For example, there is significant concordance among the mouse, human, and rat genomes in hypertension QTLs[4]; and between the human and mouse genomes in HDL cholesterol QTLs.[9]

(3) Convenience of use. Unlike the human species, animal models can be easily manipulated. Breeding is done easily and is less costly. Other benefits include the low costs of maintenance and phenotype measurements, and the ease of acquiring very large samples of designed matings.

(4) Molecular marker map. Recent technological advances — such as the development of high-resolution genetic and physical linkage maps of the mouse genome, which in turn facilitate the identification and cloning of mouse disease loci — have dramatically increased our ability to create mouse models of human disease.[10] The mouse has an extensively developed and well-organized molecular marker map, consisting of over 6500 easily typed PCR-based microsatellite markers that exhibit allelic variation between lines.

3. CROSS DESIGNS

The success of QTL analysis using linked markers is contingent upon two basic components: linkage disequilibrium (LD) between QTLs and linked molecular markers, and genetically informative linked markers. LD is defined generally as the relationship between two alleles that arises more often than can be accounted for by chance, because those alleles are physically located close to one another on a chromosome and are not frequently separated from one another by recombination. It is also called gametic disequilibrium.[6]

Generally, natural and commercial populations are at linkage equilibrium for the vast majority of the genome, thus making it difficult to associate an effect of segregating QTLs with marker genotypes because QTLs and marker alleles are independently assorted. This is the scenario in outbred populations such as livestock and humans. Consequently, to detect a single QTL, LD must be generated. Contrast this to the case of inbred line crosses. LD is maximized here by crossing two inbred parental lines, P_1 and P_2, which are themselves products of repeated sibling matings, leading to the establishment of panels of well-defined strains with consistent phenotypic differences despite being raised in a common environment.

The resulting F_1 progeny are heterozygous at all loci that differ between the crossed lines, and are therefore completely informative as parents. They also have the same linkage phase, which means that marker trait associations can be averaged over all offspring, regardless of parents.

The backcross (BC) is the simplest example of an inbred line cross design, obtained by backcrossing F_1 progeny to one of the parental lines. In the F_2 design, F_2 individuals are obtained by either self-breeding among the F_1 individuals or intercrossing them. These (BC and F_2) are the two most widely used crosses. Other crosses that have been used are recombinant inbred lines (RILs), double haploid (DH) lines, and test crosses (TCs). RILs can be produced in either of two possible ways: the first way is by several generations of self-breeding individual F_2 progeny, or by brother–sister matings in situations where self-breeding (RILF) is not possible; the second way is production from single-backcross individuals or by brother–sister matings (RILB). DH lines are produced by self-fertilizing DHs derived from the F_1 progeny. Finally, TC progeny are produced by mating F_1 individuals to a third inbred line.

The choice of cross (experimental) design in QTL mapping using animal models is subject to biological, genetic, and statistical considerations. Biological considerations ensure that there is no species-specific constraint on the design (e.g. DH), because asexual reproduction is not possible in all animal species. For certain species, it is much easier to produce large numbers of F_2 individuals than with either BC or TC designs, which require cross-fertilization. Genetic considerations refer to the parameters to be estimated, e.g. dominance effects, which cannot be estimated using the BC design unless both backcrosses are available. Finally, statistical considerations refer to the choice of a design that maximizes the statistical power of QTL detection.

Crosses of outbred populations, on the other hand, traditionally rely on collecting families or extended pedigrees that are segregating for the QTL genotypes to generate the requisite amount of disequilibrium. Other outbred populations in which LD occurs include natural populations where QTL alleles are in drift–recombination equilibrium[11,12]; populations resulting from admixture between two populations with different gene frequencies for the marker and mean values of the trait[12,13]; and populations that have not reached drift–mutation equilibrium due to a recent mutation at the trait locus, or a recent founder event followed by population expansion.[6,12] Unlike inbred line crosses, outbred populations pose several challenges for QTL mapping. For example, only heterozygous parents are informative;

individuals may differ in QTL-marker linkage phase; and in the presence of genetic heterogeneity, different families may show different associations. In outbred populations, marker–trait associations are evaluated through the effects of the marker on the trait's genetic variance, which is estimated less precisely than the mean used in inbred line crosses. Consequently, the power of QTL detection is lower in the former than in the latter.

4. GENETIC ARCHITECTURE OF COMPLEX TRAITS

Conventionally, the genetic architecture of a complex trait has been described by a genetic model based on quantitative genetics theory. Most of the observed variations between individuals for traits such as diabetes, obesity, atherosclerosis, and hypertension are quantitative; and population variation tends towards a standard normal distribution. This is in contrast to qualitative traits, for which individual phenotypes fall into discrete categories. Phenotypic variation of a quantitative trait is divided into two main components: genetic variation and nongenetic or environmental variation.

Genotypes are specified by homozygous (additive) and heterozygous (dominance) effects at each locus, and pairwise and higher-order interactions (epistasis) between loci.[6] These constitute the genetic component. In contrast to the case for simple traits with Mendelian inheritance, individual genotypes cannot be inferred from observations of the phenotype; instead, estimates of heritability and variance components from correlation between relatives are often used. Moreover, responses to selection, estimates of the degree of dominance from changes in the means of inbreeding, estimates of net pleiotropic effects from genetic correlations, and estimates of the total mutation rate from phenotypic divergence between inbred lines are also useful.[14] Although this knowledge of quantitative traits is the basis of the marked genetic progress achieved, particularly in livestock and crop production, there still remains the need to identify and determine the properties of the individual genes underlying variation in complex traits by advancing from statistics to biology.

With the discovery and rapid proliferation of highly polymorphic genetic markers due to advances in genotyping technology, coupled with refined experimental approaches and better statistical methods, thorough genomic screening for individual genes or QTLs that control measurable polygenic traits in model organisms is now possible. Subsequently, there has been a major shift in emphasis to the dissection of the different complex traits in a variety of animal models into their constituent components.

4.1. Components of Genetic Architecture

The components of the genetic architecture of a quantitative trait have been summarized by Mackay.[14,15] Overall, genetic architecture integrates all branches of genetics, such as molecular genetics, evolutionary genetics, and classical genetics. A description of the complete genetic architecture has not yet been accomplished for any single quantitative trait. However, attempts have been made through whole-genome sequencing and improved technologies for polymorphism detection. In this chapter, we concentrate on obesity as an example of a quantitative trait as studied in the mouse model. We also focus on one main component of genetic architecture: epistasis. This is in recognition of the extensive use of animal models for studies designed to dissect quantitative traits due to the aforementioned reasons. The mouse model offers excellent genetic and genomic resources critical for understanding the subject under review.[2,5,8] Besides obesity and body size, other traits that have been studied in mice using genome scans include acute alcohol withdrawal[16] and emotionality.[17]

As of the 2004 update of the genetic map of obesity,[5] the number of QTLs reported from animal models currently stands at 221. Of these, 204 QTLs have been confirmed in humans from 50 genomewide scans. A total of 38 genomic regions harbor QTLs replicated among two to four studies. The number of studies reporting associations between DNA sequence variation in specific genes and obesity phenotypes has also increased considerably, with 358 findings of positive associations with 113 candidate genes. Among them, 18 genes are supported by at least five positive studies. Overall, 600 genes, markers, and chromosomal regions have been associated or linked with human obesity phenotypes. This is a marked increase from the modest number of 22 reported 5 years earlier.[2]

4.2. Epistasis

Epistasis, defined in classical Mendelian genetics as the masking of genotypic effects at one locus by the genotype of another locus, is much broader in quantitative genetics. It refers to any statistical interaction between genotypes at two (or more) loci,[18–20] resulting in modification of the homozygous or heterozygous effects of the interacting loci. It is further referred to as synergistic if the phenotype due to one locus is enhanced by the genotype at another locus, or antagonistic if the phenotype due to one locus is suppressed by the genotype at the second locus.[14]

In QTL mapping, single-locus QTL effects have been found to vary according to epistasis. Failure to account for epistasis, when present, leads to low power of detection and decreased accuracy with which QTL locations and effects are estimated. Furthermore, ignoring epistasis precludes finding QTLs with nonsignificant main effects but significant epistatic effects. Therefore, modeling epistasis improves the chances of finding QTLs and also enhances the accuracy of estimated QTL positions and effects.[21] The power to detect epistasis can be low for a number of reasons: first, small numbers of individuals in the rarer two-locus genotype classes, even in large mapping populations; second, interference from other segregating QTLs; and lastly, multiple testing adjustments that leave only QTLs with extremely large interaction signals.[14]

The existence of epistasis, however, has been accepted by many investigators, though comprehensive empirical data is still lacking.[21-24] The challenge is mainly in the detection of epistasis using available designs. Interaction terms tend to have very high sampling variances and, unless large sample sizes are available, the power of detection is low.

Table 1 summarizes QTLs with significant epistatic effects on obesity. This summary should be considered a fraction of potential epistatic QTLs because most obesity studies do not examine the data appropriately for epistasis.[21] It is evident from Table 1 that all findings have been reported within the last 5 years, indicating that epistasis as a component of genetic architecture has recently acquired greater interest.

The proportion of phenotypic variance explained by epistasis has consistently been estimated at around one third of the phenotypic variance[25,26] in mice intercrosses. This figure is within the range of 25%–75% reported for body mass index in human family and adoption studies. Elsewhere, studies of twins place the estimate even higher, at 50%–90% of variance.[27] Segal and

Table 1   Summary of Epistatic QTLs for Obesity

Cross	Positions	References
BSB: (C57BL/6J × *Mus spretus*)	6, 7, 12, & 15	30
× C57BL/6J	3 & 7	31
	2 & 12	32
DU6i × DBA/2	2, 9, 10, 11, 14, 17, 18, X	25
F$_2$: Large (LG/J) × Small (SM/J)	1, 6, 7, 9, 12, 13, & 18	28
F$_2$: MRL/MPJ × SJL/J	6, 7, & 14	26
Intercross: CAST × C57BL/6J	10 & 12	33, 34
BC: (M16i × CAST/Ei) × M16i	1, 2, 13, 15, 18, & 19	29

Allison[27] argued that these estimates are most likely correct because they reflect the nonadditive nature of the genetic effects due to the combination of dominance and epistasis on obesity. Moreover, in twin studies, common environmental factors also influence the estimation of these genetic effects.

In an F_2 cross of DU6i (bred for extreme high growth) and DBA/2 (inbred line) mice, the *Lepq1* locus on chromosome 14 interacted with several other QTLs located on multiple chromosomes to affect obesity.[25] In an intercross of LG/J and SM/J inbred mice, eight QTLs (*Adip1–Adip8*) were identified on as many chromosomes (1, 6–9, 12, 13, and 18), one of which (*Adip8* on chromosome 18) interacted with the other seven to influence adiposity.[28] LG/J alleles were found to result in greater adiposity than SM/J, except for *Adip5* and *Adip6*, for which the reverse was observed.[28] The SM/J allele was dominant to the LG/J allele at *Adip1*, *Adip2*, and *Adip4*; while the converse was observed for the QTLs *Adip3* and *Adip5*. *Adip6* and *Adip8* acted additively, whereas *Adip7* was overdominant.[28]

Most recently, in a comprehensive genome scan, Yi *et al.*[29] detected several epistatic QTLs for body weight at different ages and body composition in a backcross mouse population. QTLs at six chromosomes with main effects (QTLs on chromosomes 2, 13, and 15) and others without (1, 18, and 19) interacted amongst themselves to cause positive and sometimes negative effects on the traits.[29] Of all QTLs, two on chromosomes 1 and 18 had pleiotropic effects on adiposity and body weight.[29] These findings were in general agreement with those reported elsewhere (Table 1).

4.3. *Limitations of QTL Mapping to Resolve Genetic Architectures of Complex Traits*

Although substantial progress has been achieved in the mapping of QTLs that influence quantitative traits, we are still far from discovering the identity and properties of individual genes, without which an understanding of epistasis is even more remote. Four reasons for this drawback are outlined below.

4.3.1. *Size of a QTL*

A QTL, as defined by a genome scan, is a relatively large chromosomal region containing one or potentially more loci affecting the trait. Recent findings indicate that, in *Drosophila* and many other species, the average size of intervals containing a significant QTL ranges from 0.1 cM to 44.7 cM and 98 kb to 19 284 kb, which corresponds to an average of 8.9 cM and

4459 kb, respectively.[14] Given that there are $\sim$13 600 genes in the 120 Mb of *Drosophila* eukaryotic DNA,[35] it follows that the average gene size is 8.8 kb, implying that a QTL may contain between 11 and 2191 genes. Methods designed to narrow the DNA region enough to conduct practical searches for a specific gene have been developed.[36] The use of congenic strains of mice is one such strategy, where the QTL is transferred into the genome of another strain, followed by inbreeding.[37] This is a powerful tool for the study of epistasis.[21] Several strains are produced that carry different or overlapping parts of the original QTL. Strains with similar phenotypes can be identified and the regions of the original QTL they contain can be determined, allowing for the discrimination of DNA regions most likely to contain the relevant gene(s). The expectation is that the use of multiple congenic strains may lead to more precise mapping.

4.3.2. *Failure to Replicate Positive Findings in Other Populations*

Besides a QTL being a potential home for tens to hundreds of genes, another challenge of QTL mapping is the replication of the findings in other populations.[38] The reasons given for low replicability are varied; they include locus heterogeneity, epistasis, low penetrance, variable expressivity and pleiotropy, and limited statistical power. Although the prospects for success have improved markedly in the recent past following the development of extensive genome resources and technologies, claims of gene discovery in complex traits still require additional evidence. Glazier *et al.*[39] have proposed standards of evidence that together establish the formal burden of proof which can be used to evaluate the evidence for gene discovery in complex traits in a wide variety of organisms. As a result of their application, the molecular bases of a number of complex trait genes localized initially in genomewide linkage studies for various species have been confirmed.[39]

4.3.3. *Identification of Genes Using Positional Cloning*

Identifying genes that underlie QTL peaks by positional cloning has proved elusive for several reasons: (i) a lack of sufficient recombination events in each QTL interval to map the QTL to successively smaller intervals; (ii) QTLs with small effect sizes that are often environmentally sensitive, leading to unreliability of the phenotype of a single individual as an indicator of QTL genotype; and (iii) determination of the causative gene from a pool of several positional candidate genes. Guessing (the "candidate gene" approach) has worked in cases where the genetic basis underlying the trait phenotype was well understood. The availability of complete genome sequences for many organisms means that it is possible to peruse the gene list in a candidate QTL region. Within this list are very likely many predicted genes of unknown function and known genes with no *a priori* relationship to the trait.

4.3.4. *Statistical and Practical Pitfalls*

In most cases, a single experiment is designed to detect only one QTL with a relatively large effect size on the trait under study. Unfortunately, most quantitative traits are influenced by many QTLs, each with a small effect size, that may be missed. Since the power to detect QTLs with small effect sizes is low, the result is an increased type II error rate. Moreover, since a large number of markers are often tested for correlation with the trait of interest, pitfalls related to multiple testing may result. Some correlations will appear significant purely by chance, thereby inflating the type I error rate. Minimizing type I errors by employing stringent criteria for declaring significant linkage may result in missing QTLs with small effects, thereby inflating the type II error rate. To achieve a balance, the research community prefers to insist on the replication of the potential QTL in follow-up populations and to relax the initial criteria for screening significant correlations.[40]

5. BASIC PRINCIPLES OF QTL MAPPING

QTL mapping seeks to identify specific regions of the genome that contribute to variation in the trait of interest. This knowledge is essential in biomedical science because it leads to an understanding of the biochemical bases of disease traits, with the potential for identification of new drug targets. QTL mapping is based on two components: a highly polymorphic linkage marker map, and a population that is genetically variable for the quantitative trait.[6]

5.1. *Linkage Marker Map*

A linkage marker map represents the ordering of marker loci along a chromosome according to recombination fractions.[41,42] Historically, little progress was possible in the construction of genetic maps because of insufficient markers. Allozymes, which were the first molecular markers, lacked sufficient protein variation for high-resolution mapping. With the discovery of DNA-based markers approximately two decades ago, variation could be scored directly at the DNA level. Starting from restriction fragment length polymorphism (RFLP) analysis using Southern blots,[43,44] methods of detecting molecular variation have since evolved to high throughput for both polymorphism discovery and genotyping.[45]

The dense genetic maps of mammalian genomes facilitate the mapping of polygenic traits, the mapping of quantitative or qualitative trait loci, and marker association, among other biological studies. DNA genetic markers provide the foundation for mammalian genetic and physical mapping[46-48] as well as the mapping of quantitative and qualitative trait loci.[49-52] Single nucleotide polymorphisms (SNPs) have currently superseded all other markers as the single most abundant class of genetic variation in mammals. SNPs have become the marker of choice because of their frequency, density, and compatibility with multiple high-throughput technology platforms.[53] The genotyping of SNPs is especially suited for the genetic analysis of model organisms, such as the mouse, because biallelic markers remain fully informative when used to characterize crosses between inbred strains. Currently, over two million SNPs in the human genome have been discovered,[54,55] which is expected to spur further development of rapid, high-throughput, accurate, and economical methods for genotyping molecular markers.

5.2. *Population*

Populations with genetically variable quantitative traits refer to the aforementioned cross designs. Knowledge of the marker map and the population creates an opportunity to test marker differences in trait mean between marker genotypes for each marker in turn. The marker in a local region that expresses the greatest difference in the mean value of the trait is thus the closest to the QTL.

Formally, the basic principle of QTL mapping can be summarized as follows. Consider two biallelic loci, an autosomal marker locus (M) and a quantitative trait locus (Q), each with two alleles, M_1 and M_2 as well as Q_1 and Q_2, respectively. Also suppose that the QTL is located r cM away from the marker locus, with additive (a) and dominance (d) effects. If individuals with the genotype $M_1M_1Q_1Q_1$ are crossed with those with $M_2M_2Q_2Q_2$, and subsequently the F_1 progeny are allowed to breed randomly, then the difference in mean values of the quantitative trait between homozygous marker classes among the F_2 progeny is $a(1 - 2r)$. Similarly, the difference between the average mean phenotype of the homozygous marker classes and the heterozygote is $d(1 - 2r)^2$. If the QTL and the marker locus are unlinked (i.e. $r = 0.5$), then the mean value of the quantitative trait will be the same for each of the marker genotypes. More details are presented in the mapping methods.

6. SINGLE QTL MODELS

Attendant to the recent increase in the availability of abundant polymorphic molecular markers is the improvement in statistical methods for mapping QTLs as well as the development of guidelines for experimental design and interpretation. Several reviews have been written on this subject in the past decade alone.[6,56–59] In this section, we summarize the key points from the literature on traditional QTL mapping methods in line crosses based on single QTL models.

6.1. *Analysis of Variance*

Analysis of variance (ANOVA) is also referred to as marker regression or a linear model. It is the simplest of all statistical methods used for mapping QTLs for the following reasons: ease of application, widespread implementation by commonly used statistical packages, and ability to analytically compute significance and power. We shall use the BC design to illustrate how the various methods work; the statistical methods and the issues that arise are the same for the different types of cross designs.

Table 2 shows how the BC design is analyzed using a linear model or ANOVA. Suppose that F_1 progeny with marker genotype $M_1 M_2$ are backcrossed to the $M_2 M_2$ parental line, and the resulting BC progeny are divided into two groups based on their genotypes at this marker locus ($M_1 M_2$ and $M_2 M_2$). In both of these marker classes, the corresponding probabilities of getting progeny with QTL genotypes $Q_1 Q_2$ and $Q_2 Q_2$ are $(1 - r, r)$ and $(r, 1-r)$, respectively, in which r is the recombination fraction. The product of trait value and conditional probability of the QTL given a marker for both marker classes is then summed to obtain the marker genotype trait expectation, which is used to compute the contrast between means. For $r = 0.5$, the contrast is zero; this means that the marker and the QTL are unlinked, so the two marker genotypes will have equal phenotypic means. Otherwise, for $r < 0.5$, the contrast will always be nonzero, indicating that the two loci are linked.

Table 2   Backcross Design Genotype Probabilities and Quantitative Trait Expectations

Marker Genotype	QTL Genotype	Conditional Probability	Trait Value	Trait × Conditional Probability	Expectation
M_1M_2	Q_1Q_2	$1-r$	d	$d(1-r)$	$d - r(d+a)$
	Q_2Q_2	r	$-a$	$-ar$	
M_2M_2	Q_2Q_2	$1-r$	$-a$	$-a(1-r)$	$-a + r(d+a)$
	Q_2Q_1	r	d	dr	
Contrast (δ) $(M_1M_2 - M_2M_2)$					$(1-2r)(d+a)$

The effect size of a QTL is often measured as the proportion of phenotypic variance explained by that QTL. Given that the variance induced by that QTL only (i.e. in the absence of environmental variation, measurement error, and other QTLs) is

$$\frac{\delta^2}{4},$$

then the proportion of phenotypic variance will be computed as $\frac{\delta^2\tau^2}{4}$, in which τ^2 is the total phenotypic variance in the BC generation.

Significance testing of the marker genotype effect has been performed using either the ratio of the marker mean squares to the residual mean squares or a t test. Under the null hypothesis (i.e. no segregating QTLs), the former follows a central F distribution while the latter follows a central t distribution. A significant deviation of these statistics indicates the presence of segregating QTLs.

The main disadvantages of this method are that (1) estimation of QTL locations and effect sizes is biased due to confounding by the effect of recombination between the marker and the QTL, leading to inconsistency across sample sizes; (2) missing data cannot be handled; (3) residuals are assumed to be normally distributed with equal variance; (4) there is no distinction between pleiotropic and linked QTLs; and (5) in the case of widely spaced markers, the QTL may be too far from all markers, leading to low power.

6.2. *Interval Mapping*

Interval mapping was developed to overcome most of the disadvantages of ANOVA under single marker analysis.[60] Interval mapping, also called flanking marker mapping, has become the most popular approach to QTL mapping in experimental crosses. It assumes, like the linear model, the presence of single putative QTLs, and it uses markers on either side of the putative QTLs (i.e. left and right or M and N, respectively) that flank a specific region of the chromosome. Two opposing approaches have been reported in the literature. According to Weller,[59] it is not possible to construct a linear model that accurately describes the relationship between the observations and the QTL parameters. On the other hand, expected marker contrasts have been created for the BC design.[58,61]

We present here the linear model approach. If a QTL exists between markers M and N with recombination frequencies r_L (between M and Q) and r_R (between Q and N), and assuming that there is no interference so that $r_L + r_R = r$, then the expected marker class means are given in Table 3. When F_1 progeny are backcrossed to the M_2N_2 parental line, two marker-class contrasts result as shown in Eqs. (1) and (2):

$$M_1N_1/M_2N_2 - M_2N_2/M_2N_2 = d + a \tag{1}$$

$$M_1N_2/M_2N_2 - M_2N_1/M_2N_2 = d(r_R - r_L) + a(r_R - r_L) + 2ar_Rr_L. \tag{2}$$

If the frequency of double recombinants is very low,[61] the term $2ar_Rr_L$ goes to zero and the contrast M_1N_2/M_2N_2 vs. M_2N_1/M_2N_2 is approximately $(d + a)(r_R - r_L)$. These contrasts will give estimates of the effects and the map positions of QTLs relative to the flanking markers.

The strength of evidence for the presence of a QTL at a particular location compared to the lack of segregating QTLs thereat is measured using the likelihood of odds (LOD) score. Larger LOD scores correspond to greater evidence for the presence of a QTL. To calculate the LOD score, we assume that, given a QTL genotype, the phenotype is distributed normally with a mean μ_{Q1} or μ_{Q2}, according to whether the genotype is Q_1Q_2 or Q_2Q_2, respectively, with a common standard deviation σ. A likelihood ratio (LR) is calculated at each position of the genome. At position z,

$$\text{LR} = -2\ln\left(\frac{\max[\Pr(Data \mid Presence\ of\ QTL)]}{\max[\Pr(Data \mid Absence\ of\ QTL)]}\right). \tag{3}$$

The LOD score is then calculated as $\text{LOD} = \frac{\text{LR}}{2\ln(10)}$.

The F test has been used in a nonlinear least squares approach to test for the presence of segregating QTLs.[59] In this case, the ratio of model mean squares to residual mean squares is used. Under the null hypothesis, this ratio follows an approximately central F distribution. Proponents of this approach argue that nonlinear regression, unlike maximum likelihood, can be performed by more statistical packages and that significance testing by

Table 3   Conditional Probabilities for the QTL Genotypes Given the Genotypes at Two Flanking Markers

Marker Genotype		QTL Genotype	Conditional Probability	Trait Value	Conditional Probability × Trait Value	Marker Trait Expectation
Left	Right					
M_1M_2	N_1N_2	Q_1Q_2	$(1-r_L)(1-r_R)/(1-r)$	d	$d(1-r-r_Lr_R)/(1-r)$ $+[-ar_Lr_R/(1-r)]$	d
		Q_2Q_2	$r_Lr_R/(1-r)$	$-a$		
	N_2N_2	Q_1Q_2	$(1-r_L)r_R/r$	d	$(1-r_L)dr_R/r+$ $[-ar_L(1-r_R)/r]$	$dr_R - ar_L-$ $r_Lr_R(d-a)/r$
		Q_2Q_2	$r_L(1-r_R)/r$	$-a$		
M_2M_2	N_1N_2	Q_1Q_2	$r_L(1-r_R)/r$	d	$dr_L(1-r_R)/r+$ $[-(r_L-1)ar_R/r]$	$dr_L + ar_R-$ $r_Lr_R(d+a)/r$
		Q_2Q_2	$(r_L-1)r_R/r$	$-a$		
	N_2N_2	Q_1Q_2	$r_Lr_R/(1-r)$	d	$dr_Lr_R/(1-r)+$ $[d(1-r-r_Lr_R)/(1-r)]$	$-a$
		Q_2Q_2	$(1-r_L)$ $(1-r_R)/(1-r)$	$-a$		

an F test is more familiar to investigators than a likelihood ratio test (LRT). The main disadvantage of nonlinear regression is its limited applicability. For instance, it cannot be applied to estimate the recombination frequency between a QTL and a single marker or to estimate QTL variance effects.

The advantages of maximum-likelihood interval mapping over ANOVA include (1) evidence for the presence of QTLs depicted as a continuous curve of LOD scores as a function of the location on a chromosome; (2) inference of QTLs to positions between markers; (3) improvement in the estimation of QTL effects due to an attenuation of apparent effects at marker loci resulting from recombination between the markers and the QTLs; and (4) proper allowance for incomplete marker genotype data by moving to the next flanking marker, for which genotype data is present if the immediate one is missing. The presence of genotyping errors can also be accommodated in maximum-likelihood interval mapping.

6.3. *Least Squares Regression*

Least squares regression has been used to approximate the likelihood map.[62,63] Least squares regression suffers from the problem that the residual variance contains part of the QTL variance not explained by the markers, and is therefore biased.[64] The iteratively reweighted least squares method[65,66] addresses this problem, but still ignores the mixture distribution of the residual error. Maximum likelihood (ML) mapping — developed by Lander and Botstein[60] and improved by Jansen,[67] Zeng,[68] and others — takes into full consideration the mixture distribution of the residual error. However, ML mapping is computationally more intensive than these least squares methods. Consequently, least squares methods are still commonly used in QTL mapping.

7. MULTIPLE QTL MODELS

Interval mapping can lead to a biased identification and estimation of QTLs when multiple QTLs are located within the same linkage group.[60,62,68] To address this problem, Jansen[67] and Zeng[68,69] independently proposed a combination of interval mapping and multiple regression. This combination, termed composite interval mapping, uses markers outside of the interval that are being tested as covariates to control other QTLs and reduce residual variance.

All of the methods discussed thus far test the alternative hypothesis that there is one QTL in the interval being tested. If there is more than one QTL in an interval, then the estimates of effect sizes and locations are subject to error. On the one hand, if two QTLs within an interval have effects with the same sign, then there is potential for a type I error. This is because the two QTLs will be combined into one QTL within one effect size that is inflated, a peak location somewhere between the two loci, and a wide interval surrounding the peak location. On the other hand, if two QTLs within an interval have effects of opposing sign, then there is potential for a type II error. This is because the two loci to some degree cancel each other. If the difference in effect sizes is small, then it is likely that no QTL will be detected; if the difference in effect sizes is large, then it is likely that the QTL with the larger effect size will be detected, but with an underestimate of its effect size and a location that is shifted toward the second, undetected locus.

Kao et al.[70] developed the multiple interval mapping method to directly control the presence of multiple QTLs. The likelihood for m QTLs is a mixture of 2^m normal distributions, with at least $2m+2$ parameters to estimate (the overall phenotypic mean, the residual error, and the position and effect for each QTL). The expectation-maximization (EM) algorithm is used to obtain the estimates of the genetic parameters, using the phenotypes and markers as observed data and the QTLs as missing data. Kao et al.[70] proposed a stepwise selection technique based on the LRT statistic for identifying QTLs. Their proposal consists of three steps. First, critical values for including or deleting a QTL are specified. Second, for each interval, LRT statistics are calculated; the position with the largest statistic exceeding the critical value for inclusion is added to the model. Third, the model with one additional QTL is considered; if the partial LRT statistic for a QTL is not significant, then the QTL is deleted from the model. These steps are iterated because the inclusion of QTLs may reduce residual error, leading to significance for inclusion from partial LRT statistics for currently excluded QTLs. After the model with all included QTLs is chosen, positions and effects are re-estimated.

Other possible model selection procedures besides stepwise selection include forward selection, backward selection, and chunk selection. Other criteria besides the LRT statistic include the Akaike information criterion,[71] cross-validation,[72] predictive sample reuse,[73] the Bayes information criterion,[74] the risk inflation criterion,[75] the minimum posterior predictive loss,[76] and the deviance information criterion.[77]

These approaches provide only point estimates for the numbers, locations, and effects of QTLs. Other procedures for establishing critical values for significance tests and interval estimates of the parameters, besides assuming asymptotic χ^2 distributions for LRT statistics, include permutation testing[78] and bootstrapping.[79]

8. BAYESIAN MODEL SELECTION FOR MAPPING MULTIPLE QTLs

Disadvantages of the ML methods include the ignoring of uncertainty in the model, a hierarchical testing strategy that includes a dynamic null hypothesis, and the dependence of selection criteria on the quantity of data. In the Bayesian approach, a likelihood function for the phenotype is constructed and prior distributions are assigned to all unknowns. The proposed Bayesian methods simultaneously estimate the number, positions, and genetic effects of QTLs, and thus avoid problems that can arise from misspecification of the number of QTLs. Inferences about particular parameters of interest are obtained using Markov chain Monte Carlo (MCMC) algorithms[80,81] conditional on the observed data, but not on the particular values of any of the unknowns. Therefore, Bayesian methods can provide more robust inferences than non-Bayesian methods. In addition to point estimates, posterior confidence intervals and variance estimates for any parameter of interest can be obtained from the marginal posterior distribution. In this way, we can avoid difficult problems concerning the critical values for testing multiple QTL hypotheses. Finally, Bayesian methods can incorporate biologically meaningful prior information, which can improve statistical power.

Bayesian methods can treat the unknown number of QTLs as a random variable. This advantage over ML methods results in the complication of the search space having a variable dimensionality. The reversible jump MCMC algorithm can explore posterior distributions in this setting.[82] However, the ability to jump among models of different dimensionality requires the careful construction of proposal distributions. Despite the challenges of implementing the reversible jump MCMC algorithm, several groups have developed approaches using it to map multiple, noninteracting QTLs.[83-90] Bayesian approaches to map epistatic interactions have also been developed using the reversible jump MCMC algorithm.[31,32,91-93]

Yi[94] proposed a composite space approach to circumvent the difficulties of implementing the reversible jump MCMC algorithm. The composite space approach fixes the dimensionality of search space.[95,96] Yi[94] proposed to augment variable dimensionality to fixed dimensionality by constraining the number of QTLs to an upper bound. Latent binary variables are then used to indicate the inclusion or exclusion status for each putative

QTL. Yi *et al.*[97] extended this model to include epistasis. Analogous to QTLs with main effects, putative epistatic interactions have latent binary variables that are used to indicate inclusion or exclusion.

9. MAPPING QtLs FOR BINARY AND ORDINAL TRAITS

A complex binary trait has a dichotomous expression with a polygenic background. Many human diseases are treated as binary traits, and are scored as present or absent. Mapping QTLs for such traits is difficult because of the discrete nature and reduced variation in the phenotypic distribution. Binary traits are usually described by a threshold function, which links the binary phenotype with a latent continuous variable called the liability. Liability can be modeled as an unobservable quantitative trait. Let s_i and $y_i (i = 1, \ldots, n)$ be the binary phenotype and the underlying liability, respectively, of the ith individual. The threshold model assumes that there is a fixed threshold in the scale of liability, t, that determines the binary phenotype of an individual by comparing y_i with t. If $y_i > t$, we assign $s_i = 1$; otherwise, $s_i = 0$.

An ordinal trait can be considered as an extension of a binary trait. An ordinal trait has a discrete number of categories that are ordered. The threshold function used for modeling the liability for binary traits can be extended to include multiple categories. For ordinal traits, we observe an ordinal response w_i on the ith individual, with w_i belonging to one of J ordered categories, $1, \ldots, J$, for $i = 1, 2, \ldots, n$. Let y_i^* represent the value of the liability associated with the ith individual. The relationship between the unobservable liability y_i^* and the observed phenotype w_i is $w_i = j \Leftrightarrow t_{j-1}^* \leq y_i^* < t_j^*$, $j \in \{1, 2, \ldots, J\}$; $t_0^* = -\infty$, $t_J^* = +\infty$, and $t_1^*, \ldots, t_{J-1}^*$ are unknown threshold values that divide the real line into J intervals. Thus, if the realized value of y_i^* belongs to the jth interval, the ordinal response w_i is j.

We use the probit model to analyze ordinal traits under the Bayesian framework. The main point of the method is that by introducing liability into the problem, the probit model on the ordinal trait is linked to the normal linear model on the continuous liability and a set of unknown threshold values. Therefore, we are able to turn the problem of discrete traits into a missing value problem in continuous traits. The missing values are easily augmented with the MCMC algorithm. With the generated liability values, MCMC algorithms for the number, locations, and genetic effects of QTLs are the same as those for normally distributed traits.

We describe our mapping method for an F_2 design; the method can be applied to other experimental designs such as backcrosses, recombination inbred lines, and four-way crosses. If no epistasis is assumed, liability follows the linear model

$$y_i^* = \mu^* + \sum_{q=1}^{l} (u_{iq} a_q^* + v_{iq} d_q^*) + e_i^*, \tag{4}$$

where μ^* is the overall mean; l is the number of QTLs affecting liability; a_q^* and d_q^* are the additive and dominance effects of the qth QTL, respectively; u_{iq} and v_{iq} are the indicator variables for the genotype of the qth QTL for the ith individual, defined as

$$u_{iq} = \begin{cases} +1 & \text{for } Q_1 Q_1 \\ 0 & \text{for } Q_1 Q_2 \\ -1 & \text{for } Q_2 Q_2 \end{cases} \quad \text{and } v_{iq} = \begin{cases} +\frac{1}{2} & \text{for } Q_1 Q_2 \\ -\frac{1}{2} & \text{for } Q_1 Q_1 \text{ or } Q_2 Q_2 \end{cases},$$

where $Q_1 Q_1$, $Q_1 Q_2$, and $Q_2 Q_2$ denote three genotypes at the qth QTL; and e_i^* is the residual error assumed to follow $N(0, \sigma^{*2})$.

Note that the inequality $t_{j-1}^* \le y_i^* < t_j^*$ can be re-expressed as

$$(t_{j-1}^* - t_1^*)/\sigma^* \le (y_i^* - t_1^*)/\sigma^* < (t_j^* - t_1^*)/\sigma^*.$$

Therefore, we can take $(y_i^* - t_1^*)/\sigma^*$ as the underlying variable, and the model in Eq. (4) can be rewritten as

$$\frac{y_i^* - t_1^*}{\sigma^*} = \frac{\mu^* - t_1^*}{\sigma^*} + \sum_{q=1}^{l} \left(u_{iq} \frac{a_q^*}{\sigma^*} + v_{iq} \frac{d_q^*}{\sigma^*} \right) + \frac{e_i^*}{\sigma^*}. \tag{5}$$

Thus, the threshold model can be reformulated in terms of a reparameterized threshold model in which underlying continuous variables correspond to $(y_i^* - t_1^*)/\sigma^*$, $i = 1, \dots, n$; threshold values correspond to $(t_j^* - t_1^*)/\sigma^*$, $j = 1, \dots, J$; overall mean, additive, and dominance effects correspond to $\mu^* - t_1^*/\sigma^*$, a_q^*/σ^*, and d_q^*/σ^*, $q = 1, \dots, l$, respectively; and residual variance corresponds to e_i^*/σ^*.

In this reparameterized threshold model, residual variance equals one and the first threshold value equals zero. For binary traits, $J = 2$; there are no unknown threshold values. For ordinal traits, we further consider the following reparameterization:

$$\sigma = \frac{\sigma^*}{t_{J-1}^* - t_1^*}, \quad t_j = \sigma \frac{t_j^* - t_1^*}{\sigma^*}, \quad \mu = \sigma \frac{\mu^* - t_1^*}{\sigma^*}, \quad a_q = \sigma \frac{a_q^*}{\sigma^*}, \quad d_q = \sigma \frac{d_q^*}{\sigma^*},$$

$$e_i = \sigma \frac{e_i^*}{\sigma^*}, \quad \text{and } y_i = \sigma \frac{y_i^* - t_1^*}{\sigma^*}, \tag{6}$$

for all i, j, and q. With the reparameterization in Eq. (6), the threshold model in Eq. (4) becomes

$$y_i = \mu + \sum_{q=1}^{l}(u_{iq}a_q + v_{iq}d_q) + e_i. \tag{7}$$

Here, y_i is the liability of the ith individual; μ is the overall mean; a_q and d_q are the additive and dominance effects of the qth QTL for the liability y_i, respectively; and e_i is the residual error distributed as $N(0, \sigma^2)$. The relationship between the liability y_i and the observed phenotype w_i is

$$w_i = j, \quad \text{if } t_{j-1} \le y_i < t_j,$$

with the reparameterized thresholds

$$-\infty = t_0 < t_1 = 0 < t_2 < \cdots < t_{J-2} < t_{J-1} = 1 < t_J = +\infty.$$

Suppose we observe the ordinal responses $\mathbf{w} = \{w_i\}_{i=1}^{n}$ and a set of marker genotypes $\mathrm{M} = \{M_{jk}\}_{j=1,k=1}^{n,K}$ for n individuals and K markers. For an F_2 population, M_{jk} takes one of three values denoting three different genotypes. Our goal is to make an inference about the number of QTLs l, their locations $\boldsymbol{\lambda} = \{\lambda_q\}_{q=1}^{l}$, and their additive effects $\mathbf{a} = \{a_q\}_{q=1}^{l}$ and dominance effects $\mathbf{d} = \{d_q\}_{q=1}^{l}$, in which λ_q is the location of the qth QTL represented by the distance of the QTL from one end of the corresponding chromosome. We denote the vector of all model parameters by $\boldsymbol{\theta} = (\mathbf{a}, \mathbf{d}, \mu, \sigma^2)$. The genotypes of putative QTLs are usually unobserved; thus, the coefficients of the model in Eq. (7), $\mathbf{u} = \{u_{iq}\}$ and $\mathbf{v} = \{v_{iq}\}$, are missing values. Note that v_{iq} is determined by u_{iq}, and thus can be suppressed from the list of unknowns.

9.1. *Prior and Posterior Distributions*

For the multiple threshold model, the joint posterior distribution of all unknowns, given the observed data and prior information for unknowns, can be written as

$$p(l, \boldsymbol{\lambda}, \boldsymbol{\theta}, \mathbf{u}, \mathbf{t}, \mathbf{y} \mid \mathbf{w}, \mathrm{M}) \propto p(\mathbf{w} \mid \mathbf{y}, \mathbf{t})p(\mathbf{y} \mid l, \boldsymbol{\theta}, \mathbf{u})p(\mathbf{u} \mid l, \boldsymbol{\lambda}, \mathrm{M}) \\ p(l, \boldsymbol{\lambda}, \boldsymbol{\theta}, \mathbf{t}). \tag{8}$$

The first term on the right is

$$p(\mathbf{w} \mid \mathbf{y}, \mathbf{t}) = \prod_{i=1}^{n} p(w_i \mid y_i, \mathbf{t}) = \prod_{i=1}^{n}\left\{\sum_{j=1}^{J} 1(t_{j-1} \le y_i < t_j)1(w_i = j)\right\},$$

where $1(X \in A)$ is an indicator function, taking a value of one if $X \in A$ is true and zero otherwise. The second term is the conditional distribution of the liability given all unknowns, and has the following form:

$$p(\mathbf{y} \mid l, \boldsymbol{\theta}, \mathbf{u}) = \prod_{i=1}^{n} p(y_i \mid l, \boldsymbol{\theta}, \mathbf{u}) = \prod_{i=1}^{n} (2\pi\sigma^2)^{-\frac{n}{2}}$$

$$\times \exp\left\{ -\frac{[y_i - \mu - \sum_{q=1}^{l}(u_{iq}a_q + v_{iq}d_q)]^2}{2\sigma^2} \right\}.$$

The third term is the conditional distribution of the genotypes of the putative QTLs. Assuming that there is at most one QTL on any marker interval, $p(\mathbf{u} \mid l, \boldsymbol{\lambda}, \mathbf{M})$ can be factorized as

$$p(\mathbf{u} \mid l, \boldsymbol{\lambda}, \mathbf{M}) = \prod_{i=1}^{n} \prod_{q=1}^{l} p(u_{iq} \mid \lambda_{iq}, m_{iq}^{\mathrm{L}}, m_{iq}^{\mathrm{R}}),$$

where m_{iq}^{L} and m_{iq}^{R} represent the left and the right flanking marker genotypes, respectively.

To implement Bayesian analysis, we need to specify the prior distributions for $l, \boldsymbol{\lambda}, \boldsymbol{\theta}$, and $\mathbf{t}$. Assuming prior conditional independence of the parameters, we can factorize the joint prior distribution into the following products:

$$p(l, \boldsymbol{\lambda}, \boldsymbol{\theta}, \mathbf{t}) = p(l)p(\boldsymbol{\lambda} \mid l)p(\mu)p(\sigma^2) \prod_{q=1}^{l} [p(a_q)p(d_q)]p(\mathbf{t}).$$

The prior distribution for l is chosen to be a uniform distribution between 0 and a prespecified integer L. A common choice for the prior of $\boldsymbol{\lambda}$, when no information regarding the locations is available, is uniform over the entire genome. The priors for the overall mean μ and the QTL effects $a_q, d_q (q = 1, 2, \ldots, l)$ are assumed to be independently normal, i.e. $\mu \sim N(\eta_0, \tau_0^2)$ and $a_q, d_q \sim N(\eta, \tau^2)$, with prespecified prior means η_0 and η as well as variances τ_0^2 and τ^2. The prior for σ^2 is assumed to be a scaled, inverted χ^2 distribution with known hyperparameter values ν_0 and σ_0^2, so that $\sigma^2 \sim \mathrm{Inv}\text{-}\chi^2(\nu_0, \sigma_0^2)$. Finally, we assume a uniform prior on

$$\mathbf{t} = (t_2, \ldots, t_{J-1}), \text{ i.e. } p(\mathbf{t}) \propto 1, \quad \text{for } 0 < t_2 < \cdots < t_{J-2} < 1.$$

9.2. *The Markov Chain Monte Carlo Algorithm*

In QTL mapping problems, the calculation of the joint posterior distribution is analytically intractable, and thus a Markov chain Monte Carlo

(MCMC) approach is required to obtain observations from the joint posterior distribution. To sample from the posterior distribution, we need to generate l, $\boldsymbol{\lambda}$, $\boldsymbol{\theta}$, $\mathbf{u}$, $\mathbf{t}$, and $\mathbf{y}$ from their respective conditional distributions. Conditional on l, $\mathbf{u}$, $\boldsymbol{\lambda}$, and $\mathbf{y}$, the model in Eq. (4) is a conventional linear model, and thus $\boldsymbol{\theta}$ can be sampled using a Gibbs sampler.[89,91,98] QTL locations depend strongly on the QTL genotypes; hence, we adopt the Metropolis–Hastings algorithm to jointly update the locations and genotypes of QTLs.[91,99,100] The dimension of the parameter space is determined by the number of QTLs l. For the multiple threshold model used here, we need an algorithm to generate $\mathbf{y}$ and $\mathbf{t}$.

Because of the strong correlation between the liability and the threshold values, it is desirable to jointly generate $\mathbf{t}$ and $\mathbf{y}$. To draw $\mathbf{t}$ and $\mathbf{y}$ jointly from the conditional distribution

$$p(\mathbf{t}, \mathbf{y} \mid \mathbf{w}, l, \boldsymbol{\lambda}, \boldsymbol{\theta}, \mathbf{u}) = p(\mathbf{t}|\mathbf{w}, l, \boldsymbol{\theta}, \mathbf{u})p(\mathbf{y} \mid \mathbf{w}, l, \boldsymbol{\theta}, \mathbf{u}, \mathbf{t}),$$

we first draw the threshold values $\mathbf{t}$ from $p(\mathbf{t}|\mathbf{w}, l, \boldsymbol{\theta}, \mathbf{u})$, and then draw the liability values $\mathbf{y}$ from $p(\mathbf{y} \mid \mathbf{w}, l, \boldsymbol{\theta}, \mathbf{u}, \mathbf{t})$.

Given l, $\boldsymbol{\theta}$, $\mathbf{u}$, and $\mathbf{t}$, $w_1, w_2, \ldots, w_n$ are independent. Therefore, the conditional distribution $(\mathbf{t} \mid \mathbf{w}, l, \boldsymbol{\theta}, \mathbf{u})$ is

$$p(\mathbf{t} \mid \mathbf{w}, l, \boldsymbol{\theta}, \mathbf{u}) \propto p(\mathbf{w} \mid \mathbf{t}, l, \boldsymbol{\theta}, \mathbf{u})$$

$$= \prod_{i=1}^{n} p(w_i \mid \mathbf{t}, l, \boldsymbol{\theta}, \mathbf{u})$$

$$= \prod_{i=1}^{n} \sum_{j=1}^{J} 1(w_i = j)p(t_{j-1} \leq y_i < t_j \mid \mathbf{t}, l, \boldsymbol{\theta}, \mathbf{u})$$

$$= \prod_{i=1}^{n} \sum_{j=1}^{J} 1(w_i = j)\left\{ \Phi\left(\frac{t_j - \mu - \sum_q^l (u_{iq}a_q + v_{iq}d_q)}{\sigma}\right)\right.$$

$$\left. - \Phi\left(\frac{t_{j-1} - \mu - \sum_q^l (u_{iq}a_q + v_{iq}d_q)}{\sigma}\right)\right\}, \quad (9)$$

where $1(w_i = j)$ is an indicator function equal to one if $w_i = j$ and zero otherwise, and $\Phi(\cdot)$ is the standardized normal distribution function.

Apparently, this distribution has a nonstandard form. Therefore, the Metropolis–Hastings algorithm is used to generate samples from this distribution. To implement the Metropolis–Hastings algorithm, we first sample new threshold values $t_2^*, t_3^*, \ldots, t_{J-2}^*$ uniformly from the intervals $[\max\{0, t_2 - d\}, \min\{t_3, t_2 + d\}]$, $[\max\{t_2^*, t_3 - d\}, \min\{t_4, t_3 + d\}], \ldots,$ $[\max\{t_{J-3}^*, t_{J-2} - d\}, \min\{1, t_{J-1} + d\}]$, respectively, where d is a predetermined tuning parameter. The proposal $\mathbf{t}^* = (t_2^*, t_3^*, \ldots, t_{J-2}^*)$ is then accepted with probability $\min\{1, r\}$, where

$$r = \frac{p(\mathbf{w} \mid \mathbf{t}, l, \boldsymbol{\theta}, \mathbf{u})}{p(\mathbf{w} \mid \mathbf{t}^*, l, \boldsymbol{\theta}, \mathbf{u})}, \tag{10}$$

and $p(\mathbf{w} \mid \mathbf{t}, l, \boldsymbol{\theta}, \mathbf{u})$ and $p(\mathbf{w} \mid \mathbf{t}^*, l, \boldsymbol{\theta}, \mathbf{u})$ are calculated by Eq. (9).

Note that $p(\mathbf{y}|\mathbf{w}, l, \boldsymbol{\theta}, \mathbf{u}, \mathbf{t}) = \prod_{i=1}^{n} p(y_i|w_i, l, \boldsymbol{\theta}, \mathbf{u}, \mathbf{t})$. Therefore, the liability values can be generated individual by individual. Given w_i, l, $\boldsymbol{\theta}$, $\mathbf{u}$, and $\mathbf{t}$, the conditional posterior of y_i is a truncated normal distribution, i.e.

$$p(y_i|w_i, l, \boldsymbol{\theta}, \mathbf{u}, \mathbf{t})$$

$$= \frac{\varphi[\mu + \sum_q^l (u_{iq}a_q + v_{iq}d_q), \sigma^2]}{\Phi\left(\frac{t_j - \mu - \sum_q^l (u_{iq}a_q + v_{iq}d_q)}{\sigma}\right) - \Phi\left(\frac{t_{j-1} - \mu - \sum_q^l (u_{iq}a_q + v_{iq}d_q)}{\sigma}\right)},$$

$$\text{if } w_i = j, \tag{11}$$

where $\varphi(x, \sigma^2)$ stands for the normal density with mean x and variance σ^2.

We adopt the inverse transformation method[101] to sample from the doubly truncated normal distribution in Eq. (11). With this method, we first simulate U uniformly from the interval $[0, 1]$. Then, the draw from the truncated normal distribution is

$$y_i = \mu + \sum_q^l (u_{iq}a_q + v_{iq}d_q) + \sigma\Phi^{-1}[p_1 + U(p_2 - p_1)],$$

where Φ^{-1} is the inverse c.d.f. of the standard normal distribution,

$$p_1 = \Phi\left(\frac{t_j - \mu - \sum_{q=1}^{l}(u_{iq}a_q + v_{iq}d_q)}{\sigma}\right) \quad \text{and}$$

$$p_2 = \Phi\left(\frac{t_{j-1} - \mu - \sum_{q=1}^{l}(u_{iq}a_q + v_{iq}d_q)}{\sigma}\right).$$

The reparameterized threshold model has several attractive features. First, the model has only J-3 unknown threshold values; therefore, when $J = 3$, there are no unknown thresholds. Second, the number of the threshold values is reduced by one at the expense of increasing a model parameter, i.e. the residual variance σ^2; however, the posterior distribution of σ^2 is a standard distribution, i.e. an inverse χ^2 distribution, and thus is easily sampled. Finally, all unknown threshold values are between 0 and 1, i.e. $0 \le t_j < 1$ for $j = 2, 3, \ldots, J - 2$; hence, it may be easier to generate these thresholds using the MCMC algorithm.

10. MAPPING QTLs FOR CONTINUOUS TRAITS

We now describe the composite space model.[94,97] The observed data consist of phenotypic trait values $\mathbf{y}$ and marker genotypes $\mathbf{m}$. We assume that markers are organized into a linkage map, and restrict our attention to models with, at most, pairwise interactions. The entire genome is partitioned into H loci, $\boldsymbol{\zeta} = \{\zeta_1, \ldots, \zeta_H\}$; we assume that the possible QTLs occur at these fixed positions. When the markers are densely and regularly spaced, we set $\boldsymbol{\zeta}$ to the marker positions; otherwise, $\boldsymbol{\zeta}$ includes not only the marker positions, but also a grid of points between markers (pseudomarkers). The tuning parameter sets the spacing of points between pseudomarkers; we typically use 1–10 cM. In general, the genotypes $\mathbf{g}$ at loci $\boldsymbol{\zeta}$ are unobservable except at completely informative markers, but their probability distribution $p(\mathbf{g} \,|\, \boldsymbol{\zeta}, \mathbf{m})$ can be inferred from the observed marker data using the multipoint method.[102]

Although a complex trait may be influenced by many loci, our emphasis is on a set of, at most, L QTLs with detectable effects. Typically, L will be much smaller than H. Let $\boldsymbol{\lambda} = \{\lambda_1, \ldots, \lambda_L\}$ $(\in \{\zeta_1, \ldots, \zeta_H\})$ be the current positions of L putative QTLs. Each locus may affect the trait through its marginal (main) effects and/or interactions with other loci (epistasis). The phenotype distribution is assumed to follow the linear model

$$\mathbf{y} = \boldsymbol{\mu} + \mathbf{X}\boldsymbol{\beta} + \mathbf{e}, \tag{12}$$

where $\boldsymbol{\mu}$ is the overall mean, $\boldsymbol{\beta}$ denotes the vector of all possible main effects and pairwise interactions of L potential QTLs, $\mathbf{X}$ is the design matrix, and $\mathbf{e}$ is the vector of independent normal errors with mean zero and variance σ^2. The number of genetic effects depends on the experimental design, and the design matrix $\mathbf{X}$ is determined from those genotypes $\mathbf{g}$ at the current loci $\boldsymbol{\lambda}$ by using a particular genetic model, e.g. the Cockerham model.[18,103]

As in the Bayesian variable selection for linear regression,[104–106] we introduce a binary variable γ for each effect, indicating that the corresponding effect is included $(\gamma = 1)$ or excluded $(\gamma = 0)$ from the model. Letting $\Gamma = \mathrm{diag}(\gamma)$, the model becomes

$$\mathbf{y} = \boldsymbol{\mu} + \mathbf{X}\Gamma\boldsymbol{\beta} + e. \tag{13}$$

This linear model defines the likelihood $p(\mathbf{y} \,|\, \boldsymbol{\gamma}, \mathbf{X}, \boldsymbol{\theta})$ with $\boldsymbol{\theta} = (\mu, \boldsymbol{\beta}, \sigma^2)$, and the full posterior can be written as

$$p(\boldsymbol{\gamma}, \boldsymbol{\lambda}, \mathbf{g}, \boldsymbol{\theta} \,|\, \mathbf{y}, \mathbf{m}) \propto p(\mathbf{y} \,|\, \boldsymbol{\gamma}, \mathbf{X}, \boldsymbol{\theta}) p(\boldsymbol{\gamma}, \boldsymbol{\lambda}, \mathbf{g}, \boldsymbol{\theta} \,|\, \mathbf{m}). \tag{14}$$

The vector γ determines the number of QTLs. Hereafter, we denote the included positions of QTLs as λ_γ. The vector (γ, λ_γ) comprises a model index that identifies the genetic architecture of the trait. A natural model selection strategy is to choose the most probable model (γ, λ_γ) based on its marginal posterior, $p(\gamma, \lambda_\gamma \mid \mathbf{y}, \mathbf{m})$.[107] For genomewide epistatic analysis, however, no single model may stand out; hence, we average over possible models when assessing the characteristics of genetic architecture, with the various models weighted by their posterior probability.[108–110]

10.1. *Prior Distributions*

We suggest specifying the prior expected number of QTLs, l_0, based on initial investigations with traditional methods, and then determining a reasonably large upper bound, L. If the number of markers is small and the number of loci H is large, a reasonable value for L is the number of markers. As an extreme case, we could take the total number of loci (H) as the upper bound. Since the number of detectable QTLs is usually much less than H, such a choice is unlikely optimal. In practice, one could experiment with several values of the expected number of QTLs and investigate their impact on posterior inference.

For the indicator vector γ, we use an independence prior of the form

$$p(\gamma) = \prod w_j^{\gamma_j} (1 - w_j)^{1-\gamma_j}, \tag{15}$$

where $w_j = p(\gamma_j = 1)$ is the prior inclusion probability for the jth effect. We assume that w_j equals the predetermined hyperparameter w_m or w_e, depending on the jth effect being the main effect or epistatic effect, respectively. Under this prior, the importance of any effect is independent of the importance of any other effect, and the prior inclusion probability of a main effect is different from that of an epistatic effect.

The hyperparameters w_m and w_e control the expected numbers of main and epistatic effects included in the model, respectively. Instead of directly specifying w_m and w_e, it may be better to first determine the prior expected numbers of main-effect QTLs l_m and all QTLs $l_0 \geq l_\mathrm{m}$ (i.e. main-effect and epistatic QTLs), and then solve w_m and w_e from the expressions of the prior expected numbers. It is reasonable to require that $w_\mathrm{m} \geq w_\mathrm{e}$, which requires some adjustment when $l_\mathrm{m} = 0$.

The prior expected number of main-effect QTLs can be expressed as

$$l_\mathrm{m} = L[1 - (1 - w_\mathrm{m})^K], \tag{16}$$

and the prior expected number of all QTLs as

$$l_0 = L[1 - (1 - w_{\mathrm{m}})^K (1 - w_{\mathrm{e}})^{K^2(L-1)}], \tag{17}$$

where K is the number of possible main effects for each QTL and K^2 is the number of possible epistatic effects for any two QTLs. From Eqs. (16) and (17), we obtain

$$w_{\mathrm{m}} = 1 - \left[1 - \frac{l_{\mathrm{m}}}{L}\right]^{\frac{1}{K}} \tag{18}$$

and

$$w_{\mathrm{e}} = 1 - \left[\frac{1 - \frac{l_0}{L}}{(1 - w_{\mathrm{m}})^K}\right]^{\frac{1}{K^2(L-1)}}. \tag{19}$$

Here, we have fixed w_{m} and w_{e}, but we could relax this by treating w_{m} and w_{e} as unknown model parameters and assigning priors.[111] We note that if no main-effect QTLs are detected by traditional nonepistatic mapping methods and $l_{\mathrm{m}} = 0$, then $w_{\mathrm{m}} = 0$. In this case, we suggest making all weights equal and using Eq. (17) to obtain

$$w = 1 - \left(1 - \frac{l_0}{L}\right)^{\frac{1}{K + K^2(L-1)}}. \tag{20}$$

If there is no prior information concerning QTL locations, these could be assumed to be independent and uniformly distributed over H possible loci. Thus, given l_0, the prior probability that any locus is included becomes l_0/H. In practice, it may be reasonable to assume that any interval of a given length (e.g. 1 cM or 10 cM) contains at most one QTL. Although this assumption is not necessary, it can substantially reduce model space and thus accelerate the search procedure.

Yi[94] discussed the choice of the prior distribution for the genetic effects $\boldsymbol{\beta}$. First, we could use a normal prior for each vector of the genetic effects, i.e. $\boldsymbol{\beta}_j \sim N(0, \Sigma)$, $j = 1, \ldots, K$, where the prior mean of zero reflects indifference between the positive and negative values, and Σ is the prior covariance matrix. The covariance matrix Σ could be chosen to be diagonal. In this prior specification, the prior distribution for each QTL is identical and is independent of γ_j. Most Bayesian mapping methods use this type of prior distribution. This prior has been used in Bayesian variable selection for the linear regression model.[105] Second, we could use the prior $p(\boldsymbol{\beta}_j | \gamma_j) = (1 - \gamma_j)N(0, \Sigma) + \gamma_j N(0, c^2\Sigma)$, where c^2 is a predetermined constant.[112,113] Third, we could use $p(\boldsymbol{\beta}_\gamma | \boldsymbol{\gamma}, \mathbf{x}_\gamma) \sim N[0, c^2(\mathbf{x}_\gamma'\mathbf{x}_\gamma)^{-1}\sigma^2]$, in which c^2 is a hyperparameter. This prior has been extensively used in Bayesian variable selection for the conventional linear model.

The prior distributions on the model index γ and the genetic effects $\boldsymbol{\beta}$ may be the most critical factors influencing the performance of the algorithms, and thus deserve careful attention. The hyperparameter w or L in the prior of γ controls the expected proportion of genetic effects and the number of QTLs included in the model. The prior covariance matrix or the hyperparameter c in the prior of $\boldsymbol{\beta}$ controls the expected size of genetic effects included in the model. Small w or L and large prior variance or c concentrate the prior on parsimonious models with large effects, while large w or L and small prior variance or c concentrate on saturated models with small effects. The reasonable choices of c and w would account for the sample size n, the marker information, and the upper bound of QTLs L. For the conventional linear model, Fernández *et al.*[114] recommended $c = \max\{n, L^2\}$. George and Foster[107] proposed to treat c and w as unknown parameters, and use empirical Bayes estimates of c and w based on the data. We could consider hyperprior distributions on w or L and Σ or c.[90,106] Gaffney[90] proposed to adjust the prior variances of genetic effects as the number of QTLs changes, and showed that such a strategy can reduce sensitivity to the priors of genetic effects.

Yi *et al.*[97] implemented a two-component mixture prior distribution, with effect sizes assigned a point mass at zero for excluded QTLs and normally distributed for included QTLs (a point-normal prior). This prior distribution reflects indifference between the negative and positive effects. Zhang *et al.*[115] proposed a Bayesian framework consisting of three categories of genetic effects: a positive-effect category for all QTLs with detectable positive effects on the phenotypic value, a negative-effect category for all QTLs with detectable negative effects on the phenotypic value, and a negligible-effect category for all non-QTL markers and all nondetectable QTLs. They used two truncated normal distributions to model the positive-effect and negative-effect categories, an inverse gamma distribution for the variance parameters, and a point mass at zero for negligible effects. Rather than using binary indicator variables, Zhang *et al.*[115] modeled the probabilities that a locus belongs to one of the three categories. This approach tests each marker in the full model, and as such may not be optimally efficient. *A priori* information regarding the probability of a locus belonging to one of the three categories can be incorporated into a Dirichlet conjugate prior distribution. A problem with the point-normal prior is that, with enough data, the posterior will favor the inclusion of effects with estimated nonzero sizes, regardless of how small the sizes are.[106]

We propose the following hierarchical mixture prior for each genetic effect:

$$\beta_j | (\gamma_j, \sigma^2, \mathbf{x}_{\cdot j}) \sim N[0, \gamma_j c \sigma^2 (\mathbf{x}_{\cdot j}^T \mathbf{x}_{\cdot j})^{-1}], \tag{21}$$

where $\mathbf{x}_{\cdot j} = (x_{1j}, \ldots, x_{nj})^T$ is the vector of the coefficients of β_j, and c is a positive scale factor. We take $c = n$, which is a popular choice and yields the BIC criterion[74] if the prior inclusion probability for each effect equals 0.5.[106,107]

In this prior setup, a point mass prior at 0 is used for the genetic effect β_j if $\gamma_j = 0$, effectively removing β_j from the model. If $\gamma_j = 1$, the prior variances reflect the precision of each β_j and are invariant to scale changes in the phenotype and the coefficients. The value $(\mathbf{x}_{\cdot j}^T \mathbf{x}_{\cdot j})^{-1}$ varies for different types of genetic effects. For a large backcross population with no segregation distortion, for example, $(\mathbf{x}_{\cdot j}^T \mathbf{x}_{\cdot j})^{-1}/n \approx 1/4$ for marginal effects and $[1 - (1 - 2r)^2]/16$ for epistatic effects, with r as the recombination fraction between two QTLs, under Cockerham's model.[103]

The prior for the overall mean μ is $N(\eta_0, \tau_0^2)$. We could empirically set

$$\eta_0 = \bar{y} = \frac{1}{n} \sum_{i=1}^n y_i, \text{ and } \tau_0^2 = s_y^2 = \frac{1}{n-1} \sum_{i=1}^n (y_i - \bar{y})^2.$$

We take the noninformative prior for the residual variance, $p(\sigma^2) \propto 1/\sigma^2$.[98] Although this prior is improper, it yields a proper posterior distribution for the unknowns and so can be used formally.[106]

10.2. *The Markov Chain Monte Carlo Algorithm*

To develop our MCMC algorithm, we first partition the vector of unknowns $(\boldsymbol{\lambda}, \mathbf{g}, \boldsymbol{\theta})$ into $(\boldsymbol{\lambda}_\gamma, \mathbf{g}_\gamma, \boldsymbol{\theta}_\gamma)$ and $(\boldsymbol{\lambda}_{-\gamma}, \mathbf{g}_{-\gamma}, \boldsymbol{\theta}_{-\gamma})$, representing the unknowns included and excluded from the model, respectively, in which $\boldsymbol{\lambda}_\gamma$ and $\mathbf{g}_\gamma$ ($\boldsymbol{\lambda}_{-\gamma}$ and $\mathbf{g}_{-\gamma}$) are the positions and the genotypes of QTLs included (excluded), respectively; $\boldsymbol{\beta}_\gamma (\boldsymbol{\beta}_{-\gamma})$ represent the genetic effects included (excluded); $\boldsymbol{\theta} = (\boldsymbol{\beta}, \mu, \sigma^2)$; $\boldsymbol{\theta}_\gamma = (\boldsymbol{\beta}_\gamma, \mu, \sigma^2)$; and $\boldsymbol{\theta}_{-\gamma} = \boldsymbol{\beta}_{-\gamma}$. Similarly, $\mathbf{X}_\gamma (\mathbf{X}_{-\gamma})$ represents the model coefficients included (excluded), which are determined by $\mathbf{g}$ and $\boldsymbol{\gamma}$.

We suppress the dependence on the observed marker data below. For a particular $\boldsymbol{\gamma}$, the likelihood function depends only upon the parameters

$(\mathbf{X}_\gamma, \boldsymbol{\theta}_\gamma)$ used by that model, i.e.

$$p(\mathbf{y} \mid \gamma, \mathbf{X}, \boldsymbol{\theta}) = p(\mathbf{y} \mid \gamma, \mathbf{X}_\gamma, \boldsymbol{\theta}_\gamma). \tag{22}$$

The prior distribution of $(\boldsymbol{\lambda}, \gamma, \mathbf{g}, \boldsymbol{\theta})$ can be partitioned as

$$p(\gamma, \boldsymbol{\lambda}, \mathbf{g}, \boldsymbol{\theta}) = p(\gamma)p(\boldsymbol{\lambda}_\gamma, \mathbf{g}_\gamma, \boldsymbol{\theta}_\gamma \mid \gamma)p(\boldsymbol{\lambda}_{-\gamma}, \mathbf{g}_{-\gamma}, \boldsymbol{\theta}_{-\gamma} \mid \gamma). \tag{23}$$

The full posterior distribution for $(\gamma, \boldsymbol{\lambda}, \mathbf{g}, \boldsymbol{\theta})$ can now be expressed as

$$p(\gamma, \boldsymbol{\lambda}, \mathbf{g}, \boldsymbol{\theta} \mid \mathbf{y}) \propto p(\mathbf{y} \mid \gamma, \mathbf{X}_\gamma, \boldsymbol{\theta}_\gamma)p(\gamma)p(\boldsymbol{\lambda}_\gamma, \mathbf{g}_\gamma, \boldsymbol{\theta}_\gamma \mid \gamma)$$
$$\times p(\boldsymbol{\lambda}_{-\gamma}, \mathbf{g}_{-\gamma}, \boldsymbol{\theta}_{-\gamma} \mid \gamma). \tag{24}$$

From Eq. (24), we can derive the following conditional posterior distributions:

$$p(\boldsymbol{\lambda}_\gamma, \mathbf{g}_\gamma, \boldsymbol{\theta}_\gamma \mid \gamma, \mathbf{y}) \propto p(\mathbf{y} \mid \gamma, \mathbf{X}_\gamma, \boldsymbol{\theta}_\gamma)p(\boldsymbol{\lambda}_\gamma, \mathbf{g}_\gamma, \boldsymbol{\theta}_\gamma \mid \gamma) \tag{25}$$

$$p(\boldsymbol{\lambda}_{-\gamma}, \mathbf{g}_{-\gamma}, \boldsymbol{\theta}_{-\gamma} \mid \gamma, \mathbf{y}) \propto p(\boldsymbol{\lambda}_{-\gamma}, \mathbf{g}_{-\gamma}, \boldsymbol{\theta}_{-\gamma} \mid \gamma) \tag{26}$$

$$p(\gamma \mid \boldsymbol{\lambda}, \mathbf{g}, \boldsymbol{\theta}, \mathbf{y}) \propto p(\mathbf{y} \mid \gamma, \mathbf{X}_\gamma, \boldsymbol{\theta}_\gamma)p(\gamma)p(\boldsymbol{\lambda}_\gamma, \mathbf{g}_\gamma, \boldsymbol{\theta}_\gamma \mid \gamma)$$
$$p(\boldsymbol{\lambda}_{-\gamma}, \mathbf{g}_{-\gamma}, \boldsymbol{\theta}_{-\gamma} \mid \gamma). \tag{27}$$

It can be seen that the unused parameters do not affect the conditional posterior of $(\boldsymbol{\lambda}_\gamma, \mathbf{g}_\gamma, \boldsymbol{\theta}_\gamma)$, and thus do not need to be updated conditional on γ. Since the unused parameters do not contribute to the likelihood, the posterior of $(\boldsymbol{\lambda}_{-\gamma}, \mathbf{g}_{-\gamma}, \boldsymbol{\theta}_{-\gamma})$ is identical to its prior. From Eq. (27), the conditional posterior of γ depends on $(\boldsymbol{\lambda}_{-\gamma}, \mathbf{g}_{-\gamma}, \boldsymbol{\theta}_{-\gamma})$, and thus the update of γ requires generation of the corresponding unused parameters in the current model. These properties led us to develop MCMC algorithms as described below. We first briefly describe the algorithms for updating $\boldsymbol{\theta}_\gamma$, $\mathbf{g}_\gamma$, and $\boldsymbol{\lambda}_\gamma$; and then develop a novel Gibbs sampler and Metropolis–Hastings algorithm to update the indicator variables for main and epistatic effects, respectively.

Conditional on γ, $\mathbf{X}_\gamma$, and $\boldsymbol{\lambda}_\gamma$, the parameters μ, σ^2, and $\boldsymbol{\beta}_\gamma$ can be sampled directly from their posterior distributions, which have standard form.[98] Conditional on γ, $\boldsymbol{\lambda}_\gamma$, and $\boldsymbol{\theta}_\gamma$, the posterior distribution of each element of $\mathbf{g}_\gamma$ is multinomial, and thus can be sampled directly as well.[91] We adapt the algorithm of Yi *et al.*[92] to our model to update locations $\boldsymbol{\lambda}_\gamma$: (1) $\boldsymbol{\lambda}$ is restricted to the discrete space $\boldsymbol{\zeta} = \{\zeta_1, \ldots, \zeta_H\}$, and (2) any interval of some length δ includes at most one QTL. To update λ_q, therefore, we propose a new location λ_q^* for the qth QTL uniformly from $2d$ most flanking loci of λ_q, where d is a predetermined integer (e.g. $d = 2$), and then generate genotypes at the new location for all individuals. The proposals for the new location and the genotypes are then jointly accepted or rejected using the Metropolis–Hastings algorithm.

At each iteration of the MCMC simulation, we update all elements of $\boldsymbol{\gamma}$ in some fixed or random order. For the indicator variable of a main effect, we need to consider two different cases: a QTL is currently (1) in or (2) out of the model. For (1), the QTL position and genotypes are generated at the preceding iteration. For (2), we sample a new QTL position from its prior distribution and generate its genotypes for all individuals. An epistatic effect involves two QTLs — hence, three different cases: (1) both QTLs are in, (2) only one QTL is in, and (3) both QTLs are out of the model. Again, the new QTL position(s) and genotypes are sampled as needed.

We update γ_j, the indicator variable for an effect, using its conditional posterior distribution of γ_j, which is Bernoulli:

$$p(\gamma_j = 1 \mid \boldsymbol{\gamma}_{-\gamma_j}, \mathbf{X}, \boldsymbol{\theta}_{-\beta_j}, \mathbf{y}) = 1 - p(\gamma_j = 0 \mid \boldsymbol{\gamma}_{-\gamma_j}, \mathbf{X}, \boldsymbol{\theta}_{-\beta_j}, \mathbf{y})$$

$$= \frac{wR}{(1-w) + wR} \qquad (28)$$

$$R = \frac{p(\mathbf{y} \mid \gamma_j = 1, \boldsymbol{\gamma}_{-\gamma_j}, \mathbf{X}, \boldsymbol{\theta}_{-\beta_j})}{p(\mathbf{y} \mid \gamma_j = 0, \boldsymbol{\gamma}_{-\gamma_j}, \mathbf{X}, \boldsymbol{\theta}_{-\beta_j})} = \left(\frac{\sigma_{\beta_j}^{-2} + \sigma^{-2} \sum_{i=1}^{n} x_{ij}^2}{\sigma_{\beta_j}^{-2}} \right)^{-0.5}$$

$$\times \exp\left(\frac{1}{2} \frac{[\sum_{i=1}^{n} x_{ij}(y_i - \mu - \mathbf{x}_{i.}\boldsymbol{\beta} + x_{ij}\beta_j)\sigma^{-2}]^2}{\sigma_{\beta_j}^{-2} + \sigma^{-2} \sum_{i=1}^{n} x_{ij}^2} \right),$$

where $\mathbf{x}_{i.}$ is the vector of the coefficients of $\boldsymbol{\beta}$ for the ith individual, $w = \Pr(\gamma_j = 1)$ is the prior probability that β_j appears in the model, $\sigma_{\beta_j}^2$ is the prior variance of β_j, $\boldsymbol{\gamma}_{-\gamma_j}$ means all of the elements of $\boldsymbol{\gamma}$ except for γ_j, and $\boldsymbol{\theta}_{-\beta_j}$ represents all of the elements of $\boldsymbol{\theta}$ except for β_j. We can sample γ_j directly from Eq. (28) or update γ_j with probability $\min\{1, r\}$, where $r = \left(\frac{w}{1-w} R \right)^{1-2\gamma_j}$. The effect β_j is integrated from Eq. (28). We can generate β_j as follows: if γ_j is sampled to be zero, $\beta_j = 0$; otherwise, β_j is generated from its conditional posterior

$$p(\beta_j | \gamma_j = 1, \boldsymbol{\gamma}_{-\gamma_j}, \mathbf{X}, \boldsymbol{\theta}_{-\beta_j}, \mathbf{y}) = N(\tilde{\mu}_j, \tilde{\sigma}_j^2), \qquad (29)$$

where $\tilde{\mu}_j = \left(\sigma^2 \sigma_{\beta_j}^{-2} + \sum_{i=1}^{n} x_{ij}^2 \right)^{-1} \sum_{i=1}^{n} x_{ij}(y_i - \mu - \mathbf{x}_{i.}\boldsymbol{\beta} + x_{ij}\beta_j)$ and

$$\tilde{\sigma}_j^{-2} = \sigma_{\beta_j}^{-2} + \sigma^{-2} \sum_{i=1}^{n} x_{ij}^2.$$

11. POSTERIOR ANALYSIS

The MCMC algorithm starts from initial values and updates each group of unknowns in turn. Initial iterations are typically discarded as "burn-in". We have observed with many real data sets that the MCMC algorithm tends to converge quickly enough such that no initial iterations need to be discarded. To reduce autocorrelation, we thin the subsequent samples by keeping every kth simulation draw and discarding the rest. In our experience, values of k from 20 to 100 are adequate to make autocorrelation insignificant. The MCMC sampler sequence $\{(\boldsymbol{\gamma}^{(t)}, \boldsymbol{\lambda}_{\gamma}^{(t)}, \mathbf{g}_{\gamma}^{(t)}, \boldsymbol{\theta}_{\gamma}^{(t)}); \ t = 1, \dots, N\}$ is a random draw from the joint posterior distribution $p(\boldsymbol{\gamma}, \boldsymbol{\lambda}_{\gamma}, \mathbf{g}_{\gamma}, \boldsymbol{\theta}_{\gamma} \mid \mathbf{y})$; hence, the embedded subsequence $\{(\boldsymbol{\gamma}^{(t)}, \boldsymbol{\lambda}_{\gamma}^{(t)}); \ t = 1, \dots, N\}$ is a random sample from its marginal posterior distribution $p(\boldsymbol{\gamma}, \boldsymbol{\lambda}_{\gamma} \mid \mathbf{y})$, which is used to infer the genetic architecture of the complex trait.

The most important characteristic may be the posterior inclusion probability of each possible locus ζ_h, estimated as

$$
p(\zeta_h \mid \mathbf{y}) = \frac{1}{N} \sum_{t=1}^{N} \sum_{q=1}^{L} 1(\lambda_q^{(t)} = \zeta_h, \xi_q^{(t)} = 1), \ h = 0, 1, \dots, H, \qquad (30)
$$

where ξ_q is the binary indicator that QTL q is included or excluded from the model. Commonly used summaries include the posterior probability that a chromosomal region contains QTLs, the most likely position of QTLs (the mode of QTL positions), and the region of highest posterior density (the HPD region).[98] To account for prior specifications, $p(\zeta_h)$, we can use the Bayes factor (BF) to show evidence for inclusion of ζ_h against exclusion of ζ_h:[116]

$$
\mathrm{BF}(\zeta_h) = \frac{p(\zeta_h \mid \mathbf{y})}{1 - p(\zeta_h \mid \mathbf{y})} \cdot \frac{1 - p(\zeta_h)}{p(\zeta_h)}. \qquad (31)
$$

We can estimate the main effects at any locus or chromosomal interval Δ:

$$
\beta_k(\Delta) = \frac{1}{N} \sum_{t=1}^{N} \sum_{q=1}^{L} 1(\lambda_q^{(t)} \in \Delta, \xi_q^{(t)} = 1) \beta_{qk}^{(t)}, \ k = 1, 2, \dots, K. \qquad (32)
$$

Heritabilities explained by the main effects can also be estimated. In epistatic analysis, we need to estimate two types of additional parameters, the posterior inclusion probability and the size of epistatic effects, both involving pairs of loci. These two types of unknowns can be estimated with natural extensions of Eqs. (30) and (32), respectively.

12. CONCLUSIONS

Returning to the topic of the components of a complex genetic architecture, there are several avenues to explore in expanding the linear regression model of phenotypes and genotypes. In this chapter, we focused our discussion on the genetic component of complex genetic architectures. We started with main effects and then added in pairwise epistatic interactions. The extension to higher-order epistatic interactions is fairly straightforward, although one would expect less power to detect such interactions. The extension to multiple phenotypes and pleiotropy requires modeling of the covariance matrix. The linear regression model can also be expanded to include terms accounting for the environmental component of complex genetic architectures. To this end, we have already incorporated fixed and random effects into the model. We have also included the interaction between genotypes and fixed effects. Other major directions include modeling longitudinal phenotypic data or function-valued traits,[117,118] monitoring the MCMC algorithm using convergence diagnostics, and addressing multiple testing.

13. SOFTWARE

MAPMAKER, available at http://linkage.rockefeller.edu/soft/mapmaker/, is used to construct genetic linkage maps. QTL Cartographer, available at http://statgen.ncsu.edu/qtlcart/, is a suite of programs for mapping quantitative traits based on maximum likelihood and the EM algorithm. qtl, available at http://www.biostat.jhsph.edu/~kbroman/qtl, is a package written in R for mapping quantitative traits. Multimapper, available at http://www.rni.helsinki.fi/~mjs/, is used for Bayesian QTL mapping. The Bayesian methods described in this chapter are available as a package R/bmqtl at http://www.stat.wisc.edu/~yandell/qtl/software/bmqtl/.

ACKNOWLEDGMENTS

This research was supported in part by the National Institutes of Health grants DK062710 and GM069430.

REFERENCES

1. Hedrich HJ. (2000) History, strains and models. In: *The Laboratory Rat*, Krinke GJ (ed.), Academic Press, San Diego, CA, pp. 3–16.

2. Moore KJ, Nagle DL. (2000) Complex trait analysis in the mouse: the strengths, the limitations and the promise yet to come. *Annu Rev Genet* **34**:653–686.

3. Korstanje R, Paigen BJ. (2002) Mapping of quantitative trait loci involved in atherosclerosis in the mouse. In: *Atherosclerosis: Risk Factors, Diagnosis, and Treatment*, Kostner GM, Kostner KM, Kostner B (eds.), Medimond, Bologna, Italy.

4. Sugiyama F, Churchill GA, Higgins DC, *et al.* (2001) Concordance of murine quantitative trait loci for salt-induced hypertension with rat and human loci. *Genomics* **71**:70–77.

5. Pérusse L, Rankinen T, Zuberi A, *et al.* (2005) The human obesity gene map: the 2004 update. *Obes Res* **13**:381–490.

6. Lynch M, Walsh B. (1998) *Genetics and Analysis of Quantitative Traits*. Sinauer Associates, Sunderland, MA.

7. DeBry RW, Seldin MF. (1996) Human/mouse homology relationships. *Genomics* **33**:337–351.

8. Pomp D. (1997) Genetic dissection of obesity in polygenic animal models. *Behav Genet* **27**:285–306.

9. Wang X, Paigen B. (2002) Quantitative trait loci and candidate genes regulating HDL cholesterol. *Arterioscler Thromb Vasc Biol* **22**:1390–1401.

10. Bedell MA, Jenkins NA, Copeland NG. (1997) Mouse models of human disease. Part I: techniques and resources for genetic analysis in mice. *Genes Dev* **11**:1–10.

11. Hill WG, Robertson A. (1968) The effects of inbreeding at loci with heterozygote advantage. *Genetics* **60**:615–628.

12. Weir BS. (1996) *Genetic Data Analysis II*. Sinauer Associates, Sunderland, MA.

13. Hartl DL, Clark AG. (1997) *Principles of Population Genetics*, 3rd ed. Sinauer Associates, Sunderland, MA.

14. Mackay TFC. (2001) The genetic architecture of quantitative traits. *Annu Rev Genet* **35**:303–339.

15. Mackay TFC. (2004) The genetic architecture of quantitative traits: lessons from *Drosophila*. *Curr Opin Genet Dev* **14**:253–257.

16. Buck KJ, Metten P, Belknap JK, Crabbe JC. (1997) Quantitative trait loci involved in genetic predisposition to acute alcohol withdrawal in mice. *J Neurosci* **17**:3946–3955.

17. Flint J, Corley R, DeFries JC, *et al.* (1995) A simple genetic basis for a complex psychological trait in laboratory mice. *Science* **269**:1432–1435.

18. Cockerham CC. (1954) An extension of the concept of partitioning hereditary variance for analysis of covariances among relatives when epistasis is present. *Genetics* **39**:859–882.

19. Kempthorne O. (1954) The correlation between relatives in a random mating population. *Proc R Soc Lond B Biol Sci* **143**:102–113.

20. Mather K, Jinks JL. (1977) *Introduction to Biometrical Genetics*. Chapman and Hall, London, England.

21. Warden CH, Yi N, Fisler J. (2004) Epistasis among genes is a universal phenomenon in obesity: evidence from rodent models. *Nutrition* **20**:74–77.
22. Templeton AR. (2000) Epistasis and complex traits. In: *Epistasis and the Evolutionary Process*, Wolf JB, Brodie ED III, Wade MJ (eds.), Oxford University Press, New York, NY, pp. 41–57.
23. Moore JH, Williams SM. (2005) Traversing the conceptual divide between biological and statistical epistasis: systems biology and a more modern synthesis. *Bioessays* **27**:637–646.
24. Moore JH. (2003) The ubiquitous nature of epistasis in determining susceptibility to common human diseases. *Hum Hered* **56**:73–82.
25. Brockmann GA, Kratzsch J, Haley CS, *et al.* (2000) Single QTL effects, epistasis, and pleiotropy account for two-thirds of the phenotypic F_2 variance of growth and obesity in DU6i × DBA/2 mice. *Genome Res* **10**:1941–1957.
26. Masinde GL, Li X, Gu W, *et al.* (2002) Quantitative trait loci (QTL) for lean body mass and body length in MRL/MPJ and SJL/J F_2 mice. *Funct Integr Genomics* **2**:98–104.
27. Segal NL, Allison DB. (2003) Twins and virtual twins: bases of relative body weight revisited. *Int J Obes* **26**:437–441.
28. Cheverud JM, Vaughn TT, Pletscher LS, *et al.* (2001) Genetic architecture of adiposity in the cross of LG/J and SM/J inbred mice. *Mamm Genome* **12**:3–12.
29. Yi N, Zinniel DK, Kim K, *et al.* (2006) Bayesian analyses of multiple epistatic QTL models for body weight and body composition in mice. *Genet Res* **87**:45–60.
30. Warden CH, Fisler JS, Shoemaker SM, *et al.* (1995) Identification of four chromosomal loci determining obesity in a multifactorial mouse model. *J Clin Invest* **95**:1545–1552.
31. Yi N, Chiu S, Allison DB, *et al.* (2004) Epistatic interaction between two nonstructural loci on chromosomes 7 and 3 influences hepatic lipase activity in BSB mice. *J Lipid Res* **45**:2063–2070.
32. Yi N, Diament A, Chiu S, *et al.* (2004) Characterization of epistasis influencing complex spontaneous obesity in the BSB model. *Genetics* **167**:399–409.
33. Corva PM, Horvat S, Medrano JF. (2001) Quantitative trait loci affecting growth in high growth (hg) mice. *Mamm Genome* **12**:284–290.
34. Corva PM, Medrano JF. (2001) Quantitative trait loci (QTLs) mapping for growth traits in the mouse: a review. *Genet Sel Evol* **33**:105–132.
35. Adams MD, Celniker SE, Holt RA, *et al.* (2000) The genome sequence of *Drosophila melanogaster*. *Science* **287**:2185–2195.
36. Darvasi A. (1998) Experimental strategies for the genetic dissection of complex traits in animal models. *Nat Genet* **18**:19–24.
37. Crabbe JC, Phillips TJ, Buck KJ, *et al.* (1999) Identifying genes for alcohol and drug sensitivity: recent progress and future directions. *Trends Neurosci* **22**:173–179.
38. Cardon LR, Bell JI. (2001) Association study designs for complex diseases. *Nat Rev Genet* **2**:91–99.

39. Glazier AM, Nadeau JH, Aitman TJ. (2002) Finding genes that underlie complex traits. *Science* **298**:2345–2349.
40. Lander E, Kruglyak L. (1995) Genetic dissection of complex traits: guidelines for interpreting and reporting linkage results. *Nat Genet* **11**:241–247.
41. Haldane JBS. (1919) The combination of linkage values and the calculation of distances between the loci of linked factors. *J Genet* **8**:299–309.
42. Kosambi DD. (1944) The estimation of map distances from recombination values. *Ann Eugen* **12**:171–175.
43. Langley CH, Montgomery E, Quattlebaum WF. (1982) Restriction map variation in the *Adh* region of *Drosophila*. *Proc Natl Acad Sci USA* **79**:5631–5635.
44. Leigh Brown AJ. (1983) Variation at the 87A heat shock locus in *Drosophila melanogaster*. *Proc Natl Acad Sci USA* **80**:5350–5354.
45. Kristensen VN, Kelefiotis D, Kristensen T, Børresen-Dale A-L. (2001) High-throughput methods for detection of genetic variation. *Biotechniques* **30**:318–332.
46. Copeland NG, Jenkins NA, Gilbert DJ, *et al.* (1993) A genetic linkage map of the mouse: current applications and future prospects. *Science* **262**:57–66.
47. Dietrich WF, Copeland NG, Gilbert DJ, *et al.* (1995) Mapping the mouse genome: current status and future prospects. *Proc Natl Acad Sci USA* **92**:10849–10853.
48. Dietrich WF, Miller J, Steen R, *et al.* (1996) A comprehensive genetic map of the mouse genome. *Nature* **380**:149–152.
49. Cargill M, Altshuler D, Ireland J, *et al.* (1999) Characterization of single-nucleotide polymorphisms in coding regions of human genes. *Nat Genet* **22**:231–238.
50. Halushka MK, Fan J-B, Bentley K, *et al.* (1999) Patterns of single-nucleotide polymorphisms in candidate genes for blood-pressure homeostasis. *Nat Genet* **22**:239–247.
51. Talbot CJ, Nicod A, Cherny SS, *et al.* (1999) High-resolution mapping of quantitative trait loci in outbred mice. *Nat Genet* **21**:305–308.
52. Barton NH, Keightley PD. (2002) Understanding quantitative genetic variation. *Nat Rev Genet* **3**:11–21.
53. Tsang S, Sun Z, Luke B, *et al.* (2005) A comprehensive SNP-based genetic analysis of inbred mouse strains. *Mamm Genome* **16**:476–480.
54. The International SNP Map Working Group. (2001) A map of human genome sequence variation containing 1.42 million single nucleotide polymorphisms. *Nature* **409**:928–933.
55. Venter JC, Adams MD, Myers EW, *et al.* (2001) The sequence of the human genome. *Science* **291**:1304–1351.
56. Doerge RW, Zeng Z-B, Weir BS. (1997) Statistical issues in the search for genes affecting quantitative traits in experimental populations. *Stat Sci* **12**:195–219.
57. Bovenhuis H, van Arendonk JAM, Davis G, *et al.* (1997) Detection and mapping of quantitative trait loci in farm animals. *Livest Prod Sci* **52**:135–144.

58. Broman KW. (2001) Review of statistical methods for QTL mapping in experimental crosses. *Lab Anim* **30**:44–52.
59. Weller JI. (2001) *Quantitative Trait Loci Analysis in Animals.* CABI Publishing, London, England.
60. Lander ES, Botstein D. (1989) Mapping Mendelian factors underlying quantitative traits using RFLP linkage maps. *Genetics* **121**:185–199.
61. Falconer DS, Mackay TFC. (1996) *Introduction to Quantitative Genetics*, 4th ed. Addison-Wesley Longman, Harlow, Essex, England.
62. Haley CS, Knott SA. (1992) A simple regression method for mapping quantitative trait loci in line crosses using flanking markers. *Heredity* **69**:315–324.
63. Martínez O, Curnow RN. (1992) Estimating the locations and sizes of the effects of quantitative trait loci using flanking markers. *Theor Appl Genet* **85**:480–488.
64. Xu S. (1995) A comment on the simple regression method for interval mapping. *Genetics* **141**:1657–1659.
65. Xu S. (1998) Iteratively reweighted least squares mapping of quantitative trait loci. *Behav Genet* **28**:341–355.
66. Xu S. (1998) Further investigation on the regression method of mapping quantitative trait loci. *Heredity* **80**:364–373.
67. Jansen RC. (1993) Interval mapping of multiple quantitative trait loci. *Genetics* **135**:205–211.
68. Zeng Z-B. (1994) Precision mapping of quantitative trait loci. *Genetics* **136**:1457–1468.
69. Zeng Z-B. (1993) Theoretical basis for separation of multiple linked gene effects in mapping of quantitative trait loci. *Proc Natl Acad Sci USA* **90**:10972–10976.
70. Kao CH, Zeng Z-B, Teasdale RD. (1999) Multiple interval mapping for quantitative trait loci. *Genetics* **152**:1203–1216.
71. Akaike H. (1974) A new look at the statistical model identification. *IEEE Trans Auto Control* **19**:716–723.
72. Stone M. (1974) Cross-validatory choice and assessment of statistical predictions. *J R Stat Soc Ser B* **36**:111–147.
73. Geisser S. (1975) The predictive sample reuse method with applications. *J Am Stat Assoc* **70**:320–328.
74. Schwarz G. (1978) Estimating the dimension of a model. *Ann Stat* **6**:461–464.
75. Foster DP, George EI. (1994) The risk inflation criterion for multiple regression. *Ann Stat* **22**:1947–1975.
76. Gelfand AE, Ghosh SK. (1998) Model choice: a minimum posterior predictive loss approach. *Biometrika* **85**:1–11.
77. Spiegelhalter DJ, Best NG, Carlin BP, van der Linde A. (2002) Bayesian measures of model complexity and fit. *J R Stat Soc Ser B* **64**:583–639.
78. Churchill GA, Doerge RW. (1994) Empirical threshold values for quantitative trait mapping. *Genetics* **138**:963–971.
79. Visscher PM, Thomson R, Haley CS. (1996) Confidence intervals in QTL mapping by bootstrapping. *Genetics* **143**:1013–1020.

80. Metropolis N, Rosenbluth AW, Rosenbluth MN, *et al.* (1953) Equation of state calculations by fast computing machines. *J Chem Phys* **21**:1087–1092.

81. Hastings WK. (1970) Monte Carlo sampling methods using Markov chains and their applications. *Biometrika* **57**:97–109.

82. Green PJ. (1995) Reversible jump Markov chain Monte Carlo computation and Bayesian model determination. *Biometrika* **82**:711–732.

83. Satagopan JM, Yandell BS. (1996) Estimating the number of quantitative trait loci via Bayesian model determination. Special Contributed Paper Session on Genetic Analysis of Quantitative Traits and Complex Diseases, Biometric Section, Joint Statistical Meeting, Chicago, IL.

84. Heath SC. (1997) Markov chain Monte Carlo segregation and linkage analysis for oligogenic models. *Am J Hum Genet* **61**:748–760.

85. Thomas DC, Richardson S, Gauderman J, Pitkaniemi J. (1997) A Bayesian approach to multipoint mapping in nuclear families. *Genet Epidemiol* **14**:903–908.

86. Uimari P, Heoschele I. (1997) Mapping linked quantitative trait loci using Bayesian method analysis and Markov chain Monte Carlo algorithms. *Genetics* **146**:735–743.

87. Sillanpää MJ, Arjas E. (1998) Bayesian mapping of multiple quantitative trait loci from incomplete inbred line cross data. *Genetics* **148**:1373–1388.

88. Stephens DA, Fisch RD. (1998) Bayesian analysis of quantitative trait locus data using reversible jump Markov chain Monte Carlo. *Biometrics* **54**:1334–1347.

89. Yi N, Xu S. (2000) Bayesian mapping of quantitative trait loci for complex binary traits. *Genetics* **155**:1391–1403.

90. Gaffney PJ. (2001) An efficient reversible jump Markov chain Monte Carlo approach to detect multiple loci and their effects in inbred crosses. PhD thesis, Department of Statistics, University of Wisconsin–Madison, Madison, WI.

91. Yi N, Xu S. (2002) Mapping quantitative trait loci with epistatic effects. *Genet Res* **79**:185–198.

92. Yi N, Xu S, Allison DB. (2003) Bayesian model choice and search strategies for mapping interacting quantitative trait loci. *Genetics* **165**:867–883.

93. Narita A, Sasaki Y. (2004) Detection of multiple QTL with epistatic effects under a mixed inheritance model in an outbred population. *Genet Sel Evol* **36**:415–433.

94. Yi N. (2004) A unified Markov chain Monte Carlo framework for mapping multiple quantitative trait loci. *Genetics* **167**:967–975.

95. Carlin BP, Chib S. (1995) Bayesian model choice via Markov chain Monte Carlo. *J Am Stat Assoc* **88**:881–889.

96. Godsill SJ. (2001) On the relationship between MCMC model uncertainty methods. *J Comput Graph Stat* **10**:230–248.

97. Yi N, Yandell BS, Churchill GA, *et al.* (2005) Bayesian model selection for genome-wide epistatic quantitative trait loci analysis. *Genetics* **170**:1333–1344.

98. Gelman A, Carlin JB, Stern HS, Rubin DB. (2004) *Bayesian Data Analysis*, 2nd ed. Chapman & Hall, New York, NY.

99. Xu S, Yi N. (2000) Mixed model analysis of quantitative trait loci. *Proc Natl Acad Sci USA* **97**:14542–14547.

100. Uimari P, Sillanpää MJ. (2001) Bayesian oligogenic analysis of quantitative and qualitative traits in general pedigrees. *Genet Epidemiol* **21**:224–242.

101. Devroye L. (1986) *Non-uniform Random Variable Generation*. Springer-Verlag, New York, NY.

102. Jiang C, Zeng Z-B. (1997) Mapping quantitative trait loci with dominant and missing markers in various crosses from two inbred lines. *Genetica* **101**:47–58.

103. Kao C-H, Zeng Z-B. (2002) Modeling epistasis of quantitative trait loci using Cockerham's model. *Genetics* **160**:1243–1261.

104. George EI, McCulloch RE. (1997) Approaches for Bayesian variable selection. *Stat Sin* **7**:339–373.

105. Kuo L, Mallick BK. (1998) Variable selection for regression models. *Sankhyā Ser B* **60**:65–81.

106. Chipman H, George EI, McCulloch RE. (2001) The practical implementation of Bayesian model selection. In: *Model Selection*, Lahiri P (ed.), Institute of Mathematical Statistics, Beachwood, OH, pp. 65–116.

107. George EI, Foster DP. (2000) Calibration and empirical Bayes variable selection. *Biometrika* **87**:731–747.

108. Raftery AE, Madigan D, Hoeting JA. (1997) Bayesian model averaging for linear regression models. *J Am Stat Assoc* **92**:179–191.

109. Ball RD. (2001) Bayesian methods for quantitative trait loci mapping based on model selection: approximate analysis using the Bayesian information criterion. *Genetics* **159**:1351–1364.

110. Sillanpää MJ, Corander J. (2002) Model choice in gene mapping: what and why. *Trends Genet* **18**:301–307.

111. Kohn R, Smith M, Chan D. (2001) Nonparametric regression using linear combinations of basis functions. *Stat Comput* **11**:313–322.

112. George EI, McCulloch RE. (1993) Variable selection via Gibbs sampling. *J Am Stat Assoc* **88**:881–889.

113. Dellaportas P, Forster JJ, Ntzoufras I. (2002) On Bayesian model and variable selection using MCMC. *Stat Comput* **12**:27–36.

114. Fernández C, Ley E, Steel MFJ. (2001) Benchmark priors for Bayesian model averaging. *J Econom* **100**:381–427.

115. Zhang M, Montooth KL, Wells MT, *et al.* (2005) Mapping multiple quantitative trait loci by Bayesian classification. *Genetics* **169**:2305–2318.

116. Kass RE, Raftery AE. (1995) Bayes factors. *J Am Stat Assoc* **90**:773–795.

117. Ma C-X, Casella G, Wu R. (2002) Functional mapping of quantitative trait loci underlying the character process: a theoretical framework. *Genetics* **161**:1751–1762.

118. Wu R, Ma C-X, Lin M, Casella G. (2004) A general framework for analyzing the genetic architecture of developmental characteristics. *Genetics* **166**:1541–1551.

Chapter 12

Animal Models in the Study of Genetics of Complex Human Diseases

Weikuan Gu[*,‡], Jian Yan[†] and Yan Jiao[*]

*Center of Genomics and Bioinformatics &
Center of Diseases of Connective Tissues
Department of Orthopedic Surgery–Campbell Clinic
University of Tennessee Health Science Center, Memphis, TN 38163, USA*

[†]*Department of Medicine, University of Tennessee Health Science Center
Memphis, TN 38163, USA*

1. INTRODUCTION: ADVANTAGES AND LIMITATIONS OF USING ANIMAL MODELS

This chapter focuses on the application of animal models in genetic mapping and transgenic animals. An animal model usually refers to a nonhuman animal with a disease or phenotype that is similar to a human condition. There are many reasons for using animal models in genetic studies; however, in this chapter, we will not deal with economic and social issues. The very reason for using animal models in genetic research is to obtain data that cannot be obtained from humans. In a broad sense, we can apply many procedures on animals that we are not able to use on humans. A very obvious example is, in gene therapy tests, a genetic material that cannot be tested on humans but can be tested on animals. We can also obtain materials from animals that are not possible to obtain from humans. For instance, genetic study needs heritable materials such as DNA and RNA; we can easily obtain tissues such as the liver, heart, and brain from animals to extract RNA, but it is difficult to get them from humans.

[‡] Correspondence author.

Advantages of using animal models include the following:

(1) Carrying out of procedures not possible to conduct on humans. The use of animal models allows researchers to investigate disease states in ways that would be inaccessible in a human patient. There are several kinds of experimental procedures that cannot be applied to humans before being tested on animal models. Surgery is the most obvious one of these procedures. To test a gene therapy on the treatment of broken bones, a bone from an animal model needs to be broken, while such a procedure should not be applied to humans. To assess the carcinogenicity of a substance, an animal model needs to be chosen. To test the toxicity of drugs, an animal model is also needed. Even for the test of alcoholism, an animal is needed.

(2) Short generation time. The short generation time is one of the advantages of animal models in pedigree analysis and genetic mapping. It is common knowledge that to analyze a genetic locus of a disease or a useful trait, the parents (father, mother), F1 (son or daughter), and F2 (grandsons or granddaughters) are needed. The short generation time of animal models will allow researchers to obtain samples from each of three generations sooner than from humans.

(3) Short life span. The short life span of animal models allows a quick completion of experiments, especially for studies on aging or age-related diseases.

(4) Large populations. Most animals have not only short life spans and generations, but also a high reproduction rate. Therefore, it is easy to produce a large population of animals for genetic analysis. This is important to generate a large F2 population for fine mapping of a genetic locus. For instance, many disease loci mapped on a mouse genetic map have been generated from a population of several hundreds of F2 mice from a single pair of parents within a year or so. However, in humans, mapping a genetic locus from a family tree to a small region (less than 5 cM) has been extremely difficult.

(5) Creation of human diseases or conditions. Animals can be used to create diseases or conditions by drug treatment or genetic manipulation. In mice, classic collagen-induced arthritis (CIA) has been used extensively as a model for rheumatoid arthritis. Thousands of gene knockout mice have a variety of disease phenotypes. Those models are extremely important for the investigation of molecular mechanisms and the treatment of diseases.

(6) Manipulation of genetic materials. Animals can be genetically manipulated to change their genetic makeup so as to fit the genetic study. Unlike humans, animals can be made as a cross between brothers and sisters. Animals can also be used for backcross to the same parent for several generations. In The Jackson Laboratory, a large number of recombinant inbred strains have been self-crossed for hundreds of generations; as a result, each mouse strain has a uniform genetic background — the genetic makeup of every mouse in the same inbred strain is exactly the same. It is much easier to recognize the phenotype of a mutation from such a genetically uniform background, and it is easy to track the individual animal that has the mutation.

(7) Manipulation of environmental conditions. Animals can be put under fixed conditions according to the requirements of the experiment. For instance, mice treated with different drugs are grown under the same day–night length, given the same food and water, placed under the same room temperature, in the same cage. As they also have the same genetic makeup, the only difference they have (if any) is due to the different drug treatment. Compared to the extremely diverse environment and activity of humans, the advantages of an animal model are very obvious.

However, for each advantage, there is also a disadvantage to animal models. Animals provide a simpler system under a unified genome background with which to work; however, if the system is too simple, it will be too far from real complex human diseases. Hence, such an animal model becomes irrelevant to our human population. As animals have shorter generation times and shorter life spans, we can apply selective genetic pressure to develop animal models, but we lose the ability to study long-term drug effects. The biggest problem is that because animals are not humans, their biological reaction to mutations or gene therapies are somewhat different from humans. Depending on different situations, sometimes the differences between humans and animals are so large that the results from animal models are not applicable to humans.

It is not very rare for spontaneous or induced mouse models of human disease to exhibit phenotypes considerably different from the counterpart human disorders. For example, gene targeting to inactivate the tumor suppressor genes *TP53* and *RB1* has often produced disappointing mouse models. The reasons are multiple. First, there may be technical problems in achieving the desired type of mutation as mentioned above. Second, there may be a strain-specific effect of genes. A good example is the mouse model of rheumatoid arthritis (RA): *Il1rn*-deficient mice. Spontaneous arthritis only happens in BALB/c-based *IL1rn* knockout mice, not in other strains such as DBA/1- and C57BL/6-based *Il1rn* mice. Third, murine physiology, anatomy, and life span differ significantly from those of humans, making the mouse model inappropriate for some human diseases. In this case, it might be advantageous to produce animal models in livestock species. The rabbit is an alternative model for this disease and other disorders, including cardiovascular diseases. The rabbit is phylogenetically closer to primates than are rodents, and is large enough to enable nonlethal monitoring of physiological and pathophysiological changes. A transgenic pig model for the rare human eye disease retinitis pigmentosa has been developed to study this disease, particularly the phase of cone degeneration that cannot be modeled in mice. Fortunately, results from most animal models are similar to or the same as in humans.

Accordingly, the selection of a species for animal models should not be based on availability, familiarity, or cost. Rather, the selection should be based on the biology, the suitability for the disease or trait, and the impact of the study on humans. There are many animal models that have been well developed. Mammals have been widely used because of their obvious similarities in both structure and function to those of humans. Rats, mice, guinea pigs, and hamsters are favored because of their small size, short life span, ease of handling, and high reproduction rate. Vertebrates, especially mammals, provide essential one-to-one models for many specific human disease processes. However, large animals such as goats, dogs, and horses are used in many special disease models such as eye disease. The monkey, unlike other animal models, has its special advantage because of its close relationship to human beings.

The mouse is an excellent model system for understanding mammalian development, in particular human development. As the development and physiology of mice and humans are basically similar, direct comparisons can be made between the two systems. A great deal is known about the human and mouse genomes. For instance, over large regions of the genome, the order of genes (and therefore linkage of the genes) is conserved between mice and humans. These regions of homology or synteny have propelled both the human and mouse genome projects forward, since information regarding one system can be directly related to the other. In addition, because the human and mouse genomes have been largely sequenced, direct comparisons between the sequences can now be made.

2. ANIMAL MODELS IN THE APPLICATION FOR QTL MAPPING

One important use of animal models is the mapping of quantitative trait loci (QTLs). The term QTLs has been used for multiple genetic loci that control the same complex trait — a complex trait is one that is influenced by multiple genes. Generally, quantitative traits are multifactorial and are influenced by several polymorphic genes and environmental conditions. There can be one or many QTLs influencing a trait or phenotype.[1,2] A typical example is bone density, for which a number of quantitative trait loci (QTLs) have been identified.[1-4]

It is particularly important to employ animal models in QTL mapping. In general, the contribution of an individual QTL to a trait of interest is small: as with bone density, a QTL usually regulates 2%–10% of the variation of total bone density. Obviously, in order to detect such changes at a statistically significant level, a uniform and large population is needed. Because the genetic heterogeneity of the human population is so profound, unreasonably large numbers of subjects would be required for an appropriate assessment of the genetic regulation of such kinds of traits. As we do not deal with the statistic aspect of QTLs in this chapter, readers can obtain more information from other chapters or population genetics textbooks such as *Introduction to Quantitative Genetics* (4th ed., Falconer DS and Mackay TFC, Prentice Hall, Harlow, England, 1996). The more genes that control a trait, the smaller the differences among the individuals within a population in which the QT genes are segregating. Thus, because the power of linkage requires an extremely large population size for detectable QTLs, most QTLs are identified through animal models.

Mouse models have gained popularity in QTL mapping for several reasons, including small size, rapid gestation period (21 days), large litter size (~4–8 mice), and relatively low maintenance costs (~US$1 per cage of 4–5 mice). Moreover, the mouse genome has been extensively characterized, its whole-genome sequence has been completed, and gene-targeted knockout (of every gene) and transgenic overexpression experiments are performed using mice. Therefore, in this chapter, we will use the mouse model to discuss the methodology and application of animal models in QTL mapping.

2.1. *Principle of QTL Mapping*

QTL mapping developed from modern Mendelian linkage analysis. In a very simplified example, Fig. 1 shows how a trait is linked to a gene. In the

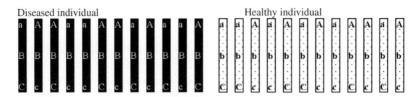

Fig. 1. Linkage between gene/genotype and phenotype/disease. On the left side, there are 12 diseased individuals represented by black bars. On the right side, there are 12 healthy individuals represented by white bars. Three molecular loci of genotypes/markers are listed for each individual. Each marker has two polymorphic types, e.g. Aa, Bb, or Cc. While A, a, C, and c are randomly segregated among the total population, every individual bearing the B genotype has the disease (on the left half) and every b genotype individual is normal (on the right half).

figure, three molecular markers are listed for each individual. The A, a, C, and c genotypes are randomly segregated among all of the individuals. Only the individual with B polymorphism has the disease, while the b individual is healthy. Obviously, it is not a difficult question if one is asked which molecular marker is linked to the disease.

From Fig. 1, we learn that in order to link a disease to a molecular marker or gene, we have to have (1) the molecular marker with polymorphism so that we can genotype or fingerprint each individual, and (2) a standard scale for a disease or any other phenotype to distinguish the individuals.

In reality, the issue of molecular markers in human and animal models is not as easy as shown in Fig. 1. Both human and mouse have about three billion nucleotides that contain about 30 000 genes in their genome. In the mouse, 20 pairs of chromosomes host those nucleotides and genes; accordingly, there should be 20 groups of molecular markers or genotypes, each group having thousands of markers or genotypes. In order to find out which gene controls a trait, several steps have to be taken by researchers. (1) First, they must group molecular markers or genes, thus establishing which genes belong to which group or chromosome. (2) Then, relative locations of markers are assigned to each group — i.e. linkage mapping. It is pointless to attempt to detect linkages between markers and QTLs if one cannot properly identify linkages between the markers themselves. (3) When the linkage map is ready, a population in which a disease or trait of interest is segregating is produced. (4) The genotype or the type of polymorphism of markers and the phenotype or the disease status of each individual in the population is then investigated and recorded. (5) Finally, an association analysis (as shown in Fig. 1, but with statistical tools) between markers and phenotypic

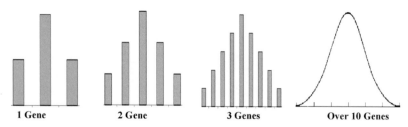

Fig. 2.   Phenotypic distribution of traits controlled by different pairs of genes. Assuming the effect of each genotype on a trait is in an additive manner, the *aa* genotype will have one phenotype, *Aa* an intermediate phenotype, and *AA* another phenotype, for instance. The number of variations in phenotype is from three with one gene to an extremely large number with the increase in gene numbers.

traits is performed. The closer the association of a marker to a trait, the more likely a gene in that location may regulate the trait.

A QTL is not as easily recognized as a particular disease as shown in Fig. 1. As we mentioned above, a QTL refers to a locus that influences the variation of a trait. A quantitative trait is usually influenced by many QTLs and by the environment. In a large population, the distribution of a quantitative trait appears like a normal curve (not as black and white), as shown in Fig. 2.

QTL mapping is basically performed to find the association between genes and the variation of a trait. The difference between the mapping of QTLs and a simple trait is that the former needs to conduct the correlations between many genes and many small changes in the trait — a very difficult thing to do. As the number of genes for a trait becomes larger, the effect of each gene on the variation of the trait becomes smaller; the influence of the environment thus becomes important, at least in terms of the measurement of the influence of a genetic locus. A 2% variation of a trait can be caused by a genetic factor as well as by an environmental factor, while a 50% variation caused by a genetic factor will not be easily influenced by the environment. This is particularly important when using animal models that can be produced within a uniformly controlled environment. One such QTL mapping using a mouse model is the QTL mapping of bone mineral density (BMD) (Table 1). Because BMD is controlled by multiple genetic loci, the effect of each individual QTL on a QT is so small that it cannot be distinguished from the environmental influence if the environment is not controlled. Obviously, it is very difficult to study a large human population under a controlled environment. Fine mapping of the QTL for

Table 1   QTL Positions of Peak Bone Density Identified from Mouse Model and Comparison to That from Humans

Chr/ Mouse	Position (cM)	LOD/ P-Value*	Genomic Sequence** (bp)	Chr/ Human	Human Homologous Regions***	References
1	82–106	8.8	159161492–195694164	1	1q21–25; 1q42–q44	3, 5, 12
1	64–82	24	123137140–159201852	2, 1	2q37–q21; 1q25–32	4
2	2.2–23	4.05	3374309–35049497	10, 9, 2	10p15–p13; 9q34–q34.1; 2q14.2	6
2	35–45	3.5	62721162–67146921	2	2q24–q32	6
2	45–55	$P = 0.002$	67146921–104378449	11, 2	11p13–p12; 2q31–2q32	3, 9, 10
4	35–54	16.3	61124566–113400698	9, 1	9p24–p21; 1p31–34	4
4	60–80	9	130853280–154299058	1	1p33–p36.3	3, 5
5	25–36	3.6	42713374–71142400	4	4p16.3–p14	12
6	51.0	4.56	~117334206	3	3p26.2–p24; 3q21–q24	4
7	11–13	$P = 0.001$	25522212–29345204	19	19q13–p13.1	6
7	44	$P = 0.0007$	~79815461	15, 11,	15q23–q26; 11q13–q21	8, 5
11	31.0	6.76	~51857154	5, 17	5q21–q23.3; 5q31–q35; 17p11–p13	4, 11
11	49–60	5.4	87424373–100790282	17	17q12–q24	3, 5, 13
11	58–72	10.8	97586515–117908416	17	17q11.2–q25.3	8
13	0–10	6.1	0–21072268	1, 7, 6	1q42–q43; 7p15–13; 6p23–p21	12
13	35	7.73	~54537729	5, 9	5q31–q35.3; 9q21.3–q22.3	4
13	22	5.5	~40945185	6	6p24–p22.3;	13
13	10–30	5.8	21072268–63274779	7, 9, 6, 17	7p15–p13; 9q21–q22; 6p25.1–p21; 17q22–q24	8, 14
14	40	4.3	~63201604	8, 13	8p21–p11.2; 13q14.1–q21.1	4
14	2.0	$P = 0.0007$	~5168925	10, 3	10q22.1–q23; 3p21–p14	8

(Continued)

Table 1 (Continued)

Chr/ Mouse	Position (cM)	LOD/ P-Value*	Genomic Sequence** (bp)	Chr/ Human	Human Homologous Regions***	References
15	24–55	3.2	59453579–10034855	8, 12, 22	8q21–q24.3; 12q12–14; 22q13–q13.3	12
16	27.6	4.07	~38896865	3	3q12–q13.2; 3q21–q23; 3q28–q29	5
16	9–28.4	$P = 0.01$	15488977–38208380	8, 22, 3	8q11–q11.2; 22q11–q11.2; 3q21–q29	4, 5
18	24	13.67	~45399525	5	5q21–q23; 5q31–q32	4
Total: 15		ND			ND	ND

* LOD score or *P*-value for each QTL is listed according to the availability of information provided from the published reference. ND: not determined.

** Mouse genomic accession numbers within each QTL are listed according to the information from the Mouse Genome Sequence Project.

*** The human homologous regions corresponding to mouse QTL positions are assigned according to the database of the Human–Mouse Homology Map: www.ncbi.nlm.nih.gov/Homology/. The "q" and "p" represent the short and long arms of human chromosomes, respectively.

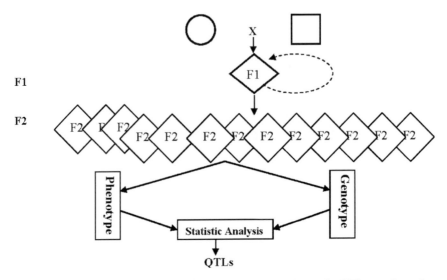

Fig. 3.   Schematic procedure of the creation of an F2 population for QTL mapping using a mouse model.

the purpose of positional cloning using a human population is even more difficult.[2] Alternatively, the mouse model has been widely used to identify the QTLs of peak bone density.[3,4]

2.2. *QTL Mapping Using F2 Populations of Mice*

Most QTLs mapped using mouse models have been achieved using F2 populations. One of the advantages of mouse models is that there are a large number of recombinant inbred or pure strains available. They are pure strains because they have been inbred for many generations and genetically unified; therefore, within the same strain, all brothers and sisters have the same genomic background. Figure 3 shows how an F2 population is created, and the QTL mapping procedure using F2 mice is described below.

(1) Parental selection. In general, the two parents should be different at the trait of study as much as possible, thus having separate sets of genetic factors that regulate the trait of interest. A good example is the strain selection of bone mineral density (BMD). Before producing an F2 population, Beamer and colleagues[1] first evaluated 11 potential mouse strains and then selected two pairs of mouse strains for the mapping of BMD.[3,4] Each of those two pairs of mouse strains had one

strain of high BMD and the other of low BMD, selected from the 11 strains.

(2) F2 population. The purpose of creating an F2 population is to have the genes/genotypes and the phenotype/variation segregates among populations so that a correlation between genotype and phenotype can be analyzed. An F2 generation is created by a self-cross of F1 mice. Theoretically, the larger the population, the more detailed a genetic locus can be localized. In practice, several hundred up to a thousand mice have been used.[3-5]

(3) Phenotyping and genotyping. From each individual mouse, two kinds of data can be obtained for the mapping purpose. One is the phenotype, such as BMD, for which many instruments such as micro-CT and PIXImus can be used; the other is the genotype. Microsatellite markers have been widely used for this purpose.

(4) QTL analysis. After both the phenotype and genotype are obtained, the correlation or linkage between phenotype (in this case, BMD) and genotype (the polymorphism of microsatellite markers) is analyzed. As there is much variation in the phenotype and there are hundreds of molecular markers existing in several hundreds to thousands of mice, statistical software is used. These include WebQTL (http://webqtl.org/), MapMaker/QTL (http://www.broad.mit.edu/genome_software), and Map Manager QTX (http://www.mapmanager.org/mmQTX.html).

2.3. *QTL Mapping Using Recombinant Inbred Lines*

Recombinant inbred (RI) strains are an important resource for mapping complex traits in many species. The basic principle is straightforward: two mouse strains are crossed, and mating pairs from the F2 generation are used to establish a new set of inbred lines by repeated generations of brother–sister matings. This approach was initially carried forward extensively by Ben Taylor in 1978[a]; since then, it has been widely adopted. Recently, a research group at Robert Williams' laboratory generated 46 RI strains — by far the largest number of RI strains of mice — from two progenitor strains, C57BL/6J (B6) and DBA/2J (D2).[7] Figure 4 shows a simplified schematic procedure of the IR strain breeding in their laboratory.

---

[a]For more detail, please see Taylor BA (1978), Recombinant inbred strains: use in gene mapping, in: *Origins of Inbred Mice*, Morse III HC (ed.), Academic Press, New York, NY, pp. 423–438.

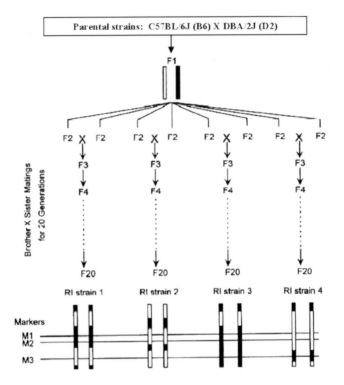

Fig. 4.   Procedure of RI strain breeding.

The RI strains have two features. First, like the F2 progeny, each strain contains different chromosomal fragments randomly chosen from two progenitors; thus, each of the new strains has a randomly different mosaic arrangement. Second, unlike the F2 progeny, within each strain, each pair of chromosomes is homozygous for a mosaic of alternating chromosomal stretches derived from the two progenitors.

The advantage of this set-of-RI-strains approach to mapping becomes apparent when we realize that each of these new inbred strains can be maintained indefinitely as the equivalent of a never-ending segregant in the cross B6 × D2. A new gene tested for its segregation pattern can now be compared for genetic linkage with every gene that was ever scored by any laboratory in the same set of strains, and it can eventually be compared for linkage with any gene that is tested in the future.

RI strains are particularly useful in the detection of QTLs because of their homozygous status. For any phenotype, we can replicate as many times as we like using the same RI strain. Therefore, statistical significance can be reached for a small variation, or the phenotype of a QTL can be confirmed by replication using the same strain. Furthermore, the QTL region on a chromosome is easily defined by comparison of the chromosomal segments among the RI strains.

Today, RI resources still have a powerful utility in mapping the genes' underlying phenotypes whose molecular basis is still unknown, and are especially useful for complex phenotypes (e.g. developmental processes, regulatory phenomena, disease progression) that cannot be determined on a single mouse. The completion of the mouse genome sequence and the extensive genotyping of RI strains have greatly enhanced QTL mapping and candidate gene discovery in QTL studies. The combination of RI strains and genomic information will enable rapid progress in the gene discovery of QTLs.

2.4. Tanscriptome-QTL Mapping

Transcriptome-QTL mapping was developed by a group of investigators at the University of Tennessee Health Science Center. This method exploits the isogenic lines of mice that can be studied using a battery of computational, statistical, molecular, and even morphometric methods. Databases on variations in gene expression are coupled with image and brain behavioral databases. A researcher interested in a particular behavioral phenotype in mice (e.g. BMD) can search for transcriptional regulators that may influence that trait, thus identifying major causes that underlie the variation.

In transcriptome-QTL mapping analysis, the microarray data (expression levels of genes) are treated as the phenotype instead of the disease phenotype. By its principle, the traditional linkage analysis is an association between the genotype of a particular location of the chromosome/genome and a particular phenotype (e.g. disease). In transcriptome-QTL mapping, an association is sought between genotypes (markers on the chromosome) and the levels of expression of genes. In this case, the procedure of transcriptome-QTL mapping is essentially the same as that of genetic QTL mapping. However, unlike the disease phenotype, there are thousands of genes with different levels of expression. Thus, the computation in transcriptome-QTL mapping will repeat thousands of more cycles than that of disease mapping. In reality, transcriptome-QTL mapping has much more of a statistical task than that of genetic QTL mapping, but the QTL mapping program can be easily used for transcriptome-QTL mapping. This group has successfully accomplished transcriptome-QTL maps using the adult brain mRNA levels, adult hematopoietic stem cell mRNA levels, and other published phenotypes of the mouse model. The data are

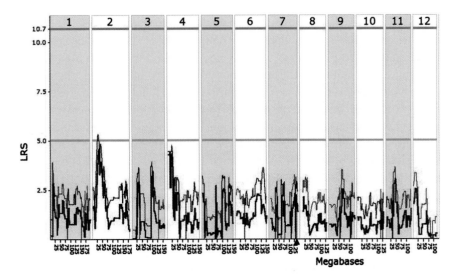

Fig. 5.   Partial picture of a transcriptome QTL mapping using an M430 chip. A positive additive coefficient (green line) indicates that DBA/2J alleles increase trait values. In contrast, a negative additive coefficient (red line) indicates that AKR/J alleles increase trait values. Source: WebQTL (http://www.genenetwork.org/cgi-bin/ WebQTL.py?sid=3b13218b8dc0bb6b7139a2d075dab5237594a274).

posted on the webpages http://headmaster.utmem.edu/search.html/ and http://webqtl.roswellpark.org/search.html/. Similar analysis on bone studies is expected in the near future.

Figure 5 shows part of a picture of transcriptome-QTL mapping using an Affymetrix mouse 430 chip. On the left are the LRS scores, while the bottom indicates the microsatellite markers along the chromosome. The peaks of the QTL mapping reflect the degree of the association of expression levels with the microsatellite markers listed on the X-axis (in this case, IGF regulation). Clicking any QTL point will give the information of the gene that is mapped to that point. Readers are encouraged to view the webpage using the addresses mentioned above.

2.5. *Transgenic and Knockout Mice*

Prior to the current revolution in applied molecular genetics, the only practical way of studying the function of mammalian genes was to utilize spontaneous mutants. The major problem with this method was that a large number of other genes flanking the mutant genetic locus were invariably

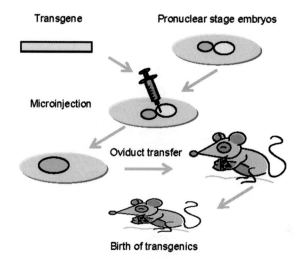

Transgene

Pronuclear stage embryos

Microinjection

Oviduct transfer

Birth of transgenics

Fig. 6. Schematic of the generation of a transgenic mouse by pronuclear microinjection.

transferred from animal to animal during meiotic recombination along with the gene(s) in question. In the early 1980s, a breakthrough technology known as transgenics or gene transfer was developed by Richard Palmiter of the University of Washington in collaboration with Ralph Brinster of the University of Pennsylvania.[15] This new technology involved the process of pronuclear microinjection — a method involving the injection of genetic material (transgene) into the nuclei of fertilized eggs — followed by implanting the modified fertilized eggs into pregnant females, who were then brought to term (Fig. 6).

A major downfall of this technique was that researchers could neither predict nor control where in the genome the foreign genetic material would be inserted. Since a gene's location in the genome is important for its expression pattern, mouse lines carrying the same transgene could display wildly varying phenotypes. An innovative solution that resolved this drawback was originated by a team of scientists led by Martin Evans, Oliver Smithies, and Mario Capecchi, who created what is known as a "knockout" through targeting genes by a process called homologous recombination in mouse embryonic stem (ES) cells[16] (Fig. 7). By creating a knockout, they proved that it was possible to aim the inserted gene at a precise location in the mouse genome. This gave scientists the ability to replace, or knock out, a specific gene with an inactive or mutated allele. The technique has been used to make several thousand different knockout mice. Knockout mice

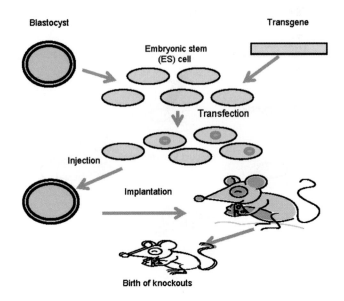

Fig. 7.   Schematic of the generation of a knockout by gene targeting.

have become one of the most useful scientific tools in helping to understand the human genome and its roles in disease. In this chapter, we will introduce the development of transgenic and knockout technologies and applications.

2.5.1. *Transgenic Mice*

Although many types of animals such as rats, rabbits, sheep, cattle, pigs, birds, and fish have been engineered to generate biological or disease models, mice have remained the main species used in the field of transgenic study for the following reasons: (1) the mouse is a mammal, and its development, physiology, and diseases are considerably similar to those of humans; (2) almost all mouse genes (~99%) have homologs in humans; and (3) it has a relatively low cost of maintenance.

There are three major methods used for the production of transgenic animals: retrovirus-mediated gene transfer,[17] DNA microinjection, and ES cell–mediated gene transfer. In the retroviral vector–based method, retroviruses are used to infect the cells of an early-stage embryo before implantation. The basic design of all retroviral vector systems is similar: the *gag, pol,* and *enu* genes that encode essential viral proteins are removed and replaced with the transgene of interest. This method is relatively quick, but includes the risk that the DNA may insert itself into a critical locus, thus causing an unexpected, detrimental genetic mutation. Alternatively, the transgene may insert into a locus that is subject to gene silencing. For these reasons, several independent lines of mice containing the same transgene must be created and studied to ensure that any resulting phenotype is not due to toxic gene dosing or to the mutations created at the site of transgene insertion. Another disadvantage of this method is the limitation of the size of the transferred gene.

DNA microinjection is the first robust method that was developed for the generation of transgenic animals, and provided the underlying concept for the other two methods. Although it is considered reliable, the success rate of this method is low ($\sim$3%–5%).[18] In addition, this technique results in random incorporation of transgenes into the host chromosomes and variable expression of the transgenes in the body.[19]

ES cells refer to cells from early mouse embryos at the blastocyst stage that have the ability to differentiate into various types of cells. ES cells have the advantage that they can be genetically modified by means of homologous recombination, a process by which a fragment of genomic DNA introduced into a mammalian cell recombines with the endogenous homologous sequence, prior to being injected into the blastocyst. This process is known as "gene targeting". In theory, using this technology, any type of mutation (e.g. deletion, point mutation, inversion, or translocation) can now be modeled in mice. Therefore, this method has been widely used to generate knockout and knockin mice (see below).

Depending on the goal of the experiment, the transgenic mouse will display overexpression of a nonmutated protein, expression of a dominant-negative protein, or expression of a fluorescent-tagged protein. By definition, transgenesis is the introduction of DNA from one species into the genome of another species. Many first transgenic mice fit this description well, as they were generated for the study of overexpression of a human protein, often an oncogene.[20] Currently, the phrase "transgenic mouse" generally refers to any mouse whose genome contains an inserted piece of DNA originating from the mouse genome or other species; this term includes the traditional transgenic mouse as well as a knockin or knockout mouse.

To avoid the problems of nonspecific insertion by a traditional transgenic approach, many researchers now rely on gene targeting technology to produce knockin mice in order to study the function of a gene. A knockin mouse is generated by targeted insertion of the transgene at a selected locus. The insert is flanked by DNA from a noncritical locus, and homologous recombination allows the transgene to be targeted to that specific, noncritical integration site. In this way, site-specific knockins result in a more consistent level of expression of the transgene from generation to generation. Also, because a targeted transgene does not interfere with a critical locus, the researcher can be more certain that any resulting phenotype is due to the exogenous expression of the protein. Although the generation of a knockin mouse does avoid many of the problems of a traditional transgenic mouse, this procedure requires more time to design the vector and to identify ES cells that have undergone homologous recombination.

Since the cloned sheep Dolly was born, nuclear transfer technology has been exploited as a novel method for the production of transgenic animals. The process of nuclear transfer in mammals involves replacing the nucleus of an egg with the nucleus from a genetically modified donor cell.[21] The generation of transgenic animals by nuclear transfer has now been achieved in sheep, cattle, mice, pigs, goats, rabbits, and cats.[22]

2.5.2. *Knockout Mice*

While traditional transgenic and knockin mice are generated to elucidate gene function by expressing a gene product, much information about gene function can also be learned from the elimination of a gene or a functional domain of the protein. This can be achieved through many approaches such as random mutation using chemical mutagenesis, gene trapping, or gene targeting to generate a knockout mouse. Homologous recombination technology allows a researcher to completely remove one or more exons from a gene, resulting in the production of a mutated or truncated protein or, more often, no protein at all. Knockout mice have become one of the most useful scientific tools.

Gene targeting in mouse ES cells has become the "gold standard" technology for determining gene function.[23] This technique involves two steps: first, a vector is constructed; then, the vector is put into ES cells and a new mouse is made from these cells. The vector consists of two regions of homology at either end, the disrupted gene of interest, and two additional genes (i.e. neomycin resistance gene and thymidine kinase gene) for quality control during this process. Conventional knockout technology has become so well established that its protocols are described in a cookbook-style manner.

The conventional knockout technology aims to knock out both alleles so that the gene is entirely absent from all cells. Recently, novel technologies have been developed for conditional knockouts, which aim to delete a gene in a particular organ, cell type, or stage of development. Conditional knockout mice have several advantages over the conventional type, such as longer survival of mice and more precise targeted location. There are several different ways of making conditional knockout models, but the most widely used approach is the Cre–loxP recombinase system.[24]

Gene targeting has been hampered by the fact that homologous recombination in mammalian cells is a rare event and varies greatly from gene to gene, making it very difficult to target some genes. Despite much effort toward generating mouse knockouts, only a very small proportion ($\sim$10%) of mouse genes ($n = \sim 25\,000$) have been described in the literature. To address this problem, two other complementary methods — gene trapping and RNA interference (RNAi) — are recommended for use in combination with gene targeting technology to create large numbers of knockout alleles in mice.[25]

Gene trapping is a high-throughput method of random knockout mutagenesis in which the insertion of a DNA element into endogenous genes leads to their transcriptional disruption.[26] Gene trapping constructs are usually delivered to the target cell via electroporation or via replication-deficient self-inactivating retroviruses. Unlike other random mutagenesis methods, such as chemical or radiation-induced mutagenesis, gene trapping allows the rapid identification of the mutated gene.[27] However, this trapping is not entirely random. Generally, larger transcription units and more highly expressed genes in ES cells are trapped more readily.

RNAi is a method of silencing gene expression using double-stranded RNA (dsRNA).[28] It offers vast promise for gene function annotation in mice, but is not well developed for the large-scale generation of gene modifications capable of reliably producing true null alleles.

2.5.3. *N-Ethyl-N-Nitrosourea (ENU) Mutagenesis Mouse Models*

Gene targeting through homologous recombination in ES cells allows the systematic production of mouse mutants for known genes. Gene trap strategies have been designed to interrupt even unknown genes that are tagged by the inserted vector. Complementary to these gene-driven approaches, phenotype-driven approaches are necessary to identify new genes or gene products through a search for mutants with specific defects, thus enabling the elucidation of the function of genes. Mutagenesis using the alkylating agent *N*-ethyl-*N*-nitrosourea (ENU) is a powerful approach for the generation of such mouse mutants.

ENU is the most potent chemical mutagen in mice, with a mutagenesis frequency of 1.5×10^{-3} per locus.[29,30] In contrast to other chemical agents, ENU primarily produces point mutations in germ cells, particularly spermatogonial stem cells.[31] In this sense, ENU-induced mutant mice are very similar to the situation of single nucleotide polymorphisms (SNPs), and will be a good animal model for the human SNP project. In addition to loss-of-function mutations, ENU mutagenesis can also result in gain-of-function mutations. In this method, the male mice are mutagenized by an intraperitoneal injection of ENU; then, the mutagenized males are bred to wild-type females, and the resulting F1 offspring are examined for dominant or semidominant phenotypes. For recessive phenotype screens, F1 offspring are bred to wild-type females, the resulting female F2 offspring are backcrossed to the mutagenized male, and F3 offspring are examined for phenotypes. Following the isolation of an animal with the desired trait, a candidate gene approach may be used to identify the mutated gene, which is confirmed to be heritable by test matings.

The main advantage of the ENU approach is that it is unbiased and does not rely on prior knowledge of gene function. It is a forward genetics approach based entirely on phenotype. Another advantage is that ENU typically produces point mutations, and the process can relate phenotypes to subtle mutations that would never be chosen and produced by gene targeting.

The major challenge in such a phenotype-driven strategy is the establishment of appropriate procedures to assess the mutant phenotypes of interest and to identify the point mutations responsible for any specific inherited phenotype. The phenotype of interest might result from the mutation of a previously characterized gene or, most likely, a majority of phenotypes of interest might result from the mutation of groups of genes. To date, the large ENU efforts have resulted in the identification of very few genes.

A comprehensive three-stage protocol for phenotype assessment called SHIRPA has recently been proposed by Rogers *et al.*[32] The first stage of screening utilizes 40 rapid standard tests to provide a behavioral and functional profile by observational assessment. The second stage of screening involves a detailed behavioral assessment and pathological analysis. The last screening stage uses more sophisticated approaches to assess existing or potential models of neurological disease and phenotypic variability that may be the result of unknown genetic influences. To evaluate the sensitivity of the protocol in identifying mutants with neuromuscular abnormalities, Rafael *et al.*[33] tested two dystrophin-deficient mutants Dmd^{mdx} and

Dmd^{mdx3cv}, both of which are indistinguishable from wild-type by a simple visual exam, using the primary screen of the SHIRPA protocol. They found that the SHIRPA primary screen is effective not only in identifying subtle neuromuscular mutants, but also in distinguishing qualitative differences between mutants with neuromuscular abnormalities. Masuya *et al.*[34] modified the first stage of the protocol by integrating new morphologic observations into the initial assays. Using the modified SHIRPA protocol, they screened the dominant phenotypes of more than 10 000 F1 progeny generated by crossing DBA/2J females with ENU-treated C57BL/6J males, and obtained 136 hereditary-confirmed mutants that exhibit behavioral and morphologic defects.

2.5.4. *Applications of Genetically Engineered Mice*

There are two major applications of genetically engineered mice: biomedical research and healthcare.[35] Many potential applications of transgenic animal technology relevant to human healthcare that were suggested when the traditional transgenic methods were developed in the 1980s have become a reality. One is to use transgenic animals to produce therapeutic proteins. Another is to treat degenerative diseases with replacement organs from other species, i.e. xenotransplantation. Many biotechnology companies have made efforts on protein production in milk and eggs as well as blood, urine, and seminal plasma. The first recombinant human protein expressed in the milk of a transgenic dairy animal was factor IX in sheep in 1988.[36] Several different proteins have since been successfully expressed in the milk of different livestock species, and some of them are subsequently used in clinical trials. Examples of these proteins are insulin-like growth factor 1 in rabbits, α1-antitrypsin in sheep, antithrombin III in goats, and protein C in pigs.[37] The product α-glucosidase from the milk of transgenic rabbits has been successfully used for the treatment of Pompe's disease in infants.[38]

The shortage of human organ supply for severe diseases has stimulated the investigation of xenotransplantation. Current research is focused on the genetic modification of pigs to avoid the hyperacute rejection response.[39] The α1,3-galactose (α1,3-Gal) epitope is the major antigen that stimulates this response. It is synthesized by an enzyme named 1,3-galactosyltransferase (a1,3GT), which is the first target for the genetic engineering of pigs as organ donors. Much work needs to be done to evaluate the results of this strategy.

The transfer of disease from farm animals to humans is a major challenge to public health. Genetically engineering livestock or poultry to make them resistant to specific zoonoses is an attractive strategy to address this issue. The major suggested approach is to engineer animals that express molecules blocking productive infection. This may be achieved by expressing a dominant-negative protein.

Genetically engineered mice induced by the abovementioned techniques allow researchers to test the specific functions of particular genes and to observe the processes that these particular genes could regulate *in vivo*. Therefore, they are widely used in biomedical research.[40–42] Understanding gene structure and functions is of great help in developing gene therapy, which is an approach to treat certain disorders by introducing specific engineered genes into a patient's cells.

The number of genetically engineered mice available for study is rapidly increasing. Unfortunately, many induced mutant mice are not commercially available, and there is no single comprehensive repository for all genetically altered mice. In the fall of 2003, an international meeting was convened at the Banbury Center of the Cold Spring Harbor Laboratory, New York, to discuss the construction of a public resource consisting of a comprehensive collection of mouse knockouts, i.e. a null mutation in every gene in the mouse genome. The meeting attendees agreed that such a resource would greatly benefit biomedical research.

Since the Banbury meeting, a significant amount of effort has been made to organize mouse knockout works around the world, such as in Europe, Canada, Australia, and Japan. The National Human Genome Research Institute (NHGRI) organized a working group of National Institutes of Health (NIH) institute and center representatives to develop an NIH Knockout Mouse Project (KOMP) plan in order to generate a comprehensive and public resource comprised of mice containing a null mutation in every gene in the mouse genome. On October 5, 2005, the NIH announced that contracts had been signed with two companies — Deltagen, Inc., and Lexicon Genetics, Inc. — to obtain 251 lines of knockout mice and extensive relevant phenotyping data from them. The NIH will make these mouse lines and data as widely available to the research community as possible. The contracts allow the acquisition of more lines and phenotypic data from over 1500 additional lines under the same terms for the next 3 years.

To facilitate the model search, we summarize several major resources of induced mutant mice as follows:

(1) The Induced Mutant Resource (IMR), sponsored by The Jackson Laboratory

This database lists over 1000 strains of mutant mice induced by various types of techniques such as transgenics, gene targeting, gene trapping, chemical mutagenesis, and irradiation. These strains can be used in numerous research areas, including apoptosis, cancer, cardiovascular system, cell biology, dermatology, developmental biology, diabetes, obesity, endocrine deficiency, hematology, immunology and inflammation, neurology, reproductive biology, sensorineural hearing loss, and virology (http://www.jax.org/imr/imr_info.html).

(2) Mouse Models of Human Cancers Consortium (MMHCC) Mouse Repository

This repository/distribution resource is located at the National Cancer Institute (NCI)–Frederick Cancer Research Center. The center can now provide about 90 mouse cancer models to the scientific community (http://web.ncifcrf.gov/researchresources/mmhcc/default.asp). Information on over 500 mouse models of human cancers can be obtained by searching another MMHCC website: http://cancermodels.nci.nih.gov/mmhcc/index.jsp.

(3) The Type 1 Diabetes Repository (T1DR)

The National Center for Research Resources (NCRR) at the NIH has funded this resource at The Jackson Laboratory. The purpose of this repository is to collect and cryopreserve about 150 mouse stocks important to research in type 1 diabetes. Current repository holdings can be found at the T1DR website (http://www.jax.org/t1dr). The website includes information about stock availability as well as an online submission form for investigators who wish to donate a new stock.

(4) The International Mouse Mutagenesis Consortium (IMMC)

The IMMC was established to encourage the exchange of information and resources among participating mutagenesis laboratories and centers for systematically and comprehensively annotating the mouse genome and compiling data on mouse mutations (http://www.informatics.jax.org/mgihome/other/phenoallele_commun_resource.shtml#pointer1).

(5) The International Gene Trap Consortium (IGTC)

The IGTC was established as a subgroup of the IMMC. It provides all publicly available resources of embryonic stem cells with gene trap insertions in any or most of the genes in the mouse genome (http://www.genetrap.org/).

3. CONCLUSIONS AND FUTURE DIRECTIONS

There are no real substitutes for the use of laboratory animals. Studies with bacteria, tissue culture, and computer simulation can provide useful information, but the complexity of living organisms requires research and testing on animals that are similar to humans so as to attain reliable and effective results. Unforeseen side-effects of inadequately tested new drugs or biomedical materials have proven the need for extensive animal tests before such drugs or materials can meet the Food and Drug Administration (FDA) licensing requirements for human application.

As a model organism, the mouse has much to offer in the better understanding of human biology and disease. Technologies for the production of induced mutant mice have advanced considerably in recent years, and will undoubtedly have an important impact on biomedical research and human healthcare. In an era where biologists are focusing on how a gene is involved in a disease rather than which gene is involved in a disease, the mouse is more than ever the organism that is able to provide the answers. A combination of genotype- and phenotype-driven approaches will greatly speed up the establishment of a catalog of mutant mice and mouse phenomes. It will provide a powerful resource to uncover novel pathways and annotate gene function.

REFERENCES

1. Beamer WG, Donahue LR, Rosen CJ, Baylink DJ. (1996) Genetic variability in adult bone density among inbred strains of mice. *Bone* **18**(5):397–403.
2. Beamer WG, Donahue LR, Rosen CJ. (2002) Genetics and bone. Using the mouse to understand man. *J Musculoskelet Neuronal Interact* **2**(3): 225–231.
3. Beamer WG, Shultz KL, Churchill GA, *et al.* (1999) Quantitative trait loci for bone density in C57BL/6J and CAST/EiJ inbred mice. *Mamm Genome* **10**(11):1043–1049.
4. Beamer WG, Shultz KL, Donahue LR, *et al.* (2001) Quantitative trait loci for femoral and lumbar vertebral bone mineral density in C57BL/6J and C3H/HeJ inbred strains of mice. *J Bone Miner Res* **16**(7):1195–1206.
5. Bouxsein ML, Uchiyama T, Rosen CJ, *et al.* (2004) Mapping quantitative trait loci for vertebral trabecular bone volume fraction and microarchitecture in mice. *J Bone Miner Res* **19**(4):587–599.

6. Benes H, Weinstein RS, Zheng W, *et al.* (2000) Chromosomal mapping of osteopenia-associated quantitative trait loci using closely related mouse strains. *J Bone Miner Res* **15**(4):626–633.

7. Williams RW, Bennett B, Lu L, *et al.* (2004) Genetic structure of the LXS panel of recombinant inbred mouse strains: a powerful resource for complex trait analysis. *Mamm Genome* **15**(8):637–647.

8. Shimizu M, Higuchi K, Bennett B, *et al.* (1999) Identification of peak bone mass QTL in a spontaneously osteoporotic mouse strain. *Mamm Genome* **10**(2):81–87.

9. Klein RF, Mitchell SR, Phillips TJ, *et al.* (1998) Quantitative trait loci affecting peak bone mineral density in mice. *J Bone Miner Res* **13**(11):1648–1656.

10. Koller DL, Liu G, Econs MJ, *et al.* (2001) Genome screen for quantitative trait loci underlying normal variation in femoral structure. *J Bone Miner Res* **16**(6):985–991.

11. Koller DL, Rodriguez LA, Christian JC, *et al.* (1998) Linkage of a QTL contributing to normal variation in bone mineral density to chromosome 11q12–13. *J Bone Miner Res* **13**(12):1903–1908.

12. Klein OF, Carlos AS, Vartanian KA, *et al.* (2001) Confirmation and fine mapping of chromosomal regions influencing peak bone mass in mice. *J Bone Miner Res* **16**(11):1953-1961.

13. Orwoll ES, Belknap JK, Klein RF. (2001) Gender specificity in the genetic determinants of peak bone mass. *J Bone Miner Res* **16**(11):1962–1971.

14. Shimizu M, Higuchi K, Kasai S, *et al.* (2001) Chromosome 13 locus, Pbd2, regulates bone density in mice. *J Bone Miner Res* **16**(11):1972–1982.

15. Brinster RL, Braun RE, Lo D, *et al.* (1989) Targeted correction of a major histocompatibility class II E alpha gene by DNA microinjected into mouse eggs. *Proc Natl Acad Sci USA* **86**(18):7087–7091.

16. Wilkie TM, Braun RE, Ehrman WJ, *et al.* (1991) Germ-line intrachromosomal recombination restores fertility in transgenic MyK-103 male mice. *Genes Dev* **5**(1):38–48.

17. Barquinero J, Eixarch H, Perez-Melgosa M. (2004) Retroviral vectors: new applications for an old tool. *Gene Ther* **11**(Suppl 1):S3–S9.

18. Nottle MB, Haskard KA, Verma PJ, *et al.* (2001) Effect of DNA concentration on transgenesis rates in mice and pigs. *Transgenic Res* **10**(6):523–531.

19. Chicas A, Macino G. (2001) Characteristics of post-transcriptional gene silencing. *EMBO Rep* **2**(11):992–996.

20. Robertson E, Bradley A, Kuehn M, Evans M. Germ-line transmission of genes introduced into cultured pluripotent cells by retroviral vector. *Nature* **323**:445–448.

21. Wilmut I, Schnieke AE, McWhir J, *et al.* (1997) Viable offspring derived from fetal and adult mammalian cells. *Nature* **385**(6619):810–813.

22. Wilmut I, Beaujean N, de Sousa PA, *et al.* (2002) Somatic cell nuclear transfer. *Nature* **419**(6907):583–586.

23. Capecchi MR. (2005) Gene targeting in mice: functional analysis of the mammalian genome for the twenty-first century. *Nat Rev Genet* **6**(6):507–512.

24. Utomo AR, Nikitin AY, Lee WH. (1999) Temporal, spatial, and cell type–specific control of Cre-mediated DNA recombination in transgenic mice. *Nat Biotechnol* **17**(11):1091–1096.

25. Austin CP, Battey JF, Bradley A, *et al.* (2004) The Knockout Mouse Project. *Nat Genet* **36**(9):921–924.
26. Durick K, Mendlein J, Xanthopoulos KG. (1999) Hunting with traps: genome-wide strategies for gene discovery and functional analysis. *Genome Res* **9**(11):1019–1025.
27. Abuin A, Holt KH, Platt KA, *et al.* (2002) Full-speed mammalian genetics: *in vivo* target validation in the drug discovery process. *Trends Biotechnol* **20**(1):36–42.
28. Sandy P, Ventura A, Jacks T. (2005) Mammalian RNAi: a practical guide. *Biotechniques* **39**(2):215–224.
29. Russell WL, Kelly PR, Hunsicker PR, *et al.* (1979) Specific-locus test shows ethylnitrosourea to be the most potent mutagen in the mouse. *Proc Natl Acad Sci USA* **76**:5918–5922.
30. Stanford WL, Cohn JB, Cordes SP. (2001) Gene-trap mutagenesis: past, present and beyond. *Nat Rev Genet* **2**:756–768.
31. Abuin A, Holt KH, Platt KA, *et al.* (2002) Full-speed mammalian genetics: *in vivo* target validation in the drug discovery process. *Trends Biotechnol* **20**(1):36–42.
32. Rogers DC, Fisher EM, Brown SD, *et al.* (1997) Behavioral and functional analysis of mouse phenotype: SHIRPA, a proposed protocol for comprehensive phenotype assessment. *Mamm Genome* **8**(10):711–713.
33. Rafael JA, Nitta Y, Peters J, Davies KE. (2000) Testing of SHIRPA, a mouse phenotypic assessment protocol, on Dmd^{mdx} and Dmd^{mdx3cv} dystrophin-deficient mice. *Mamm Genome* **11**(9):725–728.
34. Masuya H, Inoue M, Wada Y, *et al.* (2005) Implementation of the modified-SHIRPA protocol for screening of dominant phenotypes in a large-scale ENU mutagenesis program. *Mamm Genome* **16**(11):829–837.
35. Hunter CV, Tiley LS, Sang HM. (2005) Developments in transgenic technology: applications for medicine. *Trends Mol Med* **11**(6):293–298.
36. Simons JP, Wilmut I, Clark AJ, *et al.* (1988) Gene transfer into sheep. *Biotechnology* **6**:179–183.
37. Rudolph NS. (1999) Biopharmaceutical production in transgenic livestock. *Trends Biotechnol* **17**(9):367–374.
38. Van den Hout JM, Kamphoven JH, Winkel LP, *et al.* (2004) Long-term intravenous treatment of Pompe disease with recombinant human α-glucosidase from milk. *Pediatrics* **113**:e448–e457.
39. Cox A, Zhong R. (2005) Current advances in xenotransplantation. *Hepatobiliary Pancreat Dis Int* **4**(4):490–494.
40. Watase K, Zoghbi HY. (2003) Modelling brain diseases in mice: the challenges of design and analysis. *Nat Rev Genet* **4**(4):296–307.
41. Laustsen PG, Michael MD, Crute BE, *et al.* (2002) Lipoatrophic diabetes in $Irs1^{-/-}/Irs3^{-/-}$ double knockout mice. *Genes Dev* **16**:3213–3222.
42. McPherson JP, Lemmers B, Chahwan R, *et al.* (2004) Involvement of mammalian Mus81 in genome integrity and tumour suppression. *Science* **304**:1822–1826.

Chapter 13

Acquisition of Accurate Gene Expression Information from Microarray Measurements

Shuang Jia, Shouguo Gao, Martin J. Hessner and Xujing Wang*

Max McGee National Research Center for Juvenile Diabetes
Department of Pediatrics, Medical College of Wisconsin and
Children's Hospital of Wisconsin, 8701 Watertown Plank Road
Milwaukee, WI 53226, USA

Human and Molecular Genetics Center, Medical College of Wisconsin
8701 Watertown Plank Road, Milwaukee, WI 53226, USA

Gene expression profiling using microarray technology has become an important genetic tool in the study of complex diseases. However, this technology is prone to noise and the accuracy of its measurements is often in question. Here, we describe the technological and analytical advancements that we have made in microarrays to overcome this problem. We have extended the conventional dual-color spotted microarray technology to a novel three-color microarray platform, where the probes on printed arrays are labeled with a third dye for quality control (QC) in array fabrication. We have also developed a microarray image analysis package called Matarray, which achieves quantitative QC of data acquisition through the definition of a set of quality scores. With these advances, a better dissection of the sources of data variability and more efficient QC are achieved. In addition, our data QC approach has led to a new weighted statistical procedure for evaluating the significance of microarray findings that can more sensitively detect changes in gene expression. This resolves the missing value problem that has often plagued microarray data analysis. Finally and most importantly, we show that with our analytical and technical advancements, where a comprehensive and efficient QC procedure is in place, accurate gene expression measurements comparable in quality to those by quantitative reverse

---

* Correspondence author.

transcription–polymerase chain reaction (RT-PCR) can be achieved with microarrays fabricated in academic laboratories.

1. MICROARRAY TECHNOLOGY: POTENTIALS AND PROBLEMS

Since its introduction nearly a decade ago,[1] microarray technology — which allows the investigation of a whole transcriptome in a single hybridization — has become a widely used genetic tool. It has great potential in dissecting the complex regulatory networks at the genome level, and hence the genetic mechanisms that underlie complex diseases.[2,3] During this time, the technology has rapidly evolved into a number of widely used high-density platforms that are either commercially fabricated or prepared within research laboratories, mainly based on one of two general fabrication formats: *in situ* synthesized oligonucleotide arrays,[4,5] or spotted arrays. The immobilized probe type can vary considerably and includes ~300–1500 bp of spotted polymerase chain reaction (PCR) products amplified from cDNA clones, *in situ* synthesized 25-mer oligonucleotides (Affymetrix GeneChip), and spotted or *in situ* synthesized 60-mer to 70-mer oligonucleotides (Agilent). New platforms are still being introduced, such as the Illumina BeadArray (San Diego, CA)[6] and the Universal Hexamer Array from Agilix (New Haven, CT).[7] Furthermore, the varying platforms differ in terms of experimental design as well as target labeling and hybridization protocols. For example, the Affymetrix platform analyzes a single *in vitro* transcribed, biotinylated cRNA per array, which is visualized after hybridization with a streptavidin–phycoerythrin conjugate; whereas other platforms typically utilize cohybridization of two cDNA targets directly labeled during reverse transcription with different fluorescent dyes.

Each format offers unique advantages and disadvantages in terms of cost, study design considerations, array content flexibility, and required hardware and analytical protocols. cDNA arrays can be cost-effectively fabricated in-house, possess high hybridization stringency, and are not highly susceptible to single nucleotide polymorphisms (SNPs). However, they require laborious and error-prone clone library management, PCR amplification and purification, and cDNA clone sequence validation[8,9] (since clone misidentification rates within libraries have been estimated to be as high as 30%[10]). Oligonucleotide arrays exhibit a number of advantages, including the fact that they can be designed to exclude homologous sequences between genes, thereby enhancing specificity. In addition, a given gene can be represented by a set of different oligonucleotides targeting different regions or exons, allowing for the detection of splice variants or the discrimination of closely related genes. Oligonucleotide probe designs are typically based upon deposited sequence information, and are therefore dependent on the quality of the submitted information and annotation. Highlighting this reality is a recent report which found that, among Affymetrix mammalian arrays, more than 19% of the probes on any given array type did not correspond to their appropriate mRNA reference sequence defined by the highly curated, publicly available RefSeq database.[11]

Microarray technology has great potential in the study of complex human diseases,[12] systems biology,[13] and clinical diagnostics/prognostics.[14] However, improved reliability and accuracy in measurements are prerequisites for these applications to truly blossom. The technology is known to be prone to noise and low reproducibility.[15] Furthermore, correlations with other platforms, including reverse transcription–PCR (RT-PCR),[12,15] and between different microarray platforms are often unsatisfactory.[16–19] On the other hand, many disease processes may involve subtle gene perturbations that require highly accurate signal quantification. A general data quality control (QC) scheme is therefore needed, so that each platform can be systematically assessed[20,21] and data from different platforms can be utilized and integrated.

Among microarray platforms, the spotted arrays fabricated in academic laboratories are usually more subject to QC problems than commercial arrays.[16–18] Over the last few years, we have made an effort to develop a quantitative data QC procedure for cDNA arrays that can be standardized and generalized for other microarray platforms. We have developed a microarray image processing software called Matarray, which assesses the reliability of measurements acquired from each microarray spot utilizing a set of quality scores[22] and offers quality-dependent filtering and normalization.[23] We have also developed a novel three-color microarray platform, where the spotted probes are labeled with a third dye and the quality of each printed array can be evaluated prior to its use in hybridization.[24,25] Combined, these advancements have resulted in an academic cDNA microarray platform that is able to generate highly reliable gene expression measurements rivaling commercial systems in their reproducibility and correlating well with quantitative RT-PCR.[26]

In this chapter, we will summarize these developments. We will show that our quantitative QC approach can lead to significant improvements in the statistical evaluation and data mining of microarray data. We will end with a discussion of the application of our methods to other microarray platforms, including spotted and Affymetrix oligonucleotide arrays.

2. QUALITY CONTROL IN cDNA ARRAYS

Efficient QC requires a clear understanding of the sources of data variability so that the contributing factors can be appropriately dissected and most efficiently controlled. We have found that several major sources of noise generated during array fabrication and hybridization can be assessed from array images, and we have developed procedures for their QC. Much of our algorithmic investigations of microarray bioinformatics was made possible by our in-house software, Matarray.[22]

2.1. *Matarray Package*

There are several novel and unique features of Matarray that we will briefly describe here. The most critical step in image quantification is to correctly locate all spots (grid alignments).[22,27] Matarray possesses two novel algorithms to carry out this task: (1) it utilizes both spatial and intensity information in signal–background segmentation, which is more accurate and robust than when either is used alone[22]; and (2) an iterative procedure is applied to improve spot detection until a satisfactory result is obtained.[22] The accuracy of Matarray has been validated using spiked-in control clones of known ratios. The most unique feature of Matarray is its novel quantitative QC scheme for information retrieved from every spot (gene) on the array.[22] As far as we are aware, it is the first in the field to possess this feature. For each spot, nonredundant factors affecting data quality are identified, including spot size, signal-to-noise ratio, background level and uniformity, and saturation status; and individual quality scores are defined for each:

$$q_{\text{size}} = \exp\left(-\frac{|A - A_0|}{A_0}\right), \quad q_{\text{sig-noise}} = \frac{sig}{sig + bkg_1}$$

$$q_{\text{bkg1}} = f_1/\text{CV}_{\text{bkg}}, \qquad q_{\text{bkg2}} = f_2\left(1 - \frac{bkg_1}{bkg_1 + bkg_0}\right) \qquad (1)$$

$$q_{\text{sat}} = \begin{cases} 1, & \text{if \% of saturated pixels} < 10\% \\ 0, & \text{if \% of saturated pixels} \geq 10\%, \end{cases}$$

where f_i are normalization factors; sig stands for signal; and bkg_1 and bkg_0 are local and global background levels, respectively. Several other factors — including the intensity level and the variation of the signal pixels (corresponding score $q_{sig} = f_3/CV_{sig}$) — have also been considered, but were found to be not independent of the above five scores, and hence were excluded from further consideration. Based on these, a composite score q_{com} is defined to give an overall assessment of quality[22]:

$$q_{com} = (q_{size} \times q_{sig\text{-}noise} \times q_{bkg1} \times q_{bkg2})^{1/4} \times q_{sat}. \quad (2)$$

Numerous microarray image analysis softwares are available. We have compared the performance of Matarray with more than 20 of these, including ImaGene, SpotFinder, Gleams, ArrayVision, and ScanAlyze, using both synthetic and real images. Differences in ratio estimates were observed, highlighting a general source of variability in the different information acquisition algorithms and suggesting that QC measures are necessary. Profiling the difference against q_{com} revealed that different applications give highly correlative results for spots possessing high q_{com}, whereas the results can be different for low-q_{com} spots.[22] When judged by the correlation of the measured-to-input (expected) ratios of control clones and the consistency among replicate spots, Matarray was usually able to perform better for over 90% of the spots highest in q_{com}. This demonstrates the importance of having a quantitative quality measure for microarray information acquisition.

2.2. Array Fabrication and TDAV

A substantial limitation in the utilization of spotted arrays fabricated in research laboratories is their susceptibility to QC issues, resulting largely from variable DNA probe deposition and retention on the solid support surface that is difficult to control for each and every array.[25,28,29] This has been a motivation for many investigators, despite the potentially higher costs, to use commercial array systems, where highly controlled fabrication methodologies have evolved, resulting in systems that offer high intraplatform reproducibility. The generation of microarray slides involves coating the glass slides, printing up to tens of thousands of amplified cDNA or oligonucleotide probes, and fixing/blocking the slides. During this process, variable amounts of material can be deposited and/or retained on the activated glass surface. When the amount of immobilized probes is inadequate, the measurements made on such arrays can be unreliable.[24,25,28–30] Noise and artifacts introduced to the arrays at this stage will also directly affect the quality of hybridization.

Until recently, such problems were difficult to quantitatively evaluate and control for each and every array, since the array was typically "invisible" prior to hybridization.[24,25,29] To overcome this difficulty, we have made a significant technological advancement in microarray QC by conceiving and developing a three-color microarray platform,[24,25,28] which we have termed as third-dye array visualization (TDAV) technology.[31] The approach labels the cDNA probes printed on the array slides with the cyanine dye–compatible third dye (TD) fluorescein[24] and makes prehybridization quantitative assessment of array quality possible, so that precious samples as well as laboratory and analytical efforts will not be wasted over poor-quality slides.[24,25,28]

In our TDAV, cDNA array platform probes are amplified with oligonucleotide primers labeled with a third dye, fluorescein, as illustrated in Fig. 1(a). Products are then purified to remove unincorporated primer and PCR reaction components, so that fluorescein intensity detected on the probes is proportional to the amount of cDNA ultimately retained on the solid surface. Figure 1(b) shows the fluorescein TD image in the left column and the corresponding cyanine image in the right column. The TD image is highly predictive of hybridized array performance. The excitation and emission spectra of the fluorescein dye are compatible with Cy3 and Cy5 when using confocal laser scanners with narrow bandwidths.[24] We have also experimentally confirmed that there is no data contamination from the third dye.[24,25,32] Rigorous evaluation of experimental data with or without TDAV using ANOVA (analysis of variance) further indicated that the additional variation introduced by TDAV is insignificant.[32] Setting up quantitative QC rules requires a standard scanner calibration method to ensure consistent

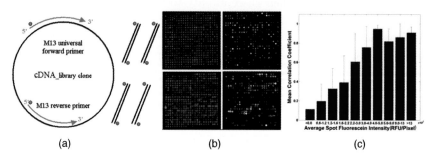

(a)   (b)   (c)

Fig. 1.   TDAV technology in cDNA array platform. (a) Amplification of library clone inserts using fluoresceinated primers. The fluorescein intensity from every spot will directly reflect the amount of bound probes. (b) The fluorescein TD image is predictive of hybridization data quality. The left column shows TD images, and the right column shows the corresponding images after hybridization. A noisy printed slide will readily affect hybridized array performance (lower panels). (c) Relationship between replicate consistency and the amount of bound probes, showing that below a threshold value of 5000 RFU/pixel, data quality is increasingly compromised.

image collection. To accomplish this, we have implemented a scanner calibration method utilizing FluorIS (CLONDIAG, Jena, Germany), a nonbleaching and reusable calibration/standardization tool.[24,25,28]

In developing the TDAV technology, we have performed microarray experiments designed to establish criteria for array and spot QC. This has included the use of control clones spiked in at known ratios as well as homotypic (self–self) and competitive hybridizations. For example, we performed replicated hybridizations between human UACC903 and Jurkat RNA on 80 cDNA arrays printed in the same batch.[24,25] The influences limiting the amount of probes and noise on the data reproducibility were then investigated. In Fig. 1(c), we give the replicate correlation vs. the TD intensity for all data from this experiment. Through these studies, we have established the following quantitative QC criteria to preselect printed cDNA arrays suitable for hybridization[24,25]:

(1) TD intensity mean > 5000 RFU/pixel, and CV < 10%.
(2) TD mean $q_{\text{sig-noise}} > 0.90$.
(3) TD mean of spot diameter/spot-to-spot distance < 0.8, and CV of spot diameter < 20%.

Up to now, we have carried out several major biological experiments using TDAV slides.[32] On average, approximately 85% of arrays printed at our facility meet these standards and are used in our microarray experiments. We have found that the benefit of being able to always use high-quality slides for hybridization is enormous.

2.3. Efficiency of TDAV in Spot-Level QC

TDAV technology has also led to valuable algorithms in spot-level data QC, since even on high-quality arrays there may be a low percentage of low-quality spots that can yield compromised data. The data in Fig. 2 show how this is possible. In this example, we profiled the gene expression in the thymus of diabetic-prone BB DR$^{lyp/lyp}$ ("DP") rats and diabetic-resistant DR$^{+/+}$ ("DR") rats at two time points (day 40 and day 65). This analysis utilized four animal pairs for each comparison, and four replicate hybridizations were performed for each pair, with two hybridizations reverse-labeled to control for dye bias, totaling 32 hybridizations. For the 16 hybridizations (8 for each way of dye labeling) that compared day 65 and day 40 DP rats, the labeling reactions of total thymus RNA were spiked with four *Arabidopsis in vitro* transcripts (cellulose synthase, chlorophyll a/b binding protein,

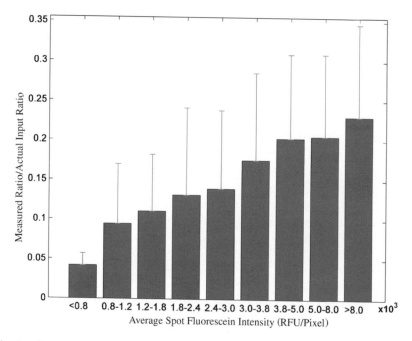

Fig. 2.   Compression and variation in data measurements. Data presented are from the *Arabidopsis* clone spiked in at a ratio of 30:1. Significant further data compression occurs when spot TD intensity falls below 5000 RFU/pixel.

ribulose-1,5-bisphosphate, and triosphosphate isomerase) at known input ratios (30:1, 10:1, 1:1, and 1:0, respectively), giving rise to a total of 1216 data points. These clones enabled an evaluation of microarray measurement accuracy through the comparison of measured output ratios to the known RNA input ratios.

We found that TD intensity, which reflects the amount of bound probes available for hybridization, is a major source of variation in measurements. In Fig. 2, the measured ratio vs. actual input ratio is presented for all spots spiked in at 30:1; evidently, the spots with insufficient support-bound probes exhibit higher compression and data variability. In a real experiment where the transcript abundance for different genes spans a wide range and their folds of change vary, gene-dependent artifacts in measurements will occur. We have also found that TD intensity is a major contributing factor to spatial-dependent heterogeneity arising from different printing pins.[33]

These results suggest that TD intensity is a major factor causing spot-level data variability. In addition, other artifacts measurable by the TD image can also influence the accuracy of expression measurements made from posthybridization images. These include noise, spot size, and shape irregularities.[24,25] Based on these observations, we formulated a quality measure for every spot from the TD image by defining

$$q_{\text{TD}} = q_{\text{int}} * q_{\text{com,TD}}, \tag{3}$$

where $q_{\text{com,TD}}$ is the composite TD image quality score, defined according to the signal-to-noise ratio, spot size, and background levels and variation, as similarly given in Eqs. (1) and (2).[22] q_{int} is given by:

$$q_{\text{int}} = \begin{cases} 1, & \text{intensity} \geq \text{threshold} \\ \text{intensity/threshold}, & \text{intensity} < \text{threshold}. \end{cases} \tag{4}$$

In Matarray, the default threshold is 5000 RFU/pixel.

2.4. *Data Filtering and Quality-Dependent LOWESS Normalization*

Through numerous experiments, we have found that both our quality scores q_{com} and q_{TD} capture very well the inherent variability in microarray measurements. High-score spots will generate less variability, and removing spots with low scores can dramatically improve the reliability of data.[22,23] There is no significant correlation between q_{TD} and q_{com},[33] thus proving that they are two nonredundant quality measures, each capturing a different major source of data variability; QC by each is necessary. In fact, we have optimized our quality score definitions such that the error variance exhibits a log-linear dependence over the quality scores, as shown in Fig. 3. This not only served to validate our quality score definitions, but also led to novel normalization and statistical algorithms, which are described next.

Based on the Matarray work, we have developed an original quality-dependent normalization and data filtering procedure. For each spot, we define a Z-score in place of the commonly used log ratio $\log R$,[23]

$$Z = \frac{\log R(q) - \text{mean\_} \log R(q)}{\text{SD\_} \log R(q)}, \tag{5}$$

and define the Z-method normalized log ratio as $\log R_Z = \log R(q) - \text{mean\_} \log R(q)$, where q is q_{com}. The local mean of log ratio $\text{mean\_} \log R(q)$ is obtained using a LOWESS scheme as described by Wang *et al.*[23] The local standard deviation (SD) $\text{SD\_} \log R(q)$ is obtained using a moving window LOWESS approach. Briefly, SD in log ratio local to a spot is calculated using f proportion of its neighboring spots, where f is the fraction of data

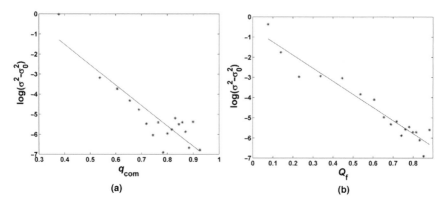

Fig. 3.   The error variance drops exponentially with increasing spot quality. Plotted are the logarithm-transformed error variance $\log(\sigma^2 - \sigma_0^2)$ vs. (a) q_{com} and (b) Q_f. The constant σ_0 is the asymptotic value of $\sigma : \sigma_0 = \lim_{q \to 1} \sigma$.[22]

used for smoothing in the LOWESS fit for mean (default value in Matarray is $f = 0.05$).[23] After that, LOWESS is performed on the SD and the fitted result is defined as $SD\_\log R(q)$.[23] Our work represents the first publication that applies the LOWESS algorithm to obtain both the mean and standard deviation of the ratio distribution, and to perform normalization at both levels.

Figure 4 shows the advantage of Z-normalization, where dashed lines are $mean\_\log R(q) \pm 3SD\_\log R(q)$. Also shown are the three SD lines from a global normalization scheme (dotted lines). If we were to use three SDs as the threshold for statistically significant outliers, then obviously the Z-method would have the potential for more sensitive detection at the high-quality region and the generation of far less false-positives in the low-quality region. Recently, we demonstrated that this approach is more efficient than the commonly used local MA LOWESS normalization.[23] These results demonstrate the advantages and efficiency of utilizing the ratio $-q$ plot for data filtering and normalization.[22,23]

2.5. *QC Pipeline*

Integrating these results, we have developed the following filtering and normalization procedure (Fig. 5):

(1) Evaluate the $\log R - q_{TD}$ plot and decide on a filtering threshold for q_{TD}. Normally, this will be the value where there is an abrupt increase in data variability, in the same fashion as shown in Fig. 4 (which describes

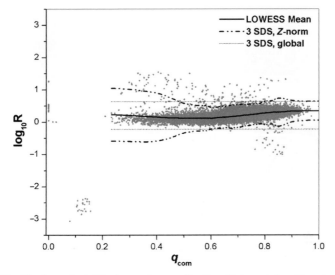

Fig. 4. Quality-dependent filtering and normalization. Data points with extremely low quality, exhibiting a discontinuous distribution for others ($q_{\mathrm{com}} < 0.22$ in this example), will be filtered through reassigning their $q_{\mathrm{com}} = 0$. The remaining data points will be normalized. The localized Z-normalization will lead to more sensitive detection of changes in gene expression with less false-positives.

the $\log R - q_{\mathrm{com}}$ filtering and normalization). The quality score for all spots below this threshold will be reset to $q_{\mathrm{TD}} = 0$.

(2) Perform a local q_{TD}-dependent normalization for all data points with $q_{\mathrm{TD}} > 0$, utilizing the robust scatterplot smoother LOWESS.[23,34,35]

(3) Evaluate the $\log R - q_{\mathrm{com}}$ plot and decide on a filtering threshold for q_{com}. The quality score for all spots below this threshold will be reset to $q_{\mathrm{com}} = 0$, as described in Fig. 4.

(4) Perform a local q_{com}-dependent LOWESS normalization for all data points with $q_{\mathrm{com}} > 0$. The LOWESS fit for SD will also be determined, and the Z-score will be calculated for every spot by

$$Z = \frac{\text{normalized } \log R}{\text{local SD}}.$$

After all normalizations, the final quality score will be defined as

$$Q_{\mathrm{f}} = q_{\mathrm{TD}} * q_{\mathrm{com}}. \tag{6}$$

Only spots with $Q_{\mathrm{f}} > 0$ will be retained for further data mining and modeling.

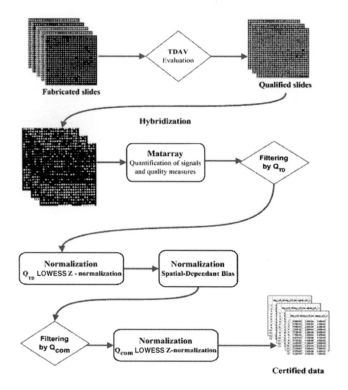

Fig. 5. The Matarray QC pipeline.

2.6. *Improvement in Measurement Quality*

The advantage of our approach is evident when we examine its efficiency at noise reduction indirectly[22,23] through (1) the reduction in the variance of log-ratio distribution from homotypic experiments, (2) the improvement in correlation and concordance for differentially expressed genes among replicate slides in a two-sample competitive hybridization, and (3) the reduction in mean coefficient of variation (CV) for differentially expressed genes among replicate hybridizations.

Recently, a set of experiments was carried out in our laboratory to directly demonstrate the efficiency of our application. In the same rat thymus experiment that we described before, we processed the data with our normalization pipeline, dropped the spots possessing low quality scores ($Q_f = 0$), and compared the measured ratios with the actual ratios. Twenty-two of the genes showing differential expression in the experiment were selected for quantitative RT-PCR for their biological significance, and the

results were compared with the microarray measurements. In both cases, high linear relationships were observed over up to 300 folds of change, with $R^2 > 0.94$. In addition, we observed much less data compression in the microarray measurements than those previously reported by others,[36] indicating the advantage of our microarray setup at the Medical College of Wisconsin. For example, if we were to calibrate microarray measurements using RT-PCR results as a standard, we would have $R_{corrected} = (R_{measured})^q$, with a correction factor of $q \sim 1.17$. This represents a big improvement over the $q \sim 1.88$ previously reported by Yuen *et al.*[36]

Furthermore, we recently compared the measurements from cDNA microarrays with those derived from commercial arrays by Affymetrix and Agilent. We observed a high correlation among the three platforms, with no significant differences in terms of data quality. Briefly, two pooled rat liver RNA samples were hybridized to Affymetrix's RG-U34A arrays, Agilent's G4130A arrays, and our in-house cDNA arrays. The Affymetrix arrays were processed with MAS5.0, and both Agilent and our in-house arrays were processed with Matarray. The three platforms shared 2824 UniGene unique genes. After data filtering, 865 genes passed QC on all platforms. Using the Affymetrix default settings, more than 50% of the probe sets were labeled "absent" on at least one array and were therefore filtered; these criteria were then used to set the filtering stringency in Matarray for the other two arrays. We then calculated (1) the correlation between each pair of platforms for the remaining 865 genes, and (2) the correlation and concordance rate for genes showing differential expression (DE) at $p = 0.05$ in at least one platform. The correlations between Affymetrix and Agilent oligonucleotide arrays, Affymetrix and our in-house cDNA arrays, and Agilent oligonucleotide arrays and our cDNA arrays were 0.91, 0.88, and 0.92, respectively.[26] The concordances were found to be 61%, 50%, and 57%, respectively.[26] Using ANOVA, we found that the different platforms did not cause more variation than the replicate hybridizations within each individual platform.[26] These results indicate that the quality of the data generated from our cDNA arrays is comparable to that from the commercial arrays.

3. QUALITY-WEIGHTED STATISTICS AND DATA MINING

In the statistical inference of differentially expressed genes, as well as in clustering and data mining, we have found the concept of relative spot weights by their quality scores to be valuable in these downstream analyses.

This is somewhat expected, as our quality score measures correlate well with the error variance in data (Fig. 3). Statistically, it is known that utilizing the error variance as a weight can lead to improved performance. For example, the maximum likelihood estimation (MLE) of the mean of a normal distribution is the error variance–weighted average of the experimental measurements.[37] Several groups have explored this possibility by replacing the mean and standard deviation (SD) calculations in statistical tests with their variance-weighted counterparts.[38–40] Limited effort has also been made to improve clustering performance by incorporating the error variance information.[41] However, these methods are all based on a variance calculation from replicates, which requires that an adequate number of replicate hybridizations be made in order to derive a reliable estimation of the variance. In addition, such approaches are not sensitive to quality issues that affect all replicates equally.

Utilizing our quality score definitions, we have recently developed quality-weighted statistical and clustering algorithms. We have found that the weighted test leads to improved sensitivity and specificity. In Fig. 6(a), the p-value distribution according to the weighted and nonweighted t-tests is plotted for the positive and negative control clones of the rat thymus analysis described in Sec. 2.3. Evidently, the type II error rate was significantly reduced without compromising the type I error rate. We have further extended our weighted approach to clustering algorithms by incorporating the quality scores for each gene in the distance metrics. We found again that the weighted clustering generally leads to more sensitive detection of groupings among samples, and is a more robust algorithm against noise. Figures 6(b) and 6(c) show an example of the hierarchical clustering of samples from an experiment conducted to profile gene expression changes in pancreatic islet β-cells during apoptosis progression initiated by the protein kinase C inhibitor staurosporine. Clearly, the weighted algorithm can recover the groupings among the samples much better than the nonweighted algorithm.

These quality-weighted approaches lead to not only improved performance, but also great convenience. In microarray experiments, it is crucial to remove unreliable data before further analysis. However, such data filtering creates the problem of missing values,[42] which makes the combining of data from replicates and the downstream statistical evaluation and data mining extremely cumbersome. Data from microarray experiments are generally formatted as large matrices of gene expression measurements or log ratios (between the target sample and the control, for example).

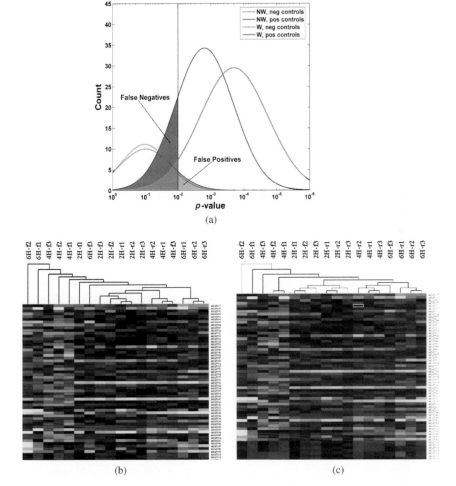

(a)

(b)   (c)

Fig. 6.   (a) Comparison between the weighted and nonweighted *t*-test demonstrates the advantage of the weighted approach. In an experiment that profiled rat thymus, we have 40 data points corresponding to a control clone spiked in at known Cy3:Cy5 input ratios of 1:1, which serve as negative controls; and 120 data points corresponding to spike-in ratios of 5:1, 10:1, and 30:1, which serve as positive controls. The relative frequencies are plotted against the *p*-values derived using the weighted (W) and nonweighted (NW) *t*-tests. (b,c) Weighted clustering can better recover the sample relationships. In an experiment, we profiled rat islet cells at 2 h, 4 h, and 6 h after staurosporine treatment. At each time point, three forward-labeling (f1, f2, f3) and three reverse-labeling (r1, r2, r3) hybridizations were made. Presented are the results of hierarchical clustering of the 18 arrays using (b) nonweighted and (c) weighted algorithms.

Normally, each row corresponds to a gene and each column corresponds to a condition. It is difficult to set up automated statistical tests where the data matrix is incomplete and the sample size varies from gene to gene due to missing values. Many pattern-finding methods, including principal component analysis and singular value decomposition, need complete data sets. Clustering methods such as hierarchical clustering[43] can handle missing values by ignoring them when calculating cluster distance; however, doing so can lead to spurious results.[44]

To deal with the missing values in a data set, many investigators would simply remove the entire rows or columns possessing missing values. This is not practical for large data sets with multiple conditions, as there are often too many genes possessing missing values.[45] In experiments with Affymetrix GeneChips, for example, normally ~50% or more of probe sets on each array will have their signal detection level labeled "M" or "A". Data from such probe sets are typically excluded from further analysis. When there are multiple conditions or samples within an experiment, the total number of data points excluded becomes significantly higher; situations where over 90% of genes (i.e. rows) possess missing values have been reported.[45] It is possible to impute the missing values by replacing them with zeros or row means, but this can lead to high deviations from true values.[42,44] More sophisticated approaches that utilize the information from the whole data set to estimate the missing values have also been proposed. Examples include methods that utilize measurements from other genes with similar or correlated expression patterns,[42,46] methods that utilize measurements from principal components of the gene expression matrix,[42,44] and model-based approaches such as Gaussian mixture[45] and Bayesian[44,47] models. However, the performance of different algorithms varies, and the accuracy and robustness of the estimation often depend on data characteristics including data size, data quality, correlation among data from different conditions, and experimental designs. There is no single algorithm that has been deemed the best under all conditions; to laboratory investigators, these issues add even more complexity to the already technically challenging task of microarray data analysis.

In our quality score–based approach to filtering or normalization, statistics, and data mining, the contribution from low-quality spots is automatically minimized through the reduction in weight. Therefore, it eliminates the need to manually flag bad data points and circumvents the complex missing value issue.[48] For large data sets, the advantages are more evident; to laboratory investigators, this means high convenience in analysis.

4. DISCUSSION: GENERALIZATIONS OF OUR APPROACH

In this chapter, we have presented our strategies for quantitative data QC in spotted cDNA microarrays through the definition of quality scores. Our approach allows a comprehensive dissection of the different noise-causing factors, and we have demonstrated its advantage in improving data reliability as well as in data analysis and mining. We have found the $\log R - q$ plot very revealing with respect to data structure and possible artifacts, and very insightful for data quality evaluation and for deciding data-filtering stringencies. In addition, we found it useful in the design and optimization of new protocols and algorithms, as it can differentiate the effect on good spots vs. bad spots and thus point out the means for improvement.

The field of microarray data analysis recognizes the necessity of a universal, quantitative data QC scheme, and the Microarray Gene Expression Data (MGED) Society is now vigorously advocating a standardized quality metric (http://www.mged.org/; the MGED Data Transformation and Normalization Working Group). Our approach offers a practical solution toward this goal. Since the publication of our work,[22,23,49] several similar approaches aimed at defining spot-level quality metrics for quantitative QC have been proposed by others.[50-53] These include the mean–median correlation measure by Tran *et al.*,[53] the coefficient of variation (CV) parameter by Sauer *et al.*,[51] and quality score definitions that were directly based on our designs.[51] The quality score–based approach has also been adopted by and benefited other fields including sequencing (e.g. the popular sequence analysis package phredPhrap,[54] in which phred assigns quality values to each base call it makes). Although our approach was initially developed for spotted cDNA arrays, the generalization to other microarray platforms is straightforward.

4.1. *Spotted Oligonucleotide Arrays*

By definition, our approach is directly applicable to other spotted microarray platforms, such as spotted oligonucleotide arrays. We have successfully utilized Matarray to process oligonucleotide arrays and apply quality score–based data QC procedures.[26,32] There is a difference, however, in terms of the utilization of TDAV. In oligonucleotide-based microarrays, synthesis costs make the approach described in Fig. 1(a) unfeasible; instead, we introduced a third-color "tracking" oligonucleotide (which, by design, will not cross-hybridize with any target in the tissue being analyzed) into the printing buffer, thus allowing the printing fidelity of a target-specific oligonucleotide library to be indirectly monitored.[32]

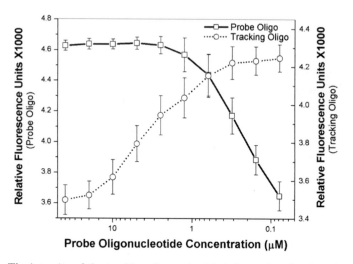

Fig. 7. The intensity of the tracking oligonucleotide is inversely related to the concentration of the probes; thus, it can be utilized in the QC of the probe concentration. Data were generated from 100 dilution series over 25 arrays.

In a preliminary study, a set of arrays was printed with a dilution series of two Cy5-labeled *Mycobacterium tuberculosis* target-specific 70-mers. A fluoresceinated 70-mer generated to a *Staphylococcus aureus* gene was used as the tracking oligonucleotide. The signal intensities derived from the tracking and the probe oligonucleotide were observed to be inversely related (Fig. 7); this indicated that as the molarity of the probe decreased, the tracking oligonucleotide was able to compete more effectively for open sites on the glass surface, resulting in increased fluorescein fluorescence. Utilizing this relationship, the probe condition can be evaluated.[32] In our experiment, no significant difference in hybridization signal was observed between elements possessing or not tracking oligonucleotides, indicating that the technology will not affect the accuracy of gene expression measurements.[32] We are currently finalizing the quantitative QC rules for our three-color oligonucleotide array platform.

4.2. *Affymetrix Microarray Platform*

Recently, we examined factors including signal intensity level, specificity in hybridization (how different perfect match [PM] signals are from mismatch [MM] signals), number of probe pairs available for signal quantification, spatial location of the probe sets, local background noise, etc. Some of

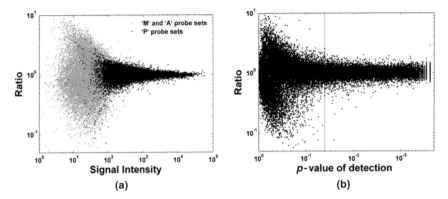

Fig. 8.   Signal intensity and signal-to-noise levels are critical QC parameters for data from the Affymetrix microarray platform. (a) The ratio to reference is plotted against the signal level for all probe sets on one array. The variation in data depends on intensity, even for the spots labeled "P" (black dots), and is particularly high in the low-intensity region. (b) The ratio to reference from the same array is plotted against the p-value of detection, again showing dependence with complex characteristics. The dotted line corresponds to $p = 0.04$, the default cut-off value for making "present" calls.

these factors (intensity, for example) have been studied by others.[55,56] In an experiment we did recently, the thymuses of BB-DP and BB-DR rats were evaluated at day 40 and day 65 using the Affymetrix U34 arrays. At each time point, six animals were sacrificed and the thymus from each was profiled, totaling 24 arrays. In each of the four groups of samples, a reference date was made by averaging the log signal values over the six replicates, and the ratio in signal levels between each sample and the reference was determined.

We found that data variability depends considerably on both the signal intensities and the specificity of hybridization for the probe set (Fig. 8). In the Affymetrix analysis package MAS5.0 or GCOS, the detection calls of "P" (present), "M" (marginal), or "A" (absent) are assigned, and typically only "P" data points are retained for further analysis. We can see from Fig. 8(a) that, even if we were to keep only "P" data points, there is still significant nonlinear data variability. We have characterized the degree of contribution from these factors to data variability, and found that quantitative measures in the form of quality scores can be defined accordingly for their QC. We are currently investigating the quality metric definitions and their optimizations for the Affymetrix oligonucleotide arrays.[48]

4.3. Generalization in Quality Score Definition

Many of the QC issues addressed here are relevant and applicable to the gene expression studies of other investigators. In addition, our approach can be adopted and optimized by other investigators for their own laboratory setting. New developments to improve the definition of q_{com} can be made through two means:

(1) More individual quality measures can be defined. There are factors that we have not yet considered or explored in detail; for example, the correlation between median and mean pixel intensity values has been indicated by others to also be a good indicator of data quality.[53]

(2) A more optimal definition of q_{com} can be obtained. Our definition uses the geometric mean of all individual scores; however, the individual factors may affect data quality at different degrees depending on the setup in a certain laboratory. Therefore, a weighted mean approach such as defining $q_{com} = \prod_i q_i^{w_i}$ may be able to capture the error profile better.

There are several possible ways to assign the individual weights w_i. One promising approach is to utilize the $\log R - q_i$ plot. We have found that the variation σ^2 in $\log R_i$ often drops exponentially with q_i[22] (Fig. 3); i.e. $\sigma^2 = \sigma_0^2 + \sigma_1^2 e^{-s_i q_i}$, where σ_0^2 and σ_1^2 are two constants. In a logarithmic scale, we have $\log(\sigma^2 - \sigma_0^2) = \log \sigma_1^2 - s_i q_i$. The slope s_i, being a good measure of the degree of variability that factor i causes, will be a good candidate for q_i's weight: $w_i = s_i / \sum_{j=1}^{k} s_j$. It is worth pointing out that this approach will allow users to optimize the definition of q_{com} for their own microarray setup.

ACKNOWLEDGMENTS

This work was supported in part by an NIH/NIBIB grant (EB001421) awarded to M. J. Hessner and X. Wang, an NIH/NIAID grant (U19-AI62627) awarded to M. J. Hessner, and a special fund from the Children's Hospital of Wisconsin Foundation (2201377).

REFERENCES

1. Schena M, Shalon D, Davis RW, Brown PO. (1995) Quantitative monitoring of gene expression patterns with complementary DNA microarray. *Science* **270**:467–470.

2. Diehn M, Alizadeh AA, Brown PO. (2000) Examining the living genome in health and disease with DNA microarrays. *JAMA* **283**(17):2298–2299.
3. Debouck C, Goodfellow PN. (1999) DNA microarrays in drug discovery and development. *Nat Genet* **21**(1 Suppl):48–50.
4. Fodor SP, Read JL, Pirrung MC, *et al.* (1991) Light-directed, spatially addressable parallel chemical synthesis. *Science* **251**(4995):767–773.
5. Fodor SP, Rava RP, Huang XC, *et al.* (1993) Multiplexed biochemical assays with biological chips. *Nature* **364**(6437):555–556.
6. Fan JB, Yeakley JM, Bibikova M, *et al.* (2004) A versatile assay for high-throughput gene expression profiling on universal array matrices. *Genome Res* **14**(5):878–885.
7. Roth ME, Feng L, McConnell KJ, *et al.* (2004) Expression profiling using a hexamer-based universal microarray. *Nat Biotechnol* **22**(4):418–426.
8. Barczak A, Rodriguez MW, Hanspers K, *et al.* (2003) Spotted long oligonucleotide arrays for human gene expression analysis. *Genome Res* **13**(7):1775–1785.
9. Taylor E, Cogdell D, Coombes K, *et al.* (2001) Sequence verification as quality-control step for production of cDNA microarrays. *Biotechniques* **31**(1):62–65.
10. Watson A, Mazumder A, Stewart M, Balasubramanian S. (1998) Technology for microarray analysis of gene expression. *Curr Opin Biotechnol* **9**(6):609–614.
11. Mecham BH, Wetmore DZ, Szallasi Z, *et al.* (2004) Increased measurement accuracy for sequence-verified microarray probes. *Physiol Genomics* **18**(3):308–315.
12. Miklos GL, Maleszka R. (2004) Microarray reality checks in the context of a complex disease. *Nat Biotechnol* **22**(5):615–621.
13. Ideker T, Thorsson V, Ranish JA, *et al.* (2001) Integrated genomic and proteomic analyses of a systematically perturbed metabolic network. *Science* **292**(5518):929–934.
14. van de Vijver MJ, He YD, van 't Veer LJ, *et al.* (2002) A gene-expression signature as a predictor of survival in breast cancer. *N Engl J Med* **347**(25):1999–2009.
15. Chuaqui RF, Bonner RF, Best CJ, *et al.* (2002) Post-analysis follow-up and validation of microarray experiments. *Nat Genet* **32**(Suppl):509–514.
16. Bammler T, Beyer RP, Battacharya S, *et al.* (2005) Standardizing global gene expression analysis between laboratories and across platforms. *Nat Methods* **2**(5):351–356.
17. Irizarry RA, Warren D, Spencer F, *et al.* (2005) Multiple-laboratory comparison of microarray platforms. *Nat Methods* **2**(5):345–350.
18. Larkin JE, Frank BC, Gavras H, *et al.* (2005) Independence and reproducibility across microarray platforms. *Nat Methods* **2**(5):337–344.
19. Yauk CL, Berndt ML, Williams A, Douglas GR. (2004) Comprehensive comparison of six microarray technologies. *Nucleic Acids Res* **32**(15):e124.
20. Johnson K, Lin S. (2003) QA/QC as a pressing need for microarray analysis: meeting report from CAMDA'02. *Biotechniques* (Suppl):62–63.

Current Topics in Human Genetics

21. van Bakel H, Holstege FC. (2004) In control: systematic assessment of microarray performance. *EMBO Rep* **5**(10):964–969.
22. Wang X, Ghosh S, Guo S-W. (2001) Quantitative quality control in microarray image processing and data acquisition. *Nucleic Acids Res* **29**:E75–E82.
23. Wang X, Hessner MJ, Wu Y, *et al.* (2003) Quantitative quality control in microarray experiments and the application in data filtering, normalization and false positive rate prediction. *Bioinformatics* **19**:1341–1347.
24. Hessner MJ, Wang X, Hulse K, *et al.* (2003) Three color cDNA microarrays: quantitative assessment through the use of fluorescein-labeled probes. *Nucleic Acids Res* **31**:e14.
25. Hessner MJ, Wang X, Khan S, *et al.* (2003) Use of a three-color cDNA microarray platform to measure and control support-bound probe for improved data quality and reproducibility. *Nucleic Acids Res* **31**:e60.
26. Hessner MJ, Xiang B, Jia S, *et al.* (2006) Three-color cDNA microarrays with prehybridization quality control yield gene expression data comparable to that of commercial platforms. *Physiol Genomics* **25**(1):166–178.
27. Chen Y, Dougherty ER, Bittner ML. (1997) Ratio-based decision and the quantitative analysis of cDNA microarray images. *J Biomed Opt* **2**:364–374.
28. Hessner MJ, Meyer L, Tackes J, *et al.* (2004) Immobilized probe and glass surface chemistry as variables in microarray fabrication. *BMC Genomics* **5**(1):53.
29. Yue H, Eastman PS, Wang BB, *et al.* (2001) An evaluation of the performance of cDNA microarrays for detecting changes in global mRNA expression. *Nucleic Acids Res* **29**(8):E41.
30. Wang Y, Wang X, Guo SW, Ghosh S. (2002) Conditions to ensure competitive hybridization in two-color microarray: a theoretical and experimental analysis. *Biotechniques* **32**(6):1342–1346.
31. Wang X, Jiang N, Feng X, *et al.* (2003) A novel approach for high quality microarray processing using third-dye array visualization technology. *IEEE Trans Nanobioscience* **2**(4):193–201.
32. Hessner MJ, Singh VK, Wang X, *et al.* (2004) Visualization and quality control of spotted 70-mer arrays using a labeled tracking oligonucleotide. *BMC Genomics* **5**:12.
33. Wang X, Jia S, Meyer L, *et al.* (2006) Comparehensive quality control utilizing the prehybridization third-dye image leads to accurate gene expression measurements by cDNA microarrays. *BMC Bioinformatics* **7**:378.
34. Cleveland WS, Devlin SJ. (1988) Locally weighted regression: an approach to regression analysis by local fitting. *J Am Stat Assoc* **83**(403):596–610.
35. Yang YH, Dudoit S, Luu P, *et al.* (2002) Normalization for cDNA microarray data: a robust composite method addressing single and multiple slide systematic variation. *Nucleic Acids Res* **30**(4):e15.
36. Yuen T, Wurmbach E, Pfeffer RL, *et al.* (2002) Accuracy and calibration of commercial oligonucleotide and custom cDNA microarrays. *Nucleic Acids Res* **30**(10):e48.
37. Bakewell DJ, Wit E. (2005) Weighted analysis of microarray gene expression using maximum-likelihood. *Bioinformatics* **21**(6):723–729.

38. Baldi P, Long AD. (2001) A Bayesian framework for the analysis of microarray expression data: regularized *t*-test and statistical inferences of gene changes. *Bioinformatics* **17**(6):509–519.

39. Lonnstedt I, Speed TP. (2002) Replicated microarray data. *Stat Sin* **12**:31–46.

40. Fan J, Tam P, Vande Woude G, Ren Y. (2004) Normalization and analysis of cDNA microarrays using within-array replications applied to neuroblastoma cell response to a cytokine. *Proc Natl Acad Sci USA* **101**(5):1135–1140.

41. Yeung KY, Medvedovic M, Bumgarner RE. (2003) Clustering gene-expression data with repeated measurements. *Genome Biol* **4**(5):R34.

42. Troyanskaya O, Cantor M, Sherlock G, *et al.* (2001) Missing value estimation methods for DNA microarrays. *Bioinformatics* **17**(6):520–525.

43. Eisen MB, Spellman PT, Brown PO, Botstein D. (1998) Cluster analysis and display of genome-wide expression patterns. *Proc Natl Acad Sci USA* **95**(25):14863–14868.

44. Oba S, Sato M, Takemasa I, *et al.* (2003) A Bayesian missing value estimation method for gene expression profile data. *Bioinformatics* **19**(16):2088–2096.

45. Ouyang M, Welsh WJ, Georgopoulos P. (2004) Gaussian mixture clustering and imputation of microarray data. *Bioinformatics* **20**(6):917–923.

46. Bo TH, Dysvik B, Jonassen I. (2004) LSimpute: accurate estimation of missing values in microarray data with least squares methods. *Nucleic Acids Res* **32**(3):e34.

47. Zhou X, Wang X, Dougherty ER. (2003) Missing-value estimation using linear and non-linear regression with Bayesian gene selection. *Bioinformatics* **19**(17):2302–2307.

48. Jia S, *et al.* (2007) Quality weighted mean and *t*-test in microarray analysis lead to improved accuracy in gene expression measurements and circumvent the missing value problem (submitted).

49. Wang X, Hessner MJ. (2006) Quantitative quality control of microarray experiments: toward accurate gene expression measurements. In: *Gene Expression Profiling by Microarrays — Clinical Implications*, Hofmann WK (ed.), Cambridge University Press, Cambridge, England, pp. 27–46.

50. Cutler DJ, Zwick ME, Carrasquillo MM, *et al.* (2001) High-throughput variation detection and genotyping using microarrays. *Genome Res* **11**(11):1913–1925.

51. Sauer U, Preininger C, Hany-Schmatzberger R. (2005) Quick and simple: quality control of microarray data. *Bioinformatics* **21**(8):1572–1578.

52. Raffelsberger W, Dembélé D, Neubauer MG, *et al.* (2002) Quality indicators increase the reliability of microarray data. *Genomics* **80**(4):385–394.

53. Tran PH, Pfeiffer DA, Shin Y, *et al.* (2002) Microarray optimizations: increasing spot accuracy and automated identification of true microarray signals. *Nucleic Acids Res* **30**(12):e54.

54. Ewing B, Hillier L, Wendl MC, Green P. (1998) Base-calling of automated sequencer traces using phred. I. Accuracy assessment. *Genome Res* **8**(3):175–185.

55. Irizarry RA, Hobbs B, Collin F, *et al.* (2003) Exploration, normalization, and summaries of high density oligonucleotide array probe level data. *Biostatistics* **4**(2):249-264.
56. Bolstad BM, Irizarry RA, Astrand M, Speed TP. (2003) A comparison of normalization methods for high density oligonucleotide array data based on variance and bias. *Bioinformatics* **19**(2):185–193.

Chapter 14

Application of DNA Microarray
Technology in Genetics

Jian Yan[*], Weikuan Gu[†,‡] and Yan Jiao[†]

[*]*Department of Medicine, University of Tennessee Health Science Center*
Memphis, TN 38163, USA

[†]*Center of Genomics and Bioinformatics &*
Center of Diseases of Connective Tissues
Department of Orthopedic Surgery–Campbell Clinic
University of Tennessee Health Science Center
Memphis, TN 38163, USA

DNA microarrays are a unique cost-effective method for simultaneously assessing the expression levels of thousands of genes. Microarrays can be fabricated by robotic spotting of gene-specific cDNAs or oligonucleotides and by *in situ* synthesis of oligonucleotides. More recent approaches include piezoelectric inkjets for noncontact printing and maskless light-directed synthesis of oligonucleotides. Although Affymetrix GeneChip arrays occupy the majority of the microarray marketplace, their drawbacks of less flexibility and high cost leave much room for the development of robotic spotting and other novel technologies. Here, we review recent technological advances in spotted arrays, *in situ* synthesized arrays, and their applications in genetics and other life science areas.

1. INTRODUCTION

By definition, a microarray is a piece of glass or plastic on which single-stranded pieces of DNA are affixed in a microscopic array. The unique

---

[‡] Correspondence author.

feature of microarray is that it has hundreds of thousands of cDNA or genetic sequences on one chip, thus enabling it to screen a biological sample for the presence of many genes or the genes of a whole genome at once.

As a result of the Human Genome Project, there has been an explosion in the amount of information available on the DNA sequence of the human and model organism genomes. Genome sequencing confronts us with the unequivocal fact that we know very little about the function of most genes. Traditionally, the "one-gene-at-a-time" strategy through generating loss-of-function or gain-of-function mutations followed by detailed phenotypic analysis has been used to assess gene function. However, this is not always possible or informative because not all gene mutations have obvious phenotypes. In such instances, the identification of a gene expression pattern will indirectly provide evidence for the possible functions of a gene. Gene expression pattern analysis can be used to suggest more appropriate genetic, molecular, or biochemical assays.

Several techniques for the analysis of gene expression at the mRNA level are available, such as Northern blotting, real-time polymerase chain reaction (PCR), differential display,[1] serial analysis of gene expression (SAGE),[2] and dot blot analysis.[3] However, each of these methods has its disadvantages. For example, Northern blot and real-time PCR only allow limited numbers of genes to be studied at one time. Differential display does enable the simultaneous detection of multiple genes; however, only a limited number of different conditions can be compared, the method is not quantitative, and screening is based on the differences in mRNA length rather than identity. SAGE is laborious, as it involves complex sample preparation procedures and requires extensive DNA sequencing, and is not very sensitive. Finally, dot blot analysis requires a relatively large amount of material due to the size of the filters. DNA microarrays are unique in offering a cost-effective and efficient method for measuring genomewide gene expression. Microarrays exploit the specificity of nucleic acid base pairing during hybridization to simultaneously assess the expression of tens of thousands of genes.

A DNA microarray study is composed of at least six different components (Fig. 1): the nucleic acid extraction and quality control system, the microarray itself, a hybridization oven, a fluidic system for array washing and staining, a scanner to read the arrays, and sophisticated data analysis softwares to quantify and interpret the results. Special equipment is now commercially available for each of these components. The various applications offer microarray-interested biomedical researchers a choice of several

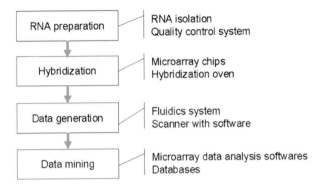

Fig. 1. Schematic diagram of the major procedures of a DNA microarray assay. A typical DNA microarray study consists of four major steps: RNA preparation, hybridization, data generation, and data mining. Different components are applied in these consecutive processes.

options, especially for gene expression profiling. At the heart of microarray work are the microarrays used and data analysis.

2. HISTORICAL PERSPECTIVE

DNA microarray technology is the assembly and convergence of several technologies, including automated DNA sequencing, DNA amplification by PCR, highly efficient oligonucleotide synthesis, and nucleic acid labeling chemistries. The use of fluorescent tags to measure nucleic acid hybridization signals is the mainstay of most array platforms; therefore, DNA microarray technology is also regarded as a logical extension of earlier immunoassay technology.[4]

In the mid- to late 1980s, Roger Ekins of University College London Medical School started to conceive and develop microarray-based assays. In the late 1980s, Edwin Southern of Oxford University developed a way to use inkjet printing to build oligo-based microarrays on glass slides. At the same time, Stephen Fodor at a little-known company called Affymax tried another way to fabricate microarrays using photolithography. The theory behind their work was that semiconductor manufacturing techniques could be united with advances in combinatorial chemistry to build vast amounts of biological data on a small glass chip. By 1991, Fodor and his colleagues were able to build arrays of peptides and dinucleotides. This technology became the basis of a new company, Affymetrix, formed as a division of Affymax in 1991. Affymetrix began operating independently in 1992.

Meanwhile, Stanford University biochemist Patrick Brown and colleagues pursued a distinct approach to microarray fabrication. Instead of building oligonucleotides on the array *in situ*, they proposed spotting PCR-amplified DNA at regular intervals on a surface. In 1995, the first gene expression microarray article using their two-color spotted arrays was published in *Science*.[5] The word "microarray" was used for the first time in this paper.

3. ADVANCES IN *IN SITU* SYNTHESIS ARRAY TECHNOLOGIES

3.1. *Affymetrix Technology*

The earliest and leading *in situ* synthesis array technology, developed by Affymetrix, combines photolithography and combinatorial chemistry to manufacture high-density GeneChip brand arrays.[6] In this approach, a quartz wafer is coated with a light-sensitive chemical compound that prevents coupling between the wafer and the first nucleotide of the DNA probe being created. Lithographic masks are used to either block or transmit light onto predetermined locations of the wafer surface. The surface is then covered with a solution containing adenine, thymine, cytosine, or guanine; and coupling occurs only in those ultraviolet (UV)-catalyzed base deprotected regions on the glass. The cycle is repeated until the desired sequence and length of oligonucleotides are synthesized. Eleven to 20 different oligonucleotides are made for each transcript to cover a portion of the 3′ end of the mRNA. A transcript is represented as a probe set, which is made up of probe pairs comprised of perfect match (PM) and mismatch (MM) probe cells. The MM oligonucleotides have a one-base mismatch in the center position, and are used as a control to detect background noise and cross-hybridization from unrelated probes. The Affymetrix method involves selective illumination of the substrate surface to fabricate a microarray and requires the use of photomasks and precise mask aligners, resulting in high costs for array fabrication.

In 1998, Affymetrix released the first commercially available array set, Hu6800, which was comprised of four arrays probing approximately 8000 full-length genes. These 50-micron arrays had 1700–1800 probe sets on each array. To increase the information content on each array, Affymetrix has been implementing innovations in its array manufacturing process, labeling, hybridization, scanning, washing, staining, image analysis, and data analysis. In 2003, Affymetrix became the first company to enable researchers to conduct whole-genome expression analysis on single arrays for human, mouse, and rat studies.

An advanced Affymetrix manufacturing process uses an improved mask design and wafers with a permanent antireflective coating (ARC) as the starting point. These improvements have increased array feature definition, enabling the generation of 11-micron arrays with higher feature density and enhanced performance. The most recent generation of 5-micron arrays has six million features, and the company envisions the release of 1-micron arrays in the near future. To improve the array scanning technology, Affymetrix produced the GeneChip Scanner 3000, which incorporates a "flying objective" confocal microscope that is implemented with a full 16-bit resolution capable of resolving >65 000 different levels of fluorescence. This new design provides a fourfold improvement in collection efficiency and a 41% improvement in data precision over the previous GeneArray 2500. It also employs an auto-set laser power feature to ensure greater consistency in scanning over time (www.affymetrix.com).

3.2. Inkjet Printing Technology

The Affymetrix method is very efficient in producing thousands or millions of identical arrays, but is less flexible and more expensive where the creation of new arrays with added or different gene content is concerned. The piezoelectric inkjet oligoarray synthesis technology as proposed by Blanchard et al.[7] addresses this point. It is a particularly flexible method that allows the rapid construction of oligonucleotide arrays containing any desired sequence. With this method, a modified inkjet printing process is applied to deliver phosphoramidites to a hydrophobic surface containing chemically active hydroxyl groups for the *in situ* synthesis of oligonucleotides. The advantages of this method are that no expensive masks are required, synthesis is faster compared to Affymetrix photolithography, and there is more flexibility in creating new arrays. Work by Rosetta Inpharmatics (Seattle, WA) and Agilent (Palo Alto, CA) has resulted in the generation of an industrial-scale inkjet oligoarraying system and a catalog of commercially available arrays.

Inkjet printing of microarrays is similar to printing a document with an inkjet printer. The printer consists of an inkjet head with computer-controlled nozzles through which ink can flow onto the paper. The printer also contains cartridges that hold a supply of ink to print with. Agilent SurePrint technology makes use of similar components to print nucleic acids (instead of ink) onto an activated glass surface (instead of paper) via noncontact inkjet deposition. The process of printing oligonucleotide microarrays is nearly identical to the process described for spotted microarrays. However, instead of printing the presynthesized oligos onto the microarray surface, it synthesizes oligos base by base in repetitive print layers using standard phosphoramidite chemistry.

3.3. *Maskless Array Synthesizer (MAS)*

The current Affymetrix method has two major drawbacks: the masks are expensive and require a significant amount of time to synthesize. For example, an array of 25-mers may require 100 different masks, leading to high costs and a long fabrication time. A maskless approach would circumvent the need for chrome/glass photolithographic masks, thereby reducing the cost and turnaround time for making custom arrays and greatly increasing their versatility. Recently, new photolithographic oligoarray synthesizers that replace mask sets with a dynamic micromirror device (DMD) have been introduced.[8,9]

NimbleGen (Madison, WI) manufactures custom, high-density DNA arrays based on its proprietary maskless array synthesizer (MAS) technology. The MAS system is comprised of a maskless light projector, a reaction chamber, a personal computer, and a DNA synthesizer. At the center of the system is a digital micromirror device (DMD), which generates virtual masks that replace the physical photolithographic masks used in traditional arrays. These virtual masks reflect the desired UV light pattern with individually addressable aluminum mirrors controlled by the computer. The DMD controls the UV light pattern projected on the microscope slide in the reaction chamber, which is coupled to the DNA synthesizer. The UV light selectively cleaves a UV-labile protecting group at the precise location where the next nucleotide will be coupled. The patterns are coordinated with the DNA synthesis chemistry in a parallel, combinatorial manner such that 390 000 unique probe features are synthesized in a single array.

The NimbleGen System's MAS technology is very similar to traditional oligonucleotide synthesis, but with some important exceptions. Unlike conventional oligo synthesis, arrays are synthesized on glass slides rather than on controlled pore glass supports. Another key difference is that the deprotection steps are performed by photodeprotection rather than by acid deprotection. However, the synthesis efficiency of photoreactive nucleotide groups is not as high as conventional oligonucleotide chemistries, and the exposure of oligonucleotides to the illuminating UV light source is undesirable.

NimbleGen Systems is now a leading supplier of flexible high-density microarray products and services. NimbleGen uniquely produces high-density arrays of isothermal long oligos that provide superior results for advanced genomic analysis methods such as comparative genomic hybridization (CGH), chromatin immunoprecipitation (ChIP), microbial whole-genome resequencing, and expression tiling. NimbleGen and Affymetrix are collaborating to provide customers with high-density, made-to-order arrays on the Affymetrix GeneChip® platform.

3.4. *XeoChip®* *Technology*

Another maskless alternative is the approach developed by Invitrogen, also known as XeoChip® technology. It is the first to introduce digital projection photolithography for parallel chemical synthesis. It provides another flexible and cost-effective process to produce DNA chips of any design. Its proprietary technologies are composed of photogenerated acid (PGA) chemistry, digital photolithography, and parallel microfluidics. The core of the technology is the *in situ* parallel combinatorial synthesis in three-dimensional nanochambers of microfluidic chips. The digital photolithography process avoids the use of expensive and time-consuming photomasks and, more importantly, enables flexibility and enhances efficiency for oligonucleotide array synthesis. Instead of deprotecting photolabile monomers, this technology uses a DMD to photogenerate detritylating acids. This has the advantage of better yields through higher coupling efficiency, but the feature density is lower.

3.5. *Electrochemical Array Technology*

CombiMatrix (Mukilteo, WA) developed an electrochemical approach to fabricate 30-mer to 50-mer oligonucleotide arrays based on the porous reaction layer (PRL) and virtual flask technologies. The synthesis protocol and chemistries are exactly the same as those used by other commercial DNA manufacturing companies, except that the acid used in the detritylation reaction is electrochemically produced.

4. ADVANCES IN SPOTTED ARRAY TECHNOLOGIES

Spotted microarrays are fabricated by dipping pins into probe DNA dissolved in spotting buffer and then depositing each probe on solid supports, which can be glass microscope slides, silicon chips, or nylon membranes. Arrays built on membranes (called filter arrays) have the advantage of being relatively affordable and not needing any special equipment, except that large format phosphorimager screens may be required with larger (e.g. $22\,cm^2$) filters. They are also useful for scarce RNA, as only approximately 50 ng of total RNA is required for a single experiment. However, the sensitivity of filter arrays is reported to be limited to high- and medium-abundance genes.[10] They can be regarded as complementary rather than competing technologies in the microarray field. We will not cover them here.

The pin-based robotic printing technique uses DNA fragments generated from cDNA or synthesized long oligonucleotides as probes. A glass slide or silicon chip is coated with a substrate that gives the slide an even binding surface, which is positively charged. The probe DNA dissolved in spotting buffer is deposited onto the prepared slide by an arrayer by lowering the printing pins until they just barely touch the slide surface. These arrays are then hybridized with cDNAs from two different sources (e.g. samples from normal and diseased tissues), which are labeled with different fluorescent dyes. The scanned image allows one to detect the relative expression levels of genes on the array.

The first glass slide arrays were produced in Patrick Brown's laboratory at Stanford in 1995. Since then, this technique has been widely applied in gene expression studies. Generally, during the printing of arrays, pins are filled with probe DNA by capillary action, and the surface tension between the spotting buffer and surface substrate acts to deposit the probe spots. This process often produces signal variation because of perturbations in spot structure. In addition, variations can be induced by labeling and hybridization. Although spotted microarrays can provide accurate, large-scale measurements of gene expression, similar to other microarray platforms, their sensitivity is limited by high levels of experimental variability. Therefore, much effort has been made to improve the spotting process.[11]

4.1. *Glass Slide Substrates*

Commercial aminosilane-coated glass slides have surpassed in-house polylysine-coated slides in terms of spot morphology and consistency. To reduce background fluorescence, a succinic anhydride wash is suggested to prevent nonspecific hybridization to the substrate surface. Other available substrates include dendrimers, epoxysilane aminosilane composites, and self-absorbing polymers.

4.2. Source of Probe DNA

Initially, PCR-amplified cDNA clones were used as probes for spotting arrays. This is still the source of choice when interrogating samples from organisms with an unknown genome sequence. Since PCR amplification can suffer frequent failures, variable DNA yields, and cross-contamination, gel electrophoresis and resequencing of PCR amplicons before printing are highly recommended. As probes, PCR amplicons are highly sensitive; however, they have inherent tolerance to small sequence variations, thus reducing their ability to discriminate similar sequences within an organism. Many microarray users have, therefore, started to spot single-stranded long oligonucleotide (35–80-mer) probes to overcome the limitations of PCR amplification and increase target sequence discrimination. The design of oligonucleotide probes in general is a challenge. Modern oligonucleotide design tools use sophisticated algorithms to predict hybridization behavior. Public and commercial sources have proprietary algorithms for oligonucleotide design. In principle, the oligonucleotides should have very similar melting temperatures or G-C (guanosine–cytosine) content, have very little homology with other oligonucleotides, be fully contained within an exon, and have no repetitive or hairpin sequences.

A recent study indicates that a long (70-mer) oligonucleotide platform is highly suitable for expression analysis, and compares favorably with the cDNA and short oligonucleotide arrays. It will probably become the platform of choice for gene expression analysis, replacing the cDNA type entirely.

4.3. Spotting Pins

Traditionally, a robotic printing pin is made from stainless steel or titanium through electrical discharge machining or laser cutting. This process has inherent inconsistencies that may result in pin–pin difference. Ceramic pins have recently been tried for robotic printing. Since these novel pin types are not generally available, no comparisons have been made. However, they are cheaper to produce and could, in theory, be used to print up to 225 000 spots on a standard microarray slide.

Microarrays made in-house are generally coated by poly-L-lysine, while commercial arrays use aminosilane substrates for coating to improve spot morphology and consistency. Modified probes and reactive substrates have also been introduced to enhance sequence discrimination. Other substrates have been optimized to produce higher spot signals with reduced amounts of probe material by coating glass slides with dendrimers, epoxysilane aminosilane composites, or self-absorbing polymers.

5. ALTERNATIVE DNA MICROARRAY TECHNOLOGIES

In the field of DNA microarrays, the abovementioned platforms are most widely used. However, there are several companies developing alternative arraying strategies, such as bead-based and optical fiber–based arrays. Bead arrays are fabricated either by impregnating beads with different concentrations of fluorescent dye or by some type of barcoding technology. BD Biosciences (San Diego, CA) produces a cytometric bead array (CBA) that can be viewed as a low-density microarray system. The BD CBA generates data that are comparable to data generated by ELISA-based assays in a multiplexed or simultaneous fashion.

Illumina (San Diego, CA) has developed a novel bead array technology. The BeadArray[TM] technology is built around specially prepared beads that self-assemble into microwells etched into an array substrate. Illumina deploys these arrays in two formats: Sentrix Array Matrix and Sentrix BeadChip. These two array platforms provide great flexibility to life science researchers who want to scale from smaller experimental throughput on the BeadChip to production-scale throughput on the Array Matrix.

SmartBead Technologies (Cambridge, England) uses barcoded microparticles as bead microarrays. Microparticles are produced using traditional optical lithography, and then unique barcodes are patterned into these particles. This process generates microparticles with a series of different-sized holes that define the barcodes. The identification of beads occurs via a binary system. Various proteins and DNA molecules can be immobilized on these particles, and can be used as an array.

The method of Nanoplex (Mountain View, CA) involves the use of cylindrical particles that are similar to conventional barcodes. These nanoparticles are prepared from inert metals such as gold, silver, nickel, or platinum. Nanobarcodes are encoded on these particles, which are used by immobilizing DNA or other biomolecules; the binding event of complementary probes is then measured by fluorescence. The barcode on particles enables the identification of those particles that have fluorescent probes attached to them.

Another distinct bead-based microarray technique is massively parallel signature sequencing (MPSS) technology, invented by Sydney Brenner and colleagues in 2000.[12] MPSS is based on Megaclone, which is Lynx's technology for cloning DNA molecules onto microbeads. Millions of DNA-signatured microbeads — each carrying a different cDNA fragment attached by *in vitro* cloning — are repeatedly cycled between restriction type II cleavage, ligation steps, and hybridization reactions to add decoder probes for detecting the signatures. The number of microbeads carrying identical cDNAs is then counted by a digital approach with a charge-coupled device camera using a flow cell.

MPSS has a routine sensitivity of a few molecules of mRNA per cell and the results are similar to SAGE expression analysis, although the tags are longer. The number of tags obtained per library is extremely large (>1 000 000), making the technology sensitive to genes expressed at low levels.

Like the sequence data contained in ESTs, data derived from MPSS experiments have many uses. The expression level of particular genes can be quantitatively determined; the counted frequency of tags is representative of the expression level of the gene in the analyzed tissue. The completion of genomes such as those of yeast, *C. elegans*, and rice permits the direct comparison of tags to the genomic sequence, and further extends the utility of MPSS data. The identification of genes and the assessment of transcriptional activity are performed by aligning the tags to the genomic sequence. The location of the polyadenylation site for each transcript can be determined within ∼256 bp (the tag is derived from a 4-bp restriction site immediately 5′ to the poly-A site). Several distinct tags matching different sites within a single gene are indicative of alternative 3′ termination.

With MPSS, differential expression may be detected simply by sequencing entire libraries and comparing them — without hybridization or the sorting of beads. Libraries are derived from distinct tissues or treatments. Each library of cDNAs-on-beads is sequenced to such depth that the transcript count for a given gene is compared among libraries. Basic statistics are then used to determine genes that are present in significantly different amounts in two libraries. Quantitative methods for the analysis of tag frequencies and the detection of differences among libraries have been published and incorporated into public databases for SAGE data.

6. CURRENT APPLICATIONS OF DNA MICROARRAYS

6.1. *Gene Expression Profiling*

This is the classical and primary application of DNA microarrays. In the past 10 years, microarrays have been widely used in most biomedical research areas, especially in cancer research.[13,14] Profiling the gene expression of various human tumors has led to the identification of gene expression patterns or signatures relating to tumor classification, disease outcome, and response to therapy. It is unquestionable that this technology will have a great impact on the management of cancer, and its applications will range from the discovery of new drug targets to new molecular tools for diagnosis and prognosis as well as for personalized treatment.

6.2. *Genomewide SNP Genotyping* [15]

It has been estimated that the human genome contains more than 10 million nucleotide positions which vary among individuals in a population. There is a lot of evidence for the association between gene polymorphisms and complex disorders or traits. With the whole-genome sequence available, large-scale screening of gene polymorphisms using SNP markers is attractive to genetics researchers. As a result of the International HapMap Project (www.hapmap.org) and the effort by Perlegen Sciences (www.perlegen.com), more than two million SNP markers with verified allele frequencies have been identified for public use. Linkage disequilibrium (LD) studies suggest that genomewide association studies will require the genotyping of several hundred thousands of SNPs in each individual, and that successful association studies will require the analysis of thousands of samples. Therefore, microarray-based SNP genotyping design needs to balance the SNP multiplexing levels and the sample number. It is not feasible, so far, to combine a high SNP multiplexing level with a high sample throughput.

6.3. *Array CGH* [16]

CGH was the first efficient approach to scanning genomewide variations in DNA copy number. The assay was first developed and described by Kallioniemi *et al.* in 1992.[17] In a traditional CGH assay, total genomic DNA is isolated from test and reference cell populations, differentially labeled, and hybridized to metaphase chromosomes. The ratio of the two fluorochrome intensities is then calculated, and regions where the test DNAs are amplified or deleted are readily detected on the metaphase spread from a normal reference.

A major development of CGH technology was heralded in 1997 by Solinas-Toldo *et al.*[18] They developed a technique by which immobilized target DNAs on glass slides were hybridized with biotinylated tumor DNA and digoxygenin-labeled reference DNA. These slides were then labeled with two different fluorochromes, and the ratios of the two dyes on the slides were analyzed by confocal microscopy. While these experiments were laborious and primitive, they provided a framework from which the current high-throughput array CGH would develop.

In the last decade, microarray-based methods have been successfully adapted to the analysis of genomic copy-number alterations by hybridization to probes spanning large chromosomal regions. Pollack *et al.*[19] were the first to use a cDNA microarray to provide full genomewide analysis of a number of cancer cell lines.[19] The advent of array platforms as probes for CGH had important implications for clinical investigation and provided a lot of possibilities for choosing probe materials. The initial approaches used arrays produced by spotting DNA obtained directly from large-insert genomic clones such as bacterial artificial chromosomes (BACs). Since the process of producing good quality BAC DNA for arrays is labor-intensive, new techniques for amplifying small amounts of starting material have been developed. The current probe options include BAC/yeast artificial chromosome (YAC) clones, cDNAs, genomic representations, oligonucleotides, and specific target regions of chromosomes or gene sets. The YAC/BAC and cDNA arrays enable higher resolution CGH analysis; while the genomic representation arrays, oligonucleotide arrays, and specific target arrays provide even greater resolution CGH platforms.

6.4. *RNA interference (RNAi) Cell Microarrays*[20]

Recent advances in genome sequencing have challenged cell biologists with the task of characterizing the functions of the products encoded by tens of thousands of genes. Two classical ways to investigate gene function in the context of the entire organism are gain-of-function and loss-of-function studies using transgenic and knockout animal technologies, respectively. However, their drawbacks include the enormous cost and time required for generating genetically modified animals, and the lack of any distinct phenotype change that would allow for its unambiguous linkage to the mutated gene in the majority of genetically modified animals. Therefore, there is a clear need for a corresponding high-throughput method in this context.

The development of RNAi technology provides a novel tool to address this issue by making loss-of-function genetic studies more tractable in many organisms. RNAi is a posttranscriptional method of gene silencing, in which mRNAs are degraded by double-stranded RNA (dsRNA) in a sequence-specific fashion. The combination of RNAi and microarray technologies has enabled the development of the transfected-cell array (TCA) technique, which is seen as a breakthrough for high-throughput functional genomics in cell biology. In this method, full-length open reading frames of genes cloned in expression vectors are printed at a high density on a glass slide along with a lipid transfection reagent. The microarray is then covered with a layer of cells. Cells growing on top of the DNA spots are transfected, resulting in the expression of specific proteins in spatially distinctive groups of cells. The phenotypic effects of hundreds or thousands of gene products can be detected using specific cell-based bioassays. Although this technique is still in its infancy, it offers an efficient platform for carrying out high-throughput loss-of-function studies. Currently, the key issues that must be resolved before RNAi cell microarrays become a routine tool are high-resolution imaging platforms which are compatible with cell microarrays and the availability of verified mammalian genomewide RNAi libraries.

6.5. *Regulatory Network Mapping*[21]

The expression of thousands of genes in an organism is regulated by the concerted action of hundreds of transcription factors and chromatin proteins, as well as by epigenetic mechanisms such as DNA methylation and histone acetylation, methylation, or phosphorylation. During the past 5 years, new microarray-based approaches have been developed that allow us to obtain broader views of gene regulation. In particular, methods have been developed for the genomic mapping of *in vivo* binding sites of regulatory proteins and for the distributions of histone modification and DNA methylation.

In terms of genomewide mapping of *in vivo* protein–genome interactions, two methods have contributed considerably to our current understanding of gene regulatory networks. One of these is ChIP microarray (ChIP-chip). In this technique, cells are treated with a cross-linking reagent, which covalently links protein complexes *in situ* to DNA, followed by immunoprecipitation of chromatin fragments using an antibody against the protein of interest. To identify the DNA fragments linked to chromatin, the cross-links are reversed, and the DNA fragments are labeled with a fluorescent dye and hybridized to microarrays with probes corresponding to genomic regions of interest. Another method is DamID, which is based on the creation of a fusion protein consisting of *Escherichia coli* adenine methyltransferase (Dam) and the chromatin protein of interest. Dam methylates adenines in the sequence GATC. Adenine methylation does not occur endogenously in the DNA of most eukaryotes. When this fusion protein is expressed *in vivo*,

Dam will be targeted to the native binding sites of the chromatin protein, resulting in local methylation of adenine residues. Hence, the sequences near a binding site of the protein will be marked with a unique methylation tag, which can be detected using microarray-based assays. DamID appears to work well for proteins that interact with target DNA sequences either directly or indirectly. ChIP-chip and DamID have distinct advantages and disadvantages. ChIP-chip requires a good antibody against the protein of interest, whereas DamID does not; however, DamID is not suitable for the detection of posttranslational modifications.

DNA methylation is one of the most important regulators of gene activity. It is altered in many diseases, and is associated with response to medicines and other factors like aging. Genomewide mapping of DNA methylation will provide a crucial link between genetics, the environment, and health. There are several microarray-based methods that have been developed to map DNA methylation in genomes.

6.6. *Diagnostics and Personalized Medicine*

DNA arrays have tremendous potential for diagnostics and personalized medicine. In 2004, Affymetrix released the first microarray instrument, the GeneChip System 3000Dx (GCS 3000Dx), for diagnostic use in the European Union. With the Roche AmpliChip CYP450 Test based on this system, diagnostic laboratories can identify certain naturally occurring variations in the drug metabolism genes *CYP2D6* and *CYP2C19*. During the past 30 years, genetic variation in infectious agents and in humans has been shown to be related to many of the interindividual differences in drug response. These variations affect the rate at which an individual metabolizes many common drugs used to treat diseases including depression, schizophrenia, bipolar disorder, and cardiovascular disease. Knowledge of these variations can help a physician select the best drug and the right dosage for a patient, as well as avoid drugs that may cause the patient to suffer adverse side-effects.

One alternative approach is the microelectronic array technology developed by Nanogen (San Diego, CA). Each electronic array contains 100 test sites, which can be controlled electronically from the system's on-board computer. Nanogen's patented technology utilizes the natural positive or negative charge of most biological molecules including DNA and RNA. Applying an electric current to individual test sites on the NanoChip® microarray enables rapid movement and concentration of the target molecules. Through electronics, molecular binding onto the NanoChip®

microarray is accelerated up to 1000 times faster than traditional passive methods. The NanoChip® microarray technology provides an open platform that allows customers to easily run common assays as well as customize their own assays.

Current applications performed on the NanoChip® array focus on genetic analysis, including single nucleotide polymorphisms (SNPs), short tandem repeats (STRs), insertions, deletions, and other mutation analyses. This technology is likely to be helpful in diagnostics applications.

6.7. *Drug Discovery and Development*[22]

DNA microarrays promise to provide powerful tools for drug discovery and development. For example, gene expression microarrays are now playing a valuable role in all phases of the cancer drug discovery process. Microarrays are providing new insights into the molecular mechanisms of human cancers, and are helping to identify many new additional targets for drug discovery. First, for target identification, we can simply compare gene expression patterns in the normal and disease tissues. Second, during the drug discovery process, gene expression microarrays can be used to profile the pharmacological effects of lead compounds on a genomewide basis. This facilitates the discovery of prognostic and pharmacodynamic markers of drug response, helps in the understanding of the molecular mechanism of drug actions, and identifies undesirable expression signatures related to toxicity that can be dealt with during the chemical optimization process. Third, in the clinical trial, expression profiling can be used to define the molecular mode of action of drugs and to predict which patient is most likely to benefit from which particular drug, aiding in personalized treatment. Thus, the ability to obtain transcriptome information represents an exceptionally powerful means to explore basic biology, diagnose the disease, facilitate drug development, and tailor therapeutics to specific pathologies.

6.8. *Special Applications of DNA Microarrays in the Identification of Genetic Mutations*

By definition, a genetic mutation is a permanent change or a structural alteration in DNA or RNA. In general, mutations occur in DNA in humans and many other organisms; however, in retroviruses like HIV, mutations occur in RNA, which is the genetic material of retroviruses. In this chapter, the genetic mutation refers to the change of nucleotides in DNA. A mutation can be inherited, such as from parents; or it can occur by natural causes, such as radiation or toxic chemicals, that induce the change or damage the DNA. However, not every mutation causes diseases. A change in the nucleotide of DNA may or may not lead to a change in function of a gene. The change in function of a gene may cause a disease or may bring a benefit. But, in this chapter, we refer to those mutations that cause disease.

How does the microarray detect mutations? When we talk about the detection of mutation by microarray, we refer to the detection of either changes in nucleotides or changes in expression/transcription of a gene. A cDNA microarray contains cDNA sequences of thousands of genes. Hybridization of cRNA from a subject to an array chip basically allows researchers to examine how much a gene is hybridized onto the cDNA in the chip. If a gene expresses at a high level, it produces a large number of copies of mRNA; the large number of copies of a gene transcript is reflected by the high level of hybridization to the cDNA on the chip. If a mutation leads to nontranscription of a gene, then no hybridization will happen to the cDNA of this gene on the chip. It is in this regard we say that microarray can be used to detect the mutation of genes. The power of detection of mutation by the microarray itself, however, is limited. In the case of the detection of changes in nucleotides, there is a question of polymorphism, or real mutation, when a difference is detected between subject and control. There are thousands of polymorphisms (almost one in every kb) existing in human populations. In terms of gene expression, the mutated gene as well as the downstream and upstream genes in the same pathway will change the expression levels. Therefore, in most cases, we actually detect a list of candidate genes rather than the real mutated gene.

Nevertheless, in some cases, microarray may detect the real mutation:

(1) Sometimes, a mutation results in the deletion of a certain coding region of a gene. The deleted region may contain the sequence of the probe in the cDNA array. When the gene is highly expressed in a tissue, the microarray will detect a high level of signal from the normal control, while there will be no signal detected from the disease subject. When the data of the microarray are combined with the mapping information, the real mutated gene can be easily singled out.

(2) Some mutations lead to the large deletion of a chromosome region that contains an entire sequence of a gene or several genes. The microarray will not detect the deleted gene(s) at all.

(3) When a mutation is in the regulatory region, it usually causes either a decrease or an increase in the expression level of a gene. However,

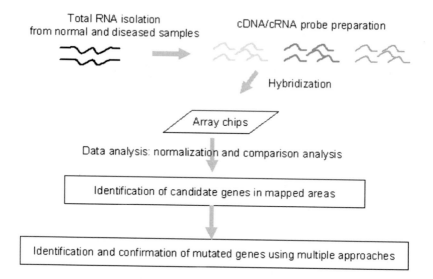

Fig. 2.   Steps for mutation identification using microarrays.

in some cases, such a mutation will completely shut off or turn on the expression of a gene at all times. Microarray data will point to such a gene and will eventually lead to the identification of the mutation.

An important note to the reader is that in this case, although the microarray identifies the mutated gene, it does not identify the exact DNA sequence of the mutation. A complete sequence comparison between normal controls and disease subjects has to be done to search for the mutated sequence. The procedure for mutation identification using microarrays is depicted in Fig. 2, and is described in detail below:

(1) RNA extraction. There are standard procedures from websites and textbooks for RNA extraction. However, at least two issues need close attention. First, due to the high sensitivity of the microarray, high-quality RNA is essential to the quality of the data. A concern is that a low-quality RNA sample will also produce data from the microarray; consequently, the data may mislead the research direction. With respect to gene identification, low-quality RNA may identify the wrong gene. We usually extract total RNA with a Trizol reagent (Invitrogen) and measure the integrity of total RNA integrity with the Agilent Bioanalyzer 2100. Second, the selection of tissue is key to the success of mutation identification. Not every gene is expressed in all tissues, and not every tissue expresses all of the genes. Therefore, a researcher needs to make

sure that every candidate gene is expressed at a certain level in the microarray data. If some genes are not expressed at the mRNA level in a tissue, microarray analysis from other tissues may be needed.

(2) Microarray chips. At present, microarray chips from major companies contain almost every gene in the human and mouse genomes. However, it is necessary for a researcher to check whether every candidate gene in his/her list is on the chips. If one uses chips that contain cDNA of fewer genes or chips made in-house, particular attention should be paid to the gene content.

(3) Microarray process. As in the case of RNA handling, an error in the microarray process will also produce bad data. At present, an automatic processor for microarray has not yet been developed. A technologist, not a technician, is needed for the process of microarray. Clear notes for every step will help to identify mistakes in the process.

(4) Signal comparison. The expression level of a gene is represented by its hybridization signal to its cDNA on the chips. When doing data analysis, normalization is needed so that signals from different chips are comparable.

(5) Confirmation of mutation. A variety of approaches can be used to confirm the mutation identified by microarrays. Several commonly used techniques are relative quantitative real-time PCR, DNA sequencing, and expression of proteins *in vitro*. To compare the protein sequences from normal and mutant genes, one needs to insert the corresponding nucleotide sequences into expression vectors. In general, protein products are analyzed using sodium dodecyl sulfate polyacrylamide gel electrophoresis (SDS-PAGE) and/or native PAGE electrophoresis and known enzymes to confirm predicted differences between normal and mutant proteins. Other techniques include *in situ* hybridization as well as antibody generation and immunolocalization. The generation of one or more high-quality antibodies to encoded proteins allows for subcellular localization studies and, in addition, corroborates the *in situ* hybridization results. Furthermore, antibodies may be needed for protein–protein interaction and other biochemical studies. Western blot analysis can then be used to analyze the specificity of the antibody and the expression of protein products of a mutated gene.

7. SUMMARY

DNA microarray technology has attained maturity and has become a crucial tool in biomedical research and drug discovery. Currently, different types of microarray platforms provide interested researchers with many options. In the area of *in situ* synthesized microarrays, Affymetrix is the dominant player, but its premier position is being challenged by NimbleGen (which is also developing high-density microarrays) and Invitrogen.

A second competitor to the *in situ* synthesis-based technology is array fabrication by spotting methods. Dozens of companies have mushroomed to produce various types of arrays, especially customized arrays. These companies offer everything from polylysine-coated slides to specialty slides containing various types of polymer and metal coatings. Recent developments show that the experimental variations of spotted microarrays will decrease as new spotting materials and systems are applied and further improved in the coming years.

Because microarray technologies provide a relative measurement of gene expression levels, it is extremely difficult to compare multiple experiments. Furthermore, the results of expression levels obtained from microarray experiments are also software-dependent. Therefore, standardized procedures for data analysis will be crucial for comparing microarray results from multiple experiments. The need for precise descriptions of microarray experiments for data archival and subsequent analysis has led to the Minimum Information About a Microarray Experiment (MIAME) standard.[23]

It is clear that DNA microarray, which was developed for gene expression profiling, is now widely used in the majority of life science research including genetics studies. The availability of various types of ready-to-use and customized microarrays meets the different needs of researchers. More importantly, the concept of personalized medicine is beginning to make its way to the marketplace, and is likely to largely increase the use of DNA microarray technologies.

In summary, microarrays are moving towards the promise of providing calibrated, truly quantitative gene expression measurements on a large scale. DNA microarray technology thus remains at the forefront of modern molecular biological methods that shape the way we address complex biological systems. It can be anticipated that microarray-based genetics will greatly speed up mutation identification in the near future.

REFERENCES

1. Liang P, Pardee AB. (1992) Differential display of eukaryotic messenger RNA by means of the polymerase chain reaction. *Science* **257**:967–971.
2. Velculescu VE, Zhang L, Vogelstein B, Kinzler KW. (1995) Serial analysis of gene expression. *Science* **270**:484–487.

3. Kafatos FC, Jones CW, Efstratiadis A. (1979) Determination of nucleic acid sequence homologies and relative concentrations by a dot hybridization procedure. *Nucleic Acids Res* **24**:1541–1552.

4. Ekins R, Chu FW. (1999) Microarrays: their origins and applications. *Trends Biotechnol* **17**(6):217–218.

5. Schena M, Shalon D, Davis RW, Brown PO. (1995) Quantitative monitoring of gene expression patterns with a complementary DNA microarray. *Science* **270**(5235):467–470.

6. Lipshutz RJ, Fodor SP, Gingeras TR, Lockhart DJ. (1999) High density synthetic oligonucleotide arrays. *Nat Genet* **21**(1 Suppl):20–24.

7. Blanchard AP, Kaiser RJ, Hood LE. (1996) High-density oligonucleotide arrays. *Biosens Bioelectron* **11**:687–690.

8. Gao X, LeProust E, Zhang H, *et al.* (2001) A flexible light-directed DNA chip synthesis gated by deprotection using solution photogenerated acids. *Nucleic Acids Res* **29**:4744–4750.

9. Singh-Gasson S, Green RD, Yue Y, *et al.* (1999) Maskless fabrication of light-directed oligonucleotide microarrays using a digital micromirror array. *Nat Biotechnol* **17**:974–978.

10. Bowtell DD. (1999) Options available — from start to finish — for obtaining expression data by microarray. *Nat Genet* **21**:25–32.

11. Auburn RP, Kreil DP, Meadows LA, *et al.* (2005) Robotic spotting of cDNA and oligonucleotide microarrays. *Trends Biotechnol* **23**(7):374–379.

12. Brenner S, Johnson M, Bridgham J, *et al.* (2000) Gene expression analysis by massively parallel signature sequencing (MPSS) on microbead arrays. *Nat Biotechnol* **18**:630–634.

13. Rhodes DR, Chinnaiyan AM. (2005) Integrative analysis of the cancer transcriptome. *Nat Genet* **37**:S31–S37.

14. Segal E, Friedman N, Kaminski N, *et al.* (2005) From signatures to models: understanding cancer using microarrays. *Nat Genet* **37**:S38–S45.

15. Syvänen A. (2005) Toward genome-wide SNP genotyping. *Nat Genet* **37**:S5–S10.

16. Pinkel D, Albertson DG. (2005) Array comparative genomic hybridization and its applications in cancer. *Nat Genet* **37**:S11–S17.

17. Kallioniemi A, Kallioniemi OP, Sudar D, *et al.* (1992) Comparative genomic hybridization for molecular cytogenetic analysis of solid tumors. *Science* **258**(5083):818–821.

18. Solinas-Toldo S, Lampel S, Stilgenbauer S, *et al.* (1997) Matrix-based comparative genomic hybridization: biochips to screen for genomic imbalances. *Genes Chromosomes Cancer* **20**(4):399–407.

19. Pollack JR, Sorlie T, Perou CM, *et al.* (2002) Microarray analysis reveals a major direct role of DNA copy number alteration in the transcriptional program of human breast tumors. *Proc Natl Acad Sci USA* **99**(20):12963–12968.

20. Wheeler DB, Carpenter AE, Sabatini DM. (2005) Cell microarrays and RNA interference chip away at gene function. *Nat Genet* **37**:S25–S30.

21. van Steensel B. (2005) Mapping of genetic and epigenetic regulatory networks using microarrays. *Nat Genet* **37**:S18–S24.

22. Clarke PA, Poele R, Workman P. (2004) Gene expression microarray technologies in the development of new therapeutic agents. *Eur J Cancer* **40**(17):2560–2591.
23. Brazma A, Hingamp P, Quackenbush J, *et al.* (2001) Minimum Information About a Microarray Experiment (MIAME) — toward standards for microarray data. *Nat Genet* **29**:365–371.

Chapter 15

Expression Quantitative Trait Locus (eQTL) Mapping

Song Huang[*], David Ballard[*], Zheyang Wu[†] and Hongyu Zhao[†,‡]

[*]*Program of Computational Biology and Bioinformatics*
Yale University, New Haven, CT 06520, USA

[†]*Department of Epidemiology and Public Health, Yale University*
New Haven, CT 06520, USA

[‡]*Department of Genetics, Yale University, New Haven*
CT 06520, USA

With the recent advances in genomewide expression microarray technology, combining the power of gene expression profiling and genetics is a natural step forward. Jansen and Nap[5] first formally proposed a new research area termed "genetical genomics", which describes the combined study of expression variations and DNA variations in segregating populations. The gene expression levels (i.e. mRNA transcript abundance) are treated as quantitative traits potentially affected by multiple genes and other factors. Traditional methods for detecting quantitative trait loci (QTLs) could be utilized to detect chromosomal regions affecting expression levels; these regions are referred to as expression quantitative trait loci (eQTLs). Recent studies have demonstrated the utility of this approach in unraveling many features of the genetic basis of variation in gene expression. Despite its great potential, there are many limitations to the current eQTL methods that demand statistical and computational novelties. Some of the issues are inherited from the traditional multiple-trait methods in QTL analysis and microarray technology. More importantly, the unique challenge is the joint consideration of tens of thousands of correlated phenotypes (i.e. transcription levels) with hundreds or thousands of genotypes. In this context, the issue of multiple testing needs to be better addressed not only to control the overall

---

[‡] Correspondence author.

rate of false discoveries, but also to fully take advantage of the correlated expression patterns. In this chapter, we will summarize statistical methods that have been applied to eQTL studies, review the knowledge and patterns emerging from these studies, and discuss future research directions.

1. INTRODUCTION

Quantitative trait locus (QTL) mapping methods have been extensively used in model organisms, plants, livestock, and humans to identify chromosomal regions associated with quantitative traits of interest. These regions, once identified as QTLs, can be targeted for positional cloning of the genes involved. In general, the regions identified through QTL methods are quite large and a considerable amount of effort is required to clone functional genes. Fortunately, during the past two decades, numerous advances in both technology and methodology have greatly improved our ability to collect, measure, and analyze the data necessary to clone genes and to understand how these genes interact with one another to manifest a quantitative phenotype. Nevertheless, these advances also raise many new problems that must be addressed in order to fully utilize the information found in the data.

The recent advances in QTL methodology are rooted in the development of high-throughput technologies, exemplified by the releases of the draft sequence of the human genome in 2001.[1,2] Its completion was due in part to the increase in computational power that had become available.[3] In parallel with the data accumulation of DNA sequences, other high-throughput molecular technologies also advanced. In 1995, Schena *et al.*[4] published the first DNA microarray paper in *Science*, describing how gene expression was simultaneously measured for 45 *Arabidopsis* genes; these technologies were quickly applied to more organisms. The fields of bioinformatics, genomics, and proteomics exploded due to the accumulation of large amounts of diverse types of data that are potentially useful for answering fundamental biological questions, e.g. the identification and evolutionary analysis of homologous genes across species and the comparison of gene expression profiles between different organisms and cell types.

In order to utilize these technologies to understand how genes affect complex disease phenotypes, Jansen and Nap[5] outlined a procedure pairing microarray data with genetics to elucidate biological pathways. The procedure, termed as "genetical genomics", identifies chromosomal regions associated with gene expression data (which are treated as quantitative phenotypes) through genomewide linkage analysis. The genomic loci controlling the mRNA abundances are called expression quantitative trait loci (eQTLs). Although most eQTL studies employ traditional QTL methods in data analysis, eQTL mapping is not simply a repeat of single-trait QTL analysis. In this chapter, we will first review the methods that have been applied in genetical genomics, along with some of the computational and statistical challenges it faces. Next, we will illustrate the type of questions and findings that have arisen from this new field. We will conclude with future directions in this rapidly evolving field.

2. CHALLENGES AND METHODS

The field of genetical genomics is gaining momentum, but what do these new studies provide in terms of biological information? As stated above, the key interest is to understand how genes manifest into a phenotype, i.e. why are some people susceptible to a disease when others are not? Is it due to a mutation in the DNA sequence? If so, where does the mutation occur — in the exons of the gene, or in a regulatory element? The regulation of a gene's transcription can be controlled by nearby genetic elements (*cis*-acting elements — the gene's own promoter, for example) or by genetic elements located further away (*trans*-acting elements).[6] It is believed that the regulation of a gene(s) leads to the susceptibility of an individual.[7] Current work is focused on reliably identifying *cis*- and *trans*-acting elements, and on understanding their impact on a phenotype to disease. Understanding the effect of *cis*- or *trans*-acting elements depends on the investigator's ability to measure the impact that these elements have on a gene's expression. This is not an easy task.

In general, genetic studies can be broadly defined as either linkage-based or association-based. Linkage studies have been successful in identifying the genes underlying simple Mendelian disorders by tracing the segregation of the disease and a genetic marker within families. In contrast, association studies are commonly used to identify alleles associated with diseases through the co-occurrence of marker alleles and disease phenotypes across families at the population level. Both approaches allow researchers to target candidate regions of the genome where disease-causing genes reside. The regions identified from linkage studies are, however, generally large and require considerable effort to systematically search for the actual disease-causing variant(s) (a process called positional cloning); whereas association studies need to first identify candidate regions for screening, except for genomewide association studies. However, both approaches have been less successful with complex disorders, which are the result of complex interactions among multiple genes and environmental risk factors. There-fore, using DNA sequences and gene expression data in QTL analysis (i.e. eQTL analysis) may allow us to get closer to the actual biological mecha-nisms controlling disease susceptibility.[8]

eQTL mapping is essentially a genomewide linkage analysis with tens of thousands of gene expression values as multiple quantitative traits. The issues existing in traditional QTL analysis still need to be addressed in eQTL studies. More importantly, due to the high dimensionality of microar-ray data, we are facing more challenges. In this section, we will first review the issues in microarray data acquisition and processing (Sec. 2.1), and then we will summarize the methods aimed at meeting the challenges in eQTL mapping, such as how to effectively integrate knowledge from multi-ple correlated traits using appropriate genetic models (Secs. 2.2–2.4), how to efficiently conduct computations (Secs. 2.5 and 2.7), and how to appro-priately control false-positive findings (Secs. 2.6 and 2.7). Lastly, we will describe the current approaches used to infer gene regulatory networks as one of the applications of eQTL analysis (Sec. 2.8).

2.1. *Microarray*

In most eQTL studies, gene expression is measured using microarrays. A microarray consists of many (e.g. up to millions of) specifically designed DNA probes bound to a slide to measure mRNA concentrations. The con-centration is quantified by the amount of intensity emitted when a fluores-cently labeled sample is put onto the microarray.[9] In the ideal case, this intensity is proportional to the level of activity of the gene.[10] Comparison of the intensity of a spot between samples taken from different conditions or individuals allows researchers to conclude whether a gene's activity has changed. Unfortunately, the process of obtaining these intensity values is rather complex and many types of noises can contribute to the observed signals. For example, intensity values may be missing from different probes due to experimental error or mishandling of the slide. It is also known that the intensity measures from the probes emitting high intensity are more variable than the probes emitting low intensity.[11] The results may also vary across different experimental conditions (e.g. slide effects), which could cause low repeatability of the experiments, so more replicates and samples may be required to keep the results consistent.

The published studies adopted plants and animals as model organisms, which can be used as homologous results to study large livestock animals at a much higher experimental cost.[12] In microarray gene expression data, we can use established preprocessing procedures, such as data transformation, normalization, and clustering, but these methods are not perfect. For example, it remains a statistically challenging problem to identify differentially expressed genes across a set of samples from tens of thousands of genes. More recently, microarray has been advocated as a platform for genomewide genotyping, presenting new statistical problems in genotype calling. Therefore, eQTL analysis needs to address and incorporate the issues arising from microarray data for both phenotypic and genotypic data.

The challenges discussed above demand thorough statistical treatment; indeed, work in these areas has provided insights and tools for data analysis. For instance, multiple imputation methods developed by Rubin[13] have been applied to microarray data to reliably predict missing data. Some of the more popular imputation methods are K-nearest neighbors,[14] Bayesian principal component analysis (PCA),[15] and local least squares.[16] Various statistical error models have been developed for the accurate inference of expression levels and the appropriate control of false-positive results, e.g. false discovery rate (FDR) control in multiple testing.

2.2. *Multiple Trait Analysis*

Statistical methods that simultaneously consider multiple traits were initially proposed in a traditional QTL setting. These methods, which can be potentially adopted in eQTL studies, are based on the joint analysis of multiple traits while considering the correlation structure among them. Such joint analysis may provide more accurate and informative results, even if the traits considered are not correlated under some circumstances (e.g. when pleiotropic effects exist).[17] The theoretical reasoning is as follows. The expected logarithmic odds (LOD) score (or equivalently, power) is an increasing function of trait heritability (H^2), which in turn is an increasing function of the discrepancy between the average trait values across different genotype groups at a putative locus.[18] When the traits are correlated, or when there is a pleiotropic effect of the putative locus on the traits (while the correlation coefficient between the traits is zero), the joint distribution of traits is more informative than the separate consideration of each trait in describing the discrepancy across different genotype groups. Under this situation, an increase in the overall LOD score will lead to a higher power (even with the increased number of parameters) to identify the QTLs and to make more precise parameter estimation.

Korol *et al.*[17,19] rationalized this intuition mathematically and offered a multiple-trait interval-mapping model based on the joint distribution of traits. By defining multiple-trait heritability, we can incorporate the variance–covariance effect of eQTLs, by which different genotypes cause different phenotypic variances of traits. For parameter estimation and hypothesis tests, they applied the maximal likelihood approach to the joint distribution of traits (multivariate normal mixture distribution). When the genetic effect is only related to the mean expression level without effect on the variance–covariance structure, this model can actually be covered by the regression-based model by Jiang and Zeng,[20] which we will introduce later.

In most cases, the joint analysis of traits has the most benefit when the traits are correlated. Correlation may result from genetic and/or environmental sources. In eQTL studies, correlations among genes at both the observed transcript level and the regulation level need to be considered, preferably at the model setup stage. In addition to increasing detection power, another advantage of taking into account the correlated structure of multiple traits is the reduction of the search space by clustering correlated gene expression traits. The information in highly correlated traits can be extracted into lower-dimension representative trait data, which may be sufficient for QTL detection and parameter estimation. Furthermore, the efficiency of mapping and the power of hypothesis tests could be increased. For example, the FDR could be decreased by (1) reducing the number of tests, or (2) reducing the noise while strengthening the signal and weakening the environmental variation after clustering.[21,22]

By considering the correlation structure in the QTL mapping model, we may also test biologically interesting hypotheses for joint QTL mapping, pleiotropy, and QTLs by environmental interaction, as well as the means to differentiate for associations among traits (e.g. due to pleiotropy or close linkage).

Here, we introduce a representative model setup by Jiang and Zeng[20] that considers the correlation structure among traits. They extended the composite interval mapping (CIM) method, which is one of the multiple QTL mapping methods, to a general multivariate linear model that handles multiple traits as response variables. Similar to the CIM method, one QTL is considered each time without epistatic effects.[20,22] The model considered is

$$\mathbf{Y} = \mathbf{x}^*\mathbf{b}^* + \mathbf{z}^*\mathbf{d}^* + \mathbf{X}\mathbf{B} + \mathbf{E}, \text{ or}$$

$$y_{j1} = b_{01} + b_1^* x_j^* + d_1^* z_j^* + \sum_l^t (b_{l1} x_{jl} + d_{l1} z_{jl}) + e_{j1}$$

$$y_{j2} = b_{02} + b_2^* x_j^* + d_2^* z_j^* + \sum_l^t (b_{l2} x_{jl} + d_{l2} z_{jl}) + e_{j2}$$

$$\vdots$$

$$y_{jm} = b_{0m} + b_m^* x_j^* + d_m^* z_j^* + \sum_l^t (b_{lm} x_{jl} + d_{lm} z_{jl}) + e_{jm},$$

where y_{jk} is the phenotypic value for trait k in individual j, b_{0k} is the mean effect of the putative QTL on trait k, b_k^* is the additive effect of the putative QTL on trait k, x_j^* is the number of effective alleles at the putative QTL in individual j, d_k^* is the dominant effect of the putative QTL on trait k, and z_j^* is an indicator variable of the heterozygosity at the QTL in individual j. x_{jl} and z_{jl} are the corresponding numbers of the effective allele and indicators for heterozygosity for marker l, coming from t markers that are selected for controlling residual genetic variation when m traits are analyzed. e_{jk} is the error term (residual effect on trait k for individual j), where we assume $\mathrm{Cov}(e_{jk}, e_{jl}) = \sigma_{kl}$ and $\mathrm{Cov}(e_{jk}, e_{j'l}) = 0$. The parameters are estimated via the expectation-maximization (EM) algorithm or, more efficiently, via the expectation–conditional maximization (ECM)[23] algorithm. The significance of the results can be assessed by the likelihood ratio test, followed by multiple testing adjustments (e.g. Bonferroni correction, permutation, or approximation of chi-square distribution).

The above model for joint analysis can significantly increase the detection power in suitable situations. However, one of the problems is that the number of parameters to be estimated is at least $O(n^2)$ of the number of traits.[17] Furthermore, if we want to integrate multiple QTL analysis into the multiple-trait model, the computational burden would be much heavier. It may be worthwhile to cluster the highly correlated traits and apply joint trait analysis to relatively small groups of combined traits.

In addition to grouping traits, information from other sources can be incorporated into the testing procedure. For example, transcripts that are grouped together with physiological traits (e.g. complex diseases) are believed to be closely relevant because physiological traits are likely to be associated with mRNA levels. In order to include the physiological information in the eQTL analysis, clustering methods can be used to group the traits of primary interest with the gene expression data based on the correlation among them.[21] This approach can reduce the model search space and potentially increase the power of detecting eQTLs.

Principal component analysis (PCA) may also be used to reduce transcript dimensions in the mapping stage. "Supergenes" can be defined as the first several principal components of the transcript matrix, each of which is a linear combination of the original correlated expression trait values. They can capture the majority of variations in gene expression traits, and thus contain most of the relevant information.[21] This methodology may elucidate a biologically meaningful interpretation for the linked trait loci, and could be helpful when constructing regulatory networks. In order to reduce the computational burden, individual genes composing the super-genes could be mapped and investigated with more emphasis. For instance, genes within a cluster can be mapped by multiple-trait interval mapping.[20] This method would enable us to detect which transcript contributes to the composite linkage signal; at the same time, we may avoid multiple trait analysis, which is counterproductive when too many transcripts are fitted into the model.

However, we should be cautious when using the clustering methods. First, the signal from a loosely linked gene could be diluted when combined with the others; second, the control of multiple testing errors should be derived from the grouping method. Specifically, the limitations of the PCA method include the potential loss of information, the lack of power to detect *cis*-acting QTLs, and the interpretation of supergenes.[21] Because of these problems, a comprehensive study should consider data reduction combined with individual trait analysis.[24]

Instead of using the correlation structure of transcript traits, Carlborg *et al.*[22] suggested a weighted least squares estimation for QTL parameters, based on the repeatability of the trait. The goal is to integrate useful information and increase the power of QTL detection. Repeatability is defined as the ratio of line variance over total variance:

$$r = \frac{\sigma_{\mathrm{b}}^2}{\sigma_{\mathrm{w}}^2 + \sigma_{\mathrm{b}}^2},$$

where σ_{b}^2 is the variance between lines (genetic variation) and σ_{w}^2 is the variance within lines (environmental variation).

2.3. *Joint Linkage Probability of eQTL*

In multiple trait analysis, the QTL detection process is also a model selection process: the best identified model leads to the identification of putative QTLs. Support for the presence of QTLs is usually based on the goodness of fit measured at different search stages. Another method is to evaluate a set of loci by the joint probability when all loci in the set are linked, indicating the joint significance of the set of loci.[22,25] For simplicity, here we use two loci as an example to illustrate the procedure of the joint linkage method:

(1) The joint probability is obtained by calculating the conditional probability of the linkage of a locus given the other locus is linked, i.e.

$$\Pr(loci\ 1\ and\ 2\ are\ linked\mid Data)$$
$$= \Pr(locus\ 1\ linked\mid Data)$$
$$\times \Pr(locus\ 2\ linked\mid locus\ 1\ linked,\ Data).$$

In order to simplify the calculation, an estimate of the joint probability of linkage is made with a nonparametric method. This method is based on the standard F-statistic to test the significance of a newly selected locus, using a forward sequential model selection strategy:

(M0)trait = baseline level + error

(M1)trait = baseline level + locus 1 + error

(M2)trait = baseline level + locus 1 + locus 2 + locus 1 × locus 2

+ error

For the ith trait, let F_{ij} be the maximal F-statistic corresponding to the jth sequentially added locus among all candidate loci. To estimate the null distribution of F_{ij}, we simulate the null observation of F_{ij}, $F_{ij}^b (b = 1, \ldots, B)$, by the bth permutation. Specifically, F_{ij}^b is the maximal F-statistic obtained in the bth permutation of trait values taking place within each segregating group of all the $(j - 1)$ previously added loci. Note that the null distribution on the jth locus is formed given the condition that all of the previously determined loci are truly linked.[26]

To represent the linkage, let $l_{ij} = 1$ if the jth marker is truly linked to the ith trait, and $l_{ij} = 0$ otherwise. Under certain assumptions, replace $\Pr(l_{ij} = 1 \mid l_{i1} = 1, \ldots, l_{i,j-1} = 1, Data)$ with $\Pr(l_{ij} = 1 \mid l_{i1} = 1, \ldots, l_{i,j-1} = 1, F_{ij})$ (note that this may lose information).

Calculations for the primary locus $\Pr(l_{i1} = 1 \mid Data) = \Pr(l_{i1} = 1 \mid F_{i1})$ are as follows. Let g_0 represent the probability density function (PDF) of the null distribution of the F-statistic, i.e. the observations under g_0 are F_{i1}^b; g_1 represent the PDF of the alternative distribution of the F-statistic; π_0 denote the prior probability that the primary locus is not linked; and π_1 denote the prior probability that the primary locus is linked. The observed F at a random locus is a randomly selected value from the mixture distribution of linkage and no linkage, with density $g = \pi_0 g_0 + \pi_1 g_1$ (note that if $F_{i1} \mid l_{i1}$ differs between the traits, one can view g_0 and g_1 as the average density over these traits). Then, the posterior probability of linkage for the primary locus is

$$\Pr(l_{i1} = 1 \mid F_{i1}) = 1 - \frac{\pi_0 g_0(F_{i1})}{\pi_0 g_0(F_{i1}) + \pi_1 g_1(F_{i1})}.$$

To estimate the posterior probability, we first estimate

$$R(F) = \frac{g_0(F)}{g(F)} \text{ by } \frac{\Pr(l_{i1} = 1 \mid F)}{\Pr(l_{i1} = 0 \mid F)}$$

through logistic regression[27]:

$$\log \frac{\Pr(l_{i1} = 1 \mid F)}{\Pr(l_{i1} = 0 \mid F)} = f(F),$$

where $l_{i1} = 1$ when $F = F_{i1}$, and $l_{i1} = 0$ when $F = F_{i1}^b$; $f(.)$ is a natural cubic spline.[28] Secondly, a conservative estimate of π_0 is given by

$$\hat{\pi}_0(c) = \frac{\{F_{i1} \le c; \forall c\}}{\{F_{i1}^b \le c; \forall c, \forall b\}/B},$$

where c can be automatically chosen.[29] Now, we have $\Pr(l_{il} = 1 \mid Data) = 1 - \hat{\pi}_0 \hat{R}(F_{i1})$.

The probability for the conditional linkage of the second- and higher-order loci can be estimated analogously. Thus, the estimated joint probability is

$$\hat{\Pr}(l_{i1} = 1, \dots, l_{iL} = 1 \mid Data)$$
$$= \hat{\Pr}(l_{i1} = 1 \mid Data)$$
$$\times \cdots \times \Pr(l_{iL} = 1 \mid l_{i1} = 1, \dots, l_{iL-1} = 1, Data).$$

(2) The criterion of linkage is decided by controlling the FDR. Let S_λ be the set of traits called significant, in which the corresponding joint linkage probabilities exceed a predetermined threshold, i.e. $\hat{\Pr}(l_{i1} = 1, \dots, l_{iL} = 1 \mid Data) \ge \lambda$. To decide a reasonable threshold, we can control the FDR:

$$\hat{\mathrm{FDR}}(S_\lambda) = \frac{\text{estimated number of false discoveries}}{\text{total number of significant traits}}$$
$$= \frac{\sum_{i \in S_\lambda} [1 - \hat{\Pr}(l_{i1} = 1, \dots, l_{iL} = 1 \mid Data)]}{|S_\lambda|}.$$

If $\lambda = 0.9$, then the estimated FDR is at most 10%. By controlling λ, the number of highly possible linked loci does not have to be fixed in advance.

By considering the overall linkage, this method partially overcomes the bias caused by loci that explain a large part of the trait variation, but are not actually linked. For instance, a large marginal effect in the model for a forward search does not mean that a locus is truly linked. Using joint probability as the criterion, we can exclude the situation where the selected set of loci is linked, but not every one of them truly represents a QTL.

By assuming that all of the primary loci, secondary loci, etc. have the same distribution, the above procedure tries to capture information shared across traits for the gene expression traits (under both null and alternative hypotheses). The argument for the validity is that the "F-statistic is an asymptotically pivotal statistic and the number of observations is reasonably large".[25] In other words, the F-test statistics are asymptotically independent of the unknown parameters with respect to different expression traits. This argument is subject to the condition that the test statistics for different traits are from the same distribution family given the linkage status of a marker; however, this may not be true in practice. When the assumption is not satisfied, the authors suggest using the average (or mixture) of the distribution over the traits.

2.4. *Bayesian Methods Based on Markers*

We can consider the mapping methods in Secs. 2.2 and 2.3 as transcript-based approaches because the basic idea is to repeatedly apply phenotype mapping methods to gene expression data. Alternatively, one could fix the marker(s) and test the significance of linkage across genes; this is equivalent to repeatedly identifying differential expression (DE) across genes between the groups defined by marker genotypes. Many DE detection approaches used in microarray analysis can be applied here. One of them is an empirical Bayesian approach under the hierarchical modeling framework.[30] Bayesian approaches that capture the shared information across genes may potentially lead to more accurate QTL identification.

Here, we briefly introduce the model setup. For notation simplicity, we assume a backcross experiment where the genotype of an F2 generation is either 0 or 1. Let $y_t = (y_{t,1}, \ldots, y_{t,n})$ represent the n expression values of the tth transcript. Given the underlying mean $\mu_{t,\bullet}$, the conditional distribution of expression $y_{t,\bullet}$ is independent across t with density $f_{\text{obs}}(y_{t,\bullet} \mid \mu_{t,\bullet})$. We consider the underlying means as random variables with priors $\pi(\cdot)$. In the case of equivalently expressed (EE) transcripts, the underlying means are the same for all observed expression values. So, the likelihood for equivalent expression (EE) is

$$f_0(y_t) = \int \left[\prod_{k=1}^{n} f_{\text{obs}}(y_{t,k} \mid \mu) \right] \pi(\mu) d\mu.$$

In the case of differentially expressed (DE) transcripts, where the genotype is equal to 0 (genotype AA) or 1 (genotype AB), the underlying means $\mu_{t,0}$ and $\mu_{t,1}$ are independent random variables from the same density $\pi(\cdot)$. Decomposing $y_t = (y_t^0, y_t^1)$, the likelihood for differential expression (DE) is

$$f_1(y_t) = f_0(y_t^0) f_0(y_t^1)$$

$$= \int \left[\prod_{\{\text{genotype}=0\}} f_{\text{obs}}(y_{t,k} \mid \mu_{t,0}) \right] \pi(\mu_{t,0}) d\mu_{t,0}$$

$$\int \left[\prod_{\{\text{genotype}=1\}} f_{\text{obs}}(y_{t,k} \mid \mu_{t,1}) \right] \pi(\mu_{t,1}) d\mu_{t,1}.$$

The marginal density of y_t is a mixture:

$$f(y_t) = p f_1(y_t) + (1-p) f_0(y_t),$$

where p is the probability of DE.

Based on the above formulation, we can estimate the parameters by maximizing the likelihood over all observed transcripts:

$$L = \prod_t f(y_t).$$

The EM algorithm can be used[31] by considering the expression pattern (DE or EE) at each transcript as the missing value:

E-step: Calculate the expectation of the pattern based on the observed data and the previously calculated (or initially assumed) parameter values. This is equivalent to updating the posterior probability:

$$P(DE \mid y_t) = \frac{p f_1(y_t)}{p f_1(y_t) + (1-p) f_0(y_t)}.$$

M-step: Update the parameter estimates by maximizing the conditional likelihood given the observations and the expected patterns in the E-step.

A linkage is significant for a gene if the posterior probability of DE is larger than a predefined threshold. To set the threshold, we can first order the estimated posterior probability from all transcripts at a specific marker, then include the largest values into a set one by one until the mean of this set is less than or equal to $1 - \alpha$. The value included last is the threshold for the current marker. This method gives a posterior expected FDR of at most $100\alpha\%$.[32]

Specifically, the log-normal model can be used to define $f_{\text{obs}}(\cdot)$ and $\pi(\cdot)$: $y_{t,k} \sim N(\mu_{t,\bullet}, \sigma^2)$ and $\mu_{t,\bullet} \sim N(\mu_0, \tau_0^2)$ ($y_{t,k}$ is a log-transformed expression measure).

A gamma–gamma model can also be used where both distributions are gamma. The argument for these models is based on the common features of microarray data, such as a constant coefficient of variation, and dependencies between intensity ratio variation and magnitude.[31]

The mixture-over-marker (MOM) model is an extended version of the empirical Bayes method.[30] Besides the probability of no linkage and the probability of linkage to the current marker, this model also addresses the probability that a specific transcript is linked to other markers. Let p_0 be the probability of no linkage to any marker and p_m be the probability of linkage to the mth marker. The marginal density for the tth transcript is

$$f(y_t) = p_0 f_0(y_t) + \sum_{m=1}^{M} p_m f_m(y_t),$$

where $f_m(y_t)$ is the density of the transcript given it links to the mth marker:

$$f_m(y_t) = \int \left[\prod_{\{\text{marker } m=0\}} f_{\text{obs}}(y_{t,k} \mid \mu_{t,0}) \right] \pi(\mu_{t,0}) d\mu_{t,0}$$

$$\int \left[\prod_{\{\text{marker } m=1\}} f_{\text{obs}}(y_{t,k} \mid \mu_{t,1}) \right] \pi(\mu_{t,1}) d\mu_{t,1}.$$

The distribution assumption for $f_{\text{obs}}(\cdot)$ and $\pi(\cdot)$ is Gaussian, as before. However, the variance σ^2 depends on the clusters of transcripts, with the cluster membership determined by K-means.

A transcript is DE if the posterior probability of EE is smaller than the threshold setup by controlling the posterior FDR. If a transcript is DE, its putative controlling loci are those markers within the Bayesian confidence interval.[33]

Through simulations, Kendziorski *et al.*[30] claimed that the MOM approach is able to simultaneously control the FDR and keep high power. They believe that this model improves the specificity of eQTL identification when compared to other methods (trait-based methods and the older version of the empirical Bayesian method). As a Bayesian method, the MOM model also permits the further incorporation of prior information to improve the power and accuracy of eQTL detection. For example, the physical location of a transcription factor and the functional categories of transcripts could be used to inform the mixing proportions or to identify gene clusters.

2.5. *Automated and Efficient Mapping Strategies*

To address the large-scale and high-dimensional problem in eQTL analysis, we will discuss the methods initially designed to alleviate the heavy computational burden in multiple-QTL mapping procedures. Since eQTL mapping involves tens of thousands of gene expression traits, computational strategies are even more relevant and critical in eQTL analysis.

Within the traditional regression-based multiple QTL analysis framework, there are two components in the computation design: (1) the definition and calculation of a target function (e.g. a kernel), and (2) the development of a model selection strategy that can efficiently find the loci which maximizes the kernel over the search space throughout the genome. The definition of a kernel can be based on traditional QTL mapping methods such as the LOD score in trait-based mapping methods. The calculation of the LOD score in interval mapping can be accomplished through the maximum likelihood (ML) method, and it can be well approximated by the least squares regression method advocated by Haley and Knott.[34] The EM algorithm and the ECM algorithm can be applied for the ML estimation.[23,35]

As for model selection, several methods can be applied in this setting. The first is to use an exhaustive search procedure,[36] but with a sparser grid for QTLs. In fact, single marker mapping does not lose much power compared to interval mapping when markers are highly informative in the linkage context.[22,37] A two-stage strategy can use refined local mapping in the second stage, thus improving the resolution of the QTL location. However, exhaustive searching is still not feasible in high-dimensional space created by having more than three traits and eQTLs in the model. The second method is to use the forward model selection strategy.[38] However, since the selection procedure can only pick loci with relatively large marginal effects first, this approach is less powerful for finding loci that have a large epistatic effect but individually small marginal effects for balancing. Furthermore, the forward model selection strategy is sensitive to the type of epistatic interactions and the size of QTL effects; for example, the detection power for the forward model selection strategy was only 21% for a duplicate epistatic model in a simulation study.[39]

Carlborg and colleagues[39,41–43] applied a genetic algorithm[40] for the simultaneous mapping of interacting loci, balancing the detection power and the computational performance. Depending on the genetic length of the genome and the resolution of QTL mapping, the computational time for this method is approximately threefold to fivefold of the forward selection method for the mapping of two to four interacting QTLs (the computational burden does not exponentially increase on QTL dimension as it does in exhaustive searching), while the detecting power is 90%–100% depending on the epistasis types. Although this algorithm has a weakness in finding the precise QTL locations, Carlborg *et al.*[39] claimed that local exhaustive searching might overcome this limitation.

The model selection problem can also be seen as a global optimization problem, especially when fixing the number of loci within a QTL model. Algorithms for global optimization can be used in mapping procedures to find the best model that maximizes the kernel over the whole search space. For example, the DIRECT algorithm,[44] which can be used to maximize a Lipschitz continuous function, has been applied to QTL mapping in a two-dimensional mapping space.[42] With some preciseness, this method reduces the computational time two to four orders of magnitude compared to an exhaustive search. Strategies with such good computational performance and higher statistical power in QTL detection are important areas for future developments.

The implementation of computation algorithms for eQTL mapping should also be emphasized. Parallel computing using computer clusters can dramatically increase the efficiency in multiple-QTL interval mapping.[45] With only minor modification of a software for regression interval mapping in outbred line crosses developed by Haley *et al.*,[46] Carlborg *et al.*[45] implemented a parallel processing program using message passing interface (MPI) and distributed the computation task to a maximum of 18 processors by assigning one or several linkage groups to each processor. The performance of QTL mapping and permutation analyses, defined as the computational time of a single process divided by that of multiple processors, increased about seven times. A higher performance can be achieved by distributing the computational load to each processor by even division of the genetic map segment instead of by linkage group.[45] As suggested and tested by Carlborg *et al.*,[45] parallel computing can also be implemented for other computationally intensive tasks (e.g. permutations) or for other QTL mapping methods based on ML and Bayesian theory, with possibly more modifications to the existing programs. In the eQTL scenario, where tens of thousands of expression traits are involved, tasks like permutation tests, model searching, and large-scale simulation can be even more computationally demanding. The similar parallel computing approach can be applied to distribute the heavy computational load to multiple processors evenly across traits to increase the efficiency in eQTL analysis.[22] Moreover, parallelization in the different types of tasks, such as partitioning traits, genetic mapping, and permutations, can be used at the same time to perform computationally feasible analyses.

2.6. Test Threshold Problem

The problem of multiple testing is naturally an issue in eQTL mapping because thousands of expression traits are simultaneously tested on hundreds or thousands of loci on the whole genome. The FDR may be preferred in this context to control false-positive results. In some of the above methods, we have discussed the setup of FDR calculation. However, many issues remain to be addressed more carefully and thoroughly; for example, whether it is safe to consider the distribution of genomewide significance thresholds the same for different traits. Some pioneering papers used the same theoretically derived LOD score (or p-value) as the testing threshold to detect putative QTLs, or to build the FDR assuming a common distribution.[47] However, some researchers argued that the thresholds should be considered empirically by randomization tests trait by trait[26,48] because thresholds can vary substantially, which is not simply due to sampling variation.[22] Further theoretical research and empirical evidence are needed to find a better choice that balances preciseness and computational simplicity.

2.7. QTL Detection Refinement

In eQTL mapping, due to the large number of traits considered, it is likely that the total number of detected QTLs is approximately equal to the expected number of false discoveries if the threshold is not carefully selected (total number of traits × false-positive rate, where the overall FDR is based on a moderate genomewide type I error, say, $\alpha = 0.05$). One way to obtain more confident putative eQTLs is to restrict the p-value in order to decrease the FDR. However, this will reduce the detection power (finding much fewer significant results) and increase the calculation burden, especially when we use permutation tests to get a trustable empirical randomization threshold. A better way to address this problem is to add information from gene expression traits to distinguish which detected loci are more likely to be true.

Repeatability calculated from the replicated transcripts has been found to be positively related to the detection of significant QTLs.[22] Based on this, we can classify traits into groups according to their repeatability. The proportion of traits with significant QTLs is higher in a group with higher repeatability (the higher the proportion, the lower the corresponding group's FDR). In the group with high repeatability, the number of detected QTLs was found to be much higher than the number of expected QTLs calculated based on the overall FDR, indicating that most of the detected QTLs are not likely to be false-positive due to chance. In addition, in order to reduce the computation burden while keeping the most reliable results, we can put more focus on transcripts with a high repeatability level (note that the detection of QTLs for a gene was also proved to be unrelated to the sample variance of the trait/gene expression values[22]).

Other useful information that could be used are the relative locations of studied genes and loci on the genome, based on which the *cis*-acting QTLs can be detected with greater confidence. By chance, the expected number of linked QTLs located within a range around a gene is

$$p_{cis} = \frac{\alpha}{n_b},$$

where α is the genomewide significant level and $n_b = \frac{\text{total genome length}}{\text{range width}}$ is the number of genomic bins. p_{cis} is small by definition of *cis*-acting QTLs. If more significant QTLs are detected as *cis*-acting around their genes, such as the output shown by Carlborg *et al.*,[22] most of them are detected due to a true underlying *cis*-acting control/connect mechanism. Considering the fact of repeatability, the detected *cis*-acting QTLs in a high-repeatability group are very likely to be true.[22] Actually, in the paper, the authors found that most detected QTLs had higher confidence and were also *cis*-acting. This coincidence suggests that either most linked QTLs are *cis*-acting in their study or *cis*-acting QTLs could have a relatively stronger effect to transcripts causing higher repeatability.[22]

The incorporation of information from multiple sources is a potential way of increasing detection ability, but we should be careful about which information is used. There is discussion in the literature on this issue. Jansen and Stam[49] suggested using data on the parental lines and the F1 population in QTL mapping; Franken *et al.*[50] used repeated measurements on BXD lines as independent observations in a QTL study for sleep regulation. However, Carlborg *et al.*[22] showed by simulation that incorporating this might increase the type I error in eQTL mapping.

2.8. Gene Regulatory Network Reconstruction

A biological network can be viewed as a complex system involving a large number of components and activities, such as DNA, protein, messenger RNA, etc. Gene regulatory networks consist of genes that interact with their regulatory elements and other genes in a cell. To study these complex interactions and reconstruct a network, systematical perturbations of genes are needed. Some of the current techniques for creating these perturbations, such as chemical or siRNA inhibition, can introduce additional off-target effects and therefore need additional experiments for validation.[51] Current genetical genomics studies[47,52] suggested that naturally occurring genetic variations derived from segregating populations are good candidates for uncovering the underlying gene regulatory networks.

Microarray-based regulatory network reconstruction approaches are based only on expression data obtained from perturbation experiments designed for detecting gene–gene interactions or coregulation of multiple genes.[53] Unfortunately, microarray data sets themselves may not provide adequate information to infer network structure.[54,55] In 2001, when Jansen and Nap first introduced genetical genomics, they also proposed the basic regulatory network reconstruction procedures using integrative information.[5] Recently, several regulatory network reconstruction studies employing QTL analysis and eQTL analysis to identify candidate genes and regulatory elements have been published.[54,56,57] The main idea of these studies is to use graphical representation to integrate genotypes, gene expression, and clinical trait data in order to elucidate the causal relationships in gene networks. The basic strategy is as follows: (1) perform eQTL mapping and QTL mapping of the interested traits; (2) construct a confidence interval for each significant eQTL/QTL locus, and do fine mapping if necessary; (3) with the help of gene annotation data, identify the candidate genes by looking at the DE genes localized within the common eQTL and QTL intervals, or use other comparable approaches to reduce the number of candidate genes and regulatory elements (see below); and (4) infer regulatory networks exploiting the dependencies among the candidate genes using existing network reconstruction algorithms, e.g. infer a probabilistic network model using Bayesian networks.[58]

The genetical genomics approach provides a unique way to infer associations between genes and traits (i.e. transcription levels). The DNA variations causing the variations of traits imply a causal relationship between them, thus reducing the ambiguity regarding the flow of information.[55] To discover the causal relationship between genes and traits is not only the first necessary step, but also one of the major challenges in building gene networks. Several recent studies suggested using *cis*-acting eQTL trait genes, which are colocalized within the QTL region associated with other complex traits (such as disease), as causal candidate genes for these traits.[5,47,52,55]

Due to the strong linkage disequilibrium (LD) in experimental populations, the chromosomal regions identified from QTL or eQTL mapping analysis are usually very large and could harbor hundreds of genes.[55] Different criteria have been used to reduce the number of identified genes within a putative chromosomal region.[54,56,57,59] Schadt *et al.*[59] developed a likelihood-based causality model selection (LCMS) test to infer relationships (causal, reactive, and independent) between genes and traits. Other challenges also exist in gene network reconstruction. The popularly used Bayesian network modeling approach has some limitations, such as not allowing cyclic edges, while negative feedback is one of the well-known mechanisms in biological systems.[55,58] Although the single-gene perturbation method has been used to validate the resulting networks,[55,59] the development of systematic approaches to reconstruct and test gene regulation networks, both statistically and biologically, remains a major challenge.

3. CASE STUDIES

In order to substantiate the power of the genetical genomics approach, researchers have conducted QTL and eQTL analyses in a number of organisms (including humans) to address many fundamental biological questions. For example, it is well known that genes are inherited genetic material and that they control traits such as height and weight. But, is gene expression level also inherited and therefore determined by genotype? If so, what is the mechanism that determines a gene's expression level? Additional questions to be answered revolve around the interpretation of results from the analyses. If standard QTL analysis is performed with gene expression as the phenotypic trait, what may we conclude about the gene from the analysis? How does eQTL analysis uncover gene regulatory networks and inform us on the biological mechanisms underlying complex disorders? And, lastly, what type of population is suitable for eQTL analysis?

In this section, we will review some recent studies aimed at addressing these biologically meaningful questions. The eQTL analyses have been done in many organisms, mostly in yeast, mouse, and human, with one common primary goal: to dissect complex quantitative traits (i.e. transcription levels) using a genetic approach. First, we will start with one of the initial eQTL studies in yeast by Brem *et al.*[47] and one of their follow-up studies addressing interactions between polymorphisms.[60] Next, we will review some analyses done in human by Cheung *et al.*[61] (who used both linkage and association approaches) and Monks *et al.*[62] (who used genetic correlation rather than Pearson correlation to investigate the relationship among genes). Then, we will use the analyses done in maize, mouse, and human by Schadt *et al.*[52] to explain an approach using integrated clinical trait information to facilitate eQTL mapping. Lastly, we will review a couple of gene regulatory network reconstruction studies in mouse and yeast, as one of the most important applications in this field.[56,57,59]

The first comprehensive eQTL study appeared to be reported by Brem *et al.*,[47] using yeast as the model organism. Based on evidence that an organism's genotype affects its gene expression,[63,64] Brem *et al.*[47] compared the gene expression profiles of two strains of *Saccharomyces cerevisiae*, a standard laboratory strain and a wild strain from a California vineyard, to investigate the underlying genetic causes. The authors then crossed these two strains and compared the expression profiles of the offspring to the parents in order to inspect the heritability of the transcript abundances. The parental strains showed that nearly half of all 6215 genes in the yeast genome were differentially expressed, and that approximately 84% of the change in parental expression was inherited by the resulting cross.

To discover what controls this variation in gene expression, Brem *et al.*[47] carried out genomewide linkage analysis between the markers and the transcript levels. A total of 570 transcripts had significant linkage to at least one locus ($p < 5 \times 10^{-6}$). Interestingly, approximately half of these linkages were found for loci that were DE and half for those that were not DE. The authors suggested that this discrepancy could be explained by the expected false-positives in the analysis, the study design, and the inheritance of parental alleles with opposite effect. On the other hand, 1220 genes that showed DE in the parental strains did not show linkage in the cross; this is quite different from the expected number of 22 false-positives. To understand this difference, the authors performed simulation studies to predict the number of linkages that would be detectable if a single locus or multiple loci caused the differences in gene expression. Their results showed that 97% of linkages would be found if differences in expression were caused by a single locus, but only 29% would be found if multiple loci were involved. Therefore, given that only 20% (208/1528) were identified in the experimental data, they concluded that most genes are affected by multiple loci.

Further simulations were performed to find the most likely number of equipotent loci that fit their data. They varied the number of loci from one to five and calculated the ability to detect linkage. The ability varied from 82% for a two-loci model to 39% for a five-loci model. Hence, after considering the variability of expression differences and gene effects, they concluded that a multilocus model of at least five loci may be responsible for affecting each DE gene, though the number is certainly variable across genes.

To understand the mechanism of the loci identified by the linkage results, the authors plotted the linkages to their respective genome locations. This method allows determining whether the loci are *cis-* or *trans-*acting ones. If the linkages are located in the same 10-kb window of the genome as the gene itself, then it is said that the gene is linked to itself — hence, *cis*-acting — and is evidence that a polymorphism within the gene affects the gene's expression. Thirty-two percent of the linkages were *cis*-acting. To identify the *trans*-acting regulators of expression, the authors then divided the genome into 20-kb regions and mapped the linkages as described above. Statistically, if the linkages are independent of each other, there should not be a bin with more than five members. The authors' data found eight regions with more than five members and evidence of *trans*-acting regulators in each of the eight groups. They were able to propose six genes as putative *trans*-acting regulators from related biological functions. Therefore, they concluded that these regions have a widespread effect on the expression of genes located distally to them and are *trans*-acting.[47]

This study was the first to offer a quantification of the number of loci involved in controlling gene expression. The surprising number of significant linkages for non-DE genes hints at the statistical issues that must be addressed when dealing with large numbers of statistical tests. While they did not provide biological examples of identified *cis*-acting loci, many examples of *trans*-acting loci were provided.

Interactions have been shown to exist among QTLs or between genes and environmental factors; they can affect the power to dissect complex quantitative traits in many organisms.[65] Brem and Kruglyak[66] used the genetical genomics approach to explore the naturally occurring genetic epistasis effect between a pair of loci in a cross between two strains of *Saccharomyces cerevisiae*. The data set consisted of 112 segregants, 2957 markers, and 5727 genes. Since using all possible pairs of loci to test interactions brought few significant results due to the multiple-testing adjustment, they used a two-stage search strategy to improve the statistical power. For each quantitative trait (transcript level), they first scanned the genome to obtain a primary locus with the most significant individual effect using Wilcoxon rank-sum tests at each locus; then, they partitioned the segregants into two subgroups based on the inheritance (genotype) at the primary locus. The locus with the highest test statistic among either subgroup could be considered as the secondary locus when tested together with the primary locus fitting model $t = ax + by + cxy + d$ across all segregants, where t is the $\log_2$ of the ratio of expression levels between the interested strain and the reference sample, and x and y are the inheritance at the primary locus and secondary locus, respectively. The above full model was compared to the reduced model $t = ax + by + d$ to test the hypothesis that $c = 0$ by the standard F-test for each locus.

To assess the test significance cut-off and estimate the FDR, they randomly permuted the expression traits to form the pooled null statistics across all transcripts. The observed and null statistics were used to estimate the probability that a transcript is false-positive, given that it is called significant with either primary or secondary loci (or both). With a 5% FDR, significant loci pairs were identified for 225 expression traits. If only testing with individual effects, 67% of the secondary loci of those significant traits could not be detected.

Cheung *et al.*[61] conducted a genomewide scan of gene expression traits based on cells obtained from the lymphoblastoid cells of 35 unrelated CEPH individuals. Their goal was to investigate the role of DNA polymorphisms in affecting gene expression. They compared the gene ontology (GO) categories of genes with the highest and lowest variable gene expression, and found that the category classification differed between the two types. Next, they looked at the variation of five highly variant genes in three groups (unrelated individuals, siblings, and monozygotic twins) in order to assess their heritabilities. They found that gene expression became more variable as the degree of relatedness decreased. Therefore, there is a heritable component to gene expression that can be exploited with genetic analysis.

Given this result, Morley *et al.*[67] performed genomewide linkage analysis on 94 unrelated individuals from the CEPH population to find regulators of the inherited gene expression. The regulators were defined as regions of the genome linked to the expression level, and were determined to be *cis-* or *trans*-acting depending on their distance to their target gene (greater or less than 5 Mb). The majority of regulators were found to be *trans*-acting. Looking at the distribution of the regulators across the genome, they found two main hotspots (master regulators): one with 31 genes, and the other with 25 genes. The common location suggests a coregulated expression of genes linked to the regulators. The authors then clustered the genes linked to these hotspots based on their correlation. They found that the hotspot with 31 genes contained a cluster of 14 genes with similar expression; the hotspot with 25 genes contained three clusters — one with four genes, and two with two genes each. They concluded that their similar correlation is a sign that they share common transcriptional regulators, but the region identified by linkage mapping was too large to rule out potential finergrain regulation.[67] The authors also confirmed the presence of *cis* linkages by typing additional single nucleotide polymorphisms (SNPs) in the area of the *cis* linkage and performing the quantitative transmission disequilibrium test (QTDT).[68] The test confirmed combined linkage of the SNP and gene expression, and allelic association for 82% of the genes followed up. This further supports the idea of *cis*-acting regulation due to allele effect, and provides evidence for the use of the QTDT as a means of identification.

Cheung *et al.*[61] continued the work by conducting an association study for the *cis*-linked genes in 57 individuals from the CEPH population. They looked to confirm the previous linkage with association. Note, however, that linkage is determined by the cosegregation of a marker and a locus, and does not rely on specific alleles of the gene to be associated with an SNP; whereas association study does consider specific alleles shared — therefore, the results may differ between linkage and association studies. Of the 27 strongest linkages found by Morley *et al.*,[67] Cheung *et al.*,[61] using genomewide association analysis, found that only 14 showed significant association. This difference can be attributed to allelic heterogeneity or the small sample size. They predicted that at least 500 samples would be needed to identify a locus responsible for the small phenotypic variance associated with complex diseases. The authors did identify a trend that more significant linkages tend to show association compared to less significant linkages. An advantage of association studies is that the candidate regions identified by association are smaller than those from linkage.

In a separate study, Monks *et al.*[62] measured gene expression in 23 499 genes in 15 CEPH families with a total of 210 individuals to study the inheritance of gene expression in humans.[68] The increase in sample size allowed for enough power to identify major loci that affect gene expression in a population of nonrelated individuals. Of the 23 499 genes, 2430 were DE, of which 31% were shown to be heritable. The annotation of these genes showed an enrichment in immune-related functions. Linkage analysis of the 2430 DE genes showed that 20 had significant eQTLs. But, contrary to mouse and yeast results, when linkage results were mapped to their genomic positions, no hotspot was found; the authors suggested that this may be due to the random population structure of the samples. The mice strains used by Schadt *et al.*[52] were bred to manifest certain phenotypes; their resulting cross was therefore a mixture of two extremes, potentially leading to more significant QTLs than found in a random population. In contrast to other studies, when determining which genes should be followed up in their analysis, Monks *et al.*[62] chose genes based on DE, significant linkage, and heritability measures (determined from variance components methods). The goal of their follow-up analysis was to determine if Pearson correlation or genetic correlation was more informative as a means to measure the common genetic determinants between genes. They found that cluster analysis based on genetic correlation rather than Pearson correlation resulted in clusters which more closely matched the known pathways in KEGG.

Schadt *et al.*[52] applied the eQTL approach in mice to elucidate expressions associated with a disease state. Their study concentrated on the heritability of gene expression profiles of liver tissue from the F2 mice from two inbred mouse strains (one known to be obese, the other lean) that had been fed a high-fat diet. To identify genes of interest, they examined the number of DE and non-DE genes as well as the distribution of LOD scores between them. The analysis showed that 33% (7861/23 574) of the genes showed DE in either the parental or 10% of F2 mice, and 9% (2123) of the genes had QTLs with an LOD score greater than 4.3. Interestingly, when looking at all of the genes, not just the DE ones, 18% (4339) of the genes had LOD scores greater than 4.3; while twice as many genes that were not DE had LOD scores greater than 7.0. Therefore, using all of the genes in the analysis provides more information about the biological processes being investigated; non-DE genes should not be ignored. Genes associated with high LOD scores but nonsignificant expression equate to genes that are strictly regulated. The authors explained that DE is not needed in order for a gene to be informative of an underlying process; rather, similar expression of a gene by members of the same population with a similar genotype represents common underlying biological processes.

To distinguish between *cis*- and *trans*-acting loci, they, too, plotted the eQTLs to their locations on the genome. They used 2-cM bins and found similar genomic regions with significantly more eQTLs than would have been predicted by chance. Interestingly, eQTLs with higher LOD scores were more likely to be mapped to the physical location of the corresponding gene. Their conclusion was that eQTLs with highly significant LOD scores (greater than 7.0) are more likely to be *cis*-acting than those with lower, yet still significant, LOD scores (greater than 4.3). Additionally, Schadt *et al.*[52] were able to validate the predicted *cis*-acting loci by locating known expression-affecting polymorphisms in the predicted *cis*-acting loci. Hence, high LOD scores were predictive of the location of the loci affecting the gene. This information can also be used to better identify regions for positional cloning.

With an understanding of the expression differences and loci involved, Schadt *et al.*[52] integrated the expression data and genotypic data with phenotypic data (e.g. measures of obesity) from the same mice. Using the expression and clinical phenotypic data (subcutaneous fat pad mass [FPM]), they were able to categorize genes via two-dimensional clustering into three sets: two sets associated with high FPM, but with different expression patterns; and a low-FPM set. Furthermore, those genes that made up the DE genes were highly represented in the regions of the genome containing greater than expected loci. This overlap of highly expressed genes in genomic hotspots allowed greater insight into the identification of the genes responsible for the FPM trait, several of which the authors validated experimentally. In the same paper, Schadt *et al.* demonstrated a similar inheritance of gene expression in maize and humans. The human samples were lymphoid cell lines from four pedigrees from the CEPH population. DE was found in 2726 genes between family founders; but the sample, which contained only 56 individuals, was not enough for QTL analysis.

The discovery of gene regulatory networks is one of the primary goals in biological research. The genetical genomics approach, which integrates gene expression data, genotypic data, and physiological traits, is a powerful and efficient approach to network reconstruction. Zhu *et al.*[57] introduced an extended Bayesian approach that incorporates genetic information into the gene regulatory network reconstruction. Integrating the gene expression data and genotypic data from the study of Schadt *et al.*[52] described above, they were able to demonstrate the ability of the derived network to capture causal associations between genes with higher discriminating power compared to the reconstructed networks using classical algorithms based on only gene expression data. In their approach, they first used the extent of genetic overlap to compute the correlation between any two trait genes (i.e. genes whose transcript levels are used as quantitative traits), and then they selected significant potential parent nodes for each gene. If two genes are highly correlated and controlled by a similar set of loci, their eQTLs should overlap. Using the extent of overlap between vectors of LOD scores associated with the identified eQTLs over entire chromosomes for each trait gene, and assuming no epistasis, a similarity measure of gene relatedness could be defined as $r = \sum w(c) * r(c)$, where $r(c)$ is the correlation coefficient for chromosome c and $w(c)$ is the chromosome-specific weight. Only genes above a certain threshold were retained for further analysis.

To further assess the association among gene expression and differentiate pleiotropy vs. multiple closely linked QTLs, they defined a mutual information (MI) measure as

$$\text{MI}(X, Y) = \sum_{i,j} p(x_i, y_j) * \log\left(\frac{p(x_i, y_j)}{[p(x_i) * p(y_j)]}\right),$$

where $p(x)$ is the probability density function for the expression level of gene X. Similar to the genetic relatedness test, only genes above a certain threshold can be chosen as potential candidate genes. Then, they used the information from eQTL overlap to infer causality by defining the prior for a candidate relationship as

$$p(X \rightarrow Y) = r(X, Y) * \frac{N(Y)}{N(X) + N(Y)}.$$

The posterior probability of a graphical model, M, given a gene expression data set, D, can be defined as

$$P(M \mid D) \propto P(D \mid M) P(M),$$

where the prior probability is

$$P(M) = \prod_{X \rightarrow Y} p(X \rightarrow Y).$$

Finally, they used Friedman's local maximum search algorithm to reconstruct Bayesian networks.[57,58] Based on the 1088 genes, they reconstructed 1000 networks. A consensus network was obtained by identifying the links existing in more than 40% of the 1000 networks. Links in the consensus networks were associated with confidence values, which were derived from the number of times they appeared. To further simplify the relationship among genes and reduce the overfit, some links can be removed by using conditional MI, defined as

$$\mathrm{MI}(X, Y \mid Z) = \sum_{i,j,k} p(x_i, y_j, z_k) * \log \left(\frac{p(x_i, y_j \mid z_k)}{p(x_i \mid z_k) * p(y_j \mid z_k)} \right).$$

The link between X and Y should be removed if the conditional MI is not significantly different from 0.

Li *et al.*[54] also extended Bayesian network reconstruction to infer gene transcriptional modulatory relations using a mouse data set. They started by building the correlation networks using only microarray data, and then used QTL intervals to construct potential candidate networks. To further reduce the complexity of networks, the authors considered eliminating those genes within QTL intervals that were identical by descent based on the density and distribution of SNPs, which distinguished the two progenitor strains. The resulting networks were called "QTL-SNP–derived candidate networks". Finally, they also used Bayesian methods to model the networks.[54]

Bing and Hoeschele[56] proposed another approach to reconstructing regulatory networks in a yeast data set containing 40 segregants, 6215 genes, and 3312 markers.[47] They first performed genomewide QTL analysis of all expression profiles to identify eQTL confidence regions using a bootstrap resampling method, followed by fine mapping for some identified eQTL intervals. Next, they identified potential regulatory candidate genes in each eQTL region. For each gene included in the confidence interval of a given trait gene, the Spearman correlation coefficient was computed. Then, they constructed regulatory networks by drawing directed edges from each candidate gene to the trait gene using a structural equation model that allowed cyclic edges.[56] There were 768 putative regulatory links (including 721 genes) with different types of interactions identified. Many biological function–related genes, such as those involved in protein synthesis, were found in the same network and loosely connected to other networks. Most of the *trans*-acting regulators found in the previous study[47] were also identified to regulate the corresponding groups of genes.[56]

4. FUTURE DIRECTIONS

As a novel approach to the dissection of complex traits, genetical genomics, even in its early phase, has shown success and has been applied to many model organisms to unravel causal relationships among genes, expression traits, and phenotypes. We believe that the directions of potential growth depend on developments in the following related areas: (1) high-throughput genotyping technologies and the integration of multiple data resources such as quantitative traits; (2) network reconstruction; and (3) new statistical methods and corresponding software applications.

First, gene expression profiling data using microarray (transcriptomics) technology have been used as quantitative phenotypes in genetical genomics analysis. Theoretically, the methods developed for multiple quantitative traits can be similarly applied to other quantitative trait measurements, such as proteomics and metabolomics data, which provide more measures of the physiological state of the cell. The integration of transcriptomics, proteomics, and metabolomics data, along with other pathology data, may provide more information and be helpful in dissecting complex disease traits. Note that the -omics technologies mentioned above take only snapshots of the dynamic activities in a cell. The measurements could vary with the different living conditions of a cell and different tissue types, which may pose some difficulties in repeating the experiments. To account for these fluctuations, more replicates and study designs under various experimental conditions will be needed to obtain consistent results. It is then that we will be able to capture the detailed picture of living organisms. Meanwhile, genotyping technology has advanced dramatically in the past few years, and a number of protocols are now available.[69–72] These array-based genotyping platforms can simultaneously handle tens of thousands of SNPs with a very low error rate, which brings down the experimental cost significantly compared to the traditional gel-based genotyping approaches. In addition, other types of data, such as protein–DNA interaction data,[73] may also provide useful information on regulatory networks. Therefore, physiological (clinical) traits and information from high-throughput profiling approaches, such as data generated by the -omics technologies, can be integrated together to conduct genomewide genetic analysis for an individual at both gene and gene product levels with the availability of high-density marker data. This integrative approach is promising in identifying the key drivers underlying complex traits and in unraveling the associated biological pathways.[8,57,59]

The use of a graphical model inferred from gene expression data to study gene regulatory networks has been demonstrated by many published studies.[58,74,75] The genetical genomics approach has the potential power to uncover the relationship between causal genetic elements and physiological (clinical) traits. As suggested in several studies,[47,52,59,76] naturally occurring genetic variations in segregating populations can be used as good candidates for the systematic perturbation of gene regulatory networks, which is especially meaningful for dissecting disease and drug response traits.[76] However, the gene regulatory networks discussed here are only a graphical representation of the complex dynamic cell activities based on gene transcription data. Different types of networks, such as metabolic and signaling networks, exist and interact with one another. It is obvious that gene regulatory networks by themselves are not sufficient to explain a living system completely. Integrating multiple data resources and considering the interactions among different networks could provide more information to elucidate the complex dynamic systems. Most of the existing network reconstruction approaches in the genetical genomics field are extensions of classical algorithms.[54,57] No matter what the algorithm is, identifying the optimum network from experimental data is often computationally intensive. Therefore, the number of genes identified within a QTL or eQTL region needs to be minimized to maximally simplify the graphic structure searching process. Approaches such as examining the correlation among identified genes and colocalizing the genes with QTLs associated with physiological traits are often used. Furthermore, using other biological knowledge extracted from a variety of existing databases could facilitate network searching and the construction process. All of these procedures aim to further narrow down the identified genes to make the network reconstruction procedures feasible. The validation of identified candidate genes is also recommended after network reconstruction. Schadt *et al.*[59] used single-gene perturbation experiments and validated the constructed network biologically. Meanwhile, more work needs to be done to assess network confidence using statistical methods.[77]

Some computer programs have been developed for eQTL data visualization and analysis, such as eQTL Explorer,[78] and they have been used in eQTL analysis. Some widely used analysis tools for traditional QTL analysis, such as R/qtl, QTL Express, MAPMAKER/QTL, QTL Cartographer, QTL Express, and WebQTL, have been either directly applied to or extended with an eQTL analysis feature as well.[79–84] However, the development of new methodologies is still in great demand due to the statistical and computational challenges. Note that the methods mentioned in Sec. 2 have not been widely applied in biological studies yet. Most of the published papers used only simple methods to identify QTLs for expression traits.[47] In order to recommend the appropriate methods for use in different scenarios, the presented methods (as well as any novel method yet to be developed) need to be systematically compared and evaluated. An even bigger challenge is to understand the role of the environmental factor in regulating expression levels. An environmental effect, as well as gene–environment interactions or gene–gene interactions, is now recognized as a rule instead of an exception in complex disease etiology,[65] and therefore should be considered in the development of new statistical models. Another challenge that all genetic fields are facing, but is most prominent in the eQTL field, is the multiple testing problem. Although the FDR approach, which has been applied in some studies, seems to be most appropriate, there are still many related issues that need to be addressed in new implementations, as summarized in the previous section. Last but not least, to lessen the computation load that is often involved in eQTL analysis, parallel computing using computer clusters — which has been widely used in many scientific fields — could be utilized. Implementing parallel programs and distributing the heavy computing load to clustered processors can significantly decrease computing time and make tasks like large-scale simulations and permutation tests feasible.[22] When new efficient computational methods are developed, the resulting performance needs to be systematically tested through intensive simulations and applied to real data.

ACKNOWLEDGMENTS

This research was supported in part by the NIH grant GM 59507 and the NSF grant DMS-0241160. D. Ballard was supported by the NIH Institutional Training Grants for Informatics Research.

REFERENCES

1. Lander ES, Linton LM, Birren B, *et al.* (2001) Initial sequencing and analysis of the human genome. *Nature* **409**(6822):860–921.
2. Venter JC, Adams MD, Myers EW, *et al.* (2001) The sequence of the human genome. *Science* **291**(5507):1304–1351.
3. Sinsheimer RL. (1989) The Santa Cruz Workshop — May 1985. *Genomics* **5**(4):954–956.
4. Schena M, Shalon D, Davis RW, Brown PO. (1995) Quantitative monitoring of gene expression patterns with a complementary DNA microarray. *Science* **270**(5235):467–470.
5. Jansen RC, Nap JP. (2001) Genetical genomics: the added value from segregation. *Trends Genet* **17**(7):388–391.
6. Alberts R, Terpstra P, Bystrykh LV, *et al.* (2005) A statistical multiprobe model for analyzing *cis* and *trans* genes in genetical genomics experiments with short-oligonucleotide arrays. *Genetics* **171**(3):1437–1439.
7. King MC, Wilson AC. (1975) Evolution at two levels in humans and chimpanzees. *Science* **188**(4184):107–116.

8. Schadt EE, Monks SA, Friend SH. (2003) A new paradigm for drug discovery: integrating clinical, genetic, genomic and molecular phenotype data to identify drug targets. *Biochem Soc Trans* **31**(2):437–443.
9. Wu TD. (2001) Analysing gene expression data from DNA microarrays to identify candidate genes. *J Pathol* **195**(1):53–65.
10. Lockhart DJ, Dong H, Byrne MC, *et al.* (1996) Expression monitoring by hybridization to high-density oligonucleotide arrays. *Nat Biotechnol* **14**(13): 1675–1680.
11. Rocke DM, Durbin B. (2001) A model for measurement error for gene expression arrays. *J Comput Biol* **8**(6):557–569.
12. Kadarmideen HN, von Rohr P, Janss LL. (2006) From genetical genomics to systems genetics: potential applications in quantitative genomics and animal breeding. *Mamm Genome* **17**(6):548–564.
13. Rubin DB. (1987) *Multiple Imputation for Nonresponse in Surveys.* Wiley, New York, NY.
14. Troyanskaya O, Cantor M, Sherlock G, *et al.* (2001) Missing value estimation methods for DNA microarrays. *Bioinformatics* **17**(6):520–525.
15. Oba S, Sato MA, Takemasa I, *et al.* (2003) A Bayesian missing value estimation method for gene expression profile data. *Bioinformatics* **19**(6):2088–2096.
16. Kim H, Golub GH, Park H. (2005) Missing value estimation for DNA microarray gene expression data: local least squares imputation. *Bioinformatics* **21**(2):187–198.
17. Korol AB, Ronin YI, Itskovich AM, *et al.* (2001) Enhanced efficiency of quantitative trait loci mapping analysis based on multivariate complexes of quantitative traits. *Genetics* **157**(4):1789–1803.
18. Lander ES, Botstein D. (1989) Mapping Mendelian factors underlying quantitative traits using RFLP linkage maps. *Genetics* **121**(1):185–199.
19. Korol AB, Ronin YI, Kirzhner VM. (1995) Interval mapping of quantitative trait loci employing correlated trait complexes. *Genetics* **140**(3):1137–1147.
20. Jiang C, Zeng ZB. (1995) Multiple trait analysis of genetic mapping for quantitative trait loci. *Genetics* **140**(3):1111–1127.
21. Lan H, Stoehr JP, Nadler ST, *et al.* (2003) Dimension reduction for mapping mRNA abundance as quantitative traits. *Genetics* **164**(4):1607–1614.
22. Carlborg O, De Koning DJ, Manly KF, *et al.* (2005) Methodological aspects of the genetic dissection of gene expression. *Bioinformatics* **21**(10):2383–2393.
23. Zeng ZB. (1994) Precision mapping of quantitative trait loci. *Genetics* **136**(4):1457–1468.
24. Mahler M, Most C, Schmidtke S, *et al.* (2002) Genetics of colitis susceptibility in IL-10–deficient mice: backcross versus F2 results contrasted by principle component analysis. *Genomics* **80**:274–282.
25. Storey JD, Akey JM, Kruglyak L. (2005) Multiple locus linkage analysis of genomewide expression in yeast. *PLoS Biol* **3**(8):e267.
26. Churchill GA, Doerge RW. (1994) Empirical threshold values for quantitative trait mapping. *Genetics* **138**(3):963–971.

27. Anderson JA, Blair V. (1982) Penalized maximum likelihood estimation in logistic regression and discrimination. *Biometrika* **69**(1):123.

28. Green PJ. (1994) *Nonparametric Regression and Generalized Linear Models*. Chapman & Hall/CRC, New York, NY.

29. Storey JD, Tibshirani R. (2003) Statistical significance for genome-wide studies. *Proc Natl Acad Sci USA* **100**:9440–9445.

30. Kendziorski C, Chen M, Yuan M, *et al.* (2004) Statistical methods for expression trait loci (ETL) mapping. Technical Report No. 184, Department of Biostatistics and Medical Informatics, University of Wisconsin–Madison, Madison, WI.

31. Kendziorski CM, Newton MA, Lan H, Gould MN. (2003) On parametric empirical Bayes methods for comparing multiple groups using replicated gene expression profiles. *Stat Med* **22**:3899–3914.

32. Newton MA, Noueiry A, Sarkar D, Ahlquist P. (2004) Detecting differential gene expression with a semiparametric hierarchical mixture method. *Biostatistics* **5**(2):155–176.

33. Carlin B, Louis T. (1998) *Bayes and Empirical Bayes Methods for Data Analysis*. Chapman & Hall, New York, NY.

34. Haley CS, Knott SA. (1992) A simple regression method for mapping quantitative trait loci in line crosses using flanking markers. *Heredity* **69**(4): 315–324.

35. Meng XLI, Rubin DB. (1993) Maximum likelihood estimation via the ECM algorithm: a general framework. *Biometrika* **80**(2):267.

36. Sen S, Churchill GA. (2001) A statistical framework for quantitative trait mapping. *Genetics* **159**:371–387.

37. Coffman CJ, Doerge RW, Wayne ML, McIntyre LM. (2003) Intersection test for single marker QTL analysis can be more powerful than two marker QTL analysis. *BMC Genet* **138**:963–971.

38. Kao CH, Zeng ZB, Teasdale RD. (1999) Multiple interval mapping for quantitative trait loci. *Genetics* **152**(3):1203–1216.

39. Carlborg O, Andersson L, Kinghorn B. (2000) The use of a genetic algorithm for simultaneous mapping of multiple interacting quantitative trait loci. *Genetics* **155**(4):2003–2010.

40. Goldberg DE. (1989) *Genetic Algorithms in Search, Optimization and Machine Learning*. Addison-Wesley Longman, Boston, MA.

41. Carlborg O. (2002) New methods for mapping quantitative trait loci. PhD dissertation, Department of Animal Breeding and Genetics, Swedish University of Agricultural Sciences, Uppsala, Sweden.

42. Ljungberg K, Holmgren S, Carlborg O. (2004) Global optimization in QTL analysis. *The 8th Annual International Conference on Research in Computational Molecular Biology*, March 27–31, 2004, San Diego, CA.

43. Ljungberg K, Holmgren S, Carlborg O. (2004) Simultaneous search for multiple QTL using the global optimization algorithm DIRECT. *Bioinformatics* **20**(12):1887–1895.

44. Jones DR, Perttunen CD, Stuckman BE. (1993) Lipschitzian optimization without the Lipschitz constant. *J Optim Theory Appl* **79**(1):157–181.

45. Carlborg O, Andersson-Eklund L, Andersson L. (2001) Parallel computing in interval mapping of quantitative trait loci. *J Hered* **92**(5):449–451.
46. Haley CS, Knott SA, Elsen JM. (1994) Mapping quantitative trait loci in crosses between outbred lines using least squares. *Genetics* **136**(3): 1195–1207.
47. Brem RB, Yvert G, Clinton R, Kruglyak L. (2002) Genetic dissection of transcriptional regulation in budding yeast. *Science* **296**(5568):752–755.
48. Doerge RW, Churchill GA. (1996) Permutation tests for multiple loci affecting a quantitative character. *Genetics* **142**(1):285–294.
49. Jansen RC, Stam P. (1994) High resolution of quantitative traits into multiple loci via interval mapping. *Genetics* **136**(4):1447–1455.
50. Franken P, Chollet D, Tafti M. (2001) The homeostatic regulation of sleep need is under genetic control. *J Neurosci* **21**(8):2610–2621.
51. Jackson AL, Bartz SR, Schelter J, *et al.* (2003) Expression profiling reveals off-target gene regulation by RNAi. *Nat Biotechnol* **21**(6):635–637.
52. Schadt EE, Monks SA, Drake TA, *et al.* (2003) Genetics of gene expression surveyed in maize, mouse and man. *Nature* **422**(6929):297–302.
53. de Jong H. (2002) Modeling and simulation of genetic regulatory systems: a literature review. *J Comput Biol* **9**(1):67–103.
54. Li H, Lu L, Manly KF, *et al.* (2005) Inferring gene transcriptional modulatory relations: a genetical genomics approach. *Hum Mol Genet* **14**(9):1119–1125.
55. Lum PY, Chen Y, Zhu J, *et al.* (2006) Elucidating the murine brain transcriptional network in a segregating mouse population to identify core functional modules for obesity and diabetes. *J Neurochem* **97**(Suppl 1):50–62.
56. Bing N, Hoeschele I. (2005) Genetical genomics analysis of a yeast segregant population for transcription network inference. *Genetics* **170**(2):533–542.
57. Zhu J, Lum PY, Lamb J, *et al.* (2004) An integrative genomics approach to the reconstruction of gene networks in segregating populations. *Cytogenet Genome Res* **105**(2–4):363–374.
58. Friedman N, Linial M, Nachman I, Pe'er D. (2000) Using Bayesian networks to analyze expression data. *J Comput Biol* **7**(3–4):601–620.
59. Schadt EE, Lamb J, Yang X, *et al.* (2005) An integrative genomics approach to infer causal associations between gene expression and disease. *Nat Genet* **37**(7):710–717.
60. Brem RB, Storey JD, Whittle J, Kruglyak L. (2005) Genetic interactions between polymorphisms that affect gene expression in yeast. *Nature* **436**(7051):701–703.
61. Cheung VG, Conlin LK, Weber TM, *et al.* (2003) Natural variation in human gene expression assessed in lymphoblastoid cells. *Nat Genet* **33**(3): 422–425.
62. Monks SA, Leonardson A, Zhu H, *et al.* (2004) Genetic inheritance of gene expression in human cell lines. *Am J Hum Genet* **75**(6):1094–1105.
63. Primig M, Williams RM, Winzeler EA, *et al.* (2000) The core meiotic transcriptome in budding yeasts. *Nat Genet* **26**(4):415–423.

64. Sandberg R, Yasuda R, Pankratz DG, *et al.* (2000) Regional and strain-specific gene expression mapping in the adult mouse brain. *Proc Natl Acad Sci USA* **97**(20):11038–11043.

65. Carlborg O, Haley CS. (2004) Epistasis: too often neglected in complex trait studies? *Nat Rev Genet* **5**(8):618–625.

66. Brem RB, Kruglyak L. (2005) The landscape of genetic complexity across 5,700 gene expression traits in yeast. *Proc Natl Acad Sci USA* **102**(5): 1572–1577.

67. Morley M, Molony CM, Weber TM, *et al.* (2004) Genetic analysis of genome-wide variation in human gene expression. *Nature* **430**(7001):743–747.

68. Abecasis GR, Cardon LR, Cookson WO. (2000) A general test of association for quantitative traits in nuclear families. *Am J Hum Genet* **66**(1):279–292.

69. Craig DW, Stephan DA. (2005) Applications of whole-genome high-density SNP genotyping. *Expert Rev Mol Diagn* **5**(2):159–170.

70. Fornage M, Doris PA. (2005) Single-nucleotide polymorphism genotyping for disease association studies. *Methods Mol Med* **108**:159–172.

71. De La Vega FM, Dailey D, Ziegle J, *et al.* (2002) New generation pharmacogenomic tools: a SNP linkage disequilibrium map, validated SNP assay resource, and high-throughput instrumentation system for large-scale genetic studies. *Biotechniques* (Suppl):48–50, 52, 54.

72. Wang WY, Todd JA. (2003) The usefulness of different density SNP maps for disease association studies of common variants. *Hum Mol Genet* **12**(23): 3145–3149.

73. Sun N, Carroll RJ, Zhao H. (2006) Bayesian error analysis model for reconstructing transcriptional regulatory networks. *Proc Natl Acad Sci USA* **103**(21):7988–7993.

74. Sun N, Zhao H. (2004) Genomic approaches in dissecting complex biological pathways. *Pharmacogenomics* **5**(2):163–179.

75. Schafer J, Strimmer K. (2005) An empirical Bayes approach to inferring large-scale gene association networks. *Bioinformatics* **21**(6):754–764.

76. Schadt EE. (2005) Exploiting naturally occurring DNA variation and molecular profiling data to dissect disease and drug response traits. *Curr Opin Biotechnol* **16**(6):647–654.

77. Kendziorski C, Wang P. (2006) A review of statistical methods for expression quantitative trait loci mapping. *Mamm Genome* **17**(6):509–517.

78. Mueller M, Goel A, Thimma M, *et al.* (2006) eQTL Explorer: integrated mining of combined genetic linkage and expression experiments. *Bioinformatics* **22**(4):509–511.

79. Basten CJ, Weir BS, Zeng Z-B. (1994) Zmap — a QTL cartographer. In: *The 5th World Congress on Genetics Applied to Livestock Production: Computing Strategies and Software*, Smith C, Benkel B, Chesnais J, *et al.* (eds.), Guelph, Ontario, Canada, pp. 65–66.

80. Broman KW, Wu H, Sen S, Churchill GA. (2003) R/qtl: QTL mapping in experimental crosses. *Bioinformatics* **19**(7):889–890.

81. Wang S, Basten CJ, Zeng Z-B. (2006) Windows QTL Cartographer 2.5. Department of Statistics, North Carolina State University, Raleigh, NC (http://statgen.ncsu.edu/qtlcart/WQTLCart.htm).

82. Lander ES, Green P, Abrahamson J, *et al.* (1987) MAPMAKER: an interactive computer package for constructing primary genetic linkage maps of experimental and natural populations. *Genomics* **1**(2):174–181.
83. Paterson AH, Lander ES, Hewitt JD, *et al.* (1988) Resolution of quantitative traits into Mendelian factors by using a complete linkage map of restriction fragment length polymorphisms. *Nature* **335**(6192):721–726.
84. Chesler EJ, Lu L, Wang J, *et al.* (2004) WebQTL: rapid exploratory analysis of gene expression and genetic networks for brain and behavior. *Nat Neurosci* **7**(5):485–486.

Chapter 16

Bridging Genotype and Phenotype: Causal Pathways from DNA to Complex Traits

Yan Cui

Department of Molecular Sciences
Center of Genomics and Bioinformatics
University of Tennessee Health Science Center
Memphis, TN 38163, USA

High-throughput molecular profiling technologies such as microarrays have recently been used in the genetic analysis of complex traits. Genetic loci influencing gene expression can be identified using quantitative trait locus (QTL) mapping. These genetic loci may also influence higher-order traits. The coregulation of gene expression traits and higher-order traits indicates that some of the gene expression traits may be involved in the causal pathways connecting the genetic loci and higher-order traits. Strong associations between the modules of gene expression traits, higher-order traits, and the genetic loci regulating them have been found, indicating that genetic loci may influence higher-order traits by affecting gene expression traits. In this chapter, I discuss the recent trend in combining genetic and genomic methods to analyze complex traits, and introduce a computational approach to studying the causal pathways from DNA to complex traits.

1. INTRODUCTION

Many genomic regions harboring disease-causing DNA variants have been identified using quantitative trait locus (QTL) mapping. However, the causal pathways through which the DNA variants influence disease phenotypes remain largely unknown. Recent advances in molecular profiling technologies have provided the tools to fill this gap.[1] Several research groups used microarrays to detect variations in gene expression across populations of human and model organisms.[2−10] The variations in gene expression are often heritable and can therefore be mapped as quantitative traits. This approach is called expression QTL (eQTL) mapping or transcriptome QTL mapping. The combination of genetic and genomic methods has emerged as a new field, genetical genomics,[11−13] which holds the promise of uncovering the causal pathways connecting genotype and phenotype.

For decades, recombinant inbred (RI) mice with fixed genomes have been used for complex trait analysis.[14,15] Hundreds of phenotypes of these RI mouse strains have been characterized, published, and recently collected in a database.[16] These phenotypes can be readily integrated with the newly available gene expression traits of the same RI strains. In this chapter, I illustrate the study of causal pathways using RI mice. A schematic demonstration of the causal pathway is shown in Fig. 1. In this scenario, a QTL regulates gene expression trait 1 (GET1, i.e. the expression level of gene 1), which in turn regulates gene expression traits 2 and 3 (GET2 and GET3); GET2 and GET3 regulate a higher-order trait (HT). Beside each node, there is a table showing the conditional probability distribution (CPD)

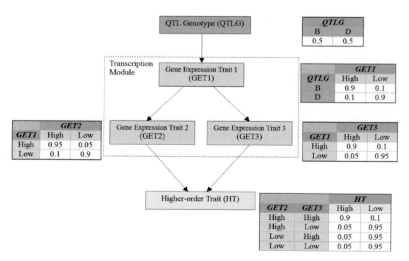

Fig. 1.   Schematic demonstration of a causal pathway. The genotype of a QTL (QTLG) influences the expression level of gene 1 (GET1), which in turn influences the expression levels of genes 2 and 3 (GET2 and GET3). Finally, the expression levels of genes 2 and 3 influence the value of a higher-order trait (HT). Beside each node, there is a table showing the conditional probability distribution (CPD) associated with the node. GET1, GET2, and GET3 belong to a transcription module.

associated with the node. Using Bayesian network methods, we can learn the structure of the causal pathway and the parameters of the CPDs from the genotype, phenotype, and gene expression data of a population. The directed edges in the causal pathway may represent direct interactions, (e.g. the binding of a transcription factor to its target gene) or indirect influences acting through a series of molecular events.

2. INFERRING CAUSAL PATHWAY MODEL FROM GENOTYPE, PHENOTYPE, AND GENE EXPRESSION DATA OF A POPULATION

2.1. *Transcription Modules Defined by Genetic Variations — Bridging Genotype and Phenotype*

A transcription module consists of a group of genes and a set of experimental conditions — the genes are coexpressed under these experimental conditions. The modular architecture of gene expression programs has been found in several model organisms.[17-20] In a recent work, we developed a mathematical model for the transcription modules caused by genetic variations.[1] Here, a transcriptional module is defined as a set of genes coregulated by allelic variations in QTLs. Twenty-nine transcription modules have been identified in mouse brain, using the microarray data of 32 BXD RI strains. The BXD strains were originally derived from two progenitor strains: C57BL/6J and DBA/2J. The genome of each BXD strain is a near-random recombination of chromosome intervals from the two progenitor strains.

The iterative signature algorithm (ISA) was developed by Ihmels and colleagues[18,21] to discover transcription modules from microarray data. In the application to gene expression traits, the ISA starts with a set of input genes, and then iteratively optimizes a gene score vector and a strain score vector calculated using two normalized expression matrices, respectively, to identify a submatrix of coexpression where the ISA converges. The two normalized expression matrices E_G and E_S are derived from the original gene expression matrix E, in which each row represents a gene and each column represents an RI strain. The row vectors of E_G and the column vectors of E_S are normalized to have zero mean and unit length. Two scores are used for optimizing the module iteratively. A randomly selected set of genes is used as the initial genes of a module. The strain score vector S_{strain} is defined as the average of the row vectors of E_G that correspond to the genes of the module. The strain score vector S_{strain} is used to select the strains of the module: strain i belongs to the module if, and only if,

$(S_{\text{strain}})_i - \langle S_{\text{strain}} \rangle > t_S \sigma_S$, where $(S_{\text{strain}})_i$ is the ith score in the strain score vector, $\langle S_{\text{strain}} \rangle$ is the average of all the scores in the strain score vector, and σ_S is the standard deviation of the strain scores. The selected set of strains is then used to calculate a gene score vector. The gene score vector S_{gene} is defined as the average of the column vectors of E_S that correspond to the strains of the module. The gene score vector S_{gene} is used to select a new set of genes: gene i belongs to the module if, and only if, $(S_{\text{gene}})_i - \langle S_{\text{gene}} \rangle > t_G \sigma_G$, where $(S_{\text{gene}})_i$ is the ith score in the strain score vector, $\langle S_{\text{gene}} \rangle$ is the average of all the scores in the gene score vector, and σ_G is the standard deviation of the gene scores. The parameters t_G and t_S determine the "compactness" of the modules. Larger t_G and t_S lead to smaller modules with more stringent coexpression, while smaller t_G and t_S lead to larger modules with less stringent coexpression.[21] The new set of genes is then used to calculate a new strain score, which in turn is used to select a new set of strains. The iteration finishes when it converges to a fixed point where the genes and strains of the module no longer change: each fixed point corresponds to a transcription module. The strain score vector S_{strain} is considered as a module signature representing the module, and can be mapped as a quantitative trait.

The transcription modules found in the gene expression data of mouse brain showed strong association with a wide range of physiological and behavioral phenotypes[1] including individual differences in response to cocaine[22] and ethanol,[23,24] acoustic startle response,[25] high-pressure seizure susceptibility,[26,27] immune responses,[28,29] age-related hearing loss,[30] and vulnerability to drug abuse.[31] A few examples are shown in Table 1. More information about the transcription modules, the associated phenotypes, and the QTLs regulating them can be found in the supplementary material of an article by Li *et al.*[1] or at the website http://moduleqtl.utmem.edu/.

Transcription modules in mouse brain associate with many neurobehavioral phenotypes, and they are often coregulated by the same QTLs. Association maps can be constructed based on the coregulation of transcription modules and physiological/behavioral phenotypes. For example, module 2 (containing 13 genes) and module 19 (containing 14 genes),[1] although not overlapping in terms of genes, are both associated with behavioral responses to cocaine[22] (Table 1). Genomewide QTL mapping showed that module 2 and the cocaine-related behavioral trait are coregulated by a QTL locating at chromosome 2 (155 Mb), and that module 19 and the cocaine activity trait are coregulated by another QTL locating at chromosome 7 (79 Mb). The coregulation indicates that the two QTLs regulate the cocaine-related

Table 1   Selected Transcription Modules and the Associated Neurobehavioral Phenotypes, and the QTLs Regulating Both the Transcription Modules and the Neurobehavioral Phenotypes

Module #	QTL Position	LOD*	Associated Phenotypes
2	Chr 2, 155 Mb	11.1	Cocaine-related exploratory behavior[22]
9	Chr 6, 146 Mb	7.2	Prepulse inhibition of the acoustic startle response[25]
10	Chr 17, 43 Mb	8.5	Seizure susceptibility to high atmospheric pressure at 1000 atmospheres [measured at atm where seizure occurred][27]
			Seizure threshold pressure at compression rate of 1000 atm per hour (1000 iPc) [atm][26]
17	Chr 6, 17 Mb	3.5	Quinine consumption — two-bottle choice[31]
19	Chr 7, 79 Mb	8.9	Dopamine transporter expression in frontal cortex[22]
			Cocaine-related exploratory behavior[22]

*LOD: log of odds.

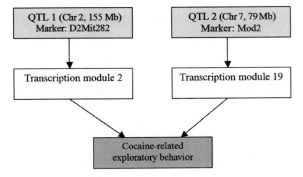

Fig. 2.   The map of association based on coregulation of the transcription modules and the cocaine-related behavioral trait.

behavioral trait through the two transcription modules, respectively. This association map (Fig. 2) can provide structural constraints for the construction of causal pathway models.

2.2. *Inferring Causal Pathways from QTLs to Phenotypes with Bayesian Network*

The structure of a Bayesian network can be learned from observational data. Thus, we can use a Bayesian network to infer causal pathways from the genotype, phenotype, and gene expression data of a population.[32–36]

2.2.1. *Bayesian Score*

A Bayesian network is a graphic probabilistic model representing the structure of dependency among multiple interacting variables.[37-39] In a Bayesian network model of a causal pathway, the variables include the values of gene expression traits and higher-order traits (e.g. disease phenotypes) as well as the genotype of the QTLs regulating them. In order to discover the causal pathway model that best matches the data set (i.e. the genotype, phenotype, and gene expression data of a population), we use a score to evaluate the posterior probability of a causal pathway G given dataset D, $S(G:D) = \log P(G \mid D)$, where P is the posterior probability.[40] In the case of a discrete Bayesian network with multinomial local CPDs, the score can be computed using a closed form equation[41]:

$$S(G:D) = \log P_0 + \sum_{i=1}^{n} \sum_{j=1}^{q_i} \log \left[\frac{\Gamma(\alpha_{ij})}{\Gamma(\alpha_{ij} + N_{ij})} \cdot \prod_{k=1}^{r_i} \frac{\Gamma(\alpha_{ijk} + N_{ijk})}{\Gamma(\alpha_{ijk})} \right], \quad (1)$$

where r_i is the number of states that variable i can assume; q_i denotes the number of joint states that the parents of variable i can have; α_{ijk} is the parameter of Dirichlet prior distribution; N_{ijk} is the number of occurrences of variable i in state k given parent configuration j,

$$N_{ij} = \sum_{k=1}^{r_i} N_{ijk} \quad \text{and} \quad \alpha_{ij} = \sum_{k=1}^{r_i} \alpha_{ijk};$$

$\Gamma(\cdot)$ is the gamma function; and P_0 is the structure prior.[33,35]

2.2.2. *Structural Constraints*

The number of possible structures of a causal pathway is superexponential in the number of variables. It is usually impractical to evaluate every possible structure. The constraints provided by an association map greatly reduce the space of structures that need to be searched. The structural constraints include the following: (1) a QTL genotype node has no parent node; (2) the parents of a gene expression trait node must be either QTL genotype nodes or other gene expression trait nodes of the same transcription module; and (3) the parents of a higher-order trait node must be either QTL genotype nodes or gene expression trait nodes of the associated transcription modules.

2.2.3. *Model Discovery*

The next step is to search for the best-scored model under the structural constraints provided by the association map. Commonly used model discovery algorithms include greedy search with random starts, Markov chain Monte Carlo (MCMC) simulation, and genetic algorithm (GA). Here, I introduce one of these model discovery algorithms — genetic algorithm.

Genetic algorithms are powerful search algorithms inspired by genetics and natural selection.[42,43] The most important phases in GAs are crossover, mutation, fitness evaluation, and selection.

(1) Encoding causal pathways. GA begins with a population of individuals. In this case, the individuals are the structures of causal pathway models. The set of potential parents of each node is determined by the constraints provided by the association map. The most common method of encoding is in a binary string. Each bit in the binary string represents a potential parent of a node; "0" or "1" represents the presence or absence of the directed edge from the potential parent to the node, respectively:

$$\underbrace{10010001000101}_{\text{node 1}}\,\underbrace{0100101011}_{\text{node 2}}\dots\dots\dots\dots\underbrace{011010110001}_{\text{node } N}.$$

Thus, a binary string can represent the structure of a causal pathway model.

(2) Fitness. We use the Bayesian score defined in Eq. (1) as the fitness function.

(3) Selection. Selection is an operation used to decide which individuals to use for crossover. Individuals are selected according to their fitness; individuals with higher fitness will have more offspring.

(4) Crossover. The individuals exchange part of their chromosomes to create offspring. Crossover can be illustrated as follows:

Parent Pathway 1	11011	00100110110
Parent Pathway 2	11011	11000011110
Offspring Pathway 1	11011	11000011110
Offspring Pathway 2	11011	00100110110

(5) Mutation. Mutation is applied to randomly selected offspring resulting from crossover. In the case of binary encoding, we can switch a few randomly chosen bits from 1 to 0 or from 0 to 1.

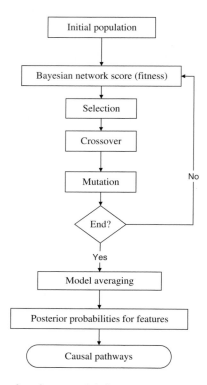

Fig. 3.   Causal pathway model discovery with genetic algorithm.

Two criteria are used to determine the ending of GA: (1) convergence, i.e. the population no longer changes; and (2) reaching a predetermined limit for the maximum number of generations. A flowchart of model discovery with GA is shown in Fig. 3.

By encouraging the best structures generated and throwing away the worst ones ("survival of the fittest"), the population keeps improving as a whole. According to the Schema Theorem, the building blocks of the good genes will spread in the population.[43] Thus, more and more high-fitness offspring will be generated. In this way, GAs search for causal pathway models of high posterior probability.

2.2.4. *Model Averaging*

Selecting a single best pathway model and ignoring all of the other models may lead to overfitting of the data. Model averaging can be used to reduce this risk.[44] An indicator function f is defined as follows: if a pathway G has the feature (here, the feature is a directed edge representing a causal relationship), then $f(G) = 1$; otherwise, $f(G) = 0$. The posterior probability of a feature is $P[f(G) \mid D] = \sum_G f(G)P(G \mid D)$. This probability reflects our confidence in the feature f. The posterior probability of a feature can be estimated by averaging over the highest scoring models visited during the search. The features (causal relationships) with high posterior probabilities are then used for constructing causal pathways.

2.2.5. *Parameter Estimation*

Given the pathway structures, we can find maximum likelihood estimates of the parameters of the CPDs. The parameter values maximize the likelihood of the observational data. The normalized log-likelihood of the data decomposes according to the pathway structure,[45]

$$L = \frac{1}{M} \sum_{i=1}^{n} \sum_{m=1}^{M} \log P[X_i \mid \mathrm{Pa}(X_i), D_m],$$

where $\mathrm{Pa}(X_i)$ are the parents of X_i, and M is the number of strains (samples). For example, in the causal pathway model shown in Fig. 1, the maximum likelihood estimate of $P(NT = high \mid GET2 = high, GET3 = low)$ is

$$\frac{P(NT = high, GET2 = high, GET3 = low)}{P(GET2 = high, GET3 = low)} \approx \frac{N_1}{N_2},$$

where N_1 is the number of individuals with $HT = high$, $GET2 = high$, $GET3 = low$; and N_2 is the number of individuals with $GET2 = high$, $GET3 = low$.

3. PREDICTING THE EFFECTS OF GENETIC AND TRANSCRIPTIONAL INTERVENTIONS

With the causal pathway models, we can compute the probability distribution of the gene expression and phenotype values, conditional on known evidence (i.e. the QTL genotype and/or the expression states of the interfered genes). For example, using the causal pathway model shown in Fig. 1, we can calculate the probability distributions of the higher-order trait in each BXD strain and the two parental strains (B6 and D2) given the QTL genotype (QTLG):

$$P(HT \mid QTLG = B) = \sum_{GET2, GET3} P(HT \mid GET2, GET3)$$
$$P(GET2 \mid QTLG = B) P(GET3 \mid QTLG = B)$$
$$P(HT \mid QTLG = D) = \sum_{GET2, GET3} P(HT \mid GET2, GET3)$$
$$P(GET2 \mid QTLG = D) P(GET3 \mid QTLG = D),$$

where

$$P(GET2 \mid QTLG = B) = \sum_{GET1} P(GET2 \mid GET1) P(GET1 \mid QTLG = B)$$
$$P(GET3 \mid QTLG = B) = \sum_{GET1} P(GET3 \mid GET1) P(GET1 \mid QTLG = B)$$
$$P(GET2 \mid QTLG = D) = \sum_{GET1} P(GET3 \mid GET1) P(GET1 \mid QTLG = D)$$
$$P(GET3 \mid QTLG = D) = \sum_{GET1} P(GET3 \mid GET1) P(GET1 \mid QTLG = D).$$

Using the conditional probability tables shown in Fig. 1, we can derive the following [Figs. 4(a) and 4(b)]:

$$P(HT = high \mid QTLG = B) = 0.705$$
$$P(HT = low \mid QTLG = B) = 0.296$$
$$P(HT = high \mid QTLG = D) = 0.126$$
$$P(HT = low \mid QTLG = D) = 0.873.$$

If gene 2 is knocked out or silenced by siRNA (i.e. GET2 = low) in B6 (QTLG = B) or D2 (QTLG = D) mouse, then the probability distribution of the higher-order trait can be calculated as

$$P(HT \mid QTLG = B, GET2 = low)$$
$$= \sum_{GET3} P(HT \mid GET2 = low, GET3) P(GET3 \mid QTLG = B)$$
$$P(HT \mid QTLG = D, GET2 = low)$$
$$= \sum_{GET3} P(HT \mid GET2 = low, GET3) P(GET3 \mid QTLG = D),$$

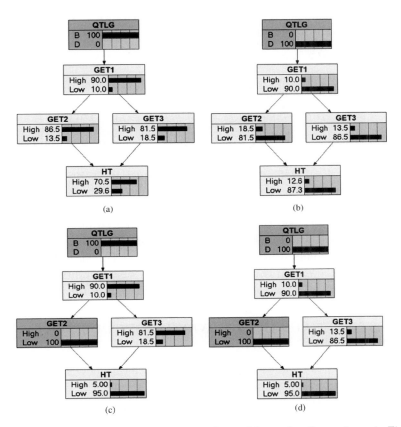

Fig. 4. The distributions of the variables in the model causal pathway shown in Fig. 1. The numbers are percentages. The values of shadowed nodes are fixed. (a) The probability distributions of the values of the variables in C57BL/6J (B6) mouse (i.e. QTLG = B). (b) The probability distributions of the values of the variables in DBA/2J (D2) mouse (i.e. QTLG = D). (c) The probability distributions of the values of the variables in gene 2–null B6 mouse (i.e. gene 2 is knocked out, therefore GET2 = low). (d) The probability distributions of the values of the variables in gene 2–null D2 mouse (i.e. gene 2 is knocked out, therefore GET2 = low).

where

$$P(GET3 \mid QTLG = B)$$
$$= \sum_{GET1} P(GET3 \mid GET1) \, P(GET1 \mid QTLG = B)$$
$$P(GET3 \mid QTLG = D)$$
$$= \sum_{GET1} P(GET3 \mid GET1) \, P(GET1 \mid QTLG = D).$$

Using the conditional probability tables shown in Fig. 1, we can derive the following [Fig. 4(c) and 4(d)]:

$$P(HT = high \mid QTLG = B, GET2 = low) = 0.05$$
$$P(HT = low \mid QTLG = B, GET2 = low) = 0.95$$
$$P(HT = high \mid QTLG = D, GET2 = low) = 0.05$$
$$P(HT = low \mid QTLG = D, GET2 = low) = 0.95.$$

Comparing the probability distributions, we can see that the probability of observing high values of the higher-order trait is 0.705 in wild-type B6 mice [Fig. 4(a)], while the probability is only 0.05 in B6 mice with the low expression of gene 2 [Fig. 4(c)]. Suppose the higher-order trait is a disease phenotype and the low expression of gene 2 is caused by a therapeutic intervention (e.g. the gene expression is silenced by an RNAi-based therapy) — the model shows that the susceptibility to the disease is affected by the genetic profiles (0.705 in B6 mice vs. 0.126 in D2 mice); however, the therapeutic intervention can reduce the disease susceptibility to the same level (0.05), regardless of the different genetic profiles of the two strains.

4. TESTING CAUSAL PATHWAY MODELS USING CROSS-VALIDATION

Cross-validation can be used to test causal pathway models. Suppose we have the data of N BXD strains. Each time, we use the data of $N - 1$ strains (training set) to construct causal pathway models; we then use the causal pathway models to predict the distributions of the higher-order trait values of the other strain (the one left out of the training set) given the QTL genotype data of the strain, and compare the predictions with the real trait value of that strain. The process is called "leave-one-out" cross-validation.

To evaluate the causal pathway models, we compute the logarithmic loss:

$$S = -\frac{1}{N} \sum_{i=1}^{N} \ln[P_i(X \mid QTLG = g_i)],$$

where X is the real value of the higher-order trait of strain i, $P_i(X \mid QTLG = g_i)$ is the probability of observing the trait value X given the QTL genotype in strain i, and $g_i = B$ or D. A small logarithmic loss indicates that the causal pathway model gives high probability to the observed trait values, thereby indicating that the performance of the causal pathway model is good.

5. DISCUSSION

It is generally believed that future medicine will be personalized and preventive medicine, which requires the capability of predicting which diseases a person is likely to develop and what medical treatments work well with that person's genetic profile.[46-50] Knowledge of the causal pathways is essential to understanding how the genetic variations (underlying the QTLs) influence phenotypes (e.g. susceptibility to a disease or response to a drug). Causal pathway models may provide many potential targets for therapeutic interventions and can also help us to predict the outcomes of therapeutic interventions, such as how the silence (e.g. by siRNA) of a gene's expression may affect the disease phenotype of interest.

6. ACKNOWLEDGMENTS

The author thanks Robert Williams, Lu Lu, Kenneth Manly, Elissa Chesler, Lei Bao, Biao Li, Hao Chen, Hongqiang Li, and Mi Zhou for their helpful discussions. This work was partly supported by a PhRMA Foundation grant.

REFERENCES

1. Li H, Chen H, Bao L, *et al.* (2006) Integrative genetic analysis of transcription modules: towards filling the gap between genetic loci and inherited traits. *Hum Mol Genet* **15**:481–492.
2. Bystrykh L, Weersing E, Dontje B, *et al.* (2005) Uncovering regulatory pathways that affect hematopoietic stem cell function using 'genetical genomics'. *Nat Genet* **37**:225–232.
3. Hubner N, Wallace CA, Zimdahl H, *et al.* (2005) Integrated transcriptional profiling and linkage analysis for identification of genes underlying disease. *Nat Genet* **37**:243–253.
4. Chesler EJ, Lu L, Shou S, *et al.* (2005) Complex trait analysis of gene expression uncovers polygenic and pleiotropic networks that modulate nervous system function. *Nat Genet* **37**:233–242.
5. Monks SA, Leonardson A, Zhu H, *et al.* (2004) Genetic inheritance of gene expression in human cell lines. *Am J Hum Genet* **75**:1094–1105.
6. Morley M, Molony CM, Weber TM, *et al.* (2004) Genetic analysis of genome-wide variation in human gene expression. *Nature* **430**:743–774.
7. Schadt EE, Monks SA, Drake TA, *et al.* (2003) Genetics of gene expression surveyed in maize, mouse and man. *Nature* **422**:297–302.
8. Brem RB, Yvert G, Clinton R, Kruglyak L. (2002) Genetic dissection of transcriptional regulation in budding yeast. *Science* **296**:752–755.
9. Lan H, Chen M, Flowers JB, *et al.* (2006) Combined expression trait correlations and expression quantitative trait locus mapping. *PLoS Genet* **2**:e6.

10. Wang S, Yehya N, Schadt EE, *et al.* (2006) Genetic and genomic analysis of a fat mass trait with complex inheritance reveals marked sex specificity. *PLoS Genet* **2**:e15.

11. Jansen RC, Nap JP. (2001) Genetical genomics: the added value from segregation. *Trends Genet* **17**:388–391.

12. Li J, Burmeister M. (2005) Genetical genomics: combining genetics with gene expression analysis. *Hum Mol Genet* **14**:R163–R169.

13. de Koning DJ, Haley CS. (2005) Genetical genomics in humans and model organisms. *Trends Genet* **21**:377–381.

14. Peirce JL, Lu L, Gu J, *et al.* (2004) A new set of BXD recombinant inbred lines from advanced intercross populations in mice. *BMC Genet* **5**:7.

15. Taylor BA. (1989) Recombinant inbred strains. In: *Genetic Variants and Strains of the Laboratory Mouse*, 2nd ed., Lyon ML, Searle AG (eds.), Oxford University Press, Oxford, England, pp. 773–796.

16. Chesler EJ, Wang J, Lu L, *et al.* (2003) Genetic correlates of gene expression in recombinant inbred strains: a relational model system to explore neurobehavioral phenotypes. *Neuroinformatics* **1**:343–357.

17. Bergmann S, Ihmels J, Barkai N. (2004) Similarities and differences in genome-wide expression data of six organisms. *PLoS Biol* **2**:E9.

18. Ihmels J, Friedlander G, Bergmann S, *et al.* (2002) Revealing modular organization in the yeast transcriptional network. *Nat Genet* **31**:370–377.

19. Segal E, Shapira M, Regev A, *et al.* (2003) Module networks: identifying regulatory modules and their condition-specific regulators from gene expression data. *Nat Genet* **34**:166–176.

20. Tanay A, Sharan R, Kupiec M, Shamir R. (2004) Revealing modularity and organization in the yeast molecular network by integrated analysis of highly heterogeneous genomewide data. *Proc Natl Acad Sci USA* **101**:2981–2986.

21. Bergmann S, Ihmels J, Barkai N. (2003) Iterative signature algorithm for the analysis of large-scale gene expression data. *Phys Rev* **67**:031902.

22. Jones BC, Tarantino LM, Rodriguez LA, *et al.* (1999) Quantitative-trait loci analysis of cocaine-related behaviours and neurochemistry. *Pharmacogenetics* **9**:607–617.

23. Crabbe JC. (1998) Provisional mapping of quantitative trait loci for chronic ethanol withdrawal severity in BXD recombinant inbred mice. *J Pharmacol Exp Ther* **286**:263–271.

24. Risinger FO. (2003) Genetic analyses of ethanol-induced hyperglycemia. *Alcohol Clin Exp Res* **27**:756–764.

25. McCaughran J Jr, Bell J, Hitzemann R. (1999) On the relationships of high-frequency hearing loss and cochlear pathology to the acoustic startle response (ASR) and prepulse inhibition of the ASR in the BXD recombinant inbred series. *Behav Genet* **29**:21–30.

26. McCall RD, Frierson D Jr. (1981) Evidence that two loci predominantly determine the difference in susceptibility to the high pressure neurologic syndrome type I seizure in mice. *Genetics* **99**:285–307.

27. Plomin R, McClearn GE, Gora-Maslak G, Neiderhiser JM. (1991) Use of recombinant inbred strains to detect quantitative trait loci associated with behavior. *Behav Genet* **21**:99–116.

28. Schrier DJ, Sternick JL, Allen EM, Moore VL. (1982) Immunogenetics of BCG-induced anergy in mice: control by genes linked to the Igh complex. *J Immunol* **128**:1466–1469.

29. Pereira P, Lafaille JJ, Gerber D, Tonegawa S. (1997) The T cell receptor repertoire of intestinal intraepithelial gammadelta T lymphocytes is influenced by genes linked to the major histocompatibility complex and to the T cell receptor loci. *Proc Natl Acad Sci USA* **94**:5761–5766.

30. Willott JF, Erway LC. (1998) Genetics of age-related hearing loss in mice. IV. Cochlear pathology and hearing loss in 25 BXD recombinant inbred mouse strains. *Hear Res* **119**:27–36.

31. Phillips TJ, Belknap JK, Crabbe JC. (1991) Use of recombinant inbred strains to assess vulnerability to drug abuse at the genetic level. *J Addict Dis* **10**:73–87.

32. Zhu J, Lum PY, Lamb J, *et al.* (2004) An integrative genomics approach to the reconstruction of gene networks in segregating populations. *Cytogenet Genome Res* **105**:363–374.

33. Li H, Lu L, Manly KF, *et al.* (2005) Inferring gene transcriptional modulatory relations: a genetical genomics approach. *Hum Mol Genet* **14**:1119–1125.

34. Schadt EE, Lamb J, Yang X, *et al.* (2005) An integrative genomics approach to infer causal associations between gene expression and disease. *Nat Genet* **37**:710–717.

35. Cui Y. (2006) Elucidating gene regulatory networks underlying complex phenotypes: genetical genomics and Bayesian network. In: *Microarrays and Transcription Networks*, Shannon F (ed.), Landes Bioscience, Georgetown, TX, pp. 114–126.

36. Schadt EE. (2005) Exploiting naturally occurring DNA variation and molecular profiling data to dissect disease and drug response traits. *Curr Opin Biotechnol* **16**:647–654.

37. Pearl J. (1988) *Probabilistic Reasoning in Intelligent Systems.* Morgan Kaufmann, San Francisco, CA.

38. Pearl J. (2000) *Causality: Models, Reasoning, and Inference.* Cambridge University Press, Cambridge, England.

39. Neapolitan RE. (2003) *Learning Bayesian Networks.* Prentice Hall, Upper Saddle River, NJ.

40. Pe'er, D, Regev A, Elidan G, Friedman N. (2001) Inferring subnetworks from perturbed expression profiles. *Bioinformatics* **17**(Suppl 1):S215–S224.

41. Heckerman D. (1999) A tutorial on learning with Bayesian networks. In: *Learning in Graphical Models*, Jordan M (ed.), MIT Press, Cambridge, MA, pp. 301–354.

42. Holland J. (1975) *Adaptation in Natural and Artificial Systems.* University of Michigan Press, Ann Arbor, MI.

43. Goldberg DE. (1989) *Genetic Algorithms in Search, Optimization and Machine Learning.* Addison-Wesley Longman, Boston, MA.

44. Hartemink AJ, Gifford DK, Jaakkola TS, Young RA. (2001) Using graphical models and genomic expression data to statistically validate models of genetic regulatory networks. *Pac Symp Biocomput*: 422–433.

45. Murphy K. (2001) An introduction to graphical models. Available at http://www.cs.ubc.ca/~murphyk/Papers/intro_gm.pdf/

46. Hood L, Heath JR, Phelps ME, Lin B. (2004) Systems biology and new technologies enable predictive and preventative medicine. *Science* **306**: 640–643.

47. Meadows M. (2005) Genomics and personalized medicine. *FDA Consum* **39**:12–17.

48. Bell J. (2004) Predicting disease using genomics. *Nature* **429**:453–456.

49. Snyderman R, Langheier J. (2006) Prospective health care: the second transformation of medicine. *Genome Biol* **7**:104.

50. Snyderman R, Yoediono Z. (2006) Prospective care: a personalized, preventative approach to medicine. *Pharmacogenomics* **7**:5–9.

Chapter 17

Functional Proteomics and Its Application in Biomedical Sciences

Guishan Xiao[*,†,§] and Teri Zhang[†,‡]

*College of Life Sciences, Hunan Normal University
Changsha, Hunan 410081, China*

†*College of Medicine, University of California, Irvine
CA 92697, USA*

‡*University of California, Berkeley
CA 94720, USA*

Proteomics, viewed as postgenomics, is the large-scale study of proteins, especially their structures and functions. It is much more complicated than genomics: whereas genomics is a rather constant entity, the proteome differs from cell to cell and is constantly changing through its biochemical interactions with the genome and the environment. Over the last several years, the emphasis on genomics has shifted to proteomics, since transcriptional regulation is often difficult to reconcile with protein abundance and the transcriptome has poor correlation with proteome in a cell.[1] Proteomics has now gained more interest in directly analyzing protein expression at the posttranslational level because it permits the qualitative and quantitative assessment of a broad spectrum of proteins that can be related to specific cellular responses, including the response to oxidative stress.[2–4] Moreover, identified protein markers can offer more effective validation protocols (e.g. tissue microarray). Analysis on proteome provides an idea of biological processes happening at their level of occurrence, allowing the comparison of physiological and pathological states of a cell line or tissue.

§ Correspondence author.

1. INTRODUCTION

Since the completion of the human genome sequence, which is one of the most important scientific achievements of this century, substantial progress has been made in the fundamental understanding of human biology, ranging from DNA structure to identification of diseases associated with genetic abnormalities. Proteomics is the large-scale study of proteins, particularly their structures and functions. This term was coined to make an analogy with genomics, and while it is often viewed as the "next step", proteomics is much more complicated than genomics. Most importantly, while the genome is a rather constant entity, the proteome differs from cell to cell and is constantly changing through its biochemical interactions with the genome and the environment. One organism has radically different protein expression in different parts of its body, in different stages of its life cycle, and in different environmental conditions. The entirety of proteins existent in an organism throughout its life cycle or, on a smaller scale, the entirety of proteins found in a particular cell type under a particular type of stimulation is referred to as the *proteome* of the organism or cell type, respectively.

With the completion of a rough draft of the human genome, many researchers are now looking at how genes and proteins interact to form other proteins. A surprising finding of the Human Genome Project is that there are far fewer protein-coding genes in the human genome than there are proteins in the human proteome ($\sim$22 000 genes vs. $\sim$400 000 proteins). The large increase in protein diversity is thought to be due to the alternative splicing and posttranslational modification (PTM) of proteins. This discrepancy implies that protein diversity cannot be fully characterized by gene expression analysis alone, making proteomics a useful tool for characterizing cells and tissues of interest.

To catalog all human proteins and ascertain their functions and interactions presents a daunting challenge for scientists. An international collaboration to achieve these goals is being coordinated by the Human Proteome Organisation (HUPO).

Proteomics is to analyze all proteins in a living system, including the description of cotranslationally and posttranslationally modified proteins and alternatively spliced variants. It includes their covalent and noncovalent associations, spatial and temporal distributions within cells, and how all of these are affected by changes in the extracellular and intracellular conditions.

Nowadays, proteomics in general has been developed into several branches, including protein fractionation, protein separation, protein

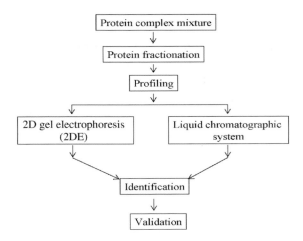

Fig. 1.   Flowchart of general proteomics.

identification, protein quantification, protein sequence analysis, structural proteomics, interaction proteomics, protein modification, and cellular proteomics. In practice, proteomics experiments include all aspects of proteome profiling (e.g. protein fractionation and separation), protein identification, and protein validation (Fig. 1).

Mass spectrometry (MS) is at the heart of virtually all proteomics experiments, as it provides the key tools for the analysis of proteins. Developments in technology and methodology — which are used for the detection and identification of proteins at different abundant levels — in the field of proteomics have been rapid in the last 6 years, and are providing improved and novel strategies for global understanding of cellular function. In this chapter, we will provide a brief summary of novel proteomics technologies that have been developed and the future direction in proteomics.

2. PROTEOME PROFILING TECHNIQUES IN PROTEOMICS

2.1. *Protein Fractionation*

The effective study of low-abundance or membrane proteins usually requires a fractionation step to reduce overall sample complexity and to elevate the concentration of those proteins relative to the original sample. Formerly undetectable proteins may be enriched to levels that allow downstream analysis by two-dimensional gel electrophoresis coupled with mass spectrometry (2DE/MS) or two-dimensional liquid chromatography/mass spectrometry (2D LC/MS), the two methods most commonly used in proteomics for the separation and identification of proteins. Reduction in sample complexity also minimizes signal suppression effects that may occur in MS analysis of complex samples.[5] Several new techniques have been developed in recent years, including ultrafiltration, liquid-phase isoelectric focusing, immunoaffinity liquid chromatography (LC), and chemical conjugated bead assays.

2.1.1. *Ultrafiltration*

Ultrafiltration is a type of filtration in which hydrostatic pressure forces a liquid against a semipermeable membrane. Water and low-molecular-weight solutes are able to pass through the membrane, but solutes of high molecular weight remain on one side of the membrane. This filtration process is used in industry and research for purifying and concentrating macromolecular (10^3–10^6 Da) solutions, especially protein solutions.

Advances in proteomics are continuing to expand the ability to analyze the serum proteome. It has now been realized that, in addition to the circulating proteins, human serum also contains a large number of peptides. In a recent study of human serum, Zheng and colleagues[6] studied human serum using ultrafiltration and a hybrid ion trap–Fourier transform mass spectrometer to identify low-molecular-weight serum peptidome. This research focused on peptides with a nominal molecular weight of less than 10 kDa. In order to study these low-molecular-weight (LMW) peptides, the researchers modified previous protocols for using centrifugal ultrafiltration[7] and did not digest the filtrate with trypsin, so endogenous peptides could be identified. It is believed that many of these LMW peptides are fragments of larger proteins that have been at least partially degraded by various enzymes such as metalloproteases. In this fractionation protocol, 500 μL of human serum from each sample was diluted in 1500 μL of distilled water. Then, the serum was centrifuged and the filtrate was collected. The amount of peptide/protein material in the filtrate was determined by Bradford-compatible assays (BCAs). To check whether the high-molecular-weight (HMW) proteins had been successfully depleted, 200 μL of the serum filtrate was concentrated either by cold acetonitrile precipitation or with a speed vacuum and analyzed with NuPAGE. A concern with the use of an ultrafiltration step to remove HMW proteins from serum is the potential loss of peptides that are strongly bound to carrier proteins, such as albumin. Next, salt was removed from the human serum before the sample was analyzed by LC-MS/MS. Protein/Peptide identifications were obtained using the SEQUEST algorithm from Thermo Electron.

2.1.2. *Liquid-Phase Isoelectric Focusing*

Isoelectric focusing (IEF), an electrophoretic technique that is used as the first-dimension separation in a traditional 2D gel electrophoresis workflow, is also applied as a fractionation technique upstream of both 2DE/MS[8,9] and LC/MS[10] workflows. For 2DE/MS, sample fractionation by IEF can result in a more effective analysis by removing the proteins that are outside the pH range of the selected immobilized pH gradient (IPG) strip. This limits protein precipitation and smearing, which are often the consequences of higher protein loads, and enables the enrichment of proteins in the isoelectric point (pI) range of interest.

Liquid-phase IEF can perform the separation of proteins entirely in free solution. Using this approach, larger volumes/amounts of samples can be loaded, yielding sufficient amounts of low-abundance proteins for further characterization. Since proteins remain in liquid phase during the entire procedure, extra steps such as electroelution, extraction, or transfer to membranes from the gels prior to mass spectrometric analysis are obviated. This method is applicable to a wide range of sample types, such as cerebrospinal fluid,[11-14] serum, tissue extracts, cell media, whole cells, and bacterial lysates.

Liquid-phase IEF provides a unique method of sample fractionation and enrichment of low-abundance proteins as well as a robust complementary separation strategy to conventional 2DE/MS and LC/MS. This prefractionation strategy prior to conventional 2DE enables higher sample loads and greater resolution for the pI fractions of interest.[15,16] The use of these types of electrophoretic fractionations in proteomic studies can also increase sequence coverage, and provides proteins of high purity and in yields sufficient for characterization by MS.[11,12,17-20]

Liquid-phase IEF fractionations of proteins coupled online to an electrospray ionization time-of-flight (ESI-TOF) mass spectrometer (MS) or to a matrix-assisted laser desorption/ionization TOF (MALDI-TOF) MS can improve the resolution of low-mass and basic proteins, and make the entire procedure highly amenable to automation and high throughput.[21-25]

For fractionation of protein using liquid-phase IEF, the reduced and alkylated sample is diluted to a concentration of 0.6 mg/mL protein in IEF buffer (7 m urea, 2 M thiourea, 4% CHAPS, 2 mM tributylphosphine, 0.001% bromophenol blue, 2% w/v Biolyte 3/10 ampholytes), and a 2.7-mL sample is loaded into the focusing chamber of a MicroRotofor cell (Bio-Rad, Hercules, CA). The sample can be focused at 1 W (constant) for 2.5 h. The pH, volume (calculated by weight/density, 1.1 g/mL), and protein concentration can be measured by the RC DC Protein Assay (Bio-Rad, Hercules, CA) for each of the 10 fractions collected. Prior to the 2DE experiment, the fractions are required to be treated with the ReadyPrep 2D Cleanup Kit (Bio-Rad, Hercules, CA) and resuspended in IEF buffer, which contains 0.2% (w/v) ampholytes matching the pH of the IPG strip to be used.

2.1.3. *Depletion of High-Abundance Proteins*

Proteins in plasma/serum and other body fluids may often serve as indicators of disease and are a rich source for biomarker discovery. However, the intrinsic large dynamic range of plasma proteins makes the analysis very challenging because a large number of low-abundance proteins are often masked by a few high-abundance proteins. The use of prefractionation methods, such as depletion of higher-abundance proteins before protein profiling, can assist in the discovery and detection of less abundant proteins that may ultimately prove to be informative biomarkers.

There are now two high-throughput methods commercially available for the fractionation of plasma/serum proteins: the Multiple Affinity Removal System (MARS) (Agilent Technologies, CA) and the Seppro MIXED 12 (GenWay Biotech, Inc.). MARS was designed for simultaneous binding of albumin, IgG, IgA, haptoglobin, transferrin, and α1-antitrypsin; while Seppro was designed for simultaneous binding of albumin, IgG, transferrin, α1-antitrypsin, IgA, IgM, α2-macroglobulin, haptoglobin, apolipoproteins A-I and A-II, orosomucoid (α1-acid glycoprotein), and fibrinogen.

Practically, for MARS, crude plasma is required to be diluted five-fold with buffer A (product no. 5185-5987; Agilent Technologies) and then passed through a 0.45-μm filter by spinning at 12 000 rpm at room temperature. Each aliquot of the sample (equal to 25–40 μL original plasma) can be injected on the MARS column in 100% buffer A at a flow rate of 0.5 mL/min for 10 min. After collection of the flow-through fraction, the column is washed and the bound proteins can then be eluted with 100% buffer B (product no. 5185-5988; Agilent Technologies) at a flow rate of 1.0 mL/min for 8 min; the bound fraction is subjected to be collected for additional verification. The column is then regenerated by equilibrating it with 100% buffer A for 10 min and is ready for the next run. The standard spectrum is shown in Fig. 2.

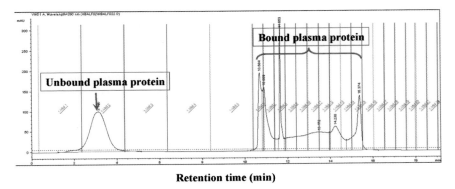

Retention time (min)

Fig. 2. The standard spectrum of whole plasma using MARS (Agilent Technologies, CA).

For Seppro, plasma can be diluted with five volumes of Dilution Buffer (GenWay Biotech, Inc.) and centrifuged through a 0.45-μm filter at 12 000 rpm. Using a Beckman Proteome Lab PF 2D system, each aliquot of the sample (equal to 25–40 μL original plasma) is injected on the Seppro column. The method starts at a flow rate of 0.1 mL/min for 10 min, and the column is then washed at a flow rate of 0.2 mL/min for 7 min; then, the flow rate is changed to 1.0 mL/min to continue the wash for 5 min and the flow-through fractions collected. Bound proteins are eluted from the column with Stripping Buffer (GenWay Biotech, Inc.) at a flow rate of 1.0 mL/min for 12 min with bound fractions collected for additional verification, and the column is then neutralized with Neutralizing Buffer (GenWay Biotech, Inc.) at a flow rate of 1.0 mL/min for 6 min. The column can be regenerated by equilibrating it with Dilution Buffer at a flow rate of 1.0 mL/min for 6 min.

In general, the fractions collected from columns must be desalted and concentrated before performing further downstream analysis (e.g. 2DE, 2D LC). A Centriplus centrifugal concentrator (YM-3, molecular weight cutoff 3 kDa; Millipore, MA) can be used for desalting and concentrating the collected fractions by centrifuging at 4500 rpm at 4°C. The protein content is measured using a revised Bio-Rad RC/DC method.

In a recent study of human plasma proteome, the performances of two different immunoaffinity fractionation columns for the top 6 or top 12 proteins in plasma were investigated, and both the proteins in column-bound and flow-through fractions were subsequently analyzed utilizing chromatographic separation techniques combined with tandem mass spectrometry (MS/MS).[26] Based on the recommended HUPO Plasma Proteome Project (PPP) criteria — namely, $X_{corr} \geq 1.9$ with charge state 1+, $X_{corr} \geq 2.2$ with charge state 2+, $X_{corr} \geq 3.75$ with charge state 3+, DeltCn ≥ 0.1, and Rsp ≤ 4 — a total of 2401 unique plasma proteins were identified in this group. MARS yielded 921 and 725 unique proteins from the flow-through and bound fractions, respectively; whereas the Seppro MIXED 12 column yielded the identification of 897 and 730 unique proteins from the flow-through and bound fractions, respectively. This study suggested that MARS may perform a little better than Seppro, even though this difference is not of statistical significance.[26]

2.2. *Protein Separation*

Proteomics as a scientific discipline emerged with the purpose to identify rapidly complex protein patterns in a comprehensive manner.[27–29] The methodology required to achieve this ambitious goal must also be powerful enough to detect subtle quantitative and qualitative differences of a protein profile in order to finally identify target proteins and modified variants.[30,31] Such comprehensive proteome characterization will give new insight into cellular responses for disease pathogeneses such as carcinogenesis, for development, as well as for aging, drug action, and environmental damage. In order to identify proteins from a complex mixture of 5×10^3 to 5×10^4 compounds with a dynamic range of at least 10,[5,32] it is crucial to develop technologies with extremely good resolving power on the one hand and extraordinary sensitivity on the other hand. It is obvious that these challenging tasks will not be achieved with a single analytical technology, but through the combination of separation and detection techniques.

Proteomics involves several complementary technologies. Protein separation is the key step for obtaining quality data in proteomic experiments. It can be largely classified into two categories: gel-based separation and gel-free separation (also liquid phase). The former includes one-dimensional polyacrylamide gel electrophoresis (1D PAGE) and 2DE, while the latter includes liquid-phase separations (e.g. 1D, 2D, or multidimensional LC).

2.2.1. *Gel-based Techniques — Polyacrylamide Gel Electrophoresis (PAGE)*

PAGE is the most common technique for the separation of complex protein mixtures before identification by MS. This includes 1D PAGE, 2DE, and the newly developed 2D difference gel electrophoresis (DIGE).

a. *One-dimensional polyacrylamide gel electrophoresis (1D PAGE)*

1D PAGE is the most widely used and least expensive analytical proteomic analysis method. In 1D PAGE, the protein sample is first dissolved in a loading buffer containing a reductant. Next, separation occurs based on molecular weight in the presence of an electrical field. Once the proteins are separated, they can be visualized by staining the gel. The degree of protein resolution by 1D PAGE is low, making the procedure sufficient for reducing protein complexity in samples containing limited amounts of proteins, but rather insufficient in samples containing complex protein mixtures.[33]

b. *Two-dimensional gel electrophoresis (2DE)*

2DE, first introduced by O'Farrell in 1975, is a method used in the analysis of complex protein mixtures extracted from cells, tissues, or other biological samples. In this technique, proteins are separated according to two parameters. First, they are separated by their pI by IEF. In the second-dimension step, sodium dodecyl sulfate–polyacrylamide gel electrophoresis (SDS-PAGE) separates proteins according to their molecular weights. Each spot on the resulting 2D gel potentially corresponds to a single protein species in the sample.

In the original technique, the first-dimension separation was performed in carrier-ampholyte-containing polyacrylamide gels cast in narrow tubes.

A large and growing application of 2DE is within the field of proteomics. The analysis involves the systematic separation, identification, and quantification of many proteins simultaneously from a single sample. 2DE is used in this field due to its unique ability to separate thousands of proteins simultaneously. The technique is also able to detect posttranslational and cotranslational modifications, which cannot be predicted from the genome sequence. Applications of 2DE include proteome analysis, cell differentiation, detection of disease markers, therapy monitoring, drug discovery, cancer research, purity checks, and microscale protein purification.

One example of a successful application of this technique is an environmental pollutant study.[2] A 2DE-based proteomics approach was used to study diesel exhaust particle (DEP)-induced responses in the macrophage cell line RAW 264.7 (Fig. 3). In this study, using 2DE coupled with MALDI-TOF MS and ESI-LC/MS/MS analysis, 32 newly induced/NAC-suppressed proteins were identified. These data suggested that DEPs induce a hierarchical oxidative stress response in which some of these proteins may serve as markers for oxidative stress during particulate matter (PM) exposure.[2]

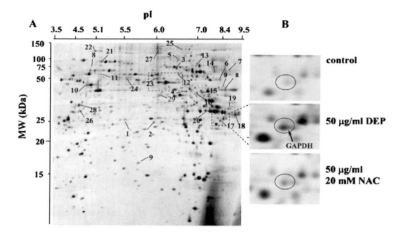

Fig. 3.   A 2DE-based proteomics approach was used to study diesel exhaust particle (DEP)-induced responses in the macrophage cell line RAW 264.7.

Currently, the World Wide Web provides direct access to spot-pattern databases for the comparison of electrophoresis results and genome sequence databases for assignment of sequence information. The HUPO is currently attempting to coordinate proteome analysis between many countries towards a common goal.

This technique has been extensively applied in studies on protein expressional profiling,[2,3] protein posttranslational modifications,[34,35] and functional proteomics[2,3,33] to gain insights into disease mechanisms. One example shown is that the 2D PAGE technique has been the main method for the study of protein expression in breast biopsies and for proteomic phenotyping of metastatic and invasive breast cancer.[33]

c. *Two-dimensional difference gel electrophoresis (2D DIGE)*

2D DIGE (Fig. 4) is different from the classic 2DE in its CyDye technology, which allows multiplexing proteome display in one gel. This technology is not solely a detection technique; it offers a complete concept for accurate quantitative proteomics. Using CyDye technology, different protein samples can be prelabeled with dyes of different excitation and emission wavelengths, then mixed and run together in a single gel (i.e. multiplexing). Three dyes with different excitation and emission wavelengths are available: Cy2, Cy3, and Cy5. The gels are usually prepared in nonfluorescent glass

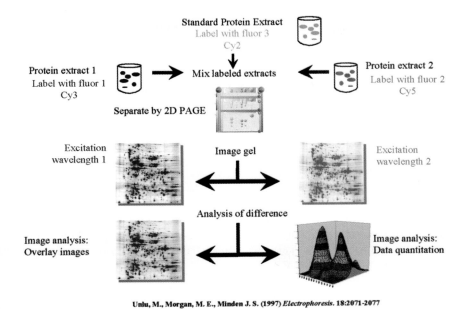

Unlu, M., Morgan, M. E., Minden J. S. (1997) *Electrophoresis.* **18:2071-2077**

Fig. 4.   Flowchart of the standard 2D DIGE approach.

cassettes and scanned with them. The key factors for this technique are sample labeling and separation on the gel. There are two labeling concepts that exist for scarce samples: lysine minimum labeling and cysteine saturation labeling.

(1) Minimum labeling

The Cye dyes bind to the ε-amino group of the lysine. Only a limited amount of dye can be added to the protein mixture; practically about 3% of protein can be effectively labeled. Hydrophobicity of proteins can affect the label efficiency. To perform effective minimum labeling, the exactly controlled conditions below must be followed:

- The pH must be adjusted to 8.5.
- No ampholytes and reductants are suggested.
- Add a small amount of lysine after labeling to stop the reaction.
- The reaction time may be restricted to half an hour in ice water.

(2) Saturation labeling

Only Cye3 and Cye5 can be modified for saturation labeling. This approach is subjected to label all available thiol groups of proteins. However, the reaction must be strictly followed:

- One-hour reduction at 37°C at pH 8.0 should be taken with a strong reductant: Tris-carboxyethyl phosphine (TCEP).
- Half-hour labeling at 37°C at pH 8.0 may be taken with the Cy dye derivative.

(3) The internal standard

A benefit of using this multiplexing technique is the use of an internal standard for every protein so that gel-to-gel variations can be eliminated. For multisample experiments, the standard is run in each gel. This allows normalization between the gels according to the spot positions and the spot volumes of the standard.[36,37]

In summary, compared to traditional 2DE, 2D DIGE offers several advantages:

- DIGE improves traditional 2DE.
- DIGE makes protein quantification more statistically relevant by reducing the effects of gel-to-gel variation with its internal standard.
- DIGE reduces the number of gels necessary due to multiple available covalent labels.
- DIGE offers a large dynamic range due to typhoon imaging and Cy dyes.

2.2.2. *Gel-Free System — High-Performance Liquid Chromatography (HPLC)*

High-performance liquid chromatography (HPLC) is a chemistry-based tool for quantifying and analyzing mixtures of chemical compounds. It is used to find the amount of a chemical compound within a mixture of other chemicals. Nowadays, HPLC plays a key role in the core technological development of proteomics. Today, 2D or multidimensional online or offline LC coupled with MS is state of the art for the identification of proteins from complex proteome samples in many laboratories.

a. *One-dimensional liquid chromatography (1D LC)*

This is a simple way to reduce protein complexity. Based on the purpose of the research, a size-exclusion (strong cation exchange, SCX) column — in which protein is separated according to its molecular weight — or a reverse-phase (RP) column — in which protein is separated according to its hydrophobicity — can be chosen to make the final separation of protein or peptides prior to MS identification.

b. *Multidimensional liquid chromatography (MDLC)*

Analysis of complex protein samples (e.g. plasma) is troublesome due to the wide dynamic range of protein expression levels and the different hydrophobicities of proteins. To overcome this problem, multiple efficient steps are required to reduce the complexity of protein mixture. Although gel-based 2DE has been widely used for protein separation, it suffers from several significant shortcomings such as lack of throughput potential and reproducibility as well as difficulties in resolving proteins that are highly basic, of high molecular weight, or in low abundance. Separation of highly basic proteins by 2DE has been a challenge despite the limited improvement achieved by changing IPG strip composition,[38,39] using strips of narrower pH range[40] or cup-loading methods.[41] To circumvent problems associated with 2DE, the development of several liquid-phase separation methods has been attempted[42] using size-exclusion chromatography,[43] affinity chromatography,[44] and ion-exchange chromatography.[45]

Any combination of the methods above can form MDLCs that may be used for the separation of whole cellular proteome and plasma proteome. In a recent study, newly developed liquid-based 2D liquid column chromatographic separation systems (2D LC) were attempted using a combination of chromatofocusing (CF)[46,47] and nonporous RP column chromatography (NPRPC) in order to separate proteins by both pI and hydrophobicity, and to provide greater throughput potential for the reproducible separation of complex mixtures of proteins present in mammalian cells such as mouse macrophage cell lines[48–50] (Fig. 5).

2.3. *Protein Detection*

The ideal detection technique should have a wide linear dynamic range; should be very sensitive, quantitative, compatible with further analysis using MS, quick, nontoxic, and reasonably priced; and should not require

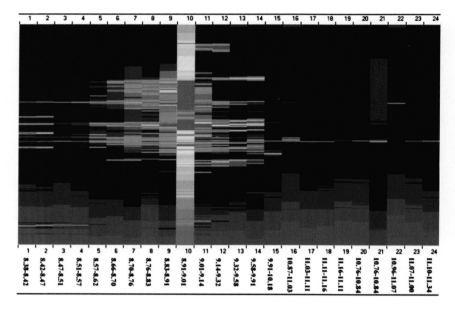

Fig. 5.   Representative 2D liquid protein map of macrophages (Refs. 49–51).

living cells for labeling. Unfortunately, none of the existing techniques combines all of these features.[51]

2.3.1. *Radioactive Labeling*

This is the most sensitive and quantitatively reliable method to detect cellular protein dynamics. Basically, a protein is labeled with ^{35}S or ^{32}P isotopes, separated on a 2D gel, labeled as a protein in the gel, and then exposed to a storage phosphor screen, which is subsequently scanned with a laser. The detection limit is less than 1 pg protein. However, there is a trend to replace radioactivity in the laboratories wherever possible.

2.3.2. *Gel Staining*

a. *Coomassie Brilliant Blue (CBB) staining*

It stains almost all proteins and peptides with good quantitative linearity and is also compatible with MS, but the sensitivity is not the same for all proteins. Its sensitivity is about 0.1 µg of protein.

b. *Negative staining*

Its combination with imidazole zinc provides a more sensitive detection limit ($\leq 15\,\text{ng}$)[52] than CBB. It has good compatibility with MS; however, it cannot be used for quantification.

c. *Silver staining*

It was introduced by Merril *et al.*[53] This method can pick up a protein amount less than 0.2 ng. There are two main technical methods being used for 2DE: silver nitrate and silver diamine procedures.[53] Silver diamine has better sensitivity for basic proteins; however, it contains caustic solutions and has limited application in Tricine buffer gel because of the development of a silver mirror on the gel surface. The advantage of the silver nitrate method is that it can be modified for MS compatibility, since the silver is more weakly bound to the proteins. To avoid keratin contamination, the experiment should be performed in closed trays.

d. *Fluorescence staining*

These methods are less sensitive than silver staining: their sensitivity is down to 2–8 ng protein. However, they have very wide linear dynamic ranges and are also compatible with subsequent MS analysis. Among them, Deep Purple is the most sensitive dye,[54] followed by SYPRO Ruby.[55]

New fluorescent dyes for specific detection of posttranslational modifications have recently been introduced: Pro-Q Diamond dye for the detection of phosphorylated proteins, and Pro-Q Emerald for glycosylated proteins. It is possible to stain gel sequentially with the different dyes and finally with a total stain (SYPRO Ruby).

Pro-Q Diamond phosphoprotein stain can detect phosphate groups attached to tyrosine, serine, or threonine residues. It is ideal for the identification of kinase targets in signal transduction pathways and for phosphoproteomic studies. Signal intensity is linear over three orders of magnitude and correlates with the number of protein phosphates. Stained proteins can be accurately identified by MS.

The Pro-Q Diamond phosphoprotein gel stain is particularly useful when used in conjunction with SYPRO® Ruby protein gel stain. The SYPRO® Ruby dye quantitatively stains total proteins. Determining the ratio of Pro-Q Diamond dye to SYPRO® Ruby dye signals provides a measure of the phosphorylation level normalized to the total amount of protein (Fig. 6). Using both stains in combination, it is possible to distinguish a

Fig. 6.   Visualization of total protein (green) and phosphoproteins (red) on a single 2D gel.

lightly phosphorylated, high-abundance protein from a heavily phosphorylated, low-abundance protein.

The Pro-Q Emerald 300 glycoprotein stain provides the most advanced technology available for direct detection of glycoproteins in gels.[56] In less than 3 h, it is possible to detect as little as 500 pg of glycoprotein per band depending on the degree of glycosylation, making the stain at least 50-fold more sensitive than standard fuchsin staining. The combination of Pro-Q Emerald glycoprotein stain with SYPRO Ruby protein gel stain provides a picture of both the proteome and glycoproteome in 2D gels.

3. PROTEIN IDENTIFICATION

3.1. *Introduction*

Proteins have become central to the identification, understanding, diagnosis, and treatment of diseases. With this new emphasis on proteins comes the need for solutions that allow researchers to quickly and easily determine protein identities, modifications, and expression levels. Thus far, many unique technologies have been developed to meet the specific challenges of preparing, separating, and identifying proteins. The integration of these technologies can help researchers discover more proteins more quickly and with less effort. In this section, we will illustrate these new techniques in detail.

3.2. *Mass Spectrometry Techniques*

3.2.1. *Isotope-Coded Affinity Tag (ICAT) Technology*

ICAT is an innovative method of protein profiling technology that utilizes stable isotope labeling of protein samples from two different sources, which are chemically identical in all aspects other than isotope composition. ICAT analysis profiles the relative amounts of peptides containing cysteine that are derived from tryptic digests of protein extracts. Cells extracted from the two samples are labeled with either light or heavy ICAT reagents, and react via cysteinyl thiols on the proteins. Peptides are recovered by avidin affinity chromatography and are then analyzed by LC-MS-MS. This produces a full-scan spectrum that displays the abundance of light and heavy peptide ions and their relative proteins.

The significance of ICAT technology is that it can be used to identify 300–400 proteins per sample without the need for 2D gel.[57] Also, enrichment of low-abundance proteins can be performed before the analysis through cell lysate fractionation.[58] ICAT technology has been used for protein identification and quantification in mammalian, liver, and breast tumor cells.[57] Disadvantages of ICAT analysis are that it is only applicable to proteins containing cysteine; identifies far less proteins than 2D PAGE; and contains a large label that makes database searching more difficult, especially for short peptides.[58]

3.2.2. *Immobilized Metal Ion Affinity Chromatography (IMAC)*

Immobilized metal ion affinity chromatography (IMAC) was first introduced by Porath *et al.*[59] IMAC interaction relies on specific binding between an analyte and an immobilized metal ion. Initially, immobilized metal ions (e.g. Ni^{2+}, Co^{2+}, Zn^{2+}, Mn^{2+}) were shown to bind strongly to proteins with a high density of histidines. The immobilized metal ions of Fe^{3+}, Ga^{3+}, and Al^{3+}, however, have shown strong binding characteristics with phosphopeptides, thus opening roads for the development of methods that can selectively enrich phosphopeptides from complex protein and peptide mixtures. IMAC is now a commonly used process for reversibly capturing phosphopeptides; allowing for preconcentration and selective retention of phosphopeptides; and removing salts, detergents, and nonspecific contaminants not compatible with MS analysis.

IMAC quickly established itself as a highly reliable purification procedure, showing rapid expansion in the number of preparative and analytical applications while not remaining confined to protein separation. It was soon applied to protein refolding (matrix-assisted refolding), evaluation of protein folding status, protein surface topography studies, and biosensor development. One example of using this technique is to identify phosphorylation sites in protein. As is well known, phosphopeptides are acidic by virtue of the phosphate group and bind preferentially to chelated metal ions; however, other peptides, particularly those containing strings of acidic amino acids, are also coenriched. In general, binding strength is dependent on numerous factors, such as the degree of phosphorylation, the pH of solutions, salt and peptide concentrations, chelated metal ions, temperature, and the degree of exposure of chelated ions interacting with the peptide side chains.

Because the enrichment of phosphopeptides prior to MS reduces ion suppression effects that would otherwise occur with untreated complex mixtures, enrichment allows for a higher success rate in the assignment of site-specific phosphorylation. The use of IMAC either online or offline is followed by MS analysis directly or coupled to other capillary electrophoresis (CE) or liquid chromatography (LC) systems.

3.2.3. *Multidimensional Protein Identification Technology (MudPIT)*

The innovative protein identification technology MudPIT was first introduced by Yates and coworkers.[60] This method was developed based on this group's online method of coupling 2D LC to MS/MS.[61] In brief, this method is made up of HPLC, MS/MS, and database searching. It focuses on the use of two columns interfaced back to back in silica capillary in order to permit 2D HPLC: one column contains an SCX material, and the other contains RP materials. After loading the complex peptide mixture into the microcapillary column, the column is inserted into the instrumental setup. Xcalibur software, HPLC, and MS are controlled simultaneously by means of the user interface of the mass spectrometer. As chromatography proceeds, an increase in salt concentration gradient causes the release of proteins from the SCX resin to the RP resin. Then, an RP reagent is introduced to eject peptides from the RP resin into a mass spectrometer. The tandem mass spectra generated are correlated to theoretical mass spectra generated from protein or DNA databases by the SEQUEST algorithm.

MudPIT has been extensively used to study cytokinesis proteins and membrane proteins, and is capable of identifying low-abundance proteins.[57] The advantage of MudPIT over 2DE is that MudPIT applies LC, allowing it to digest protein that is soluble in a solvent; whereas 2DE is much more rigid and will only break down proteins that can be dissolved in the gel.

3.2.4. *Surface-Enhanced Laser Desorption/Ionization Time-of-Flight (SELDI-TOF) Technology*

The SELDI method analyzes changes at the protein level and uses protein chips with an affinity matrix. Various wash buffers allow the differential binding of proteins to the surface of the proteins or to the surface of the chip, based on the stringency of their binding in various conditions.

SELDI-MS technology can be used without special skills and further enables the analysis of more than 100 samples within a reasonable time. This is a major step forward, since alternative approaches like 2DE followed by MS or HPLC-MS can be operated only by skilled personnel and are time-consuming. Unfortunately, SELDI-MS is limited in that only a very small fraction of polypeptides are detectable on the chip surface. Although SELDI-MS technology has severe limitations, it nevertheless enabled the start of "practical clinical proteomics", which would not have been possible using the rather complicated alternative technologies.[62]

3.2.5. *Nanoelectrospray Ionization Liquid Chromatography/Tandem Mass Spectrometry (nanoESI LC-MS/MS)*

As we know, the characterization of proteins and peptides originating from complex biological samples often requires sample isolation, cleanup, and separation using HPLC and 2DE before MS or MS/MS analysis. The amount of material available from such samples is usually limited, and sample concentrations are very low. Therefore, low limits of detection are necessary for the analysis of these samples and can be obtained by using nanoESI LC-MS/MS.

What is electrospray ionization (ESI)? ESI is a method of generating a very fine liquid aerosol through electrostatic charging. In electrospray, a liquid passes through a nozzle. The plume of droplets is generated by electrically charging the liquid to a very high voltage. The charged liquid in the nozzle becomes unstable as it is forced to hold more and more charge. Soon, the liquid reaches a critical point at which it can hold no more electrical charge, and at the tip of the nozzle it blows apart into a cloud of tiny, highly charged droplets.

What is the advantage of ESI-MS? ESI-MS is a proteomic technique that uses high voltage to generate ions from an aerosol of charged liquid droplets. Classically, commercial ESI mass spectrometers utilize flow rates from tens of microliters per minute $(10 \mu L/min)$ to milliliters per minute $(1 \ mL/min)$. Since a relatively large volume of liquid exits the emitter, aerosol formation must be assisted by pneumatic nebulization and/or by thermal heating in an effort to obtain a stable spray. This requirement is especially pronounced for highly aqueous liquids.[63]

As the flow rate is reduced to nanoliters per minute (nL/min), droplet formation occurs more readily, requiring only the applied voltage to generate spray. No sheath gas or additional heat is required. Consequently, the stability of spray, and therefore signal, at lower flow rates is typically improved for aqueous or "salty" mobile phases. The term "nanospray" means that it is working at lower flow rates of nL/min, and has now become a popular method employed in protein analysis. Low-flow ESI is especially tolerant to a wide range of liquid compositions, and can even spray "pure" water with a high degree of stability.[63]

The efficiency of ionization improves as the flow rate is lowered because less volume of mobile phase passes through the emitter, producing smaller aerosol droplets. The lower flow rates in the nanospray technique also allow for a longer length of analysis time. This provides ample time to perform novel mass spectrometer scan functions so as to obtain structural information of an analyte. Nanospray also provides for the direct coupling of nanoscale chromatographic methods; therefore, signal-robbing dilution by a sheath or makeup liquid is eliminated.[63]

Tandem mass spectrometers (MS/MS) are usually made of two mass analyzers separated by a collision cell. The first analyzer is used as a mass filter to isolate one single molecular ion that will enter the collision cell in which its fragmentation will be induced. The resulting fragments are directly analyzed in the second mass analyzer, where their molecular mass is measured, leading to highly valuable structural information such as amino acid sequences.

In a recent research done by Zheng *et al.*,[64] as an example, human serum was studied to identify peptides through the use of nanoLC-MS/MS analysis. To assess reproducibility, the researchers measured the average coefficient of variance (CV) of the retention time for five different peptides, and the value was found to be 0.15% for three runs. Peptides were identified in this study by MS/MS fragmentation in a hybrid linear ion trap–Fourier transform mass spectrometer (LTQ-FTMS).

It has been noted that low-level peptides may not be detected in an LC-MS/MS analysis due to the time constraints of the measurement, and thus repeated analyses are required. In addition, many correctly identified peptides, particularly the lower-abundance fragments, are not expected to be observed in replicate analyses because of the limitations of LC-MS.

In the study, a total of 804 unique peptides representing 359 unique proteins in human serum filtrate were identified using the LTQ-FT system.[64]

3.2.6. *Matrix-Assisted Laser Desorption/Ionization–Tandem Time-of-Flight Mass Spectrometry (MALDI-TOF/TOF MS)*

The ability to use laser energy to convert molecules into gas-phase ions was a long-standing goal. Laser light has a specific energy level, but attempts to use this energy to effect thermal desorption of molecules into gas-phase ions were ineffective. Finally, it was found that ions could be desorbed from a glycerol-based liquid laced with a finely powdered metal. Rather than tuning the laser energy to absorption wavelengths of the analyte, a "matrix" that absorbs the energy of the laser could be used to assist and promote thermal desorption. The glycerol/metal powder method quickly gave way to an approach that is based on cocrystallization of the analyte with an organic matrix — matrix-assisted laser desorption/ionization (MALDI) — as a more practical and reliable method to produce ions.[60]

MALDI is a laser-based soft ionization method used in MS of large molecules. In this technique, a sample is mixed with a molar excess of matrix, typically an aromatic acid with a chromophore that strongly absorbs the laser wavelength. Next, a laser hits the chemical matrix with short pulses of light for desorption and ionization. The matrix absorbs its light energy and causes a small part of the target substrate to vaporize. After the laser ionizes and vaporizes all of the sample molecules, the sample is transferred into a mass spectrometer where it is separated from the matrix ions and individually detected, usually by TOF MS.

The typical laser used for MALDI applications has been a 20-Hz nitrogen laser (314 nm). Recently, 200-Hz lasers that increase the rate of analysis by a factor of 10 have appeared.

However, MALDI has certain disadvantages such as low shot-to-shot reproducibility, short sample life time, and strong dependence on the sample preparation method. Polypeptides above ca. 3 kDa generally give unsatisfactory data; this problem can be solved by using FT0ICR methods, which enable the identification of polypeptides even larger than 10 kDa.[62]

3.2.7. *Tandem Time-of-Flight Mass Spectrometry (TOF/TOF MS)*

The TOF mass spectrometer is a type of mass analyzer used in proteomics. In this technique, a mass spectrum is measured by determining the flight time of ions down a field-free flight tube. The TOF of ions is related to their m/x values; therefore, a mass spectrum can be acquired. The major limitation of a TOF mass analyzer is its inability to perform true MS/MS.

In recent technology, Medzihradszky *et al.*[65] developed a tandem TOF mass spectrometer, using the high-speed capabilities of the mass analyzer to create a high-throughput tandem mass spectrometer. The TOF/TOF instrument is equipped with a MALDI source and a high-speed laser to create ions, making m/z measurement fast. A TOF mass analyzer is used in the ion selection process, and the selected ions are then transferred into a collision cell. Analysis of product ions occurs in a second TOF mass spectrometer.

The use of a MALDI source to create ions for the TOF/TOF mass spectrometer allows separation of the peptide fractionation process from the mass analysis, creating two separate workflows. In theory, this could enable several separation stations to create samples for one mass spectrometer, but this only improves efficiency if the MS analysis of each fractionated sample is faster than the separation.

3.2.8. *Fourier Transform Ion Cyclotron Resonance (FT-ICR)*

FT-ICR is a mass analyzer that provides mass spectra with a high content of information (MS/MS spectra). In FT-ICR MS, mass accuracy is strongly affected by the total number of ions trapped in the ICR cell. A high number of ions in an ICR cell will cause space charge, which is the reason for mass shifts of up to several hundred ppm. This problem is resolved by the Finnigan LTQ FT, which uses automatic gain control to regulate the number of ions for an FT-ICR MS experiment. Ions are generated in the electrospray ion source and then guided through transfer ion optics into the linear ion trap, which determines the total ion current being generated. After calculating the optimal injection time, ions are injected with a very reproducible ion population. Finally, ions are ejected from the linear trap into the FT-ICR part of the instrument, excited, and detected.

3.2.9. *Quadrupole Time-of-Flight Mass Spectrometry (Q-TOF MS)*

In this instrument, a collision cell is placed between a quadrupole mass filter and a TOF analyzer. Ions of a particular m/z ratio are selected in the quadrupole mass filter and fragmented in the collision cell, and then the fragment ion masses are read out by a TOF analyzer.[66] Q-TOF instruments can be used interchangeably with a MALDI and ESI source. Q-TOF instruments have a high sensitivity, resolution, and mass accuracy. The resulting fragment ion spectra are often more extensive and informative than those generated by trapping instruments.

4. PROTEIN MICROARRAY (PMA)

Protein microarrays (PMAs) are tools that can be used in many different areas of research, including basic and translational research. Because most drug targets are proteins, the pharmaceutical industry has a vested interest in high-throughput methods to perform protein expression profiling. While DNA microarrays effectively measure the levels of mRNA expressed in a cell, they cannot directly measure the amounts or functions of the proteins that these messengers produce. This has led to the use of PMAs. The miniaturization, parallelism, and high throughput that microarrays provide reduce the cost and speed up the collection of high-quality protein expression information.

Analyzing protein expression brings to light the molecular basis of disease. Microarrays can be used to study disease susceptibility, perform diagnosis, monitor progression, and discover potential points of therapeutic interference faster and more accurately than ever before. For example, the quantitative measurement of many serum proteins in parallel is more descriptive of a disease than a single biomarker, and should allow us to identify new biomarkers (or patterns of biomarkers) for improved diagnosis and/or prognosis.

PMAs can take on many different formats, and can be used to do more than simple expression profiling of samples. Recent publications have demonstrated that PMAs can be used to phenotype leukemia cells, identify novel protein–protein interactions, screen entire proteomes for new proteins, and profile hundreds of patient samples simultaneously.[67,68]

Proteomic arrays are typically high-density arrays (>1000 elements/array) that are used to identify novel proteins or protein/protein interactions. The library that is arrayed can come from any one of many possible sources, including expression libraries, and can contain known as well as unknown elements. The sample to probe the array can come from virtually any source. To detect proteins that are bound to the array, the samples must be labeled directly with a fluorophore or hapten; alternatively, in some applications, antibodies can be used to detect binding events. One common use is for antibody screening.

Protein microarrays are classified into three different groups according to their functionality:

(1) MicroSpot enzyme-linked immunosorbent assay (ELISA) and antibody arrays

MicroSpot ELISA and antibody arrays are used for quantitative profiling of protein expression in cell cultures or clinical specimens. Typically, these arrays are of low density (9–100 elements/array). In these arrays, known antibodies are arrayed and used to capture antigens from unknown samples. To detect an antigen that is bound to the array, either the antigen needs to be labeled directly with a fluorophore or a second binder/antibody can be used. The latter option creates a sandwich assay similar to a traditional ELISA, only in a microspot format.[69]

(2) Reverse arrays

Reverse arrays are used to profile dozens or hundreds of samples (research or clinical) for the presence of a small number of antigens.[70] Cell lysates, material from laser capture microdissection, or serum samples are arrayed. This creates an array of "unknowns" that can be probed with a small number of antibodies. Visualization can be performed with a detection or "top" antibody linked to a fluorophore or color detection reagent.

(3) Protein binder arrays

Protein arrays can be used to identify novel protein-binding motifs or protein/protein interactions. Engineered or synthetic proteins or peptides with various binding motifs are arrayed, and the array is probed with complex protein samples. Detection with a known antibody allows the researcher to identify previously unknown binding events.[71]

ProMAT is a software tool that statistically analyzes data from ELISA microarray experiments. The software estimates standard curves, sample protein concentrations, and their uncertainties for multiple assays. ProMAT generates a set of comprehensive figures for assessing results and diagnosing process quality. The tool is available for Windows and Mac, and is distributed as open-source Java and R code.[72]

Expansion towards experimentally complex systems will be an important direction of future development. Given that surfaces have such a crucial role in biology, microarrays might be the archetypal platform for an eventual model of a cell.[73]

5. PROTEOMIC DATA ANALYSIS

Proteomic studies involve the identification as well as the qualitative and quantitative comparison of proteins expressed under different conditions, and the elucidation of their properties and functions, usually in a large-scale, high-throughput format. The high dimensionality of data generated from these studies requires the development of improved bioinformatics tools and data-mining approaches for an efficient and accurate data analysis of biological specimens from healthy and diseased individuals. Mining large proteomics data sets provides a better understanding of the complexities between the normal and abnormal cell proteomes of various biological systems, including environmental hazards, infectious agents (bioterrorism), and cancers.

Proteomic analysis inherently involves the handling and interpretation of a huge amount of data. Originally, proteome analysis was aimed at collecting comprehensive information on all proteins present in a specified sample. Proteome inventories developed to be highly sophisticated databases collecting all of the different kinds of data formats, including 2D gel images, mass spectra, protein sequences, and posttranslational modifications (PTMs). These databases laid the foundations for identifying proteins in proteomic studies aimed at finding differentially expressed proteins, thus promoting proteomics from constructing "descriptive" databases to the design of "functional" experiments.

In order to find differentially expressed proteins, data have to be compared between two or more experimental groups. Therefore, a new field of bioinformatic tools had to be developed for the proteomic high-throughput technologies or had to be adapted from other applications, such as genomic and transcriptomic analysis based on nucleotide microarrays. In principle, these types of analyses — Kruskal–Wallis test, Fisher's exact test, t-test, significance analysis of microarrays (SAM),[74] weighted gene analysis (WGA),[75] and the mutual-information scoring method (InfoScore)[75] — would be applied to multidimensional gel electrophoresis or MDLC analysis.

In this section, the strategies for comparing protein concentration and functional activity by different statistical means, as well as the methodology of comparing whole sets of genes or proteins, will be evaluated. Comments on the statistical algorithms incorporated in 2D gel analysis software and discussions on alternatives for data comparison are incorporated. The use of supervised and unsupervised data analyses and their application in proteomic experiments, including the use of hierarchical clustering for identification of functional pathways in proteome analysis, will also be reviewed.

2DE is a powerful technique to examine PTMs of complexly modulated proteins. Currently, spot detection is a necessary step to assess the relations between spots and biological variables. This often proves time-consuming and difficult when working with imperfect gels. An analysis technique to measure the correlation between 2DE images and biological variables on a pixel-by-pixel basis has recently been developed.[76] In brief, after image alignment and normalization, the biological parameters and pixel values are replaced by their specific rank; these rank-adjusted images and parameters are then put into a standard linear Pearson correlation, and further tested for significance and variance. This analysis method measures the relations between 2DE images and external variables without requiring spot detection, thereby enabling the exploration of biosignatures of complex signaling networks in biological systems.[76]

6. VALIDATION TECHNOLOGIES

Nowadays, the majority of computational protein annotation methods used are based on scoring during MS protein identification.[77] In general, when testing whether a protein should be given a certain annotation, a score threshold is set, and proteins that score higher than the threshold are given the annotation. Obviously, some annotation mistakes may occur. Such mistakes can be divided into false-positives (FPs) and false-negatives (FNs). FPs (or false hits) are annotations that were mistakenly assigned to a protein (type I error). FNs (or misses) are annotations that should have been assigned to a protein, but were not (type II error).

To further confirm the validity of the proteins identified by MS, a certain validation strategy must be required. Currently, the available techniques that are used for this purpose are tissue microarray of immunohistochemistry (TMA IHC), ELISA, Western blot, and reverse transcription–polymerase chain reaction (RT-PCR). Quantitative MS can now also be an approach used for validation of the proteins identified.[78] In this section, we will highlight these commonly used techniques.

6.1. *Tissue Microarray of Immunohistochemistry (TMA IHC)*

A large gap currently exists between the ability to discover potential biomarkers and the ability to assess the real value of these proteins for cancer screening. One major challenge in biomarker validation is the inherent heterogeneity of biomarker expression. An effective validation procedure, therefore, seems to be most important for biomarker validation. Cancer tissue microarray (TMA) and ELISA as well as Western blot assay have been demonstrated to be effective and reliable antibody-based methods used for biomarker validation. There are several TMA companies available, including US Biomax (Rockville, MD; http://www.biomax.us/), Cybrdi Inc. (Frederick, MD; http://www.cybrdi.com/), and BioCat GmbH (Heidelberg, Germany; http://www.biocat.de/):

- US Biomax offers more than 298 types of human tissue arrays and microarrays.
- BioCat GmbH offers a variety of human cancer TMAs with detailed information (e.g. Cat# LC2001-BX: 200 cores, single core for each case, 192 cases of cancer in various stages and 8 cases of normal tissue [matched or unmatched]).
- Cybrdi manufactures hundreds of predeveloped human tissue arrays that can be used to survey hundreds or even thousands of clinical specimens in a single experiment using common probes such as DNA, RNA, peptide, protein, and antibodies.

For immunohistological staining of the newly induced proteins of TMA slides, a standard two-step indirect avidin–biotin complex (ABC) method is used (Vector Laboratories, Burlingame, CA). Four-micrometer-thick tissue array sections are cut prior to staining. They are first heated to $56°C$ for 20 min, followed by deparaffinization in xylene. The sections are then rehydrated in graded alcohols, and endogenous peroxidase is quenched with 3% hydrogen peroxide in methanol at room temperature. The sections are placed in a $95°C$ solution of 0.01 M sodium citrate buffer (pH 6.0) for antigen retrieval. Protein blocking is accomplished through the application of 5% normal horse serum for 30 min. Endogenous biotin is then blocked with sequential application of avidin D and biotin (A/B blocking system). Primary antibody is applied at a 1:750 dilution for 60 min at room temperature. After washing, biotinylated secondary antibody is applied for 30 min at room temperature. Next, the ABC complex is applied for 25 min and diaminobenzidine (DAB) is used as the chromagen. Ten millimolar phosphate buffered saline (PBS) at pH 7.4 is used for all intermediate wash steps, and a moist humidity chamber is used for prolonged incubations. The sections are counterstained with Harris' hematoxylin, followed by dehydration and mounting.

Semiquantitative assessment of antibody staining is performed blinded to clinicopathological variables. The intensity of DAB brown chromagen staining is graded using a 0–2 scale (0 = negative; 1 = weak staining; 2 = strong staining). The median value of four repeated core spots for each individual sample is used for the final analysis.

6.2. *Enzyme-Linked Immunosorbent Assay (ELISA)*

ELISA can be readily utilized in large-scale case/control studies to validate the proteomics-discovered protein biomarkers. ELISA approaches are targeted and selective. They are also high in throughput, reproducible, and sensitive.

Alexander *et al.*[79] analyzed nipple aspirate fluid (NAF) proteome and found a limited number of proteins. The researchers studied three protein spots that were upregulated in three or more cancer samples: a1-acid glycoprotein (AAG), gross cystic disease fluid protein (GCDFP)-15, and apolipoprotein D (apoD). Since the preliminary association of these biomarkers with breast cancer was made in a small sample set, the proteins were investigated by ELISA in 105 NAF samples. This technique is a well-established method to quantitate the level of antigen in a sample. Proteomic analysis by 2D PAGE and ELISA showed that the expression of AAG and the nonexpression of GCDFP-15 correlated with disease presence and stage.

The same antibody concentration was used throughout the experiment in all of the wells, and the results were scored at the point where 50% of the antibody was bound. Through this, the researchers were able to determine how much each NAF sample needed to be diluted in order to bind 50% of the antibody.

6.3. *MS-based Validation*

New configurations of mass spectrometers have been developed to isolate ions, fragment them, and then measure the mass-to-charge ratio of the fragments. The linear 2D ion trap is a device based on the 3D quadrupole, which traps ions in the center of the device and then scans the ions from the trap to a detector. There were several issues that limited the performance of 3D traps. First, there was a limit to the number of ions that could be trapped in the device. Second, when ions were scanned from the trap, half of them exited in the direction of the detector and the other half exited in the opposite direction. Third, there was a limitation in mass accuracy and resolution, although a narrow mass range scan could be employed to obtain high-resolution data with improved mass accuracy. Two-dimensional quadrupole ion traps or linear ion traps can hold almost 10 times more ions than 3D traps.

Collecting ions with two detectors doubles the ion current collected during a scan of the m/z range. A second feature of the linear ion trap is the ability to scan at a much faster speed, thus increasing the number of scans that can be acquired over the course of an LC analysis. Linear ion traps have limits to the mass resolution or accuracy that can be obtained. At normal scan speeds, a unit resolution of ~2000 is obtained, but slowing the scan speed can yield much higher resolutions of 15 000 over a 10-amu window.

There is a new LC-MS technology called Extended Range Proteomic Analysis (ERPA),[80] which is able to achieve very high sequence coverage and comprehensive characterization of PTMs in complex proteins. This novel platform combines protein digestion with an enzyme, such as Lys-C, which cuts less frequently than trypsin, leading to an average higher-molecular-weight peptide size. It also leads to HPLC separation of the resulting fragments, a new data acquisition strategy using LTQ-FTMS — a hybrid mass spectrometer that couples a linear ion trap with an FT-ICR cell — for the analysis of peptides in the range of 0.5–10 kDa, and new data analysis methods for assigning large peptide structures and determining the site of attachment of PTMs as well as structural features from the accurate precursor mass together with MS/MS and MS/MS/MS fragmentations.

7. PROTEOMIC APPLICATION

The field of proteomics has now gained more interest in directly analyzing protein expression at the posttranslational level because it permits the qualitative and quantitative assessment of a broad spectrum of proteins that can be related to the response to oxidative stress.[81–83] Consequently, there have been many innovations in the applications of proteomics. Proteomics research has been utilized to gain a better understanding of diseases and is significant in the fields of biomarker or drug target discovery. In this section, we will highlight the related research and novel applications for each of the specific areas of proteomics.

7.1. *Blood*

7.1.1. *Red Blood Cells*

The topic of red blood cells is fairly new to proteomics research. In one study, red blood cell membrane proteins were separated by SDS-PAGE and analyzed by MS; the researchers Low *et al.*[84] were able to identify 44 polypeptides, of which only 19 were also found on 2DE gels. The most complete study describing the proteome of red blood cells has been reported by Kakhniashvili *et al.*[85]: they identified about 100 membrane proteins, including several proteins of the cytoskeletal membrane such as spectrin and ankyrin. A few red blood cell antigens were also characterized, such as Rh blood group D, glycophroin A, aquaporin, and Lutheran blood group protein. The study of the cytoplasmic fraction of erythrocytes allowed the identification of many proteins, including hemoglobin α, β, and γ chains.

The analysis of phosphoprotein profiling of erythropoietin receptor–dependent pathways has also offered substantial information on red blood cell biomarkers. Research done by Korbel *et al.*[86] combined the proteomics approaches 2DE/MALDI-TOF and 1D LC-MS/MS; this method offered the detection of low-level–expressed signaling molecules because an effective enrichment of phosphoproteins was achieved. Proteomics was also successfully employed to analyze the invasion of malaria parasites into human red blood cells, detecting two novel surface proteins on *Plasmodium faciparum*–infected erythrocytes.[87] Red blood cell membrane proteins such as flotillin 1, syntaxin 1C, and arginase also appear to be dysregulated in type 2 diabetes patients.[87]

7.1.2. *Platelets*

Platelets, also known as thrombocytes, are the smallest circulating blood particles. They derive from megakaryocytes located in the bone marrow, and play critical roles in primary and secondary hemostasis (blood coagulation) as well as contribute to the formation of vascular plugs.[88] Unwanted platelet activation and arterial thrombus formation are implicated in the onset of myocardial infarction, stroke, and other cardiovascular diseases. The significance of platelets has revolutionized the pharmacological treatment of cardiovascular diseases, and aspirin is now an essential antiplatelet drug.[89] Proteomics is the best way to approach the biochemistry of platelets because platelets do not have a nucleus.[88]

Over the last few years, several research groups have applied proteomics to platelet research, establishing a detailed analysis of the general proteome and signaling cascades in human platelets. This knowledge of platelet function will aid in the development of new therapeutic agents to treat thrombotic diseases. Recently, research groups have focused on a detailed analysis of the human platelet proteome, especially with platelets in their basal state. Researchers use 2DE technology to separate cytosolic platelet proteins, and analyze them by MALDI-TOF MS. Reports from the Glycobiology Institute at Oxford University (England) provided a high-resolution 2DE platelet proteome map comprising more than 2000 different protein features.[90,91] The platelets were isolated by a sensitive method that minimized contaminations from other vascular cells. Proteins were separated by 2DE by using narrow pH gradients during IEF, and 9%–16% SDS-PAGE gradient gels in the second dimension (18×18 cm). Gels were stained with a highly sensitive fluorescent dye, and the corresponding protein spots were excised, in-gel trypsinized, and analyzed by LC-MS/MS. The studies reported the identification of 284 protein features that correspond to 123 different genes in the pI 4–5 region, and 760 protein features that correspond to 311 different genes in the pI 5–11 region.

However, it is known that very basic and/or hydrophobic proteins, such as membrane proteins, escape the 2DE approach of protein analysis. To overcome this problem, nongel techniques must be improved to analyze the proteome. These techniques rely on mass spectrometric analysis of enzymatically digested protein mixtures. In 2002, Geveart and colleagues[92] reported a novel gel-free proteomics technology called combined fractional diagonal chromatography (COFRADIC™). This technology alters the retention behavior of specific peptides on RP columns. It has been used to identify 264 proteins in a cytosolic and membrane skeleton fraction of human platelets, and later another 163 different platelet proteins with a broad range of functions and abundance.

7.1.3. *Plasma*

Blood plasma is a complex body fluid containing an estimate of more than 10 000 different proteins, with concentrations ranging over at least 15 orders of magnitude. Albumin is the main plasma protein, and represents 50% of the plasma protein in high concentration; albumin is followed by immunoglobulin (Ig), fibrinogen, transferrin, haptoglobin, and lipoprotein in abundance. Current peptidome studies have revealed about 5000 peptides in human serum.[87] However, a major challenge of proteomics analysis of plasma is the low-abundance proteins such as (1) proteins or peptides resulting from tissue leakage, (2) proteins released from normal cells as a result of cell death or damage, and (3) proteins released from tumor cells as aberrant secretions. All of these proteins may be significant as potential biomarkers for diseases.

Several advancements have been achieved in the field of plasma proteomics in the last 5 years. In 2004, Anderson and Anderson[93] published one of the most precise available lists of human plasma proteins. Chan *et al.*[94] analyzed the serum proteome by combining multidimensional peptide separation strategies and MS/MS; they identified 1444 unique proteins and constructed a public database of human serum proteome.

An important technique for the study of plasma proteomics is 2DE because it allows analysis of the microheterogeneity of proteins, due to genetic polymorphisms or PTMs. In specific studies, Watanabe *et al.*[95] analyzed serum samples from pregnant women by 2DE and identified a group of overexpressed spots corresponding to clusterin in women with preclampsia. Also using 2DE and MS analysis, Kwak *et al.*[96] showed that the serum proteome of patients with acute myeloid leukemia differed from that of controls: patients with acute myeloid leukemia expressed upregulated levels of haptoglobin-1. Two-dimensional electrophoresis studies have also led to the discovery of a new peptide called hSpα or CD5L. This peptide is able to bind to different cells of the immune system (monocytes and lymphocytes), thus suggesting that it may play an important role in the regulation of the immune system. CD5L is a relatively abundant serum protein with a concentration of 60 μg/mL and circulates in association with other serum proteins, particularly immunoglobulin M (IgM). Chen *et al.*[97] studied four patients with severe acute respiratory syndrome. After 2DE, a total of 38 different spots were selected for protein identification, and most of them corresponded to acute phase proteins; the researchers also identified proteins not detected before on plasma 2DE, such as peroxiredoxin II. Comparative proteome analysis of human plasma and LC-MS/MS analysis allowed Qian *et al.*[98] to identify 32 proteins that were significantly increased after lipopolysaccharide administration.

Various non-2DE approaches have also been successful in the study of the proteome of biological samples and the discovery of new potential plasma biomarkers. This principally includes the combination of protein chip arrays with high-resolution MS, such as SELDI-TOF MS. This technology has been used in biomarker discovery and clinical diagnostics, especially in oncology. For example, SELDI-TOF MS patterns have shown prostate-specific antigen levels of 2.5–15.0 ng/mL in patients presenting with either prostate cancer or benign prostate pathologies; artificial intelligence based on pattern recognition algorithms was able to distinguish between the two categories. SELDI-TOF MS has also uncovered three biomarkers of early-stage ovarian cancer: the identification of apolipoprotein A1 (downregulated in cancer), a truncated form of transthyretin (downregulated in cancer), and a cleavage fragment of inter-α-trypsin inhibitor heavy chain H4 (upregulated in cancer). Heart fatty acid–binding protein as well as apolipoproteins C-I and C-III have been proposed as potential biological markers for the diagnosis of stroke; interestingly, apolipoproteins C-I and C-III appear to be the first reported plasmatic biomarkers able to accurately distinguish between ischemic and hemorrhagic strokes within a small number of patients. However, a setback of SELDI-TOF MS approaches in proteomics is its poor reproducibility.

7.2. *Body Fluid*

7.2.1. *Saliva*

The ability to utilize saliva to monitor the health and disease state of an individual is a highly desirable goal. However, there has only recently been a growing appreciation of saliva as a reflection of the body that can virtually show the entire spectrum of normal and disease states.[99] These include tissue levels of natural substances and a large variety of molecules introduced for therapeutic, dependency, or recreational purposes; emotional, hormonal, or immunological status; neurological effects; and nutritional and metabolic influences. However, a disadvantage of the analysis of saliva as a diagnostic fluid is that informative analytes are generally present in lower amounts in saliva than in serum.[100] With new and very sensitive techniques, the lower level of analytes in saliva is no longer a limitation.

Saliva is useful in clinical applications because it is inexpensive, noninvasive, and displays easy-to-use diagnostic methods. Research done by Li *et al.*[101] has led to the discovery that discriminatory and diagnostic human mRNAs are present in the saliva of both normal and diseased individuals. The salivary transcriptome presents an additional valuable resource for disease diagnostics. The normal salivary transcriptome consists of $\sim$3000 mRNAs.[101] Of particular value is that, of these 3000 mRNAs, 180 are common between different normal subjects, constituting the normal salivary transcriptome core (NSTC). To demonstrate the diagnostic and translation potential of the salivary transcriptome, investigators have profiled and analyzed saliva from head and neck cancer patients. Based on four mRNAs from the NSTC (IL8, OAZ1, SAT, and IL1B), they were able to discriminate and predict whether a saliva sample was from a cancer or normal subject with a combined sensitivity and specificity of 95%.[102]

While head and neck cancer was used in the first analysis of salivary transcriptome diagnostics, data are now available for systemic disease applications. Analysis of these data provides sufficient confidence and highlights the necessity to fully explore salivary transcriptome diagnostics fully for major human disease translational applications.[103]

7.2.2. *Urine*

For several years, proteomics research has been expected to lead to the finding of new markers that will translate into clinical tests applicable to samples such as serum, plasma, and urine — the so-called *in vitro* diagnostics (IVDs).[104] Attempts to implement technologies applied in proteomics, particularly protein arrays and SELDI-TOF MS, as IVD instruments have initiated constructive discussions on the opportunities and challenges inherent in such a translation process with respect to the use of multimarker profiling approaches and pattern signatures in IVDs.

Different diagnostic applications of disease-specific proteins have been comprehensively outlined by Zolg and Langen.[105] For example, screening markers and disease-staging markers are used to detect proteins in urine. Screening markers identify disease preferentially in an asymptomatic stage within a population to start treatment as soon as possible, while disease-staging or classification markers differentiate disease states.

7.2.3. *Bronchial Alveolar Lavage Fluid (BALF)*

The bronchial alveolar lavage technique is one approach to evaluate cellular and protein components in the lower respiratory tract of lungs. It is a diagnostic and therapeutic procedure conducted by placing a suction catheter into the lung of a patient and injecting sterile saline into the lung. The bronchial alveolar lavage fluid (BALF) is acquired from the terminal bronchi. The first 2D map providing the major soluble proteins present in BALF was published in 1979.[106] Currently, research in BALF is directed towards establishing an exhaustive 2DE reference database of BALF proteins. Protein maps of BALF in various disease states using 2DE and MS have been constructed,[107–109] including cystic fibrosis,[110] pulmonary fibrosis,[111] hypersensitivity pneumonitis,[112] and immunosuppression.[113]

Unfortunately, the use of BALF has a disadvantage in that soluble proteins are very diverse and can originate from a broad range of sources, both endogenous and exogenous. Therefore, detected differences in the amount of lung-specific proteins in BALF may result from different kinds of sources. BALF proteomics also possesses two problems, low protein concentration and high salt concentration, which require special sample-handling procedure analysis.

7.2.4. *Nipple Aspirate Fluid (NAF)*

Nipple aspirate fluid (NAF) contains proteins that are secreted and represents the final processed form of a protein. Alexander *et al.*[114] analyzed NAF as a noninvasive method to identify candidate biomarkers of breast cancer. Analysis of NAF is more advantageous than proteomics analysis of cells from breast cancer tissue because the protein comparison is less complex.[114]

A recent study of the interstitial fluid of breast tumors identified more than 1000 proteins, while NAF analysis under 2DE and MALDI-TOF MS identified 41 different proteins. In the study, NAF proteins from a single subject without breast cancer were separated by 2D PAGE. First-dimension separation was done by focusing the proteins at a total of 80 000 volt-hours, with a 6000-V programmable power supply; second-dimension separation was performed by PAGE on 8%–18% gradient gel chips. A total of 41 different proteins were identified, 25 of which were known to be secreted. Then, the samples were subjected to MALDI-TOF MS to identify prolactin-induced proteins, and three proteins were found to be upregulated in NAF from a cancer-containing breast compared with a normal breast from the same woman: AAG, GCDFP-15, and apoD.

Next, ELISA was used to quantitate the level of antigen in a sample. ELISAs revealed that the median GCDFP-15 expression was sixfold lower in samples from breasts with cancer than in those without ($P < 0.001$).[114] AAG levels in the NAF samples were higher in women with breast cancer ($P = 0.001$), when considering both all subjects and premenopausal subjects; however, the levels of AAG were not related to breast cancer in postmenopausal women. ApoD levels were not associated with the presence of breast cancer, whether considering all subjects or subjects divided by menopausal status.[114]

Also, when determining whether GCDFP-15, apoD, or AAG levels in NAF were related to disease stage, it was found that the lowest levels of GCDFP-15 and the highest levels of AAG were in breasts from women with ductal carcinoma *in situ* (DCIS); apoD levels were not significantly different based on disease stage.[114] Breasts with invasive cancer had higher levels of AAG than breasts with benign disease, but lower levels than in breasts with DCIS; apoD levels were not significantly different based on disease stage. Women with invasive cancer had higher levels of AAG than breasts with benign disease, but lower levels than in breasts with DCIS.[114]

In regards to the effects of menopause, GCDFP-15 levels were higher in the breasts of premenopausal women, both overall and in breasts without cancer; apoD expression was not significantly influenced by menopausal status; and AAG levels were significantly higher in postmenopausal than premenopausal women, both overall and in breasts without cancer.[114] Nonetheless, the researchers suggested that all of the identified proteins be measured in a large number of NAF samples by using ELISAs to further correlate their levels with various disease stages, due to the variability of expression of biomarkers within the population.

7.3. *Tissue*

Proteomic approaches have been applied to describe the proteomics profiling of epithelial cells and the epithelial origin of cancer and stroma cells, monitor molecular targeted therapy, measure intracellular signaling pathways in primary and metastatic cancer, and provide implications for diagnosis of the diseases. Proteomic analysis can be useful for screening and selecting highly sensitive, specific proteins as biomarkers of drug-associated interstitial lung diseases. Most published papers and reviews have demonstrated the changes of protein profiles in epithelial cells in a certain organ, tissue, or disease.[115]

Tissue proteomics has also been implemented in the study of heart failure. Proteomic investigations of heart failure have been concentrated on dilated cardiomyopathy (DCM),[116] a disease of unknown etiology for which contributory factors include genetic factors, prior viral infections, cardiac specific autoantibodies, and toxic agents, among others. Furthermore, heart failure is caused not only by DCM, but also by ischemia and/or hypertension or congenital heart defects.[117] The causes of contractile dysfunction in heart failure are still largely unknown, but they are likely to result from underlying alterations in gene and protein expression level, including modifications.[118] Therefore, proteomic studies are likely to give new insights into the cellular mechanisms involved in cardiac dysfunction, and may also provide new diagnostic and therapeutic markers. Because cardiac myocytes are mainly responsible for contractility and hypertrophic growth, the use of *in vitro* models of more or less homogeneous cardiomyocytes for protein profiling purposes may be advantageous.

Researchers have characterized alterations in DCM protein patterns using conventional 2DE and silver staining.[119–121] These studies revealed few significantly altered proteins, of which only a few proteins could be identified; the latter was limited by the lack of MS identification. Investigators have also studied (chamber-specific) DCM protein patterns and presented them in a Web-accessible database.[121,122] Another study used 2DE to determine quantitative and qualitative changes in protein expression in heart tissue from patients with DCM compared to patients with ischemic heart disease and undiseased controls[123]; among the 88 spots that were found to be downregulated in DCM compared with ischemic cardiopathy, multiple desmin-containing spots were found as well as several metabolic enzymes and stress-related proteins. The aforementioned studies are examples of the implementation of proteomics in studying heart disease. The limited use of MS in these studies still impedes proper identification of proteins and their PTMs.

7.4. *Population Proteomics*

Population proteomics is the quantitative and qualitative study of protein diversity in human populations. High-throughput, top-down mass spectrometric approaches are employed to investigate, define, and understand protein diversity and modulations across and within populations. Concentration ranges of human plasma have been determined for a large number of proteins via immunoassays and similar methods of protein quantification.

The MS method of detection is widely used in proteomics because it incorporates a second structural dimension to protein analysis, enabling rapid assessment of PTMs and point mutations. MS methods that can be utilized in population proteomics must be capable of analyzing hundreds, if not thousands, of samples per day with high reproducibility and sensitivity. Existing LC-MS/MS proteomics methods fail in this area due to their difficulties in reproducibility, throughput, and protein sequence coverage.[124] Existing methods use the bottom-up method of protein detection, which is disadvantageous in population proteomics because only a small part of the sequence is generally detected for each protein when analyzed from a complex mixture such as plasma.

Therefore, population proteomics relies on a few top-down MS methods to perform peptide mapping characterization and sequence verification. MS arrays based on SELDI technology offer the capability to selectively screen for groups of proteins that share similar characteristics such as hydrophobicity. Other MALDI MS approaches that may be useful in population proteomics include 2D gels coupled to MS,[125] bead-based methods,[126,127] MS tissue imaging,[128] and whole-fluid MS for detection of abundant proteins. The main setback of these wide-specificity MS approaches is that sequence-based protein verification is difficult due to the simultaneous analyses of tens (if not hundreds) of proteins and the limitations in throughput, particularly in gel-based methods.[124]

One technology that seems very effective in achieving high throughput and reproducibility, as well as utilizes MS as a method of detection, was perfected and utilized in Nedelkov's laboratory. It is a hybrid methodology that combines protein affinity extraction with rigorous characterization using MALDI-TOF MS. Protein affinity extraction is achieved with the help of affinity pipettes, and is followed by elution and MALDI-TOF MS of the eluted proteins. Specificity and sensitivity are dictated by the antibodies, i.e. the affinity-capture reagents. A second measure of specificity is incorporated in the resulting mass spectra. High-throughput analysis is achieved via the use of robotics that enables parallel manipulation of 96 affinity pipettes, similar to the 96-well plate workstations and readers in ELISA. However, before disseminating the protein into peptide chains, it is important to evaluate the mass of the intact protein. Once this mass value is confirmed, the presence of protein modifications can be noted by the appearance of other signals in the mass spectra. Modifications can be identified through accurate measurement of the observed mass shifts and through knowledge of the protein sequence and possible modifications. The identify of the modification is then verified using proteolytic digestion and mass mapping approaches in combination with high-performance MS. This will ultimately assemble a catalog of the modifications and their frequencies.[124]

8. FUTURE DIRECTION OF PROTEOMICS

2D PAGE is the current main method of separating complex proteomic samples, due to its high resolving power for proteins. However, this technique still has major setbacks when analyzing low-abundance proteins, membrane proteins, proteins with extreme pI values, and very large or very small proteins. In order to overcome these obstacles, LC-based separation techniques directly coupled with MS detection have been developed to obtain a comparable resolution as in 2D PAGE.

Today, online or offline 2D LC-MS is state of the art for the identification of proteins from complex proteome samples in many laboratories.[129] Usually, online 2D LC-MS/MS is performed using an SCX column in series with an RP column. In the course of analysis, tryptic peptides are eluted stepwise by injecting salt plugs of increasing ionic strength from the SCX column in the first dimension; in the second dimension, these peptides are first trapped on an RP enrichment column and then separated on an analytical RP column.[130] This methodology is capable of resolving large proteomes as well as identifying a subset of proteins expressed under special conditions.

In contrast, offline 2D LC-MS/MS methodology significantly increases chromatographic resolution of highly complex proteome samples. This technology is contributed to continuous linear gradient SCX chromatography in the first dimension without intermitting RP chromatography intervals. As a consequence, higher resolution and peak capacity and, therefore, a higher number of identified proteins are associated with the offline 2D LC-MS/MS approach. For fraction collection in the first dimension (SCX), a microfraction collection device must be used. In the second dimension, nanoRP-LC/MS is applied to the separation of the eluent obtained from the first dimension.

The offline approach has several advantages over the online SCX methodology. In the offline approach, peptide separation is superior in a linear gradient and more organic solvents can be used to improve chromatography. Also, more peptide fractions can be collected and reanalyzed or chemically modified.[131] Chromatography and the corresponding fraction collection can also be more interactively optimized for the individual sample in the offline approach by using the ultraviolet trace of the chromatogram and adjusting the fraction width accordingly. The offline SCX approach has been very successful in elucidating the proteomes from yeast[131] and human serum.[132]

The advantage of using 2D LC-MS is a robust and fully automated separation: it is able to perform the complete analysis in an unattended fashion. However, the disadvantage is that the SCX column is working far away from its optimum under these conditions; resolution and the number of salt steps are also rather limited with this approach. To overcome these drawbacks in an online 2D LC system, a more complex instrument setup capable of working with a pumped semicontinuous gradient has been developed. This enhanced system enables the investigator to obtain more resolution out of a given sample without compromising automation. For the analysis of proteome samples of the highest complexity, a HPLC instrument setup for offline 2D MS has been developed; this system separates the two dimensions in two different HPLC systems. The advantages of the system are high resolution, user-defined fraction width and number, easy optimization by using ultraviolet light detection, chemical manipulation possibility, and a reanalysis option.

In the future, the critical issue will be to use multiple techniques in order to balance each other and support their weaknesses.[129] This is important because there is no universal technique that is superior in protein identification. Therefore, both 2DE and 2D LC-MS separation techniques will contribute complementarily to the progress in proteomic research and drug development. The offline 2D LC-MS approach provides high resolution and increased flexibility over the online 2D LC approach; this will promote the offline 2D LC methodology as the best choice for future needs in proteomic separations.[129]

ACKNOWLEDGMENTS

This research was supported by a setup fund to G. Xiao at Hunan Normal University, Changsha, China; and by a new faculty startup fund to G. Xiao at the Department of Medicine and Chao Family Comprehensive Cancer Center, University of California–Irvine, Irvine, CA 92697.

REFERENCES

1. Crameri R, Schulz-Knappe P, Zucht HD. (2005) The future of post-genomic biology at the proteomic level: an outlook. *Comb Chem High Throughput Screen* **8**:807–810.
2. Xiao G, Wang M, Li N, *et al.* (2003) Use of proteomics to demonstrate a hierarchical oxidative stress macrophage cell line. *J Biol Chem* **278**:50781–50790.
3. Xiao GG, Nel AE, Loo JA. (2005) Nitrotyrosine-modified proteins and oxidative stress induced by diesel exhaust particles. *Electrophoresis* **26**:280–292.
4. Alaiya A, Al-Mohanna M, Linder S. (2005) Clinical cancer proteomics: promises and pitfalls. *J Proteome Res* **4**:1213–1222.

5. Wang MZ, Howard B, Campa MJ, *et al.* (2003) Analysis of human serum proteins by liquid phase isoelectric focusing and matrix-assisted laser desorption/ionization–mass spectrometry. *Proteomics* **3**:1661–1666.

6. Zheng X, Baker H, Hancock WS. (2006) Analysis of the low molecular weight serum peptidome using ultrafiltration and a hybrid ion trap–Fourier transform mass spectrometer. *J Chromatogr A* **1120**:173–184.

7. Magueijo V, Semiao V, Norberta de Pinho M. (2005) Fluid flow and mass transfer modelling in lysozyme ultrafiltration. *Int J Heat Mass Transf* **48**:1716–1726.

8. Hansson SF, Puchades M, Blennow K, *et al.* (2004) Validation of a prefractionation method followed by two-dimensional electrophoresis — applied to cerebrospinal fluid proteins from frontotemporal dementia patients. *Proteome Sci* **2**:7.

9. Puchades M, Hansson SF, Nilsson CL, *et al.* (2003) Proteomic studies of potential cerebrospinal fluid protein markers for Alzheimer's disease. *Brain Res Mol Brain Res* **118**:140–146.

10. Harper RG, Workman SR, Schuetzner S, *et al.* (2004) Low-molecular-weight human serum proteome using ultrafiltration, isoelectric focusing, and mass spectrometry. *Electrophoresis* **25**:1299–1306.

11. Hesse C, Nilsson CL, Blennow K, Davidsson P. (2001) Identification of the apolipoprotein E4 isoform in cerebrospinal fluid with preparative two-dimensional electrophoresis and matrix assisted laser desorption/ionization–time of flight–mass spectrometry. *Electrophoresis* **22**:1834–1837.

12. Davidsson P, Paulson L, Hesse C, *et al.* (2001) Proteome studies of human cerebrospinal fluid and brain tissue using a preparative two-dimensional electrophoresis approach prior to mass spectrometry. *Proteomics* **1**:444–452.

13. Davidsson P, Westman A, Puchades M, *et al.* (1999) Characterization of proteins from human cerebrospinal fluid by a combination of preparative two-dimensional liquid-phase electrophoresis and matrix-assisted laser desorption/ionization time-of-flight mass spectrometry. *Anal Chem* **71**:642–647.

14. Puchades M, Westman A, Blennow K, Davidsson P. (1999) Analysis of intact proteins from cerebrospinal fluid by matrix-assisted laser desorption/ionization mass spectrometry after two-dimensional liquid-phase electrophoresis. *Rapid Commun Mass Spectrom* **13**:2450–2455.

15. Davidsson P, Folkesson S, Christiansson M, *et al.* (2002) Identification of proteins in human cerebrospinal fluid using liquid-phase isoelectric focusing as a prefractionation step followed by two-dimensional gel electrophoresis and matrix-assisted laser desorption/ionisation mass spectrometry. *Rapid Commun Mass Spectrom* **16**:2083–2088.

16. Westman-Brinkmalm A, Davidsson P. (2002) Comparison of preparative and analytical two-dimensional electrophoresis for isolation and matrix-assisted laser desorption/ionization–time-of-flight mass spectrometric analysis of transthyretin in cerebrospinal fluid. *Anal Biochem* **301**:161-167.

17. Thoren K, Gustafsson E, Clevnert A, *et al.* (2002) Proteomic study of non-typable *Haemophilus influenzae. J Chromatogr B Analyt Technol Biomed Life Sci* **782**:219–226.

18. Covert BA, Spencer JS, Orme IM, Belisle JT. (2001) The application of proteomics in defining the T cell antigens of *Mycobacterium tuberculosis. Proteomics* **1**:574–586.

19. Gustafsson E, Thoren K, Larsson T, *et al.* (2001) Identification of proteins from *Escherichia coli* using two-dimensional semi-preparative electrophoresis and mass spectrometry. *Rapid Commun Mass Spectrom* **15**:428–432.

20. Nilsson CL, Larsson T, Gustafsson E, *et al.* (2000) Identification of protein vaccine candidates from *Helicobacter pylori* using a preparative two-dimensional electrophoretic procedure and mass spectrometry. *Anal Chem* **72**:2148-2153.

21. Janini GM, Conrads TP, Veenstra TD, Issaq HJ. (2003) Development of a two-dimensional protein–peptide separation protocol for comprehensive proteome measurements. *J Chromatogr B* **787**:43–51.

22. Lubman DM, Kachman MT, Wang H, *et al.* (2002) Two-dimensional liquid separations — mass mapping of proteins from human cancer lysates. *J Chromatogr B* **782**:183–196.

23. Wang H, Kachman MT, Schwartz DR, *et al.* (2002) A protein molecular weight map of ES2 clear cell ovarian carcinoma cells using a two-dimensional liquid separations/mass mapping technique. *Electrophoresis* **23**:3168–3181.

24. Wall DB, Parus SJ, Lubman DM. (2002) Three-dimensional protein map according to pI, hydrophobicity and molecular mass. *J Chromatogr B* **774**:53–58.

25. Kachman MT, Wang H, Schwartz DR, *et al.* (2002) A 2-D liquid separations/mass mapping method for interlysate comparison of ovarian cancers. *Anal Chem* **74**:1779–1791.

26. Gong Y, Li X, Yang B, *et al.* (2006) Different immunoaffinity fractionation strategies to characterize the human plasma proteome. *J Proteome Res* **5**:1379–1387.

27. Gorg A, Weiss W, Dunn MJ. (2004) Current two-dimensional electrophoresis technology for proteomics. *Proteomics* **4**:3665–3685.

28. Rabilloud T. (2002) Two-dimensional gel electrophoresis in proteomics: old, old fashioned, but it still climbs up the mountains. *Proteomics* **2**:3–10.

29. O'Farrell PH. (1975) High resolution two-dimensional electrophoresis of proteins. *J Biol Chem* **250**:4007–4021.

30. Klose J, Kobalz U. (1995) Two-dimensional electrophoresis of proteins: an updated protocol and implications for a functional analysis of the genome. *Electrophoresis* **16**:1034–1059.

31. Park OK. (2004) Proteomic studies in plants. *J Biochem Mol Biol* **37**:133–138.

32. Gorg A, Obermaier C, Boguth G, *et al.* (1997) Very alkaline immobilized pH gradients for two-dimensional electrophoresis of ribosomal and nuclear proteins. *Electrophoresis* **18**:328–337.

33. Somiari RI, Somiari S, Russell S, Shriver CD. (2005) Proteomics of breast carcinoma. *J Chromatogr B* **815**:215–225.
34. Boyle WJ, Smeal T, Defize LH, *et al.* (1991) Activation of protein kinase C decreases phosphorylation of c-Jun at sites that negatively regulate its DNA-binding activity. *Cell* **64**:573–584.
35. Corthals GL, Aebersold R, Goodlett DR. (2005) Identification of phosphorylation sites using microimmobilized metal affinity chromatography. *Methods Enzymol* **405**:66–81.
36. Friedman DB, Hill S, Keller JW, *et al.* (2004) Proteome analysis of human colon cancer by two-dimensional difference gel electrophoresis and mass spectrometry. *Proteomics* **4**:793–811.
37. Alban A, David SO, Bjorkesten L, *et al.* (2003) A novel experimental design for comparative two-dimensional gel analysis: two-dimensional difference gel electrophoresis incorporating a pooled internal standard. *Proteomics* **3**: 36–44.
38. Chiari M, Micheletti C, Nesi M, *et al.* (1994) Towards new formulations for polyacrylamide matrices: N-acryloylaminoethoxyethanol, a novel monomer combining high hydrophilicity with extreme hydrolytic stability. *Electrophoresis* **15**:177–186.
39. Pennington K, McGregor E, Beasley CL, *et al.* (2004) Optimization of the first dimension for separation by two-dimensional gel electrophoresis of basic proteins from human brain tissue. *Proteomics* **4**:27–30.
40. Bae SH, Harris AG, Hains PG, *et al.* (2003) Strategies for the enrichment and identification of basic proteins in proteome projects. *Proteomics* **3**:569–579.
41. Barry RC, Alsaker BL, Robison-Cox JF, Dratz EA. (2003) Quantitative evaluation of sample application methods for semipreparative separations of basic proteins by two-dimensional gel electrophoresis. *Electrophoresis* **24**:3390–3404.
42. Zhu K, Miller FR, Barder TJ, Lubman DM. (2004) Identification of low molecular weight proteins isolated by 2-D liquid separations. *J Mass Spectrom* **39**:770–780.
43. Zhang Z, Smith DL, Smith JB. (2001) Multiple separations facilitate identification of protein variants by mass spectrometry. *Proteomics* **1**: 1001–1009.
44. Davis MT, Beierle J, Bures ET, *et al.* (2001) Automated LC-LC–MS-MS platform using binary ion-exchange and gradient reversed-phase chromatography for improved proteomic analyses. *J Chromatogr B Biomed Sci Appl* **752**:281–291.
45. Lubman DM, Kachman MT, Wang H, *et al.* (2002) Two-dimensional liquid separations — mass mapping of proteins from human cancer cell lysates. *J Chromatogr B Analyt Technol Biomed Life Sci* **782**:183–196.
46. Shin YK, Lee HJ, Lee JS, Paik YK. (2006) Proteomic analysis of mammalian basic proteins by liquid-based two-dimensional column chromatography. *Proteomics* **6**:1143–1150.

47. Chong BE, Yan F, Lubman DM, Miller FR. (2001) Chromatofocusing nonporous reversed-phase high-performance liquid chromatography/ electrospray ionization time-of-flight mass spectrometry of proteins from human breast cancer whole cell lysates: a novel two-dimensional liquid chromatography/mass spectrometry method. *Rapid Commun Mass Spectrom* **15**:291–296.

48. Wall DB, Kachman MT, Gong S, *et al.* (2000) Isoelectric focusing nonporous RP HPLC: a two-dimensional liquid-phase separation method for mapping of cellular proteins with identification using MALDI-TOF mass spectrometry. *Anal Chem* **72**:1099–1111.

49. Wall DB, Kachman MT, Gong SS, *et al.* (2001) Isoelectric focusing nonporous silica reversed-phase high-performance liquid chromatography/electrospray ionization time-of-flight mass spectrometry: a three-dimensional liquid-phase protein separation method as applied to the human erythroleukemia cell-line. *Rapid Commun Mass Spectrom* **15**:1649–1661.

50. Shin YK, Lee HJ, Lee JS, Paik YK. (2006) Proteomic analysis of mammalian basic proteins by liquid-based two-dimensional column chromatography. *Proteomics* **6**:1143-1150.

51. Westermeier R, Gronau S, Becket P, *et al.* (2005) A guide to methods and applications of DNA and protein separations. In: *Electrophoresis in Practice,* 4th ed., Westermeier R (ed.), Wiley-VCH Verlag, Weinheim, Germany.

52. Hardy E, Santana H, Sosa A, *et al.* (1996) Recovery of biologically active proteins detected with imidazole–sodium dodecyl sulfate–zinc (reverse stain) on sodium dodecyl sulfate gels. *Anal Biochem* **240**:150–152.

53. Merril CR, Switzer RC, Van Keuren ML. (1979) Trace polypeptides in cellular extracts and human body fluids detected by two-dimensional electrophoresis and a highly sensitive silver stain. *Proc Natl Acad Sci USA* **76**:4335–4339.

54. Mackintosh JA, Choi HY, Bae SH, *et al.* (2003) A fluorescent natural product for ultra sensitive detection of proteins in one-dimensional and two-dimensional gel electrophoresis. *Proteomics* **3**:2273–2288.

55. Berggren KN, Schulenberg B, Lopez MF, *et al.* (2002) An improved formulation of SYPRO Ruby protein gel stain: comparison with the original formulation and with a ruthenium II tris (bathophenanthroline disulfonate) formulation. *Proteomics* **2**:486–498.

56. Steinberg TH, Pretty On Top K, Berggren KN, *et al.* (2001) Rapid and simple single nanogram detection of glycoproteins in polyacrylamide gels and on electroblots. *Proteomics* **1**:841–855.

57. Somiari R. (2005) Proteomics of breast carcinoma. *J Chromatogr B* **815**:215–225.

58. Graves PR, Haystead TAJ. (2002) Molecular biologist's guide to proteomics. *Microbiol Mol Biol Rev* **66**:39–63.

59. Porath J, Carlsson J, Olsson I, Belfrage G. (1975) Metal chelate affinity chromatography, a new approach to protein fractionation. *Nature* **258**:598–599.

60. Washburn MP, Wolters D, Yates JR 3rd. (2001) Large-scale analysis of the yeast proteome by multidimensional protein identification technology. *Nat Biotechnol* **19**:242–247.

61. Link AJ, Eng J, Schieltz DM, *et al.* (1999) Direct analysis of protein complexes using mass spectrometry. *Nat Biotechnol* **17**:676–682.

62. Fliser D, Wittke S, Mischak H. (2005) Capillary electrophoresis coupled to mass spectrometry for clinical diagnostic purposes. *Electrophoresis* **26**:2708–2716.

63. Covey TR, Devanand P. (2002) Nanospray electrospray ionization development: LC/MS, CE/MS application. In: *Applied Electrospray Mass Spectrometry*, Practical Spectroscopy Series, Vol. 32, Pramanik BN, Ganguly AK, Gross ML (eds.), Marcel Dekker, New York, NY.

64. Zheng X, Baker H, Hancock WS. (2006) Analysis of the low molecular weight serum peptidome using ultrafiltration and a hybrid ion trap–Fourier transform mass spectrometer. *J Chromatogr A* **1120**:173–184.

65. Medzihradszky KF, Campbell JM, Baldwin MA, *et al.* (2000) The characteristics of peptide collision-induced dissociation using a high-performance MALDI-TOF/TOF tandem mass spectrometer. *Anal Chem* **72**:552–558.

66. Guerrara I, Kleiner O. (2005) Application of mass spectrometry in proteomics. *Biosci Rep* **25**:71–93.

67. Schweitzer B, Predki P, Snyder M. (2003) Microarrays to characterize protein interactions on a whole-proteome scale. *Proteomics* **3**:2190–2199.

68. Kersten B, Feilner T, Kramer A, *et al.* (2003) Generation of *Arabidopsis* protein chips for antibody and serum screening. *Plant Mol Biol* **52**:999–1010.

69. Tonkinson JL, Osborn DS, Zhao WW, Stillman BA. (2003) Development of microscale immunoassays for parallel analysis of multiple analytes. *IVD Technol* **2**:29–34.

70. Wulfkuhle JD, Aquino JA, Calverts VS, *et al.* (2003) Signal pathway profiling of ovarian cancer from human tissue specimens using reverse-phase protein microarrays. *Proteomics* **3**:2085–2090.

71. Espejo A, Cote J, Bednarek A, *et al.* (2002) A protein-domain microarray identifies novel protein–protein interactions. *Biochem J* **367**:697–702.

72. White AM, Daly DS, Varnum SM, *et al.* (2006) ProMAT: protein microarray analysis tool. *Bioinformatics* **22**:1278–1279.

73. Hoheisel JD. (2006) Microarray technology: beyond transcript profiling and genome type analysis. *Nat Rev Genet* **7**:200–210.

74. Tusher VG, Tibshirani R, Chu G. (2001) Significance analysis of microarrays applied to the ionizing radiation response. *Proc Natl Acad Sci USA* **98**:5116–5121.

75. Hedenfalk I, Duggan D, Chen Y, *et al.* (2001) Gene-expression profiles in hereditary breast cancer. *N Engl J Med* **344**:539–548.

76. Van Belle W, Anensen N, Haaland I, *et al.* (2006) Correlation analysis of two-dimensional gel electrophoretic protein patterns and biological variables. *BMC Bioinformatics* **7**:198.

77. Kaplan N, Linial M. (2005) Automatic detection of false annotations via binary property clustering. *BMC Bioinformatics* **6**:46.
78. Duncan MW, Hunsucker SW. (2005) Proteomics as a tool for clinically relevant biomarker discovery and validation. *Exp Biol Med* **230**: 808–817.
79. Alexander H, Stegner AL, Wagner-Mann C, *et al.* (2004) Proteomic analysis to identify breast cancer biomarkers in nipple aspirate fluid. *Clin Cancer Res* **10**:7500–7510.
80. Wu S, Kim J, Hancock WS, Karger S. (2005) Extended Range Proteomic Analysis (ERPA): a new and sensitive LC-MS platform for high sequence coverage of complex proteins with extensive post-translational modifications — comprehensive analysis of beta-casein and epidermal growth factor receptor (EGFR). *J Proteome Res* **4**:1155–1170.
81. Bergman AC, Benjamin T, Alaiya A, *et al.* (2000) Identification of gel-separated tumor marker proteins by mass spectrometry. *Electrophoresis* **21**:679–686.
82. Shannon WD, Watson MA, Perry A, Rich K. (2002) Mantel statistics to correlate gene expression levels from microarrays with clinical covariates. *Genet Epidemiol* **23**:87–96.
83. Bafna V, Edwards N. (2001) SCOPE: a probabilistic model for scoring tandem mass spectra against a peptide database. *Bioinformatics* **17**(Suppl 1): S13–S21.
84. Low TY, Seow TK, Chung MC. (2002) Separation of human erythrocyte membrane associated proteins with one-dimensional and two-dimensional gel electrophoresis followed by identification with matrix-assisted laser desorption/ionization–time of flight mass spectrometry. *Proteomics* **2**:1229–1239.
85. Kakhniashvili DG, Bulla LA Jr, Goodman SR. (2004) The human erythrocyte proteome: analysis by ion trap mass spectrometry. *Mol Cell Proteomics* **3**:501–509.
86. Korbel S, Buchse T, Prietzch H, *et al.* (2005) Phosphoprotein profiling of erythropoietin receptor–dependent pathways using different proteomic strategies. *Proteomics* **5**:91–100.
87. Thadikkaran L, Siegenthaler M, Crettaz D, *et al.* (2005) Recent advances in blood-related proteomics. *Proteomics* **5**:3019–3034.
88. Garcia A, Watson S, Dwek R, Zitzmann N. (2005) Applying proteomics technology to platelet research. *Mass Spectrom Rev* **24**:918–930.
89. Von Bruchhausen F, Walter U. (1997) *Platelets and Their Factors.* Springer-Verlag, Berlin, Germany.
90. O'Neill EE, Brock CJ, von Kriegsheim AF, *et al.* (2002) Towards complete analysis of the platelet proteome. *Proteomics* **2**:288–305.
91. Garcia A, Zitzmann N, Watson SP. (2004) Analyzing the platelet proteome. *Semin Thromb Hemost* **30**:485–489.
92. Geveart K, Van Damme J, Goethals M, *et al.* (2002) Chromatographic isolation of methionine-containing peptides for gel-free proteome analysis. *Mol Cell Proteomics* **1**:896–903.

93. Anderson NL, Anderson NG. (2002) The human plasma proteome: history, character, and diagnostic prospects. *Mol Cell Proteomics* **1**:845–867.

94. Chan KC, Lucas DA, Hise D, *et al.* (2004) Analysis of the human serum proteome. *Clin Proteomics* **1**:101–226.

95. Watanabe H, Hamada H, Yamada N, *et al.* (2004) Proteome analysis reveals elevated serum levels of clusterin in patients with preeclampsia. *Proteomics* **4**:537–543.

96. Kwak JY, Ma TZ, Yoo MJ, *et al.* (2004) The comparative analysis of serum proteomes for the discovery of biomarkers for acute myeloid leukemia. *Exp Hematol* **32**:836–842.

97. Chen JH, Chang YW, Yaho CW, *et al.* (2004) Plasma proteome of severe acute respiratory syndrome analyzed by two-dimensional gel electrophoresis and mass spectrometry. *Proc Natl Acad Sci USA* **101**:17039–17044.

98. Qian WJ, Jacobs JM, Camp DG, *et al.* (2005) Comparative proteome analyses of human plasma following *in vivo* lipopolysaccharide administration using multidimensional separations coupled with tandem mass spectrometry. *Proteomics* **5**:572–584.

99. Mandel ID. (1993) Salivary diagnosis: more than a lick and a promise. *J Am Dent Assoc* **124**:85–87; erratum 20–21.

100. Miller SM. (1994) Saliva testing — a nontraditional diagnostic tool. *Clin Lab Sci* **7**:39–44.

101. Li Y, Zhou X, St John MA, Wong DT. (2004) RNA profiling of cell-free saliva using microarray technology. *J Dent Res* **83**:199–203.

102. Li Y, St John MA, Zhou X, *et al.* (2004) Salivary transcriptome diagnostics for oral cancer detection. *Clin Cancer Res* **10**:8442–8450.

103. Li Y, Denny P, Ho C, *et al.* (2005) The Oral Fluid MEMS/NEMS Chip (OFMNC): diagnostic and translational applications. *Adv Dent Res* **18**:3–5.

104. Vitzthum F, Behrens F, Anderson L, Shaw J. (2005) Proteomics: from basic research to diagnostic application. A review of requirements and needs. *J Proteome Res* **4**:1086–1097.

105. Zolg JW, Langen H. (2004) How industry is approaching the search for new diagnostic markers and biomarkers. *Mol Cell Proteomics* **3**:345–354.

106. Bell DY, Hook GE. (1979) Pulmonary alveolar proteinosis: analysis of airway and alveolar proteins. *Am Rev Respir Dis* **119**:979–990.

107. Bowler RP, Duda B, Chan ED, *et al.* (2004) Proteomic analysis of pulmonary edema fluid and plasma in patients with acute lung injury. *Am J Physiol Lung Cell Mol Physiol* **286**:L1086–L1087.

108. Noel-Georis I, Beman A, Falmagne P, Wattiez R. (2002) Database of bronchoalveolar lavage fluid proteins. *J Chromatogr B Analyt Technol Biomed Life Sci* **771**:221–236.

109. Noel-Georis I, Bernard A, Falmagne P, Wattiez R. (2001) Proteomics as the tool to search for lung disease markers in bronchoalveolar lavage. *Dis Markers* **17**:271–284.

110. von Bredow C, Birrer P, Griese M. (2001) Surfactant protein A and other bronchoalveolar lavage fluid proteins are altered in cystic fibrosis. *Eur Respir J* **17**:716–722.

111. Lenz AG, Meyer B, Costabel U, Maier K. (1993) Bronchoalveolar lavage fluid proteins in human lung disease: analysis by two-dimensional electrophoresis. *Electrophoresis* **14**:242–244.

112. Wattiez R, Hermans C, Cruyt C, *et al.* (2000) Human bronchoalveolar lavage fluid protein two-dimensional database: study of interstitial lung diseases. *Electrophoresis* **21**:2703–2712.

113. Neumann M, von Bredow C, Ratjen F, Griese M. (2002) Bronchoalveolar lavage protein patterns in children with malignancies, immunosuppression, fever and pulmonary infiltrates. *Proteomics* **2**:683–689.

114. Alexander H, Steger A, Wagner-Mann C, *et al.* (2004) Proteomic analysis to identify breast cancer biomarkers in nipple aspirate fluid. *Clin Cancer Res* **10**:7500–7510.

115. Zhao H, Adler K, Bai C, *et al.* (2006) Epithelial proteomics in multiple organs and tissues: similarities and variations between cells, organs, and diseases. *J Proteome Res* **5**:743–755.

116. Kondo T, Seike M, Mori Y, *et al.* (2003) Application of sensitive fluorescent dyes in linkage of laser microdissection and two-dimensional gel electrophoresis as a cancer proteomic study tool. *Proteomics* **3**:1758–1766.

117. Faber MJ, Agnetti G, Bezstarosti K, *et al.* (2006) Recent developments in proteomics: implications for the study of cardiac hypertrophy and failure. *Cell Biochem Biophys* **44**:11–29.

118. Gey GO, Coffman WD, Kubicek MT. (1952) Tissue culture studies of the proliferative capacity of cervical carcinoma and normal epithelium. *Cancer Res* **12**:264–265.

119. Diaz JI, Cazares LH, Corica A, John Semmes O. (2004) Selective capture of prostatic basal cells and secretory epithelial cells for proteomic and genomic analysis. *Urol Oncol* **22**:329–336.

120. Yingling JM, Blanchard KL, Sawyer JS. (2004) Development of TGF-beta signaling inhibitors for cancer therapy. *Nat Rev Drug Discov* **3**:1011–1022.

121. Schmidt-Weber CB, Blaser K. (2004) Regulation and role of transforming growth factor-beta in immune tolerance induction and inflammation. *Curr Opin Immunol* **16**:709–716.

122. Kanamoto T, Hellman U, Heldin CH, Souchelnytskyi S. (2002) Functional proteomics of transforming growth factor-beta1–stimulated Mv1Lu epithelial cells: Rad51 as a target of TGFbeta1-dependent regulation of DNA repair. *EMBO J* **21**:1219–1230.

123. Paron I, D'Elia A, D'Ambrosio C, *et al.* (2004) A proteomic approach to identify early molecular targets of oxidative stress in human epithelial lens cells. *Biochem J* **378**:929–937.

124. Nedelkov D. (2005) Population proteomics: addressing protein diversity in humans. *Expert Rev Proteomics* **2**:315–324.

125. Steel LF, Haab BB, Hanash SM. (2005) Methods of comparative proteomic profiling for disease diagnostics. *J Chromatogr B Analyt Technol Biomed Life Sci* **815**:275–283.

126. Villanueva J, Philip J, Entenberg D, *et al.* (2004) Serum peptide profiling by magnetic particle-assisted, automated sample processing and MALDI-TOF mass spectrometry. *Anal Chem* **76**:1560–1570.
127. Zhang X, Leung SM, Morris CR, Shigenaga MK. (2004) Evaluation of a novel, integrated approach using functionalized magnetic beads, bench-top MALDI-TOF-MS with prestructured sample supports, and pattern recognition software for profiling potential biomarkers in human plasma. *J Biomol Tech* **15**:167–175.
128. Stoeckli M, Chaurand P, Hallahan DE, Caprioli RM. (2001) Imagining mass spectrometry: a new technology for the analysis of protein expression in mammalian tissues. *Nat Med* **7**:493–496.
129. Nägele E, Vollmer M, Hörth P, Vad C. (2004) 2D-LC/MS techniques for the identification of proteins in highly complex mixtures. *Expert Rev Proteomics* **1**:37–46.
130. Davis MT, Beierle J, Bures ET, *et al.* (2001) Automated LC/LC-MS/MS platform using binary ion-exchange and gradient reversed-phase chromatography for improved proteomic analyses. *J Chromatogr B Biomed Sci Appl* **752**:281–291.
131. Peng J, Elias JE, Thoreen CC, *et al.* (2003) Evaluation of multidimensional chromatography coupled with tandem mass spectrometry (LC/LC-MS/MS) for large-scale protein analysis: the yeast proteome. *J Proteome Res* **2**:43–50.
132. Adkins JN, Varnum SM, Auberry KJ, *et al.* (2002) Towards a human blood proteome. *Mol Cell Proteomics* **1**:947–955.

Chapter 18

An Introduction to Biomedical Informatics

Hai Hu* and Michael N. Liebman

*Biomedical Informatics, Windber Research Institute, 620 7th St.
Windber, PA 15963, USA*

Biomedical informatics can be defined as a multidisciplinary subject applying information-based, i.e. computational and statistical, technologies to clinical, genomic, and proteomic studies. One of its major components involves healthcare informatics, as it focuses on advancing the practice of medicine. The other component focuses on basic and applied research, integrating clinical perspectives into genomic and proteomic studies to address issues of clinical significance. Biomedical informatics is a comprehensive subject in the study of human disease, encompassing a cycle of from bedside to bench and then from bench to bedside. This chapter will give an overview of human genetic studies in the genomic era from the biomedical informatics perspective. Included topics are carefully selected to minimize overlap with other chapters, while at the same time presenting a relatively comprehensive picture. It is expected that, after reading this chapter and others on bioinformatics in this book, readers will have a deeper understanding of the importance of bioinformatics and biomedical informatics in the study of human genetics as well as the differences in their approach. Conscientiously keeping up with and open-mindedly adopting new technologies developed in bioinformatics and biomedical informatics has proven invaluable to modern human genetic studies.

The previous chapter provided an excellent overview of the field of bioinformatics, i.e. the use of informatic and statistical technologies to solve biological problems associated with large quantities of data. It focuses on data

---

* Correspondence author.

analysis, data mining, and algorithm development for genomic and proteomic studies. Biomedical informatics, on the other hand, can be defined as the use of informatic and statistical technologies to solve clinical problems utilizing genomic and proteomic data.

Despite being a relatively new multidisciplinary subject, and in fact probably because it is young, biomedical informatics bears many different definitions. One popular definition is more from the perspective of "informatics for healthcare", and thus includes hospital informatics for demographics and general clinical data integration; clinical informatics for physical examinations, lab work, and disease diagnosis; clinical trial data tracking, focusing on information associated with clinical trials; and bioimaging, including X-ray, MRI, ultrasound images, etc. A good discussion on biomedical informatics that focuses on this perspective has been written by Shortliffe and Cimino.[1]

Since the purpose of this book is to give an overview of the methods for genetic and genomic research, the focus here will be on the basic and applied research nature of biomedical informatics. The following is our definition of biomedical informatics: "Biomedical Informatics is an integrative multi-disciplinary subject applying computational and statistical technologies to clinical, genomic, and proteomic studies, to meet the needs of data collection, tracking, storage, visualization, analysis, and knowledge discovery, with a goal that the knowledge will be eventually applied in the clinic."[2] Biomedical informatics typically adopts a systems-oriented, top-down approach that complements the traditional bottom-up approach of systems biology.[3,4]

There is a clear overlap between bioinformatics and biomedical informatics in genomic and proteomic studies, as well as in the informatic and statistical technologies used for data tracking, storage, analysis, etc. A clear distinction between the two is that the latter maintains a heavy clinical perspective; therefore, genomic and proteomic studies are primarily focused on human biospecimens. This emphasis on the analysis of human biospecimens demands that biomedical informatics research be conducted under strict compliance with regulations which protect the privacy and rights of human subjects. However, the power of biomedical informatics approaches resides in their direct targeting of clinically important questions, making it possible to transfer newly discovered knowledge quickly into clinical applications so as to aid in clinical decision making by healthcare providers. Figure 1 shows the major components in biomedical informatics as defined above.

Human genetic and genomic studies on diseases certainly require collaboration among clinicians, experimental molecular scientists, and biomedical

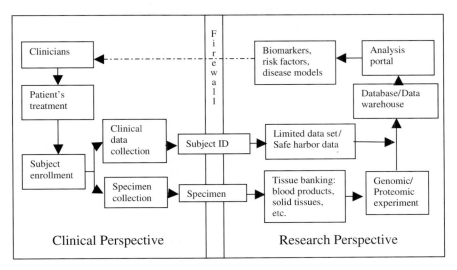

Fig. 1. Major components of biomedical informatics. Clinically, patients receive treatment, subjects are enrolled in the study, and clinical data as well as specimens are collected. To protect the privacy of human subjects, deidentified subject IDs and specimen IDs are created and properly mapped before being transferred to the research side with the corresponding clinical data and specimens. On the research side, clinical data are properly stored, tissues properly banked, and genomic and proteomic studies conducted. All data are then warehoused, analyzed, and mined for biomarkers, risk factors, and disease models. Newly obtained knowledge is fed back to the clinic to aid in clinical decision making.

informaticians. This chapter presents an overview of human genetic studies in the genomic era from the biomedical informatics perspective. The topics are carefully selected to minimize overlap with other chapters, while developing a relatively comprehensive picture of the subject. This chapter includes the clinical perspective, genomic and proteomic perspectives, data tracking (laboratory information management system), data centralization, application development and integration, and examples of the application of biomedical informatics for translational research.

1. CLINICAL PERSPECTIVE

Biomedical informatics directly studies human subjects. Typically, comparative studies are conducted for nondiseased (control) and diseased subjects. Enrollment of the subjects is usually done through clinics; therefore, discussion of biomedical informatics should start with the clinical perspective.

1.1. *The Role of the Institutional Review Board*

Before any research project involving human subjects begins, the research protocol must first be approved by the local institutional review board (IRB). An IRB is typically composed of people from different backgrounds, such as physicians, researchers, patient advocates, and administrators. The responsibility of the IRB is to examine the research protocol from multiple angles, including but not limited to the following: the design of the project is scientifically reasonable; the use of human subjects is appropriate for the study; all risks to the subject are minimized; the privacy of the human subject is protected; and the informed consent form properly informs the prospective human subject of the goal, risks, and benefits of the study.

A more comprehensive guide to IRBs is provided by Penslar.[5] If a research project involves the use of animals, then the protocol should be approved by the local institutional animal care and use committee (IACUC). The latter is not the focus of this chapter.

1.2. *Handling of Protected Health Information*

Biomedical informatics research uses human subjects (e.g. clinical data and clinical specimens) as the object of study. In doing so, such research needs to abide by special rules and regulations developed for the protection of the privacy of the human subject. Specifically, protected health information (PHI) — information that may lead to the identification of a human subject — cannot be released to researchers.

Over the last several years, regulations to protect PHI have been developed and are currently enforced across the world in developed countries. These regulations include the Personal Information Protection and Electronic Document Act (PIPEDA) in Canada as well as the European Union (EU) Privacy Directive in the EU. In the United States, as of April 14, 2003, any and every research project involving human subjects needs to be compliant with the Health Insurance Portability and Accountability Act of 1996 (HIPAA) Privacy Rule, the first comprehensive federal protection for the privacy of personal health information in the US. HIPAA regulations are for the general protection of PHI in any application where the transfer of clinical data is involved. In the following paragraphs, the application of HIPAA rules to biomedical informatics research will be discussed.

The HIPAA Privacy Rule allows deidentified clinical data to be released for biomedical informatics research purposes. A clinical data set is considered deidentified when all of the 18 data elements that could be used to identify the individual are removed. The following are the listed 18 data elements:

(1) Names
(2) All geographic subdivisions smaller than a state, including street address, city, county, precinct, ZIP code, and their equivalent geographical codes, except for the initial three digits of a ZIP code if, according to the current publicly available data from the Bureau of the Census:

  (a) The geographic unit formed by combining all ZIP codes with the same three initial digits contains more than 20 000 people
  (b) The initial three digits of the ZIP code for all such geographic units containing 20 000 or fewer people are changed to 000

(3) All elements of dates (except year) for dates directly related to an individual, including birth date, admission date, discharge date, and date of death; all ages over 89 and all elements of dates (including year) indicative of such age, except if such ages and elements may be aggregated into a single category of age 90 or older
(4) Telephone numbers
(5) Facsimile numbers
(6) Electronic mail addresses
(7) Social security numbers
(8) Medical record numbers
(9) Health plan beneficiary numbers
(10) Account numbers
(11) Certificate/License numbers
(12) Vehicle identifiers and serial numbers, including license plate numbers
(13) Device identifiers and serial numbers
(14) Web universal resource locators (URLs)
(15) Internet protocol (IP) address numbers
(16) Biometric identifiers, including fingerprints and voiceprints
(17) Full-face photographic images and any comparable images
(18) Any other unique identifying number, characteristic, or code, unless otherwise permitted by the Privacy Rule for reidentification

The second way to establish deidentification is to have a statistician — "a person with appropriate knowledge . . . and experience" — use "generally accepted statistical and scientific principles and methods" to determine and certify that, even though not all of the above 18 data elements are removed, there is a "very small" risk that the information could be used by the recipient to identify the individual who is the subject of the information, alone or in combination with other reasonably available information.

Sometimes, restricted data elements, such as the date of birth or the date of disease diagnosis, are needed for research. In addition, in some studies a subject's information from different sources needs to be interpreted. Thus, a unique identification number is required for the subject. This leads to the use of a "limited data set", which is a clinical data set after removal of the following 16 categories of direct identifiers of a subject[6]:

(1) Names
(2) Postal address information, other than town or city, state, and ZIP code
(3) Telephone numbers
(4) Fax numbers
(5) Electronic mail addresses
(6) Social security numbers
(7) Medical record numbers
(8) Health plan beneficiary numbers
(9) Account numbers
(10) Certificate/License numbers
(11) Vehicle identifiers and serial numbers, including license plate numbers
(12) Device identifiers and serial numbers
(13) Web universal resource locators (URLs)
(14) Internet protocol (IP) address numbers
(15) Biometric identifiers, including fingerprints and voiceprints
(16) Full-face photographic images and any comparable images

In transferring a limited data set from a clinical premise to a research entity, a Data Use Agreement must be executed between the data supplier and the recipient, whereby the recipient gives satisfactory assurances that the data will only be used for specifically specified purposes and will be properly protected following HIPAA regulations.

In practice, in some research projects, PHI is strictly held at the healthcare premise accessible only to the clinical principal investigator (PI) or designees of the PI. Each human subject is assigned a unique, anonymous ID, and the mapping of the ID to the subject identifications is maintained by the PI. This is practically the same exercise as the use of a master patient index (MPI).

The limited data set satisfying HIPAA regulations and any biological specimens (i.e. tissue or blood samples) properly mapped to the anonymous

ID are then transferred to the research entity. At the research entity, this ID is the only reference to the corresponding human subject; thus, the privacy of the human subject is protected. In this practice, the clinical PI serves as an "honest broker". Decoding the ID back to the human subject for identification can only be done by the PI under predefined conditions following predefined procedures in the protocol previously approved by the IRB. For example, decoding may be necessary when the study scope expands such that additional data from the human subject is needed. In some cases, clinical diagnostic mistakes may be identified during the research and some IRB-approved protocols may allow such information to be fed back to the clinical site for error correction.

In the HIPAA regulations, there is no specification on where PHI should be stored. In some projects, research data are stored in one database, but PHI is stored in a completely different database or data storage system. However, it is acceptable to store PHI and research data in the same database as long as it is secured such that PHI is only accessible to clinicians, while the researchers only have access to the research data. This can be achieved through, for example, a role-based privilege structure.

One problem in this PHI-removed research model is that some research projects enroll subjects at multiple clinical sites, and one subject may visit more than one clinical site and end up being enrolled in a study more than once. If the subject remembers being previously enrolled at another site, a remedy can potentially be developed to accommodate this situation depending on how the clinical data are collected. However, if the subject does not remember the previous enrollment, then the same subject will be enrolled multiple times as different subjects in the study. This situation can happen when a subject is aged or has a deteriorated memory. Some subjects participate in multiple research projects, and so it is difficult to count on them to remember all of the studies they have participated in.

A major obstacle to solving this problem in the US is that the different healthcare providers adopt competing hospital information systems (HISs) or clinical information systems (CISs) which do not communicate with each other. A notable effort is being made by the National Health Information Infrastructure (NHII) initiative, currently led by the US Department of Health and Human Services, to harmonize data exchange standards.[7] If the clinical data for research are directly derived from HIS and/or CIS, then they are required to be stripped of PHI at the healthcare site before being released to the researchers; thus, there is no definitive way to prevent duplicate enrollment. However, if the clinical data are collected through a questionnaire mechanism, PHI is maintained at the healthcare premises, and the clinical PI is permitted to oversee all such information, then it is possible to develop a mechanism that allows for a centralized subject ID assignment and therefore potentially prevents a subject from being enrolled multiple times at multiple clinical sites.

1.3. *Clinical Data Collection*

As discussed, there are two major approaches to collect clinical data for biomedical informatics research. One approach is to directly access the data in HIS and/or CIS, and then deidentify the information to meet HIPAA regulations. Academic deidentification computer programs are available,[8,9] and there are companies dedicated to this service.[10,11] Good deidentification software should be able to remove PHI as much as possible, while at the same time disturb non-PHI information as little as possible.

Another approach to collect clinical data is through the completion of questionnaires. Issues involved here include questionnaire design, questionnaire completion, data entry, and quality assurance. The design of the questionnaire(s) requires involvement from domain experts such as clinicians specialized in the disease of study, biomedical informatics researchers, and experts on questionnaire design and implementation. It is important to balance the scope of the questions: too few questions will not enable the collection of sufficient data for research, whereas too many questions will increase the operational cost and may discourage subjects from participating in the study. Another important issue is the language of the question. The question should be composed to precisely describe the question to be answered without ambiguity. Sometimes, repetitive questions are used for the collection of important data elements (e.g. both date of birth and age may be listed) to improve the quality and consistency of the data collected.

Questionnaires can be completed by the subject alone or with the help of an experienced nurse, or by the research nurse interviewing the subject or reviewing the medical record. The nurses should be properly trained so that they can accurately and consistently interpret the questions and clarify any language ambiguity. The questionnaire can be in either a hardcopy form or an electronic version. The electronic version of the questionnaire eliminates the later data entry step required for the hardcopy form, and thus eliminates another step of error generation. Current tablet technology makes it very convenient to implement an electronic questionnaire such that the completion of questionnaires can be done via a touch screen.

Irrespective of the questionnaire format, a quality assurance (QA) step needs to be performed. This step includes a visual review of the answers to all of the questions for missing values or obvious errors, and a built-in program that deploys established QA metrics of rules to examine the validity of the value entered for each field and the consistency of the data between different questions.

1.4. *Clinical Specimen Collection and Biobanking*

In addition to clinical data, biological specimens are often also collected for molecular studies in biomedical informatics research. It is important that a specimen is accompanied with detailed annotations possibly including (but not limited to) specimen type, quantity, location, diagnosis, collection date, and other key data points. Specimens should be mapped to the donor subjects via the anonymous ID. Examples of specimen types are peripheral circulatory blood and blood-based products (e.g. plasma, white blood cells, serum, blood clots), saliva, urine, and solid tissues that are disease-specific (e.g. breast tissue and lymph nodes in breast disease studies).

Solid tissues can be preserved as flash-frozen, formalin-fixed paraffin-embedded (FFPE), or as optimal cutting temperature compound–embedded (OCT-embedded). When stored in vapor-phase liquid nitrogen freezers, specimens can be preserved for months or years and remain suitable for genomic and proteomic studies. However, the precise length of time before mRNAs and protein molecules are no longer suitable for analysis has yet to be established. In addition, tissue preservation methods may affect the condition of the specimen and thus affect its viability for molecular studies. For example, in FFPE tissues, DNA is not damaged and thus FFPE samples remain suitable for DNA-based studies. However, mRNA and protein molecules are damaged by the FFPE procedure; as a consequence, these tissues are not suitable for gene expression (mRNA-based) or proteomic studies.

Long-term preservation of human specimens is done through tissue banking or biobanking. Many research institutions have their own small-scale tissue banking facilities to support internal biomedical informatics research. Nationwide biobanking efforts are also underway. The first such effort in the world was started in Iceland by a company called deCODE Genetics. Authorized by the government of Iceland, the primary focus of this effort is to identify the genetic causes of common diseases and to apply this information to developing new drugs and DNA-based diagnostics.[12] To date, more than 100 000 volunteers with blood samples have been enrolled in more than 50 disease projects.

Encouraged by the success of this effort (which will be discussed later in Sec. 6), other countries have also started nationwide biobanking projects.

In 2002, the Swedish National Biobank Program was started.[13,14] The UK Biobanks project has been in development for several years and started to recruit participants in March 2006, aiming at gathering information on the health and lifestyle of 500 000 volunteers aged between 40 and 69 with blood and urine samples.[15] In the US, the National Cancer Institute (NCI) of the National Institutes of Health (NIH) established the Office of Biorepositories and Biospecimen Research, which is dedicated to providing leadership for biobanking activities that support all types of cancer research funded by the NCI.[16] The blueprint of a national biospecimen network has been developed, funded by the NCI and the nonprofit institute National Dialogue on Cancer.[17,18] Taiwan is also starting a biobanking pilot project to support genomic and proteomic studies in the country.

A more systematic description of national-level biobanks has been written by Austin *et al.*,[19] comparing the then-known (in development, proposed, or terminated) eight international biobanks in Iceland, the United Kingdom, Estonia, Latvia, Sweden, Singapore, Tonga, and Canada with respect to the purpose of biobanking, role of the government, commercial involvement, consent and confidentiality procedures, opposition to biobanking, and the then-current progress of the biobanks.

As a last note, in tissue banking, it is important to reduce or avoid repeated freeze-and-thaws, i.e. taking specimens out to thaw for experiments and then sending the remaining tissue back to the freezer for storage. This practice deteriorates mRNAs and protein molecules in the tissues. One way to avoid this practice is to aliquot the specimen properly after sample collection and before placing in the freezer, so that only aliquots of the specimen are thawed for experiments when needed.

2. GENOMIC AND PROTEOMIC PERSPECTIVES

Several preceding chapters have detailed the available genomic and proteomic technologies from the wet lab point of view. These reviews have laid a solid background for us to offer a genomic and proteomic perspective of biomedical informatics research. In this section, we will focus on several issues or problems that are considered important in applying these technologies.

Genomic studies are conducted at the DNA and RNA levels. Several DNA analysis technologies — such as array comparative genomic hybridization (aCGH), fluorescent *in situ* hybridization (FISH), single nucleotide polymorphism (SNP) analysis, and genotyping — can be applied to study

DNA copy number changes (genes or chromosome regions), nucleotide change in genes, and loss of heterozygosity (LOH). Such genomic changes may impact disease susceptibility.

For mRNA-level gene expression analysis, reverse transcription–polymerase chain reaction (RT-PCR) was the main technology before the introduction of microarray technology. Many researchers still consider RT-PCR as the gold standard to validate microarray results. Gene expression microarray technology, on the other hand, is quickly maturing. Arrays can be oligonucleotide-based or cDNA-based, and the current dominating commercial arrays include Affymetrix's GeneChip, Agilent's oligo microarray, Applied Biosystems' TaqMan Gene Expression Assays, GE Healthcare's CodeLink, Illumina's Sentrix BeadChips, and NimbleGen's gene expression arrays. The number of probes on a microarray chip has steadily increased, and is currently at 50 000.

The rapid development of microarray technology has greatly benefited basic research, but at the same time has created a major challenge in biomedical research, especially for long-term projects in which samples are accumulated over time. The challenge is this: how can results from new arrays be compared to those from earlier arrays? Newer arrays contain higher probe density, and special probes may have been designed for improved sensitivity. Thus, long-term research faces a dilemma: since it is best that experiments for one project be conducted on one type of array, which array type should be used? Waiting until all of the samples are collected to start the experiments using the newest arrays with the densest probe coverage is not an option because of funding and time constraints. However, there is also difficulty in using older arrays from when the project was started, since one would certainly want to optimize the data from the precious human specimens. In addition, the vendor may stop supporting or even providing an earlier version of the microarray chip.

In general, proteomic technologies are not as mature as genomic technologies. Analytical technologies have been developed for protein separation, identification, and interaction studies. For protein separation, a major problem is the wide dynamic range of protein concentrations in a specimen; for example, in human serum or plasma, the dynamic range is more than 10 orders of magnitude.[20] Gel electrophoresis–based technology cannot resolve very low protein concentrations. In addition, in order to study proteins with low concentrations, high-concentration (also known as high-abundance) proteins need to first be depleted; for example, in the study of human serum samples, it is almost a routine practice to deplete albumin and several other high-abundance proteins. Proteins with extreme molecular weights also cannot be studied using this technology, as they cannot be well separated on the 2D gel.

The lack of reproducibility in the 2D gels was largely relieved by the introduction of two-dimensional difference gel electrophoresis (2D DIGE), whereby one can simultaneously apply two samples and one standard labeled with different dyes to the same gel, with the expectation that the same protein from different samples will migrate to the same position on the gel and thus allow for differential analysis.[21] Technologies for comparison between gels and batches of experiments have been developed, though there are still many hurdles to overcome before these technologies can be readily applied to the study of human specimens.

Proteomic analyses also include protein interaction and structural analysis. In the next chapter, the available experimental methods for protein interaction studies are reviewed; and previously we have also reported the mapping of the human WW domain family interactions using a protein array technology based on functional protein domain and peptide ligand binding.[22] Structural analysis often involves direct or indirect imaging technologies such as magnetic resonance imaging (MRI) or X-ray crystallography of proteins; these technologies are not as often applied to biomedical informatics studies, so we will not elaborate on them here.

The selection of the type of sample for study is dependent on the goal of the project and the technology selected. For diagnostic biomarker discovery, a minimally invasive specimen is preferred, for example, saliva, urine, or blood. It should be cautioned that when conducting microarray gene expression studies using peripheral circulatory blood to identify biomarkers for solid tissue cancers, the results most likely reflect the immunological response to the developed lesion, since the mRNA molecules are almost all from the white blood cells. Although it is possible that some invasive cancer cells are released into the circulatory blood, the sheer volume of the white blood cells will likely dilute the mRNA signal from cancer cells to a level where it is undetectable by the current microarray technology. The serum or plasma sample, however, may be suitable for identifying peptide-level biomarkers from lysing cancer cells shed into the bloodstream, where proteomic technologies are capable of detecting such weak signals.

Genomic and proteomic experiments can be done either sample by sample or on a comparative basis in which two samples are studied in one experiment (e.g. a diseased sample is compared to a control sample). For gene expression analysis by microarray, which is relatively mature, it has been a consensus for several years that biological replicates are more important than technical replicates.

In a recent review focused on microarray technology,[23] several consensus points were discussed that can be generalized to other technologies in genomic and proteomic studies. When the goal is to identify differential expression of genes or proteins in samples from different groups of subjects, if biological variability is high relative to measurement error and samples are inexpensive relative to experimental cost, then samples can be pooled so that by conducting limited experiments conclusions can be obtained from a larger number of samples. In this practice, it is expected that random effects — called "noise" — are averaged and reduced, but that signals are accumulated and intensified, so that the signal-to-noise ratio is enhanced and signals are easier to detect. Virtually pooled experiments may also be conducted, in which case samples are tested individually and results are pooled for analysis with the same expectation that the signal-to-noise ratio is enhanced. It is understood that results from virtually pooled experiments may be different from physically pooled experiments.

The purity of the sample is an issue that needs to be taken into account when designing a biomedical informatics research project. For example, to study molecular expression changes in cancer tissues, diseased tissues are assayed against normal tissues. However, sometimes a cancer lesion is small compared to the diseased tissue block dissected through surgery (since surgeons always purposely take out "normal" tissues surrounding the cancer to make sure that cancer cells are removed). When using such tissues where the diseased sample contains a nondominant cancer cell component, the cancer signal may be too weak to be detected through experiments. Under such a condition, laser capture microdissection technology is often applied to dissect cancer cells of consecutive cryosections from the cancer tissue block. Cancer cells from the original tissue block are then repooled to create a diseased sample for analysis.[24]

Human beings are complex systems, and the development of human disease may be due to many contributing factors. For example, it is now widely accepted that breast cancer is a complex disease which may result from multiple disease progression pathways; one single breast cancer pathology slide may contain disease lesions of invasive ductal carcinoma, ductal carcinoma *in situ*, ductal hyperplasia, and other breast pathology types. For the study of biomarkers for ductal invasive carcinoma, it is better to use laser capture microdissection technology in order to obtain invasive carcinoma cells for study. However, if the sample type is whole blood for gene expression microarray studies or serum/plasma for protein expression studies, then the researcher needs to understand that this cancer subject may also have other diseases. It would be ideal to use specimens from subjects who do not have complicated co-occurring diseases.

Genomic and proteomic studies generate a large amount of data, and these data are generated by different experimental platforms. Properly keeping track of such data is a challenge. Additionally, it is often necessary for researchers to compare the results from different experimental platforms. The following sections will discuss how these problems can be solved.

3. LABORATORY INFORMATION MANAGEMENT SYSTEMS FOR DATA TRACKING

Traditionally, genetic research was done on a small scale such that one scientist would study one gene at a time. Spending one's whole life to study a single gene was not rare at all. Experimental procedures and the observed results used to be recorded manually in a paper notebook.

In contrast, in the genomic era, many experiments are done on a large scale, making it close to impossible to manually track the experiment and the data. Hence, electronic data tracking systems to record all of the necessary information and to allow rapid retrieval of needed information have become necessary. A laboratory information management system (LIMS) is specialized computer software that meets such a need.

3.1. *Background*

There are hundreds of commercial LIMS providers. The website http://www.limsource.com/ alone lists more than 150 LIMS companies. In addition, there are dozens of companies providing LIMSs specialized for healthcare-related industries. However, most of these companies do not offer LIMSs for integrative biomedical informatics research, high-throughput genomic or proteomic research, or life science research in general. Sometimes, the more general term "life science" actually better describes the current status of LIMSs, since the term "biomedical informatics" may not be broad enough. Therefore, in the following section, we will sometimes refer to "life science" instead of "biomedical informatics" where appropriate when describing LIMSs.

An estimated less than 10% of LIMS companies offer a system for life sciences, and the leading players are young companies. Traditional LIMS companies, starting from the early 1980s, are mostly focused on existing industries such as chemistry, pharmaceuticals, healthcare, petroleum, and quality assurance/quality control (QA/QC) in general. At the turn of this century, when new LIMSs for life sciences became necessary due to the development of high-throughput genomic and proteomic platforms, some traditional LIMS companies such as LabVantage (http://www.labvantage.com/) made a strategic decision to diversify into life sciences. At the same time, new LIMS companies were incorporated to specifically serve this field; these companies often start with a specific platform such as sequencing or mass spectrometry protein identification, and then expand to other platforms for coverage of different needs in life science research. Since these new companies did not have an existing LIMS structure, they are often able to design a new structure based on the new needs of life science research; thus, they are in general better for life science research. Such new companies include Cimarron Software (http://www.cimsoft.com), Genologics Life Sciences Software (http://www.genologics.com), and Ocimum Biosolutions (http://www.ocimumbio.com/web/default.asp).

Supplementary to commercial software providers, academic LIMS systems have also gradually become available. It is probably fair to say that until the turn of this century, there was no LIMS present in academic labs, even in chemistry, one of the oldest fields of research.[25] Currently, the sheer volume of data generated in the lab makes it impossible for an academic life science lab to rely on paper lab notebooks to keep track of experiments such as microarray or protein expression studies, and to access such experimental data for analysis without an electronic system. Realizing such a need, the Cancer Biomedical Informatics Grid (caBIG[TM]) — initiated by the US National Cancer Institute's Center for Bioinformatics (NCICB) of the NIH — has developed an open source web-based cancer LIMS[26] that can be adapted for specific lab needs, such as ABI 3700 sequencing equipment.

While experiments in most academic labs and some biotech companies are relatively confined to a couple of platforms, in an integrative biomedical informatics research setting multiple experimental platforms are deployed. For example, the following operations are conducted at our institution, the Windber Research Institute:

- Clinical operations — patient enrollment, questionnaire filling, blood and solid tissue collection, data entry and QA, and tissue banking
- Genomic studies — microarray gene expression, genotyping, array comparative genomic hybridization (aCGH), polymerase chain

reaction (PCR), immunohistochemistry (IHC), and fluorescent *in situ* hybridization (FISH)

- Proteomic studies — two-dimensional difference gel electrophoresis (2D DIGE) for protein separation, liquid chromatography (LC) for protein separation, and several mass spectrometry analysis systems for protein identification (MALDI-TOF MS, Q-TOF MS, LCQ MS, LTQ-FT MS, etc.)

3.2. *To Develop or to Buy?*

There are two approaches to address the need for LIMSs: to buy off the shelf or to develop in-house. Both solutions have advantages and disadvantages.

Given the breadth of biomedical informatics research, it is very difficult for one commercial LIMS to cover all the needs of a comprehensive research entity. Identifying a commercial LIMS supplier to meet your specific needs can be a painstaking process, and many with such an experience have said that commercial LIMSs are expensive and too rigid. Some organizations, after trials and failures, ended up developing a system of their own. Debate is ongoing with regard to which is better: to buy or to develop.[27] No doubt, an in-house developed system can serve your specific needs better, and is easier to maintain and expand for new needs; however, developing a LIMS requires experience, and it can be costly since you are developing one complete system for your own use. It is certainly possible to develop a LIMS in-house if you only need to cover one or two experimental platforms.

To buy, you have to identify critical factors to establish a good product. From our own experiences, we think the following factors are very important in evaluating a LIMS supplier:

(1) Front end — Is the workflow interface user-friendly? Is the workflow flexible? Is the coverage complete with regard to the breadth of your institutional operations? Is the coverage complete with regard to the depth of your specific experimental platform needs? Is the reporting utility satisfactory?

(2) Back end — Is the data model sound and able to support the front end features? Can the infrastructure support the required functionalities? Is the LIMS readily expandable for new experiments, and well designed for modification of existing workflows?

(3) Corporate strengths — Are the sales and technical teams competent and eager to help? Is the company going to be able to support your institute for long?

(4) Cost — Is the price of the product reasonable? Is the price for customization reasonable? What is the cost of the preferred database for the candidate LIMS?

(5) Reference checking and test driving of the candidate LIMS.

Of the five categories of factors listed above, the front end issues need input from lab operators — after all, it is the lab scientists who will use the LIMS on a daily basis. If the front end is appealing to them, some of the challenges discussed in the next section will be less of a problem. The back end infrastructure of the LIMS needs to be sound, and this is an issue that informatics/bioinformatics professionals can help with. The last three items on the list are the issues that the administrators need to handle.

After due diligence, you either identify a good LIMS product, customize it for implementation, and the system is in operation in your laboratories; or you develop a LIMS in-house, after extensive communication between your developers and your end users, and the system is up and running. You may think you are done. However, there are still many challenges awaiting you.

3.3. *Challenges*

One major challenge in LIMS implementation is to convince laboratory users that a LIMS is essential to their research. There is often resistance from end users in adopting it. Adopting a LIMS means changing the way the lab experiment is done, as the end user has to find time to type in the needed information in the LIMS. In addition, there is a learning curve, and no matter how well the system is designed it needs to be debugged in the first weeks or months of operation. Resistance from end users is common when a LIMS is newly deployed, and it takes effort (sometimes hardline commands) from the executive of the organization to enforce the adoption of the system.

Another challenge is LIMS training. Sometimes, the LIMS design is good, but the training is insufficient such that good features are not known to the user, with the features being wasted or even becoming a burden to the user.

Mismatches between designed operations and real practice may also happen; when such mismatches occur at a fundamental level, it may pose a serious challenge. Often, a commercial LIMS is designed based on the needs of the initial customers, but not all of the subsequent customers follow the same operations as the initial customers. For example, one LIMS for microarray asked for a barcode for every step of the experiment, from mRNA preparation to cDNA synthesis to cRNA synthesis. However, in the lab, there were no barcodes attached to the vials used so the lab scientists had to punch in arbitrary barcodes. Eventually, a sheet of sequential barcodes was printed for users to scan in one after another, pretending that such barcodes were attached to the vials used in the experiments.

The last challenge discussed here is that life science research is very dynamic. New technologies emerge all the time. This is less severe a problem to labs in established industries (i.e. a chemistry lab). A LIMS can work well in a stable environment, but when the environment changes the LIMS has to change to meet the evolving needs. Thus, a LIMS for life sciences needs to have a built-in feature such that the system is flexible enough to allow it to grow with the field it serves.

3.4. *Additional Notes*

The LIMS database is a transactional database, mostly with "insert" but little "update" and "delete" operations, and all such data-altering operations are audited. Given that the nature of the LIMS is to track experimental operations, the "delete" operation is rarely available from the user interface, but can be done at the back end with a strict auditing track.

Analytical capability is typically not a requirement for LIMSs, although it is normal that the analytical results from an interface instrument are parsed and captured in the LIMS. For example, in proteomics labs, the mass spectrometry identification of proteins through MALDI-TOF, Q-TOF, or MudPIT analysis should be captured by the LIMS. The logic for why LIMSs should not also conduct data analysis is that the LIMS is supposed to track the experimental data generated in the lab, with an audit trail such that any modifications of the captured data are tracked. If the experimental data are analyzed in the LIMS, where should the results be stored? In addition, data analysis is often a recursive process involving error corrections, but the LIMS is really not for trial and failure. Moreover, if the LIMS is serving for a large laboratory system, competition for central processing unit (CPU) time and CPU bus traffic may become a problem.

4. DATA FORMAT STANDARDS AND DATA INTEGRATION OR FEDERATION

The clinical, genomic, and proteomic data discussed in previous sections are generated or collected on multiple platforms. These data must be physically combined before integrative data analysis and data mining can be done to identify risk factors for diseases of study, to discover biomarkers to aid in disease diagnosis or as drug targets for drug development, and to develop disease models for better understanding of disease development.

To achieve this goal, the data generated on different platforms by different organizations need to be in certain common formats so as to (1) allow data to be exchanged between research entities and (2) allow for easy integration of the data needed for a research project. In this section, we will first summarize the types of biomedical informatics data discussed so far and give an overview of the relevant public databases. This will be followed by a discussion on data format standards, and how such data can be integrated or federated.

4.1. *Clinically Collected and Laboratory-Generated Data*

In organizations where there is no LIMS, the data are scattered in instrument computers in different formats such as Oracle database, MySQL, MS Access, MS Excel spreadsheets, or simply flat files. Even in organizations implementing a LIMS, although most of the data may be in the LIMS database there are still data stored in other formats. The following is a list of data types described in the previous three sections:

(1) Clinical data

Clinical data may be collected through the clinical information system (CIS), hospital information system (HIS), and specially designed questionnaires. The following categories of data may be included:

- Demographics — including age, sex, ethnicity, body mass index, family structure, etc.
- Medical exam results and medical history — including lab blood test results, previous and current known diseases, current medications, etc.
- Medical images — including mammograms, ultrasound images, MRIs, PET/CT scans, images of pathology slides, etc.
- Family history — including primary and secondary family members' history of diseases. Diseases relevant to the disease in study should also be included.
- Lifestyle — including exercise pattern as well as smoking and alcohol intake habits along the temporal dimension.
- Risk factors — disease-specific risk factors. For example, use of hormone replacement therapy may be a risk factor for breast cancer; consumption of fried foods may be a risk factor for obesity and cardiovascular diseases.
- Specimen annotations — including tissue type, sample date, location, diagnosis, and preservation type.

- Treatment and outcome data — including surgical treatment and drug treatment details and results. Some of them are recorded during follow-up visits.

(2) Genomic data

Genomic data may be collected through microarray gene expression, aCGH, genotyping, sequencing, FISH, RT-PCR, etc. For the first three array-based analyses, a data format standard called Minimum Information About a Microarray Experiment (MIAME) has been proposed and is being gradually adopted, while the standard itself is being improved.[28] For genotyping, RT-PCR, and sequencing, the platform and the results are different from array-based technology (gene expression and aCGH microarray).

(3) Proteomic data

Studies include protein expression, interactions, and structural analysis. Qualitative protein expression can be studied using antibody arrays, while quantitative protein expression often involves protein separation (2D gel, HPLC) and identification (mass spectrometry).

4.2. Clinical, Genomic, and Proteomic Data in the Public Domain

OMIM is a human disease database focused on genes and genetic disorders.[29] It can be searched using a disease name or a gene name. With records written in free text, OMIM is designed for use by physicians and genetic researchers. Its unstructured format makes parsing of the data a major task, thus making it difficult to apply OMIM to automated data analysis and research. Improvements in text and mining technology will make the information in OMIM easier to access.

For nucleotide sequences and annotations, the most widely used public databases are GenBank maintained by the National Center for Biotechnology Information (NCBI) in the USA,[30] the EMBL Nucleotide Sequence Database maintained by the European Molecular Biology Laboratory (EMBL),[31] and the DNA DataBank of Japan (DDBJ) maintained by the National Institute of Genetics of Japan,[32] all of which belong to the International Nucleotide Sequence Database Collaboration. These three databases update and exchange data on a daily basis. Since this international collaboration accepts submissions from research scientists almost automatically with minimal verification, the databases are plagued with redundant information and data entry errors. To solve this problem, the NCBI initiated

RefSeq, aiming to provide a nonredundant database of sequences for DNA, RNA, and proteins.[33]

For protein sequences and their annotations, the most prominent databases are Universal Protein Resource (UniProt),[34] which is the result of the merger of SwissProt/Trembl between the Swiss Institute of Bioinformatics (SIB) in Switzerland and the European Bioinformatics Institute (EBI) in the UK, and Protein Information Resources (PIR) developed and maintained by the National Biomedical Research Foundation at Georgetown University in the USA. Trembl is a computer translation of the genes in EMBL with annotation provided by software program. Entries in SwissProt are based on those in Trembl, but are manually annotated and examined by experts in the field to maintain high data quality.

Gene Ontology (GO) is a consortium effort that aims to develop three structured and controlled vocabularies (ontologies) to describe gene products in terms of their associated biological processes, cellular components, and molecular functions in a species-independent manner.[35] Started as a collaboration between three model organism databases — FlyBase (*Drosophila*), the *Saccharomyces* Genome Database (SGD), and the Mouse Genome Database (MGD) — the consortium has now expanded to include about 20 leading organizations working on many organisms, including the EBI Gene Ontology Annotation (GOA) efforts in annotating UniProt. Applying these ontologies to gene products provides consistent descriptions across different databases, thus greatly facilitating scientific research and development.

Previously known as LocusLink, Entrez Gene provides a single query interface to curated sequence and descriptive information about genetic loci.[36] It presents information on official nomenclature, aliases, sequence accessions, phenotypes, EC numbers, OMIM numbers, UniGene clusters, homologies, map locations, and related websites. It is an indispensable source of information to integrative biomedical informatics research.

PubMed[37] is a major source of literature covering biomedical articles dating back to the 1950s. Abstracts for most of the articles are available. Access to full papers is limited because of the copyright issue. As the open access policy becomes more popular, limitations on accessing the full paper will gradually ease.

Pathway information is very important to the genomic and proteomic study of the mechanisms of human diseases. There are three types of biological pathways: metabolic, protein synthesis, and signal transduction (including protein–protein interactions). Well-known pathway databases include

BIND,[38] DIP,[39] BioCarta,[40] KEGG,[41] and STKE.[42] Integration of pathway databases is being worked on; for example, Krishnamurthy and colleagues[43] presented and integrated a system called the Pathway Database System, with a set of software tools for modeling, storing, analyzing, visualizing, and querying biological pathway data.

There are also ongoing public efforts to collect experimental microarray gene expression data, SNPs, 2D gels, and mass spectrometric protein identification data, which we will not discuss in detail here. However, caution needs to be exercised in using such data. It is not uncommon that the results from different studies are not comparable, possibly due to differences in the handling of the tissue, the experimental platform and procedures, and the data analysis methods.

The total number of public databases has dramatically increased in the last few years. The journal *Nucleic Acids Research* provides a good index to all of these databases in its January special issue on databases every year. In 1994, when the first database issue was published, there were only about 30 databases; whereas the 2005 database issue covered 719 databases[44] in 13 categories such as "nucleotide sequence databases", "protein sequence databases", and "human genes and diseases". All of the databases described above are covered in greater detail in this database issue.[44]

4.3. *Data Format Standards*

Biomedical informatics data generated by different organizations need to be interchangeable in order to make the best use of the data. With a data format standard, integration of the data becomes easier. Toward this goal, different data standardization consortia have formed to standardize different types of data. Some standards are relatively mature and thus better adopted by the clinical and research communities, while others are still at the infant stage or even at the stage of being conceived.

For clinical data, there is a widely adopted standard, Health Level Seven (HL7), currently in version 3.[45] HL7 refers to the nonprofit organization Health Level Seven, accredited by the American National Standards Institute (ANSI) for developing standards to support clinical practice and healthcare service.[45] Oracle also adopted this standard in its database specialized for healthcare, the Health Transaction Base (HTB). For digital imaging data, the Digital Imaging Communications in Medicine (DICOM) standard is dominant. Standards for digital mammogram MRI ultrasound images etc. are developed by specialized working groups appointed by the DICOM committee.

For microarray-based technologies, a standard called MIAME (Minimum Information About a Microarray Experiment) has been developed,[28] requiring that the following information be included: sample selection, platform (type of chip), results, and information about randomization. Another effort in standardizing the microarray data model and data exchange format is by the MicroArray and Gene Expression (MAGE) group.[46] The two groups coordinate in their respective standards development.

For proteomics data, the Human Proteome Organisation Proteomics Standard Initiative (HUPO PSI) is leading the development of data format standards, including MIAPE (Minimum Information About a Proteomics Experiment), the gel electrophoresis data format GelML, the PSI Object Model incorporating the mzData model for mass spectrometry, a standard data model for the representation and exchange of protein interaction data MIF, the PSI ontology, associated tools for data entry and format conversion, and reference implementations in the form of an XML interchange format and a repository.[47,48]

In addition to the efforts by the HUPO PSI, other organizations and consortia are also developing standard formats for proteomics data exchange. For proteomics mass spectrometry data, the Institute for Systems Biology in the USA developed an XML-based common data format — mzXML — that is widely adopted by the proteomic research community. In June 2006, the HUPO PSI announced that it would combine the mzData format with the mzXML format to form the newly named dataXML format, which was mostly complete by the end of 2006.

For molecular pathways, a computer-readable format for representing models of biochemical reaction networks called Systems Biology Markup Language (SBML) has been developed in Japan. SBML is applicable to metabolic networks, cell signaling pathways, and regulatory networks.[49] A consortium named BioPAX, which was formed in 2002 with the goal to develop a common exchange format for biological pathway data, has released its Level 2 ontology files.[50] A recent publication in the journal *Bioinformatics* compared and contrasted these two efforts with the HUPO PSI's molecular interaction format (MIF),[51] and concluded that the main structure of the three formats is similar. However, SBML is designed to aid the development of simulation models for molecular pathways, while PSI MI is more suitable for representing details about particular interactions and experiments. BioPAX is the most general and expressive of the formats.

Today, it appears that, for biomedical informatics research, different data formats will coexist even after the standards are unified to meet the different levels of need. A semantic metadata model such as the

Resource Description Framework (RDF) has also been developed to facilitate information exchange through the internet. Across many applications, the object-oriented data model is probably dominant. No matter what the data model is, a controlled vocabulary or a defined ontology is necessary to lay the foundation for seamless data exchange.

4.4. *Approaches for Data Integration or Federation*

In conducting integrative biomedical informatics research, it is necessary that the data from different sources, as described above, are accessible when needed. One way to achieve this goal is through data warehousing. Data warehousing will need to consider the data of different formats, including flat files, spreadsheets, and XML files; different databases; object data models; and the RDF. Although, for data exchange purposes, the object data model is probably a good model, at present the standards for different data types are still generally evolving. Thus, it is very difficult to develop a general method for data warehouse development for biomedical informatics research. These difficulties aside, the following paragraphs will discuss the three major data warehousing models: integration, federation, and a hybrid of the two.[52,53]

(1) Integration
In this model, data elements from different sources are integrated into one database with a comprehensive table structure. The advantage of this model is its efficiency in operation, but the disadvantage is its high maintenance cost. If one data source changes, it is likely that the data warehouse table structure also needs to be modified.

(2) Federation
In this model, all data are retained in their original data source structure. Middleware wrappers are developed to access each data source for needed data elements, and required data elements are then integrated at the time of execution. The advantage of this model is the low maintenance cost: modifying a middleware wrapper is typically easier than modifying a data warehouse table structure, since the latter has a global impact. The disadvantage of the federated model lies in the operational cost: it takes time to access data sources through the internet, and extracting and integrating needed data elements may also take time.

To partially improve the operation, an aggregation model can be used whereby the remote sources of the data are mirrored on a local server. Access to different data sources can be done via an intranet instead of the

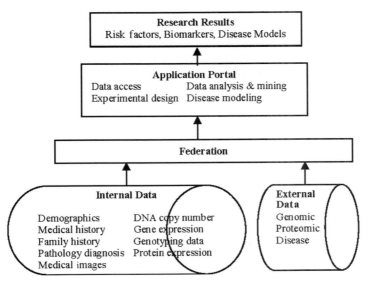

Fig. 2. Illustration of a diagram of a hybrid data warehouse. The internal data collected or obtained from clinical, genomic, and proteomic platforms are mostly integrated, and are then federated with external data of a variety of types so that researchers can readily access not only the internal data, but also the external data. Researchers can then use the application tools on the application portal for needed data analysis and mining to derive results and new knowledge from the composite data.

more time-intensive internet. Some people call this a different model of data warehouse, but we consider it an improved version of the federated model.

(3) Hybrid

In this model, data from internal sources that are controllable by the organization are integrated, whereas data from sources that are external to the organization and are therefore not controlled by the organization are federated. All internal data do not have to be integrated. Data types with a large file size (e.g. image data) that may impede the data warehouse performance typically should not be integrated; instead, data of this nature should be stored in a separate media from the data warehouse. Only a pointer to the file location and the annotation of the image need be integrated. Figure 2 illustrates an example of a hybrid data warehouse.

No matter which data warehouse model is selected, it is important that data from the same human subject is properly mapped so that the genomic and proteomic data can be properly linked to the correct subject's specimen used in the experiment. These data must be further mapped to the proper date of specimen collection and the subject ID, so that the detailed tissue annotation and the subject's clinical data are readily available for data analysis and mining. In addition, the internal experimental genomic and proteomic data can be cross-examined with the information in the public domain for better knowledge development.

Despite the data warehouse model, an application layer is needed to access and analyze the data. In Fig. 2, this layer is termed "application portal". In the next section, we will further elaborate what this portal is and how it can be developed. A concrete development environment example and a couple of tool examples will also be provided.

5. ANALYTICAL APPLICATION DEVELOPMENT AND INTEGRATION

Given the breadth of biomedical informatics research, a wide variety of applications are needed for analysis and mining of the generated or collected clinical, genomic, and proteomic data. Also, since biomedical informatics research is relatively new, not all research can be conducted using existing applications, so novel tools are still in the process of being developed. The application tools for biomedical informatics research include the following:

- General statistical functions for experimental design, sample size estimate, time series analysis and outcome prediction, data mining and knowledge discovery tools, and other data analysis tools (i.e. sample randomization and t-test, ANOVA, correlation, linear or logistic regression, power analysis, and nonparametric approaches)
- Visualization tools for generating heat maps, the results of clustering analysis, and scatter plots
- Missing value estimation methods such as k-nearest neighbor algorithm
- Sequence-based analysis tools including BLAST, FASTA, and ClusterW
- Tools for genotyping analysis such as linkage and association analysis
- A family of tools for microarray differential gene expression analysis, including normalization and false-positive rate control
- A family of tools for mass spectrometry data analysis and protein identification
- A family of tools for protein structure analysis
- A set of tools for text mining
- A set of tools for image processing applied to chip-based and gel-based technologies, as well as medical images

- Other tools for artificial neural network analysis, Bayesian analysis, and disease and biological process modeling
- Public database query tools such as Entrez (NCBI), Sequence Retrieval System or SRS (EBI), and Ensembl (jointly developed by EMBL–EBI and Sanger Centre, UK)
- Specialized platform-specific applications such as DeCyder for DIGE analysis, among many other applications too numerous to list

These analytical tools may be shareware, open source code, or licensed commercial packages. Examples include BioPerl, BioJava, BioPHP, BioPython, and BioLinux, which are dedicated to providing bioinformatics tools developed in different programming languages; R and R-based Bioconductor, SPSS of SPSS Inc., SAS of SAS Inc., Statistica of Statsoft, and Matlab of The Mathworks for statistical analysis and data mining; GCG of Accelrys and the European Molecular Biology Open Software Suite (EMBOSS) for sequence-based analysis; Expert Protein Analysis System (ExPASy) developed by the Swiss Institute of Bioinformatics (EBI) for protein sequence and structure analysis; Vector Xpression of Invitrogen, Spotfire of Spotfire Inc., and Gene Spring of Agilent for gene expression analysis and visualization; Genetic Profiler of GE Healthcare for microsatellite genotyping analysis; DeCyder of GE Healthcare for DIGE analysis; and MASCOT of Matrix Science and SEQUEST distributed by Thermo Finnigan for protein identification using mass spectrometry data. The tools listed above are only intended to provide examples in several application fields. We do not intend to endorse these specific tools, nor do we imply that these fields are more important than others.

Given the increasing number of tools and websites being released to the public, as a complementary effort to the annual database issue, the journal *Nucleic Acids Research* published a web server issue in 2002 and thereafter every year midyear. The 2006 issue highlighted 149 bioinformatics and molecular biology servers that are open to the public.[54] All of them and the tools published in previous special web server issues, as well as other useful tools and resources for bioinformatics and molecular biology research, are compiled in the Bioinformatics Link Directory.[55] This directory currently lists more than 1000 different servers and databases hosted in over 35 different countries. Servers are grouped by intended use (e.g. DNA-, RNA-, protein-, expression-, or computer-related), and each group includes a brief summary and relevant PubMed citations.[54] Most of the open-source, open-development tools described earlier are listed in this Bioinformatics Link Directory.

The biomedical informatics analytical tools operate in one or more of the three common operating systems: Microsoft Windows, Apple Macintosh, and Unix/Linux. In an integrative biomedical informatics research environment such as the Windber Research Institute (WRI), a variety of tools are needed to analyze many types of data; it would be ideal if all of the portable tools could be launched from the same environment. We believe that an analytical application portal for tool integration is the best approach to meet this need. Developing such an application portal for application integration on top of the data warehouse can effectively make use of the warehoused data, and thus lay a good structural foundation for result sharing and, as a consequence, knowledge sharing.

There are three key features that this application portal must provide. First, this portal should have a user-friendly interface, as both computer-literate and relatively computer-illiterate researchers need to use the tools on the portal. Second, this portal should already have integrated a large number of available analytical tools so that, when researchers need to analyze, mine, or model the data, all common tools or functions are readily accessible to them. Third, and very critically, this portal should be in a dynamic development environment such that, when the need to develop a new tool arises, a sophisticated user can develop, test, and publish this tool for average users to use it through the interface.

Given the sheer number of available tools that are applicable to biomedical informatics research and the fact that new tools are being developed everyday, it is probably impractical and also unnecessary to integrate all of the application tools onto one application portal. However, a partial integration of the tools that is focused on one application field is practical. In fact, many software packages can be considered examples of such an integration. Spotfire is focused on data analysis and visualization with many built-in tools and functions; SPSS for statistical analysis and data management is another example. However, almost all of these commercially available software packages do not allow users to modify the tools/functions supplied. Clementine, a data mining software package from SPSS Inc., does allow users to use supplied tool nodes to develop a workflow in order to meet the dynamic data mining needs, but it does not offer a dynamic programming development environment.

Currently, the WRI considers the Knowledge Discovery Environment (KDE) developed by InforSense Ltd. (London, UK) as the leading development environment that possesses the three key features described above. In the KDE, each application node is a tool performing a specific function. A user can not only build a workflow using these nodes to carry out a complicated task, but also convert this workflow into a new node for repeated future use. New nodes can also be directly developed through programming language.

Before closing this section, we would like to provide real examples of our tool integration effort. The WRI has teamed up with InforSense in the development of a patient-centric, object-oriented data warehouse and an application portal. Currently, we have developed and deployed a questionnaire-based clinical component of the data warehouse and a new

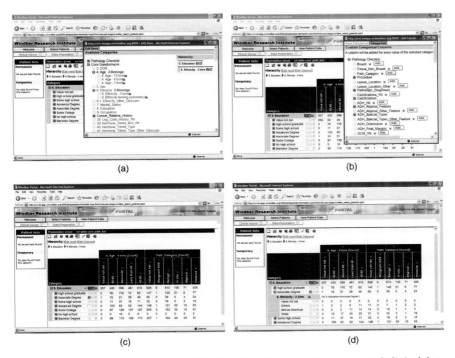

(a)　　　　　　(b)

(c)　　　　　　(d)

Fig. 3.　Screenshots of our relational OLAP tool to show how the aggregated clinical data collected from questionnaires can be viewed. (a) Rows can be selected and viewed in a hierarchical structure. In this example, the fields of education and prebinned "Ethnicity" from the core questionnaire are selected. (b) Columns can be selected and viewed in a flat structure. In this example, the "Path_Category" from the pathology checklist is selected together with the prebinned "Age" already selected from the core questionnaire. (c) The resulting view shows the aggregated data with "Education" displayed in rows, and "Age" and "Path_Category" displayed as columns. (d) The resulting view is the same as in (c), but with the row value of "Associate Degree" being expanded to show the hierarchical results including prebinned "Ethnicity".

relational online analytical processing (OLAP) tool to directly access the data in the data warehouse.[56] Figure 3 shows screenshots of the tool when composing a data view to analyze the aggregated patients' data. Shown here are data from the Clinical Breast Care Project (CBCP). Note that the data in Fig. 3 are from two questionnaires: the core questionnaire, which is completed by a nurse while interviewing the patient; and the pathology checklist, which is completed by a pathologist after reviewing the patient's pathology slide(s).

This OLAP tool enables us to easily access the aggregated clinical data for patient and sample selection, and this tool sits on the application portal and uses the KDE as its framework.[56] This integrated environment allows us to rapidly develop and deploy new application tools. For example, the WRI biomedical informatics team is responsible for identifying subjects from the collected clinical data and samples for research projects. Although subject selection criteria vary from project to project, there are common criteria for control subject selection and specimen selection. For example, since an invasive surgery may impact the genomic and proteomic properties of the blood, blood products should be chosen such that the blood-draw date is before any surgical date.

In addition, data from the questionnaire, which contains more than 500 data fields, are often requested by researchers in Excel spreadsheet format. Therefore, we have developed a small Perl program to format the data queried from the data warehouse into multiple spreadsheets, each limited to no more than 256 columns. This tool has been developed and implemented in the KDE. Figure 4 shows screenshots of application workflow nodes that take the patient CBCP numbers (anonymous ID), query the data warehouse for the report, and convert the report to spreadsheets as a saved Excel file. The nodes for data reformatting are condensed into one node, which is very convenient for repeated future use.

6. EXAMPLES: BIOMEDICAL INFORMATICS FOR TRANSLATIONAL RESEARCH

With multiplatform data properly warehoused and the needed analytical tools readily available, biomedical informatics can facilitate translational research. One key goal in biomedical informatics, as previously stated, is to discover new knowledge that will eventually be applied to clinical practices. To achieve this goal, researchers must focus on clinically important problems. In the following, we will provide three real examples on the application

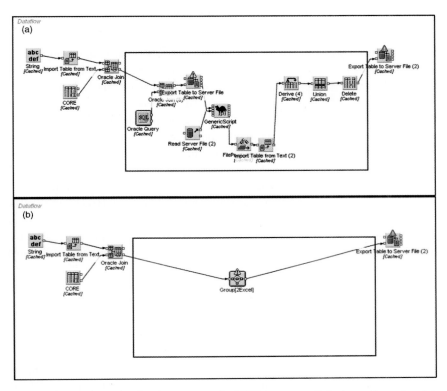

Fig. 4. Screenshots showing how to use the KDE to query clinical questionnaire data and format data for Excel spreadsheets. (a) Nodes showing that CBCP numbers are received as strings and are then converted to a table, which is joined and queried in the Oracle database with the core questionnaire table for the needed questionnaire data. The resulting flat file is saved, and called by a Perl program that has already been developed in-house to format the flat file into spreadsheets of about 250 columns each (spreadsheet column limit is 256, which is less than the number of the data fields in the core questionnaire) in one Excel file. The file is further processed and saved to a hard disk in a specified subdirectory. (b) Same as A, but here the nodes in the shaded area have been grouped to form a new node, which can be repeatedly used in the future when querying the data in other questionnaires that contain more than 256 data elements.

of biomedical informatics for translational research, focusing on infrastructure development for information and tool sharing, biobanking for genomic study and drug discovery, and statistical modeling for progressive disease pathway analysis.

6.1. *Cancer Biomedical Informatics Grid*

Cancer is a group of complex diseases and is the second leading cause of death in Western countries. According to the American Cancer Society,[57] cancer has claimed more than half a million lives in the US since the 1990s. While death rates from many other major diseases have dramatically decreased during the last 50 years, the death rate from cancer has remained almost unchanged at nearly 2 deaths per 1000 people. In the long battle against cancer, researchers have accumulated a large amount of information about the disease; and in the current genomic and proteomic era, an explosive amount of data has been generated by the cancer research community.

In 2003, an ambitious project for integrative biomedical informatics research called the cancer Biomedical Informatics Grid (caBIG^{TM})[58,59] was launched by the NCICB. The caBIG^{TM} is a US\$60 million 3-year pilot project, and its goal is to create the World Wide Web of cancer research by developing standards for data and applications so that cancer researchers can work together and share data more easily.

By 2006, the caBIG^{TM} community included more than 800 individuals at 50 NCI-designated cancer centers and 30 other organizations from the public and private sectors, including industry and patient advocacy groups.[59] The main technical fields of focus are clinical trial management systems, tissue banks and pathology tools, integrative cancer research, *in vivo* imaging, and vocabularies and common data elements. Fundamental to the caBIG^{TM}'s effort in unifying the infrastructure of cancer research are controlled vocabularies and object-oriented data models. Any data a researcher collects is required to be described using metadata that are already registered in the form of common data elements registered in the cancer Data Standards Repository (caDSR), and the information model is required to be expressed in the Unified Modeling Language (UML).

Currently, dozens of products have been released by the caBIG^{TM}. These products include caTISSUE Core for tissue banking; caTIES (cancer Text Information Extraction System); caTISSUE CAE as a Clinical Annotation Engine; caArray for microarray data services; rProteomics for proteomics analytical services; and an earlier product before caBIG^{TM} was launched, caCORE (cancer Common Ontological Reference Environment) with its major component caBIO (cancer Bioinformatics Infrastructure Objects) UML model.[60]

The caBIG^{TM} is built on the principles of open source, open access, open development, and federation.[58] The NCICB believes that this initiative will help to achieve the NCI's goal of eliminating suffering and death due to cancer by 2015. The systems designed through this effort will benefit not only the current cancer research field, but also the application of biomedical informatics research in general to other diseases. However, one of the challenges faced by the caBIG[TM] is that, although the caBIG[TM] community has steadily increased, most researchers still do not know of it at present.[61]

6.2. *The Icelandic Biobanking Effort by deCODE Genetics*

The Icelandic biobanking effort by deCODE Genetics has achieved unparalleled success, largely because of its population's relatively homogeneous genetics. The company's scientists, working with its collaborators, primarily used genotyping technology and linkage analysis to identify genes associated with diseases of interest.[62–67] The identification of such genes has led to the discovery of drug targets, and compounds have been developed for the potential treatment or prevention of these diseases. Since the project started 10 years ago, the company now has eight lead programs in drug discovery and development (including three in clinical trials) for cardiovascular diseases, asthma, pain, type 2 diabetes, and obesity. This successful approach to biobanking (i.e. genomic study) for drug discovery has set a good example of biomedical informatics research, and a number of other countries and regions are following suit.

It is worth noting that, even though the Icelandic population is relatively isolated, the population substructure still needs to be taken into account.[68] This will be even more important when similar studies are to be carried out in the USA, since the US is known as a "melting pot" in terms of both culture and ethnicity. However, even in the US, there are still regions which are relatively isolated with populations mostly from the same ethnic origin that are potentially suitable for association studies.

6.3. *The Clinical Breast Care Project: Hypothesis-Generating Analysis Using the Co-occurrence of Pathology Diagnoses Made on the Same Breast*

Earlier in this chapter, we mentioned the Clinical Breast Care Project (CBCP) without describing what it is. The CBCP is a US Department of Defense–funded project, with its clinical component headed by the Walter Reed Army Medical Center located in Washington, DC, and its research component headed by the Windber Research Institute (WRI) located in

Windber, PA. Launched in 2000, the CBCP had enrolled 3000 subjects by the fall of 2006. Comprehensive clinical data are collected from participants in the form of up to four questionnaires, each containing hundreds of data elements. Specimens, including blood and blood products as well as breast tissues and lymph nodes where applicable, are collected and stored at the WRI tissue banking facility.[69] A data warehouse has been developed and is evolving with the expansion of the research scope of the WRI.[53,56] Molecular studies using these specimens have resulted in important findings.[70–74] Several application tools have been developed to aid in data analysis and quality assurance of the data.[75–77] Here, a specific overview of a study analyzing the co-occurrence of pathology diagnoses from the same breast will be given.

Breast cancer is a heterogeneous disease, and multiple pathology diagnoses can often be made on the same diseased breast tissue. In normal clinical practice, a pathologist usually records the most severe diagnosis. In the CBCP, however, our pathologists diligently record all of the diagnoses that can be made; and in practice up to seven subpathologies may be recorded from a list of more than 130 possible diagnoses for each breast biopsy. Thus, the CBCP provides a rich data source for studying the co-occurrence of breast pathologies typically found in breast biopsies.

Given that this is a relatively new field of study, researchers at the WRI first developed a heat map–like tool for visualization of the co-occurrence frequencies between different pathologies.[78] In the initial study, the definition of the co-occurrence score was based on the Jaccard coefficient, and statistically significant differences in pathology co-occurrence patterns were identified for premenopausal and postmenopausal women. As the project developed, a Bayesian network learning algorithm was used by deploying the commercial software FasterAnalytics (DecisionQ Corp., Washington, DC), and this study resulted in the proposition of new breast cancer progression pathways.[79]

Such derived pathways are based on the assumption that, if breast cancer develops from a precursor disease, then when the cancer is detected the precursor disease may also be diagnosed in the tissue surrounding the cancer. These new pathways hypothesize that different grades of ductal carcinoma *in situ* (DCIS) develop into corresponding grades of infiltrating ductal carcinoma from different precursor pathologies.[79] This suggests that different grades of DCIS should probably be considered as different diseases instead of different grades of the same disease. Interestingly, a parallel genotyping study by the WRI also implied that the different grades of DCIS are separate genomic entities.[80] New genomic experiments have been proposed to verify the hypotheses. If these hypotheses prove to be true, it will likely change the clinical treatment of DCIS patients of different grades.

There are many examples of the successful application of integrative biomedical informatics technologies to genetic, genomic, and biological research in general. Because of space limitations, we have only discussed three specific cases here, focusing on tool development, tissue-based molecular studies, and hypothesis generation through data mining, respectively. Although we cannot call these examples representative, these three examples are indeed active research and development areas in integrative biomedical informatics research.

7. SUMMARY

Biomedical informatics is a relatively young multidisciplinary subject. It focuses on analyzing biological specimens and clinical information from human beings, as opposed to using animal or cell culture models of disease. From the disciplinary point of view, it involves many fields in medicine, such as physiology, surgery, pathology, and oncology. It involves life sciences including biology, genomics, and proteomics. It involves information technology for data tracking, data warehousing, and application tool development. It also involves applied mathematics and statistics for data analysis, modeling, and mining.

From the content point of view, in this chapter we have provided a definition of biomedical informatics and have elaborated on it from clinical, genomic, and proteomic perspectives. The issues of data tracking (LIMS), data format standards and warehousing, and data analysis and tool integration were discussed. At the end, several biomedical informatics application examples were provided. We are confident that, as technology develops, biomedical informatics will play a more important role in basic and applied research on human genetic diseases.

ACKNOWLEDGMENTS

We thank the previous and current leaders of the Windber Research Institute (WRI) and the Clinical Breast Care Project (CBCP) for the opportunity to develop pioneering application systems in the relatively new field of biomedical informatics. We thank Dr Richard Mural and the WRI biomedical informatics team for useful discussions. We specially thank Dr Susan

Maskery for critically commenting on the manuscript. The comments and opinions made in this chapter should be considered as personal, and should not be interpreted as the official stance of the WRI or the CBCP.

REFERENCES

1. Shortliffe EH, Cimino JJ. (2006) *Biomedical Informatics: Computer Applications in Health Care and Biomedicine*, 3rd ed., Health Informatics Series. Springer, New York, NY.
2. Liebman MN, Hu H, Mural R. Biomedical informatics: problems and opportunities. In: *Biomedical Informatics — A Translational Approach*, Hu H, Mural R, Liebman MN (eds.), Artech House, Norwood, MA (in preparation).
3. Hood L, Heath JR, Phelps ME, *et al.* (2004) Systems biology and new technologies enable predictive and preventative medicine. *Science* **306**:640–643.
4. Ideker T, Galitski T, Hood L. (2001) A new approach to decoding life: systems biology. *Annu Rev Genomics Hum Genet* **2**:343–372.
5. Penslar RL. (1993) *Protecting Human Research Subjects: Institutional Review Board Guidebook*. US Government Printing Office, Washington, DC.
6. Health Insurance Portability and Accountability Act (HIPAA). Protecting personal health information in research: understanding the HIPAA Privacy Rule. Available at http://privacyruleandresearch.nih.gov/pdf/HIPAA_Booklet_4-14-2003.pdf/
7. US Department of Health and Human Services. The National Health Information Infrastructure (NHII). Available at http://aspe.hhs.gov/sp/NHII/
8. Beckwith BA, Mahaadevan R, Balis UJ, *et al.* (2006) Development and evaluation of an open source software tool for deidentification of pathology reports. *BMC Med Inform Decis Mak* **6**:12.
9. Gupta D, Saul M, Gilbertson J. (2004) Evaluation of a deidentification (De-Id) software engine to share pathology reports and clinical documents for research. *Am J Clin Pathol* **121**:176–186
10. Claredi. Available at http://www.claredi.com/
11. De-ID. Available at http://www.de-id.com/
12. deCODE Genetics. Available at http://www.reykjavikresources.com/displayer2.asp?cat_id=397/
13. Dillner J. (2002) The new biobank law accepted: a unique chance of improvement. *Lakartidningen* **99**:2774–2776.
14. Swedish National Biobank Program. Available at http://www.biobanks.se/
15. UK Biobank. Available at http://www.ukbiobank.ac.uk/about/overview.php/
16. National Cancer Institute (NCI) Office of Biorepositories and Biospecimen Research. Available at http:// biospecimens.cancer.gov/
17. National Dialogue on Cancer (NDC), National Cancer Institue (NCI). (2003) *National Biospecimen Network Blueprint*. Constella Group, Durham, NC.
18. National Biospecimen Network Blueprint. Available at http://biospecimens.cancer.gov/nbn/blueprint.asp/

19. Austin MA, Harding S, McElroy C. (2003) Genebanks: a comparison of eight proposed international genetic databases. *Community Genet* **6**:37–45.
20. Anderson NL, Anderson NG. (2002) The human plasma proteome: history, character, and diagnostic prospects. *Mol Cell Proteomics* **1**:845–867.
21. Unlu M, Morgan ME, Minden JS. (1997) Difference gel electrophoresis: a single gel method for detecting changes in protein extracts. *Electrophoresis* **18**:2071–2077.
22. Hu H, Columbus J, Zhang Y, *et al.* (2004) A map of WW domain family interactions. *Proteomics* **4**:643–655.
23. Allison DB, Cui X, Page GP, *et al.* (2006) Microarray data analysis: from disarray to consolidation and consensus. *Nat Rev Genet* **7**:55–65.
24. Ellsworth DL, Shriver CD, Ellsworth RE, *et al.* (2003) Laser capture microdissection of paraffin-embedded tissues. *Biotechniques* **34**:42–44, 46.
25. Perry D. (2002) LIMS in the academic world. *Todays Chem Work* **11**: 15–16, 19.
26. National Cancer Institute (NCI). Cancer laboratory information management system. Available at http://calims.nci.nih.gov/caLIMS/
27. Salamone S. (2004) LIMS: to buy or not to buy?, *Bio-IT World*, June 27.
28. Brazma A, Hingamp P, Quackenbush J, *et al.* (2001) Minimum Information About a Microarray Experiment (MIAME) — toward standards for microarray data. *Nat Genet* **29**:365–371.
29. Online Mendelian Inheritance in Man (OMIM). Available at http://www3.ncbi.nlm.nih.gov/omim/
30. Benson DA, Karsch-Mizrachi I, Lipman DJ, *et al.* (2005) GenBank. *Nucleic Acids Res* **33**:D34–D38.
31. Kanz C, Aldebert P, Althorpe N, *et al.* (2005) The EMBL Nucleotide Sequence Database. *Nucleic Acids Res* **33**:D29–D33.
32. Okubo K, Sugawara H, Gojobori T, *et al.* (2006) DDBJ in preparation for overview of research activities behind data submissions. *Nucleic Acids Res* **34**:D6–D9.
33. Pruitt KD, Tatusova T, Maglott DR. (2005) NCBI Reference Sequence (RefSeq): a curated non-redundant sequence database of genomes, transcripts and proteins. *Nucleic Acids Res* **33**:D501–D504.
34. Bairoch A, Apweiler R, Wu CH, *et al.* (2005) The Universal Protein Resource (UniProt). *Nucleic Acids Res* **33**:D154–D159.
35. Gene Ontology. Available at http://www.geneontology.org/
36. Maglott D, Ostell J, Pruitt KD, *et al.* (2005) Entrez Gene: gene-centered information at NCBI. *Nucleic Acids Res* **33**:D54–D58.
37. PubMed. Available at http://www.ncbi.nlm.nih.gov/entrez/query.fcgi/
38. Alfarano C, Andrade CE, Anthony K, *et al.* (2005) The Biomolecular Interaction Network Database and related tools 2005 update. *Nucleic Acids Res* **33**:D418–D424.
39. Salwinski L, Miller CS, Smith AJ, *et al.* (2004) The Database of Interacting Proteins: 2004 update. *Nucleic Acids Res* **32**:D449–D451.
40. BioCarta. Available at http://www.biocarta.com/

41. Kanehisa M, Goto S, Kawashima S, *et al.* (2004) The KEGG resource for deciphering the genome. *Nucleic Acids Res* **32**:D277–D280.
42. Signal Transduction Knowledge Environment. Available at http://stke.sciencemag.org/
43. Krishnamurthy L, Nadeau J, Ozsoyoglu G, *et al.* (2003) Pathways database system: an integrated system for biological pathways. *Bioinformatics* **19**:930–937.
44. Galperin MY. (2005) The Molecular Biology Database Collection: 2005 update. *Nucleic Acids Res* **33**:D5–D24.
45. Health Level 7. Available at http://www.hl7.org/
46. MicroArray and Gene Expression (MAGE). Available at http://www.mged.org/Workgroups/MAGE/mage.html/
47. Hermjakob H, Montecchi-Palazzi L, Bader G, *et al.* (2004) The HUPO PSI's molecular interaction format — a community standard for the representation of protein interaction data. *Nat Biotechnol* **22**:177–183.
48. Orchard S, Hermjakob H, Binz PA, *et al.* (2005) Further steps towards data standardisation: the Proteomic Standards Initiative HUPO 3rd Annual Congress, Beijing 25–27th October, 2004. *Proteomics* **5**:337–339.
49. Systems Biology Markup Language (SBML). Available at http://sbml.org/index.psp/
50. BioPAX. Available at http://www.biopax.org/
51. Stromback L, Lambrix P. (2005) Representations of molecular pathways: an evaluation of SBML, PSI MI and BioPAX. *Bioinformatics* **21**:4401–4407.
52. Inmon WH. (2002) *Building the Data Warehouse*, 3rd ed. John Wiley & Sons, New York, NY.
53. Hu H, Brzeski H, Hutchins J, *et al.* (2004) Biomedical informatics: development of a comprehensive data warehouse for clinical and genomic breast cancer research. *Pharmacogenomics* **5**:933–941.
54. Fox JA, McMillan S, Ouellette BF. (2006) A compilation of molecular biology web servers: 2006 update on the Bioinformatics Links Directory. *Nucleic Acids Res* **34**:W3–W5.
55. Bioinformatics Links Directory. Available at http://bioinformatics.ubc.ca/resources/links_directory/
56. Hu H, Correll M, Osmond M, *et al.* (2007) Development of a data warehouse and on-line analytical processing tool for translational research (in preparation).
57. American Cancer Society. Cancer facts and figures 2006 by American Cancer Society. Available at http://www.cancer.org/downloads/STT/CAFF2006PWSecured.pdf/
58. von Eschenbach AC, Buetow KH. (2006) Cancer informatics vision: caBIG^{TM}. *Cancer Inform* **2**:22-24.
59. caBIG Fact Sheet. Available at https://cabig.nci.nih.gov/overview/caBIG_Fact_ Sheet.pdf/
60. Covitz PA, Hartel F, Schaefer C, *et al.* (2003) caCORE: a common infrastructure for cancer informatics. *Bioinformatics* **19**:2404–2412.

61. Salamone S. (2006) Is caBIG ready to bloom? *Bio-IT World* **5**:33–34.
62. Gulcher JR, Jonsson P, Kong A, *et al.* (1997) Mapping of a familial essential tremor gene, *FET1*, to chromosome 3q13. *Nat Genet* **17**:84–87.
63. Gudmundsson G, Matthiasson SE, Arason H, *et al.* (2002) Localization of a gene for peripheral arterial occlusive disease to chromosome 1p31. *Am J Hum Genet* **70**:586–592.
64. Hakonarson H, Bjornsdottir US, Halapi E, *et al.* (2002) A major susceptibility gene for asthma maps to chromosome 14q24. *Am J Hum Genet* **71**: 483–491.
65. Gretarsdottir S, Thorleifsson G, Reynisdottir ST, *et al.* (2003) The gene encoding phosphodiesterase 4D confers risk of ischemic stroke. *Nat Genet* **35**:131–138.
66. Stefansson H, Steinthorsdottir V, Thorgeirsson TE, *et al.* (2004) Neuregulin 1 and schizophrenia. *Ann Med* **36**:62–71.
67. Stacey SN, Sulem P, Johannsson OT, *et al.* (2006) The *BARD1* Cys557Ser variant and breast cancer risk in Iceland. *PLoS Med* **3**:e217.
68. Helgason A, Yngvadottir B, Hrafnkelsson B, *et al.* (2005) An Icelandic example of the impact of population structure on association studies. *Nat Genet* **37**:90–95.
69. Somiari SB, Somiari RI, Hooke J, *et al.* (2004) Establishment and management of a comprehensive biorepository for integrated high throughput genomics and proteomics research. *Trans Integr Biomed Inform Enabling Technol Symp* **1**:131–143.
70. Somiari RI, Sullivan A, Russell S, *et al.* (2003) High-throughput proteomic analysis of human infiltrating ductal carcinoma of the breast. *Proteomics* **3**:1863–1873.
71. Ellsworth DL, Ellsworth RE, Love B, *et al.* (2004) Outer breast quadrants demonstrate increased levels of genomic instability. *Ann Surg Oncol* **11**:861–868.
72. Somiari S, Shriver CD, Heckman C, *et al.* (2006) Plasma concentration and activity of matrix metalloproteinase 2 and 9 in patients with breast cancer and at risk of developing breast cancer. *Cancer Lett* **233**:98–107.
73. Ellsworth RE, Ellsworth DL, Deyarmin B, *et al.* (2005) Timing of critical genetic changes in human breast disease. *Ann Surg Oncol* **12**:1054–1060.
74. Ru QC, Zhu LA, Silberman J, *et al.* (2006) Label-free semiquantitative peptide feature profiling of human breast cancer and breast disease sera via two-dimensional liquid chromatography–mass spectrometry. *Mol Cell Proteomics* **5**:1095–1104.
75. Guo X, Liu R, Shriver CD, *et al.* (2006) Assessing semantic similarity measures for the characterization of human regulatory pathways. *Bioinformatics* **22**:967–973.
76. Yang S, Guo X, Yang YC, *et al.* (2006) Detecting outlier microarray slides by correlation and percentage of outliers spots. *Cancer Inform* **2**:351–360.
77. Zhang Y, Hu H, Shriver CD, *et al.* (2007) New peak detection and alignment algorithms for LC-MS analysis (submitted).

78. Maskery SM, Zhang Y, Jordan RM, *et al.* (2006) Co-occurrence analysis for discovery of novel breast cancer pathology patterns. *IEEE Trans Inf Technol Biomed* **10**:497–503.
79. Maskery S, Hu H, Mural R, *et al.* (2006) Separate DCIS progression pathways derived from breast disease heterogeneity data. Presented at American Society of Clinical Oncology, Atlanta, GA.
80. Ellsworth RE, Ellsworth DL, Love B, *et al.* (2007) Correlation of levels and patterns of genomic instability with histological grading of DCIS. *Ann Surg Oncol* [Epub ahead of print].

Chapter 19

Protein–Protein Interactions: Concepts, Databases, Software Tools, and Biomedical Implications

SudhaRani Mamidipalli and Jake Yue Chen*

Indiana University School of Informatics
Indianapolis, IN 46202, USA

The publication of the draft human genome consisting of 30 000 genes is merely the beginning of genome biology. A new way to understand the complexity and richness of the molecular and cellular functions of proteins in human biology is through an understanding of large collections of protein–protein interactions and the network that they form. In this chapter, we first introduce various concepts involved in the study of protein–protein interaction data. Then, we introduce different experimental methods that can be used to measure protein interactions. Next, we review public data sources that provide protein interaction data and compare their coverage. We also review the computational tools and methods used to predict, validate, and interpret protein interaction data. Finally, we present our perspectives on how systems-scale protein interaction and network analysis represent an essential step towards understanding human biology and identifying novel targets for drug development.

1. INTRODUCTION

The study of protein interactions has been vital to understanding how proteins function within the cell. The publication of the draft sequence of the human genome and proteomics-based protein profiling studies have brought

---

*Correspondence author.

forth a new era in protein interaction analysis. Understanding the characteristics of protein interactions in a given cellular proteome, often called the interactome, will be the next milestone in revolutionizing the understanding of cell biochemistry.[1]

A comprehensive collection of information related to human proteins, their features, and their functions is invaluable to biological researchers in several ways. For instance, the type of sequence domain found in proteins generally predicts the functional class or biological role of proteins. The exact subcellular localization of proteins and their tissue-specific distribution are also important to understanding the functional activities of proteins in cells.[2] In addition, it is important to know whether and how any proteins encoded by disease-associated genes play their roles in molecular complexes and biological pathways.[3]

Around 30 000 genes of the human genome are expected to give rise to one million proteins through a series of gene splicing and posttranslational modification mechanisms.[1] Posttranslational modifications such as phosphorylation and ubiquitination can influence the activity of proteins, and are generally used as regulatory mechanisms in signal transduction pathways.[4] Although a few of these proteins may function in relative isolation, the vast majority of them are expected to act in concert with other proteins in protein complexes and molecular networks in order to interweave many elements of biological processes and achieve complex cellular and molecular functions. Examples of these processes include cell cycle control, differentiation, protein folding, signaling, transcription, translation, posttranslational modification, and protein transport. Therefore, understanding the total identity and characteristic features of interacting proteins, furthered by the additional interacting details such as binding sites, can help generate promising new hypotheses on how global molecular regulations of networks implicated in human disease processes are achieved.[5]

New protein functions may also be inferred from protein–protein interactions. If the functions of an unknown protein's interaction partners are mostly uniform and known, it is not difficult to conclude that the unknown protein may have similar functions as its interacting partners.[6] This insight may be further validated using a variety of experimental methods as reviewed by Phizicky and Fields,[7] such as altering the kinetic properties of enzymes, allowing for substrate channeling, creating a new binding site, inactivating or destroying a protein, changing the specificity of a protein for its substrate through interaction with different binding partners, or serving a regulatory role in either an upstream or a downstream action.

When two proteins interact with each other, the domains (and amino acid residues) from each protein are in physical contact with each other.[8] Therefore, understanding the interaction between two proteins at the sub-protein level is essential. Domains are structural subunits of proteins and are considered as "building blocks" that are conserved during evolution. Proteins may have one domain or, more frequently, a combination of several domains. In eukaryotes, about one fifth of all proteins contain just one domain; while in prokaryotes, the fraction is about 34%.[9] Usually, each domain performs an independent function for the protein, such as binding to a small ligand, spanning the plasma membrane (transmembrane proteins), holding the catalytic site (enzymes), DNA binding (in transcription factors), or providing a protein interaction surface.[2] In some cases, a protein domain can be encoded by a distinct domain-encoding exon.

Motifs, similar to domains but smaller in scope, are also important sequence/structure elements involved in protein–protein interactions. For example, coiled-coil and nuclear localization signal motifs are well-known protein-binding motifs.[10] Motifs publicly cataloged today can be broadly viewed as either sequence motifs or folding motifs. Sequence motifs are often locally conserved regions of a sequence or short sequence patterns shared by a set of sequences. Folding motifs, on the other hand, are independent folding units or particular structures that recur in many molecules.

Assembling protein interaction data and their interaction details into multilevel molecular networks represents a grand challenge for systems biology. Systems biology is an emerging discipline that utilizes experimental techniques and bioinformatics to help understand biological systems on global scales, including cellular scales, intercellular scales, and physiological scales.[11] A protein interaction network is an example that can provide researchers with an opportunity to study the intracellular molecular functions and biological processes as a unified system. However, limited experimental protein interaction data published to date, limited tools to study complex biological networks, and limited ability to integrate genomic and proteomic data efficiently with other sources of information present practical challenges in the application of protein interaction networks into current biomedical research.[12]

In this chapter, we first discuss the different levels of observation of protein interactions and survey the various experimental methods used to detect protein interactions for mapping networks. We then summarize major public databases available to document these interactions. We also review the computational tools and methods used to predict, validate, and interpret protein interaction data from these databases. Finally, we present our perspectives on how protein–protein data analysis represents a significant initial step towards a systems approach for understanding human biology and identifying drug target candidates in the postgenomics era.

2. METHODS TO DETECT PROTEIN INTERACTIONS

Only in recent years have experimentally determined protein–protein interactions been collected to analyze novel functions of proteins.[13] This traditional source of information has been steadily accumulated in concert with the simultaneous development of high-throughput experiments aimed at probing potential interactions within entire proteomes (see Table 1 for details).

These experimental methods to detect protein interactions vary at different levels of resolution.[21] Figure 1 shows schematically different levels of observation of protein interactions (adapted from the Central Dogma drawing online from the course BIOL 103: Principles of Biology, Queens University of Charlotte, NC[22]). The first atomic-level observation refers to protein interactions detected at an atom level, for example, X-ray crystallography. The second direct interaction–level observation refers to protein interactions detected at the protein/peptide level, for example, yeast two-hybrid experiment. The third complex-level observation refers to protein interactions detected at the multiprotein complex level, for example, immunoprecipitation and mass spectrometry. The fourth cellular-level observation refers to protein interactions detected through activity bioassay, for example, proliferation assays of cells stimulated by a receptor–ligand interaction. Of these four categories of protein interaction detection methods, the complex level is the most commonly used method to detect and report protein interactions in public literature, followed by the cellular interaction, the direct interaction, and finally the atomic observation level.[21]

2.1. *Mass Spectrometry (MS)*

Mass spectrometry (MS)[15] is used for the identification of peptides (and, therefore, proteins) from protein complexes derived from either biological samples (e.g. serum, plasma, urine) or purified protein complex mixtures. MS experiments usually begin with digesting a protein mixture sample with trypsin and determining the masses of the intact peptides, producing a peptide mass fingerprint of a sample. This fingerprint can be used to search protein databases. A search algorithm that carries out virtual

Table 1   Experimental Methods to Identify Protein Interactions

Method	Features	Observation Level
NMR spectroscopy[14]	Provides insights into the dynamic interaction of proteins in solution.	Atomic and direct
Phage display[14]	Includes identification of binding motifs, followed by computational identification of potential interacting partners and a yeast two-hybrid validation step.	Direct and complex
Yeast two-hybrid[14]	A protein of interest serves as the "bait" to fish for and bind to unknown proteins, the "prey".	Direct and complex
Mass spectrometry[15]	Isolates binding partners and complexes, and identifies the component proteins.	Direct and complex
X-ray crystallography[16]	Crystallization of the interacting complex allows definition of the interaction structure.	Atomic and direct
Protein arrays/ Protein chips[17]	Antibody or bait-based arrays allow for the screening and detection of specific interactions of proteins from complex mixtures.	Direct and complex
Surface plasmon resonance[18]	Relates binding information to small changes in refractive indices of laser light reflected from gold surfaces, to which a bait protein has been attached.	Direct and complex
Pull-down assays[19]	Uses a labeled bait to create a specific affinity matrix so as to enable binding and purification of a prey protein from a lysate sample or other protein-containing mixture.	Direct and complex
Protein interaction mapping[19]	Utilizes an artificial protease on a bait protein to initiate contact-dependent cleavages in the prey protein in the presence of specific reactants. The nonspecific cleavage fragments are then analyzed to map the interface of a known protein–protein interaction.	Complex
Label transfer[19]	Upon reduction of the cross-linked complex, the prey protein gets the label from a bait protein that was first modified with the reagent. The label is typically used in the detection process to isolate or identify the unknown prey protein.	Direct and complex

(Continued)

Table 1 *(Continued)*

Method	Features	Observation Level
Far-Western analysis[19]	Similar strategy to Western blotting, except that the antibody probe is substituted with an appropriately labeled bait protein as the probe.	Direct and complex
Cross-linking reagents[19]	Strategies involve homobifunctional or heterobifunctional reagents whose chemical cross-links may or may not be reversed. Nearest neighbors (suspected to interact) *in vivo* or *in vitro* can be trapped in their complexes for further study.	Complex
Coimmunoprecipitation (Co-IP)[19]	Designed to affinity purify a bait protein antigen together with its binding partner using a specific antibody against the bait.	Complex
Fluorescence resonance energy transfer (FRET) analysis[20]	Green fluorescent protein (GFP) and its derivatives make fluorescent protein fusions inside cells to detect protein interactions in real time in living cells.	Atomic, direct, complex, and cellular

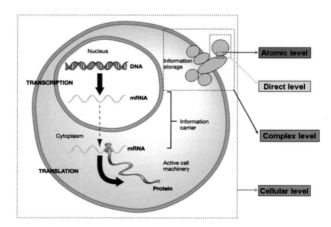

Fig. 1. Different levels of observation of protein interactions.

digestion of protein sequences based on the sequence specificity of trypsin and then calculates the masses of the predicted peptides from first principles (for example, by adding up the masses of the individual atoms) is used. If no matches are found, a different procedure called tandem mass spectrometry (MS/MS) can be used to generate random fragmentation of the peptides. The masses of these shorter fragments can be searched against further databases containing short sequences. The fragment ions can also be ordered by size, and the masses of sequential fragments used to establish which amino acids have been cleaved off. In this way, the protein sequence can be established and used to search sequence databases in order to find related rather than exact matches.[14] An example of this method is the identification of caspase-8 (FLICE) as an interactor with immunoprecipitated CD95 (Fas/APO1).

The major limitations are cost, need for high level of technical sophistication, and need for isolation of protein complexes by physical methods such as electrophoresis. Many proteins will not be detected because, for example, they are too scarce, large, small, acidic, or alkaline to be studied by two-dimensional (2D) gel analysis; or because the interaction is too weak and transient to survive affinity purification. As its sensitivity and ease of use improves, the MS method is expected to outperform other biological assays for detecting and analyzing protein interactions in protein complexes.

2.2. *Nuclear Magnetic Resonance (NMR) Spectroscopy*

Nuclear magnetic resonance (NMR) can be used to determine the structure of molecules (usually small soluble proteins) or interacting molecules.[14] NMR describes the way some atomic nuclei waver in a magnetic field when exposed to radio waves. The frequency of radio waves required to cause resonance in different nuclei depends on the way atoms are arranged in a molecule. Structures solved by NMR are generally presented as an ensemble of models, all of which satisfy the available data constraints.

2.3. *Yeast Two-Hybrid System*

The yeast two-hybrid system exploits protein interactions to assemble a functional transcription factor; the transcription factor then activates a test gene, allowing yeast cells containing interacting proteins to be identified. The principle of the method can be summarized as follows.[14] Transcription factors are proteins required for gene expression. They have at least two functional domains: a DNA-binding domain (BD) and a transcriptional activation domain (AD). The function of protein X, for example, is unknown. The DNA sequence of protein X is cloned into a vector containing a yeast transcription factor's BD and expressed in a strain of haploid yeast cells to produce a hybrid protein known as the "bait". All of the other genes in the genome are expressed as hybrid proteins cloned into a different vector containing the yeast transcription factor's BD, known as the "prey". This forms a library of potential preys. In a high-throughput screening experiment, the bait yeast strain is mated to each member of the prey library, so all possible interactions are tested in the resulting cells. In those cells where the bait interacts with the prey, a functional transcription factor is assembled and the test gene is activated.

There are three main approaches for large-scale two-hybrid studies. The matrix approach (one vs. one) systematically tests pairs of proteins for an interaction phenotype; a positive result can indicate that these particular proteins interact. Array experiments (one vs. all) examine the interactions of a single BD fusion protein against a pool of AD fusions. Pooling studies (all vs. all) involve yeast strains expressing different BD fusions mass-mated with strains expressing AD hybrids. The two-hybrid system can be scaled up so that multiple baits can be tested. The pooling search space, i.e. the total number of possible interactions to test, can be huge and therefore error-prone, even if the sensitivity is reasonable compared to low-throughput methods.

2.4. *Phage Display*

Like the two-hybrid system, phage display is used for the high-throughput screening of protein interactions. The principle of this method is summarized as follows. The function of protein X is unknown. The protein is used to coat the surface of a small plastic dish. All of the other genes in the genome are expressed as fusions with the coat protein of a bacteriophage (a virus that infects bacteria), so that they are displayed on the surface of the viral particle. This phage display library is added to the dish. After a while, the dish is washed. Phage-displaying proteins that interact with protein X remain attached to the dish, while all others are washed away. DNA extracted from interacting phage contains the sequences of interacting proteins.

2.5. *Two-Dimensional Polyacrylamide Gel Electrophoresis (2D PAGE)*

Two-dimensional polyacrylamide gel electrophoresis (2D PAGE)[14] is a technique for separating complex mixtures of proteins. Thousands of proteins can be resolved in a single experiment, allowing the major proteins in a sample to be isolated and protein levels in related samples to be compared.

Coupled with MS methods, 2D PAGE allows complex protein samples to be systematically identified.

The technique is particularly effective when comparing related samples, such as healthy tissue vs. disease tissue. For example, proteins that are more abundant in disease tissue may represent novel drug targets or diagnostic disease markers. Comparative 2D PAGE can also be used to look for proteins whose expression varies similarly under the same set of conditions (these may have related functions), and to identify proteins produced in response to drug therapy (these may be responsible for drug-related side-effects).

2.6. *X-Ray Crystallography*

Like NMR spectroscopy,[14] the main application of X-ray crystallography in biology is to determine the structure of a protein or a protein complex. X-ray crystallography produces very accurate structural models and can even show how proteins interact with other molecules. When X-rays strike a protein crystal, they are scattered by individual atoms. The way in which individual X-rays are diffracted depends on the types of atoms in the molecule and the way they are arranged. Using sophisticated mathematical techniques, an X-ray diffraction pattern can be used to work out the relative positions of different atoms in a molecule. X-ray crystallography can thus be used to determine three-dimensional (3D) protein structures. A protein structure solved by X-ray crystallography is regarded as the gold standard to which all other structures are compared.

3. DATABASE RESOURCES FOR PROTEIN INTERACTIONS

Publicly accessible databases of protein–protein interactions greatly simplify and facilitate the analysis of various protein interactions. Several databases that are currently available provide access to both experimental data and the results of different computational methods of inference. Some databases also identify the most reliable subsets of the interaction data. Further development of interaction databases is essential for the standardization of the interaction data sets and data exchange formats, as well as for the integration of the databases with other bioinformatics resources. Here, we review the different types of biological databases dedicated to protein interactions, and show a summarized comparison of selected databases in Table 2.

Table 2 Major Public Databases of Protein Interactions

Database	URL	Content	Interaction Statistics	Type of Interactions/Networks
DIP[23]	http://dip.doe-mbi.ucla.edu/	Catalog of protein–protein interactions	55 732 interactions among 19 051 proteins in 110 species	Experimental physical network
BIND[24]	http://bind.ca/	Biomolecular interaction complexes and pathways	201 896 interactions in 1528 organisms	Experimental physical network
MPPI[25]	http://mips.gsf.de/proj/ppi/	A new resource of high-quality human protein interaction data	1800 interactions among 900 proteins from 10 mammalian species	Experimental physical and genetic networks
STRING[26]	http://string.embl.de/	A database of predicted functional associations among genes or proteins	23 256 408 interactions among 736 429 proteins in 179 species	Experimental, predicted
HPRD[27]	http://hprd.org/	Comprehensive collection of protein features, posttranslational modifications, and protein–protein interactions	33 710 interactions among 20 097 proteins in human organism	Experimental physical network
HPID[29]	http://hpid.org/	Provides human protein interaction information and predicts potential interactions between proteins submitted by users	8565 interactions among 1690 proteins in human organism	Experimental, structural, and predicted
HAPPI	http://bio.informatics.iupui.edu/HAPPI/	Provides comprehensive human protein interaction information including confidence, cocitation, domain/pathway profiles, PDB structures, and feature alignments	142 956 nonredundant reliable human protein interaction pairs among 10 592 human proteins	Experimental, structural, and predicted

3.1. *DIP: Database of Interacting Proteins*

The DIP is a database that documents experimentally determined protein–protein interactions. It provides a comprehensive and integrated tool for browsing and retrieving information about pairwise interactions between proteins, signaling pathways, and complex systems. The DIP allows the visual representation and navigation of protein interaction networks as well as the integration of a diverse body of information onto a protein interaction network, such as the predominance of certain domains or the different subcellular compartments in which a protein can be found. So far, around 55 732 interactions among 19 051 proteins are available. For *Homo sapiens*, 1407 interactions among 916 proteins from 2061 experiments have been documented. All of these features make the DIP a starting point for comparing and assessing the reliability of different experimental methodologies (including high-throughput interaction screening) for the development of prediction methods, as well as in studies of the properties of protein interaction networks.[23]

3.2. *BIND: Biomolecular Interaction Network Database*

The BIND is a database that archives biomolecular interaction, reaction, complex, and pathway information. It encompasses the growing network of protein and other biomolecular interactions such as protein–protein, protein–RNA, protein–DNA, and protein–small-molecule interactions. It presents protein interactions from the molecular level to the pathway level. The contents of the BIND include high-throughput data submissions and hand-curated information gathered from the scientific literature. At present, there are around 201 896 interaction records, out of which 82 164 are protein–protein interactions and 35 191 are human protein interactions.[24]

3.3. *MPPI: MIPS (Munich Information Center for Protein Sequences) Mammalian Protein–Protein Interaction Database*

The MPPI database is a new resource of high-quality experimental protein interaction data in mammals. Its content is based on published experimental evidence that has been processed by human expert curators. Currently, this data set contains more than 1800 evidence entries for protein–protein interactions among more than 900 proteins from 10 mammalian species. On average, each protein in this database is involved in 1.92 interactions and each interaction is supported by 1.98 evidence entries.[25]

3.4. STRING: Search Tool for the Retrieval of Interacting Genes/Proteins

The STRING[26] is a database of known and predicted protein–protein interactions. These interactions include direct or physical and indirect or functional associations. The associations are obtained from high-throughput experiments, mining of databases and literature, and predictions based on genomic context analysis. It integrates and ranks these associations by comparing them against a common reference set, and presents evidence in a consistent web interface. Importantly, the associations are extended beyond the organism in which they were originally described by automatic transfer to orthologous protein pairs in other organisms, where applicable. The database currently contains 736 429 proteins from 179 species.

3.5. HPRD: Human Protein Reference Database

The HPRD represents a centralized platform to visually depict and integrate information pertaining to domain architecture, posttranslational modifications (PTMs), interaction networks, and disease association for each protein in the human proteome. All of the information in the HPRD has been manually extracted from the literature by expert biologists who read, interpret, and analyze the published data. The HPRD was created using an object-oriented database in Zope, an open source web application server that provides versatility in query functions and allows data to be displayed dynamically. With over 20 000 protein entries, this database has become the largest database for literature-derived protein–protein interactions (more than 30 000) and PTMs (more than 8000) for human proteins.[7,27]

3.6. HPID: Human Protein Interaction Database

The HPID[29] was designed to provide human protein interaction information precomputed from existing experimental and structural data, to predict potential interactions between proteins submitted by users, and to provide a depository for new human protein interaction data from users. Structural interactions between human proteins are predicted by homology-based assignment of domain structures to the whole genome. The "homologous interaction" concept is applied at the SCOP superfamily level. Interactions are shown at the protein superfamily level and at those transferred from the interactions of yeast proteins. It also provides Web Interviewer for visualizing large-scale protein interaction networks in 3D space. Currently, the refined network contains 8565 interactions among 1690 human proteins with an average node degree of 5.

3.7. *OPHID: Online Predicted Human Interaction Database*

The OPHID[28] is an online database of human protein–protein interactions. It explores known and predicted protein–protein interactions, and facilitates bioinformatics initiatives in exploring protein interaction networks. It has been built by mapping high-throughput model organism (yeast, mouse, *Drosophila* and *C. elegans*) data to human proteins. The database currently contains 47 656 interactions with 10 652 proteins.

3.8. *HAPPI: Human Annotated and Predicted Protein Interaction Database*

The HAPPI database is the latest development of an open-access comprehensive collection of computer-annotated human protein–protein interactions from public data sources and computational predictions. The database was developed by exhaustively integrating publicly available human protein interaction data from BIND, OPHID, MINT, IntAct, HPRD, and STRING databases into a data warehouse powered by an Oracle 10g relational database server. In the data warehouse, various types of sequence, structure, pathway, and literature annotation data from established bioinformatics resources (e.g. NCBI, PubMed, UniProt, HUGO, EBI, PDB) were also integrated. The HAPPI database represents a new developmental trend, in which the protein interaction database serves as the primary resource for biomedical scientists who are interested in evaluating biologically significant protein interactions, developing disease pathway models, and identifying disease drug targets or diagnostic biomarkers. As of August 2006, the database contained 142 956 nonredundant reliable human protein interaction pairs among 10 592 human proteins.

4. COMPUTATIONAL METHODS AND PROTEIN INTERACTIONS

Computational methods play an important role in the study of interactions between proteins. At minimum, they allow efficient querying, browsing, and visualizing of protein interactions stored in databases. They can help predict potential interactions when the experimental data set is scarce. They can also help validate the results of high-throughput interaction screens by correlating multiple evidences together. Moreover, they can help analyze protein interaction and signaling networks extracted from databases. Many software tools are now available (for a summary, refer to Table 3).

Table 3 Software Tools to Explore Protein Interactions

Web Tool	URL	Features
ADVICE[34]	http://advice.i2r-a-star.edu.sg/	Prediction and validation of protein–protein interactions using observed coevolution of interacting proteins
BioLayout Java[35]	http://cgg.ebi.ac.uk/services/biolayout/	Versatile network visualization of structural and functional relationships
eF-site[36]	http://ef-site.hgc.jp/eF-site/	A database and viewer for molecular surfaces of protein functional sites
EP:PPI[37]	http://ep.ebi.ac.uk/EP/PPI/	Explores protein interaction data using expression data
IntAct[38]	http://www.ebi.ac.uk/intact/index.html/	Database and toolkit for the storage, presentation, and analysis of protein interactions
InterWeaver[39]	http://interweaver.i2r-a-star.edu.sg/	A web server of interaction reports
InterPreTS[40]	http://speedy.embl-heidelberg.de/people/patrick/interprets/index.html/	Predicts the potential interaction of two proteins from 3D information of protein complexes
Medusa[41]	http://www.bork.embl-heidelberg.de/medusa/	An interface to the STRING protein interaction database and a general graph visualization tool
SPIN-PP[42]	http://trantor.bioc.columbia.edu/cgi-bin/SPIN/	A database of all protein–protein interfaces for protein–protein interactions in the Protein Data Bank
Protein–Protein Interaction Server[43]	http://www.biochem.ucl.ac.uk/bsm/PP/server/server_help.html	A means of calculating a series of descriptive parameters for the interface between any two proteins in a 3D protein structure
PIVOT[44]	http://www.cs.tau.ac.il/~rshamir/pivot/	A Java-based tool for visualizing protein–protein interactions
PIMWalker[45]	http://pim.hybrigenics.com/pimriderext/pimwalker/	An interactive tool for displaying protein interaction networks
PathBLAST[46]	http://chianti.ucsd.edu/pathblast/	Network alignment and search tool for cross-species comparison of protein interaction networks for evolutionary patterns of conservation

(Continued)

Table 3 *(Continued)*

Web Tool	URL	Features
iSPOT[47]	http://cbm.bio.uniroma2.it/ispot/	Prediction of protein–protein interactions mediated by families of peptide recognition modules
InterViewer[48]	http://165.246.44.45/hpid/webforms/Visualization.aspx/	Visualizes and analyzes large-scale protein interaction networks in 3D space
iPPI	http://www.bioinfo.cu/iPPI/	Infers protein–protein interactions through homology search
iPfam[49]	http://www.sanger.ac.uk/Software/Pfam/iPfam/	Visualization of protein–protein interactions at domain and amino acid resolutions
Virtual Ligand Screening[50]	http://www.molsoft.com/vls.html/	A computer technique simulating the interaction between proteins and small molecules that might be a good lead to potential new drugs
ProteoLens[51]	http://bio.informatics.iupui.edu/proteolens/	A Java-based tool for visualizing protein–protein interaction networks from relational databases

4.1. *Prediction of Protein–Protein Interactions*

Several computational approaches are available today for predicting interactions between proteins. Protein–protein interactions can be extracted on the basis of functional relationships between proteins, such as patterns of domain fusion and protein occurrence, sequence and structural analysis, correlation of functional genomic features, and existence of conserved interactions in other organisms. These interactions can also be inferred from the literature in an automated way.

The function of a protein can be analyzed as its position within the cellular interaction network. Therefore, the understanding and prediction of a protein's interaction partners are an important step towards the identification of its role within a cell. Recently developed methods for the inference of protein–protein interactions can be broadly classified into physical and functional linkages.[30]

The physical linkage methods include the following types:

(1) Interspecies interaction transfer based on the interacting sequence motif pairs identified in Y2H screens.[31,32]
(2) Interactions inferred from correlated mutations.[33]
(3) Co-occurrence of sequence domains.[8,52]
(4) Structure assignment followed by threading-based interaction energy evaluation.[53]
(5) Ortholog-based transfer of interactions between species followed by experimental validation.[54]

The functional linkage methods include the following types:

(1) Network topology–based functional annotation. It uses the topology of a protein network, using expression-distance measurements to annotate the nodes on the shortest paths connecting proteins of the same function. In other words, a function is transferred to proteins that form the shortest path connecting two proteins of the same, known function.[55]
(2) Phylogenetic profile method. Functional links are created between proteins with a similar evolutionary history, as guided by the similar pattern of their presence across multiple genomes.[56]
(3) Phylogenetic profile enhancements. Measures of phylogenetic profile distance that reflect the detailed evolutionary history of the species improve the performance of the method.[57–59]
(4) Rosetta stone or gene fusion method.[60,61]
(5) Conventional sequence-based method. Prediction is based on primary sequence and associated physicochemical properties.

4.2. *Validation Method of Protein Interactions*

In order to evaluate the reliability of individual interactions, a number of tests are used to identify the most reliable core subset of the interactions. The tests range from simple evaluation methods based on the reliability of individual experiments to the sophisticated analysis of interaction patterns between analogous proteins using the paralogous verification method (PVM).[62]

Confidence in a reported interaction is increased by observation of the same interaction using other methods. For example, if an interaction is measured with two distinct experiments, one using the two-hybrid system and another using immunoprecipitation, the joint observation increases our confidence in this particular interaction. Another approach to validating a given interaction is the use of subcellular colocalization. If an interaction occurs between two proteins that are both known to be localized to the same subcellular compartment, the likelihood of physiological interaction is increased. This type of validation might be useful in increasing confidence levels in protein interactions detected by error-prone methods.

An example can be found in the DIP, where evaluation methods are implemented as publicly available services (http://dip.doe-mbi.ucla.edu/dip/Services.cgi) that can be used to evaluate the reliability of new experimental and predicted interactions. These services include the PVM and expression profile reliability (EPR) method[37] as well as the domain pair verification (DPV) method, which analyzes domain–domain interaction preferences as described by Deng *et al.*[8] The EPR index estimates the biologically relevant fraction of protein interactions detected in a high-throughput screen, whereas the PVM judges an interaction likely if the putatively interacting pair has paralogs that also interact. In contrast to the EPR index, which evaluates data sets of interactions, PVM scores individual interactions.

4.3. *Assessment of Protein Interaction Data Using Network Properties*

The global properties of protein–protein interactions are commonly analyzed by graph theory. According to this method, individual proteins are modeled as graph vertices connected by edges that correspond to binary protein–protein interactions. Despite its lack of temporal/spatial resolution and its weakness in representing multiprotein complexes, graph theoretical analysis has provided interesting leads into the structure of the protein interaction network. For example, Jeong and colleagues[63] described the scale-free topology of protein interaction networks. The most characteristic feature of scale-free networks is the presence of a few highly connected nodes well separated within the network.[64] It was postulated that such topology is responsible for the robustness of the scale-free networks[65]; the finding that the essential protein-encoding genes within the protein–protein interaction network coincide with the highly connected nodes seems to confirm this interpretation of robustness. However, current models of network growth cannot explain all features of biological networks. For example, some of the characteristic parameters of the metabolic networks, such as the degree of clustering, deviate from the expected values of those in a scale-free model.

In general, the results indicate that a significant fraction of high-throughput data sets may contain errors. Methods introduced by Mrowka *et al.* [66] and Deane *et al.* [62] analyzed the collective properties of the interaction data sets, such as the distribution of the expression distances between interacting partners; statistical analysis and comparison of these properties to those of a trusted reference set result in quantitative estimation of the accuracy of the high-throughput data.[62] In addition to these evaluations of the overall total quality of the interaction data sets, attempts have been made to identify the most reliable subsets of high-throughput data. These attempts usually involve combining different sources of experimental information; however, as there is only a small overlap between data sets, the number of interactions that may be validated this way is very marginal. The number of validated interactions increases if one also takes into consideration known interactions between paralogs of the putative interacting pair. This approach, as demonstrated by Deane *et al.*,[62] allows one to identify roughly half of the true interactions within a typical high-throughput data set. Recently, another method of quality evaluation has been proposed by Bader *et al.*[67] At its root, this method exploits the observation, recently made by Ravasz *et al.*,[68] that interacting proteins tend to form highly connected clusters within interaction networks; it is therefore possible to assess the quality of a prospective interaction by examining the length of the shortest path that connects the potential interactors.

5. BIOLOGICAL SIGNIFICANCE OF PROTEIN INTERACTIONS

As postgenome biological studies move from -omics data collection to systems biology data analysis, biologists are now more interested in understanding how proteins function in molecular and cellular network contexts. By interacting with each other, proteins relay intracellular signals in intricate signaling networks that are capable of well-tuned and highly adaptive responses to environmental stimuli, such as in apoptosis. With the advent of protein interactomics, i.e. a collective study of proteins and protein interactions, interactions between proteins can be studied on a large global scale. The excitement lies in the fact that information from protein–protein interaction data may help find new drug targets or diagnostic biomarkers. Understanding the proteins and their interacting partners in a network is essential in postgenome biomedical research.

Protein–protein interactions take place in cells through a number of different mechanisms. The most represented protein–protein interactions are antibody–antigen interactions. Other proteins act as enzymes that modify the structure of other proteins; for example, the γ-secretase enzyme complex separates the amyloid precursor protein into two fragments, and errors in this process seem to play a pivotal role in the development of Alzheimer's disease. Protein–protein interactions also control the localization of proteins, their substrate-processing activity, and even their tagging for destruction or recycling. A large majority of proteins undergo modifications during or after translation; these cotranslational or posttranslational modifications are important for protein stability, sorting, and function. Many modifications are directly related to diseases, and information about these modifications is an important resource to study the function of proteins.

Protein–protein interactions also play roles in extracellular signaling. Most cells react to extracellular molecular signals that control their production of enzymes, their growth, or their metabolic activity. For example, the arrival of a growth factor hormone at a cell membrane may lead to a pairing of receptor proteins, which then sets in motion a signaling cascade within the cell that eventually leads to cell growth–related responses.

A prominent feature of protein–protein interactions is their variety. Proteins interact in different ways as their structures are so vastly complex. Amino acid side chains that come out from the body of a molecule create pits or bumps of different shapes and sizes. Proteins use this structural diversity to the maximum, producing binding pockets and recognition sites with varying degrees of specificity and subtlety of interaction. This variability gives protein–protein interactions an exciting future in the search for new drug targets.

The therapeutic applications of protein–protein interactions are very extensive. Most current drugs target the important binding site of a protein, typically affecting its entire area of operation. This new generation of drugs can act as competitive antagonists, but can also make much more obscure alterations through allosteric inhibition, only disrupting the way in which a protein interacts with other specific proteins. The drug colchicine, for example, actually works by disrupting the interaction between α and β tubulins in microtubules that is crucial for cell division processes.

There are a few obstacles to taking advantage of protein interactions in drug design. First, there is usually an unacceptably large range of concentrations over which proteins might interact. Second, protein–protein interactions are often transient and subtle, making them difficult to study. Third is the size. Although large- or medium-sized peptides have often been used to modulate protein–protein interactions, therapeutic drugs must be small enough to get to the site of target protein interaction, which often takes place inside cells. The interacting surfaces of the proteins are many times larger than a small molecule. Instead of revealing the binding sites, X-ray structures of protein–protein pairs often do not reveal the deep pockets that have these binding sites. On top of that, proteins are also very flexible structures, being in constant motion between different conformational states with similar energies; and important fluctuations in the binding area would not show up in X-ray crystal structures. Although protein–protein interactions occur over a large surface area, X-ray crystallography and site-directed mutagenesis have shown that many protein–protein binding sites contain closely packed, centralized regions of residues called "hot spots" that are important for interactions. Many proteins function by binding to multiple proteins, and these partner proteins seem to reuse the same hot spot, which adjusts to present the same residues in different structural contexts.

Targeting protein–protein interactions is clearly more challenging than traditional approaches for the identification of small-molecule inhibitors of protein targets. However, researchers are making progress and some of the difficulties are gradually being lessened. Molecules have been identified that inhibit the function of inducible nitric oxide synthase by binding to the heme cofactor in the protein active site, which disrupts protein dimerization. Recently, small-molecule inhibitors of the MDM2–p53 tumor suppressor protein interaction have generated excitement in the field. MDM2 diminishes the ability of p53 to repair potential cancer-causing breakages in genes, and disturbing this interaction could be a novel strategy of cancer therapy.[69] It is hoped that such achievements will encourage more interest and research in this area, and that what seemed impossible only a few years ago might now become true.

Peptidomimetics represents another promising area for designing new drugs to disrupt protein interactions. Short, synthesized peptide fragments that simulate the most common peptide motifs α-helix or β-sheet can be used in lieu of small molecules to interfere with protein–protein interactions. The fragment discovery approach can also be used to screen a number of small organic compounds in order to find the fragment that binds to a protein, and then string the fragments together to find the most effective molecule. The advantage of this approach lies in the lead compounds as they are less hydrophobic, thus laying a good foundation for drug discovery.

In summary, protein interaction study is at the core of biomedical research in the postgenome era. Many public databases and software tools are available for the management and analysis of protein–protein interactions. The collection of protein interaction data and protein interaction networks overall may serve as a model of the molecular network context, in which different types of biomedical data can be integrated and interpreted. Although many experimental methods may be used to generate this data for human biomedical research studies, the scope and quality of many large-scale data sets still need to be thoroughly assessed. With combined experimental and computational efforts, the complete mapping and continued mining of the protein interaction network in humans promise to shed new light on how to develop new drugs to treat human diseases.

REFERENCES

1. Golemis E. (2002) Toward an understanding of protein interactions. In: *Protein–Protein Interactions: A Molecular Cloning Manual*, Cold Spring Harbor Laboratory Press, New York, NY, pp. 1–5.
2. Nakai K. (2000) Protein sorting signals and prediction of subcellular localization. *Adv Protein Chem* **54**:277–344.
3. Hanash S. (2003) Disease proteomics. *Nature* **422**:226–232.
4. Karin M, Ben-Neriah Y. (2000) Phosphorylation meets ubiquitination: the control of NF-κB activity. *Annu Rev Immunol* **18**:621–663.
5. Pawson T, Nash P. (2003) Assembly of cell regulatory systems through protein interaction domains. *Science* **300**:445–452.
6. Albert R, Jeong H, Barabasi AL. (2000) Error and attack tolerance of complex networks. *Nature* **406**:378–382.
7. Phizicky EM, Fields S. (1995) Protein–protein interactions: methods for detection and analysis. *Microbiol Rev* **59**:94–123.
8. Deng M, Mehta S, Sun F, Chen T. (2002) Inferring domain–domain interactions from protein–protein interactions. *Genome Res* **12**:1540–1548.
9. Apic G, Gough J, Teichmann SA. (2001) Domain combinations in archaeal, eubacterial and eukaryotic proteomes. *J Mol Biol* **310**:311–325.

10. Peri S, Navarro JD, Amanchy R, *et al.* (2003) Development of Human Protein Reference Database as an initial platform for approaching systems biology in humans. *Genome Res* **13**:2363–2371.
11. Kitano H. (2002) Computational systems biology. *Nature* **420**:206–210.
12. Birney E, Clamp M, Hubbard T. (2002) Databases and tools for browsing genomes. *Annu Rev Genomics Hum Genet* **3**:293–310.
13. Golemis E. (2002) *Protein–Protein Interactions: A Molecular Cloning Manual.* Cold Spring Harbor Laboratory Press, New York, NY.
14. http://genome.wellcome.ac.uk/
15. McLafferty FW, Turecek F. (1993) *Interpretation of Mass Spectra*, 4th ed. University Science Books, Mill Valley, CA.
16. http://en.wikipedia.org/wiki/X-ray_crystallography/
17. http://www.gene-chips.com/
18. http://www.uksaf.org/tech/spr.html/
19. http://www.piercenet.com/
20. http://www.andor.com/biology/?app=92/
21. Xenarios I, Eisenberg D. Protein interaction databases. (2001) *Curr Opin Biotechnol* **12**:334–339.
22. http://campus.queens.edu/faculty/jannr/bio103/helpPages/c11DNA.htm/
23. Xenarios I, Salwinski L, Duan XJ, *et al.* (2002) DIP, the Database of Interacting Proteins: a research tool for studying cellular networks of protein interactions. *Nucleic Acids Res* **30**:303–305.
24. Alfarano C, Andrade CE, Anthony K, *et al.* (2005) The Biomolecular Interaction Network Database and related tools 2005 update. *Nucleic Acids Res* **33**(database issue):D418–D424.
25. Pagel P, Kovac S, Oesterheld M, *et al.* (2005) The MIPS Mammalian Protein–Protein Interaction database. *Bioinformatics* **21**:832–834.
26. von Mering C, Jensen LJ, Snel B, *et al.* (2005) STRING: known and predicted protein–protein associations integrated and transferred across organisms. *Nucleic Acids Res* **33**:D433–D437.
27. Albert R, Jeong H, Barabasi AL. (2000) Error and attack tolerance of complex networks. *Nature* **406**:378–382.
28. Brown KR, Jurisica I. (2005) Online Predicted Human Interaction Database. *Bioinformatics* **21**:2076–2082.
29. Han K, Park B, Kim H, *et al.* (2004) HPID: the Human Protein Interaction Database. *Bioinformatics* **20**:2466–2470.
30. Salwinski L, Eisenberg D. (2003) Computational methods of analysis of protein–protein interactions. *Curr Opin Struct Biol* **13**:377–382.
31. Wojcik J, Schachter V. (2001) Protein–protein interaction map inference using interacting domain profile pairs. *Bioinformatics* **17**(Suppl 1):S296–S305.
32. Wojcik J, Boneca IG, Legrain P. (2002) Prediction, assessment and validation of protein interaction maps in bacteria. *J Mol Biol* **323**:763–770.
33. Pazos F, Valencia A. (2002) *In silico* two-hybrid system for the selection of physically interacting protein pairs. *Proteins* **47**:219–227.

34. Tan SH, Zhang Z, Ng SK. (2004) ADVICE: Automated Detection and Validation of Interaction by Co-Evolution. *Nucleic Acids Res* **32**: W69–W72.
35. Goldovsky L, Cases I, Enright AJ, Ouzounis CA. (2005) BioLayout(Java): versatile network visualization of structural and functional relationships. *Appl Bioinformatics* **4**:71–74.
36. Kinoshita K, Nakamura H. (2004) eF-site and PDBjViewer: database and viewer for protein functional sites. *Bioinformatics* **20**:1329–1330.
37. Kapushesky M, Kemmeren P, Culhane AC, *et al.* (2004) Expression Profiler: next generation — an online platform for analysis of microarray data. *Nucleic Acids Res* **32**(web server issue):W465–W470.
38. Hermjakob H, Montecchi-Palazzi L, Lewington C, *et al.* (2004) IntAct: an open source molecular interaction database. *Nucleic Acids Res* **32**:D452–D455.
39. Zhang Z, Ng SK. (2004) InterWeaver: interaction reports for discovering potential protein interaction partners with online evidence. *Nucleic Acids Res* **32**(web server issue):W73–W75.
40. Aloy P, Russell RB. (2003) InterPreTS: protein interaction prediction through tertiary structure. *Bioinformatics* **19**:161–162.
41. Hooper SD, Bork P. (2005) Medusa: a simple tool for interaction graph analysis. *Bioinformatics* **21**:4432–4433.
42. Surface Properties of Interfaces — Protein Protein Interfaces (SPIN-PP). Available at http://honiglab.cpmc.columbia.edu/SPIN/main.html/
43. Jones S, Thornton JM. (1996) Principles of protein–protein interactions. *Proc Natl Acad Sci USA* **93**:13–20.
44. Orlev N, Shamir R, Shiloh Y. (2004) PIVOT: Protein Interactions Visualization Tool. *Bioinformatics* **3**:424–425.
45. Meil A, Durand P, Wojcik J. (2005) PIMWalker: visualizing protein interaction networks using the HUPO PSI molecular interaction format. *Appl Bioinformatics* **4**:137–139.
46. Kelley BP, Yuan B, Lewitter F, *et al.* (2004) PathBLAST: a tool for alignment of protein interaction networks. *Nucleic Acids Res* **32**:W83–W88.
47. Brannetti B, Helmer-Citterich M. (2003) iSPOT: a web tool to infer the interaction specificity of families of protein modules. *Nucleic Acids Res* **13**:3709–3711.
48. Han K, Ju BH, Jung H. (2004) WebInterViewer: visualizing and analyzing molecular interaction networks. *Nucleic Acids Res* **32**:W89–W95.
49. Finn RD, Marshall M, Bateman A. (2005) iPfam: visualization of protein–protein interactions in PDB at domain and amino acid resolutions. *Bioinformatics* **21**:410–412.
50. Bernacki K, Kalyanaraman C, Jacobson MP. (2005) Virtual ligand screening against *Escherichia coli* dihydrofolate reductase: improving docking enrichment using physics-based methods. *J Biomol Screen* **10**:675–681.
51. Sivachenko A, Chen JY. ProteoLens: a software tool for visual data mining of protein–protein networks. Available at http://bio.informatics.iupui.edu/proteolens/

52. Sprinzak E, Margalit H. (2001) Correlated sequence-signatures as markers of protein–protein interaction. *J Mol Biol* **311**:681–692.
53. Lu L, Lu H, Skolnick J. (2002) MULTIPROSPECTOR: an algorithm for the prediction of protein–protein interactions by multimeric threading. *Proteins* **49**:350–364.
54. Matthews LR, Vaglio P, Reboul J, *et al.* (2001) Identification of potential interaction networks using sequence-based searches for conserved protein–protein interactions or "interologs". *Genome Res* **11**:2120–2126.
55. Zhou X, Kao MC, Wong WH. (2002) Transitive functional annotation by shortest-path analysis of gene expression data. *Proc Natl Acad Sci USA* **99**:12783–12788.
56. Pellegrini M, Marcotte EM, Thompson MJ, *et al.* (1999) Assigning protein functions by comparative genome analysis: protein phylogenetic profiles. *Proc Natl Acad Sci USA* **96**:4285–4288.
57. Pazos F, Valencia A. (2001) Similarity of phylogenetic trees as indicator of protein–protein interaction. *Protein Eng* **14**:609–614.
58. Liberles DA, Thoren A, Heijne GV, Elofsson A. (2002) The use of phylogenetic profiles for gene predictions. *Curr Genomics* **3**:131–137.
59. Vert JP. (2002) A tree kernel to analyse phylogenetic profiles. *Bioinformatics* **18**(Suppl 1):S276–S284.
60. Enright A, Illioupolos I, Kyrpides NC, Ouzounis CA. (1999) Protein interaction maps for complete genomes based on gene fusion events. *Nature* **402**:86–90.
61. Clark BF. (1981) Towards a total human protein map. *Nature* **292**:491–492.
62. Deane CM, Salwinski L, Xenarios I, Eisenberg D. (2002) Protein interactions: two methods for assessment of the reliability of high throughput observations. *Mol Cell Proteomics* **1**:349–356.
63. Jeong H, Mason SP, Barabasi AL, Oltvai ZN. (2001) Lethality and centrality in protein networks. *Nature* **411**:41–42.
64. Maslov S, Sneppen K. (2002) Specificity and stability in topology of protein networks. *Science* **296**:910–913.
65. Albert R, Jeong H, Barabasi AL. (2000) Error and attack tolerance of complex networks. *Nature* **406**:378–382.
66. Mrowka R, Patzak A, Herzel H. (2001) Is there a bias in proteome research? *Genome Res* **11**:1971–1973.
67. Bader JS, Chaudhuri A, Rothberg JM, *et al.* (2004) Gaining confidence in high-throughput protein interaction networks. *Nat Biotechnol* **22**:78–85.
68. Ravasz E, Somera AL, Mongru DA, *et al.* (2002) Hierarchical organization of modularity in metabolic networks. *Science* **297**:1551–1555.
69. Buckingham S. (2004) Picking the pockets of protein–protein interactions. *Horizon Symposia — Charting Chemical Space* (April):1–4.

Chapter 20

Genomic Cloning in the Postgenome Era

Yan Jiao and Weikuan Gu[*]

Center of Genomics and Bioinformatics &
Center of Diseases of Connective Tissues
Department of Orthopedic Surgery–Campbell Clinic
University of Tennessee Health Science Center
Memphis, TN 38163, USA

Genomic cloning is a relatively new term that was developed after the completion of the human and mouse genome sequences. It refers to the identification of a gene responsible for a trait of interest by localizing its position based on the genomic sequences. Positional cloning, a classical term for gene cloning, refers to the identification of a gene responsible for a trait of interest according to its location in a genetic map. Thus, our current term of genomic cloning is derived from positional cloning. The difference between genomic and positional cloning is the substance that the cloning procedure is based on. A gene locus on a genetic map is obtained by finding the recombinant frequency of the locus relative to other markers on the genetic map without knowing the actual genomic sequences on the locus; because the recombinant frequency is not directly correlated to the genomic sequences, it is an approximate position. In contrast, a locus on a genomic position directly links the locus to the precise locations in base pairs in the genomic sequences. The translation from classical positional cloning to genomic cloning greatly enhances our ability in gene identification and discovery.

The strategy for identifying genes using positional cloning[1,2] has been limited by the availability of genetic-based data and technology. Most procedures in traditional positional cloning are labor-intensive and

---

[*]Correspondence author.

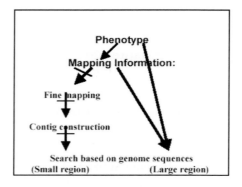

Fig. 1.   Strategy for positional cloning.

time-consuming. As shown in Fig. 1, the first step is to define a candidate region as precisely as possible, including initial linkage analysis and fine mapping. Linkage analysis requires the production of a large pedigree and polymerase chain reaction (PCR)-based analysis of microsatellite markers for a whole-genome search for linkage; fine mapping is the particularly difficult task of breaking the linkage and identifying useful markers in the targeted region. The next step is to construct a contig connecting the targeted region; this entails the identification of a large insert genomic library, either bacterial artificial chromosome (BAC) or yeast artificial chromosome (YAC) with known markers.[3] Finally, the contig is sequenced and analyzed using a technique termed chromosomal walking. Because all of these complicated procedures entail a tremendous amount of work, positional cloning has required team effort spanning a number of years.

However, the recent completion of the genome sequences of human, mouse, and a large number of other species[4,5] — along with other new technologies such as mutation analysis and microarrays — allows for dramatically faster progress in the positional cloning of genes from mutated models. The technique of positional cloning has changed in five ways: (1) We do not need to make BAC clones that contain the genomic sequences within the targeted/finely mapped region. (2) We do not need to sequence the entire region, usually 10 Mbp of the genomic sequences; the sequences are available through public (Ensembl) and private (Celera) databases. (3) We now know that the majority of the human and mouse genomes are repetitive sequences such as transposons, which are easy to be identified and can therefore be eliminated from further analysis. (4) On average, a 10-cM region may contain 300 genes[1]; assuming each gene has five exons, we need to analyze

1500 coding regions, so it is feasible for our laboratory to either sequence or analyze the expression profiles in less than half a year. (5) Recent technologies for gene expression profiling (e.g. microarray) and single nucleotide polymorphism (SNP) identification (e.g. SpectruMedix, which we use in our procedure) allow us to analyze large numbers of genes in a shorter time (less than 6 months, as in our current work).

Over the past couple of years, a new strategy of gene identification has rapidly developed.[6] The combination of genomic resources, advanced biotechnology, and creative ideas makes it possible to identify genes from multiple mutation models within a year. This chapter will present the new strategy and examples of gene identification using such a strategy. As the genome sequences of more and more species are finishing, this most updated strategy will greatly benefit studies in the genetic field and will be widely used by researchers. The new term "genomic cloning" refers to genome-based cloning. The protocol includes determining the targeted genomic region according to the genetic map of a mutation locus, identifying every gene and biologically functional element within the region, performing heteroduplex analysis of the transcripts to search for mutations in the coding region, and conducting gene expression profiling to locate mutation regions of noncoding sequences. We will describe the methodology and successful examples of genomic cloning in this chapter.

1. LOCALIZATION OF CANDIDATE GENES ON A GENOMIC MAP

Before the genomic sequence map was created, there were several kinds of genetic-based maps, including linkage, cytogenetic, and physical maps.

In the classical genetic map, a genetic locus is located on the map according to its relative position/recombinant frequency to other markers. Figure 2 is a partial genetic map that includes markers between 92.6 cM and 93.3 cM on mouse chromosome 1 (as of July 30, 2005). On the left side is the genetic distance along the chromosome. As mentioned above, the genetic distance represents the frequency of recombinant or rate of breakdown between homologous sister chromates. Next to the distance is the genetic marker, also listed along the chromosome. Genetic markers include molecular (e.g. the microsatellite D1Mit113 and the SNP of the gene *Ly9* for lymphocyte antigen 9) and phenotypic (e.g. Lsd for lymphocyte-stimulating determinant) markers.

One of the features of the markers is that many of them are located on the same genetic locus. There are several possible reasons for this. First

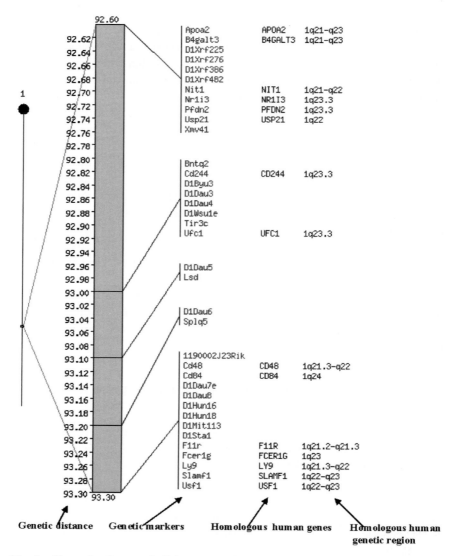

Fig. 2.   Example of a genetic linkage map. A portion of a genetic map of mouse chromosome 1 is shown above. It includes a total of 37 genetic markers within a genetic distance of 0.7 cM (from 92.6 cM to 93.3 cM). Among the genetic markers, 15 of them are known genes. Their homologous human genes and locations are also given. The figure was downloaded from the webpage for linkage mapping of The Jackson Laboratory, Bar Harbor, ME, available at http://www.informatics.jax.org/searches/linkmap.cgi/.

of all, there are not enough genetic markers that cover a detailed distance along the chromosome. For example, if two genetic markers have a distance of one million base pairs, any recombinant that happened within the one-million bp will appear as one recombinant event if it is measured by those two genetic markers. The second reason is the limited number of progenies in the mapping. In general, 1 recombinant in 100 progenies gives approximately 1 cM distance on the genetic map. A highly detailed map requires an extremely large mapping population.

Furthermore, locations of markers given by different research groups vary in precision and accuracy. The markers on the same locations in the genetic map may be far apart in the genomic map. Like human genetic maps, mouse maps also include human homologous genes. Many genes have been located on the mouse genetic map by Northern hybridization, while others were located by sequencing. The accuracy of the homologous mapping is largely unquarantined. Accordingly, the localization of your favorite gene on a genetic map is always a relative position.

After the completion of the whole-genome sequence of humans or other organisms, the same set of markers is located on genomic sequences. Figure 3 shows the corresponding genomic map (from Ensembl version 32) of the genetic map in Fig. 2. The figure includes genome information between two markers (*Apoa2* at 171 153 929–171 155 210 bp; *usf1* at 171 340 224–171 346 961 bp). The genome assembly provides all genome information in this chromosome region in the order of nucleotides (Fig. 3). The biggest advantage of the genomic map is the nucleotides in DNA contigs along the chromosome. The nucleotide information — including exons, introns, 5′ and 3′ primers, repetitive sequences of genes, and the nongene region — is assembled on each chromosome.

To localize a candidate gene or genetic locus onto the genomic map is to define a genomic region that confines the candidate gene or locus.

2. HIGH-THROUGHPUT SCREENING OF GENE MUTATIONS

High-throughput screening is a method for conducting experiments through a combination of modern robotics and other laboratory equipment. Because of the advanced robotics and high efficiency, one can process hundreds of thousands of samples at once.

In general, the cloning or identification of a gene of interest requires a large-scale screening of genes and other genetic elements along a region of

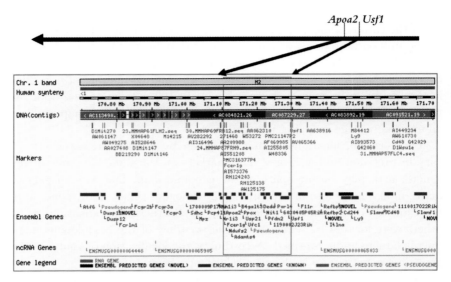

Fig. 3. Example of a genomic map. A genomic map of a genetic map between 92.6 cM and 93.3 cM on mouse chromosome 1 is shown above. It includes a total of 193 032 nucleotides. It contains molecular markers, genes, and their locations on the genome sequence assembly. The figure was downloaded from the webpage of Ensembl v32 of genome sequences, available at http://www.ensembl.org/Mus_musculus/contigview? seq_region_name=1&seq_region_right=171750944.5&click_right=695&h=1&seq_region_left=170749945.5&click_left=110&seq_region_width=193033&seq_region_strand=1&click. x=399&click.y=161/.

a chromosome. Linkage mapping usually locates a genetic or disease locus onto a region of several cM. Fine mapping to a region of less than 1 cM is time-consuming; for example, if one uses the mouse model, it takes more than 1 year for genotyping and phenotyping and a population of more than 200 mice. However, if one wants to eliminate the fine map step, a high-throughput system has to be conducted to screen a large number of genetic elements in the targeted region.[7,8] Figure 4 illustrates the current procedure for high-throughput screening of gene mutations used in many laboratories.

The first step is to identify all genetic elements in the quantitative trait locus (QTL) region. Genetic elements here include not only genes, but also noncoding regions as well as regulatory and other genome sequences that may play a role in the phenotype. Currently, the Ensembl database has complete sequence coverage of almost the entire human and mouse genomes and major parts of several other animals. Genes and transcripts notated by Ensembl and other sources/programs (EMBL mRNAs, UniGene, Genscan)

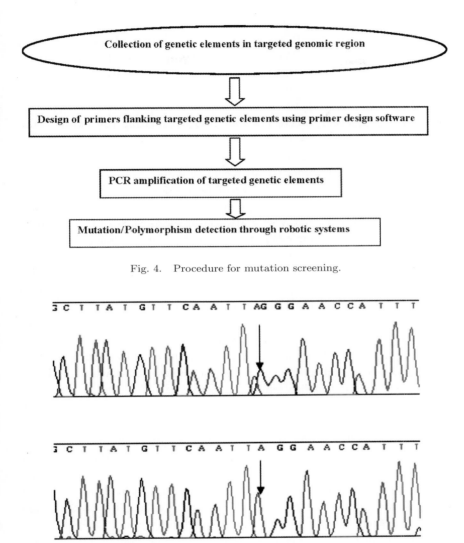

Fig. 4.   Procedure for mutation screening.

Fig. 5.   DNA sequences showing the heterozygous and homozygous statuses. The top sequence photograph shows the A/G heterozygous status of an individual, while the bottom one shows the A homozygous status of another individual.

are presented in the genome assembly. Occasionally, there are still gaps in the genome sequences, such as the genome assembly shown in Fig. 5; however, these gaps usually contain repetitive sequences, which in most cases do not have a biological role. Many software programs have been used for the identification of genetic elements in a targeted area of a chromosome, including the DeCypher II hardware-software system (Time Logic, Inc., Incline Village, NV), the PepPepSearch program of the Darwin suite at Zurich (http://cbrg.inf.ethz.ch), and Genefinder. Annotation goals for all sequences include (1) identifying genes, operons, regulatory sites, and other genetic elements and repetitive sequences such as Alu and Line elements; (2) assigning functions where possible; and (3) relating the mouse genes and sequences to human genes.

Primer design and synthesis is the next step following the identification of genetic elements. In order to allow complete detection, the sequences of each targeted region and about 100 bp on both sides should be taken for the primers designed. Many software programs can be used for computerized primer design, for example, Vector III, web freeware such as Primer3 software (http://www-genome.wi.mit.edu/cgi-bin/primer/primer3_www.cgi), and commercial ware such as Primer Premier 5 (http://www.premierbiosoft.com/primerdesign/index.html). In general, for the heteroduplex analysis, each pair of primers covers a DNA fragment of 300–400 bp in length. The size of the DNA fragment is important because a DNA fragment with too large a size (over 800 bp) may affect the accuracy of the heteroduplex analysis. The annealing temperature required for primers should be limited to a similar range, so that multiple DNA fragments can be amplified using different primers on the same PCR machine with the same parameter of amplification.

After primers are synthesized, DNA fragments are amplified with genomic DNA using PCR. The amplification of DNA fragments is performed in either 96- or 384-well plates. PCR machines for such a purpose usually contain a 0.2-mL tube (such as themocyclers from MJ Research or Applied Biosystems). The cycling protocol usually consists of 30–40 cycles of three temperatures: strand denaturation at 96°C, primer annealing at 54–60°C, and primer extension at 72°C; typically, 30–60 s, 60–90 s, and 60–120 s, respectively.[9,10]

PCR product analysis can be done with either resequencing or heteroduplex analysis methods. Several types of mutations can happen, including SNPs and insertion/deletion polymorphisms. There are several high-throughput sequencing systems available commercially. Applied Biosystems (Foster City, CA) has series of sequencers at varying levels of automated and high throughput. One that is widely used at present is the 3100 Genetic Analyzer, which is a four-capillary electrophoresis instrument for automated DNA sequencing and genotyping with 96- or 384-well plates. Other systems are also available, such as the LI-COR 4300 System from LI-COR Bioscience (http://www.licor.com). The 4300 System is a third-generation instrument based on LI-COR's highly sensitive, infrared fluorescence detection technology; it is a 96-well membrane comb, gel-based sequencing system. Recently, microarray has been developed as an array-based sequencing platform for rapid and high-throughput analysis of genomic DNA.[11,12] This superhigh-throughput system has great potential for rapid, large-scale sequencing.

Resequencing of the targeted genome region provides a direct comparison between sample and standard/control sequences; however, with known genomic sequences, a comparison can be done without sequencing. One widely used nonsequence comparison method is denaturing high-performance liquid chromatography (HPLC). HPLC detects polymorphisms by analyzing the DNA mobility of different heteroduplexes based on temperature gradient capillary electrophoresis (TGCE). The underlying principle of this method is that the melting characteristics of double-stranded DNA are largely defined by its sequence. Therefore, a single-base mismatch can produce conformation changes in the double helix that cause the differential migration of homoduplexes and heteroduplexes containing base mismatches during gel electrophoresis. The advantages of this method are its high accuracy and relatively low costs. One such system is the WAVE® DNA analysis system, which is a commercialized method from Transgenomic, Inc. (San Jose, CA), that contains temperature-modulated heteroduplex analysis.[13,14]

PCR products in our laboratory are analyzed on our SpectruMedix system (SpectruMedix LLC, State College, PA) (http://www.spectrumedix.com). The SpectruMedix system includes high-throughput capillary electrophoresis instruments (scalable onsite in 96 capillaries), specialized separation polymers, and a suite of automated software applications. In this case, the system is powerful for mutation discovery in the coding or regulatory regions of genes for this project. The ability to detect SNPs with the SpectruMedix system has been tested in different laboratories. Our study and a group of researchers at Oak Ridge National Laboratory (Oak Ridge, TN) have tested the system in their SNP detection.[9,15] Figure 6 shows PCR products from two mouse strains and their mixture using the SpectruMedix system.

3. GENOMIC EXPRESSION

Screening the sequences of a targeted genomic region provides a way for a complete search of the mutation. However, as microarray chips containing

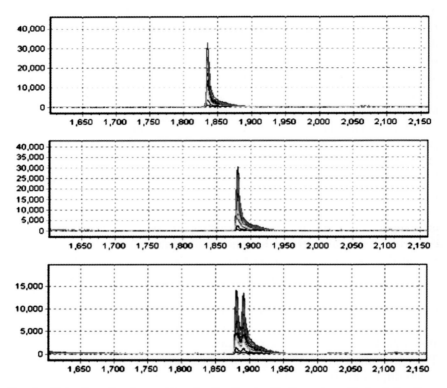

Fig. 6. Detection of polymorphism by the SpectruMedix system. Each of the three pictures shows the image of a PCR product amplified from the same genomic sequence. The left vertical line indicates the relative amount of the DNA fragments, while the horizontal number from left to right indicates the size of the DNA fragments. The top and middle pictures show the PCR products from two individuals. The bottom picture shows the mixture of the above two.

genes of whole genomes are available (see Chapter 14 for a discussion on microarray technology), gene expression provides a much quicker way to detect the mutated gene for certain mutations. The unique advantage of microarray technology is that it allows for simultaneous screening for the differential expression of thousands of genes. This approach has been made even more valuable with the availability of DNA array chips that include essentially all expressed human and mouse genes. For example, if a whole gene is deleted because of a large deletion on the chromosome, the expression of this deleted gene will be at a nondetectable level; in this case, a microarray analysis will immediately point out the candidate gene in the targeted region. Also, if a mutation is located in the regulatory region, it takes much more effort to detect such a mutation than a mutation in exons or a coding region using mutation screening by sequencing and heteroduplex analysis; however, mutations in the regulatory region usually either decrease/shut down or increase/activate the expression of a gene. An experiment using cDNA microarray will detect such changes, therefore providing information that may narrow down the candidate genes in the targeted region.

However, the application of cDNA microarray on mutation detection has several limitations.[11] One limitation is that high-quality RNA is required. Unlike Northern blotting or reverse transcription–PCR (RT-PCR), microarray analysis is very sensitive to the quality of RNA. For a comparison between diseased and normal samples, RNA with the same concentration and quality from each sample has to be used for microarray analyses. Currently, we follow the Life Technologies procedure using Trizol reagent.[9,10] We usually quantify total RNA on a spectrophotometer and ascertain its quality by analysis using Agilent Bioanalyzer.

Choosing the right tissues for microarray is also difficult. A gene does not express in every tissue. If a gene does not express in a tissue, then microarray analysis using this tissue will not be able to detect the mutation, even if the gene is mutated. One may choose the wrong tissues, even though the decision is carefully made. Furthermore, even if a gene is expressed in a tissue, if the level is very low, the gene may be categorized as nonexpressive; in this case, the microarray may miss the mutation as well.

In addition, the age of the subject or animal is critical. Some genes only express in early development or at the embryonic stage; older tissues may not detect the expression of the gene at all. Thus, the mutation will not be detected by microarray.

Because of the high sensitivity, errors occur in microarray experiments. One alternative approach is to use RT-PCR to detect the expression level of selected genes that express at a nondetectable or very low level in the microarray experiment.

4. CONFIRMATION OF CANDIDATE GENES

Although gene discovery depends heavily on the initial heteroduplex analysis and microarray profiling, direct consequences of the mutation on cDNA analysis at protein level provide further evidence to confirm the mutation and provide a solid genetic background for future studies. There are different approaches that can be used in different cases. Briefly, techniques that are widely used for mutations with more than one nucleotide change include the following:

(1) Relative quantitative real-time RT-PCR

The relative quantitative, temporal, and spatial expressions of a transcript are central to understanding the function of a given gene.[6,7] Sometimes, relative levels of a target transcript are determined at seven developmental time points (e.g. E20, P1, P7, P14, P21, P28, P60) and in different tissues (e.g. cerebral cortex, cerebellar cortex, hippocampus, striatum, thalamus, heart, liver, kidney, testis, lung). A deletion or insertion that affects a few bp or an exon may not affect the transcription of the rest of the gene. In this case, cDNA analysis will reveal the changes in the RNA level.

(2) Sequencing of the genomic DNA and/or cDNA

Sequencing can be used to verify the mutated gene and/or the flanking sequences of a deletion or an insertion, such as we did for the deletions in *sfx* and *wdl* diseases[9,10] as well as hemophilia B.[16] One good feature of the SpectruMedix system is that it can also be used for DNA sequencing. In practice, PCR products from both genomic and cDNA were purified using an AMPure PCR Purification Kit (Agencourt Beverly, MA), and the purified products were sequenced using a Big Dye V.3 Terminator Kit (Applied Biosystems, Inc., Foster City, CA). A total of 20 μL sequencing reactions — including 8 μL Big Dye (plus Half BD), 40–100 ng of purified DNA template, and 4–20 pmol of either forward or reverse universal sequencing primers — were incubated for 37 cycles at 96°C for 180 s, 50°C for 30 s, and 60°C for 180 s. Nonreacted primers were removed by ethanol–acetate precipitation (3.75% 3M NaOAc, 87.5% nondenatured 100% EtOH, and 8.75% dH$_2$O; pH 4.6). The labeled products were dissolved in 0.02M EDTA in HiDi formamide prior to electrophoretic loading onto the SpectruMedix 96-capillary sequencing system.

(3) Expression of proteins *in vitro*

To compare the protein sequences from normal and mutant genes, the corresponding nucleotide sequences can be inserted into expression vectors.[9,10] Then, the product of the mutated gene can be compared to that of a normal gene. In general, protein products are analyzed using sodium dodecyl sulfate–polyacrylamide gel electrophoresis (SDS-PAGE) and/or native PAGE as well as known enzymes to confirm predicted differences between normal and mutant proteins.

(4) *In situ* hybridization

In situ hybridization is still useful for the detection of gene expression in different tissues and at different time points. Particularly when one identifies a mutated gene that has not been studied or its function has not been known, it is necessary to detect its expression in its developmental process and in a variety of tissues.[9] In general, a large deletion of a gene will result in a complete nontranscription, such as in the case of *sfx*.[9] The cDNA will not be detected from the diseased mouse.

(5) Antibody generation and immunolocalization

The generation of one or more high-quality antibodies to encoded proteins allows for subcellular localization studies and corroboration of the *in situ* hybridization results. Furthermore, antibodies may be needed for protein–protein interaction and other biochemical studies. The predicted amino acid sequences of encoded proteins are analyzed with Vector NTI Suite (BioPlot) for antigenicity, hydrophilicity, hydrophobicity, and accessibility. In addition, basic local alignment search tool (BLAST) analyses are performed to exclude significant homology to other peptide sequences. Standard immunohistochemical techniques can be used to localize encoded proteins in sample tissues.[8−10]

(6) Western blotting to characterize antibody specificity

Western blot analysis is still a useful technique for protein analysis.[10] Usually, small tissue blocks from fresh tissues are homogenized in a chilled lysis buffer. Lysates are clarified by centrifugation. Proteins are then electrophoretically resolved on precast gels (Bio-Rad) and transferred to polyvinylidene fluoride (PVDF) membranes (Bio-Rad). Membranes are incubated with the primary antibodies. Finally, targeted proteins are visualized (with the ECL Plus detection system from Amersham).

The approaches above are useful for large mutations, but may not work for single nucleotide mutations. In fact, a major obstacle in the study of quantitative traits (QTs) has been the confirmation of genes for the traits. Because the effects of the QT genes are small and most of them are SNPs, which occur in almost every 1 kbp, identification of the real one for the trait from hundreds of SNPs in a short genome area has proven to be extremely difficult. Analyzing hundreds of SNPs for their very small effect on the trait by linkage analysis or recombination requires a population of an incredibly large size. Other approaches such as knockout and function studies provide evidence, but in most cases not the exact phenotype. In this regard, it is clear that, in the past, strategies for identifying QT genes have been limited by the availability of genetic-based data and technology. Today, this limitation still exists, but tremendous progress has been made in technology and genome resources. We will discuss several approaches that are particularly useful for single nucleotide mutations:

(1) Polymorphic testing

When multiple strains are available, a comparison between single nucleotide changes among strains is useful to eliminate some nonreal mutations.[17] For example, in the mouse model, if a study is to detect mutations of alcohol preference, then mouse strains and substrains are divided into three groups: alcohol preference, alcohol nonpreference, and unknown.

Every DNA fragment that contains polymorphisms is then amplified from each of the strains and substrains. After the genotype of each polymorphism from every strain/substrain is obtained, the correlation/cosegregationship between the genotype and the phenotype of alcohol preference is analyzed. Based on the results, polymorphisms are divided into different categories: candidate for strong correlation between polymorphism and alcohol preference, noncandidate for noncorrelation, and uncertain for intermediate correlation.

(2) Bioinformatic analysis

Function analysis is conducted at the gene and polymorphic levels.[18] At the gene level, if the candidate gene is a known gene, it is useful to search publications and examine the protein structure of the candidate gene so as to look for the possible connection between gene function and phenotype. If the candidate gene is novel, it is necessary to search the GeneBank for possible similarities between selected candidate genes and other known genes. It is also useful to use the available protein structure model to predict the possible function of a candidate gene in the behavior.

At the polymorphic level, a careful examination of the location of the polymorphism is needed. If it is in the exon region, the question will be whether the polymorphism leads to a change of amino acids or the DNA codon; much emphasis should be put on how the amino acids change and what possible impact it may have. If a polymorphism is in the noncodon region, it may be in the intron or in the $5'$ or $3'$ end or sequence between genes. The importance of such a polymorphism is difficult to evaluate, but an extensive literature search for its possible function in the regulation of gene expression should be conducted. More importantly, the decision should be made based on the combined information of the literature, gene function, and gene expression profiling.

(3) *N*-ethyl-*N*-nitrosourea (ENU)-induced point mutations

With the majority of changes occurring at A–T base pairs, ENU-induced mutations have great potential in the candidate gene confirmation of single nucleotide mutations. Assuming that QTLs are due to point mutations, it is reasonable to expect that ENU mutagenesis could affect QT genes (QTGs). Mutations can be found for almost every gene from a relatively large ENU-treated mouse population (of frozen sperm or F1 tissue). However, because this is a genotype-driven approach, gene selection is based on known genes, and many QTGs are genes either with unknown function or whose functional relationship to the quantitative phenotype is not immediately apparent. For this reason, the identification of QTG(s) in a candidate interval must include every gene in that region. The application of this strategy to biomedical research has been limited by the availability of genomic data and technology. Now, with the completion of the human and mouse genomic sequences and the development of high-throughput technology for mutation detection, it is time for a genotype-driven approach for QTG identification.

One critical question is this: can one identify an ENU-induced mutation that is identical to every polymorphism underlying the QTL? The answer is no. Some ENU mutations may be the same as the polymorphism in the QTG, but others are not. However, it is well known that multiple mutations of the same gene may cause the same problem or the same phenotype.[19] Considering the goal in research is to identify the QTGs, not the exact mutation/polymorphism, ENU-induced mutations might provide a realistic tool for the gene confirmation of single nucleotide mutations.

5. PROBLEMS AND FUTURE DIRECTIONS

Genomic cloning is based on genomic sequences. Its success depends on the completion of the sequences in the targeted genome. However, at present, sequencing gaps exist in every genome. Although genome regions that contain known genes may have been sequenced, it is unknown whether sequences in those gaps contain genetically important elements. In this regard, genomic sequences are not perfect for genomic cloning for every genetic material in the genome. However, we believe that, as time goes on and technologies are developed to sequence those gaps, the genome will eventually be completed. Therefore, genomic cloning represents the future of cloning technology.

It should be advised that errors may exist among the currently assembled genome sequences. This has been seen by the inconsistency in the number of exons between the National Center for Biotechnology Information (NCBI) and Ensembl in our first positional cloning of the *sfx* mutation.[9] Therefore, it is necessary to search different databases and obtain information from different resources, including the BAC and YAC contigs deposited in GeneBank. In most cases, for the genomic cloning of a particular gene, researchers need to search through a relatively small chromosomal region; and the number of genes within this region is not beyond the capacity of the normal laboratory. In this regard, errors in the genome sequence assembly will not affect their project much.

Although high-throughput technologies for mutation screening and sequencing are rapidly developing, their scale and accuracy have yet to completely meet the needs of research. The current platform for mutation screening allows researchers to screen 96- or 384-well plates; however, primer design, synthesis, and PCR amplification are still time-consuming. Elimination of PCR amplification will greatly enhance our screening. In this aspect, microarray resequencing may provide a future direction. The accuracy of mutation detection is the other concern of genomic cloning. It is known that heteroduplex analysis misses some single nucleotide mutations. At present, no system can claim 100% accuracy. Further developments in the detection system is another goal of biotechnology.

REFERENCES

1. Allen M, Heinzmann A, Noguchi E, *et al.* (2003) Positional cloning of a novel gene influencing asthma from chromosome 2q14. *Nat Genet* **35**(3):258–263.
2. Zhang Y, Leaves NI, Anderson GG, *et al.* (2003) Positional cloning of a quantitative trait locus on chromosome 13q14 that influences immunoglobulin E levels and asthma. *Nat Genet* **34**(2):181–186.
3. Gu WK, Li XM, Edderkaoui B, *et al.* (2002) Construction of a BAC contig for a 3 cM biologically significant region of mouse chromosome 1. *Genetica* **114**(1):1–9.
4. International Human Genome Sequencing Consortium. (2004) Finishing the euchromatic sequence of the human genome. *Nature* **431**(7011):931–945.
5. Nadeau JH, Balling R, Barsh G, *et al.*; International Mouse Mutagenesis Consortium. (2001) Sequence interpretation. Functional annotation of mouse genome sequences. *Science* **291**(5507):1251–1255.
6. Gu W-K, Li X-M, Roe BA, *et al.* (2003) Application of genomic resources and gene expression profiles to identify genes that regulate bone density. *Curr Genomics* **4**:75–102.

7. Jiao Y, Li X, Beamer WG, *et al.* (2005) Identification of a deletion causing spontaneous fracture by screening a candidate region of mouse chromosome 14. *Mamm Genome* **16**(1):20–31.

8. Jiao Y, Yan J, Zhao Y, *et al.* (2005) Carbonic anhydrase–related protein VIII deficiency is associated with a distinctive lifelong gait disorder in waddles mice. *Genetics* **171**(3):1239–1246.

9. Gu W-K, Aguirre GD, Ray K. (1998) Detection of single nucleotide polymorphisms. *Biotechniques* **24**:836–837.

10. Gu W-K, Ray K, Pearce-Kelling S, *et al.* (1999) Evaluation of apolipoprotein H *(APOH)* gene as a positional candidate gene for progressive rod-cone degeneration *(prcd)* disease. *Invest Ophthalmol Vis Sci* **40**:1229–1237.

11. Li X, Gu W, Mohan S, Baylink DJ. (2002) DNA microarrays: their use and misuse. *Microcirculation* **9**(1):13–22.

12. Gu W, Li X, Lau KH, *et al.* (2002) Gene expression between a congenic strain that contains a quantitative trait locus of high bone density from CAST/EiJ and its wild-type strain C57BL/6J. *Funct Integr Genomics* **1**(6):375–386.

13. Skopek TR, Glaab WE, Monroe JJ, *et al.* (1999) Analysis of sequence alterations in a defined DNA region: comparison of temperature-modulated heteroduplex analysis and denaturing gradient gel electrophoresis. *Mutat Res* **430**(1):13–21.

14. Kuklin A, Munson K, Taylor P, Gjerde D. (1999) Isolation and analysis of amplified cDNA fragments during detection of unknown polymorphisms with temperature modulated heteroduplex chromatography. *Mol Biotechnol* **11**(3):257–261.

15. Culiat CT, Klebig ML, Liu Z, *et al.* (2005) Identification of mutations from phenotype-driven ENU mutagenesis in mouse chromosome 7. *Mamm Genome* **16**(8):555–566.

16. Gu W-K, Brooks M, Catalfamo J, *et al.* (1999) Two distinct mutations cause severe hemophilia B in two unrelated canine pedigrees. *Thromb Haemost* **82**:1270–1275.

17. Kidd JM, Trevarthen KC, Tefft DL, *et al.* (2005) A catalog of nonsynonymous polymorphism on mouse chromosome 16. *Mamm Genome* **16**(12):925–933.

18. Hitzemann R, Malmanger B, Reed C, *et al.* (2003) A strategy for the integration of QTL, gene expression, and sequence analyses. *Mamm Genome* **14**(11):733–747.

19. Schrijver I, Oitmaa E, Metspalu A, Gardner P. (2005) Genotyping microarray for the detection of more than 200 CFTR mutations in ethnically diverse populations. *J Mol Diagn* **7**(3):375–387.

Chapter 21

RNA Metabolism and Human Diseases

Zhongwei Li* and Gayatri Kollipara

Department of Biomedical Sciences, Florida Atlantic University
777 Glades Road, Boca Raton, FL 33431, USA

Diseases related to abnormal RNA metabolism is an understudied field. RNA plays a central role in translating genetic information into proteins, and in many other catalytic and regulatory tasks. Recent advances in the study of RNA metabolism revealed complicated pathways for the generation and maintenance of functional RNA. Defects in RNA are detrimental to cells and cause diseases. This chapter aims to summarize the current understanding of various aspects of RNA metabolism, and to highlight recent findings of diseases related to defects in RNA processing and quality control.

1. INTRODUCTION

RNA plays a central role in gene expression by translating genetic information in DNA to functional proteins. A protein is made by translating a trinucleotide codon sequence in the coding region of its messenger RNA (mRNA) into a corresponding amino acid sequence. Ribosome, the machinery that performs translation, is a large microscopic complex containing four ribosomal RNA (rRNA) species and numerous proteins in eukaryotes. Transfer RNA (tRNA) is an adapter during translation, recognizing codons in mRNA and bringing the matching amino acids to the growing peptide train.

It has also been found that many RNA species carry out catalytic and regulatory activities (recently reviewed by Hall and Russell,[1] Mattick and

---

* Correspondence author.

Makunin,[2] and Costa[3]). Most of these noncoding RNAs are small and can be in complexes with proteins. Examples of such RNAs are small nuclear RNA (snRNA), small nucleolus RNA (snoRNA), and ribozymes. Genome sequencing has fermented the identification of numerous noncoding RNAs, some of which have been shown to participate in RNA processing and regulate mRNA translatability or stability.[4,5] The discovery of RNA interference and microRNAs opens up new fields of RNA function and possibly new therapies.[6,7]

Undoubtedly, RNA or RNA–protein complexes play important roles in the central part of life. It is therefore not surprising that RNA metabolism is complicated and highly regulated. This is evidenced by the fact that genome sequencing revealed large investments of genes involved in RNA metabolism.[5] Various aspects of RNA metabolism are being surprisingly discovered. While much remains unknown, it is widely recognized that abnormalities in RNA metabolism would create problems for cells and organisms.[8]

Many cellular processes are employed to make functional RNA and to maintain its level of activities (Fig. 1). RNA is synthesized as primary transcripts that undergo extensive processing reactions before becoming functional. RNA processing includes the removal of extra sequences from both 5′ and 3′ termini, splicing of introns, and modification and assembly of multimolecular complexes. mRNA undergoes 5′ capping and 3′ polyadenylation; in some cases, polyadenylation also happens on other RNA species.[9] Eukaryotic RNAs are transported from nucleus to cytoplasm, where they function; in some cases, a specific RNA is required to localize and function in a specific cytoplasmic location. Furthermore, RNA species differ widely in their stability: some are rapidly degraded minutes after being made, whereas others are quite stable passing through generations. Transient transcriptional regulations and varied stability are the major controls for the level of functional RNA. RNA can become nonfunctional due to mutation, abnormal processing, or damage. Recently, it has been shown that specific mechanisms exist to eliminate defective RNA or to correct mistakes.[8,10] Problems occurring in any of the above processes may affect the viability of cells and the health of organisms. Important processes of RNA metabolism discussed here are illustrated in Fig. 1.

Although much is learnt by studying model organisms, humans have similar pathways in every aspect of RNA metabolism. In this chapter, we will discuss how functional RNA is made, and how its intracellular levels are controlled. Abnormal RNA metabolism related to human diseases is an understudied field. We aim to summarize aspects of RNA metabolism that

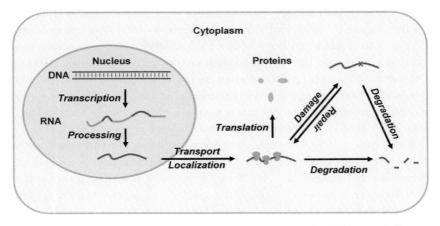

Fig. 1. The travel of a messenger RNA. Major processes of mRNA metabolism are illustrated in arrowed lines. mRNA molecules synthesized and processed in the nucleus are transported to the cytoplasm, where they are translated. Some mRNAs are localized in a specific part of the cytoplasm and are translated to make proteins of local function. mRNA undergoes degradation; each mRNA has its own characteristic half-life. Defective mRNA can be generated by gene mutation, erroneous processing, or chemical damaging agents. Damaged RNA molecules are either repaired or degraded by specific mechanisms. Each process is subject to regulations so as to assure that functional proteins are produced at the right time, location, and amount.

are related to human diseases. Emphasis will be given to processes that cause RNA dysfunction and to cellular surveillance mechanisms that deal with RNA defects. Undoubtedly, an understanding of RNA metabolism underlying human diseases will greatly enhance our ability to cope with the diseases.

2. MECHANISMS OF RNA METABOLISM

2.1. *Transcription*

Transcription is the cellular process through which RNA is made from its gene. Using nucleotide triphosphates, RNA polymerases make RNA complementary to DNA sequences. Depending on the type of RNA being made, different RNA polymerases are employed. The 18S, 5.8S, and 28S rRNAs are cotranscribed in one single transcript by RNA polymerase I. mRNA and snRNA are transcribed using RNA polymerase II. A third enzyme, RNA polymerase III, is used to make tRNA and the 5S rRNA. In addition, cytoplastic organelles such as mitochondria have their own RNA polymerases.

Numerous protein factors, and sometimes also specific RNAs, are involved in transcription. In the eukaryotic systems, there are many proteins called transcription factors, i.e. enhancers or silencers that interact with specific regions of DNA and RNA polymerase to turn transcription on or off. The rate of RNA synthesis from a particular gene depends on the availability and transient interaction of such factors with the regulatory regions of the gene. The transcription regulators are made or activated by cellular signals that sense changes in environmental stimuli or cellular activities. On the other hand, there are genes that are constitutively transcribed to make RNA that is always needed, such as housekeeping genes.

Eukaryotic mRNAs are made with two additional features that are not encoded by their gene sequences. The 5′ end triphosphate is capped with a seven-methyl guanosine. The 3′ end is released from the growing RNA strain by an endonuclease cleavage at a specific site, followed by the addition of a polyadenylate sequence of about 200 nucleotides in length. Poly(A) polymerase carries out this nontemplated polyadenylation.

RNA polymerases copy the DNA sequences faithfully; therefore, mutations in genes are normally transcribed into RNA. However, although the fidelity of RNA polymerase is high, misincorporation of nucleotides may still occur, especially when the DNA template is modified by chemicals or oxidants (see below). Mutations and nucleotide misincorporation can result in nonfunctional RNA molecules.

2.2. *RNA Processing*

Primary RNA transcripts in eukaryotic cells usually contain extra sequences that are removed by various RNA processing pathways. Messenger RNAs normally contain extra sequences in the middle of the primary transcripts called introns, which are removed by splicing reactions. The resulting fragments, exons, are joined to form the translatable form of mRNA. Ribosomal RNAs and tRNAs are transcribed with extra sequences at each end that are removed by nuclease cleavage reactions. tRNAs, and rRNAs in some organisms, also contain introns; they are removed by reactions somewhat different from mRNA splicing. In addition, various forms of modification occur in RNA on base and/or sugar. In this respect, tRNA and rRNA are modified at much higher levels than mRNA. Finally, nucleotides in RNA can also be changed at specific locations by RNA "editing" activities, so that the encoded proteins are changed. Processing events for nucleus-encoded RNA occur in the nucleus, except RNA editing which may happen in the cytoplasm. RNAs made in organelles are processed *in situ*. In conclusion, RNA processing produces RNA with designated functionalities, and sometimes makes adjustments on what RNA is made.

2.2.1. *Processing of mRNA*

The removal of introns in mRNA is carried out by the splicing machinery spliceosome, a large assembly containing various small nuclear RNAs (snRNAs) that form complexes with proteins (snRNPs). Unlike ribosomes, components in the spliceosome usually do not stick together tightly and the membership of a spliceosome changes dynamically. In the primary mRNA transcripts containing introns, there are usually conserved sequences at intron–exon junctions (5′ and 3′ splice sites) and in the middle of introns (A sites or branch point sites). Members of the spliceosome recognize splicing signals in mRNA through base pairing between snRNAs and the conserved sequences in mRNA, and remove introns.

The majority of splicing reactions start with the recognition of the A site by the branch point–binding protein (BBP) and the helper protein U2AF. This recruits the binding of the U1 snRNP to the 5′ splice site and of the U2 snRNP to the A site through base pairing. The following steps are for the recruitment of triplet snRNPs (U4/U6 × U5): The rearrangement of RNA–RNA interactions that is promoted by ATP hydrolysis allows the catalytic site of the spliceosome to be formed. The 5′ end of the intron is cleaved and joined to the 2′ of the adenosine in the A site, forming a branched circular structure known as a lariat structure. The 3′ end of the intron is subsequently cleaved off, and the lariat RNA is released. This step is coupled by the joining of the 3′ end of the upstream exon and the 5′ end of the downstream exon. The lariat RNA is degraded in the nucleus, with the exception of the formation of small functional RNAs from some introns.

In addition to the major splicing pathway described above, there are alterations in the conserved sequences of pre-mRNAs and snRNPs involved in the sliceosome. Exons from two independent transcripts can be joined together by *trans*-splicing. It should be noted that splicing happens before transcription of the pre-mRNA is complete. Some splicing factors are associated with RNA polymerase, and start working on intron sequences as soon as they are synthesized.

An important aspect of mRNA splicing is that multiple exons in a pre-mRNA may be joined in more than one way by alternative splicing, producing mature mRNAs that encode different protein products. Alternative splicing occurs when the spliceosome sometimes skips or picks up certain splicing sites. A number of proteins as well as conserved sequence features in pre-mRNA are involved in splicing site selection. The final forms of mature mRNA depend on the combination of these factors. Alternative splicing produces many more proteins than the number of genes. Moreover, alternative splicing is highly regulated so that a specific splicing product is produced when needed. The weak nature of RNA splicing determinants sometimes causes mistakes in splicing site selection, resulting in aberrant mRNA that has to be taken care of by specific surveillance mechanisms (see below).

In addition, mRNA undergoes 5′ capping and 3′ polyadenylation, which are also coupled with transcription. Capping is the addition of seven-methylguanosine to the 5′ end of mRNA through a 5′-to-5′ linkage. This is carried out by three enzymes — phosphatase, guanyl transferase, and methyl transferase — in sequential reactions using GTP. Polyadenylation occurs at the 3′ end of mRNA downstream of the encoded polyadenylation signal sequences. When such signal sequences are made in the growing RNA chain, they are recognized by specific protein factors: cleavage stimulation factor F (CstF), and cleavage and polyadenylation specificity factor (CPSF). Both proteins are associated with the C-terminal domain of the RNA polymerase, and travel with the polymerase complex along the growing pre-mRNA chain. Once the polyadenylation signal is bound by the two factors, additional proteins are recruited to cleave the RNA downstream of the signal sequence. The newly formed 3′ end is added a poly(A) tail by the enzyme poly(A) polymerase. Both capping and polyadenylation happen in the nucleus. Because of the tight coupling of RNA processing to transcription, a complete primary mRNA transcript does not normally exist.

2.2.2. *Maturation of rRNA and tRNA*

The four RNA components in ribosomes are made in two types of primary transcripts. The 18S, 5.8S, and 28S rRNAs are made in one large transcript of 45S in size by RNA polymerase I; the 5S rRNA is transcribed separately by RNA polymerase III. The functional forms of all rRNAs are generated by the removal of extra sequences at both their 5′ and 3′ termini by RNA cleavage reactions. All ribonucleases and other factors are not yet known to carry out the reactions. It is observed in yeast that the 45S primary transcript is cleaved to smaller processing intermediates by stepwise cleavage reactions, followed by the removal of the extra sequences surrounding each of the mature rRNA products.[11] Endoribonucleases and exoribonucleases are involved in the removal and degradation of the extra sequences at each step. rRNA maturation is coupled with transcription and rRNA assembly into ribosomes.

Transfer RNAs are made in primary transcripts that usually contain one tRNA each. Important features of tRNA transcripts for processing include (1) extra sequences at both the 5′ and 3′ ends of tRNA; (2) an intron that separates the tRNA sequences in the anticodon loop; and (3) the 3′ terminal CCA sequence present in all mature tRNAs that is not encoded, but added as the final step in tRNA processing. The 5′ end of tRNA is matured by the cleavage of the ribonuclease RNase P. Recently, it has been shown in many organisms that the endoribonuclease RNase Z, also known as 3′-tRNase, is responsible for removing the 3′ extra sequences from pre-tRNAs.[12,13] The CCA sequence is then added by tRNA nucleotidyl transferase (the CCA enzyme) without the need for any template. Introns in tRNAs are removed by endonuclease cleavages and ligase reactions.

2.2.3. *RNA Modification*

In addition to four normal ribonucleotides, RNA also contains nucleotides with modifications at sugar or base. The types of modifications are numerous and are present in varied abundance in different RNA species.[14] Usually, rRNA and tRNA contain more modifications than mRNA.[15] It has also been shown that snRNAs contain modified residues.[16] The most abundant modifications are pseudouridines formed from uridines, base or 2′-O-ribose methylation, and thymidine or dihydrouridine in tRNA. Although not completely known for all types and positions in RNA, modification plays a role in the optimal function of RNA.[14] In some cases, it is clear that specific modifications are needed for RNA to adopt a specific tertiary structure. Modification has also been shown to maintain the stability of RNA.

Specific enzymes carry out various types of modifications in RNA. Methyl transferases are used to add a methyl group to bases or sugars. Pseudouridine synthases convert uridine to pseudouridine. For a long time, it has been asked how the enzymes find targets at specific locations of modification in RNA sequences; it has recently been shown that there are small RNAs in the nucleolus (snoRNAs) that guide the synthesis of 2-O-methyl nucleotides and pseudouridines in eukaryotic rRNAs.[15,17] snoRNAs form a complex with the targets by base pairing, which is followed by recognition of the modification sites by the respective enzymes. The H/ACA snoRNAs guide site-directed pseudouridylation, whereas the CD-box snoRNAs guide the formation of 2-O-methyl nucleotides.

2.2.4. *RNA Editing*

One facet of RNA processing is the posttranscriptional change of the encoded nucleotides at specific locations, thus further expanding genomic encoding capability.[18] These RNA-editing reactions may add or delete a short RNA sequence, or change a nucleotide at specific locations. Commonly found RNA editing in mammalian systems include changes of adenine to inosine and cytosine to uridine, both of which alter base pairing in RNA × RNA duplexes.[19] Insertions or deletions of guanosine, adenine, or uridine residues are also found in viruses and kinetoplastids (parasites causing sleeping sickness).[20]

An example of RNA editing in humans is the conversion of a C to a U residue in the coding region of the apo-B mRNA, changing the glutamine codon to a stop codon. This editing reaction happens in the intestine, but not in the liver, resulting in the production of a shorter protein in the intestine (apo-B48) and a longer protein in the liver (apo-B100). The shorter form lacks the domain for binding of the receptor of low-density lipoprotein, which is important for its function in liver. An A-to-I change happens in the glutamate receptor GluR.[21]

Editing reactions of mammalian mRNA are carried out by posttranscriptional mRNA editing complexes or editosomes in cell nuclei. Each complex consists of multiple structural and functional proteins.[19] Editosomes contain homodimers of deaminases that convert C to U and A to I in specific locations of mRNA. The site specificity is determined by both specific RNA-binding activities of the editosomes and RNA sequences in the editing sites. The A-to-I editing of GluR mRNA is catalyzed by adenosine deaminases acting on RNA (ADARs), which have the capacity for unaided site-specific RNA binding and editing.[21] Base-pairing interaction of exon and intron in the GluR pre-mRNA is important for recognition of the editing site by ADAR. The deaminase involved in C-to-U editing of apo-B mRNA (APOBEC-1) requires RNA-binding auxiliary proteins that selectively bind to single-stranded RNA recognition element 3′ of the editing site. The multiple uridine insertions or deletions occurring in mRNA in kinetoplastid mitochondria require unique guide RNAs (gRNAs) and multiple enzyme complexes for endonucleolytic cleavage, uridine addition or deletion, and mRNA religation.[20]

2.3. *mRNA Transport and Localization*

After completion of processing in the nucleus, most RNAs (mRNA, rRNA, tRNA, and cytoplasmic regulatory RNA and ribozymes) are moved to the cytoplasm, where they function. RNA transport from nucleus to cytoplasm has been studied extensively, and many factors have been identified to play roles in this process. RNAs are also transported within the cytoplasm and localized, and function at specific sites. Large assemblies may be involved in RNA trafficking. Well-studied cases are RNA localization in *Xenopus* oocytes,[22] and RNA transport and localization in synapses of mammalian neurons.[23]

Certain RNAs in neural cells are targeted to dendrites by a specific RNA trafficking pathway mediated by the trafficking factor, heterogeneous nuclear ribonucleoprotein (hnRNP) A2. In the nucleus, hnRNP A2 recognizes RNAs containing an 11-nucleotide sequence (GCCAAGGAGCC) termed the hnRNP A2 response element (A2RE). RNAs bound by A2 are exported to the cytoplasm, where they assemble into trafficking intermediates termed granules. Granules also contain components of the translation machinery and molecular motors (cytoplasmic dynein and kinesin). RNA granules move along microtubules to the cell periphery, where they become localized and where the encoded protein is translated. Numerous trafficking factors are involved in various parts of the A2 RNA trafficking pathway.[23] Concentrations and specific interactions of these factors are regulated and govern the polarity of RNA trafficking. In some cases, membrane receptors interact with the granules, directing RNA trafficking to the organelle or vesicle membrane.[22]

In addition to motorized RNA trafficking, RNA can be localized by efficient degradation of specific RNAs within part of the cytoplasm. RNAs can randomly diffuse in cytosol, and are trapped by localized RNA-binding proteins that recognize specific RNAs. Localized mRNAs produce proteins locally when needed. For instance, a single neuron in the central nervous system may receive input signals from thousands of different cells. Such signals are used by synapses to establish memory of the stimuli and to make the synapses plastic, since their activation is influenced by their stimulation history. Such plasticity, known as long-term potentiation and long-term depression, requires newly synthesized proteins in the synapses from localized mRNAs. Therefore, synaptic plasticity could be at least partially mediated by local translational control from localized mRNAs as well as other parts of translational machinery by synaptic activation.[24]

2.4. *Translation*

Translation, the process of synthesizing proteins from mRNA coding sequences, is well characterized in various systems. Amino acids are activated by forming covalent linkages to cognate tRNAs. The ribosome subunits assemble on mRNA, and recruit aminoacyl-tRNAs based on codon–anticodon interaction. Ribosome catalyzes the formation of peptide bonds between amino acids. Proteins are released when the ribosome travels to the end of the coding region of mRNA.

An eukaryotic mRNA is rapidly translated only when it recruits ribosomes efficiently. The cap structure along with the poly(A) tail are important for efficient translation. The translation initiation factor eIF4E binds to the cap, followed by binding of the small ribosome subunit and the initiator methionyl-tRNA to the complex. The small subunit moves downstream of mRNA until a start codon is reached. The large subunit of ribosome then joins the complex to start peptide synthesis (recently reviewed by Preiss and Hentze[25]). A capless mRNA is not translated well and is degraded quickly in cells. One exception is the existence of internal ribosome entry sites (IRESs) in some viral mRNAs that allow the ribosome to bind directly at the internal position of the mRNA and start translation at a downstream start codon.[26] Aberrant translation initiation may cause human diseases such as cancer.[27]

The poly(A) tail is normally associated with poly(A)-binding proteins (PABPs). It was shown that PABPs on the poly(A) tail interact with the cap complex. Therefore, it is likely that the 5′ and 3′ termini of a translating mRNA connect to each other through RNA–protein interactions, forming a closed-loop architecture (Fig. 2). Such a structure provides a plausible explanation for the requirement of cap and poly(A) tail for efficient translation. After one round of translation, the ribosome at the 3′ end of mRNA may be transferred directly to the 5′ end, starting a new round of protein synthesis.[28,29]

As for other processes, translation is able to be highly regulated.[24] A number of *cis*-acting sequences in mRNA are involved in translational regulation. The Kozak sequence is a consensus around the start codon; its variation affects the efficiency of translation initiation. The 3′ untranslated regions (UTRs) of some mRNAs contain specific sequences that can be recognized by regulatory proteins. Binding of such sequences by cytoplasmic polyadenylation element binding protein (CPEB) and maskin causes translational inactivation. The mRNA structure may interact with the ribosome, causing frameshifting of codon reading during translation. *Trans*-acting factors regulating the translation of specific mRNAs are numerous and are controlled by cellular signals.[24,30]

2.5. *mRNA Degradation*

An important feature of mRNA is its instability. mRNAs are constantly degraded in the cytoplasm. The rates of transcription and degradation determine the level of a specific mRNA in a cell. mRNAs vary in their

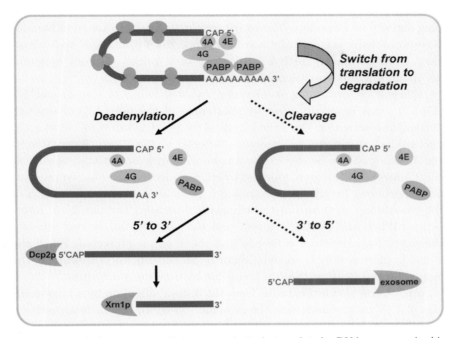

Fig. 2.   mRNA decay: a controlled process. Actively translated mRNAs are organized in a closed-circle structure involving poly(A)-binding proteins (PABPs) and translation initiation factors (4A, 4E, 4G). Such interactions may facilitate the recruiting of ribosomes (shown in orange color) after each round of translation. For reasons yet to be understood, mRNA undergoes a switch and becomes accessible to degradation activities. Shortening of the 3′ end by deadenylation or endoribonuclease cleavage is a rate-limiting step and probably precedes fast degradation. The major mRNA degradation pathway involves 5′ decapping followed by fast 5′-to-3′ exonucleolytic digestion. A decapping complex containing the decapping enzyme (Dcp2p in yeast) removes the cap structure. A 5′–3′ exoribonuclease (Xrn1p in yeast) converts RNA to nucleotide 5′ monophosphates. mRNA can be degraded from 3′ to 5′ by a complex of exoribonucleases, the exosome, independent of decapping activity.

rates of degradation, which are described as half-lives. The average half-life of mRNA in cultured human cells is about 10 h; however, the half-lives of some human mRNAs are as short as a few minutes. Fast mRNA turnover is an efficient posttranscriptional regulation of gene expression.[31] Known mechanisms of mRNA decay in eukaryotes are described in Fig. 2.

The most efficient degradation pathway of eukaryotic mRNA involves the removal of the 5′ cap followed by quick exonucleolytic digestion of the decapped RNA. A number of proteins form a complex that carries out decapping. The human protein hDcp2 is the enzyme that confers decapping activity in a complex. The exoribonuclease Xrn1p (hXrn1p in human) is responsible for the degradation of decapped RNA from the 5′-to-3′ direction. Once decapped, mRNA is digested very quickly without apparent accumulation of degradation intermediates.

Efficient 5′ decapping is preceded by shortening of the 3′ end[32,33] (Fig. 2). This is usually done by shortening of the poly(A) tail through deadenylase activities. Once the length of the poly(A) tail is too short (<30 nucleotides), decapping will proceed efficiently at the 5′ end. Alternatively, the 3′ sequence can be cleaved by an endoribonuclease activity, usually at the 3′ untranslated region, followed by efficient decapping. The requirement of 3′ shortening for efficient decapping is probably due to the formation of the closed-loop structure of the translation complex through 5′–3′ interaction. When the 3′ end is intact and the poly(A) tail is long enough, the closed-loop structure is formed and the 5′ cap is protected. When the 3′ end is shortened by deadenylation or by an endonucleolytic cleavage, the loop is broken, making the cap accessible by decapping activities.[34]

mRNA can also be degraded from the 3′-to-5′ direction by a large complex of 3′–5′ exoribonucleases, the exosome[35] (Fig. 2). Again, deadenylation or endonucleolytic cleavage is required for mRNA decay by this mechanism. Decapping at the 5′ end is not required for 3′-to-5′ degradation.

mRNA decay can be highly regulated. There are specific protein factors recognizing sequence and structural features in mRNA that will stabilize the RNA or trigger RNA degradation. Features in mRNA include the (A + U)-rich instability elements (AREs). There exist diverse families of proteins that can bind AREs, regulating the mRNA decay rate. Messenger RNA decay rates can change during development and differentiation and during aging.[36]

A curious question that has been asked for a long time is how an mRNA is switched from an actively translated state to a rapidly degraded state (Fig. 2). Apparently, access to the degradation machinery is a key step for such a switch. Recently, it has been discovered that nontranslating mRNA is largely present in a microscopic cytoplasmic structure termed the processing body (P-body).[37] Also present in the P-body are the decapping complex and Xrn1p enzyme.[38–41] It is likely that the recruitment of mRNA into the P-body initiates decapping-dependent mRNA decay.[42,43] However, what causes the distinction for an mRNA to be translated or degraded is still not clear. Sequence or structural features of mRNA may be recognized by specific proteins that determine the rate of the mRNA being sent to the

P-body. It is also likely that the mRNAs accumulate defect or damage over time during previous translation reactions, and are sent to the P-body by specific sequestering mechanisms.

2.6. *RNA Damage and Quality Control*

RNA can become damaged or defective through many different ways. Gene mutation or misincorporation of nucleotides during transcription alters the RNA sequence. Misfolding may result in RNA with nonfunctional conformation. Inappropriate processing may produce aberrant RNA. In addition, functional RNA can be modified by toxic chemicals or reactive oxygen/nitrogen species, resulting in dysfunction of the RNA. This area is currently gaining more attention, since many of the defective forms of RNA are associated with human diseases.[44] Some RNA quality control mechanisms are described in Fig. 3.

Mutations that cause insertion, deletion, or substitution of nucleotides in RNA have been well documented in various systems. It should be noted that many mutations are lethal and are not observed in survived individuals. Recessive lethal mutations may be found in heterozygotes. These are active areas in the study of genetics. Mutations affecting RNA function are discussed here. A good example is the various mutations in mitochondrial tRNA (mt tRNA) that cause human diseases due to reduced energy metabolism.[45,46] The number of mitochondria in any cell is large, making it possible for mitochondria carrying nonfunctional tRNA mutations to accumulate at various levels. Mutations in mt tRNAs may affect their maturation from precursors, modification, amino acylation, or interaction with translational machinery for peptide synthesis.[46,47] In *E. coli*, defective tRNAs are found unstable,[48,49] and can be degraded by polynucleotide phosphorylase (PNPase) and polyadenylation-dependent activities.[9,49] Yeast tRNA precursors that are hypomodified are degraded in a similar fashion in the nucleus.[50] It awaits to see if such surveillance mechanisms exist for mutant tRNAs in mitochondria.

Aberrant mRNAs are made from mutant genes or by mistakes in splicing. Alterations in mRNA with or without substitution of amino acids in proteins are active subjects of studies on single nucleotide polymorphisms (SNPs), which usually escape RNA metabolism and will not be discussed here. Aberrant mRNA can be made either with stop codons in the middle of the normal coding region (nonsense mRNA) or without any stop codon (nonstop mRNA).[51] Such mRNAs can be generated by mutations or

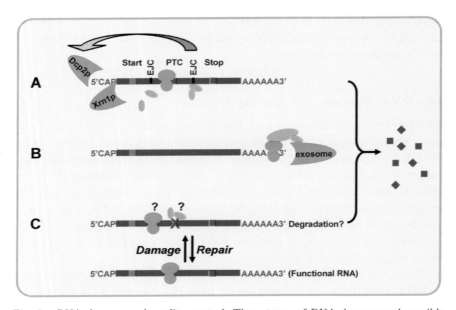

Fig. 3.   RNA damage and quality control. Three types of RNA damage and possible
mechanisms for removing damaged RNA are described. The start codon in mRNA is
shown in green. Stop codons (including the pretermination codon or PTC) are shown
in red. (A) Nonsense-mediated decay (NMD).[51] mRNA exported to the cytoplasm is
associated with proteins for processing, e.g. proteins binding the exon–junction complex
(EJC). During the first round of translation, these proteins in the coding region are
removed by ribosomes. Ribosomes may fail to remove such proteins in the presence of
a PTC, resulting in interactions between inappropriate RNA-binding proteins and the
surveillance complex as well as rapid degradation from the 5' end. In yeast, the Upf1,
Upf2, and Upf3 proteins; EJC components; and proteins for Upf phosphorylation and
dephosphorylation are involved in NMD. (B) Nonstop mRNA degradation.[55] mRNA
containing no inframe stop codon may cause the ribosome to reach the very 3' end during translation. The yeast protein Ski7p, which adopts a similar structure as translation
release factor, may recognize the empty A site of the stalling ribosome. This leads to
recruitment of the exosome and initiation of 3'-to-5' degradation. (C) Damage control.
Modification of the bases or backbone may cause RNA dysfunction. Such modifications
may be repaired by specific activities.[67] Alternatively, damaged RNA can be degraded
by specific or nonspecific mechanisms. A specific mechanism suggested here is the recognition of damaged base or ribose by specific activities, followed by recruiting mRNA
decay pathways. Such activities are yet unknown.

splicing at the wrong sites, and may direct the synthesis of aberrant proteins. Fortunately, specific surveillance mechanisms are developed to recognize and degrade nonsense or nonstop mRNAs (Fig. 3). Eukaryotic mRNAs
with nonsense mutations are degraded by a nonsense-mediated decay
(NMD) pathway (reviewed by Baker and Parker,[51] Hentze and Kulozik,[52]

and Hilleren and Parker[53]). In addition, yeast mRNAs having nonstop mutations or those failing to be exported to the cytoplasm are shown to be recognized or degraded by a large exonuclease complex, the exosome.[54,55]

An important area of study that has been much underappreciated in the past is chemical damage of RNA by toxic chemicals or radiation. Similar damages to DNA have been extensively studied and are implied in many human diseases. The same agents that damage DNA must damage RNA as well. In cases where studies have been carried out, chemical damage of RNA is observed at much higher levels than that of DNA.[56–59] Reactive oxygen species (ROS), whose levels are increased under oxidative stress, provide a major source of damage to cellular components including RNA. Among over 20 oxidized lesions in nucleic acids, an oxidized guanine species, 8-hydroxydeoxyguanine (8-oxo-dG) or 8-hydroxyguanine (8-oxo-G), appears to be the most deleterious[60–62] because it can pair with both cytosine and adenine at about the same efficiency. Oxidatively damaged mRNA may direct the synthesis of defective protein products. In addition, oxidatively damaged ribosomal and transfer RNA may be nonfunctional in translation. Despite its potential importance, little is known about how cells deal with oxidized RNA or ribonucleotides.[62–66]

Some other active chemicals can also cause RNA damage. It is known that cisplatin, a cancer chemotherapeutic agent, causes RNA modification and interferes with translation. Cigarette smoke contains over 60 chemicals that form adducts on DNA; similar adducts must also be formed on RNA from many (if not all) of them. Other agents such as nucleic acid–reacting dyes or pesticides may also modify RNA. In view of this, a large area of toxicology of RNA damage has not been explored. Damaged RNA may be removed from the total RNA population by specific RNA degradation mechanisms (Fig. 3). Remarkably, a recent study demonstrated that RNA chemical damage can even be repaired: human hABH2 and hABH3 proteins remove methyl groups from methylated bases in RNA.[67] The existence of RNA quality control activities indicates that RNA chemical damage is a challenging problem for cells (Fig. 3).

2.7. microRNA: Formation and Function

The discovery of microRNAs overlapped with the elucidation of the mechanism of RNA interference (RNAi).[68–70] It was found that the introduction of double-stranded RNA (dsRNA) into cells causes efficient degradation of cellular RNA containing the same sequence — a process termed

RNA interference. The extraneous dsRNA is converted into short dsRNAs of 21–23 nucleotides in length, termed small interfering RNAs (siRNAs), which are used to target specific RNA. The formation of siRNA is carried out by a complex of proteins including the dsRNA-specific ribonuclease Dicer. The siRNA is then assembled into an RNA-induced silencing complex (RISC), which finds and degrades target RNA that is recognized by siRNA in the complex by base pairing. Animal cells do not seem to produce siRNA from the genome, but may use this mechanism to destroy invading dsRNA viruses. siRNA is now widely used to inactivate specific mRNA in cultured cells or animals in research.

Mutations affecting the development of *C. elegans* were mapped to very short genes encoding tiny RNA products.[71] These single-stranded short RNAs, termed microRNAs (or miRNA), are 19–23 nucleotides in length and have been discovered in fruit fly and mammals.[4] It is estimated that hundreds of miRNAs are encoded in the human genome, and over 100 have been experimentally validated.[72,73] miRNAs are expressed at various stages of development or in different tissues, and may regulate ∼30% of total mRNA in human.[74] Interestingly, miRNAs are transcribed as precursors. Pre-miRNAs are probably processed similar to the generation of siRNAs.[6] miRNAs are also assembled into RISCs, which inactivate with target mRNA through base pairing. Depending on the nature of the RISC formed and on the complementarity of miRNA × mRNA, the target mRNA can be either degraded or inhibited for translation.

This miRNA-mediated mechanism of mRNA inactivation has been found to regulate protein products made during cell differentiation and development in mammals.[75] For instance, specific miRNAs are expressed during mouse hematopoiesis and help define particular cell lineages. The expression of miR-181 in hematopoietic stem cells promotes the formation of B-cells, whereas miR-142 and miR-223 define differentiation to T-cells. Disruption of miRNA function has been implicated in a number of diseases discussed below.

3. DISEASES RELATED TO RNA METABOLISM

Various processes associated with RNA metabolism are very important for the proper functioning of RNAs. Abnormalities in RNA metabolism at any stage could lead to abnormal functioning of RNAs, cause detrimental effects to the cell, and finally lead to diseases. The most important processes associated with RNA metabolism are posttranscriptional processes like splicing, editing, base modifications, etc., which give rise to functional RNAs from pre-RNAs. The mature mRNAs are then sorted with the help of RNA-binding proteins and many other factors that control the transport, localization, and local translation of mRNAs. Defects in any of these post-transcriptional mechanisms may lead to diseases. In addition, the failure to trigger mRNA surveillance mechanisms, like NMD by PTC-containing mRNAs, could be associated with some severe disease conditions. RNA damage (i.e. oxidation) is another factor that could be detrimental to the cell. Various abnormalities in RNA metabolism associated with defects in the abovementioned and other processes will be discussed in this section. Some defects in RNA metabolism that could be related to human diseases are illustrated in Fig. 4.

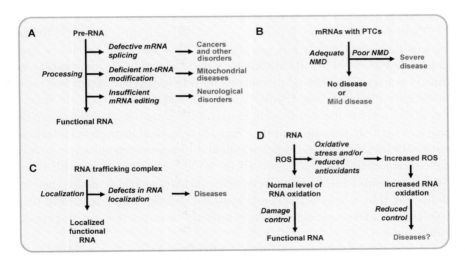

Fig. 4. Erroneous RNA metabolism and human diseases. (A) Posttranscriptional processes that give rise to mature RNAs from pre-RNAs, when they go wrong at any stage, can show association with human diseases. (B) PTC-containing mRNAs that are able to trigger adequate NMD are associated with no disease or with mild disease condition, whereas those that escape NMD can cause severe forms of disease. (C) RNA-binding proteins (RBPs), which act as part of the RNA transport and localization complex, play important roles in mRNA transport, localization, and local translation. Mutations in RBPs might lead to defects in mRNA-sorting mechanisms and might contribute to diseases. (D) Oxidation of RNA from reactive oxygen species (ROS) is prevented during normal conditions by the combined action of antioxidants that destroy ROS and RNA quality control activities that repair or clean up damaged RNA. During oxidative stress, where ROS exists in excess amounts in the body, antioxidants and repair enzymes might not be able to prevent oxidation of RNA; this might lead to excess oxidation of RNA associated with diseases.

3.1. *Human Cancers and Splicing*

Pre-mRNA splicing is an important posttranscriptional mechanism that can give rise to mature mRNAs and different protein isoforms from a single pre-mRNA by alternative splicing. Defects in this splicing are reported to play important roles in causing many diseases, most commonly cancers (Fig. 4). The most common causes of aberrant splicing in cancerous cells are mutations in genomic splice sites that usually result in exon skipping, mutations that cause inappropriate cryptic splice sites, and mutations in other *cis*- and *trans*-acting elements.[76,77] Cancer-specific alternative splicing independent of mutations has also been reported that may include the use of alternate individual splice sites, exons, and introns, which in turn results in abnormal cancer-specific splice forms.[76] Some mRNAs that show splicing defects in cancers will be discussed here, mostly along with a brief description on the association of defective splicing with diseases other than cancers.

There have been many reported cases where aberrant splicing has implications in cancer, and it has recently been observed that mutations and abnormal splicing of Smad4 mRNA have implications in thyroid cancer. As Smad4 is frequently altered in cancers, its role in thyroid cancer has been investigated by analyzing thyroid tumors of various histotypes for Smad4 mutations and for abnormalities in Smad4 mRNA expression due to alternative splicing.[78] It was observed that Smad4 has several intragenic mutations with high incidence in the linker region and tumor-associated alternatively spliced isoforms which lack portions of the linker region, along with three Smad4 isoforms which show unreported exon–exon rearrangements, thereby indicating that Smad4 is frequently mutated and aberrantly spliced in thyroid tumors. These changes might contribute to an early event in thyroid tumorigenesis.[78]

Defects in the splicing of BRCA1, AIB1, and some other mRNAs have been found to be associated with breast cancer. It has also been reported that *mdm2*, which regulates the expression of p53, is alternatively and aberrantly spliced and lacks either the entire or the majority of the binding domain for p53 in breast cancer specimens. This inappropriate splicing of *mdm2* was found to be associated with the shortening of overall patient survival, thus suggesting that altered *mdm2* expression in invasive breast cancer is associated with more aggressive disease.[79]

Although members of the superfamily cadherins (transmembrane glycoproteins which mediate calcium-dependent cell adhesion in tissues) are found to be associated with different cancers, it has recently been observed that liver intestine–cadherin (LI-cadherin) is overexpressed in hepatocellular carcinoma (HCC) patients and showed alternative splicing by lacking exon 7 in 50% of HCC patient specimens examined. This might play a role in early tumor recurrence and poor overall survival of the patients.[80]

Apart from the abovementioned examples, there are many other cases where defects in splicing due to mutations or disease-specific alternative splicing are found to show association with some other cancers like small cell lung cancer, hereditary nonpolyposis colorectal cancer, endometrial cancer, brain tumors, etc., as well as some other diseases like Frasier syndrome, Parkinson's disease, myotonic dystrophy, etc. Alternative splicing is also known to have implications in cardiovascular disease, as observed in the case of genetically altered mice that expressed SERCA2b isoform (but not SERCA2a isoform) in cardiac myocytes and that showed increased incidence of major cardiac malformations during embryonic development and cardiac hypertrophy in adulthood (reviewed by Musunuru[81]). These findings denote that defective splicing may also have implications in cardiovascular disease.

3.2. *ALS and Insufficient RNA Editing*

RNA editing is another important posttranscriptional mechanism that gives rise to mature RNA from pre-RNA. Editing of mRNA occurs in neurotransmitter receptors in the central nervous system (CNS) of the brain, particularly in ionotropic glutamate receptors (GluRs), and is therefore the molecular mechanism involved in fine modulation of neuronal transmission.[82] Any disturbances in RNA editing in the brain leads to neurological disorders, as reported in cases of epilepsy, depression, amyotropic lateral sclerosis (ALS), and malignant glioma. Here, we will briefly discuss the editing defects observed in some neurological disturbances, specifically the mRNA editing defect associated with ALS.

Seratonin 2C receptor (5HT2C) mRNAs, which undergo the editing process in a seratonin-dependent manner, showed altered editing in brain tissues of patients with schizophrenia and major depression. The most complex alterations in 5HT2C mRNAs were found in the brains of depressed suicidal victims, suggesting the defective regulation of editing by synaptic seratonin.[83] Early onset of epilepsy and premature death in mice models were found to be associated with underediting of the GluR-B subunit of AMPA receptors, showing the importance of RNA editing in mammalian brain.[84] Underediting at the Q/R site editing position of GluR-B has also been found to be associated with malignant human brain tumors, indicating the role of RNA editing in the occurrence of epileptic seizures in association with malignant gliomas.[85]

Amyotropic lateral sclerosis (ALS) is a neurodegenerative disease associated with weakness and progressive paralysis with muscle wasting. ALS was found to be associated with AMPA receptor–mediated slow neuronal death, and further investigations showed defect in editing at the Q/R site of mRNA of the GluR2 subunit of AMPA receptors in motor neurons of ALS patients.[86,87] This failure to substitute glutamine with arginine at the Q/R site in the GluR2 subunit led to high Ca^{2+} conductance by AMPA receptors, causing exocitotoxicity and death of motor neurons, thus indicating that the defect in GluR2 mRNA editing in ALS motor neurons might be the contributory cause of neuronal death in ALS patients.[86,87] These findings suggest that RNA editing is an important process in RNA metabolism; and when it goes wrong, it could be associated with diseases.

3.3. *WS4/PCWH Syndromes and NMD*

Nonsense-mediated mRNA decay (NMD) is the RNA surveillance mechanism that downregulates premature termination codon (PTC)-containing mRNAs, thereby protecting the cell. It has been estimated that one third of genetic diseases are caused by frameshift and nonsense mutations that result in premature termination codons (PTCs),[88] and the role of NMD has become increasingly evident as the mechanism behind the occurrences of distinct phenotypes from different PTC-generating nonsense or frameshift mutations in the same gene (Fig. 4). Previously, it was observed that the reason behind the occurrences of mild and more severe disease phenotypes from different nonsense mutations in the same gene — e.g. in the case of beta globulin, beta thalasemia, fibrillin-1, and Marfan's syndrome — was NMD, as observed by severely reduced mutant mRNAs in mild disease conditions and intermediate or high levels of mutant mRNA levels in severe disease conditions.[88] Here, we will briefly discuss the role of NMD in causing two different phenotypes from different mutations in the *SOX10* gene.

Recently, it has been found that some mutations in the *SOX10* gene — a gene that produces a protein required for the development of cells in the neural crest lineage and for the control of proliferation of Schwann cells and oligodendrocytes — cause a more chronic neurological disease called PCWH, while some other mutations in the same gene cause a less chronic disease called WS4. When looking for the molecular mechanism underlying the occurrence of distinct neurological phenotypes from mutations in the same *SOX10* gene, it was found that, although all of the PTC-causing mutations in the *SOX10* gene which result in PCWH and WS4 are capable of producing truncating proteins with dominant negative activity, the mutations causing WS4 are present in exon 3 and exon 4 of the *SOX10* gene and are able to trigger NMD, as observed by the reduction in mRNA levels; whereas the PCWH-causing mutations are present in exon 5 (final exon) of the *SOX10* gene. In the latter case, the mRNAs escape NMD, leading to the production of dominant negative proteins and, hence, the chronic disease condition.[89]

NMD plays a protective role as a modulator of disease conditions by downregulating the PTC-containing mRNAs, which may otherwise result in proteins with dominant negative activity as seen in the abovementioned disease conditions. However, NMD can also be hazardous, as it can eliminate PTC-containing mRNAs that produce truncated proteins, which are partially or fully functional.[90] Therefore, it is important for potential NMD therapeutics to maintain a balance between the benefits of NMD in general and its detrimental effects (as in cases of specific genetic mutations).[90] Moreover, splicing efficiency is found to influence the magnitude of PTC-mediated downregulation of mRNAs by NMD, and thus can have potential applications for the treatment of genetic diseases caused by nonsense or frameshift mutations.[91]

3.4. *Fragile X Syndrome and BC1 RNA Localization*

mRNA transport, localization, and local translation have been observed in neurons as a means of influencing development and plasticity, and many RNA-binding proteins take part in these mRNA sorting mechanisms.[92] Neurological diseases like fragile X syndrome and spinal muscular atrophy (SMA) are found to be associated with impaired function of messenger ribonucleoproteins in the assembly, localization, and translational regulation of mRNAs.[92] Here, we focus on the association of the impaired function of two RNA-binding proteins (RBPs), FMRP and SMN, with fragile X syndrome and SMA, respectively, as these RBPs play roles in RNA localization and local translation.

Fragile X syndrome is a heritable form of mental retardation that arises from the absence or mutation of the fragile X mental retardation protein (FMRP), which can bind to specific mRNAs and utilize its RNA-binding domains. FMRP was found to act as a translational repressor of specific mRNAs at synapses, as observed by increased levels of specific proteins (e.g. CaMKIIα, MAP1B, Arc) in *FMR1* knockout mice compared to wild-type ones. This suggests that the loss of FMRP results in the loss of translational repression of specific mRNAs at synapses, thus leading to the synaptic dysfunction phenotype of fragile X patients.[93] FMRP was found to associate with dendritic nontranslatable RNA, BC1, which associates with both FMRP and mRNAs regulated by FMRP independently. Blocking of BC1 was found to inhibit FMRP interaction with its target mRNAs, thereby suggesting a model in which BC1 RNA provides a link between FMRP and its target proteins.[93]

Spinal muscular atrophy (SMA) is a recessive autosomal disorder associated with the degeneration of motor neurons in the spinal cord, and is caused by mutations in the survival of motor neuron (*SMN*) gene.[94] *SMN* is localized in granules that distribute in processes and growth cones of cultured neurons. The mutant SMN protein in SMA lacks its C-terminal exon 7, which is required for the localization of *SMN* from nucleus to cytoplasm, and overexpression of mutant SMN protein that lacks exon 7 in neurons causes shorter neurites.[95] *SMN* is found to facilitate interactions between RNA-binding proteins and their target sequences, and also promotes the assembly of SnRNPs.[92] *SMN* forms a complex with hnRNP R and β-actin mRNA, and translocates to axons and growth cones of motor neurons, thereby playing a role in the localization of β-actin mRNA.[96] So, mutations in *SMN* in SMA might lead to localization defects of β-actin mRNA.

3.5. *RNA Oxidation in Association with Age-Related or Genetic Diseases*

Oxidation of nucleic acids, proteins, and lipids can occur due to reaction with reactive oxygen species (ROS). Reactive oxygen species are produced in the body as part of the body's normal metabolism, and prevail in excess amounts in the body during oxidative stress, i.e. a state of imbalance between the levels of ROS and the antioxidants that usually remove them. Excess amounts of ROS that fail to be removed can cause damage to the biomolecules, thus leading to severe pathological conditions. Although many different types of oxidatively damaged bases have been detected in nucleic acids, 8-oxo-dG in DNA and 8-oxo-G in RNA are the most potent mutagens because of their ambiguous base-pairing property. Although ROSs have the ability to oxidize both DNA and RNA in the same manner,[44] DNA oxidation, its possible repair mechanisms, and the association of oxidized DNA with numerous neurodegenerative, cardiovascular, and autoimmune diseases have been well studied compared to RNA oxidation and its biological effects.[97] However, RNA oxidation has recently started to receive attention, as increased amounts of oxidized RNA have been found to be associated with some age-related and neurodegenerative diseases. Here, we will discuss some diseases that show association with increased levels of oxidized RNA.

It has been observed that there is an increased oxidative damage to DNA and RNA in neurons of the substantia nigra (SN) of Parkinson's disease patients. This oxidative damage to nucleic acids is restricted to the cytoplasm, thus indicating both RNA and mitochondrial DNA as targets.[98] Neuronal RNA oxidation was also found to be associated with Alzheimer's disease (AD) and Down syndrome. Increased RNA oxidation was found to be a significant feature of vulnerable neurons in AD, as observed by the *in situ* immunohistochemistry approach used to identify 8OHdG and 8OHG, the potential biomarkers of oxidative damage to nucleic acids.[99,100] In addition, Shan *et al.*[101] have identified and characterized oxidized RNA species in AD. Studies done by this group showed that significant amounts of poly(A)$^+$ mRNAs are oxidized in AD brains, many of them being implicated in the pathogenesis of AD. This oxidative mRNA damage was found to be highly selective, as only some mRNA species were more susceptible to oxidative damage, thus implicating the potential contribution of RNA oxidation in the pathogenesis of AD.[101]

Apart from the abovementioned diseases, dementia with Lewy bodies was also found to be associated with neuronal RNA oxidation. This was observed by relative intensity measurements of neuronal 8-hydroxyguanosine immunoreactivity that showed a significant increase in nucleic acid oxidation in dementia with Lewy bodies compared to controls, and DNase and RNase treatment prior to immunostaining indicated that RNA is the major site of nucleic acid oxidation.[102] Increase in RNA oxidation was also found to be associated with myopathies and human atherosclerosis.[103,104]

3.6. *Interferon-Activated RNA Degradation and Prostate Cancer (RNase L)*

There are an increasing number of men being diagnosed with and dying from prostate cancer each year in the United States. *HPC1* (hereditary prostate cancer 1), a major prostate cancer susceptibility locus, is linked to chromosome 1 (1q24–25).[105] Germ line mutations in the gene encoding RNase L have been found to segregate in prostate cancer families that show

HPC1 linkage, and tumors with germ line mutations of the *RNASEL* gene showed loss of heterozygosity and loss of RNase L protein, thus showing the importance of *RNASEL* in the occurrence of prostate cancer and in developing therapies for prostate cancer patients.[106] Furthermore, genetic studies conducted by different groups suggested that mutations in the *RNASEL* gene predispose men to high incidence of prostate cancer, and in some cases may lead to more aggressive disease.[107] Here, we briefly discuss the role of RNase L in interferon-mediated antiviral pathways and in apoptosis of prostate cancer cells.

RNase L is a regulated endonuclease that functions in interferon-mediated antiviral pathways. It is converted to active dimeric form from inactive monomeric form by associating with 2-5A, which includes 2-5A synthases (enzymes that convert ATP to PPi) and a series of short $2'$-to-$5'$–linked oligoadenylates.[107] Activation of RNase L by 2-5A in virus-infected cells leads to cellular rRNA damage, followed by viral activation of c–Jun NH_2-terminal kinase (JNK) and induction of apoptosis, both of which are found to be deficient in mice lacking RNase L. This indicates that JNK and RNase L function in the integrated signaling pathway during interferon response and lead to apoptosis of virus-infected cells.[108]

Studies done to investigate the role of RNase L in apoptosis of prostate cancer cells showed that DU145 human prostate cells, which are made deficient in RNase L by targeting with siRNA, are resistant to apoptosis mediated by the RNase L activator 2-5A compared to wild-type cells.[109] Also, the induction of apoptosis, through the combination of treatments with apoptosis-inducing factors like topoisomerase (Topo) I inhibitor and tumor necrosis factor–related apoptosis-inducing ligand (TRAIL), was found to be deficient in RNase L–deficient cells; whereas cells expressing siRNA targeted against the RNase L inhibitor RL1 showed enhanced apoptosis in response to apoptosis-inducing factors.[109] Furthermore, the inhibitor of JNK was found to reduce apoptosis.[109] Taken together, all of these results implicate RNase L in mediating apoptosis of prostate cancer cells during the treatment of those cells with 2-5A, Topo I inhibitors, and TRAIL by integrating and amplifying apoptotic signals.[109]

3.7. *Diseases Caused by Mitochondrial tRNA (mt tRNA) Defects*

The circular human mitochondrial genome encodes 22 tRNAs apart from 13 proteins corresponding to respiratory chain subunits, and two rRNAs.

Thus, the minimal set of tRNAs present in human mitochondria is critical for mitochondrial protein synthesis. Considering the small size of the mitochondrial genome, it is rather surprising to know that over 150 pathogenic mutations have been associated with it, and more than half of them are located within tRNA genes.[110] Here, we will see some diseases associated with mutations in mt tRNA genes.

3.7.1. *Mutations in mt tRNA and Diseases*

Many diseases have been found to be associated with point mutations in mt tRNA genes, mostly due to mutations in tRNA^{Ile}, tRNA$^{Leu(UUR)}$, and tRNA^{Lys} genes. Individuals with maternally inherited diabetes and deafness (MIDD) were found to carry an A3243G mutation in the tRNA$^{Leu(UUR)}$ gene. Another disease that was associated with the same point mutation and the point mutation U3271C in mt tRNA$^{Leu(UUR)}$ is mitochondrial myopathy, encephalopathy, lactic acidosis, and stroke-like episodes (MELAS). Myoclonus epilepsy and ragged red fibers (MERRF), a disease associated with an A8344G point mutation in the tRNA^{Lys} gene, was found to be associated with decreased protein synthesis due to a decrease in tRNA stability and a reduction in the steady-state aminoacylation level of mutant tRNA^{Lys} compared to wild-type ones.[110] Apart from these diseases, studies on a large kindred with syndrome X or dyslepidemic hypertension that included a cluster of metabolic defects (e.g. hypertension, hypercholesterolemia, hypomagnesemia) showed that each phenotype is transmitted by mitochondrial inheritance. Further analysis of the mitochondrial genome of maternal lineage showed a T4291C mutation in the tRNA^{Ile} gene at a position immediately 5′ to the tRNA^{Ile} anticodon, thus indicating the cause for common clustering of metabolic defects in the kindred.[111]

The pathogenic effects of the point mutations in mt tRNA genes can be attributed to the loss of secondary and tertiary confirmations of the tRNA, leading to impaired function in aminoacylation, protein synthesis, posttranscriptional modification, and processing.[110]

3.7.2. *MELAS and mt tRNA Modification*

One reason for the pathologic effects of point mutations in mt tRNA genes is defect in posttranscriptional modification, as observed in the case of MELAS patients. It has been shown that MELAS patients carrying A3243G or T3271C point mutations in the mt tRNA$^{Leu(UUR)}$ gene are deficient in normal taurine, containing modification (5-taurino methyluridine) at the anticodon wobble position.[112] Further studies using mt tRNA$^{Leu(UUR)}$ lacking taurine modification and without pathogenic mutations showed severe reduction in UUG decoding and no reduction in UUA decoding, thus suggesting the role of unmodified uridine in causing the translational defect and thereby possibly causing the complex I deficiency observed in MELAS patients with mutant mt tRNA$^{Leu(UUR)}$.[113]

Wobble modification defect was also observed in mt tRNA^{Lys} with an A8344G point mutation in MERRF patients.[114] This defect is attributed to the loss of translational activity in mutant mt tRNA^{Lys} for its cognate codons.[115]

3.8. *MicroRNAs and Diseases*

MicroRNAs (miRNAs) function as regulators of gene expression in eukaryotes by binding to partly complementary regions in the 3' UTR regions of specific mRNAs and inhibiting their translation or rendering them to degradation. miRNAs are found to play important roles in different cellular processes in eukaryotes by regulating their development, differentiation, organogenesis, embryogenesis, apoptosis, etc. Recent studies suggest that miRNAs might have a role in oncogenesis, and it has also been recently observed that *let-7* miRNA negatively regulates Ras guanosine triphosphates (proto-oncogenes) in *C. elegans* and in human tumor cell lines, thus suggesting the possible role of miRNAs as tumor suppressors and regulators of oncogenes.[116] Owing to the abovementioned important roles played by miRNAs, their aberrant expression may lead to human diseases especially cancers, as indicated by the recent studies that found altered expression of miRNA to be associated with chronic lymphocytic leukemia (CLL), pediatric Burkitt's lymphoma, gastric cancer, lung cancer, etc.[117] Here, we will discuss some miRNAs that showed aberrant expression in some diseases.

miRNA expression profiles were recently analyzed in hepatocellular carcinoma (HCC) tissue, adjacent nontumorous tissue (NT), and chronic hepatitis (CH) tissue specimens. It was found that five miRNAs showed higher expression and three miRNAs showed lower expression in HCC samples when compared with NT samples; also, when compared, CH and liver cirrhosis samples showed different patterns of miRNA expression. Therefore, this suggests the role that miRNA might play in liver disease progression.[118] Altered levels of some other miRNAs have been found to be associated with other malignancies like Hodgkin's, primary mediastinal, and diffuse large

B-cell lymphomas, where upregulation of the levels of BIC primary miRNA and its derivative miR-155 has been observed.[119]

A cluster of miRNAs, mir-17-92 polycistron, located in a region of DNA, was found to be amplified in B-cell lymphomas; and the levels of primary and mature miRNAs derived from this mir-17-92 locus were found to be increased in B-cell lymphoma samples when compared to the normal tissues.[120] This increased expression of the mir-17-92 cluster was found to act in association with *c-myc* expression in promoting tumor development in a B-cell lymphoma mouse model; and tumors expressing increased levels of a subset of the mir-17-92 cluster together with *c-myc* showed absence of apoptosis, a process which is prevalent in *c-myc*-induced lymphomas, thus implicating miRNAs as modulators of tumor formation and the mir-17-92 cluster as a potential human oncogene.[120] Deregulated expression of the *Tcl1* oncogene that resulted in CLL in mice was found to be associated with abnormalities in the expression of murine miRNAs mmu-mir-15a and mmu-mir-16-1, and the human homologs of these genes — mir-15a and mir-16-1 — were also found to be deleted in CLL patients, thus demonstrating the involvement of the Tcl1 oncoprotein and miRNA genes in the pathogenesis of CLL.[121]

It has been reviewed by Alvarez-Garcia and Miska[122] that DGCR8 — which encodes Dorsha, an essential enzyme in miRNA biogenesis — maps to the chromosomal region 22q11.2. This region is found to be commonly deleted in DiGeorge syndrome, leading to haploinsufficiency in over 90% of DiGeorge syndrome patients; and if this haploinsufficiency contributes to the disease, then reduced levels of miRNA might play a role in contributing to this disease.[122]

The abovementioned examples indicate that abnormal expression of miRNAs has been associated with human diseases. Further investigation of the precise roles played by specific miRNAs in causing different diseases and the pathways associated with them might provide powerful tools in diagnosing and treating those diseases.

3.9. *Regulation of Gene Expression and Age-Related Neurodegeneration (DNA Microarray)*

Microarray technology has become a very powerful tool in biomedical research, with its potential clinical applications in disease diagnosis, prognosis, and treatment.[123] The usage and application of this technique has become diverse, as it facilitates the monitoring of thousands of genes on a single chip at a given time, which is difficult to achieve using conventional molecular biology techniques. Microarray technology is being used to study the expression levels of DNA, RNA, and proteins, depending upon the application and requirement in different research areas.

Apart from its usage in understanding the causes of many diseases, microarray technology using gene chips — which enables the study of alterations in the entire genome — has recently been used to study the age-dependent regulation of gene expression and DNA damage in human brain, as aging of the human brain is the major risk factor in Alzheimer's disease. To determine this, RNA from postmortem samples of the human frontal cortex of 30 normal individuals ranging in age from 26 to 106 years was analyzed using Affymetrix gene chips to get the transcriptional profiling, and the genes with similar age-dependent expression patterns were resolved by analyzing the hierarchical clustering of age-regulated genes.[124] Microarray analysis showed that the expression of a set of genes — which play important roles in synaptic plasticity, vesicular transport, and mitochondrial function — reduced after age 40, followed by the induction of stress response, antioxidant, and DNA repair genes, thus leading to the hypothesis that DNA damage might target these specific genes.[124] Increased DNA damage was observed in promoters of age-downregulated genes in the frontal cortex, with their promoters being selectively damaged by oxidative stress in cultured human neurons and showing reduced base excision repair. This led to the conclusion that the reduced expression of selectively vulnerable genes involved in learning, memory, and neuronal survival might be due to DNA damage and that this process starts early in adult age, thereby initiating a program of brain aging.[124]

4. CONCLUDING REMARKS

RNA, once thought to be unstable molecules that functioned mainly in protein synthesis, are now known to directly participate in many cellular processes. The complexity of RNA structures, metabolism, and interactions with other macromolecules are far beyond the initial thoughts. Organisms have a large investment of genome materials in RNA metabolism. Recent discoveries of various defects in RNA processing, localization, degradation, and chemical damage have been overwhelming. These defects are caused either by genetic mutations in genes affecting RNA or RNA metabolism, or by environmental factors that damage RNA. It has been proven that various human diseases are caused by such defects through carrying such mutations or having genetic backgrounds sensitive to RNA damage by environmental factors. A thorough understanding of the mechanisms of RNA metabolism will help us develop prevention and treatment measures of related diseases.

ACKNOWLEDGMENTS

This work was supported by the NIH Grant No. S06 GM073621 to Z. Li. G. Kollipara was supported by the Graduate Program in Biomedical Sciences at Florida Atlantic University.

REFERENCES

1. Hall PA, Russell SH. (2005) New perspectives on neoplasia and the RNA world. *Hematol Oncol* **23**:49–53.
2. Mattick JS, Makunin IV. (2005) Small regulatory RNAs in mammals. *Hum Mol Genet* **14**:R121–R132.
3. Costa FF. (2005) Non-coding RNAs: new players in eukaryotic biology. *Gene* **357**:83–94.
4. Eddy SR. (2001) Non-coding RNA genes and the modern RNA world. *Nat Rev Genet* **2**:919–929.
5. Brent MR. (2005) Genome annotation past, present, and future: how to define an ORF at each locus. *Genome Res* **15**:1777–1786.
6. Tang G. (2005) siRNA and miRNA: an insight into RISCs. *Trends Biochem Sci* **30**:106–114.
7. Leung RK, Whittaker PA. (2005) RNA interference: from gene silencing to gene-specific therapeutics. *Pharmacol Ther* **107**:222–239.
8. Weischenfeldt J, Lykke-Andersen J, Porse B. (2005) Messenger RNA surveillance: neutralizing natural nonsense. *Curr Biol* **15**:R559–R562.
9. Li Z, Pandit S, Deutscher MP. (1998) Polyadenylation of stable RNA precursors *in vivo. Proc Natl Acad Sci USA* **95**:12158–12162.
10. Deutscher MP. (2006) Degradation of RNA in bacteria: comparison of mRNA and stable RNA. *Nucleic Acids Res* **34**:659–666.
11. Gonzales FA, Zanchin NI, Luz JS, Oliveira CC. (2005) Characterization of *Saccharomyces cerevisiae* Nop17p, a novel Nop58p-interacting protein that is involved in pre-rRNA processing. *J Mol Biol* **346**:437–455.
12. Mörl M, Marchfelder A. (2001) The final cut. *EMBO Rep* **2**:17–20.
13. Hopper AK, Phizicky EM. (2003) tRNA transfers to the limelight. *Genes Dev* **17**:162–180.
14. Helm M. (2006) Post-transcriptional nucleotide modification and alternative folding of RNA. *Nucleic Acids Res* **34**:721–733.
15. Decatur WA, Fournier MJ. (2003) RNA-guided nucleotide modification of ribosomal and other RNAs. *J Biol Chem* **278**:695–698.
16. Patton JR. (1994) Pseudouridine formation in small nuclear RNAs. *Biochimie* **76**:1129–1132.

17. Tran E, Brown J, Maxwell ES. (2004) Evolutionary origins of the RNA-guided nucleotide-modification complexes: from the primitive translation apparatus? *Trends Biochem Sci* **29**:343–350.

18. Gott JM. (2003) Expanding genome capacity via RNA editing. *C R Biol* **326**:901–908.

19. Wedekind JE, Dance GS, Sowden MP, Smith HC. (2003) Messenger RNA editing in mammals: new members of the APOBEC family seeking roles in the family business. *Trends Genet* **19**:207–216.

20. Stuart KD, Schnaufer A, Ernst NL, Panigrahi AK. (2005) Complex management: RNA editing in trypanosomes. *Trends Biochem Sci* **30**:97–105.

21. Valente L, Nishikura K. (2005) ADAR gene family and A-to-I RNA editing: diverse roles in posttranscriptional gene regulation. *Prog Nucleic Acid Res Mol Biol* **79**:299–338.

22. Cohen RS. (2005) The role of membranes and membrane trafficking in RNA localization. *Biol Cell* **97**:5–18.

23. Carson JH, Barbarese E. (2005) Systems analysis of RNA trafficking in neural cells. *Biol Cell* **97**:51–62.

24. Macdonald P. (2001) Diversity in translational regulation. *Curr Opin Cell Biol* **13**:326–331.

25. Preiss T, Hentze MW. (2003) Starting the protein synthesis machine: eukaryotic translation initiation. *Bioessays* **25**:1201–1211.

26. Pisarev AV, Shirokikh NE, Hellen CU. (2005) Translation initiation by factor-independent binding of eukaryotic ribosomes to internal ribosomal entry sites. *C R Biol* **328**:589–605.

27. Stoneley M, Willis AE. (2003) Aberrant regulation of translation initiation in tumorigenesis. *Curr Mol Med* **3**:597–603.

28. Preiss T, Hentze MW. (1999) From factors to mechanisms: translation and translational control in eukaryotes. *Curr Opin Genet Dev* **9**:515–521.

29. Kahvejian A, Roy G, Sonenberg N. (2001) The mRNA closed-loop model: the function of PABP and PABP-interacting proteins in mRNA translation. *Cold Spring Harb Symp Quant Biol* **66**:293–300.

30. Kozak M. (2005) Regulation of translation via mRNA structure in prokaryotes and eukaryotes. *Gene* **361**:13–37.

31. Jacobson A. (2004) Regulation of mRNA decay: decapping goes solo. *Mol Cell* **15**:1–2.

32. Bernstein P, Ross J. (1989) Poly(A), poly(A) binding protein and the regulation of mRNA stability. *Trends Biochem Sci* **14**:373–377.

33. Mangus DA, Evans MC, Jacobson A. (2003) Poly(A)-binding proteins: multifunctional scaffolds for the post-transcriptional control of gene expression. *Genome Biol* **4**:223.

34. Jacobson A, Peltz SW. (1996) Interrelationships of the pathways of mRNA decay and translation in eukaryotic cells. *Annu Rev Biochem* **65**:693–739.

35. van Hoof A, Parker R. (1999) The exosome: a proteasome for RNA? *Cell* **99**:347–350.

36. Brewer G. (2002) Messenger RNA decay during aging and development. *Ageing Res Rev* **1**:607–625.

37. Teixeira D, Sheth U, Valencia-Sanchez MA, *et al.* (2005) Processing bodies require RNA for assembly and contain non-translating mRNAs. *RNA* **11**:371–382.
38. Bashkirov V, Scherthan H, Solinger J, *et al.* (1997) A mouse cytoplasmic exoribonuclease (mXRN1p) with preference for G4 tetraplex substrates. *J Cell Biol* **136**:761–773.
39. Ingelfinger D, Arndt-Jovin D, Luhrmann R, Achsel T. (2002) The human LSm1–7 proteins colocalize with the mRNA-degrading enzymes Dcp1/2 and Xrn1 in distinct cytoplasmic foci. *RNA* **8**:1489–1501.
40. Lykke-Andersen J. (2002) Identification of a human decapping complex associated with hUpf proteins in nonsense-mediated decay. *Mol Cell Biol* **22**:8114–8121.
41. van Dijk E, Cougot N, Meyer S, *et al.* (2002) Human Dcp2: a catalytically active mRNA decapping enzyme located in specific cytoplasmic structures. *EMBO J* **21**:6915–6924.
42. Coller J, Parker R. (2005) General translational repression by activators of mRNA decapping. *Cell* **122**:875–886.
43. Fillman C, Lykke-Andersen J. (2005) RNA decapping inside and outside of processing bodies. *Curr Opin Cell Biol* **17**:326–331.
44. Bellacosa A, Moss EG. (2003) RNA repair: damage control. *Curr Biol* **13**:R482–R484.
45. Florentz C, Sohm B, Tryoen-Toth P, *et al.* (2003) Human mitochondrial tRNAs in health and disease. *Cell Mol Life Sci* **60**:1356–1375.
46. Jacobs HT, Turnbull DM. (2005) Nuclear genes and mitochondrial translation: a new class of genetic disease. *Trends Genet* **21**:312–314.
47. Levinger L, Morl M, Florentz C. (2004) Mitochondrial tRNA 3′ end metabolism and human disease. *Nucleic Acids Res* **32**:5430–5441.
48. Tuohy TM, Li Z, Atkins JF, Deutscher MP. (1994) A functional mutant of tRNA(2Arg) with ten extra nucleotides in its TFC arm. *J Mol Biol* **235**:1369–1376.
49. Li Z, Reimers S, Pandit S, Deutscher MP. (2002) RNA quality control: degradation of defective transfer RNA. *EMBO J* **21**:1132–1138.
50. Kadaba S, Krueger A, Trice T, *et al.* (2004) Nuclear surveillance and degradation of hypomodified initiator tRNAMet in *S cerevisiae*. *Genes Dev* **18**:1227–1240.
51. Baker KE, Parker R. (2004) Nonsense-mediated mRNA decay: terminating erroneous gene expression. *Curr Opin Cell Biol* **16**:293–299.
52. Hentze MW, Kulozik AE. (1999) A perfect message: RNA surveillance and nonsense-mediated decay. *Cell* **96**:307–310.
53. Hilleren P, Parker R. (1999) Mechanisms of mRNA surveillance in eukaryotes. *Annu Rev Genet* **33**:229–260.
54. Hilleren P, McCarthy T, Rosbash M, *et al.* (2001) Quality control of mRNA 3′-end processing is linked to the nuclear exosome. *Nature* **413**:538–542.
55. van Hoof A, Frischmeyer PA, Dietz HC, Parker R. (2002) Exosome-mediated recognition and degradation of mRNAs lacking a termination codon. *Science* **295**:2262–2264.

56. Fiala ES, Conaway CC, Mathis JE. (1989) Oxidative DNA and RNA damage in the livers of Sprague–Dawley rats treated with the hepatocarcinogen 2-nitropropane. *Cancer Res* **49**:5518–5522.

57. Wamer WG, Yin JJ, Wei RR. (1997) Oxidative damage to nucleic acids photosensitized by titanium dioxide. *Free Radic Biol Med* **23**:851–858.

58. Shen Z, Wu W, Hazen SL. (2000) Activated leukocytes oxidatively damage DNA, RNA, and the nucleotide pool through halide-dependent formation of hydroxyl radical. *Biochemistry* **39**:5474–5482.

59. Hofer T, Badouard C, Bajak E, *et al.* (2005) Hydrogen peroxide causes greater oxidation in cellular RNA than in DNA. *Biol Chem* **386**:333–337.

60. Kasai H, Crain PF, Kuchino Y, *et al.* (1986) Formation of 8-hydroxyguanine moiety in cellular DNA by agents producing oxygen radicals and evidence for its repair. *Carcinogenesis* **7**:1849–1851.

61. Ames BN, Gold LS. (1991) Endogenous mutagens and the causes of aging and cancer. *Mutat Res* **250**:3–16.

62. Mo J, Maki H, Sekiguchi M. (1992) Hydrolytic elimination of a mutagenic nucleotide, 8-oxodGTP, by human 18-kilodalton protein: sanitization of nucleotide pool. *Proc Natl Acad Sci USA* **89**:11021–11025.

63. Taddei F, Hayakawa H, Bouton M, *et al.* (1997) Counteraction by MutT protein of transcriptional errors caused by oxidative damage. *Science* **278**:128–130.

64. Hayakawa H, Hofer A, Thelander L, *et al.* (1999) Metabolic fate of oxidized guanine ribonucleotides in mammalian cells. *Biochemistry* **38**:3610–3614.

65. Hayakawa H, Uchiumi T, Fukuda T, *et al.* (2002) Binding capacity of human YB-1 protein for RNA containing 8-oxoguanine. *Biochemistry* **41**:12739–12744.

66. Ishibashi T, Hayakawa H, Ito R, *et al.* (2005) Mammalian enzymes for preventing transcriptional errors caused by oxidative damage. *Nucleic Acids Res* **33**:3779–3784.

67. Aas PA, Otterlei M, Falnes PO, *et al.* (2003) Human and bacterial oxidative demethylases repair alkylation damage in both RNA and DNA. *Nature* **421**:859–863.

68. Fire A, Xu S, Montgomery MK, *et al.* (1998) Potent and specific genetic interference by double-stranded RNA in *Caenorhabditis elegans*. *Nature* **391**:806–811.

69. Tuschl T. (2001) RNA interference and small interfering RNAs. *Chem Biochem* **2**:239–245.

70. Moss EG. (2001) RNA interference: it's a small RNA world. *Curr Biol* **11**:R772–R775.

71. Lee RC, Feinbaum RL, Ambros V. (1993) The *C. elegans* heterochronic gene *lin-4* encodes small RNAs with antisense complementarity to *lin-14*. *Cell* **75**:843–854.

72. Bentwich I, Avniel A, Karov Y, *et al.* (2005) Identification of hundreds of conserved and nonconserved human microRNAs. *Nat Genet* **37**:766–770.

73. Bentwich I. (2005) Prediction and validation of microRNAs and their targets. *FEBS Lett* **579**:5904–5910.

74. Lewis BP, Burge CB, Bartel DP. (2005) Conserved seed pairing, often flanked by adenosines, indicates that thousands of human genes are microRNA targets. *Cell* **120**:15–20.

75. Pasquinelli AE, Hunter S, Bracht J. (2005) MicroRNAs: a developing story. *Curr Opin Genet Dev* **15**:200–205.

76. Venables JP. (2004) Aberrant and alternative splicing in cancer. *Cancer Res* **64**:7647–7654.

77. Kalnina Z, Zayakin P, Silina K, Line A. (2005) Alterations of pre-mRNA splicing in cancer. *Genes Chromosomes Cancer* **42**:342–357.

78. Lazzereschi D, Nardi F, Turco A, *et al.* (2005) A complex pattern of mutations and abnormal splicing of *Smad4* is present in thyroid tumours. *Oncogene* **24**:5344–5354.

79. Lukas J, Gao D, Keshmeshian M, *et al.* (2001) Alternative and aberrant messenger RNA splicing of the *mdm2* oncogene in invasive breast cancer. *Cancer Res* **61**:3212–3219.

80. Wang XQ, Luk JM, Leung PP, *et al.* (2005) Alternative mRNA splicing of liver intestine–cadherin in hepatocellular carcinoma. *Clin Cancer Res* **11**:483–489.

81. Musunuru K. (2003) Cell-specific RNA-binding proteins in human disease. *Trends Cardiovasc Med* **13**:5.

82. Barlati S, Barbon A. (2005) RNA editing: a molecular mechanism for the fine modulation of neuronal transmission. *Acta Neurochir Suppl* **93**:53–57.

83. Schmauss C. (2003) Serotonin 2C receptors: suicide, serotonin, and runaway RNA editing. *Neuroscientist* **9**:237–242.

84. Brusa R, Zimmermann F, Koh DS, *et al.* (1995) Early-onset epilepsy and postnatal lethality associated with an editing-deficient *GluR-B* allele in mice. *Science* **270**:1677–1680.

85. Maas S, Patt S, Schrey M, Rich A. (2001) Underediting of glutamate receptor GluR-B mRNA in malignant gliomas. *Proc Natl Acad Sci USA* **98**: 14687–14692.

86. Kawahara Y, Ito K, Sun H, *et al.* (2004) Glutamate receptors: RNA editing and death of motor neurons. *Nature* **427**:801.

87. Kwak S, Kawahara Y. (2005) Deficient RNA editing of GluR2 and neuronal death in amyotropic lateral sclerosis. *J Mol Med* **83**:110–120.

88. Frischmeyer PA, Dietz HC. (1999) Nonsense-mediated mRNA decay in health and disease. *Hum Mol Genet* **8**:1893–1900.

89. Inoue K, Khajavi M, Ohyama T, *et al.* (2004) Molecular mechanism for distinct neurological phenotypes conveyed by allelic truncating mutations. *Nat Genet* **36**:361–369.

90. Holbrook JA, Neu-Yilik G, Hentze MW, Kulozik AE. (2004) Nonsense-mediated decay approaches the clinic. *Nat Genet* **36**:801–808.

91. Gudikote JP, Imam JS, Garcia RF, Wilkinson MF. (2005) RNA splicing promotes translation and RNA surveillance. *Nat Struct Mol Biol* **12**:801–809.

92. Bassell GJ, Kelic S. (2004) Binding proteins for mRNA localization and local translation, and their dysfunction in genetic neurological disease. *Curr Opin Neurobiol* **14**:574–581.

93. Zalfa F, Giorgi M, Primerano B, *et al.* (2003) The fragile X syndrome protein FMRP associates with BC1 RNA and regulates the translation of specific mRNAs at synapses. *Cell* **112**:317–327.

94. Frugier T, Nicole S, Cifuentes-Diaz C, Melki J. (2002) The molecular bases of spinal muscular atrophy. *Curr Opin Genet Dev* **12**:294–298.

95. Zhang HL, Pan F, Hong D, *et al.* (2003) Active transport of the survival motor neuron protein and the role of exon-7 in cytoplasmic localization. *J Neurosci* **23**:6627–6637.

96. Rossoll W, Jablonka S, Andreassi C, *et al.* (2003) Smn, the spinal muscular atrophy–determining gene product, modulates axon growth and localization of β-actin mRNA in growth cones of motoneurons. *J Cell Biol* **163**:801–812.

97. Evans DM, Cooke SM. (2004) Factors contributing to the outcome of oxidative damage to nucleic acids. *Bioessays* **26**:533–542.

98. Zhang J, Perry G, Smith MA, *et al.* (1999) Parkinson's disease is associated with oxidative damage to cytoplasmic DNA and RNA in substantia nigra neurons. *Am J Pathol* **154**:1423–1429.

99. Nunomura A, Perry G, Hirai K, *et al.* (1999) Neuronal RNA oxidation in Alzheimer's disease and Down's syndrome. *Ann NY Acad Sci* **893**:362–364.

100. Nunomura A, Perry G, Pappolla MA, *et al.* (1999) RNA oxidation is a prominent feature of vulnerable neurons in Alzheimer's disease. *J Neurosci* **19**:1959–1964.

101. Shan X, Tashiro H, Lin CG. (2003) The identification and characterization of oxidized RNAs in Alzheimer's disease. *J Neurosci* **23**:4913–4921.

102. Nunomura A, Chiba S, Kosaka K, *et al.* (2002) Neuronal RNA oxidation is a prominent feature of dementia with Lewy bodies. *Neuroreport* **13**: 2035–2039.

103. Tateyama M, Takeda A, Onodera Y, *et al.* (2003) Oxidative stress and predominant Aβ42(43) deposition in myopathies with rimmed vacuoles. *Acta Neuropathol (Berl)* **105**:581–585.

104. Martinet W, de Meyer GR, Herman AG, Kockx MM. (2004) Reactive oxygen species induce RNA damage in human atherosclerosis. *Eur J Clin Invest* **34**:323–327.

105. Smith JR, Freije D, Carpten JD, *et al.* (1996) Major susceptibility locus for prostate cancer on chromosome 1 suggested by a genome-wide search. *Science* **274**:1371–1374.

106. Carpten J, Nupponen N, Isaacs S, *et al.* (2002) Germline mutations in the ribonuclease L gene in families showing linkage with *HPC1*. *Nat Genet* **30**:181–184.

107. Silverman RH. (2003) Implications for RNase L in prostate cancer biology. *Biochemistry* **42**:1805–1812.

108. Li G, Xiang Y, Sabapathy K, Silverman RH. (2004) An apoptotic signaling pathway in the interferon antiviral response mediated by RNase L and c-Jun NH$_2$-terminal kinase. *J Biol Chem* **279**:1123–1131.

109. Malathi K, Paranjape JM, Ganapathi R, Silverman RH. (2004) *HPC1/RNASEL* mediates apoptosis of prostate cancer cells treated with

2′,5′-oligoadenylates, topoisomerase I inhibitors, and tumor necrosis factor–related apoptosis-inducing ligand. *Cancer Res* **64**:9144–9151.

110. Wittenhagen LM, Kelly OS. (2003) Impact of disease-related mitochondrial mutations on tRNA structure and function. *Trends Biochem Sci* **28**:605–611.

111. Wilson FH, Hariri A, Farhi A, *et al.* (2004) A cluster of metabolic defects caused by mutation in mitochondrial tRNA. *Science* **306**:1190–1194.

112. Yasukawa T, Suzuki T, Suzuki T, *et al.* (2000) Modification defect at anti-codon wobble nucleotide of mitochondrial tRNAs$^{Leu(UUR)}$ with pathogenic mutations of mitochondrial myopathy, encephalopathy, lactic acidosis, and stroke-like episodes. *J Biol Chem* **275**:4251–4257.

113. Kirino Y, Yasukawa T, Ohta S, *et al.* (2004) Codon-specific translational defect caused by wobble modification deficiency in mutant tRNA from a human mitochondrial disease. *Proc Natl Acad Sci USA* **101**:15070–15075.

114. Yasukawa T, Suzuki T, Ishii N, *et al.* (2000) Defect in modification at the anticodon wobble nucleotide of mitochondrial tRNA^{Lys} with the MERRF encephalomyopathy pathogenic mutation. *FEBS Lett* **467**:175–178.

115. Yasukawa T, Suzuki T, Ishii N, *et al.* (2001) Wobble modification defect in tRNA disturbs codon–anticodon interaction in a mitochondrial disease. *EMBO J* **20**:4794–4802.

116. Morris JP 4th, McManus MT. (2005) Slowing down the Ras lane: miRNAs as tumor suppressors? *Sci STKE* **2005**:41.

117. Croce CM, Calin GA. (2005) miRNAs, cancer, and stem cell division. *Cell* **122**:6–7.

118. Murakami Y, Yasuda T, Saigo T, *et al.* (2005) Comprehensive analysis of microRNA expression patterns in hepatocellular carcinoma and non-tumorous tissues. *Oncogene* **25**:2537–2545.

119. Kluiver J, Poppema S, de Jong D, *et al.* (2005) BIC and miR-155 are highly expressed in Hodgkin, primary mediastinal and diffuse large B cell lymphomas. *J Pathol* **207**:243–249.

120. He L, Thomson JM, Hemann MT, *et al.* (2005) A microRNA polycistron as a potential human oncogene. *Nature* **435**:828–833.

121. Pekarsky Y, Calin GA, Aqeilan R. (2005) Chronic lymphocytic leukemia: molecular genetics and animal models. *Curr Top Microbiol Immunol* **294**:51–70.

122. Alvarez-Garcia I, Miska EA. (2005) MicroRNA functions in animal development and human disease. *Development* **132**:4653–4662.

123. Fadiel A, Naftolin F. (2003) Microarray applications and challenges: vast array of possibilities. *Int Arch Biosci*:1111–1121.

124. Lu T, Pan Y, Kao SY, *et al.* (2004) Gene regulation and DNA damage in the ageing human brain. *Nature* **429**:883–891.

Chapter 22

Recent Progress in the Genetics of Osteoporosis

Hui Shen[*], Yong-Jun Liu[*], Peng Xiao[†], Dong-Hai Xiong[†] and
Hong-Wen Deng[*,‡,§,¶]

[*] *Departments of Orthopedic Surgery & Basic Medical Sciences
School of Medicine, University of Missouri–Kansas City
Kansas City, MO 64108, USA*

[†] *Osteoporosis Research Center, Creighton University Medical Center
Omaha, NE 68131, USA*

[‡] *Key Laboratory of Biomedical Information Engineering of the
Ministry of Education & Institute of Molecular Genetics
School of Life Science and Technology, Xi'an Jiaotong University
Xi'an 710049, P. R. China*

[§] *Laboratory of Molecular and Statistical Genetics
College of Life Sciences, Hunan Normal University
Changsha, Hunan 410081, P. R. China*

Osteoporosis is a complex multifactorial disease, determined by genetic
and environmental factors as well as their interactions. Extensive efforts
have been made to identify the genetic determinants of osteoporosis.
Studies using multiple approaches, including linkage analysis, association
analysis, functional genomic analysis, as well as genetic and functional
studies in animal models, have greatly advanced our understanding of the
genetic basis of osteoporosis. This review summarizes recently published
important and representative molecular genetics studies of gene identification for osteoporosis. Evidence from candidate gene association studies
and genomewide linkage studies in humans as well as quantitative trait
locus (QTL) mapping in animal models are reviewed separately. We also
summarize the recent progress of functional genomic studies (including

---

[¶] Correspondence author.

DNA microarrays and proteomics) on osteogenesis and osteoporosis in light of the rapid advances and promising prospects of this field.

1. INTRODUCTION

Osteoporosis is a common disease defined as a skeletal disorder characterized by compromised bone strength predisposing a person to an increased risk of fracture. In the United States, osteoporosis is a major public health problem. Ten million individuals already have osteoporosis, and approximately 34 million more have low bone mineral density (BMD), placing them at increased risk for the disease. Osteoporosis is responsible for more than 1.3 million osteoporotic fractures (OFs) annually. OFs could cause permanent disabilities, nursing home placement, and even death. In 1997 alone, osteoporosis incurred an estimated direct cost of over US$17 billion in the US.

Osteoporosis is a complex multifactorial disease, determined by genetic and environmental factors as well as their interactions. Extensive efforts have been made to identify the genetic determinants of osteoporosis. In this article, we summarize recent advances in the genetics of osteoporosis, including association and linkage studies in human populations, QTL mapping in animal models, and functional genomic studies of osteoporosis. As the important findings made in the field of genetics of osteoporosis prior to 2002 were reviewed in our previous review article,[1] here we focus on studies published thereafter. Readers are also referred to a number of other review articles that have successfully addressed the strategies[2–4] and status[5–14] of the genetic study of osteoporosis. We recently overviewed powerful and promising methodologies in the study of complex bone disorders at the whole-genome level[15]; in addition, we systematically addressed the confounding factors that cause nonreplication in gene mapping of osteoporosis and proposed tentative remedies.[16] Therefore, a detailed discussion on these aspects will not be attempted here.

2. CANDIDATE GENE ASSOCIATION STUDIES

In the past 2 years, about 170 association studies have been published. The candidate genes involve classical genes (e.g. *VDR*, *ER-α*, *COL1A1*) that have been extensively studied as well as novel genes (e.g. *LRP5*, *SOST*) recently discovered to be important in bone and mineral metabolism. The newly investigated genes include *CYP17* (17-alpha-hydroxylase),[17] *CYP1B1* (cytochrome P450),[18] *DBP* (vitamin D–binding protein),[19] *GH1* (growth hormone 1),[20] *GnRH* (gonadotropin-releasing hormone 1),[21] *IGF-II* (insulin-like growth factor II),[22] *LEPR* (leptin receptor),[23] *LRP5* (low density lipoprotein receptor–related protein 5),[24] *BMP2* (bone morphogenetic protein 2),[25] *CCR2* (chemokine),[26] *CLCN7* (chloride channel 7),[27] *COMT* (catechol-O-methyltransferase),[28] *CTSK* (cathepsin K),[29] *DRD4* (dopamine receptor D4),[30] *I-TRAF* (TRAF family member–associated NF-kappa-B activator),[31] *LCT* (lactase),[32] *MIF* (macrophage migration inhibitory factor),[33] *MMP-1* (matrix metalloproteinase 1),[34] *MMP-9* (matrix metalloproteinase 9),[35] *NCOA3* (nuclear receptor coactivator 3),[36] *NPY* (neuropeptide Y),[37] *OSCAR* (osteoclast-associated receptor),[38] *PLOD1* (procollagen-lysine, 2-oxoglutarate 5-dioxygenase),[39] *PON1* (paraoxonase 1),[40] *RIL* (LIM domain protein RIL),[41] *SERT* (serotonin transporter),[42] *SOST* (sclerostin),[43] and *TCIRG1* (T-cell immune regulator 1).[44]

In the following, we will highlight some studies performed with a relatively large sample size. This is because statistical power is among the foremost factors for robust and replicable results, and generally at least 1000 subjects are needed to detect modest genetic effects (e.g. a QTL explaining 5% of phenotypic variation) in a population-based association study.[16] Some meta-analyses will also be addressed, as meta-analyses — by combining results across studies — are helpful to solve the problems of underpowered studies, revealing unexpected sources of heterogeneity and resolving discrepancies in genetic studies.[45] For the classical candidate genes, their potential physiological effects on bone metabolism and pathophysiological implications to osteoporosis have been elaborated elsewhere[1,14]; for the novel genes, their potential functions will be briefly outlined.

2.1. *Classical Candidate Genes*

2.1.1. *VDR*

Association between the *VDR* gene and osteoporosis-related traits has been extensively investigated.[1] The frequently studied markers include *BsmI*, *ApaI*, *TaqI*, and *FokI*, which currently have unknown functional effects.[1] A novel polymorphism at the Cdx-2 (an intestine-specific homeodomain-containing transcription factor) binding site of the promoter region was identified to be able to activate *VDR* gene transcription, and was associated with BMD variation in a Japanese population.[46]

Morita *et al.*[47] randomly selected 50 women from each of the 5-year age-stratified groups (15–79 years) in three Japanese municipalities, that is, 650 subjects for each area and 1950 in total. After excluding subjects who had medical or menstrual histories affecting BMD, 1434 women were analyzed for *ApaI*, *TaqI*, and *FokI* polymorphisms. *TaqI* was significantly associated with the distal 1/3 radius BMD in premenopausal women ($p = 0.019$). Analyzing three major combined genotypes (aaTT, AaTT, AaTt) of *ApaI* and *TaqI* revealed even stronger association for the distal 1/3 radius BMD in premenopausal women ($p = 0.009$). However, analyses on major haplotypes (AT or aT) failed to detect any significant association. In addition, they tested the relationship between the three polymorphisms and BMD change over 3 years in 976 subjects.[47] The annual percent changes in lumbar spine BMD of the *TaqI* tt subjects were different from other genotypes in women who were either premenopausal at the follow-up survey or postmenopausal at the baseline survey; however, the effect of the tt genotype on BMD change was opposite in the two groups. The authors concluded that none of the individual polymorphisms were consistently associated with baseline BMD or BMD change; hence, the effect of the VDR genotype on BMD is negligible in Japanese women.

Fang *et al.*[48] examined Cdx-2 with BMD and risk of fracture in a cohort of 2848 Dutch patients aged ≥55 years. They did not find a significant association for BMD, but they detected borderline association with vertebral fracture ($p = 0.04$) and any type of fracture ($p = 0.06$), with subjects carrying the Cdx-2 A-allele having a reduced relative risk (RR) of fracture by 20%.[48] The protective effect of the A-allele was similar in women and men.

Thakkinstian and colleagues reported two meta-analyses.[49,50] One study focused on the relationship between *VDR BsmI* and BMD/osteoporosis at the femoral neck or spine in adult women.[49] *BsmI* was associated with spine BMD in postmenopausal ($p = 0.028$), but not premenopausal, women. The association was modest and followed a recessive model, with the BB genotype having a lower BMD than the Bb/bb genotype. The magnitude of the decrease in spinal BMD by BB genotype was 2.4%, which translated into a population attributable risk of spine fracture of 1.98%.[49] They also investigated the association between *BsmI* and mean percent BMD change over time. Analyses revealed a significant effect ($p = 0.017$), with BB and Bb genotypes having greater bone loss per year than the bb genotype. The other meta-analysis was conducted with data on *BsmI*, *ApaI*, and *TaqI*, as well as on haplotypes defined by them.[50] Although none of the individual polymorphisms was associated with osteoporosis on its own, they found a significant association between spinal osteoporosis and haplotypes BAT ($p < 0.001$) and BaT ($p = 0.031$), with an odds ratio (OR) of ~4. For spine BMD, the only association was found for *BsmI* in postmenopausal

Asian women ($p < 0.001$), suggesting that the genetic effect of *VDR BsmI* could be population-, menstrual status-, and site-dependent. This was in agreement with their previous meta-analysis.

Based on the available data, Morrison[51] wrote a commentary on the association between VDR and BMD. By using the data provided in one meta-analysis mentioned above,[49] the author indicated that the effect size of the *BsmI* polymorphism on BMD was approximately 0.11–0.13 standard deviation (SD) in postmenopausal women. To detect such modest effects, large sample sizes are required. For instance, based on the estimated allele frequency, 3046 Caucasians or 4700 Japanese are required to have 80% power to detect the effect of 0.13 SD at $p = 0.01$.

In 2006, two large-scale association studies were reported for the *VDR* gene in European Caucasians.[205,206] Macdonald *et al.*[205] tested five polymorphisms (i.e. Cdx-2, *FokI*, *BsmI*, *ApaI*, and *TaqI*) in 3100 early postmenopausal British women, but found no significant association with BMD, bone loss, or fracture. In a multicenter association study involving 26 242 subjects from nine European research teams,[206] the *FokI*, *BsmI*, *ApaI*, and *TaqI VDR* polymorphisms were not associated with BMD or with fractures; but the Cdx-2 polymorphism showed borderline association ($p = 0.039$) with risk for vertebral fractures.

While most of the association studies for the *VDR* gene tested only one or a few single nucleotide polymorphisms (SNPs), two recent studies examined multiple SNPs across the *VDR* gene and assessed the linkage disequilibrium (LD) pattern across the gene.[52,53] Nejentsev *et al.*[53] resequenced the gene, and genotyped 55 common SNPs (minor allele frequency > 10%) in four European populations and one African population. LD patterns were identical (with three block-like regions) in all four European populations, but with two additional LD-breaking spots in the African population. In another study, Fang *et al.*[52] sequenced 22 kb of the gene (including the promoter, all exons, and the 3′ UTR) and identified 62 SNPs. LD analyses on common SNPs revealed four to eight haplotype blocks, which were less fragmented in Caucasians and Asians than in Africans. They subsequently tested 15 tagging SNPs with BMD and fracture risk in 6535 elderly Caucasians. Two haplotypes (containing the Cdx-2 SNP and 3′ UTR, respectively) were associated with increased risk for vertebral fractures or overall osteoporotic fractures. The combined risk alleles showed 46% increased risk for vertebral fracture ($p = 0.03$) and 34% increased risk for osteoporotic fracture ($p = 0.01$).[52] This whole-gene analysis suggested that *VDR* polymorphisms in the promoter region and 3′ UTR may interactively affect fracture risk. In another study, the same group found that tagging SNPs encompassing the 3′ UTR of the *VDR* gene interacted with an SNP in the lactase-phlorizin hydrolase (*LPH*) gene — a variant correlated to lactose intolerance — to influence height, bone geometry, and BMD.[54]

To date, more than 150 studies have been reported on associations between *VDR* gene polymorphisms and bone-related traits. Based on the available data, a tentative conclusion is that *VDR* gene polymorphisms, individually or interactively, may have effects on a number of biological endpoints, including BMD variation and bone fractures. Moreover, *VDR* may interact with nongenetic factors, i.e. calcium intake and estrogens, to modulate BMD. The interpretation of *VDR* polymorphisms is currently hindered by the fact that most studies have investigated only limited polymorphisms (e.g. *BsmI*, *ApaI*, *TaqI*, *FokI*) which have unknown effects. Whole-gene analysis, which explores all potential sequence variations within/around the gene, will be helpful to identify the functional variants. It is also imperative to clarify the molecular mechanisms underlying the observed associations by conducting functional studies using well-defined cell types and/or animal models.

2.1.2. *ER-α*

Three common *ER-α* polymorphisms have been extensively studied, namely, the *PvuII* and *XbaI* polymorphisms in intron 1 and the TA repeat polymorphism in the promoter. The TA repeat polymorphism was speculated to affect mRNA production or stability, while the functional significance of *PvuII* and *XbaI* polymorphisms remains to be elucidated. A possible mechanism is that the two polymorphisms are in LD with nearby functional variant(s), resulting in positive associations.

Yamada *et al.*[55] examined the relationship between *PvuII* and *XbaI* and BMD in ∼2230 Japanese subjects aged 40–79 years. In women ≥60 years old, *XbaI* alone or in combination with *PvuII* showed significant association with femoral neck BMD. Zhao *et al.*[56] examined eight SNPs spanning the *ER-α* gene in 405 Caucasian nuclear families (1873 subjects). Marginal evidence was observed for femoral neck BMD with rs932477 ($p = 0.015$) and rs2228480 ($p = 0.010$). The most common seven-SNP haplotype (TCGCGGG) was associated with higher lumbar spine BMD ($p = 0.015$). van Meurs *et al.*[57] investigated the association of *PvuII*, *XbaI*, and TA repeat polymorphisms with BMD, vertebral bone area, and fractures in 2042 elderly Dutch Caucasians. In women, subjects homozygous for haplotype px and a low number of TA repeats had significantly lower lumbar spine BMD ($p = 0.003$ and 0.008, respectively) and decreased vertebral bone area ($p = 0.016$) than homozygous noncarriers. They also found an increased vertebral fracture risk with an allele dose effect of OR $= 2.2$ for haplotype px, and OR $= 2.0$ for a low number of TA repeats.

A meta-analysis was reported for *Pvu*II, *Xba*I, and promoter TA repeats with BMD and fractures in 18 917 individuals from eight European research centers.[58] None of the three polymorphisms or haplotypes thereof had statistically significant association with BMD. However, there was a highly significant protection conferred by the *Xba*I XX genotype against fracture risk. In women with the XX genotype, the OR was 0.81 ($p = 0.002$) for any fracture and 0.65 ($p = 0.003$) for vertebral fractures. The observed effects on fractures were independent of BMD.

As with the *VDR* gene, the individual contributions of *ER-α* polymorphisms to osteoporosis remain to be universally confirmed. Future endeavors will be to elucidate their functional molecular relevance and their interaction with the environment in the causation of osteoporosis.

2.1.3. *COL1A1*

Thirteen association studies have been reported over the past 2 years for the *COL1A1* gene, many focusing on the Sp1 polymorphism, a G $\rightarrow$ T substitution at the first base of a consensus site in the first intron for the transcription factor. It is notable that the association of *COL1A1* Sp1 with osteoporotic fractures is among those mostly replicated.[59]

In 1044 elderly Swedish women, Gerdhem *et al.*[60] found an association between *COL1A1* Sp1 and femoral neck BMD ($p = 0.027$) and prevalent wrist fracture ($p = 0.024$). Women carrying at least one copy of the s-allele had lower femoral neck BMD and higher prevalence of wrist fracture. The ORs for prevalent wrist fracture were 2.73 and 1.4 for ss and Ss genotypes, respectively, suggesting the pronounced effect of the s-allele on increasing risk of wrist fractures. In 401 Chinese nuclear families, Zhang *et al.*[61] found that a -1997 G/T polymorphism in the *COL1A1* upstream regulatory region was associated with hip BMD, explaining 1.6% of the total hip BMD variation. By testing three *COL1A1* SNPs in 405 Caucasian nuclear families, Long *et al.*[62] reported that subjects bearing the T-allele of SNP2 had, on average, 3.05% smaller wrist size than noncarriers.

Mann and Ralston[63] performed a meta-analysis for *COL1A1* Sp1 and BMD/osteoporotic fractures that involved 7849 subjects from 26 published studies. The Ss genotype group had significantly lower lumbar spine BMD than the SS group ($p = 0.00005$), but the difference between the SS and ss groups in spine BMD did not reach a significant level ($p = 0.13$). The femoral neck BMD was lower in the Ss group ($p < 0.00001$) and ss group ($p = 0.001$) vs. the SS group. They also found increased OR for any fracture in Ss subjects (OR $= 1.26, p = 0.002$) and an even greater increase in ss subjects (OR $= 1.78, p = 0.003$). Subgroup analysis showed that the increased risk was largely attributable to vertebral fracture, where the OR was 1.37 ($p = 0.0004$) and 2.48 ($p < 0.00001$) for Ss and ss subjects, respectively. Their results suggested that the *COL1A1* Sp1 alleles contribute to a modest reduction in BMD and a significant increase in risk of osteoporotic fractures, particularly vertebral fractures. A large-scale multicenter study with 20 786 individuals also demonstrated that the *COL1A1* Sp1 polymorphism is associated with BMD at femoral neck and lumbar spine.[207] Also, this polymorphism could predispose to incident vertebral fractures in women independent of BMD.[207]

The functional importance of Sp1 has been investigated.[64] The s-allele had greater affinity for Sp1 protein compared to the S-allele. In Ss heterozygotes, the RNA transcripts derived from the s-allele were three times more abundant than those from the S-allele, suggesting allele-specific transcription or a different splicing process for the two alleles. The yield strength of bone derived from Ss individuals was reduced when compared with bone derived from SS subjects.[64] With the availability of SNP information of the human genome, together with the ongoing HapMap project,[65] further studies are necessary to identify more causative variants in the *COL1A1* gene using the LD mapping approach.

2.2. *Novel Candidate Genes*

2.2.1. *MTHFR*

MTHFR (5,10-methylenetetrahydrofolate reductase) affects the methylation of homocysteine to methionine, and high serum homocysteine concentrations have adverse effects on bone.[66,67] A *MTHFR* polymorphism, C677T, causes an alanine-to-valine substitution and gives rise to a thermolabile variant of *MTHFR* with reduced activity.[68] *MTHFR* C677T was associated with elevated levels of circulating homocysteine[69] and lumbar spine BMD.[70] In a study using 1748 healthy postmenopausal Danish women,[71] the TT genotype of the *MTHFR* C677T polymorphism was associated with lower BMD at the femoral neck, total hip, and spine ($p < 0.05$), with the effect sizes ranging from 0.1 to 0.3 SD. In consistency, the fracture incidence was increased more than twofold in subjects with the TT genotype. Although this variant did not affect the response to hormone replacement therapy (HRT), the association of the TT genotype with lower BMD was maintained at the total hip after 5 years of HRT.

Since the effect of C677T on circulating homocysteine levels is dependent on plasma folate concentration,[72] one study in 1632 Caucasians evaluated whether the folate status may modify the association between C677T and BMD or quantitative ultrasound (QUS) parameters.[73] The results showed that adjusted mean QUS parameters and BMD measurements did not significantly differ between C677T groups. However, suggestive interactions between folate status and the C677T group (CC + CT vs. TT) were found for hip BMD ($p \leq 0.05$) and for one of the QUS parameters, broadband ultrasound attenuation (BUA) ($p = 0.11$). In subjects with low folate concentration (<4 ng/mL), the TT group had lower mean BUA ($p = 0.06$) and Ward's area BMD ($p = 0.08$) compared with the CC + CT group; whereas in subjects with high folate levels (≥ 4 ng/mL), the TT group had significantly higher hip BMD ($p \leq 0.05$).

Since dietary B vitamins (folate, vitamin B12, vitamin B6, and riboflavin) can also influence circulating homocysteine levels, a study further exploited the association bewteen C677T and BMD and bone loss in relation to vitamin B intake in a cohort of perimenopausal and early postmenopausal Scottish women.[74] Although no association was observed for BMD, bone loss, or biochemical markers of bone turnover, there was a significant interaction between C677T and riboflavin intake in relation to BMD. BMD was lower for the TT group at low intakes of riboflavin compared to the other genotypes; but at high intakes, BMD was higher in the TT group. The results suggested that the association between C677T and bone phenotypes might depend on vitamin B levels, especially on folate and riboflavin status. The mechanism underlying this modification remains unclear.

2.2.2. *IGF-I*

IGF-I plays an essential role in longitudinal bone growth by stimulating the proliferation of chondrocytes in the growth plate. It is also involved in the formation of trabecular bone, and is essential for coupling matrix biosynthesis to sustain mineralization. Plasma *IGF-I* levels were associated with BMD and osteoporotic fractures.[75,76] One study examined the role of a CA repeat polymorphism in the promoter region with hip BMD in a group of elderly Dutch women and men.[77] A total of 5648 and 4134 individuals underwent examination at baseline and 2-year follow-up, respectively. Lower baseline BMD and higher BMD loss were associated with the absence of the 192-bp allele in women ($p = 0.03$), but not in men. This polymorphism only explains a minor portion of the variance in BMD (0.2%) and BMD change (0.1%) among females.

The same group conducted another study evaluating the association of the CA repeat polymorphism with incidence of nonvertebral fractures in 2799 men and 4212 women.[78] They also estimated the effects of this polymorphism on several hip bone geometry parameters, including neck width, cortical thickness, buckling ratio, and section modulus.[78] Women who were noncarriers or heterozygotes of the 192-bp allele had increased risk (1.5 and 1.2, respectively) of osteoporotic fractures compared with homozygotes for this allele ($p = 0.0007$); this effect was not observed in men. For hip geometry parameters, noncarrier males had a narrower femoral neck and lower section modulus ($p < 0.05$) than homozygote men; whilst noncarrier females had thinner cortices and higher buckling ratios ($p < 0.05$) than homozygote women, but no significant differences in femoral neck width and section modulus. The observed genotype-dependent differences in fracture risk cannot be fully explained by the genotype-dependent effects on hip bone geometry.

2.2.3. IL-6

IL-6 is a pleiotropic cytokine that promotes the differentiation of osteoclast precursor cells into mature osteoclasts.[79] Yamada *et al.*[25] examined the -634C-G polymorphism of the *IL-6* gene alone and in combination with the 298C-T polymorphism of the osteocalcin gene in ~2200 Japanese subjects (~1100 men and ~1100 women). Both polymorphisms were associated with BMD of total body and lumbar spine in postmenopausal women ($p < 0.05$). Analyses with combined genotypes suggested that the two polymorphisms exert an additive effect on BMD in postmenopausal women.

Another study in 3376 Caucasian women aged 65 or older reported association for distal and proximal radius BMD with the *IL-6* G174C polymorphism ($p = 0.016$ and 0.049, respectively).[80] The risk of wrist fractures decreased by 17% per copy of C-allele significantly ($p = 0.043$). Moreover, compared with women having the GG phenotype, women having the CC genotype also had slower rates of bone loss in the total hip and femoral neck at $\sim$3.5 years follow-up as well as 33% lower risk of wrist fractures over an average of 10.8 years.

2.2.4. *LRP5*

LRP5 was recently found to be a key regulator of osteoblast proliferation and bone formation. *LRP5* mutations resulted in high bone mass phenotypes or osteoporosis–pseudoglioma syndrome.[87,88] Ferrari *et al.*[89] examined the LD patterns using 13 previously reported SNPs in Caucasians, and then selected five informative SNPs (a minor allele frequency of $\geq$5% and one SNP for each pair of markers in nearly complete LD [$R^2 \leq 0.9$]) for further association analysis. They ruled out potential population stratification using the program STRUCTURE.[90] Significant associations were found between 2047G-A (i.e. V677M) and lumbar spine BMD ($p = 0.041$), bone mineral content or BMC ($p = 0.0032$), bone area ($p = 0.0014$), and stature ($p = 0.0062$). The observed associations between lumbar spine BMC and bone area were driven mainly by men. Haplotype analyses suggested that additional *LRP5* genetic variants might also influence lumbar spine bone mass and area. Furthermore, they also found an association between *LRP5* haplotypes with 1-year bone mass and area changes in prepubertal males, but not in females. Taken together, the results suggested that *LRP5* polymorphisms may influence lumbar spine bone mass and size in men, probably by affecting vertebral bone growth during childhood.[89] In contrast to the male-specific effects observed by Ferrari *et al.*,[89] a study in 1301 elderly Australian women showed that the *LPR5* gene was associated with hip bone mass and osteoporotic fractures.[92]

2.2.5. *SOST*

SOST is a disease-causing gene for sclerosteosis, a sclerosing bone dysplasia characterized by hyperostosis and overgrowth of normal bone tissue.[93] A study analyzed eight *SOST* polymorphisms in 1939 elderly Dutch men and women aged 55–80 years.[94] A 3-bp insertion in the promoter region (SRP3) was associated with decreased BMD in women at the femoral neck ($p = 0.05$) and lumbar spine ($p = 0.01$), with evidence of an allele-dose effect in the oldest age group ($p = 0.006$). The corresponding effect size between extreme genotypes was 0.2 SD. The polymorphism SRP9 was associated with femoral neck BMD ($p = 0.007$) and lumbar spine BMD ($p = 0.02$) in men, with an effect size between extreme genotypes of 0.2 SD. However, haplotype analyses failed to confirm the association observed in single

SNP analyses. They further studied interactions between *SOST* polymorphisms and the *VDR* 3' UTR polymorphisms as well as the *COL1A1* Sp1 polymorphism.[94] An additive effect was observed between the *SOST* and *COL1A1* Sp1 polymorphisms. The *SOST–COL1A1* additive effect increased with age, and reached 0.5 SD difference in lumbar spine BMD in the oldest age group ($p = 0.02$).

Other notable candidate genes for osteoporosis include the *TGF-β* gene,[81,82] the *ER-β* gene,[83] the *OPG* gene,[84] the *TNFRSF1B* (i.e. *TNFR2*) gene,[85] and the *TNF-α* gene.[86]

2.3. *Summary*

Association results in the bone field are currently inconsistent/inconclusive. A few articles have addressed this from perspectives of study design, analytical methodology, and others.[16,95–99] In the following, we re-emphasize important issues that undermine the validity of association studies.

First, a large portion of studies are severely underpowered, which may yield unreliable/unrepeatable results.[16,100] Meta-analysis is a powerful tool for pooling previous research when individual studies have insufficient power.[49,63] However, meta-analysis is not always a panacea, as it is prone to biases resulting from different ascertainment and diagnostic criteria and population stratification in different studies.[16] Moreover, there exists publication bias favoring positive results, which further exacerbates the situation.

Second, a limited number of polymorphisms is evaluated in the target gene(s). Given the complexity of LD and haplotype patterns across the human genome and among different populations,[101–104] this may not provide a comprehensive and accurate evaluation of the gene(s) of interest.[16,105]

Third, given multiple markers and/or phenotypes tested in a single study, it is too liberal to set the significance threshold at $p \leq 0.05$, which, however, has inappropriately been used in many studies. Traditional Bonferroni correction is overconservative, because of nonindependence of the tests. Other methods accounting for multiple testing of correlated hypotheses have been developed.[106–108]

Fourth, population stratification, a confounding factor for association analyses,[109] is not appropriately assessed and controlled. Although the actual impact of population stratification on association studies has been a matter of some debate,[110–112] modest amounts of population stratification have indeed been detected in case-control and case-cohort studies[113,114] and even in a relatively homogeneous genetic isolate.[115] A family-based association approach (e.g. transmission/disequilibrium test [TDT]), could be effective to deal with the problem.[116,117] For population-based studies, methods to control population stratification have been developed (e.g. Genomic Control[118] and Structured Association[119]) that, bearing in mind some assumptions, merit application and further development.

Finally, other factors such as heterogeneity, epistasis, and gene–environment interaction have posed special challenges to gene discovery. Endeavors to maximize sample homogeneity by recruiting subjects from the same ethnic group and rigorously controlling possible environmental or confounding factors may increase the chance of success. Studies have suggested that there exists sex-specific genetic contributions to BMD.[120,121] The genes regulating peak bone mass might differ from those regulating bone loss.[122] It is imperative for association studies to be well designed, statistically powerful, and appropriately interpreted.

A critical question to be answered is the correlation between osteoporotic fractures (OFs) and osteoporosis risk factors. The search for OF genes should start with significant heritability for OFs, and should include risk factors (e.g. BMD) that are genetically correlated with OFs. However, it is unclear to what extent the BMD variation can predict bone fractures, the endpoint of osteoporosis. Studies have suggested that OFs *per se* can be employed as a direct trait for gene hunting of osteoporosis.[123]

3. LINKAGE STUDIES

Twelve large-scale linkage studies were reported for osteoporosis during 2003–2004, including 10 whole-genome linkage scans[124–133] and 2 follow-up studies[134,135] focusing on candidate regions identified in previous genome scans. Our first review[1] summarized several whole-genome linkage scan results.[136–140] To substantiate the initial findings and to detect novel loci, several groups performed further studies in expanded samples.[124,129–131,134,135] Efforts were also made to explore confounding factors such as heterogeneity[127] and pleiotropy.[125,128] We outline below some interesting studies classified by the studied phenotypes.

3.1. *BMD and Osteoporotic Fractures*

Styrkarsdottir *et al.*[132] reported a whole-genome linkage scan in 207 Icelandic osteoporotic families (1323 individuals), using phenotypes combining BMD and osteoporotic fractures. The most significant linkage was found on chromosome 20p12.3 (multipoint logarithm of odds [LOD] = 5.10; $p =$ 6.3×10^{-7}). By saturating 30 additional markers on 20p12.3, the region was narrowed down to 6.6 cM. In the follow-up LD mapping and association analyses, they found that three variants in the *BMP2* gene — a missense polymorphism Ser37Ala and two haplotypes (hapB and hapC) — were associated with osteoporosis. Depending on the phenotypes, Ser37Ala yielded relative risks (RRs) in the range of 3.8 to 6.3; Ser37Ala and hapC were associated with low BMD and osteoporotic fractures. They further replicated their associations in a Danish cohort of postmenopausal women. However, in a large cohort of 6353 Dutch Caucasians, both of the Ser37Ala and Arg190Ser polymorphisms or haplotypes exhibited no association with BMD, bone loss, hip structural parameters, or fracture risk.[208] Thus, the importance of *BMP2* variants on BMD variation and fracture risk may vary across different ethnic groups, and their biological effects await to be explored.

In 29 Mexican-American families (664 individuals), Kammerer *et al.*[126] found a QTL affecting forearm (radius midpoint) BMD on chromosome 4p (LOD = 4.33) and suggestive linkage on 12q (LOD = 2.35). They also found suggestive linkage for trochanter BMD on chromosome 6 (LOD = 2.27). In subgroup analysis for men, they obtained linkage for femoral neck BMD on 2p (LOD = 3.98) and trochanter BMD on 13q (LOD = 3.46).

Shen *et al.*[131] performed their second whole-genome scan in a largely expanded sample (79 pedigrees, 1816 subjects). The sample contained $>$ 80 000 relative pairs informative for linkage analysis. The strongest linkage was found on Xq27, with two-point LOD scores of 4.30 for wrist BMD and 2.57 for hip BMD. Linkage was also found on 11q23 for spine BMD (LOD = 3.13), confirming the findings in two earlier independent studies.[141,142]

Another group performed extension studies[124] to confirm their previous finding on chromosome 1q in 464 premenopausal white sister pairs.[140] They reported replication on chromosome 1q in an independent sample of 254 premenopausal white sister pairs. They further fine-mapped (4 cM) the region in all white sister pairs ($n = 938$) and achieved an LOD score of 4.3. In addition, they tested the linkage for hip BMD in 570 white sister pairs and 204 African-American sister pairs[130] to compare the results with their earlier study.[140] Significant linkage was found on chromosomes 14q (LOD = 3.5 for trochanter BMD) and 15q (LOD = 4.3 for femoral BMD). However, their previous highly significant linkage on 11q12–13[91] disappeared in their follow-up efforts,[124,130] implying that underpowered studies may yield unreliable/unstable results. Another follow-up study[143] was conducted to confirm the linkage on chromosome 3p21 for spine BMD.[144] Thirty additional microsatellite markers within the 3p21 region were genotyped for a cohort of extreme discordant and concordant sib pairs (1098 individuals). The maximum evidence of linkage was LOD = 3.6 for age-adjusted spine BMD.

To explore potential pleiotropic effects, Karasik *et al.*[128] performed principal component analyses in 323 pedigrees from the Framingham Osteoporosis Study. For PC1, loci of suggestive linkage were identified on chromosomes 1q21.3 and 8q24.3 with LOD scores of 2.5 and 2.4, respectively. For PC2, the multipoint LOD score was 2.1 on 1p36. Suggestive linkage of PC_hip was found on 8q24.3 and 16p13.2 (LOD > 1.9). The study suggested that QTLs underlying bone mass variation are likely to have pleiotropic effects at different skeletal sites.

To estimate the potential heterogeneity of their earlier linkage findings,[138] Karasik *et al.*[127] performed analyses in subsamples stratified by sex, age, and body mass index (BMI). Heterogeneity was found on chromosomes 6p21.2 and 21qter, where linkage findings in the total sample were not supported in the subsample analyses. However, subsample-specific maxima were found on 4q34.1 (males), 9q22–9q31 (younger), 16p13.2 (high BMI), and 17p13.3 (older), which were not reflected by the total sample results. Their results suggested that the genetic effects on bone mass variation could be different between men and women, young and old, and lean and obese adults.

A genomewide scan involving 3691 individuals from 715 families provided further evidence for gender-specific, skeletal site–specific, and age-specific QTLs underlying BMD variation.[209] In this study, no regions of suggestive or significant linkage were identified when all of the subjects were analyzed together; whereas several potential QTLs, such as on 10q21, 18p11, and 20q13, were detected in various subgroups, which were stratified by age and gender.[209] A concern with this kind of study is that subgroup analyses may reduce the statistical power (by reducing sample sizes), although stratified samples could be more homogeneous.

3.2. Bone Structure or Bone Size

Xu *et al.*[135] saturated several previously identified regions by genotyping denser markers (∼5 cM apart) in an expanded sample of 79 pedigrees. Significant linkage was achieved for wrist bone size on 17q22 (two-point LOD = 2.27, P = 0.0006; multipoint LOD = 1.78, P = 0.002). The chromosomal region 17q22 contains *COL1A1*, a strong candidate gene associated with osteoporotic fractures.

Another extension study was performed for femoral structures in 437 white and 201 black healthy premenopausal sister pairs,[129] of which 191 white pairs overlapped with their previous sample.[139] Linkage was replicated on chromosomes 3 (LOD = 5.0 for femoral head width; LOD = 3.6 for femur shaft width), 7 (LOD = 5.0 for femoral head width), and 19 (LOD = 3.2 for femoral neck axis length) in the white sister pairs, with a new locus identified on chromosome 8 (LOD = 6.0 for femoral head width). To identify sex-specific loci, they conducted a genome scan in 257 Caucasian brother pairs (aged 18–61) to compare with their sister pairs.[145] The significant linkage was for pelvic axis length (LOD = 4.1) on chromosome 4p. There was no overlap between the linkage regions identified in men and in women, suggesting possible genetic heterogeneity for bone structure. A potential problem of the study is the small sample (257 brother pairs), which may affect the validity of the results.[16]

3.3. *Quantitative Ultrasound (QUS)*

Wilson *et al.*[133] performed a genome scan in a cohort of dizygous twin pairs to identify the QTL for bone QUS. Two specific indices, broadband ultrasound attenuation (BUA) and velocity of sound (VOS), were used as phenotypes. Linkage was found for BUA (LOD = 2.1–5.1) on 2q33–37 and for VOS (LOD = 2.2–3.4) on 4q12–21. In the Fels Longitudinal Study (453 subjects),[146] linkage was found for calcaneal QUS on chromosome 4p15 near the marker D4S419 (LOD = 2.12), a region previously linked to forearm BMD.

3.4. *Summary*

Like many other common complex diseases, the replication of linkage results for osteoporosis and related traits has been difficult. Reasons could be multiple, including complex etiology of osteoporosis (e.g. genetic heterogeneity, incomplete penetrance, epistasis, variable expressivity and pleiotropy) and poor study design.[16] However, with the increasing number of studies, the pattern of linkage that is beginning to emerge is encouraging. Not only are there strong linkage signals, but some of these are in genomic regions that harbor prominent positional candidate genes. A more encouraging fact is that we are already beginning to observe replications across studies.[124,131,135] However, caution should be taken in interpreting the replication/confirmation results. Actually, the probability that an observed linkage is true depends on the power of both initial study and replication study.[16] A high LOD score does not always imply a true linkage, especially when studies are underpowered.

So far, more than 60 QTLs have been identified from ~20 genome linkage scans. These QTLs were found on every human chromosome, except chromosome Y. The regions harboring the largest number of replications are 1p36, 1q21–24, 4q31–34, and 12q23–24 in five different studies; and 13q31–34 and 17p11–13 in at least three different studies. As the number of linkage studies continues to grow, there is strong reason to anticipate that additional replications will be brought to light. Of course, some genomic regions will eventually be proven to be false-positive.

4. QTL MAPPING IN ANIMALS

QTL mapping in animals, especially in mice, provides a powerful tool to identify human disease genes. It has greatly facilitated this daunting task facing human researchers because of the high homology (~75%) between the human and mouse genomes. In the bone field, some pilot QTL mapping studies in mice include (1) the SAMP6 model of osteopenia by Shimizu *et al.*[147] and Benes *et al.*,[148] which uncovered QTLs on chromosomes 2, 7, 11, 13, and 16 for low bone mass; (2) crosses between C57BL/6J (B6) and DBA/2J (D2) inbred strains by Klein *et al.*,[142,149] which located whole-body BMD QTLs on chromosomes 1, 2, 4, and 11; and (3) crosses between B6 and *castaneus* (CAST) or C3H/HeJ (C3H) by Beamer *et al.*,[150,151] which located femoral BMD QTLs on seven chromosomes.

During 2003–2004, 10 studies were published reporting new QTLs for BMD, bone structure and strength, bone mechanical properties, and biochemical markers. Klein *et al.*[152] reported their interesting findings by following a region on mouse chromosome 11 that strongly influenced peak BMD in their earlier investigation.[149] They generated a DGA/2 (D2) background congenic mouse with an 82-megabase (Mb) region of chromosome 11 replaced by the corresponding region of the C57BL/6 (B6) genome. The congenic mice had increased peak BMD and improved measures of femoral shaft strength (failure load and stiffness) relative to heterozygous or D2 littermates. Linkage analysis of the congenic B6D2F2 population narrowed the BMD QTL to a 31-Mb region.

Klein *et al.*[152] next analyzed gene expression in B6 and D2 mice. Microarray analysis of kidney tissue showed that *Alox15*, which is located in the middle of the identified QTL interval, was the only differentially expressed gene within this chromosomal region. Their studies on *Alox15* knockout mice confirmed the role of 12/15-lipoxygenase (12/15-LO) in skeletal development. Pharmacological inhibitors of this enzyme improved bone density and strength in two rodent models of osteoporosis. The *Alox15* gene encodes 12/15-LO, an enzyme that converts arachidonic and linoleic acids into endogenous ligands for PPAR-γ. Their *in vitro* observations suggested that genetically determined, constitutively high 12/15-LO expression might limit peak bone mass attainment by suppressing osteogenesis through activation of PPAR-γ–dependent pathways. A concern with the study is that gene expression analysis using kidney tissue revealed a gene relevant to bone metabolism.

Recently, Ichikawa *et al.*[210] performed a population- and family-based association study for two arachidonate lipoxygenase genes — *Alox15*, which is the human homology of mouse *Alox15*; and *Alox12*, which is functionally similar to *Alox 15* — in 411 men and 1291 premenopausal women. Six SNPs in the human *Alox12* gene and the haplotype thereof showed significant association with spine BMD in both men and women, whereas no association was observed in the human *Alox 15* gene.[210]

A previous interval mapping implicated 12 distinct QTLs for peak femoral BMD in (B6 × C3H)F2 and (B6 × CAST)F2 4-month-old female progeny.[150,151] To test the effect of each QTL, Shultz *et al.*[153] selected two sets of loci (six each from C3H and CAST) to make congenic strains by repeated backcrossing of donor mice carrying a given QTL-containing chromosomal region to recipient mice of the B6 progenitor strain. In addition, they selected the femoral BMD QTL region on chromosome 1 of C3H for congenic subline development to facilitate fine mapping of this locus. In 11 of 12 congenic strains, 6 B6.C3H and 5 B6.CAST, femoral BMD in mice carrying *C3H* or *CAST* alleles in the QTL regions was significantly different from that of littermates carrying *B6* alleles. Analyses of eight sublines derived from the B6.C3H-1T congenic region revealed two QTLs. Their results indicated that many QTLs identified in the F2 analyses exert independent effects when transferred and expressed in a common genetic background, and that decomposition of QTL regions by congenic sublines can reveal additional loci for phenotypes assigned to a QTL and can markedly refine QTLs. Using congenic mice, Turner *et al.*[154] revealed sex-specific genetic regulation of femoral structure.

An earlier study identified QTLs on mouse chromosomes 1, 4, 6, 13, and 18 for femoral BMD in (B6×C3H) F2 mice.[151] In a subsequent study,[155] the

999 F2 mouse progeny were phenotyped for measures of femoral biomechanics, structure, and more refined femoral midshaft bone density measures. Two novel multivariate phenotypes were derived using principal component analysis. Results of genomewide analyses provided strong evidence of pleiotropic effects on chromosome 4. Chromosomes 1, 8, 13, and 14 were found to harbor QTLs affecting phenotypes in two of the three aspects of bone properties. Principal component analysis identified pleiotropic QTLs on chromosomes 4 and 14, influencing nearly all of the bone phenotypes.

BMD is not the only risk factor for osteoporosis. Other intermediate phenotypes (e.g. bone structure or size, mechanical properties, biochemical markers) may offer additional mechanistic insights into the overall processes of peak bone mass acquisition and the determinants of bone strength. Volkman et al.[156] investigated the genetic determinants of geometric properties of cortical bone (measured by micro-CT) using a mouse population containing 487 female UM-HET3 mice derived as the progeny of (BALB/cJ × C57BL/6J) F1 females and (C3H/HeJ × DBA/2J) F1 males. Fourteen markers were associated with one or more geometric traits. Since bone strength depends on the geometric as well as material properties of the bone, they further exploited QTLs that may affect mechanical and material properties of cortical bone.[157] Femurs from 18-month-old mice were tested to failure in four-point bending in order to assess the mechanical properties of cortical bone. They found QTLs on maternal chromosomes 11 and 13 and on paternal chromosomes 2, 4, 7, 10, 11, and 17.

To exploit genetic determinants of vertebral trabecular bone traits, Bouxsein et al.[158] evaluated the fifth lumbar vertebra in 914 adult female mice from the F2 intercross of B6 and C3H progenitor strains. They found a pattern of genetic regulation derived from 13 autosomes, with 5–13 QTLs associated with each of the traits. Using 633 MRL/SJL F2 mice, Masinde et al.[159] identified nine QTLs underlying the periosteal circumference (PC), which accounted for 38.6% of phenotype variance. In addition, four epistatic interactions were found that accounted for 37.6% of phenotype variance. In another study, using a B6.C3H-4T (4T) congenic mouse strain, which is genetically 98.4% B6 and carries the C3H chromosome 4 QTL genomic DNA, Robling et al.[160] found evidence for a skeletal mechanosensitivity gene on mouse chromosome 4. Finally, two studies examined the genetic component for two biochemical markers: Srivastava et al.[161] identified three major QTLs (on chromosomes 2, 6, and 14) that influence blood levels of alkaline phosphatase (ALP) in 518 F2 female mice of MRL/MpJ and SJL progenitor strains; and using 633 F2 female mice (MRL/MpJ × SJL),

Mohan *et al.*[162] revealed six QTLs on chromosomes 1, 9, 10, and 11 for serum insulin-like growth factor–binding protein 5 levels.

4.1. *Summary*

QTL mapping in polygenic mice models continues to yield interesting findings. Among these is the discovery of *Alox15* as a negative regulator of peak bone density in mice,[152] which suggests that traditional gene mapping methods combined with functional studies may provide a novel understanding of bone mass regulation. A remarkable investment of effort over the past years has been on congenic lines, which can be created by repeated backcrossing into one of the parental strains, and the recombinants thereof. Congenic lines prove to be a useful tool to confirm the QTL existence and fine map the QTL, and to test the quantitative effect of individual QTLs.[153]

5. FUNCTIONAL GENOMIC STUDIES

Genetic epidemiology studies have provided valuable data for osteoporosis research. However, by exploring the relationship between genes and osteoporosis or related traits at the DNA level, genetic epidemiology studies cannot tell how the genes contribute to the disease. Functional genomic studies, by looking for such a relationship at the mRNA and protein levels, may help elucidate the intermediate biochemical processes of a disease. The goal of functional genomics is not simply to provide a catalog of all the genes and information about their functions, but to understand how the components work together to comprise functioning cells and organisms.[163] Moreover, functional genomic studies may infer novel candidate genes and relevant pathways for DNA-based studies, and confirm the results of DNA-based studies by providing functional evidence. DNA microarray and proteomic technologies — which systemically and quantitatively profile the mRNA and protein expression underlying the functions of a cell type, tissue, or organism at the genome level — are important aspects of functional genomics.

5.1. *DNA Microarray Studies*

Microarray technology allows simultaneous monitoring of gene expression for tens of thousands of genes. In the bone field, DNA microarray was first applied in 1997, when Heller and coworkers[164] discovered genes involved in rheumatoid arthritis. Since then, microarrays have been applied to study various aspects of osteogenesis and gene expression profiles in osteoporosis tissues.

Several studies have exploited the orchestrated gene expression during the process of osteoblast differentiation,[165–171] providing novel insights into the mechanisms of bone formation and mineralization. For example, microarrays were used to determine genes regulated by different factors, such as *BMP2* during the differentiation of human marrow stromal cells[172] and *Tbx2* in the mouse NIH3T3 cell line.[173] Changes in the expression of many genes were reported to occur during differentiation of the mouse calvarial-derived MC3T3-E1 cell line to an osteoblast-like phenotype.[174] Studies have been done to discover genes regulated in bone-related diseases: rheumatoid arthritis,[175–177] ossification of the posterior longitudinal ligament,[178] and osteosarcoma.[179] Human dental pulp stem cells and bone marrow stromal stem cells were found to have a similar level of gene expression for more than 4000 known human genes, while only a few genes were expressed differentially.[180]

The microarray technique proves to be a powerful tool to find genes for diseases of interest. For example, based on the C2C12-generated expression data set and a training set containing known members of the osteogenic, myoblastic, and adipocytic pathways, 176 new genes were found as relevant to osteogenesis.[181] Gene array analysis of osteoblast differentiation in the mouse calvarial-derived MC3T3-E1 cell line revealed changes that were not anticipated.[174] In a comparative study of gene expression between a congenic strain containing a QTL of high BMD and its wild-type strain in mice, about 40% out of the 8734 studied genes and ESTs were not documented previously.[182] Gene expression profiles during the mineralization process in bone marrow–derived human mesenchymal stem cells disclosed transcriptional stimulation of 55 genes and repression of 82 genes among more than 20 000 examined.[183] Among over 8700 genes studied in the mouse MC3T3-E1 cell line during osteoblast differentiation, 252 were found to be differentially expressed during the proliferation and mineralization phases.[184]

So far, most high-throughput gene expression studies have used cultured cell lines of humans, mice, or rats. Using cell cultures has certain advantages, such as virtually no limit to tissue supply, high purity and homogeneity of desired cell populations, possibility to easily modify experimental conditions, etc. However, it may pose potential problems to the results obtained. Even primary cultures of stromal cells from human bone marrow gradually lose their osteogenic potential,[185,186] suggesting changes in gene expression profiles. Moreover, cultured tissues lack relationships with other tissues. For complex disease research, this is a significant short-coming because it eliminates the influence of genes and/or other factors that are not expressed in the diseased tissues, but nevertheless contribute to their development. This may introduce a significant bias into gene expression profiles of cultured cells as compared to freshly isolated ones. Given that, studying fresh bone or bone-related tissues may be a promising way to obtain closer-to-reality data on expression profiles for genes that are tissue-specific and differentially expressed locally.

Bone marrow is heterogeneous in terms of cell composition, and consists of many cell types that have the potential to differentiate into various cell lineages. Some cell types of interest, such as bone marrow mesenchymal stem cells, which can differentiate into osteoblastic lineage, comprise only a minor portion of the whole bone marrow cell population. Although bone marrow cell lineages may be ideal for microarray studies, there are currently technical difficulties related to the isolation of a sufficient number of intact cells and, subsequently, RNA for the analysis. Studying gene expression profiles in monocyte/macrophage lineage cells may be of interest because these cells are early progenitors of osteoclasts and produce cytokines important to bone metabolism,[187] and because it is practical to isolate sufficient RNA from circulating monocytes for microarray analysis.

Liu *et al.*[188] reported an *in vivo* study in humans using circulating monocytes. They performed comparative gene expression studies for circulating monocytes in 10 subjects with high *BMD* vs. 9 with low BMD,[188] and identified 66 differentially expressed genes. After real-time reverse transcription–polymerase chain reaction (RT-PCR) validation, they found three most interesting genes — *CCR3* (chemokine receptor 3), *HDC* (histidine decarboxylase), and *GCR* (glucocorticoid receptor) — to be upregulated in subjects with lower BMD. The results suggested a novel pathophysiology mechanism for osteoporosis that is characterized by increased bone monocyte recruitment, increased monocyte differentiation into osteoclasts, and increased osteoclast stimulation via monocyte functional changes. The study provided helpful information for future research using fresh tissue samples such as bone marrow cells that include precursors for osteoblast lineages.

Recent developments in cell and molecular biology make it possible to obtain a genomewide expression profile of a single cell.[189] Single-cell gene expression analysis is currently used to generate data within the fundamental unit, the single cell, thereby freeing the analysis from assumptions or questions regarding cell population homogeneity. The data obtained with this approach suggest that even morphologically identical cells may have quite distinct transcriptomes, which give deeper insight into cell differentiation and functional assignment.[190,191] Given that the processes of osteogenesis and bone metabolism are multistage and involve many cell types (osteogenic precursors, various bone cellular components, T- and B-cells, etc.), including those which are able to transdifferentiate, this approach may be useful for fine characterization of the distinct bone-related cell types. Single-cell expression profiling may also offer a highly parallel view of the workings of a gene regulatory network at one specific point in time, and will hopefully provide insights that could lead to an improved ability to interpret gene expression patterns. However, its direct relevance and significance in relation to disease gene identification (e.g. osteoporosis) remain to be determined.

Advances in functional genomics and computer sciences make it possible to analyze gene expression changes in the context of known biological pathways.[166,167,169] However, many challenges remain in the use of the functional genomic approach for complex bone disorders. Questions to be answered include how various environmental, lifestyle, or inherent regulating factors contribute to the BMD variation, onset and development of osteoporosis, and risk of osteoporotic fractures. Another question is the identification of relationships and interactions between genes which have been reported to be either associated with or linked to disorders. Do their expression levels correlate with the degree of association or linkage? Do these genes interact in developing the trait?

5.2. *Proteomics Studies*

Due to complicated processes such as alternative mRNA splicing and post-translational modification, the correlation between mRNA and protein expression level[166,167,192–195] could be low. There are $\sim 30\,000$ genes in the human genome, but the estimated number of proteins in human cells is between $\sim 300\,000$ and $\sim 1\,000\,000$.[196] Thus, the proteomics approach, as a complement to DNA microarray, is an indispensable component of functional genomics.

Proteomics can be classified into three major types: expression proteomics, functional proteomics, and structural proteomics.[197] Expression proteomics quantitatively analyzes and identifies differentially expressed proteins from protein profiles between case and control groups. These differentially expressed proteins provide biomarkers or important proteins in pathways underlying different biomedical conditions. Functional proteomics involves the global understanding of protein–protein interactions. Since key proteins involved in disease development generally interact with other proteins, functional proteomics is a favorable approach for unveiling whole pathways participating in the etiology of diseases. To better understand and even predict protein function, one should determine the 3D structures of the proteome. Structural proteomics may prospectively fulfill this goal by mapping out the structures of protein complexes or proteins in a specific cellular organelle.[198]

The application of proteomics in the bone field has a relatively short history, and more results from their use are yet to come. Applying an expression proteomics approach on cultured cells, pilot studies identified some novel proteins important for the development of bone marrow hematopoietic cells,[199] mesenchymal chondroblasts,[200] and osteoblasts.[201] Combining two-dimensional gel electrophoresis (2DE) and isotope-coded affinity tag (ICAT) techniques, a study suggested novel proteins related to osteoclast differentiation.[202] A proteomics approach was used in seeking inhibitors of osteoclast-mediated bone resorption, and is currently screening for bone anabolic agents.[203] Some *in vivo* studies were performed to characterize global-scale molecular profiling of a variety of bone-related diseases.[204] For example, a study demonstrated a comparative molecular characterization at the transcriptome (microarray with 12 526 gene specificities) and proteome levels (multi-Western blot PowerBlot with 791 antibodies) of synovial tissue from rheumatoid arthritis (RA) compared to osteoarthritis (OA) patients.[204] Several new candidate molecules, e.g. cathepsin D and Stat1, displayed reproducible differences of expression in RA vs. OA patients.

Proteomics represents one of the most promising fields poised to boost our understanding of systems-level cellular behavior and the fundamental etiology of osteoporosis that can be related to specific genes. It is anticipated that a huge amount of data will be produced for years to come, and the wealth of information will undoubtedly benefit osteoporosis research communities.

6. SUMMARY

Remarkable progress has been made in revealing the molecular genetic basis of osteoporosis. The number of QTLs, genes, and other markers linked and/or associated with osteoporosis-related traits continues to expand, and becomes significantly more detailed and complex. There are now several promising chromosomal regions and candidate genes that are supported by multiple studies. However, the majority of findings are still inconclusive pending further investigation, thus calling for new approaches and strategies with both sensitivity and robustness to accommodate confounding effects from various sources. With the rapid development in human and model organism genome sequences, as well as the progress in molecular technologies, analytical tools, bioinformatics, and functional genomics, one can expect that it will be possible to define genes and mutations and their functions in the predisposition or resistance to osteoporosis.

ACKNOWLEDGMENTS

H.-W. Deng was partially supported by the NIH and the State of Nebraska (LB595). The study was also benefited by grants from CNSF, Huo Ying Dong Education Foundation, the Ministry of Education of China, and Xi'an Jiaotong University.

REFERENCES

1. Liu YZ, Liu YJ, Recker RR, Deng HW. (2003) Molecular studies of identification of genes for osteoporosis: the 2002 update. *J Endocrinol* **177**:147–196.
2. Blank RD. (2001) Breaking down bone strength: a perspective on the future of skeletal genetics. *J Bone Miner Res* **16**:1207–1211.
3. Nguyen TV, Eisman JA. (2000) Genetics of fracture: challenges and opportunities. *J Bone Miner Res* **15**:1253–1256.
4. Nguyen TV, Blangero J, Eisman JA. (2000) Genetic epidemiological approaches to the search for osteoporosis genes. *J Bone Miner Res* **15**:392–401.
5. Audi L, Garcia-Ramirez M, Carrascosa A. (1999) Genetic determinants of bone mass. *Horm Res* **51**:105–123.
6. Baldock PA, Eisman JA. (2004) Genetic determinants of bone mass. *Curr Opin Rheumatol* **16**:450–456.
7. Eisman JA. (1999) Genetics of osteoporosis. *Endocr Rev* **20**:788–804.
8. Huang QY, Recker RR, Deng HW. (2003) Searching for osteoporosis genes in the post-genome era: progress and challenges. *Osteoporos Int* **14**:701–715.
9. Peacock M, Turner CH, Econs MJ, Foroud T. (2002) Genetics of osteoporosis. *Endocr Rev* **23**:303–326.
10. Ralston SH. (2003) Genetic determinants of susceptibility to osteoporosis. *Curr Opin Pharmacol* **3**:286–290.
11. Rizzoli R, Bonjour JP, Ferrari SL. (2001) Osteoporosis, genetics and hormones. *J Mol Endocrinol* **26**:79–94.
12. Stewart TL, Ralston SH. (2000) Role of genetic factors in the pathogenesis of osteoporosis. *J Endocrinol* **166**:235–245.

13. Zmuda JM, Cauley JA, Ferrell RE. (1999) Recent progress in understanding the genetic susceptibility to osteoporosis. *Genet Epidemiol* **16**:356–367.

14. Shen H, Recker RR, Deng HW. (2003) Molecular and genetic mechanisms of osteoporosis: implication for treatment. *Curr Mol Med* **3**:737–757.

15. Dvornyk V, Xiao P, Liu YJ, *et al.* (2004) Systematic approach to the study of complex bone disorders at the whole-genome level. *Curr Genomics* **5**:93–108.

16. Shen H, Liu YJ, Liu PY, *et al.* (2005) Nonreplication in genetic studies of complex diseases — lessons learned from studies of osteoporosis and tentative remedies. *J Bone Miner Res* **20**:365–376.

17. Gorai I, Inada M, Morinaga H, *et al.* (2004) Cytochrome P450c17∞ (CYP17) gene polymorphism indirectly influences on bone density through their effects on endogenous androgen in postmenopausal Japanese women — are the effects of age and body mass index greater than those of endogenous sex steroids? *J Bone Miner Res* **19**:S382.

18. Napoli N, Mumm S, Sheik S, *et al.* (2004) Effect of CYP450 gene polymorphisms on estrogen metabolism and bone density. *J Bone Miner Res* **19**:S384.

19. Ezura Y, Nakajima T, Kajita M, *et al.* (2003) Association of molecular variants, haplotypes, and linkage disequilibrium within the human vitamin D–binding protein (DBP) gene with postmenopausal bone mineral density. *J Bone Miner Res* **18**:1642–1649.

20. Dennison EM, Syddall HE, Rodriguez S, *et al.* (2004) Polymorphism in the growth hormone gene, weight in infancy, and adult bone mass. *J Clin Endocrinol Metab* **89**:4898–4903.

21. Iwasaki H, Emi M, Ezura Y, *et al.* (2003) Association of a Trp16Ser variation in the gonadotropin releasing hormone signal peptide with bone mineral density, revealed by SNP-dependent PCR typing. *Bone* **32**:185–190.

22. Langdahl BL, Husted LB, Stenkjaer L, Carstens M. (2004) Polymorphisms in the *IGF-II* gene are associated with body weight and bone mass. *J Bone Miner Res* **19**:S385.

23. Koh JM, Kim DJ, Hong JS, *et al.* (2002) Estrogen receptor alpha gene polymorphisms *Pvu* II and *Xba* I influence association between leptin receptor gene polymorphism (Gln223Arg) and bone mineral density in young men. *Eur J Endocrinol* **147**:777–783.

24. Boot AM, Wilson SG, Dick IM, *et al.* (2004) *LRP5* gene polymorphisms predict bone mass and incident fractures in elderly Australian women. *J Bone Miner Res* **19**:S383.

25. Yamada Y, Ando F, Niino N, Shimokata H. (2003) Association of polymorphisms of interleukin-6, osteocalcin, and vitamin D receptor genes, alone or in combination, with bone mineral density in community-dwelling Japanese women and men. *J Clin Endocrinol Metab* **88**:3372–3378.

26. Yamada Y, Ando F, Niino N, Shimokata H. (2002) Association of a polymorphism of the CC chemokine receptor-2 gene with bone mineral density. *Genomics* **80**:8–12.

27. Kornak U, Branger S, Ostertag A, de Vernejoul MC. (2004) A VNTR in the *CLCN7* gene influences bone density in patients with autosomal dominant osteopetrosis (ADO) type II and in post-menopausal women. *J Bone Miner Res* **19**:S387.

28. Eriksson A, Lorentzon M, Anderson N, *et al.* (2004) The catechol-O-methyltransferase val158met polymorphism is associated with bone mineral density in young adult men. *J Bone Miner Res* **19**:S131.

29. Giraudeau FS, McGinnis RE, Gray IC, *et al.* (2004) Characterization of common genetic variants in cathepsin K and testing for association with bone mineral density in a large cohort of perimenopausal women from Scotland. *J Bone Miner Res* **19**:31–41.

30. Yamada Y, Ando F, Niino N, Shimokata H. (2003) Association of a polymorphism of the dopamine receptor D4 gene with bone mineral density in Japanese men. *J Hum Genet* **48**:629–633.

31. Ishida R, Ezura Y, Emi M, *et al.* (2003) Association of a promoter haplotype $(-1542G/-525C)$ in the tumor necrosis factor receptor associated factor–interacting protein gene with low bone mineral density in Japanese women. *Bone* **33**:237–241.

32. Enattah N, Valimaki VV, Valimaki MJ, *et al.* (2004) Molecularly defined lactose malabsorption, peak bone mass and bone turnover rate in young Finnish men. *Calcif Tissue Int* **75**:488–493.

33. Joseph C, Prestwood KM, Burleson JA, *et al.* (2004) Macrophage migration inhibitory factor (MIF) gene promoter region polymorphism and femoral neck bone density. *J Bone Miner Res* **19**:S130–S131.

34. Yamada Y, Ando F, Niino N, Shimokata H. (2002) Association of a polymorphism of the matrix metalloproteinase-1 gene with bone mineral density. *Matrix Biol* **21**:389–392.

35. Yamada Y, Ando F, Niino N, Shimokata H. (2004) Association of a polymorphism of the matrix metalloproteinase-9 gene with bone mineral density in Japanese men. *Metabolism* **53**:135–137.

36. Zmuda JM, Cauley JA, Ferrel RE. (2004) Nuclear receptor coactivator-3 (NCO A3/AIB1/SRC3) alleles are a strong correlate of bioavailable testosterone and vertebral bone mass in older men. *J Bone Miner Res* **19**:S387.

37. Heikkinen AM, Niskanen LK, Salmi JA, *et al.* (2004) Leucine7 to proline7 polymorphism in prepro-NPY gene and femoral neck bone mineral density in postmenopausal women. *Bone* **35**:589–594.

38. Koh J, Park E, Kim G, *et al.* (2004) Polymorphisms in the osteoclast-associated receptor gene are associated with risk of osteopenia and osteoporosis. *J Bone Miner Res* **19**:S384.

39. Spotila LD, Rodriguez H, Koch M, *et al.* (2003) Association analysis of bone mineral density and single nucleotide polymorphisms in two candidate genes on chromosome 1p36. *Calcif Tissue Int* **73**:140–146.

40. Yamada Y, Ando F, Niino N, *et al.* (2003) Association of polymorphisms of paraoxonase 1 and 2 genes, alone or in combination, with bone mineral density in community-dwelling Japanese. *J Hum Genet* **48**:469–475.

41. Omasu F, Ezura Y, Kajita M, *et al.* (2003) Association of genetic variation of the *RIL* gene, encoding a PDZ-LIM domain protein and localized in 5q31.1, with low bone mineral density in adult Japanese women. *J Hum Genet* **48**:342–345.

42. Haney EM, Marshall L, Lambert L, *et al.* (2004) An intron 2 polymorphism at the serotonin transporter is associated with reduced hip bone mineral density. *J Bone Miner Res* **19**:S131.

43. Balemans W, Foernzler D, Parsons C, *et al.* (2002) Lack of association between the *SOST* gene and bone mineral density in perimenopausal women: analysis of five polymorphisms. *Bone* **31**:515–519.

44. Sobacchi C, Vezzoni P, Reid DM, *et al.* (2004) Association between a polymorphism affecting an AP1 binding site in the promoter of the *TCIRG1* gene and bone mass in women. *Calcif Tissue Int* **74**:35–41.

45. Munafo MR, Flint J. (2004) Meta-analysis of genetic association studies. *Trends Genet* **20**:439–444.

46. Arai H, Miyamoto KI, Yoshida M, *et al.* (2001) The polymorphism in the caudal-related homeodomain protein Cdx-2 binding element in the human vitamin D receptor gene. *J Bone Miner Res* **16**:1256–1264.

47. Morita A, Iki M, Dohi Y, *et al.* (2004) Prediction of bone mineral density from vitamin D receptor polymorphisms is uncertain in representative samples of Japanese women. The Japanese Population-based Osteoporosis (JPOS) Study. *Int J Epidemiol* **33**:979–988.

48. Fang Y, van Meurs JB, Bergink AP, *et al.* (2003) Cdx-2 polymorphism in the promoter region of the human vitamin D receptor gene determines susceptibility to fracture in the elderly. *J Bone Miner Res* **18**:1632–1641.

49. Thakkinstian A, D'Este C, Eisman J, *et al.* (2004) Meta-analysis of molecular association studies: vitamin D receptor gene polymorphisms and BMD as a case study. *J Bone Miner Res* **19**:419–428.

50. Thakkinstian A, D'Este C, Attia J. (2004) Haplotype analysis of *VDR* gene polymorphisms: a meta-analysis. *Osteoporos Int* **15**:729–734.

51. Morrison N. (2004) Commentary: vitamin D receptor polymorphism and bone mineral density: effect size in Caucasians means detection is uncertain in small studies. *Int J Epidemiol* **33**:989–994.

52. Fang Y, van Meurs J, Zhao HY, *et al.* (2004) Intragenic interaction of *VDR* haplotype alleles determines fracture risk. *J Bone Miner Res* **19**:S385.

53. Nejentsev S, Godfrey L, Snook H, *et al.* (2004) Comparative high-resolution analysis of linkage disequilibrium and tag single nucleotide polymorphisms between populations in the vitamin D receptor gene. *Hum Mol Genet* **13**:1633–1639.

54. van Meurs J, Zillikens C, Fang Y, *et al.* (2004) A DNA variant correlated to lactose intolerance interacts with *VDR* gene variants to determine height, bone geometry and BMD. *J Bone Miner Res* **19**:S385.

55. Yamada Y, Ando F, Niino N, *et al.* (2002) Association of polymorphisms of the estrogen receptor alpha gene with bone mineral density of the femoral neck in elderly Japanese women. *J Mol Med* **80**:452–460.

56. Zhao LJ, Liu PY, Long JR, *et al.* (2004) Test of linkage and/or association between the estrogen receptor alpha gene with bone mineral density in Caucasian nuclear families. *Bone* **35**:395–402.

57. van Meurs JB, Schuit SC, Weel AE, *et al.* (2003) Association of 5′ estrogen receptor alpha gene polymorphisms with bone mineral density, vertebral bone area and fracture risk. *Hum Mol Genet* **12**:1745–1754.

58. Ioannidis J, Ralston S, Bennett S, *et al.* (2004) Large-scale evidence for differential genetic effects of *ESR1* gene polymorphisms on osteoporosis outcomes: the GENOMOS study. *J Bone Miner Res* **292**:2105–2114.

59. Lohmueller KE, Pearce CL, Pike M, *et al.* (2003) Meta-analysis of genetic association studies supports a contribution of common variants to susceptibility to common disease. *Nat Genet* **33**:177–182.

60. Gerdhem P, Brandstrom H, Stiger F, *et al.* (2004) Association of the collagen type 1 (*COL1A 1*) Sp1 binding site polymorphism to femoral neck bone mineral density and wrist fracture in 1044 elderly Swedish women. *Calcif Tissue Int* **74**:264–269.

61. Zhang YY, Lei SF, Mo XY, *et al.* (2005) The −1997 G/T polymorphism in the *COL1A1* upstream regulatory region is associated with hip bone mineral density (BMD) in Chinese nuclear families. *Calcif Tissue Int* **76**:107–112.

62. Long JR, Liu PY, Lu Y, *et al.* (2004) Association between *COL1A1* gene polymorphisms and bone size in Caucasians. *Eur J Hum Genet* **12**:383–388.

63. Mann V, Ralston SH. (2003) Meta-analysis of *COL1A1* Sp1 polymorphism in relation to bone mineral density and osteoporotic fracture. *Bone* **32**:711–717.

64. Mann V, Hobson EE, Li B, *et al.* (2001) A *COL1A1* Sp1 binding site polymorphism predisposes to osteoporotic fracture by affecting bone density and quality. *J Clin Invest* **107**:899–907.

65. The International HapMap Consortium. (2003) The International HapMap Project. *Nature* **426**:789–796.

66. Passaro A, Vanini A, Calzoni F, *et al.* (2001) Plasma homocysteine, methylenetetrahydrofolate reductase mutation and carotid damage in elderly healthy women. *Atherosclerosis* **157**:175–180.

67. Hak AE, Polderman KH, Westendorp IC, *et al.* (2000) Increased plasma homocysteine after menopause. *Atherosclerosis* **149**:163–168.

68. Goyette P, Sumner JS, Milos R, *et al.* (1994) Human methylenetetrahydrofolate reductase: isolation of cDNA mapping and mutation identification. *Nat Genet* **7**:551.

69. Frosst P, Blom HJ, Milos R, *et al.* (1995) A candidate genetic risk factor for vascular disease: a common mutation in methylenetetrahydrofolate reductase. *Nat Genet* **10**:111–113.

70. Miyao M, Morita H, Hosoi T, *et al.* (2000) Association of methylenetetrahydrofolate reductase (*MTHFR*) polymorphism with bone mineral density in postmenopausal Japanese women. *Calcif Tissue Int* **66**:190–194.

71. Abrahamsen B, Madsen JS, Tofteng CL, *et al.* (2003) A common methylenetetrahydrofolate reductase (C677T) polymorphism is associated with low bone mineral density and increased fracture incidence after menopause: longitudinal data from the Danish Osteoporosis Prevention Study. *J Bone Miner Res* **18**:723–729.

72. Jacques PF, Bostom AG, Williams RR, *et al.* (1996) Relation between folate status, a common mutation in methylenetetrahydrofolate reductase, and plasma homocysteine concentrations. *Circulation* **93**:7–9.

73. McLean RR, Karasik D, Selhub J, *et al.* (2004) Association of a common polymorphism in the methylenetetrahydrofolate reductase (*MTHFR*) gene with bone phenotypes depends on plasma folate status. *J Bone Miner Res* **19**:410–418.

74. Macdonald HM, McGuigan FE, Fraser WD, *et al.* (2004) Methylenetetrahydrofolate reductase polymorphism interacts with riboflavin intake to influence bone mineral density. *Bone* **35**:957–964.

75. Janssen JA, Burger H, Stolk RP, *et al.* (1998) Gender-specific relationship between serum free and total *IGF-I* and bone mineral density in elderly men and women. *Eur J Endocrinol* **138**:627–632.

76. Kurland ES, Rosen CJ, Cosman F, *et al.* (1997) Insulin-like growth factor-I in men with idiopathic osteoporosis. *J Clin Endocrinol Metab* **82**:2799–2805.

77. Rivadeneira F, Houwing-Duistermaat JJ, Vaessen N, *et al.* (2003) Association between an insulin-like growth factor I gene promoter polymorphism and bone mineral density in the elderly: the Rotterdam Study. *J Clin Endocrinol Metab* **88**:3878–3884.

78. Rivadeneira F, Houwing-Duistermaat JJ, Beck TJ, *et al.* (2004) The influence of an insulin-like growth factor I gene promoter polymorphism on hip bone geometry and the risk of nonvertebral fracture in the elderly: the Rotterdam Study. *J Bone Miner Res* **19**:1280–1290.

79. Ishimi Y, Miyaura C, Jin CH, *et al.* (1990) *IL-6* is produced by osteoblasts and induces bone resorption. *J Immunol* **145**:3297–3303.

80. Moffett SP, Zmuda JM, Cauley JA, *et al.* (2004) Association of the G-174C variant in the interleukin-6 promoter region with bone loss and fracture risk in older women. *J Bone Miner Res* **19**:1612–1618.

81. Bollerslev J, Wilson SG, Dick IM, *et al.* (2004) Calcium-sensing receptor gene polymorphism A986S does not predict serum calcium level, bone mineral density, calcaneal ultrasound indices, or fracture rate in a large cohort of elderly women. *Calcif Tissue Int* **74**:12–17.

82. Dick IM, Devine A, Li S, *et al.* (2003) The T869C TGF beta polymorphism is associated with fracture, bone mineral density, and calcaneal quantitative ultrasound in elderly women. *Bone* **33**:335–341.

83. Shearman AM, Karasik D, Gruenthal KM, *et al.* (2004) Estrogen receptor beta polymorphisms are associated with bone mass in women and men: the Framingham Study. *J Bone Miner Res* **19**:773–781.

84. Yamada Y, Ando F, Niino N, Shimokata H. (2003) Association of polymorphisms of the osteoprotegerin gene with bone mineral density in Japanese women but not men. *Mol Genet Metab* **80**:344–349.

85. Albagha OM, Tasker PN, McGuigan FE, *et al.* (2002) Linkage disequilibrium between polymorphisms in the human *TNFRSF1B* gene and their association with bone mass in perimenopausal women. *Hum Mol Genet* **11**:2289–2295.

86. Moffett SP, Zmuda JM, Oakley JI, *et al.* (2004) Tumor necrosis factor alpha polymorphism, bone strength phenotypes, and the risk of fracture in older women. *J Bone Miner Res* **19**:S250.

87. Johnson ML, Harnish K, Nusse R, Van Hul W. (2004) *LRP5* and *Wnt* signaling: a union made for bone. *J Bone Miner Res* **19**:1749–1757.

88. Johnson ML, Gong G, Kimberling W, *et al.* (1997) Linkage of a gene causing high bone mass to human chromosome 11 (11q12–13). *Am J Hum Genet* **60**:1326–1332.

89. Ferrari SL, Deutsch S, Choudhury U, *et al.* (2004) Polymorphisms in the low-density lipoprotein receptor–related protein 5 (*LRP5*) gene are associated with variation in vertebral bone mass, vertebral bone size, and stature in whites. *Am J Hum Genet* **74**:866–875.

90. Pritchard JK, Stephens M, Donnelly P. (2000) Inference of population structure using multilocus genotype data. *Genetics* **155**:945–959.

91. Koller DL, Rodriguez LA, Christian JC, *et al.* (1998) Linkage of a QTL contributing to normal variation in bone mineral density to chromosome 11q12–13. *J Bone Miner Res* **13**:1903–1908.

92. Bollerslev J, Wilson SG, Dick IM, *et al.* (2004) *LRP5* gene polymorphisms predict bone mass and incident fractures in elderly Australian women. *J Bone Miner Res* **19**:S383.

93. Brunkow ME, Gardner JC, Van Ness J, *et al.* (2001) Bone dysplasia sclerosteosis results from loss of the *SOST* gene product, a novel cystine knot–containing protein. *Am J Hum Genet* **68**:577–589.

94. Uitterlinden AG, Arp PP, Paeper BW, *et al.* (2004) Polymorphisms in the sclerosteosis/van Buchem disease gene (*SOST*) region are associated with bone-mineral density in elderly whites. *Am J Hum Genet* **75**:1032–1045.

95. Freimer N, Sabatti C. (2004) The use of pedigree, sib-pair and association studies of common diseases for genetic mapping and epidemiology. *Nat Genet* **36**:1045–1051.

96. Huizinga TW, Pisetsky DS, Kimberly RP. (2004) Associations, populations, and the truth: recommendations for genetic association studies in *Arthritis & Rheumatism. Arthritis Rheum* **50**:2066–2071.

97. Cardon LR, Bell JI. (2001) Association study designs for complex diseases. *Nat Rev Genet* **2**:91–99.

98. Romero R, Kuivaniemi H, Tromp G, Olson J. (2002) The design, execution, and interpretation of genetic association studies to decipher complex diseases. *Am J Obstet Gynecol* **187**:1299–1312.

99. Zondervan KT, Cardon LR. (2004) The complex interplay among factors that influence allelic association. *Nat Rev Genet* **5**:89–100.

100. Ioannidis JP, Trikalinos TA, Ntzani EE, Contopoulos-Ioannidis DG. (2003) Genetic associations in large versus small studies: an empirical assessment. *Lancet* **361**:567–571.

101. Weiss KM, Clark AG. (2002) Linkage disequilibrium and the mapping of complex human traits. *Trends Genet* **18**:19–24.
102. Ardlie KG, Kruglyak L, Seielstad M. (2002) Patterns of linkage disequilibrium in the human genome. *Nat Rev Genet* **3**:299–309.
103. Cardon LR, Abecasis GR. (2003) Using haplotype blocks to map human complex trait loci. *Trends Genet* **19**:135–140.
104. Shifman S, Kuypers J, Kokoris M, *et al.* (2003) Linkage disequilibrium patterns of the human genome across populations. *Hum Mol Genet* **12**:771–776.
105. Neale BM, Sham PC. (2004) The future of association studies: gene-based analysis and replication. *Am J Hum Genet* **75**:353–362.
106. Benjamini Y, Hochberg Y. (1995) Controlling the false discovery rate: a practical and powerful approach to multiple testing. *J R Stat Soc B* **57**:289–300.
107. Doerge RW, Churchill GA. (1996) Permutation tests for multiple loci affecting a quantitative character. *Genetics* **142**:285–294.
108. McIntyre LM, Martin ER, Simonsen KL, Kaplan NL. (2000) Circumventing multiple testing: a multilocus Monte Carlo approach to testing for association. *Genet Epidemiol* **19**:18–29.
109. Deng HW. (2001) Population admixture may appear to mask, change or reverse genetic effects of genes underlying complex traits. *Genetics* **159**: 1319–1323.
110. Thomas DC, Witte JS. (2002) Point: population stratification: a problem for case-control studies of candidate–gene associations? *Cancer Epidemiol Biomarkers Prev* **11**:505–512.
111. Wacholder S, Rothman N, Caporaso N. (2002) Counterpoint: bias from population stratification is not a major threat to the validity of conclusions from epidemiological studies of common polymorphisms and cancer. *Cancer Epidemiol Biomarkers Prev* **11**:513–520.
112. Cardon LR, Palmer LJ. (2003) Population stratification and spurious allelic association. *Lancet* **361**:598–604.
113. Freedman ML, Reich D, Penney KL, *et al.* (2004) Assessing the impact of population stratification on genetic association studies. *Nat Genet* **36**:388–393.
114. Marchini J, Cardon LR, Phillips MS, Donnelly P. (2004) The effects of human population structure on large genetic association studies. *Nat Genet* **36**:512–517.
115. Helgason A, Yngvadottir B, Hrafnkelsson B, *et al.* (2004) An Icelandic example of the impact of population structure on association studies. *Nat Genet* **37**:90–95.
116. Spielman RS, Ewens WJ. (1996) The TDT and other family-based tests for linkage disequilibrium and association. *Am J Hum Genet* **59**:983–989.
117. Spielman RS, McGinnis RE, Ewens WJ. (1993) Transmission test for linkage disequilibrium: the insulin gene region and insulin-dependent diabetes mellitus (IDDM). *Am J Hum Genet* **52**:506–516.
118. Devlin B, Roeder K. (1999) Genomic control for association studies. *Biometrics* **55**:997–1004.

119. Pritchard JK, Rosenberg NA. (1999) Use of unlinked genetic markers to detect population stratification in association studies. *Am J Hum Genet* **65**:220–228.

120. Orwoll ES, Belknap JK, Klein RF, *et al.* (2001) Gender specificity in the genetic determinants of peak bone mass site and gender specificity of inheritance of bone mineral density. *J Bone Miner Res* **16**:1962–1971.

121. Duncan EL, Cardon LR, Sinsheimer JS, *et al.* (2003) Site and gender specificity of inheritance of bone mineral density. *J Bone Miner Res* **18**:1531–1538.

122. Harris M, Nguyen TV, Howard GM, *et al.* (1998) Genetic and environmental correlations between bone formation and bone mineral density: a twin study. *J Bone Miner Res* **22**:141–145.

123. Deng HW, Mahaney MC, Williams JT, *et al.* (2002) Relevance of the genes for bone mass variation to susceptibility to osteoporotic fractures and its implications to gene search for complex human diseases. *Genet Epidemiol* **22**:12–25.

124. Econs MJ, Koller DL, Hui SL, *et al.* (2004) Confirmation of linkage to chromosome 1q for peak vertebral bone mineral density in premenopausal white women. *Am J Hum Genet* **74**:223–228.

125. Huang QY, Xu FH, Shen H, *et al.* (2004) Genome scan for QTLs underlying bone size variation at 10 refined skeletal sites: genetic heterogeneity and the significance of phenotype refinement. *Physiol Genomics* **17**:326–331.

126. Kammerer CM, Schneider JL, Cole SA, *et al.* (2003) Quantitative trait loci on chromosomes 2p, 4p, and 13q influence bone mineral density of the forearm and hip in Mexican Americans. *J Bone Miner Res* **18**:2245–2252.

127. Karasik D, Cupples LA, Hannan MT, Kiel DP. (2003) Age, gender, and body mass effects on quantitative trait loci for bone mineral density: the Framingham Study. *Bone* **33**:308–316.

128. Karasik D, Cupples LA, Hannan MT, Kiel DP. (2004) Genome screen for a combined bone phenotype using principal component analysis: the Framingham Study. *Bone* **34**:547–556.

129. Koller DL, White KE, Liu G, *et al.* (2003) Linkage of structure at the proximal femur to chromosomes 3, 7, 8, and 19. *J Bone Miner Res* **18**:1057–1065.

130. Peacock M, Koller DL, Hui S, *et al.* (2004) Peak bone mineral density at the hip is linked to chromosomes 14q and 15q. *Osteoporos Int* **15**:489–496.

131. Shen H, Zhang YY, Long JR, *et al.* (2004) A genome-wide linkage scan for bone mineral density in an extended sample: evidence for linkage on 11q23 and Xq27. *J Med Genet* **41**:743–751.

132. Styrkarsdottir U, Cazier JB, Kong A, *et al.* (2003) Linkage of osteoporosis to chromosome 20p12 and association to *BMP2*. *PLoS Biol* **1**:E69.

133. Wilson SG, Reed PW, Andrew T, *et al.* (2004) A genome-screen of a large twin cohort reveals linkage for quantitative ultrasound of the calcaneus to 2q33–37 and 4q12–21. *J Bone Miner Res* **19**:270–277.

134. Huang QY, Xu FH, Shen H, *et al.* (2004) A second-stage genome scan for QTLs influencing BMD variation. *Calcif Tissue Int* **75**:138–143.

135. Xu FH, Liu YJ, Deng H, *et al.* (2004) A follow-up linkage study for bone size variation in an extended sample. *Bone* **35**:777–784.
136. Deng HW, Shen H, Xu FH, *et al.* (2003) Several genomic regions potentially containing QTLs for bone size variation were identified in a whole-genome linkage scan. *Am J Med Genet A* **119**:121–131.
137. Deng HW, Xu FH, Huang QY, *et al.* (2002) A whole-genome linkage scan suggests several genomic regions potentially containing quantitative trait loci for osteoporosis. *J Clin Endocrinol Metab* **87**:5151–5159.
138. Karasik D, Myers RH, Cupples LA, *et al.* (2002) Genome screen for quantitative trait loci contributing to normal variation in bone mineral density: the Framingham Study. *J Bone Miner Res* **17**:1718–1727.
139. Koller DL, Liu G, Econs MJ, *et al.* (2001) Genome screen for quantitative trait loci underlying normal variation in femoral structure. *J Bone Miner Res* **16**:985–991.
140. Koller DL, Econs MJ, Morin PA, *et al.* (2000) Genome screen for QTLs contributing to normal variation in bone mineral density and osteoporosis. *J Clin Endocrinol Metab* **85**:3116–3120.
141. Devoto M, Shimoya K, Caminis J, *et al.* (1998) First-stage autosomal genome screen in extended pedigrees suggests genes predisposing to low bone mineral density on chromosomes 1p, 2p and 4q. *Eur J Hum Genet* **6**:151–157.
142. Klein RF, Mitchell SR, Phillips TJ, *et al.* (1998) Quantitative trait loci affecting peak bone mineral density in mice. *J Bone Miner Res* **13**:1648–1656.
143. Wilson SG, Reed PW, Andrew T, *et al.* (2004) Fine mapping provides further evidence of linkage for bone mineral density to 3p21. *J Bone Miner Res* **19**:S154.
144. Wilson SG, Reed PW, Bansal A, *et al.* (2003) Comparison of genome screens for two independent cohorts provides replication of suggestive linkage of bone mineral density to 3p21 and 1p36. *Am J Hum Genet* **72**:144–155.
145. Peacock M, Koller DL, Fishburn T, *et al.* (2005) Sex-specific and non–sex-specific quantitative trait loci contribute to normal variation in bone mineral density in men. *J Clin Endocrinol Metab* **90**:3060–3066.
146. Czerwinski SA, Lee M, Choh AC, *et al.* (2004) Genome-wide scan for QTL underlying normal variation in calcaneal quantitative ultrasound measures: the Fels Longitudinal Study. *J Bone Miner Res* **19**:S157.
147. Shimizu M, Higuchi K, Bennett B, *et al.* (1999) Identification of peak bone mass QTL in a spontaneously osteoporotic mouse strain. *Mamm Genome* **10**:81–87.
148. Benes H, Weinstein RS, Zheng W, *et al.* (2000) Chromosomal mapping of osteopenia-associated quantitative trait loci using closely related mouse strains. *J Bone Miner Res* **15**:626–633.
149. Klein OF, Carlos AS, Vartanian KA, *et al.* (2001) Confirmation and fine mapping of chromosomal regions influencing peak bone mass in mice. *J Bone Miner Res* **16**:1953–1961.

150. Beamer WG, Shultz KL, Churchill GA, *et al.* (1999) Quantitative trait loci for bone density in C57BL/6J and CAST/EiJ inbred mice. *Mamm Genome* **10**:1043–1049.
151. Beamer WG, Shultz KL, Donahue LR, *et al.* (2001) Quantitative trait loci for femoral and lumbar vertebral bone mineral density in C57BL/6J and C3H/HeJ inbred strains of mice. *J Bone Miner Res* **16**:1195–1206.
152. Klein RF, Allard J, Avnur Z, *et al.* (2004) Regulation of bone mass in mice by the lipoxygenase gene *Alox15*. *Science* **303**:229–232.
153. Shultz KL, Donahue LR, Bouxsein ML, *et al.* (2003) Congenic strains of mice for verification and genetic decomposition of quantitative trait loci for femoral bone mineral density. *J Bone Miner Res* **18**:175–185.
154. Turner CH, Sun Q, Schriefer J, *et al.* (2003) Congenic mice reveal sex-specific genetic regulation of femoral structure and strength. *Calcif Tissue Int* **73**:297–303.
155. Koller DL, Schriefer J, Sun Q, *et al.* (2003) Genetic effects for femoral biomechanics, structure, and density in C57BL/6J and C3H/HeJ inbred mouse strains. *J Bone Miner Res* **18**:1758–1765.
156. Volkman SK, Galecki AT, Burke DT, *et al.* (2003) Quantitative trait loci for femoral size and shape in a genetically heterogeneous mouse population. *J Bone Miner Res* **18**:1497–1505.
157. Volkman SK, Galecki AT, Burke DT, *et al.* (2004) Quantitative trait loci that modulate femoral mechanical properties in a genetically heterogeneous mouse population. *J Bone Miner Res* **19**:1497–1505.
158. Bouxsein ML, Uchiyama T, Rosen CJ, *et al.* (2004) Mapping quantitative trait loci for vertebral trabecular bone volume fraction and microarchitecture in mice. *J Bone Miner Res* **19**:587–599.
159. Masinde GL, Wergedal J, Davidson H, *et al.* (2003) Quantitative trait loci for periosteal circumference (PC): identification of single loci and epistatic effects in F2 MRL/SJL mice. *Bone* **32**:554–560.
160. Robling AG, Li J, Shultz KL, *et al.* (2003) Evidence for a skeletal mechanosensitivity gene on mouse chromosome 4. *FASEB J* **17**:324–326.
161. Srivastava AK, Masinde G, Yu H, *et al.* (2004) Mapping quantitative trait loci that influence blood levels of alkaline phosphatase in MRL/MpJ and SJL/J mice. *Bone* **35**:1086–1094.
162. Mohan S, Masinde G, Li X, Baylink DJ. (2003) Mapping quantitative trait loci that influence serum insulin-like growth factor binding protein-5 levels in F2 mice (MRL/MpJ × SJL/J). *Endocrinology* **144**:3491–3496.
163. Lockhart DJ, Winzeler EA. (2000) Genomics, gene expression and DNA arrays. *Nature* **405**:827–836.
164. Heller RA, Schena M, Chai A, *et al.* (1997) Discovery and analysis of inflammatory disease–related genes using cDNA microarrays. *Proc Natl Acad Sci USA* **94**:2150–2155.
165. Chiba S, Un-No M, Neer RM, *et al.* (2002) Parathyroid hormone induces interleukin-6 gene expression in bone stromal cells of young rats. *J Vet Med Sci* **64**:641–644.

166. Dahlquist KD, Salomonis N, Vranizan K, *et al.* (2002) GenMAPP, a new tool for viewing and analyzing microarray data on biological pathways. *Nat Genet* **31**:19–20.

167. Hughes TR, Marton MJ, Jones AR, *et al.* (2000) Functional discovery via a compendium of expression profiles. *Cell* **102**:109–126.

168. Liu F, Aubin JE, Malaval L. (2002) Expression of leukemia inhibitory factor (*LIF*)/interleukin-6 family cytokines and receptors during *in vitro* osteogenesis: differential regulation by dexamethasone and *LIF*. *Bone* **31**:212–219.

169. Power RA, Iwaniec UT, Wronski TJ. (2002) Changes in gene expression associated with the bone anabolic effects of basic fibroblast growth factor in aged ovariectomized rats. *Bone* **31**:143–148.

170. Srivastava S, Weitzmann MN, Kimble RB, *et al.* (1998) Estrogen blocks *M-CSF* gene expression and osteoclast formation by regulating phosphorylation of *Egr-1* and its interaction with *Sp-1*. *J Clin Invest* **102**:1850–1859.

171. Tomlinson JW, Bujalska I, Stewart PM, Cooper MS. (2000) The role of 11 beta-hydroxysteroid dehydrogenase in central obesity and osteoporosis. *Endocr Res* **26**:711–722.

172. Locklin RM, Riggs BL, Hicok KC, *et al.* (2001) Assessment of gene regulation by bone morphogenetic protein 2 in human marrow stromal cells using gene array technology. *J Bone Miner Res* **16**:2192–2204.

173. Chen J, Zhong Q, Wang J, *et al.* (2001) Microarray analysis of *Tbx2*-directed gene expression: a possible role in osteogenesis. *Mol Cell Endocrinol* **177**:43–54.

174. Beck GR Jr, Zerler B, Moran E. (2001) Gene array analysis of osteoblast differentiation. *Cell Growth Differ* **12**:61–83.

175. Connor JR, Kumar S, Sathe G, *et al.* (2001) Clusterin expression in adult human normal and osteoarthritic articular cartilage. *Osteoarthritis Cartilage* **9**:727–737.

176. Stokes DG, Liu G, Coimbra IB, *et al.* (2002) Assessment of the gene expression profile of differentiated and dedifferentiated human fetal chondrocytes by microarray analysis. *Arthritis Rheum* **46**:404–419.

177. Vincenti MP, Brinckerhoff CE. (2001) Early response genes induced in chondrocytes stimulated with the inflammatory cytokine interleukin-1beta. *Arthritis Res* **3**:381–388.

178. Furushima K, Shimo-Onoda K, Maeda S, *et al.* (2002) Large-scale screening for candidate genes of ossification of the posterior longitudinal ligament of the spine. *J Bone Miner Res* **17**:128–137.

179. Khanna C, Khan J, Nguyen P, *et al.* (2001) Metastasis-associated differences in gene expression in a murine model of osteosarcoma. *Cancer Res* **61**:3750–3759.

180. Shi S, Robey PG, Gronthos S. (2001) Comparison of human dental pulp and bone marrow stromal stem cells by cDNA microarray analysis. *Bone* **29**:532–539.

181. Theilhaber J, Connolly T, Roman-Roman S, *et al.* (2002) Finding genes in the C2C12 osteogenic pathway by *k*-nearest-neighbor classification of expression data. *Genome Res* **12**:165–176.

182. Gu W, Li X, Lau KH, *et al.* (2002) Gene expression between a congenic strain that contains a quantitative trait locus of high bone density from CAST/EiJ and its wild-type strain C57BL/6J. *Funct Integr Genomics* **1**:375–386.

183. Doi M, Nagano A, Nakamura Y. (2002) Genome-wide screening by cDNA microarray of genes associated with matrix mineralization by human mesenchymal stem cells *in vitro*. *Biochem Biophys Res Commun* **290**:381–390.

184. Raouf A, Seth A. (2002) Discovery of osteoblast-associated genes using cDNA microarrays. *Bone* **30**:463–471.

185. Bianco P, Riminucci M, Gronthos S, Robey PG. (2001) Bone marrow stromal stem cells: nature, biology, and potential applications. *Stem Cells* **19**:180–192.

186. Krebsbach PH, Kuznetsov SA, Satomura K, *et al.* (1997) Bone formation *in vivo*: comparison of osteogenesis by transplanted mouse and human marrow stromal fibroblasts. *Transplantation* **63**:1059–1069.

187. Takahashi N, Udagawa NT, Takami M. (2002) Cells of bone: osteoclast generation. In: *Principles of Bone Biology*, Bilezikian JP, Raisz LG, Rodan GA (eds.), Academic Press, San Diego, CA, pp. 109–126.

188. Liu YZ, Dvornyk V, Lu Y, *et al.* (2004) Microarray study of circulating monocytes in search for functional genes for osteoporosis. *J Bone Miner Res* **19**:S150.

189. Kamme F, Erlander MG. (2003) Global gene expression analysis of single cells. *Curr Opin Drug Discov Devel* **6**:231–236.

190. Kamme F, Salunga R, Yu J, *et al.* (2003) Single-cell microarray analysis in hippocampus CA1: demonstration and validation of cellular heterogeneity. *J Neurosci* **23**:3607–3615.

191. Sanz E, Alvarez-Mon M, Martinez-A C, de la Hera A. (2003) Human cord blood CD34^{+}Pax-5^{+} B-cell progenitors: single-cell analyses of their gene expression profiles. *Blood* **101**:3424–3430.

192. Abbott A. (1999) A post-genomic challenge: learning to read patterns of protein synthesis. *Nature* **402**:715–720.

193. Anderson L, Seilhamer J. (1997) A comparison of selected mRNA and protein abundances in human liver. *Electrophoresis* **18**:533–537.

194. Gygi SP, Rochon Y, Franza BR, Aebersold R. (1999) Correlation between protein and mRNA abundance in yeast. *Mol Cell Biol* **19**:1720–1730.

195. Ideker T, Thorsson V, Ranish JA, *et al.* (2001) Integrated genomic and proteomic analyses of a systematically perturbed metabolic network. *Science* **292**:929–934.

196. Banks D. (2003) Proteomics: a frontier between genomics and metabolomics. *Chance* **16**:6–7.

197. Graves PR, Haystead TA. (2002) Molecular biologist's guide to proteomics. *Microbiol Mol Biol Rev* **66**:39–63.

198. Blackstock WP, Weir MP. (1999) Proteomics: quantitative and physical mapping of cellular proteins. *Trends Biotechnol* **17**:121–127.
199. Evans CA, Tonge R, Blinco D, *et al.* (2004) Comparative proteomics of primitive hematopoietic cell populations reveals differences in expression of proteins regulating motility. *Blood* **103**:3751–3759.
200. Brown RE, Boyle JL. (2003) Mesenchymal chondrosarcoma: molecular characterization by a proteomic approach, with morphogenic and therapeutic implications. *Ann Clin Lab Sci* **33**:131–141.
201. Behnam K, Murray SS, Whitelegge JP, Brochmann EJ. (2002) Identification of the molecular chaperone alpha B–crystallin in demineralized bone powder and osteoblast-like cells. *J Orthop Res* **20**:1190–1196.
202. Kubota K, Wakabayashi K, Matsuoka T. (2003) Proteome analysis of secreted proteins during osteoclast differentiation using two different methods: two-dimensional electrophoresis and isotope-coded affinity tags analysis with two-dimensional chromatography. *Proteomics* **3**:616–626.
203. Nuttall ME. (2001) Drug discovery and target validation. *Cells Tissues Organs* **169**:265–271.
204. Lorenz P, Ruschpler P, Koczan D, *et al.* (2003) From transcriptome to proteome: differentially expressed proteins identified in synovial tissue of patients suffering from rheumatoid arthritis and osteoarthritis by an initial screen with a panel of 791 antibodies. *Proteomics* **3**:991–1002.
205. Macdonald HM, McGuigan FE, Stewart A, *et al.* (2006) Large-scale population-based study shows no evidence of association between common polymorphism of the *VDR* gene and BMD in British women. *J Bone Miner Res* **21**:151–162.
206. Uitterlinden AG, Ralston SH, Brandi ML, *et al.* (2006) The association between common vitamin D receptor gene variations and osteoporosis: a participant-level meta-analysis. *Ann Intern Med* **145**:255–264.
207. Ralston SH, Uitterlinden AG, Brandi ML, *et al.* (2006) Large-scale evidence for the effect of the *COLIA1* Sp1 polymorphism on osteoporosis outcomes: the GENOMOS Study. *PLoS Med* **3**:e90.
208. Medici M, van Meurs JB, Rivadeneira F, *et al.* (2006) *BMP-2* gene polymorphisms and osteoporosis: the Rotterdam Study. *J Bone Miner Res* **21**:845–854.
209. Ralston SH, Galwey N, MacKay I, *et al.* (2005) Loci for regulation of bone mineral density in men and women identified by genome wide linkage scan: the FAMOS Study. *Hum Mol Genet* **14**:943–951.
210. Ichikawa S, Koller DL, Johnson ML, *et al.* (2006) Human *ALOX12*, but not *ALOX15*, is associated with BMD in white men and women. *J Bone Miner Res* **21**:556–564.

Chapter 23

Critical Molecules in Bone Development and Genetic Bone Diseases

Qing Wang*, Ming Zhang*, Tianhui Zhu* and Di Chen*,†,‡

*Medical College, Nankai University
Tianjin 300070, P.R. China

†Department of Orthopaedics, Center for Musculoskeletal Research
University of Rochester, Rochester, NY 14642, USA

1. WNT/LRP5

1.1. *Wnt/β-Catenin Signaling*

Wnt proteins play critical roles in carcinogenesis and in early development, controlling mesoderm induction, patterning, cell fate determination, morphogenesis, and bone development.[1,2] Wnts form a dual-receptor complex with Frizzled and low-density-lipoprotein (LDL)-receptor–like protein 5 or 6 (LRP5/6) on cell surfaces. This triggers signaling through a large protein complex in the Wnt canonical pathway, including glycogen synthase kinase 3β (GSK3β); casein kinase I; and the scaffolding proteins, adenomatous polyposis coli (APC), Disheveled, and Axin. This complex has multiple effects on β-catenin: it promotes β-catenin phosphorylation by GSK3β at specific amino terminal residues, creates docking sites for F-box protein/E3 ligase complexes,[2–4] and enables β-catenin to be detected and destroyed by the 26S proteasome in the absence of Wnt signaling.[5] The N-terminal domain of LRP5 mediates its interaction with the Wnt–Frizzled ligand–receptor complex, and the intracellular tail of LRP5/6 binds Axins. The activation of Wnt signaling inhibits the stimulatory effect of Axins on β-catenin phosphorylation, and allows β-catenin to move to the nucleus.[6]

---

‡ Correspondence author.

Nuclear β-catenin combines with the transcription factors TCF and LEF1 to activate expression of target genes.

At the surface of cells, Wnts interact with two kinds of proteins: Frizzled receptor and LRP5/6. There are many genes encoding Frizzled proteins (10 in the human genome), and different Frizzled proteins probably have different affinities to different types of Wnt proteins. Wnt proteins form a complex with the cysteine-rich domain (CRD) of Frizzled and with LRP5/6.[1,7] The intracellular parts of the receptors turn on the pathway through β-catenin inside the cell. LRP5/6 receptors also bind to Dickkopf-1 (Dkk1), a molecule that counteracts Wnt and blocks Wnt function. Binding of Dkk1 to LRP5/6 might alter the confirmation of LRP5/6 so that it can no longer interact with Wnt and Frizzled, thus halting the intracellular signaling. Kremen-1 associates with Dkk1 and interacts with LRP5 to induce LRP5 endocytosis.[8]

β-catenin is a key molecule in the canonical Wnt signaling pathway, and plays a critical role in multiple steps during osteoblast and chondrocyte differentiation. Conditional deletion of the β-catenin gene in early mesenchymal cells shows that the differentiation of mesenchymal cells into osteoblasts is inhibited, but into chondrocyte precursors is accelerated[9,10]; this suggests that β-catenin promotes early mesenchymal cells to differentiate into osteoblasts, but inhibits mesenchymal cells to differentiate into chondrocyte precursors. However, the specific deletion of the β-catenin gene in *Col2a1*-expressing chondrocytes causes decreased chondrocyte proliferation and delayed chondrocyte maturation,[11] suggesting that β-catenin promotes chondrocyte maturation in growth plate chondrocytes.

1.2. *LRP5 and Its Homolog*

In humans, the *LRP5* gene has 23 coding exons extending over more than 100 kb and is located in chromosome 11q13.4. *LRP5* cDNA contains 4845 base pairs that encode a single-pass membrane receptor, a chain of 1615 amino acids with a calculated molecular weight of about 180 kDa. *LRP5* is highly conserved between species, and its extracellular domain contains four modules consisting of six YWTD repeats followed by an epidermal growth factor (EGF)-like motif and an LDLR-like ligand-binding domain.[7] The first and second most N-terminal modules of *LRP5* mediate its interaction with the Wnt–Frizzled seven-transmembrane ligand–receptor complex.

Even though *LRP5* and *LRP6* are highly homologous, *LRP6* (but not *LRP5*) is primarily expressed in the nervous system and deletion of the *LRP6* gene results in significant brain abnormalities, while deletion of *LRP5* results in primarily decreased bone density. Furthermore, *LRP6* (but not *LRP5*) overexpression can induce dorsal axis duplication in *Xenopus*, suggesting that these receptors control distinct functions.[12] *LRP5* deficiency does not cause developmental defects, but causes a decrease in bone mass in humans and in adult mice, thus indicating the importance of the Wnt pathway to the regulation of postnatal functions.

1.3. *LRP5 Receptor and OPPG and HBM Syndromes*

The role of *LRP5* in bone mineral density (BMD) determination was initially uncovered by genetic analyses that mapped the osteoporosis pseudoglioma (OPPG) syndrome and the high bone mass (HBM) syndrome to the human chromosome 11q12–13 locus.[13,14] Recent studies indicate that *LRP5* exerts a major effect on the entire spectrum of BMD, from osteoporosis to normal BMD to high-bone-mass phenotypes. The *LRP5* gene was shown to influence the normal population variation of BMD and diseases marked by abnormal bone mass.

1.3.1. *LRP5 and OPPG*

OPPG is an autosomal recessive childhood disorder with both skeletal and eye abnormalities. OPPG patients have normal bone growth associated with severe osteopenia without abnormal collagen synthesis or hormonal defects. They also have congenital or juvenile-onset blindness due primarily to hyperplasia of the primary vitreous. Interestingly, obligate heterozygotes for the OPPG allele have an increased incidence of osteoporotic fractures, suggesting the existence of a skeleton-restricted haploinsufficient phenotype. OPPG patients harbor inactivating mutations in the *LRP5* gene,[15] and heterozygous carriers of the mutations have reduced bone mass. Several loss-of-function mutations of the *LRP5* gene have been found in patients affected by OPPG, including W10X, R428X, E485X, D490fs, R494Q, R570W, V667M, D718X, W734X, D769fs, Q853X, and E1270fs.[15]

In addition to OPPG, other low-bone-mass–related phenotypes have also been associated with inactivating mutations in the *LRP5* gene. The 1067C > T mutation, impairing *LRP5* synthesis, and the 1364C > T mutation, inhibiting protein trafficking to the cell surface, have been related to idiopathic osteoporosis.[16] The *LRP5* 1330-valine variant was found to be associated with decreased bone mass in men, but not in women.[17] Two missense mutations (A29T and R1036Q) and one frameshift mutation (C913fs) in the *LRP5* gene were found in patients with juvenile osteoporosis.[18]

Kato *et al.*[19] generated $LRP5^{-/-}$ mice, which mimicked the *LRP5* mutation observed in human OPPG. These mice were characterized by early-onset osteoporosis, with delayed ossification from day 4 postnatally and significant reductions in bone volume and bone mineralization, all of which resulted from the reduction in osteoblast numbers. The phenotype of $LRP5^{-/-}$ mice was so severe that a significant proportion of these mice died within the first month of life as a result of multiple fractures. Moreover, these mice displayed persistent defects in embryonic eye vascularization due to the failure of macrophage-induced endothelial cell apoptosis, which implicates Wnt proteins in the postnatal control of vascular regression and bone formation.[19] Another recent *in vivo* study demonstrated that the phenotype of $LRP5^{-/-}$ mice can be reversed by introducing lithium chloride, an inhibitor of GSK3β; such treatment restored bone mass to the level close to that of wild-type mice.[20] These findings demonstrate the importance of canonical Wnt/β-catenin signaling in bone metabolism, and suggest a prospective therapy of low bone mass–related diseases through activation of canonical Wnt signaling.

1.3.2. *LRP5 and HBM Syndromes*

In contrast to OPPG, carriers of the autosomal dominant HBM trait have very high spinal BMD compared to unaffected individuals.[14] Several human HBM syndromes have been shown to have activating mutations in the *LRP5* gene, including endosteal hyperostosis, van Buchem disease, autosomal dominant osteosclerosis, craniosynostosis, macrocephaly, mild developmental delay, and a palatal developmental disorder termed "torus palatinus".[21–23] An activating mutation (Gly171Val) in *LRP5* is the most prominent mutation responsible for the HBM syndrome.[24–27] At the same time, other gain-of-function mutations of the *LRP5* gene have also been reported — such as D111Y, G171R, A214T, A214V, A242T, T253I,[23,28] and R154M[29] — all of which are located in the amino terminal part of the *LRP5* gene before the first EGF-like domain, alter the region important for *LRP5* antagonism by Dkk1, and result in a series of HBM syndromes.

In 2003, Babij *et al.*[26] generated mice bearing the G171V mutation in their *LRP5* gene, thus mimicking human HBM syndromes. These mice are now widely used as the *LRP5* gain-of-function mutant animal model. $LRP5^{G171V}$ mice have increased trabecular volumetric BMD and cortical size due to increased osteoblast activity and survival. Since the mutant receptor *LRP5* is resistant to Dkk-mediated inhibition of Wnt signaling, the phenotype occurs despite expression levels of $LRP5^{G171V}$, is evident early during mouse development, and persists in adult mice.[26,30]

Since HBM-related mutant LRP5 proteins cannot activate Wnt signaling in the absence of ligands and have reduced physical interaction with Dkk1, the likely mechanism for the HBM mutations is through reduced affinity of LRP5 to Dkk1 or inhibition of Dkk1.[31] However, this is not always necessarily the case. Recent *in vitro* studies demonstrate that the Gly171Val mutation disrupts the interaction of LRP5 with Mesd, a chaperone protein for LRP5/6 that is required for transport of the coreceptor to the cell surface, resulting in fewer LRP5 molecules on the cell surface. The reduced number of G171V LRP5 mutant molecules on the cell surface would render them less susceptible to the paracrine effects of the Dkk1 antagonist, while retaining LRP5 activation by autocrine Wnts.[27] The detailed mechanism for the constantly activated Wnt signaling by the gain-of-function mutant of *LRP5* remains to be revealed.

1.3.3. *LRP5 and BMD Variations*

Besides several severe activating and inactivating mutations of *LRP5* leading to extremely low or high bone mass phenotypes, the associations of several single nucleotide polymorphisms (SNPs) in this gene at IVS17–1677C > A, IVS17–30G > A, c.2220C > T, c.3989C > T, c.2047G > A, c.3357A > G, c.4037C > T, C171346A, C135242T, C141759T, G121513A, C135242T, G138351A, C141759T, C165215T, Q89R, N740N, V667M, and A1330V with human BMD variations have also been reported to underscore the crucial regulatory role of *LRP5* in bone formation and metabolism.[32–38] Not only SNPs, but also haplotypes in the *LRP5* gene, have been demonstrated to be associated with variation in human BMD. A most recent report on the haplotype analysis of *LRP5* indicated an association between a haplotype (C-G-C-C-A) in the *LRP5* gene and osteoarthritis (OA).[39] Interestingly, all of the polymorphisms reported to date were localized to the extracellular domain of the *LRP5* protein.

To date, two animal models have been generated to investigate the role of *LRP5* in maintaining bone mass. The $Lrp5^{-/-}$ mice generated by Kato *et al.*[11] disrupted exon 6 of the *Lrp5* gene,[19] and the other $Lrp5^{-/-}$ mice generated by Fujino *et al.*[40] disrupted exon 18 of the *Lrp5* gene that encodes a ligand-binding repeat. Both mice exhibited low bone mass throughout life and resembled human osteoporosis. Furthermore, these mice also exhibited impaired chylomicron plasma clearance, hepatic uptake, and impaired glucose-induced insulin secretion following high fat or glucose challenge. Such findings offer an additional mechanism through which *LRP5* haploinsufficiency might potentiate osteopenia or osteoporosis, because impaired lipid metabolism might disrupt vitamin D absorption and bioavailability.[19,40]

1.4. Wnt/LRP5 Signaling Pathway in Bone Metabolism

LRP5 is a key regulator of bone metabolism in the Wnt signaling pathway. Through the utility of *LRP5* gain-of-function or loss-of-function mutations, the role of *LRP5* and related Wnt signaling in bone metabolism is uncovered.

1.4.1. Osteoblast: The Major Target of Wnt Signaling

LRP5 is expressed by osteoblasts of the endosteal and trabecular bone surfaces (but not by osteoclasts), and regulates osteoblastic proliferation, survival, and activity. Human mesenchymal stem cells (MSCs) express various members of the Wnt- and Frizzled-related proteins, *LRP5* coreceptor Frizzled families, and *Lrp5* and *Dkk1* during osteogenesis.[41,42] Wnt–LRP5 stimulates the expression of alkaline phosphatase (ALP) — a marker of osteoblastic differentiation — in the pluripotent mesenchymal cell lines C3H10T1/2, C2C12, and ST2 cells, and in the osteoblast cell line MC3T3-E1 cells.[43] Effects of BMP2 on osteoblast differentiation and extracellular matrix mineralization are enhanced by Wnts.[43] In addition, BMP2 stimulates *Lrp5* and *Lrp6* expression in ST2 bone marrow stromal cells.[15,44] In the same cells, Wnt1, Wnt2, Wnt3a, and Wnt7b (but not Wnt4 and Wnt5a) induce expression of ALP. Overexpression of *Lrp5* does not enhance Wnt3a-induced ALP activity, whereas a gain-of-function mutation in this gene does induce the HBM phenotype in humans. This suggests that other Wnts (or even other ligands) may be involved in signaling through *LRP5*, or that Wnt-induced ALP stimulation may reflect only one aspect of their activity.

A recent report shows that Wnt signaling is essential in determining whether mesenchymal progenitors become osteoblasts or chondrocytes.[10] Moreover, canonical Wnt/LRP5 signaling prevents mesenchymal cells from differentiating into chondrocytes.[9] Conditional deletion of Wnt signaling in limb mesenchymal cells during early embryonic development resulted in the arrest of osteoblastic differentiation and the lack of mature osteoblasts in membranous bone.[9] Further studies showed that Wnt signaling functions downstream of Indian hedgehog (*Ihh*) during osteoblast development, and Wnt7b was identified as a potential endogenous ligand regulating osteogenesis.[45] Furthermore, a most recent report by Almeida *et al.*[46] demonstrated that Wnt proteins, irrespective of their ability to stimulate canonical Wnt signaling, prolong the survival of osteoblasts and uncommitted osteoblast progenitors via activation of the Src/ERK and PI3K/Akt signaling pathway.

1.4.2. *Osteoclast: The Promising Wnt Signaling Target*

Considering several new interesting evidences, Wnt/LRP5 signaling seems to not only regulate the level of bone formation in a given individual, but also affect osteoblast–osteoclast coupling by regulating the RANKL–RANK (receptor activator of NF-kappaB) and osteoprotegerin (OPG) pathways. Evidence correlating Wnt signaling and osteoclast activity was first presented by Hausler *et al.*,[47] showing that *Sfrp1* could bind RANKL and neutralizing antibodies against *Sfrp1*-enhanced osteoclast development in an *in vitro* coculture assay. Recent studies revealed that the Wnt signaling protein β-catenin stimulates osteoblast expression of OPG, a major inhibitor of osteoclast differentiation, thus resulting in high bone mass; while deletion of the β-catenin gene in differentiated osteoblasts leads to osteopenia.[48] These findings demonstrate that Wnt signaling promotes the ability of differentiated osteoblasts to inhibit osteoclast differentiation.

Deletion of the β-catenin gene in osteocalcin-expressing mature osteoblasts also demonstrated that Wnt signaling in osteoblasts coordinates postnatal bone acquisition by controlling the differentiation and activity of osteoclasts.[49] In these mice, the expression of RANKL was elevated, whereas the expression of OPG (the RANKL decoy receptor) was reduced. As a result, these mice exhibited severe osteopenia with a significant increase in osteoclast numbers.[49] These findings indicate the importance of Wnt/LRP5 signaling in bone metabolism.

2. RUNX2

2.1. *Runx2, a Bone-Specific Transcription Factor*

Runx2 (also named *Cbfa1*, *PEBP2αA*, *AML3*, or *Osf2*) was first discovered as one of the human members of the Runt-domain gene family and mapped to human chromosome 6p12.3–6p21.1.[50] *Runx2* is a bone-specific transcription factor essential for osteoblast differentiation, and is obligatory for the expression of many osteoblast and chondrocyte marker genes.[51–53] The expression of *Runx2* is abundant in calcified cartilage, bone tissues, thymus, and testes; while absent from the brain, heart, lung, gut, and liver.[51,54–57] *Runx2* can be initially detected during mouse embryogenesis on E9.5 in the notochord and E10.5 in the mesoderm that is destined to become shoulder bone. By E11.5, *Runx2* is strongly expressed in mesenchymal cells surrounding mesenchymal condensation and in all developing skeletal elements.[53] The expression of *Runx2* is regulated by two distinct promoters.[58–61] The two isoforms of *Runx2* produced by the two promoters, respectively, differ from each other by only 42 amino acids at their N-terminus, but have equivalent activities in transcription assays.[62] The expression and activity of *Runx2 in vivo* is controlled at multiple levels, including the production of isoforms from the two promoters,[58,60,61,63] autoregulation of transcription,[59] interaction of *Runx2* with its DNA-binding partners, and posttranslational modification.[64,65]

As one of the three members of the Runt-domain gene family of transcription factors (*Runx1/Cbfa2/PEBP2αB/AML1*, *Runx2/Cbfa1/PEBP2αA/AML3*, and *Runx3/Cbfa3/PEBP2αC/AML2*), *Runx2* shares a highly conserved DNA-binding domain called the Runt domain (128 amino acid residues) with the other two members; while *Runx2* is noticeably different from *Runx1* and *Runx3* for the presence of two unique domains, a 41-amino-acid glutamine/alanine-rich domain at the N-terminus and a 22-amino-acid region at the C-terminus. DNA-binding sites for *Runx2* have been characterized as PyGPyGGTPy, and this sequence has been identified in promoter regions of many osteoblast-specific genes. *Runx2* binds to its response elements in promoters of target genes, and activates these genes at the transcriptional level.

Multiple lines of evidence illustrate that *Runx2* is the earliest transcription factor essential for bone formation,[51,53,66–70] and Osterix is a downstream transcription factor of *Runx2*.[71,72] Homozygous *Runx2*-deficient mice die soon after birth due to an inability to breathe. The most pronounced effect is a complete lack of both endochondral and intramembranous ossification,[51,53] with an absence of mature osteoblasts throughout the body, demonstrating that *Runx2* expression is absolutely required for bone development. To investigate the function of *Runx2* in postnatal bone formation, Ducy *et al.*[73] generated transgenic mice overexpressing the *Runx2* DNA-binding domain (Δ*Runx2*) driven by the osteocalcin gene 2 (*OG2*) promoter. Δ*Runx2* was expressed in differentiated osteoblasts only postnatally, and acted in a dominant-negative fashion due to its higher affinity to bind to DNA than *Runx2* itself.[73] Skeletons of the Δ*Runx2*-transgenic mice were normal at birth, but the mice suffered from osteopenia due to decreases in bone volume and bone formation rates, evident 3 weeks after birth.[73] These results indicated that *Runx2* plays a crucial role not only in osteoblast differentiation and bone development, but also in osteoblast function and postnatal bone formation. Furthermore, overexpression of *Runx2* in nonosteoblastic cells leads to the expression of osteoblast-specific genes such as osteocalcin and bone sialoprotein.[74,75]

2.2. *Runx2 Target Genes*

Many bone-related genes are regulated by *Runx2*, such as type I collagen (*col1a1*), alkaline phosphatase (*ALP*),[76] osteopontin (*OP*),[77] osteocalcin (*OC*),[78,79] bone sialoprotein (*BSP*),[75] collagenase,[80–82] *Runx2* itself,[59] dentin sialoprotein (*DSP*),[83] type X collagen,[84,85] *C/EBPδ*,[86] *Ihh*,[87] and *VEGF*.[88] Many cytokines and their receptors are also regulated by *Runx2*, including granulocyte–macrophage colony-stimulating factor (GM-CSF),[89] interleukin-3 (IL-3),[90] macrophage inflammatory protein 1α (MIP-1α),[91] macrophage colony-stimulating factor receptor 1 (M-CSF-R1),[92] type I TGF-β receptor,[93] human vitamin D receptor (hVDR),[94] estrogen receptor α,[94,95] osteoprotegrin (OPG),[96,97] and RANKL.[96,97] *Runx2* has been reported to cooperate with growth factor signaling molecules such as Smads, MAPK, YAP, and β-catenin; and integrates a variety of signals to regulate bone metabolism.

2.3. *Runx2 and Bone Formation*

2.3.1. *Runx2 in Osteoblast Proliferation*

The regulation of cell proliferation is accomplished by a precise cell cycle control executed by different signaling pathways. Since there are stage-specific demands for the *de novo* synthesis of proteins at different stages of the cell cycle, cell cycle control is mediated at the transcriptional level of different genes where *Runx2* is involved.

To determine the role of *Runx2* in osteoblast proliferation, Pratap et al.[98] obtained calvarial osteoblasts from wild-type and *Runx2*-deficient mice. They found that calvarial cells (but not embryonic fibroblasts) from *Runx2*-deficient mice exhibited increased cell growth rates as reflected by elevations of DNA synthesis and G_1/S phase markers (e.g. cyclin E) and that reintroduction of *Ad-Runx2* into $Runx2^{-/-}$ calvarial cells inhibited osteoblast proliferation, suggesting that *Runx2* is a negative regulator of osteoblast proliferation.[98]

Recent reports indicated that *Runx2* regulates cell proliferation by affecting several cell cycle–related proteins, such as directly regulating p21/p27,[99,100] type I TGF-β receptor,[93] pRb,[101] and Twist1/2.[102] *Runx2* attenuates osteoblast growth, triggers osteoblast exit from the cell cycle, and promotes osteoblast maturation.[63,103,104]

2.3.2. *Runx2 in Osteoblast Differentiation*

Although the mechanism through which osteoblast differentiation is triggered remains unclear, the initiation of osteoblast differentiation starts from the exit from the cell cycle and then cell lineage commitment. The osteoblast phenotype development is consequently activated by specific transcription factors that promote osteoblast maturation. *Runx2* plays a critical role in every step of osteoblast differentiation.

Very low levels of marker genes for early-stage osteogenic cells can be detected in calvarial-derived cells of $Runx2^{-/-}$ mice, thus defining *Runx2* as a primary and essential factor for osteoblast differentiation.[51] *Runx2*-targeted bone-related genes have been characterized by many research groups (see Sec. 2.2). Among these gene products, many of them contribute to signal transduction, tissue-specific gene transcription, skeletal phenotype determination, and bone extracellular matrix biosynthesis and mineralization. *ALP* and *col1a1* are two marker genes in the early stage of osteoblast differentiation in addition to type I TGF-β receptor (*TGFβR1*), collagenase 3, and *OP*; while other marker genes such as *BSP*, osteonectin (*ON*), and *OC* are expressed during late-stage osteoblast differentiation.[74,76,78,82,83,96,105,106]

2.3.3. *Runx2 in Chondrocyte Hypertrophy*

During the process of chondrogenesis, *Runx2* is initially expressed in chondrogenic mesenchymal cells as well as prehypertrophic and hypertrophic chondrocytes,[87,107] suggesting the important role of *Runx2* in chondrocyte hypertrophy. *In vivo* studies demonstrated that $Runx2^{-/-}$ mice showed a severe delay in chondrocyte maturation, resulting from lack of hypertrophic chondrocytes in many bones.[108] Furthermore, $Runx2^{-/-}:Runx3^{-/-}$ mice showed a complete absence of chondrocyte maturation,[87] indicating that *Runx2* and *Runx3* are absolutely required for chondrocyte maturation. On the contrary, overexpression of *Runx2* hastens chondrocyte hypertrophy and promotes the differentiation of chondrocytes in the trachea, which normally never hypertrophies, into hypertrophic chondrocytes.[109,110]

Runx2 has been reported to directly regulate type X collagen and Indian hedgehog (*Ihh*); and indirectly regulate *MMP13*, *PTHrP*, and *VEGF*. Type X collagen knockout mice have subtle growth plate phenotypes with compressed proliferating and hypertrophic zones and altered mineral deposition,[111] which are similar to those of human chondrodysplasia.[112] *Ihh* stimulates chondrocyte proliferation and induces *PTHrP* expression, which inhibits both *Ihh* and *Runx2* expression (negative feedback regulation). *PTHrP* maintains chondrocytes in a proliferating stage.[113,114] This local negative feedback regulation contributes to a proper rate of maturation and proliferation of chondrocytes. *MMP13*-deficient mice show the phenotype of delayed endochondral ossification in the growth plate with increased length of the hypertrophic zone.[115,116] Furthermore, through activating *VEGF*, *Runx2* has been demonstrated to play a crucial role in the vascular invasion of cartilage,[88] which is required for the subsequent replacement of cartilage by bone.[117,118]

2.3.4. *Runx2 in Osteoclast Formation*

Runx2 also regulates osteoclast formation. Osteoclasts are developed from monocytic precursors of the hematopoietic lineage, and osteoclast formation is regulated by cytokines and their receptors. *Runx2* has been reported to upregulate osteoprotegrin,[96,97] RANKL,[96,97] and M-CSF[89] expression, which are key factors in osteoclast formation. After monocytes are initially committed to the osteoclast lineage, there are mainly two signaling pathways subsequently regulating downstream osteoclast-specific genes: the macrophage colony-stimulating factor (M-CSF) signaling pathway, and the RANKL/RANK signaling pathway. The absence of RANKL and reduced OPG mRNA expression were found in the cells derived from *Runx2*$^{-/-}$ embryos compared to those of *Runx2*$^{+/-}$ and *Runx2*$^{+/+}$ embryos, and a delay in osteoclast formation was observed in *Runx2*$^{-/-}$ mice. These observations indicated that *Runx2* mediates osteoclast formation through regulating RANKL/RANK and OPG signaling.[97]

Upon binding RANKL, RANK activates six major signaling pathways: NFATc1, NF-κB, Akt/PKB, JNK, ERK, and p38, all of which play distinct roles in osteoclast differentiation, function, and survival. Recent studies have revealed that several other factors, including INF-γ, IFN-β, and ITAM-activated signals, also regulate osteoclastogenesis via direct crosstalk with RANK signaling.[119–128] OPG, working as an antagonist of RANKL/RANK signaling by directly binding to RANKL, is also under the control of *Runx2*.[96,97]

2.4. *Transcriptional and Posttranslational Control of Runx2*

Like other proteins, Runx2 activity is regulated at both transcriptional and posttranscriptional levels.

2.4.1. *Transcriptional Control of Runx2*

There are two distinct isoforms of mRNAs of the Runx2 protein, namely type I and type II isoforms, which are transcribed under the control of two distinct promoters, termed P2 and P1 promoters, respectively.[58–61,129] The activities of the two isoforms are similar, and the structures of the two Runx2 isoforms are almost identical with the exception of an additional 16 amino acids at the N-terminus of the type II isoform.[129] The P2 promoter driving the type I isoform is activated during the initial development of osteoprogenitor cells, whereas the P1 promoter driving the type II isoform is activated at the late stage of osteogenic cells.[129–132] The P1 (but not P2) promoter of Runx2 has been well characterized to date, and is regulated by many transcriptional regulators.

2.4.2. *Posttranslational Control of Runx2*

Control of Runx2 activity at the posttranslational level is mainly composed of both phosphorylation and ubiquitination on the Runx2 protein posttranslationally. Phosphorylation of Runx2 contributes greatly to the regulation of Runx2 activity. TGF-β/BMP and FGF — through the signaling pathways Smad, MAPK, and PKC/PKA — activate Runx2[133–135] by phosphorylation at its Ser-125,[136] Ser-247,[137] Ser-451,[138] and Thr-341.[105] However, phosphorylation occurring on the two highly conserved serine residues (Ser-104 and Ser-451) downregulates its activity.[139] Our recent findings demonstrated that cyclin D1-CDK4 phosphorylates Runx2 through Ser-451 and induces subsequent ubiquitination and proteasome degradation of Runx2,[140] suggesting that Runx2 activity is regulated coordinately with the cell cycle machinery.

The Runx2 protein is regulated by the ubiquitin–proteasome pathway. Our previous works showed that the E3 ubiquitin ligase, Smad ubiquitin regulatory factor 1 (Smurf1), interacts with Runx2 and induces Runx2 ubiquitination and proteasomal degradation.[64,141] The degradation of Runx2 may disrupt the Runx2/CBFβ heterodimeric complex, and negatively affects Runx2 transcriptional activity. As a Hect domain E3 ligase, Smurf1 normally interacts with the PY motif of substrate proteins through its

WW domain.[142] Interestingly, Smurf1 partially maintains its activity to induce Runx2 degradation even when the PY motif of Runx2 is deleted. Our most recent findings indicated that the BMP signaling inhibitor Smad6 serves as an adaptor protein and mediates Smurf1-induced Runx2 degradation.[65]

2.5. *Runx2-Related Diseases*

2.5.1. *Human Cleidocranial Dysplasia (CCD) Syndrome*

The human cleidocranial dysplasia (CCD) syndrome — a disease featuring supernumerary tooth buds, delayed tooth eruption, patent fontanels, Wormian bones, short stature, dysplasia of the clavicles, growth retardation, and hypoplasia of the distal phalanges — is an autosomal dominant bone disease whose genetic locus has been located on chromosome 6p21, where the *Runx2* gene also maps. Heterozygous *Runx2* mutant mice have skeletal abnormalities similar to those seen in the human CCD syndrome,[52,143] including hypoplasia of the clavicle, delayed development of membranous bones,[51,53] and markedly reduced bone density[144] (Fig. 1; Table 1).

Most of the human CCD syndromes are due to mutations of *Runx2* in its highly conserved DNA-binding domain and deletion mutations on the Runx2 C-terminus that block its interaction with Smads.[155] Using the blot microarray method, Chen *et al.*[163] analyzed 226 genes in human primary dental pulp cells from a CCD patient harboring the R225Q mutation in *Runx2*, and found 25 upregulated and 17 downregulated genes; many of the downregulated genes are *Runx2* direct target genes.

2.5.2. *Association of Runx2 Polymorphisms with BMD*

So far, only a few studies have examined the relationship between polymorphisms in the *Runx2* gene and bone mineral density (BMD) variation.[164,165]

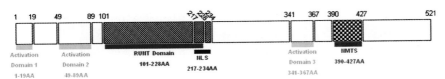

Fig. 1. Runx2 protein structure. The activation domains 1/2/3, DNA-binding RUNT domain, nuclear localization sequence (NLS), and nuclear matrix targeting signal (NMTS) of the Runx2 protein are indicated in different colors.

Table 1   *Runx2* Mutations and CCD Syndromes

Peptide Change	Nucleotide Change	Reference
Q50STOP	148C > T	145
Q53L	158A > T	146
Q64STOP	190C > T	147
Q65STOP	196C > T	148
Q66STOP	196C > T	147
Q69R	206A > G	149
S104R		139
E112STOP	334G > T	145
L113R	338T > G	149
S118R	354C > G	149
F121C	362T > G	149
C123R	366–367delGTinsTC	149
S128F	383insT	150,151
W130STOP	389G > A	152
R131G	391C > G	145
N133del	397–399delAAC	153
STOP143	178delC; 211del173	147,148
ins10AA	222–223ins30	52
R148G		154
Exon2 skipping	IVS2 + T > A; IVS2 + 1G > C	147,152
V156G	467T > G	147
STOP159	186–187ins16	52,149
STOP160	90–91insC; 134–135insCGGT(GTCC); 382–383insT	147,148,150,153
D161STOP	481–481delGA	149
V165STOP		155
G166STOP		153
R169P	506G > C	147,156
R169Q		153
STOP175	495delT	147,148
M175R	524T > G	157
STOP179	532delC	155
STOP185	539–548del10	52
R190W	568C > T	147,148
R190Q		153
S191N	572G > A	157
G192R	574G > A	158
R193STOP	577C > T	147–149,155
R193C		153
F197S	590T > C	148,155
Exon3 skipping	IVS3 + 2T > C; IVS3 + delGTAA	148
L199F		153
T200A		153
I201K	602T > A	147
T205R	614C > G	149
Q209R		153

(Continued)

Table 1 *(Continued)*

Peptide Change	Nucleotide Change	Reference
STOP217	553–554delCT	153
STOP218	522–523insA	152
K218N	654A > T	148
T220I	659C > T	148
STOP221	636delC	153
V221G	662T > G	152
R225W	673C > T	147–149,159
Exon1 skipping	IVS1 + 1G > A	145
R225Q	674G > A	147–149,153
	IVS4 + 1G > T	153,155
	IVS4 + 2T > C	147
	IVS4 + 4delAAGT	147
Q280STOP	838C > T	147,148
W297STOP	891G > A	52
STOP299	873–874delCA	152
STOP307	821delC; 821delG; 884delC; 887delC; 915delC	144,149,153
A362V	1085C > T	147,148
Exon7 skipping	IVS6–1G > C	160
R377STOP		155
STOP384	1127–1128insT	149
R391STOP	1171C > T	153,155,161
STOP410	1228insC	162
STOP477	1111–1129del19	152
STOP483	950–971del22; 1157delG	149,160
STOP489	1205–1206insC; 1215–1216insC; 1379–1390insC	149,160
STOP522S	1565G > C	146

In 495 randomly selected healthy women and 800 female fracture patients, two common polymorphisms within exon 1 of the *Runx2* gene were identified: an 18-bp deletion and a synonymous alanine codon polymorphism with alleles GCA and GCG. The former was not significantly associated with BMD variation; whereas the GCA allele of the latter variant was related to significantly greater BMD at all measured bone sites, including spine (L2–L4), femoral neck, trochanter, ultradistal forearm, whole body, etc. In addition, the GCA allele was associated with an approximately threefold protection against Colle's fracture. These results suggested that *Runx2* variants might be related to genetic effects on BMD and osteoporosis.[164,165] Another report on the genotype analysis of 991 women from a Scottish cohort revealed that the GCA allele was associated with higher femoral neck BMD within a postmenopausal subgroup of the population and that the effect of *Runx2* GCA alleles increased with increasing weight, thus suggesting that the *Runx2* alleles are associated with BMD in a menopause- and weight-dependent manner.[165] In addition to BMD, *Runx2* SNPs are also associated with femoral length; three *Runx2* SNPs (rs2819858, rs1406846, rs2819854) have been evaluated to date.[166]

Two latest reports demonstrated the functional association of different *Runx2* promoter alleles with BMD. Napierala *et al.*[167] identified promoter sequence variants unique for CCD families on the *Runx2* P1 promoter. A −334A allele and a −330T allele modified the phenotype of a CCD patient; and a −1176C allele abolished the binding of the GATA-type transcriptional repressor TRPS1 on the *Runx2* P1 promoter, thus increasing the expression of *Runx2* and causing a high BMD phenotype. Another report demonstrated that three SNPs on the *Runx2* P2 promoter are associated with increased BMD.[168]

3. PTH/PTHrP

Parathyroid hormone (PTH) is an 84-amino-acid polypeptide secreted by the parathyroid glands in response to relatively small changes in serum Ca^{2+}. PTH is the most important endocrine regulator of calcium and phosphate concentration in extracellular fluid. The primary physiological effect of PTH is to increase serum calcium, decrease serum phosphate, and increase circulating 1,25-dihydroxy-vitamin D_3 [1,25-$(OH)_2$-vitamin D_3]. Bone is one of the two major target organs for PTH action.

PTH-related peptide (PTHrP) has a similar N-terminal structure and biological function to PTH. PTHrP was initially identified as a factor responsible for the syndrome of humoral hypercalcemia of malignancy. PTHrP maintains growth plate chondrocytes in a proliferating stage and plays a critical role in cartilage development.[169]

3.1. *PTH/PTHrP and Their Receptors*

Human PTH is produced almost exclusively by parathyroid cells as a pre-propeptide containing a presequence of 25 amino acids, a prosequence of 6 amino acids, and a mature peptide of 84 amino acids.[170] PTHrP is comprised of 141, 139, or 173 amino acids due to alternative mRNA splicing.[171] PTHrP shares the N-terminal amino acid sequence homology with PTH, particularly within the first 13 residues. The classical PTH/PTHrP receptor is the type 1 PTH receptor (PTHR1), which possesses the unusual property of being bound and activated by PTH as well as PTHrP. The capacity of these two ligands to bind PTHR1 with high affinity and efficiently activate this receptor is based on the sequence similarity in their N-terminal (1–34 amino acids) regions, where 8 of the first 13 amino acids are identical, and on the substantial three-dimensional structural resemblance in the non-homologous 14–34-amino-acid region.

PTHR1 belongs to a distinct group of G protein–coupled receptors termed Family B.[172] In bone, PTHR1 is found primarily in osteoblasts.[173–175] PTH and PTHrP bind to the classical PTHR1, and activate the adenylyl cyclase (cAMP)/protein kinase A (PKA) pathway and the calcium or inositol phosphate/protein kinase C (PKC) pathway. PTHR1 also mediates the signal regulating the extracellular influx of Ca^{2+} through regulation of calcium channels.[176] While the best-characterized second messenger of PTHR1 is undeniably cAMP, activation of the phospholipase C (PLC) and PKC pathways by PTH is likely to play a significant role in renal phosphate transport[177] and osteoblast proliferation.[178] Furthermore, it has been reported that PTHR1 activates PKC through a PLC-independent pathway.[179] A second receptor, PTHR2, binds PTH (but not PTHrP) *in vitro*.[180–182] PTHR2 is closely related to PTHR1 (51% amino acid identity), while its expression appears to be limited to pancreas, brain, kidney, and testis.

In addition to its function through membrane receptor signal transduction cascades, the PTHrP signal can be initiated by the translocation of the nascent protein into the nucleus. PTHrP contains a midregion nuclear localization sequence (NLS) in its 88–106-amino-acid region that is similar to nuclear or nucleolar localization signals found in viral and mammalian transcription factors.[183] In chondrocytes and vascular smooth muscle cells, PTHrP has been found to be located in the nucleus.[184,185] In a more recent study, a second tetrabasic KKKK (147–150) motif has been proved to determine intracrine regulatory effects of PTHrP (1–173) in human chondrocytes.[186] Nuclear localization of PTHrP has been shown to regulate cell proliferation and apoptosis.

3.2. PTH: A Hormone with Dual Functions

The role of PTH in postnatal bone formation was examined in a PTH-deficient mouse model. Mice homozygous for the null mutation of PTH showed low serum $1,25(OH)_2$-vitamin D_3 levels, reduced bone turnover, and increased trabecular and cortical bone volume when they were maintained on a normal calcium intake. When mutant mice were fed with a low-calcium diet, they showed marked increases in circulating $1,25(OH)_2$-vitamin D_3 levels and bone resorption, indicating that the skeletal phenotype of PTH-null mice depends on the calcium intake and circulating $1,25(OH)_2$-vitamin D_3 levels.[187]

PTH stimulates bone resorption and bone formation, depending on the mode of administration. Continuous infusion of PTH *in vivo* results in a marked and prolonged suppression of bone formation and active bone resorption.[188,189] PTH (1–38) infusion (0.01–20 μg/100 g body weight, 1–24-hour infusion) on weanling rats, which were parathyroidectomized and fed with a calcium-free diet, resulted in a dose-dependent increase in serum-ionized calcium and osteoclast numbers in the distal femur. RANKL mRNA expression was increased, but OPG expression was decreased. The expression of osteoblast marker genes was also significantly decreased.[190] These results indicated that continuous infusion of PTH *in vivo* leads to an increased bone resorption and a net loss in bone mass. However, intermittent daily injection with the N-terminal 1–34 fragment of human PTH (hPTH 1–34) is an approved anabolic therapy for the treatment of osteoporosis.[191–197] Treatment of postmenopausal osteoporosis women with 20 μg of PTH (1–34) daily decreases the risk of vertebral and nonvertebral fractures and increases vertebral, femoral, and total-body BMD.[198]

Our knowledge about the molecular events in which PTH mediates these different biological responses in bone is very limited. It has been reported that daily injections of PTH attenuate osteoblast apoptosis and thereby increase osteoblast numbers, bone volume, and bone formation rate, but do not affect the osteoblast number.[199] In contrast, sustained elevation of PTH — achieved either by infusion or by raising endogenous hormone secretion with a calcium-deficient diet — does not affect osteoblast apoptosis, but increases osteoclast numbers.[190] Bellido *et al.*[200] found that the antiapoptotic effect of PTH requires Runt-related transcription factor 2 (*Runx2*)-mediated transcription of survival genes such as *Bcl2*. The duration of the PTH effect is short because PTH itself also stimulates Runx2 degradation through a ubiquitin–proteasome pathway.[200] The self-limiting nature of PTH-induced survival signaling provides a mechanistic explanation for the necessity of intermittent administration to elicit the bone anabolic action of this hormone.[200] Koh *et al.*[201] found that cells of the osteoclast lineage recruited to the bone marrow in response to resorptive stimuli are necessary for the anabolic response to PTH. In addition, PTH is known to stimulate the expression of IGF-I-, IGF-II-, and IGF-binding proteins. PTH also stimulates the synthesis of growth factors such as IL-6,

M-CSF, and RANKL, which promotes osteoclast formation and is likely involved in the catabolic effect of PTH.

PTH regulates bone remodeling and calcium homeostasis by acting on osteoblasts through regulating the expression of several growth factors. Latent TGFβ-binding proteins (LTBPs) are required for the proper folding and secretion of TGFβ, which plays a key role in bone remodeling. Recent data showed that PTH stimulates LTBP1 mRNA expression in osteoblasts in a PKA-dependent manner.[202] A new member of the epidermal growth factor (EGF) family protein, amphiregulin, was found to be rapidly and highly upregulated by PTH in several osteoblastic cell lines and bone tissues.[203] PTH also upregulates the mRNA expression of osteocalcin, and this effect may be mediated by multiple signaling pathways and may require OSE1 and associated nuclear proteins.[204] Osteoblast-derived membrane-type matrix metalloproteinase 1 (MT1-MMP) has been shown to play an important role in bone resorption by degrading bone matrix. PTH (1–34) reduced MT1-MMP expression in cultures of human osteoblast-like MG63 cells in a RANKL-dependent manner, suggesting that the decreased MT1-MMP expression by PTH may play an essential role in the activation of bone resorption.[205]

3.3. *PTHrP: The Bone-Specific Peptide*

PTHrP (1–36) — as a secretory form of PTHrP — is equipotent to PTH (1–34) in terms of its receptor-binding affinity and its ability to stimulate adenylyl cyclase in human bone cells via the PTH/PTHrP receptor *in vitro*, and they show the same effect when infused into human subjects: both peptides produce identical degrees of hypercalcemia, hypophosphatemia, phosphaturia, anticalciuria, nephrogenous cAMP generation, and $1,25(OH)_2$-vitamin D_3 production, suggesting that PTHrP (1–36) may be as effective as PTH (1–34) as an anabolic agent on bone.[206] Daily subcutaneous PTHrP (1–36) administration in postmenopausal women caused an increase in bone formation markers; this is in contrast to the findings in PTH-treated subjects, in whom bone resorption markers were decreased. These findings suggest that PTHrP (1–36) may uncouple bone formation from bone resorption in favor of bone formation, thus predicting that PTHrP (1–36) might be a potent anabolic agent for the treatment of osteoporosis.[207]

PTHrP mRNA and protein are also detected in intramembranous bone during the early mesenchymal stage of development. Three types of genetically modified animal models have been used to demonstrate the critical role of PTHrP in skeleton development: $PTHrP^{+/-}$ mice, $PTHrP^{-/-}$ mice, and $PTHrP$-overexpressing mice. Heterozygous $PTHrP$-null mutant mice had normal bone phenotype at birth, but developed a marked osteoporotic phenotype at 3 months of age characterized by a marked decrease in trabecular thickness and connectivity, and possessed an increased number of adipocytes in their bone marrow.[169] Mice missing both alleles of the $PTHrP$ gene $(PTHrP^{-/-})$ displayed a striking form of accelerated skeletal mineralization leading to dwarfism and a fixed, smaller-than-normal rib cage.[113] Neonatal mice died immediately after delivery from apparent respiratory failure, presumably resulting at least in part from their skeletal (rib cage) inflexibility.

Homozygous ablation of the gene encoding PTHR1 $(PTHR1^{-/-})$ led to early lethality and limited developmental defects, including an acceleration of chondrocyte differentiation. It is now known that PTHrP stimulates chondrocyte proliferation through activation of Ihh, and prevents chondrocyte differentiation. $Ihh^{-/-}$ mice showed an increase in postmitotic, hypertrophic chondrocytes due to its failure to synthesize PTHrP.[208] The actions of PTHrP on chondrocytes are mediated by PTHR1, which is most abundantly expressed in chondrocytes residing in the zone between proliferation and hypertrophy; this may explain why $PTHR1$-ablated mice $(PTHR1^{-/-})$ showed skeletal abnormalities that were similar to, but more severe than, those observed in $PTHrP^{-/-}$ mice.[209,210]

The third phenotype is also striking: $PTHrP$-overexpressing mice in which the $PTHrP$ transgene was targeted to chondrocytes using the type II collagen promoter ($col2a1$) displayed shortened limbs due to an almost complete arrest of mineralization, resulting in a persistent endochondral nonmineralized skeleton.[211] Taken together, these three mouse models reveal an unequivocal message: $PTHrP$ plays a central role in skeletal development, particularly in the regulation of chondrocyte proliferation and maturation as well as skeletal mineralization.

3.4. *Genetic Disorders Caused by PTHR1 Mutations*

The linkage between mutations in PTH/PTHrP signaling genes and human diseases has been investigated. Heterozygous $PTHR1$ missense mutations were identified in patients with Jansen's metaphyseal chondrodysplasia (JMC), a rare form of short-limb dwarfism associated with hypercalcemia and normal or undetectable levels of PTH and PTHrP.[212] All three types of mutations — I458R, H223R, and T410P — resulted in constitutive activation of the cAMP signaling pathway, and provide a plausible explanation for the abnormalities in skeletal development and mineral homeostasis.[213,214] Beier and Luvalle[225] demonstrated that JMC mutations of the *PTH/PTHrP* receptor induced activation of the cyclin D1 and cyclin A promoters through a functional cAMP response element in primary mouse chondrocytes and rat chondrosarcoma cells, suggesting that the stimulation of cell cycle gene expression and cell cycle progression by mutant *PTH/PTHrP* receptors contributes to the pathogenesis of JMC. Mice expressing constitutively active PTHR1 (caPPR) under the control of the 2.3-kb bone-specific mouse *col1a1* promoter showed increased bone formation within prospective marrow space; but delayed the transition from bone to bone marrow during growth, the formation of marrow cavities, and the appearance of stromal cell types such as marrow adipocytes and cells supporting hematopoiesis — hence, indicating that, during endochondral ossification, *PTHR1* plays an important role in regulating the timed transition between bone and marrow.[216]

Blomstrand's lethal chondrodysplasia (BLC), an autosomal recessive disorder, was first reported in 1985. Inhibition of chondrocyte proliferation and accelerated endochondral bone formation are major changes in patients with BLC.[217,218] Genetic studies led to the finding that BLC is caused by inactivating mutations in the *PTH/PTHrP* receptor. To date, three types of mutant *PTH/PTHrP* receptors have been identified. The first of these, Δ373-383-PPR, lacks a segment of the fifth transmembrane domain of *PTHR1* because of a single nucleotide change ($G \rightarrow A$ substitution at nucleotide 1176) that affects mRNA splicing[219]; this mutation causes the loss of responsiveness of PTHR to PTH/PTHrP. The second one, P132L-PPR, is a single nucleotide exchange that leads to the P132L mutation in the N-terminal extracellular domain.[220] Δ365-593-PPR — the most severe mutant of the three — lacks a 230-a.a.-long fragment including transmembrane domains 5, 6, and 7; the connecting loops; and the cytoplasmic tail. This mutant receptor loses its responsiveness to PTH.[221] COS cells transiently expressing *PTHR1* with either the P132L, Δ373-383, or Δ365-593 mutations showed a significantly lower accumulation of cyclic AMP in response to PTH than that of the cells expressing wild-type *PTHR1*. Thus, BLC is associated with compound heterozygous or homozygous mutations that lead to mutant *PTH1R* with severely impaired functional properties.

Enchondromas are common benign cartilage tumors of bone that occur as solitary lesions. In normal growth plates, differentiation of proliferative chondrocytes to postmitotic hypertrophic chondrocytes is regulated in part by a tightly coupled signaling relay involving *PTHrP* and Indian hedgehog (*IHH*).[222–224] Recent research showed that the R150C mutation of the *PTH/PTHrP* receptor constitutively activates the *Ihh* signaling pathway and leads to the formation of enchondromas.[225] More detailed molecular mechanism studies are required to fully understand *PTH/PTHrP* signaling in bone cells.

REFERENCES

1. Huelsken J, Birchmeier W. (2001) New aspects of Wnt signaling pathways in higher vertebrates. *Curr Opin Genet Dev* **11**:547–553.
2. Westendorf JJ, Kahler RA, Schroeder TM. (2004) Wnt signaling in osteoblasts and bone diseases. *Gene* **341**:19–39.
3. Behrens J, Jerchow BA, Wurtele M, *et al.* (1998) Functional interaction of an Axin homolog, Conductin, with beta-catenin, APC, and GSK3beta. *Science* **280**:596–599.
4. Jiang J, Struhl G. (1998) Regulation of the Hedgehog and Wingless signalling pathways by the F-box/WD40-repeat protein Slimb. *Nature* **391**:493–496.
5. Aberle H, Bauer A, Stappert J, *et al.* (1997) Beta-catenin is a target for the ubiquitin–proteasome pathway. *EMBO J* **16**:3797–3804.
6. Mao J, Wang J, Liu B, *et al.* (2001) Low-density lipoprotein receptor–related protein-5 binds to Axin and regulates the canonical Wnt signaling pathway. *Mol Cell* **7**:801–809.
7. Bejsovec A. (2000) Wnt signaling: an embarrassment of receptors. *Curr Biol* **10**:R919–R922.
8. Mao B, Wu W, Davidson G, *et al.* (2002) Kremen proteins are Dickkopf receptors that regulate Wnt/beta-catenin signalling. *Nature* **417**:664–667.
9. Hill TP, Spater D, Taketo MM, *et al.* (2005) Canonical Wnt/beta-catenin signaling prevents osteoblasts from differentiating into chondrocytes. *Dev Cell* **8**:727–738.
10. Day TF, Guo X, Garrett-Beal L, Yang Y. (2005) Wnt/beta-catenin signaling in mesenchymal progenitors controls osteoblast and chondrocyte differentiation during vertebrate skeletogenesis. *Dev Cell* **8**:739–750.
11. Akiyama H, Lyons JP, Mori-Akiyama Y, *et al.* (2004) Interactions between Sox9 and beta-catenin control chondrocyte differentiation. *Genes Dev* **18**:1072–1087.
12. Tamai K, Semenov M, Kato Y, *et al.* (2000) LDL-receptor–related proteins in Wnt signal transduction. *Nature* **407**:530–535.
13. Gong Y, Vikkula M, Boon L, *et al.* (1996) Osteoporosis–pseudoglioma syndrome, a disorder affecting skeletal strength and vision, is assigned to chromosome region 11q12–13. *Am J Hum Genet* **59**:146–151.
14. Johnson ML, Gong G, Kimberling W, *et al.* (1997) Linkage of a gene causing high bone mass to human chromosome 11 (11q12–13). *Am J Hum Genet* **60**:1326–1332.

15. Gong Y, Slee RB, Fukai N, *et al.* (2001) LDL receptor–related protein 5 (LRP5) affects bone accrual and eye development. *Cell* **107**:513–523.

16. Crabbe P, Balemans W, Willaert A, *et al.* (2005) Missense mutations in LRP5 are not a common cause of idiopathic osteoporosis in adult men. *J Bone Miner Res* **20**:1951–1959.

17. van Meurs JB, Rivadeneira F, Jhamai M, *et al.* (2006) Common genetic variation of the low-density lipoprotein receptor–related protein 5 and 6 genes determines fracture risk in elderly white men. *J Bone Miner Res* **21**:141–150.

18. Hartikka H, Makitie O, Mannikko M, *et al.* (2005) Heterozygous mutations in the LDL receptor–related protein 5 (LRP5) gene are associated with primary osteoporosis in children. *J Bone Miner Res* **20**:783–789.

19. Kato M, Patel MS, Levasseur R, *et al.* (2002) Cbfa1-independent decrease in osteoblast proliferation, osteopenia, and persistent embryonic eye vascularization in mice deficient in LRP5, a Wnt coreceptor. *J Cell Biol* **157**:303–314.

20. Clement-Lacroix P, Ai M, Morvan F, *et al.* (2005) LRP5-independent activation of Wnt signaling by lithium chloride increases bone formation and bone mass in mice. *Proc Natl Acad Sci USA* **102**:17406–17411.

21. Van Hul E, Gram J, Bollerslev J, *et al.* (2002) Localization of the gene causing autosomal dominant osteopetrosis type I to chromosome 11q12–13. *J Bone Miner Res* **17**:1111–1117.

22. Bollerslev J, Ueland T, Odgren PR. (2003) Serum levels of TGF-beta and fibronectin in autosomal dominant osteopetrosis in relation to underlying mutations and well-described murine counterparts. *Crit Rev Eukaryot Gene Expr* **13**:163–171.

23. Kwee ML, Balemans W, Cleiren E, *et al.* (2005) An autosomal dominant high bone mass phenotype in association with craniosynostosis in an extended family is caused by an LRP5 missense mutation. *J Bone Miner Res* **20**:1254–1260.

24. Little RD, Carulli JP, Del Mastro RG, *et al.* (2002) A mutation in the LDL receptor–related protein 5 gene results in the autosomal dominant high-bone-mass trait. *Am J Hum Genet* **70**:11–19.

25. Boyden LM, Mao J, Belsky J, *et al.* (2002) High bone density due to a mutation in LDL-receptor-related protein 5. *N Engl J Med* **346**:1513–1521.

26. Babij P, Zhao W, Small C, *et al.* (2003) High bone mass in mice expressing a mutant LRP5 gene. *J Bone Miner Res* **18**:960–974.

27. Zhang Y, Wang Y, Li X, *et al.* (2004) The LRP5 high-bone-mass G171V mutation disrupts LRP5 interaction with Mesd. *Mol Cell Biol* **24**:4677–4684.

28. Van Wesenbeeck L, Cleiren E, Gram J, *et al.* (2003) Six novel missense mutations in the LDL receptor–related protein 5 (LRP5) gene in different conditions with an increased bone density. *Am J Hum Genet* **72**:763–771.

29. Rickels MR, Zhang X, Mumm S, Whyte MP. (2005) Oropharyngeal skeletal disease accompanying high bone mass and novel LRP5 mutation. *J Bone Miner Res* **20**:878–885.

30. Akhter MP, Wells DJ, Short SJ, *et al.* (2004) Bone biomechanical properties in LRP5 mutant mice. *Bone* **35**:162–169.

31. Ai M, Holmen SL, Van Hul W, *et al.* (2005) Reduced affinity to and inhibition by Dkk1 form a common mechanism by which high bone mass–associated missense mutations in LRP5 affect canonical Wnt signaling. *Mol Cell Biol* **25**:4946–4955.

32. Urano T, Shiraki M, Ezura Y, *et al.* (2004) Association of a single-nucleotide polymorphism in low-density lipoprotein receptor–related protein 5 gene with bone mineral density. *J Bone Miner Metab* **22**:341–345.

33. Mizuguchi T, Furuta I, Watanabe Y, *et al.* (2004) LRP5, low-density-lipoprotein-receptor-related protein 5, is a determinant for bone mineral density. *J Hum Genet* **49**:80–86.

34. Ferrari SL, Deutsch S, Choudhury U, *et al.* (2004) Polymorphisms in the low-density lipoprotein receptor–related protein 5 (LRP5) gene are associated with variation in vertebral bone mass, vertebral bone size, and stature in whites. *Am J Hum Genet* **74**:866–875.

35. Koay MA, Woon PY, Zhang Y, *et al.* (2004) Influence of LRP5 polymorphisms on normal variation in BMD. *J Bone Miner Res* **19**:1619–1627.

36. Bollerslev J, Wilson SG, Dick IM, *et al.* (2005) LRP5 gene polymorphisms predict bone mass and incident fractures in elderly Australian women. *Bone* **36**:599–606.

37. Zhang ZL, Qin YJ, He JW, *et al.* (2005) Association of polymorphisms in low-density lipoprotein receptor–related protein 5 gene with bone mineral density in postmenopausal Chinese women. *Acta Pharmacol Sin* **26**: 1111–1116.

38. Ferrari SL, Deutsch S, Baudoin C, *et al.* (2005) LRP5 gene polymorphisms and idiopathic osteoporosis in men. *Bone* **37**:770–775.

39. Smith AJ, Gidley J, Sandy JR, *et al.* (2005) Haplotypes of the low-density lipoprotein receptor–related protein 5 (LRP5) gene: are they a risk factor in osteoarthritis? *Osteoarthritis Cartilage* **13**:608–613.

40. Fujino T, Asaba H, Kang MJ, *et al.* (2003) Low-density lipoprotein receptor–related protein 5 (LRP5) is essential for normal cholesterol metabolism and glucose-induced insulin secretion. *Proc Natl Acad Sci USA* **100**:229–234.

41. Etheridge SL, Spencer GJ, Heath DJ, Genever PG. (2004) Expression profiling and functional analysis of Wnt signaling mechanisms in mesenchymal stem cells. *Stem Cells* **22**:849–860.

42. Boland GM, Perkins G, Hall DJ, Tuan RS. (2004) Wnt 3a promotes proliferation and suppresses osteogenic differentiation of adult human mesenchymal stem cells. *J Cell Biochem* **93**:1210–1230.

43. Rawadi G, Vayssiere B, Dunn F, *et al.* (2003) BMP-2 controls alkaline phosphatase expression and osteoblast mineralization by a Wnt autocrine loop. *J Bone Miner Res* **18**:1842–1853.

44. Zhou S, Eid K, Glowacki J. (2004) Cooperation between TGF-beta and Wnt pathways during chondrocyte and adipocyte differentiation of human marrow stromal cells. *J Bone Miner Res* **19**:463–470.

45. Hu H, Hilton MJ, Tu X, *et al.* (2005) Sequential roles of Hedgehog and Wnt signaling in osteoblast development. *Development* **132**:49–60.
46. Almeida M, Han L, Bellido T, *et al.* (2005) Wnt proteins prevent apoptosis of both uncommitted osteoblast progenitors and differentiated osteoblasts by beta-catenin-dependent and -independent signaling cascades involving Src/ERK and phosphatidylinositol 3-kinase/AKT. *J Biol Chem* **280**: 41342–41351.
47. Hausler KD, Horwood NJ, Chuman Y, *et al.* (2004) Secreted Frizzled-related protein-1 inhibits RANKL-dependent osteoclast formation. *J Bone Miner Res* **19**:1873–1881.
48. Glass DA 2nd, Bialek P, Ahn JD, *et al.* (2005) Canonical Wnt signaling in differentiated osteoblasts controls osteoclast differentiation. *Dev Cell* **8**: 751–764.
49. Holmen SL, Zylstra CR, Mukherjee A, *et al.* (2005) Essential role of beta-catenin in postnatal bone acquisition. *J Biol Chem* **280**:21162–21168.
50. Levanon D, Negreanu V, Bernstein Y, *et al.* (1994) AML1, AML2 and AML3, the human members of the Runt domain gene-family: cDNA structure, expression, and chromosomal localization. *Genomics* **23**:425–432.
51. Komori T, Yagi H, Nomura S, *et al.* (1997) Targeted disruption of Cbfa1 results in a complete lack of bone formation owing to maturational arrest of osteoblasts. *Cell* **89**:755–764.
52. Mundlos S, Otto F, Mundlos C, *et al.* (1997) Mutations involving the transcription factor CBFA1 cause cleidocranial dysplasia. *Cell* **89**:773–779.
53. Otto F, Thornell AP, Crompton T, *et al.* (1997) Cbfa1, a candidate gene for cleidocranial dysplasia syndrome, is essential for osteoblast differentiation and bone development. *Cell* **89**:765–771.
54. Ogawa E, Maruyama M, Kagoshima H, *et al.* (1993) PEBP2/PEA2 represents a family of transcription factors homologous to the products of the *Drosophila* Runt gene and the human AML1 gene. *Proc Natl Acad Sci USA* **90**:6859–6863.
55. Satake M, Nomura S, Yamaguchi-Iwai Y, *et al.* (1995) Expression of the Runt domain-encoding PEBP2 alpha genes in T cells during thymic development. *Mol Cell Biol* **15**:1662–1670.
56. Stewart M, Terry A, Hu M, *et al.* (1997) Proviral insertions induce the expression of bone-specific isoforms of PEBP2alphaA (CBFA1): evidence for a new myc collaborating oncogene. *Proc Natl Acad Sci USA* **94**: 8646–8651.
57. Zhang YW, Bae SC, Huang G, *et al.* (1997) A novel transcript encoding an N-terminally truncated AML1/PEBP2 alphaB protein interferes with transactivation and blocks granulocytic differentiation of 32Dcl3 myeloid cells. *Mol Cell Biol* **17**:4133–4145.
58. Sudhakar S, Li Y, Katz MS, Elango N. (2001) Translational regulation is a control point in RUNX2/Cbfa1 gene expression. *Biochem Biophys Res Commun* **289**:616–622.
59. Drissi H, Luc Q, Shakoori R, *et al.* (2000) Transcriptional autoregulation of the bone related CBFA1/RUNX2 gene. *J Cell Physiol* **184**:341–350.

60. Xiao ZS, Liu SG, Hinson TK, Quarles LD. (2001) Characterization of the upstream mouse Cbfa1/Runx2 promoter. *J Cell Biochem* **82**:647–659.

61. Xiao ZS, Simpson LG, Quarles LD. (2003) IRES-dependent translational control of Cbfa1/Runx2 expression. *J Cell Biochem* **88**:493–505.

62. Javed A, Guo B, Hiebert S, *et al.* (2000) Groucho/TLE/R-esp proteins associate with the nuclear matrix and repress RUNX (CBF(alpha)/AML/PEBP2(alpha)) dependent activation of tissue-specific gene transcription. *J Cell Sci* **113**:2221–2231.

63. Franceschi RT, Xiao G. (2003) Regulation of the osteoblast-specific transcription factor, Runx2: responsiveness to multiple signal transduction pathways. *J Cell Biochem* **88**:446–454.

64. Zhao M, Qiao M, Oyajobi BO, *et al.* (2003) E3 ubiquitin ligase Smurf1 mediates core-binding factor alpha1/Runx2 degradation and plays a specific role in osteoblast differentiation. *J Biol Chem* **278**:27939–27944.

65. Shen R, Chen M, Wang YJ, *et al.* (2006) Smad6 interacts with Runx2 and mediates Smad ubiquitin regulatory factor 1–induced Runx2 degradation. *J Biol Chem* **281**:3569–3576.

66. Ryoo HM, Hoffmann HM, Beumer T, *et al.* (1997) Stage-specific expression of Dlx-5 during osteoblast differentiation: involvement in regulation of osteocalcin gene expression. *Mol Endocrinol* **11**:1681–1694.

67. Lee MH, Javed A, Kim HJ, *et al.* (1999) Transient upregulation of CBFA1 in response to bone morphogenetic protein-2 and transforming growth factor beta1 in C2C12 myogenic cells coincides with suppression of the myogenic phenotype but is not sufficient for osteoblast differentiation. *J Cell Biochem* **73**:114–125.

68. Komori T. (2002) Runx2, a multifunctional transcription factor in skeletal development. *J Cell Biochem* **87**:1–8.

69. Stricker S, Fundele R, Vortkamp A, Mundlos S. (2002) Role of Runx genes in chondrocyte differentiation. *Dev Biol* **245**:95–108.

70. Vaes BL, Dechering KJ, Feijen A, *et al.* (2002) Comprehensive microarray analysis of bone morphogenetic protein 2–induced osteoblast differentiation resulting in the identification of novel markers for bone development. *J Bone Miner Res* **17**:2106–2118.

71. Milona MA, Gough JE, Edgar AJ. (2003) Expression of alternatively spliced isoforms of human Sp7 in osteoblast-like cells. *BMC Genomics* **4**:43.

72. Nakashima K, Zhou X, Kunkel G, *et al.* (2002) The novel zinc finger–containing transcription factor osterix is required for osteoblast differentiation and bone formation. *Cell* **108**:17–29.

73. Ducy P, Starbuck M, Priemel M, *et al.* (1999) A Cbfa1-dependent genetic pathway controls bone formation beyond embryonic development. *Genes Dev* **13**:1025–1036.

74. Javed A, Gutierrez S, Montecino M, *et al.* (1999) Multiple Cbfa/AML sites in the rat osteocalcin promoter are required for basal and vitamin D–responsive transcription and contribute to chromatin organization. *Mol Cell Biol* **19**:7491–7500.

75. Javed A, Barnes GL, Jasanya BO, *et al.* (2001) Runt homology domain transcription factors (Runx, Cbfa, and AML) mediate repression of the bone

sialoprotein promoter: evidence for promoter context-dependent activity of Cbfa proteins. *Mol Cell Biol* **21**:2891–2905.

76. Harada H, Tagashira S, Fujiwara M, *et al.* (1999) Cbfa1 isoforms exert functional differences in osteoblast differentiation. *J Biol Chem* **274**: 6972–6978.

77. Sato M, Morii E, Komori T, *et al.* (1998) Transcriptional regulation of osteopontin gene *in vivo* by PEBP2alphaA/CBFA1 and ETS1 in the skeletal tissues. *Oncogene* **17**:1517–1525.

78. Ducy P, Zhang R, Geoffroy V, *et al.* (1997) Osf2/Cbfa1: a transcriptional activator of osteoblast differentiation. *Cell* **89**:747–754.

79. Merriman HL, van Wijnen AJ, Hiebert S, *et al.* (1995) The tissue-specific nuclear matrix protein, NMP-2, is a member of the AML/CBF/PEBP2/ Runt domain transcription factor family: interactions with the osteocalcin gene promoter. *Biochemistry* **34**:13125–13132.

80. Hess J, Porte D, Munz C, Angel P. (2001) AP-1 and Cbfa/Runt physically interact and regulate parathyroid hormone–dependent MMP13 expression in osteoblasts through a new osteoblast-specific element 2/AP-1 composite element. *J Biol Chem* **276**:20029–20038.

81. Jimenez MJ, Balbin M, Lopez JM, *et al.* (1999) Collagenase 3 is a target of Cbfa1, a transcription factor of the Runt gene family involved in bone formation. *Mol Cell Biol* **19**:4431–4442.

82. Selvamurugan N, Chou WY, Pearman AT, *et al.* (1998) Parathyroid hormone regulates the rat collagenase-3 promoter in osteoblastic cells through the cooperative interaction of the activator protein-1 site and the Runt domain binding sequence. *J Biol Chem* **273**:10647–10657.

83. Chen S, Gu TT, Sreenath T, *et al.* (2002) Spatial expression of Cbfa1/Runx2 isoforms in teeth and characterization of binding sites in the DSPP gene. *Connect Tissue Res* **43**:338–344.

84. Kern B, Shen J, Starbuck M, Karsenty G. (2001) Cbfa1 contributes to the osteoblast-specific expression of type I collagen genes. *J Biol Chem* **276**:7101–7107.

85. Zheng Q, Zhou G, Morello R, *et al.* (2003) Type X collagen gene regulation by Runx2 contributes directly to its hypertrophic chondrocyte–specific expression *in vivo*. *J Cell Biol* **162**:833–842.

86. McCarthy TL, Ji C, Chen Y, *et al.* (2000) Runt domain factor (Runx)-dependent effects on CCAAT/enhancer-binding protein delta expression and activity in osteoblasts. *J Biol Chem* **275**:21746–21753.

87. Yoshida CA, Yamamoto H, Fujita T, *et al.* (2004) Runx2 and Runx3 are essential for chondrocyte maturation, and Runx2 regulates limb growth through induction of Indian hedgehog. *Genes Dev* **18**:952–963.

88. Zelzer E, Olsen BR. (2001) Multiple roles of vascular endothelial growth factor (VEGF) in skeletal development, growth, and repair. *Curr Top Dev Biol* **65**:169–187.

89. Takahashi A, Satake M, Yamaguchi-Iwai Y, *et al.* (1995) Positive and negative regulation of granulocyte–macrophage colony-stimulating factor promoter activity by AML1–related transcription factor, PEBP2. *Blood* **86**:607–616.

90. Cameron S, Taylor DS, TePas EC, *et al.* (1994) Identification of a critical regulatory site in the human interleukin-3 promoter by *in vivo* footprinting. *Blood* **83**:2851–2859.

91. Bristow CA, Shore P. (2003) Transcriptional regulation of the human MIP-1alpha promoter by RUNX1 and MOZ. *Nucleic Acids Res* **31**:2735–2744.

92. Zhang DE, Fujioka K, Hetherington CJ, *et al.* (1994) Identification of a region which directs the monocytic activity of the colony-stimulating factor 1 (macrophage colony-stimulating factor) receptor promoter and binds PEBP2/CBF (AML1). *Mol Cell Biol* **14**:8085–8095.

93. Ji C, Casinghino S, Chang DJ, *et al.* (1998) CBFa(AML/PEBP2)–related elements in the TGF-beta type I receptor promoter and expression with osteoblast differentiation. *J Cell Biochem* **69**:353–363.

94. Sasaki-Iwaoka H, Maruyama K, Endhoh H, *et al.* (1999) A *trans*-acting enhancer modulates estrogen-mediated transcription of reporter genes in osteoblasts. *J Bone Miner Res* **14**:248–255.

95. Tou L, Quibria N, Alexander JM. (2001) Regulation of human cbfa1 gene transcription in osteoblasts by selective estrogen receptor modulators (SERMs). *Mol Cell Endocrinol* **183**:71–79.

96. Enomoto H, Enomoto-Iwamoto M, Iwamoto M, *et al.* (2000) Cbfa1 is a positive regulatory factor in chondrocyte maturation. *J Biol Chem* **275**: 8695–8702.

97. Thirunavukkarasu K, Halladay DL, Miles RR, *et al.* (2000) The osteoblast-specific transcription factor Cbfa1 contributes to the expression of osteoprotegerin, a potent inhibitor of osteoclast differentiation and function. *J Biol Chem* **275**:25163–25172.

98. Pratap J, Galindo M, Zaidi SK, *et al.* (2003) Cell growth regulatory role of Runx2 during proliferative expansion of preosteoblasts. *Cancer Res* **63**:5357–5362.

99. Westendorf JJ, Zaidi SK, Cascino JE, *et al.* (2002) Runx2 (Cbfa1, AML-3) interacts with histone deacetylase 6 and represses the p21(CIP1/WAF1) promoter. *Mol Cell Biol* **22**:7982–7992.

100. Thomas M, Dadgar N, Aphale A, *et al.* (2004) Androgen receptor acetylation site mutations cause trafficking defects, misfolding, and aggregation similar to expanded glutamine tracts. *J Biol Chem* **279**:8389–8395.

101. Thomas DM, Carty SA, Piscopo DM, *et al.* (2001) The retinoblastoma protein acts as a transcriptional coactivator required for osteogenic differentiation. *Mol Cell* **8**:303–316.

102. Bialek P, Kern B, Yang X, *et al.* (2004) A twist code determines the onset of osteoblast differentiation. *Dev Cell* **6**:423–435.

103. Zaidi SK, Sullivan AJ, van Wijnen AJ, *et al.* (2002) Integration of Runx and Smad regulatory signals at transcriptionally active subnuclear sites. *Proc Natl Acad Sci USA* **99**:8048–8053.

104. Zhang YW, Yasui N, Ito K, *et al.* (2000) A RUNX2/PEBP2alpha A/CBFA1 mutation displaying impaired transactivation and Smad interaction in cleidocranial dysplasia. *Proc Natl Acad Sci USA* **97**:10549–10554.

105. Selvamurugan N, Pulumati MR, Tyson DR, Partridge NC. (2000) Parathyroid hormone regulation of the rat collagenase-3 promoter by protein kinase A–dependent transactivation of core binding factor alpha1. *J Biol Chem* **275**:5037–5042.

106. Ducy P. (2000) Cbfa1: a molecular switch in osteoblast biology. *Dev Dyn* **219**:461–471.

107. Kim WY, Sieweke M, Ogawa E, *et al.* (1999) Mutual activation of Ets-1 and AML1 DNA binding by direct interaction of their autoinhibitory domains. *EMBO J* **18**:1609–1620.

108. Inada M, Yasui T, Nomura S, *et al.* (1999) Maturational disturbance of chondrocytes in Cbfa1-deficient mice. *Dev Dyn* **214**:279–290.

109. Ueta C, Iwamoto M, Kanatani N, *et al.* (2001) Skeletal malformations caused by overexpression of Cbfa1 or its dominant negative form in chondrocytes. *J Cell Biol* **153**:87–100.

110. Takeda S, Bonnamy JP, Owen MJ, *et al.* (2001) Continuous expression of Cbfa1 in nonhypertrophic chondrocytes uncovers its ability to induce hypertrophic chondrocyte differentiation and partially rescues Cbfa1-deficient mice. *Genes Dev* **15**:467–481.

111. Jacenko O, Chan D, Franklin A, *et al.* (2001) A dominant interference collagen X mutation disrupts hypertrophic chondrocyte pericellular matrix and glycosaminoglycan and proteoglycan distribution in transgenic mice. *Am J Pathol* **159**:2257–2269.

112. Gress CJ, Jacenko O. (2000) Growth plate compressions and altered hematopoiesis in collagen X null mice. *J Cell Biol* **149**:983–993.

113. Karaplis AC, Luz A, Glowacki J, *et al.* (1994) Lethal skeletal dysplasia from targeted disruption of the parathyroid hormone–related peptide gene. *Genes Dev* **8**:277–289.

114. Iwamoto M, Kitagaki J, Tamamura Y, *et al.* (2003) Runx2 expression and action in chondrocytes are regulated by retinoid signaling and parathyroid hormone–related peptide (PTHrP). *Osteoarthritis Cartilage* **11**:6–15.

115. Inada M, Wang Y, Byrne MH, *et al.* (2004) Critical roles for collagenase-3 (Mmp13) in development of growth plate cartilage and in endochondral ossification. *Proc Natl Acad Sci USA* **101**:17192–17197.

116. Stickens D, Behonick DJ, Ortega N, *et al.* (2004) Altered endochondral bone development in matrix metalloproteinase 13–deficient mice. *Development* **131**:5883–5895.

117. Colnot C, Lu C, Hu D, Helms JA. (2004) Distinguishing the contributions of the perichondrium, cartilage, and vascular endothelium to skeletal development. *Dev Biol* **269**:55–69.

118. Colnot C. (2005) Cellular and molecular interactions regulating skeletogenesis. *J Cell Biochem* **95**:688–697.

119. Hershey CL, Fisher DE. (2004) Mitf and Tfe3: members of a b-HLH-ZIP transcription factor family essential for osteoclast development and function. *Bone* **34**:689–696.

120. Steingrimsson E, Tessarollo L, Pathak B, *et al.* (2002) Mitf and Tfe3, two members of the Mitf-Tfe family of bHLH-Zip transcription factors, have

important but functionally redundant roles in osteoclast development. *Proc Natl Acad Sci USA* **99**:4477–4482.

121. Partington GA, Fuller K, Chambers TJ, Pondel M. (2004) Mitf-PU.1 interactions with the tartrate-resistant acid phosphatase gene promoter during osteoclast differentiation. *Bone* **34**:237–245.

122. Luchin A, Purdom G, Murphy K, *et al.* (2000) The microphthalmia transcription factor regulates expression of the tartrate-resistant acid phosphatase gene during terminal differentiation of osteoclasts. *J Bone Miner Res* **15**:451–460.

123. Motyckova G, Weilbaecher KN, Horstmann M, *et al.* (2001) Linking osteopetrosis and pycnodysostosis: regulation of cathepsin K expression by the microphthalmia transcription factor family. *Proc Natl Acad Sci USA* **98**:5798–5803.

124. So H, Rho J, Jeong D, *et al.* (2003) Microphthalmia transcription factor and PU.1 synergistically induce the leukocyte receptor osteoclast-associated receptor gene expression. *J Biol Chem* **278**:24209–24216.

125. Matsuo K, Galson DL, Zhao C, *et al.* (2004) Nuclear factor of activated T-cells (NFAT) rescues osteoclastogenesis in precursors lacking c-Fos. *J Biol Chem* **279**:26475–26480.

126. Cappellen D, Luong-Nguyen NH, Bongiovanni S, *et al.* (2002) Transcriptional program of mouse osteoclast differentiation governed by the macrophage colony-stimulating factor and the ligand for the receptor activator of NFκB. *J Biol Chem* **277**:21971–21982.

127. Ishida N, Hayashi K, Hoshijima M, *et al.* (2002) Large scale gene expression analysis of osteoclastogenesis *in vitro* and elucidation of NFAT2 as a key regulator. *J Biol Chem* **277**:41147–41156.

128. Takayanagi H, Kim S, Koga T, *et al.* (2002) Induction and activation of the transcription factor NFATc1 (NFAT2) integrate RANKL signaling in terminal differentiation of osteoclasts. *Dev Cell* **3**:889–901.

129. Banerjee C, Javed A, Choi JY, *et al.* (2001) Differential regulation of the two principal Runx2/Cbfa1 N-terminal isoforms in response to bone morphogenetic protein-2 during development of the osteoblast phenotype. *Endocrinology* **142**:4026–4039.

130. Choi KY, Lee SW, Park MH, *et al.* (2002) Spatio-temporal expression patterns of Runx2 isoforms in early skeletogenesis. *Exp Mol Med* **34**:426–433.

131. Park MH, Shin HI, Choi JY, *et al.* (2001) Differential expression patterns of Runx2 isoforms in cranial suture morphogenesis. *J Bone Miner Res* **16**:885–892.

132. Xiao ZS, Hinson TK, Quarles LD. (1999) Cbfa1 isoform overexpression upregulates osteocalcin gene expression in non-osteoblastic and pre-osteoblastic cells. *J Cell Biochem* **74**:596–605.

133. Kim S, Koga T, Isobe M, *et al.* (2003) Stat1 functions as a cytoplasmic attenuator of Runx2 in the transcriptional program of osteoblast differentiation. *Genes Dev* **17**:1979–1991.

134. Kim HJ, Lee MH, Park HS, *et al.* (2003) Erk pathway and activator protein 1 play crucial roles in FGF2–stimulated premature cranial suture closure. *Dev Dyn* **227**:335–346.

135. Xiao G, Jiang D, Gopalakrishnan R, Franceschi RT. (2002) Fibroblast growth factor 2 induction of the osteocalcin gene requires MAPK activity and phosphorylation of the osteoblast transcription factor, Cbfa1/Runx2. *J Biol Chem* **277**:36181–36187.

136. Phillips JE, Gersbach CA, Wojtowicz AM, Garcia AJ. (2006) Glucocorticoid-induced osteogenesis is negatively regulated by Runx2/Cbfa1 serine phosphorylation. *J Cell Sci* **119**:581–591.

137. Kim BG, Kim HJ, Park HJ, *et al.* (2006) Runx2 phosphorylation induced by fibroblast growth factor-2/protein kinase C pathways. *Proteomics* **6**:1166–1174.

138. Qiao M, Shapiro P, Fosbrink M, *et al.* (2006) Cell cycle–dependent phosphorylation of the RUNX2 transcription factor by cdc2 regulates endothelial cell proliferation. *J Biol Chem* **281**:7118–7128.

139. Wee HJ, Huang G, Shigesada K, Ito Y. (2002) Serine phosphorylation of RUNX2 with novel potential functions as negative regulatory mechanisms. *EMBO Rep* **3**:967–974.

140. Shen R, Wang X, Liu F, *et al.* (2006) Cyclin D1–Cdk4 induce Runx2 ubiquitination and degradation. *J Biol Chem* **281**:16347–16353.

141. Zhao M, Qiao M, Harris SE, *et al.* (2004) Smurf1 inhibits osteoblast differentiation and bone formation *in vitro* and *in vivo*. *J Biol Chem* **279**:12854–12859.

142. Zhu H, Kavsak P, Abdollah S, *et al.* (1999) A SMAD ubiquitin ligase targets the BMP pathway and affects embryonic pattern formation. *Nature* **400**:687–693.

143. Mundlos S, Mulliken JB, Abramson DL, *et al.* (1995) Genetic mapping of cleidocranial dysplasia and evidence of a microdeletion in one family. *Hum Mol Genet* **4**:71–75.

144. Bergwitz C, Prochnau A, Mayr B, *et al.* (2001) Identification of novel CBFA1/RUNX2 mutations causing cleidocranial dysplasia. *J Inherit Metab Dis* **24**:648–656.

145. Kim HJ, Nam SH, Kim HJ, *et al.* (2006) Four novel RUNX2 mutations including a splice donor site result in the cleidocranial dysplasia phenotype. *J Cell Physiol* **207**:114–122.

146. Machuca-Tzili L, Monroy-Jaramillo N, Gonzalez-del Angel A, Kofman-Alfaro S. (2002) New mutations in the CBFA1 gene in two Mexican patients with cleidocranial dysplasia. *Clin Genet* **61**:349–353.

147. Otto F, Kanegane H, Mundlos S. (2002) Mutations in the RUNX2 gene in patients with cleidocranial dysplasia. *Hum Mutat* **19**:209–216.

148. Yoshida T, Kanegane H, Osato M, *et al.* (2003) Functional analysis of RUNX2 mutations in cleidocranial dysplasia: novel insights into genotype–phenotype correlations. *Blood Cells Mol Dis* **30**:184–193.

149. Quack I, Vonderstrass B, Stock M, *et al.* (1999) Mutation analysis of core binding factor A1 in patients with cleidocranial dysplasia. *Am J Hum Genet* **65**:1268–1278.

150. Yokozeki M, Ohyama K, Tsuji M, *et al.* (2000) A case of Japanese cleidocranial dysplasia with a CBFA1 frameshift mutation. *J Craniofac Genet Dev Biol* **20**:121–126.

151. Goseki-Sone M, Orimo H, Watanabe A, *et al.* (2001) Identification of a novel frameshift mutation (383insT) in the RUNX2 (PEBP2 alpha/CBFA1/AML3) gene in a Japanese patient with cleidocranial dysplasia. *J Bone Miner Metab* **19**:263–266.
152. Tessa A, Salvi S, Casali C, *et al.* (2003) Six novel mutations of the RUNX2 gene in Italian patients with cleidocranial dysplasia. *Hum Mutat* **22**:104.
153. Zhou G, Chen Y, Zhou L, *et al.* (1999) CBFA1 mutation analysis and functional correlation with phenotypic variability in cleidocranial dysplasia. *Hum Mol Genet* **8**:2311–2316.
154. Golan I, Preising M, Wagener H, *et al.* (2000) A novel missense mutation of the CBFA1 gene in a family with cleidocranial dysplasia (CCD) and variable expressivity. *J Craniofac Genet Dev Biol* **20**:113–120.
155. Zhang YW, Yasui N, Kakazu N, *et al.* (2000) PEBP2alphaA/CBFA1 mutations in Japanese cleidocranial dysplasia patients. *Gene* **244**:21–28.
156. Morava E, Karteszi J, Weisenbach J, *et al.* (2002) Cleidocranial dysplasia with decreased bone density and biochemical findings of hypophosphatasia. *Eur J Pediatr* **161**:619–622.
157. Lee B, Thirunavukkarasu K, Zhou L, *et al.* (1997) Missense mutations abolishing DNA binding of the osteoblast-specific transcription factor OSF2/CBFA1 in cleidocranial dysplasia. *Nat Genet* **16**:307–310.
158. Puppin C, Pellizzari L, Fabbro D, *et al.* (2005) Functional analysis of a novel RUNX2 missense mutation found in a family with cleidocranial dysplasia. *J Hum Genet* **50**:679–683.
159. Sakai N, Hasegawa H, Yamazaki Y, *et al.* (2002) A case of a Japanese patient with cleidocranial dysplasia possessing a mutation of CBFA1 gene. *J Craniofac Surg* **13**:31–34.
160. Cunningham ML, Seto ML, Hing AV, *et al.* (2006) Cleidocranial dysplasia with severe parietal bone dysplasia: C-terminal RUNX2 mutations. *Birth Defects Res A Clin Mol Teratol* **76**:78–85.
161. Tsai FJ, Wu JY, Lin WD, Tsai CH. (2000) A stop codon mutation in the CBFA 1 gene causes cleidocranial dysplasia. *Acta Paediat* **89**:1262–1265.
162. Zheng Q, Sebald E, Zhou G, *et al.* (2005) Dysregulation of chondrogenesis in human cleidocranial dysplasia. *Am J Hum Genet* **77**:305–312.
163. Chen S, Santos L, Wu Y, *et al.* (2005) Altered gene expression in human cleidocranial dysplasia dental pulp cells. *Arch Oral Biol* **50**:227–236.
164. Vaughan T, Pasco JA, Kotowicz MA, *et al.* (2002) Alleles of RUNX2/CBFA1 gene are associated with differences in bone mineral density and risk of fracture. *J Bone Miner Res* **17**:1527–1534.
165. Vaughan T, Reid DM, Morrison NA, Ralston SH. (2004) RUNX2 alleles associated with BMD in Scottish women; interaction of RUNX2 alleles with menopausal status and body mass index. *Bone* **34**:1029–1036.
166. Ermakov S, Malkin I, Kobyliansky E, Livshits G. (2006) Variation in femoral length is associated with polymorphisms in RUNX2 gene. *Bone* **38**:199–205.
167. Napierala D, Garcia-Rojas X, Sam K, *et al.* (2005) Mutations and promoter SNPs in RUNX2, a transcriptional regulator of bone formation. *Mol Genet Metab* **86**:257–268.

168. Doecke JD, Day CJ, Stephens AS, *et al.* (2006) Association of functionally different RUNX2 P2 promoter alleles with BMD. *J Bone Miner Res* **21**: 265–273.

169. Amizuka N, Karaplis AC, Henderson JE, *et al.* (1996) Haploinsufficiency of parathyroid hormone–related peptide (PTHrP) results in abnormal postnatal bone development. *Dev Biol* **175**:166–176.

170. Habener JF, Rosenblatt M, Potts JT Jr. (1984) Parathyroid hormone: biochemical aspects of biosynthesis, secretion, action, and metabolism. *Physiol Rev* **64**:985–1053.

171. Brandt DW, Wachsman W, Deftos LJ. (1994) Parathyroid hormone–like protein: alternative messenger RNA splicing pathways in human cancer cell lines. *Cancer Res* **54**:850–853.

172. Gardella TJ, Juppner H. (2001) Molecular properties of the PTH/PTHrP receptor. *Trends Endocrinol Metab* **12**:210–217.

173. Partridge NC, Alcorn D, Michelangeli VP, *et al.* (1981) Functional properties of hormonally responsive cultured normal and malignant rat osteoblastic cells. *Endocrinology* **108**:213–219.

174. Majeska RJ, Rodan GA. (1982) Alkaline phosphatase inhibition by parathyroid hormone and isoproterenol in a clonal rat osteosarcoma cell line. *Calcif Tissue Int* **34**:59–66.

175. Suda N, Gillespie MT, Traianedes K, *et al.* (1996) Expression of parathyroid hormone–related protein in cells of osteoblast lineage. *J Cell Physiol* **166**: 94–104.

176. Castro M, Dicker F, Vilardaga JP, *et al.* (2002) Dual regulation of the parathyroid hormone (PTH)/PTH-related peptide receptor signaling by protein kinase C and β-arrestins. *Endocrinology* **143**:3854–3865.

177. Iida-Klein A, Guo J, Takemura M, *et al.* (1997) Mutations in the second cytoplasmic loop of the rat parathyroid hormone (PTH)/PTH-related protein receptor result in selective loss of PTH-stimulated phospholipase C activity. *J Biol Chem* **272**:6882–6889.

178. Carpio L, Gladu J, Goltzman D, Rabbani SA. (2001) Induction of osteoblast differentiation indexes by PTHrP in MG-63 cells involves multiple signaling pathways. *Am J Physiol Endocrinol Metab* **281**:E489–E499.

179. Whitfield JF, Isaacs RJ, Chakravarthy B, *et al.* (2001) Stimulation of protein kinase C activity in cells expressing human parathyroid hormone receptors by C- and N-terminally truncated fragments of parathyroid hormone 1–34. *J Bone Miner Res* **16**:441–447.

180. Usdin TB, Gruber C, Bonner TI. (1995) Identification and functional expression of a receptor selectively recognizing parathyroid hormone, the PTH2 receptor. *J Biol Chem* **270**:15455–15458.

181. Usdin TB, Bonner TI, Harta G, Mezey E. (1996) Distribution of parathyroid hormone-2 receptor messenger ribonucleic acid in rat. *Endocrinology* **137**: 4285–4297.

182. Juppner H. (1999) Receptors for parathyroid hormone and parathyroid hormone-related peptide: exploration of their biological importance. *Bone* **25**:87–90.

183. Nguyen MT, Karaplis AC. (1998) The nucleus: a target site for parathyroid hormone-related peptide (PTHrP) action. *J Cell Biochem* **70**: 193–199.
184. Massfelder T, Dann P, Wu TL, *et al.* (1997) Opposing mitogenic and anti-mitogenic actions of parathyroid hormone–related protein in vascular smooth muscle cells: a critical role for nuclear targeting. *Proc Natl Acad Sci USA* **94**:13630–13635.
185. Henderson JE, Amizuka N, Warshawsky H, *et al.* (1995) Nucleolar localization of parathyroid hormone–related peptide enhances survival of chondrocytes under conditions that promote apoptotic cell death. *Mol Cell Biol* **15**:4064–4075.
186. Goomer RS, Johnson KA, Burton DW, *et al.* (2000) The tetrabasic KKKK(147–150) motif determines intracrine regulatory effects of PTHrP 1–173 on chondrocyte PPi metabolism and matrix synthesis. *Endocrinology* **141**:4613–4622.
187. Miao D, He B, Lanske B, *et al.* (2004) Skeletal abnormalities in Pth-null mice are influenced by dietary calcium. *Endocrinology* **145**:2046–2053.
188. Horwitz MJ, Tedesco MB, Sereika SM, *et al.* (2005) Continuous PTH and PTHrP infusion causes suppression of bone formation and discordant effects on 1,25(OH)$_2$-vitamin D. *J Bone Miner Res* **20**:1792–1803.
189. Iida-Klein A, Lu SS, Kapadia R, *et al.* (2005) Short-term continuous infusion of human parathyroid hormone 1–34 fragment is catabolic with decreased trabecular connectivity density accompanied by hypercalcemia in C57BL/J6 mice. *J Endocrinol* **186**:549–557.
190. Ma YL, Cain RL, Halladay DL, *et al.* (2001) Catabolic effects of continuous human PTH (1–38) *in vivo* is associated with sustained stimulation of RANKL and inhibition of osteoprotegerin and gene-associated bone formation. *Endocrinology* **142**:4047–4054.
191. Lindsay R, Nieves J, Formica C, *et al.* (1997) Randomised controlled study of effect of parathyroid hormone on vertebral-bone mass and fracture incidence among postmenopausal women on oestrogen with osteoporosis. *Lancet* **350**:550–555.
192. Duan Y, Luca VD, Seeman E. (1999) Parathyroid hormone deficiency and excess: similar effects on trabecular bone but differing effects on cortical bone. *J Clin Endocrinol Metab* **84**:718–722.
193. Cosman F, Nieves J, Woelfert L, *et al.* (2001) Parathyroid hormone added to established hormone therapy: effects on vertebral fracture and maintenance of bone mass after parathyroid hormone withdrawal. *J Bone Miner Res* **16**:925–931.
194. Chen H, Frankenberg L, Goldstein S, McCauley LK. (2003) The combination of local and systemic PTH enhances fracture healing. *Clin Orthop Relat Res* **416**:291–302.
195. Finkelstein JS, Hayes A, Hunzelman JL, *et al.* (2003) The effects of parathyroid hormone, alendronate, or both in men with osteoporosis. *N Engl J Med* **349**:1216–1226.

196. Fukata S, Hagino H, Okano T, *et al.* (2004) Effect of intermittent administration of human parathyroid hormone on bone mineral density and arthritis in rats with collagen-induced arthritis. *Arthritis Rheum* **50**:4060–4069.

197. Dobnig HA. (2004) Review of teriparatide and its clinical efficacy in the treatment of osteoporosis.*Expert Opin Pharmacother* **5**:1153–1162.

198. Zanchetta JR, Bogado CE, Ferretti JL, *et al.* (2003) Effects of teriparatide [recombinant human parathyroid hormone (1–34)] on cortical bone in postmenopausal women with osteoporosis. *J Bone Miner Res* **18**:539–543.

199. Jilka RL, Weinstein RS, Bellido T, *et al.* (1999) Increased bone formation by prevention of osteoblast apoptosis with parathyroid hormone. *J Clin Invest* **104**:439–446.

200. Bellido T, Ali AA, Plotkin LI, *et al.* (2003) Proteasomal degradation of Runx2 shortens parathyroid hormone–induced anti-apoptotic signaling in osteoblasts. *J Biol Chem* **278**:50259–50272.

201. Koh AJ, Demiralp B, Neiva KG, *et al.* (2005) Cells of the osteoclast lineage as mediators of the anabolic actions of parathyroid hormone in bone. *Endocrinology* **146**:4584–4596.

202. Kwok S, Qin L, Partridge NC, Selvamurugan N. (2005) Parathyroid hormone stimulation and PKA signaling of latent transforming growth factor-β binding protein-1 (LTBP-1) mRNA expression in osteoblastic cells. *J Cell Biochem* **95**:1002–1011.

203. Qin L, Tamasi J, Raggatt L, *et al.* (2005) Amphiregulin is a novel growth factor involved in normal bone development and in the cellular response to parathyroid hormone stimulation. *J Biol Chem* **280**:3974–3981.

204. Jiang D, Franceschi RT, Boules H, Xiao G. (2004) Parathyroid hormone induction of the osteocalcin gene. Requirement for an osteoblast-specific element 1 sequence in the promoter and involvement of multiple-signaling pathways. *J Biol Chem* **279**:5329–5337.

205. Luo XH, Liao EY, Su X, Wu XP. (2004) Parathyroid hormone inhibits the expression of membrane-type matrix metalloproteinase-1 (MT1-MMP) in osteoblast-like MG-63 cells. *J Bone Miner Metab* **22**:19–25.

206. Stewart AF. (1996) PTHrP (1–36) as a skeletal anabolic agent for the treatment of osteoporosis. *Bone* **19**:303–306.

207. Plotkin H, Gundberg C, Mitnick M, Stewart AF. (1988) Dissociation of bone formation from resorption during 2-week treatment with human parathyroid hormone–related peptide-(1–36) in humans: potential as an anabolic therapy for osteoporosis. *J Clin Endocrinol Metab* **83**:2786–2791.

208. Kronenberg HM. (2003) Developmental regulation of the growth plate. *Nature* **423**:332–336.

209. Lanske B, Karaplis AC, Lee K, *et al.* (1996) PTH/PTHrP receptor in early development and Indian-hedgehog–regulated bone growth. *Science* **273**: 663–666.

210. Lanske B, Amling M, Neff L, *et al.* (1999) Ablation of PTHrP gene or PTH/PTHrP receptor gene leads to distinct abnormalities in bone development. *J Clin Invest* **104**:399–407.

211. Weir EC, Philbrick WM, Amling M, *et al.* (1996) Overexpression of parathyroid hormone–related peptide in chondrocytes causes chondrodysplasia and delayed endochondral bone formation. *Proc Natl Acad Sci USA* **93**: 10240–10245.

212. Jansen M. (1934) Ueber atypische Chondrostrophie (Acondroplasie) und ueber eine noch nicht beschriebene angeborene Wachstumsstoerung des Knochensystems: metaphysaere dysostosis. *Z Orthop Chir* **61**: 253–286.

213. Schipani E, Kruse K, Juppner H. (1995) A constitutively active mutant PTH–PTHrP receptor in Jansen-type metaphyseal chondrodysplasia. *Science* **268**:98–100.

214. Schipani E, Langman C, Hunzelman J, *et al.* (1999) A novel parathyroid hormone (PTH)/PTH-related peptide receptor mutation in Jansen's metaphyseal chondrodysplasia. *J Clin Endocrinol Metab* **84**:3052–3057.

215. Beier F, LuValle P. (2002) The cyclin D1 and cyclin A genes are targets of activated PTH/PTHrP receptors in Jansen's metaphyseal chondrodysplasia. *Mol Endocrinol* **16**:2163–2173.

216. Kuznetsov SA, Riminucci M, Ziran N, *et al.* (2004) The interplay of osteogenesis and hematopoiesis: expression of a constitutively active PTH/PTHrP receptor in osteogenic cells perturbs the establishment of hematopoiesis in bone and of skeletal stem cells in the bone marrow. *J Cell Biol* **167**: 1113–1122.

217. Blomstrand S, Claesson I, Save-Soderbergh J. (1985) A case of lethal congenital dwarfism with accelerated skeletal maturation. *Pediatr Radiol* **15**:141–143.

218. Oostra RJ, van der Harten JJ, Rijnders WP, *et al.* (2000) Blomstrand osteochondro-dysplasia: three novel cases and histological evidence for heterogeneity. *Virchows Arch* **436**:28–35.

219. Jobert AS, Zhang P, Couvineau A, *et al.* (1998) Absence of functional receptors for parathyroid hormone and parathyroid hormone–related peptide in Blomstrand chondrodysplasia. *J Clin Invest* **102**:34–40.

220. Karaplis AC, He B, Nguyen MT, *et al.* (1998) Inactivating mutation in the human parathyroid hormone receptor type 1 gene in Blomstrand chondrodysplasia. *Endocrinology* **139**:5255–5258.

221. Karperien M, van der Harten HJ, van Schooten R, *et al.* (1999) A frameshift mutation in the type I parathyroid hormone (PTH)/PTH-related peptide receptor causing Blomstrand lethal osteochondrodysplasia. *J Clin Endocrinol Metab* **84**:3713–3720.

222. Karp SJ, Schipani E, St-Jacques B, *et al.* (2000) Indian hedgehog coordinates endochondral bone growth and morphogenesis via parathyroid hormone related-protein-dependent and -independent pathways. *Development* **127**:543–548.

223. St-Jacques B, Hammerschmidt M, McMahon AP. (1999) Indian hedgehog signaling regulates proliferation and differentiation of chondrocytes and is essential for bone formation. *Genes Dev* **13**:2072–2086.

224. Chung UI, Lanske B, Lee K, *et al.* (1998) The parathyroid hormone/parathyroid hormone–related peptide receptor coordinates endochondral bone development by directly controlling chondrocyte differentiation. *Proc Natl Acad Sci USA* **95**:13030–13035.

225. Hopyan S, Gokgoz N, Poon R, *et al.* (2002) A mutant PTH/PTHrP type I receptor in enchondromatosis. *Nat Genet* **30**:306–310.

Chapter 24

Genetic Factors for Human Type 1 Diabetes

Cong-Yi Wang*, Junyan Han and Jin-Xiong She

Center for Biotechnology and Genomic Medicine
Medical College of Georgia, 1120 15th Street, CA4098
Augusta, GA 30912, USA

Type 1 diabetes (T1D) is an autoimmune disorder characterized by specific destruction of the insulin-secreting beta cells of the pancreatic islets. It is believed that susceptibility to T1D is determined by the interactions of multiple genes with unknown environmental factors. Because the onset of the autoimmune process occurs many years before the onset of clinical diabetes, it is difficult to ascertain the nature of possible environmental triggers. Therefore, for the past two decades, a great deal of research has been focused on identifying T1D susceptibility genes. To date, only two susceptibility loci, the HLA region on chromosome 6p21 (*IDDM1*) and the insulin gene (*INS*) on chromosome 11p15 (*IDDM2*), have been well characterized. These two loci only contribute a portion of the familial clustering ($\sim$40% for *IDDM1* and $\sim$10% for *INS*), suggesting that other susceptibility loci must exist. The next confirmed locus for T1D was the *PTPN22* gene, which contributes to multiple autoimmune disorders with a relative risk (RR) of 1.67–2.3. Recent studies have also provided evidence for the existence of two additional susceptibility genes, the *SUMO4* gene on chromosome 6q25 (*IDDM5*) and the *CTLA4* gene on chromosome 2q33 (*IDDM12*). In addition, extensive association studies and linkage analyses using various analytical methods have suggested a large number of putative genomic intervals that may contribute genetic susceptibility to T1D. This chapter summarizes the current state of genetic linkage and association studies in T1D, and discusses challenges and strategies for future studies.

---

* Correspondence author.

1. INTRODUCTION

Type 1 diabetes (T1D) is a multifactorial disease resulting from autoimmune destruction of the insulin-secreting beta cells within the pancreatic islets of Langerhans.[1] It occurs worldwide, and is common in childhood and adolescence. At the time of clinical diagnosis, patients have already lost a major proportion of their beta cells. Although destruction of the beta cells may proceed subclinically over a long time, overt diabetes develops when the beta cells drop to such a low level that the patients are no longer able to secret adequate levels of insulin to control their blood glucose levels. The loss of beta cells results in complete dependence on exogenous insulin for survival. Furthermore, the inability to regulate glucose levels by exogenous administration of insulin as tightly as by functioning pancreatic islets can eventually lead to devastating complications such as neuropathy, nephropathy, and retinopathy. T1D is also a major cause of cardiovascular disease and premature death in the general population.

It has long been recognized that inherited genetic factors are implicated in the pathogenesis of T1D.[2-4] There is compelling evidence that diabetes susceptibility is likely linked to a major locus and that several other minor loci may contribute to diabetes risk in an epistatic way.[3-6] The risk of developing T1D in first-degree relatives of diabetic patients is approximately 6%, which is significantly higher than the prevalence of the disease in the general population (0.4%). This increased risk in relatives compared to the general population prevalence suggests a familial aggregation. The degree of familial aggregation (λ_s) can be estimated by a ratio of the risk for siblings of patients over the general population prevalence (i.e. $\lambda_s = 6/0.4 = 15$). Therefore, relatives have a much higher risk of developing T1D, as they share genes with patients to a greater extent than do unrelated individuals.

However, the incomplete concordance for the phenotype in monozygotic twins (30%–70%) indicates that other nongenetic components, such as environmental factors, also play a major role in T1D.[7] Indeed, the environmental triggers may explain some of the differences in disease frequencies across different populations and the rapid rise in disease frequency in the last few decades.[8] The lack of a perfect correlation between genotype and phenotype complicates the identification of susceptibility genes for the disorder. Despite these difficulties, the genetic factors for T1D are probably the best known among complex diseases, and significant progress has been made in the last decade. To date, more than 20 T1D susceptibility regions have

Table 1   Human T1D Susceptibility Intervals

Locus	Region	Candidate Genes	Locus Status	Mechanism
IDDM1	6p21.3	*HLA DR, DQ*	Identified	Antigen presentation
IDDM2	11p15	*INS-VNTR*	Identified	Tolerance induction
PTPN22	1p13	*PTPN22*	Identified	T-cell activation
IDDM5	6q25	*SUMO4*	Identified	Regulation of immune response
IDDM12	2q33	*CTLA4*	Identified	Regulation of T-cell response
IDDM3	15q26	?	Suggestive	Unknown
IDDM4	11q13	*MDU1, ZMF1, RT6, ICE, LRP5, FADD, CD3*	Confirmed	Unknown
IDDM6	18q12–21	*JK (kidd), ZNF236*	Suggestive	Unknown
IDDM7	2q31–33	*NEUROD*	Suggestive	Unknown
IDDM8	6q27	*PMSB1, TBP, PDCD2*	Confirmed	Unknown
IDDM9	3q21–25	?	Suggestive	Unknown
IDDM10	10p11–q11	?	Suggestive	Unknown
IDDM11	14q24–q31	*ENSA, SEL-1L*	Suggestive	Unknown
IDDM13	2q33	*IGFBP2, IGFBP5, NERUROD, HOXD8*	Suggestive	Unknown
IDDM15	6q21	?	Significant	Unknown
IDDM16	14q32	*IGH*	Suggestive	Unknown
IDDM17	10q25	?	Significant	Unknown
IDDM18	5q31–33	*IL-12B*	Significant	Unknown
	3p13–14	?	Significant	Unknown
	9q33–34	?	Significant	Unknown
	12q12–14	?	Significant	Unknown
	16p12–q11.1	?	Significant	Unknown
	16p22–24	?	Significant	Unknown
	19p13	?	Significant	Unknown
	1q42	?	Suggestive	Unknown
	2p12	?	Suggestive	Unknown
	5p11–q32	?	Suggestive	Unknown
	7p15–p13	*GCK*	Suggestive	Unknown
	8q11–24	?	Suggestive	Unknown
	16p11–13	?	Suggestive	Unknown
	17q25	?	Suggestive	Unknown
	19q11	?	Suggestive	Unknown

been suggested (Table 1). This chapter will review and update the current progress of genetic factors for human T1D, and will discuss challenges and strategies for the characterization of more modest T1D susceptibility genes.

2. HLA-ENCODED SUSCEPTIBILITY TO T1D *(IDDM1)*

The human leukocyte antigen (HLA) region located on chromosome 6p21.3 is the major susceptibility locus (designated as *IDDM1*) for T1D, providing up to 40% of the familial clustering.[4] The HLA genes span 3.5 Mb of genomic DNA, and are structurally and functionally subdivided into three subregions (classes I, II, and III).

Class I molecules (HLA-A, HLA-B, and HLA-C) are heterodimers of a class I chain and a β2-microglobulin chain that are encoded by a monomorphic gene outside of the HLA complex (Fig. 1). They are present on the surface of all nucleated cells and are responsible for antigen presentation to CD8-positive T-lymphocytes.[9] A particular class I molecule from human cell membranes has two structural motifs: the end furthest from the membrane contains two domains with immunoglobulin folds, and the distal end has a platform of eight antiparallel beta strands topped by alpha helices. A large groove between the alpha helices provides a binding site for processed foreign antigens (Fig. 1).

Class II molecules are dimers consisting of an alpha and a beta polypeptide chain. Each chain contains an immunoglobulin-like region next to the cell membrane. The antigen-binding cleft, which is composed of two alpha helices above a beta-pleated sheet, specifically binds short peptides about 15–24 residues long (Fig. 2). There are three major class II molecules (DR, DQ, and DP) that are expressed in humans. All class II genes, with the exception of *DRA* (A gene for DR), are polymorphic. Therefore, the functional properties of DR molecules are solely determined by the polymorphic *DRB* genes, while the DQ and DP functions are determined by both alpha and beta chains. The amino acid sequence around the binding site, which specifies the antigen-binding properties, is the most variable site in the HLA molecule. Class II genes are expressed on the surface of antigen-presenting cells (macrophages, dendritic cells, and B-lymphocytes) and activated T-lymphocytes, and present antigens to CD4-positive T-lymphocytes.[9]

The class III region encodes molecules with different functions, including complement components (factors C4A, C4B, B, and C2), tumor necrosis factor (TNF), heat shock protein Hsp70, and 21-hydroxylase (CYP21).

HLA class I genes (e.g. *HLA-B8*, *HLA-B18*, and *HLA-B15*) were the first HLA alleles found to be associated with T1D.[10] Subsequent studies suggested that class I genes probably influence the age of onset[11,12] and the rate of β-cell destruction.[13–16] Further studies revealed that virtually all genes in the HLA region are associated with T1D in all studied populations; this

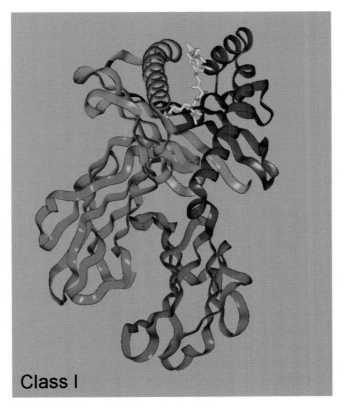

Fig. 1.   Three-dimensional structure for major histocompatibity complex (MHC) class I. MHC class I is a membrane-spanning molecule composed of two proteins. The membrane-spanning protein is approximately 350 amino acids in length, with about 75 amino acids at the carboxylic end comprising the transmembrane and cytoplasmic portions. The remaining 270 amino acids, as shown in the ribbon diagram, are divided into three globular domains labeled alpha-1 (blue-green), alpha-2 (purple), and alpha-3 prime (magenta), with alpha-1 being closest to the amino terminus and alpha-3 closest to the membrane. The second portion of the molecule is a small globular protein called beta-2 microglobulin (green). It is primarily associated with the alpha-3 prime domain and is necessary for MHC stability. The bound peptide (cream) sits within the groove.

could be a consequence of the strong linkage disequilibrium (LD) between genes within the region. The statistically strongest genetic association with T1D is conferred by the HLA class II genes (DQ, DR, and DP). Association studies also revealed evidence for the association of the HLA class III region with T1D.[17,18] It is suggested that the interval between the *TNF* and *HSP70* genes in the class III region may harbor additional susceptibility loci.[19,20]

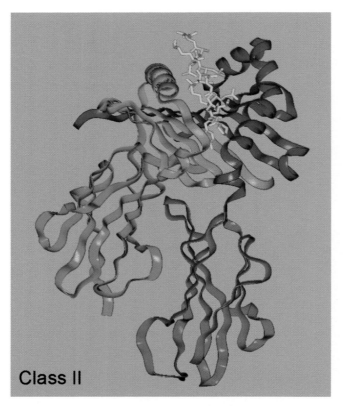

Fig. 2.   Three-dimensional structure for MHC class II. Similar to class I, the MHC class II molecule is composed of two membrane-spanning proteins. Each chain is approximately 30 kD in size, and is made of two globular domains as shown in the ribbon diagram. The domains are named alpha-1 (blue-green), alpha-2 (green), beta-1 (purple), and beta-2 (magenta). The two regions farthest from the membrane are alpha-1 and beta-1. The two chains are associated without covalent bonds. The bound peptide (cream) sits within the groove.

For example, *TNF* is a strong candidate gene since polymorphisms of this gene may affect the production of TNFα, a potent inflammatory cytokine, which in turn affects the magnitude of the immune response.[21] There is suggesting evidence that other loci within or near the HLA complex appear to modulate diabetes risk, such as the class I antigen-processing genes *LMP7* and *LMP2*,[22] thus adding additional complexity to the analysis of HLA-encoded susceptibility. Because of strong LD between these loci, it has been very difficult to study the effect of other genes that are linked to *HLA-DQ* and *HLA-DR*.

It has been well demonstrated that DQ molecules are a key suscepti-bility factor for T1D. There is strong evidence that some combinations of *HLA-DQ* genes, such as *DQ8* (i.e. *DQA1*03–DQB1*0302*) and *DQ2* (i.e. *DQA1*05–DQB1*02*), confer the strongest susceptibility to T1D. An amino acid at position 57 of DQB1 is thought to play an important role in deter-mining T1D susceptibility.[23,24] All *DQB1* alleles with an aspartic acid at DQβ residue 57 (Asp57 or D57) — including **0301*, **0303*, **0401*, **0402*, **0503*, **0601*, **0602*, and **0603* — confer neutral-to-protective effects. In particular, the *DQB1*0602* allele provides dominant protection against T1D even in the presence of HLA or non-HLA susceptibility genes. On the contrary, *DQB1* alleles with an alanine (Ala57 or A57) — including **0201* and **0302* — confer strong susceptibility to T1D in all ethnic groups. Other DQB1 alleles with substitutions for Asp57 — such as Val57 (V57) for *DQB1*0604* and **0501*, and Ser57 (S57) for *DQ*0502* — appear to confer weak susceptibility. A recent study suggested that the D57A substitution of the DQβ chain probably affects peptide binding and recognition. X-ray crystallography analysis of the 3D structure of the DQ8 molecule (encoded by *DQA1*0301/DQB1*0302*) complexed with a peptide from the insulin B chain (residues 9–23, SHLVEALYLVCGERG) revealed that the residue at position 57 is probably involved in the shaping of the P9 pocket, which is associated with peptide binding and autoantigen presentation.[25] Of note, the susceptibility effects conferred by the HLA class II genes are determined by the combination of *DRB1*, *DQB1*, and *DQA1* alleles. For example, the DR4 (*DRB1*0401* or *DRB1*0404*)–*DQB1*0302* haplotypes confer strong susceptibility (RR = 6–28), while the DR4–*DQB1*0301* haplotypes are protective (RR = 0.3) in Caucasian populations.[4]

All DQα and β chains encoded by alleles located on the same haplotype (chromosome) can form functional DQ dimers known as *cis*-complementation, which play an important role in T1D etiology. *Trans*-complementation of DQα and β chains from different chromosomes has also been demonstrated, and this could be responsible for the synergistic effect observed in certain heterozygous genotypes. For example, a great major-ity of Caucasian patients carry the *HLA-DR3* or *HLA-DR4* class II anti-gens, and approximately 30% of these patients are *DR3/DR4* heterozygotes. These DR3/4 heterozygotes (RR ≈ 20) confer a much higher risk than those of DR4/4 (RR ≈ 15) or DR3/3 (RR ≈ 7) homozygote genotypes.[3,4,23,26–28]

A similar synergistic effect was also observed for many other heterozygotes (e.g. DR3/9 in Chinese and Africans; DR8/9, DR9/13, DR4/8, and DR4/13 in Japanese).[3,27–33] A recent study with a large collection of nuclear families further demonstrated the importance of *trans*-encoded HLA DQ molecules in the determination of HLA-associated risk of T1D.[34]

Although DQ molecules are unambiguously implicated in T1D, the association between DQ and T1D is not consistent in all populations. For example, *DQB1*0302* is not positively associated with T1D in Asian populations,[3,30,31,35–37] even though it confers high risk in Caucasians and Africans.[2,3,37–40] These observations suggest a role for DR molecules in T1D. It is believed that the *DRB1* gene, tightly linked to the *DQB1* and *DQA1* genes, is responsible for the discrepant associations of DQ alleles.[3,4] Several elegant studies have convincingly demonstrated the independent effects of the DR molecules in T1D.[31,41,42] For example, in the Chinese population, the relative risk conferred by the *DRB1*0405–DQA1*0301–DQB1*0302* haplotype (RR = 34) is 168 times higher than that of the *DRB1*0406–DQA1*0301–DQB1*0302* haplotype (RR = 0.2) and 48 times higher than that of the *DRB1*0403–DQA1*0301–DQB1*0302* haplotype (RR = 0.7).[31] As the *DQA1* and *DQB1* alleles are identical for all three haplotypes, the logical explanation for this discrepancy is that DR4 subtypes contribute to the dramatic differences in T1D susceptibility between these haplotypes. Similar results are also found for these and other haplotypes in other populations,[42,43] indicating that DR molecules contribute independently to T1D susceptibility.[3]

Of note, the effect of *DR* and *DQ* alleles on T1D susceptibility varies greatly, and ranges from highly susceptible to strongly protective (Table 2). For example, some DR4 alleles (i.e. 0405, 0402, and 0401) confer high risk, while others (i.e. 0403, 0406, and 0408) confer strong protection (Table 2).[3,4] In general, all DR and DQ molecules are susceptible in a high-risk genotype, such as all DR3/4 and DR4/4 subjects containing susceptible DR4 alleles (0401, 0402, 0404, 0405, and 0407). However, the presence of a highly protective DR or DQ molecule could be sufficient to overcome the effects of all susceptibility molecules in an individual (dominant protection). For example, the *DQB1*0602* allele appears to be dominant protective, since *DQB1*0602* protects from diabetes even in the presence of high-risk HLA alleles.[5,44,45] Similarly, DR3/4 heterozygous subjects with the protective *DRB1*0403* allele are protected against diabetes, even though several highly susceptible DR and DQ molecules are present.[46] Several other alleles (*DRB1*0406*, *DRB1*0408*) also confer similar dominant protection.[31,47]

Table 2   Susceptibility Effects of HLA-DR/DQ Molecules

Effect	DRB1	DQα/β	DQA1	DQB1	DQB1 57
1. Highly susceptible (RR > 10)	0405	0301/0201	0301	0201	Ala
	0402	0301/0302		0302	Ala
2. Susceptible (RR = 2–10)	0401	0301/0604	?	0604	Val
	0301	0301/0501		0501	Val
		0301/0502		0502	Ser
3. Neutral to susceptible (RR = 1–2)	0404	0301/0303	0501?	—	—
	01, 08	0501/0201			
	09	0501/0302			
4. Neutral to protective (RR = 1–0.5)	07	Others	Others	0303	Asp
	12			0503	
	15			0601	
	16			0603	
5. Protective (RR = 0.1–0.5)	11	Others	Others	0301	Asp
	13			0401	
	14			0402	
6. Highly protective (RR < 0.1)	0403	All/0602	—	0602	Asp
	0406				
	0408				

Therefore, DR and DQ genotypes should be used instead of individual alleles to assess the risk of developing T1D.

It is suggested that protective HLA molecules may have higher affinity for one or several peptides than the predisposing molecules. Indeed, HLA molecules encoded by the predisposing *DQA1*0301–DQB1*0302* and protective *DQA1*0102–DQB1*0602* alleles appear to differ in their affinity and specificity for peptides originated from autoantigens, such as insulin, glutamic acid decarboxylase (GAD), and tyrosine phosphatase IA-2.[27,43,45,48] Therefore, allelic variations in a particular DR or DQ locus may affect the binding and functional properties of DR or DQ heterodimers, which in turn regulate the presentation of islet cell autoantigen-derived peptides to immunocompetent cells.

3. THE INSULIN GENE (*IDDM2*)

The insulin (*INS*) gene on chromosome 11p15 is the second most well-documented susceptibility gene (designated as *IDDM2*) for T1D. Autoimmunity leading to the development of T1D is highly specific to the pancreatic beta cells. The insulin gene is, therefore, a plausible candidate gene for T1D susceptibility since insulin or insulin precursors may act as autoantigens and, alternatively, the insulin levels could modulate the interaction between the immune system and beta cells.[49]

The association of the insulin gene region with T1D was mapped to a variable number of tandem repeats (VNTR) located approximately 500 bp upstream of the insulin gene.[50–52] The disease locus was localized to a 19-kb region of genomic DNA by showing absence of association with a polymorphic marker at the 3′ neighboring locus *IGF2* and a marker at the 5′ neighboring locus tyrosine hydroxylase (*TH*).[4] The disease interval was subsequently further restricted to 4.1 kb of genomic DNA, encompassing the insulin gene and its flanking regions including *INS-VNTR*.[53] More than 10 polymorphisms have been identified within this region, and all of them lie outside the coding regions. Several polymorphisms in the region are in LD, indicating that one or more of these variants might be the etiological mutation for *IDDM2*.[54,55] Each of these variants was analyzed in a large collection of diabetic families to pinpoint the etiological mutation to *INS-VNTR*.[56]

INS-VNTR is a tandem repeat unit consisting of a 14–15-bp consensus sequence [ACAGGGGT(G/C)(T/C)GGGG]. According to the number of repeat units, the alleles have been divided into three major classes: class I has 30–60 repeat units; class II, an average of 80–85 repeat units; and class III, 120–170 repeat units.[4] The intermediate-sized class II alleles are very rare in Caucasians and even rarer in African populations.[57,58] This GC-rich repeat unit tends to form unusual DNA structures, probably through the formation of G-quartets.[59,60] The frequency of the class I homozygous genotype in the general population is around 50%–60%, which is significantly lower than the frequency in diabetic patients (∼75%–85%) in most Caucasian populations (RR = 2.0–5.0).[50,51,53,56,61] These results indicate that the class I homozygous genotype can account for ∼10% of the familial clustering of type 1 diabetes[62]; whereas the longer class III alleles provide a protective effect and are rarely seen in T1D patients,[63,64] representing a threefold-to-fivefold reduction in the risk of T1D compared with class I homozygotes.[65] Of note, VNTR alleles are heterogeneous and can be further classified based on size differences; as a result, alleles within each class may not have the same degree of susceptibility effect on T1D.[56]

Based on the observation that VNTR regulates insulin expression in the thymus, the effect of *IDDM2* on T1D susceptibility is linked to the development of self-tolerance to insulin. It was found that *INS* transcripts in *cis* with protective class III VNTR alleles are transcribed in the thymus at much higher levels (on average, twofold to threefold) than those in *cis* with class I VNTR alleles, which are associated with increased susceptibility to T1D.[66,67] During negative selection, the depletion of autoreactive thymocytes depends on the expression of self-molecules (e.g. insulin) in a dose-dependent manner. As a result, higher insulin levels (associated with protective class III alleles) in the thymus may promote efficient negative selection of insulin-specific T-lymphocytes or improved selection of regulatory T-cells[9]; whereas the disease-associated class I alleles determine lower insulin levels, which may be associated with a less efficient deletion of insulin-specific autoreactive T-cells or an impaired selection of regulatory T-cells.[9]

Interestingly, this assumption was confirmed in the nonobese diabetic (NOD) mouse model. Mice have two insulin genes: *Ins1* and *Ins2* on chromosomes 7 and 19, respectively. *Ins2* is virtually the only insulin gene expressed in the thymus. *Ins2* knockout mice with an NOD background developed spontaneous diabetes faster, and the incidence in males increased to levels similar to those observed in female NOD mice[68]; in contrast, NOD mice with transgenic *Ins2* expression in the MHC class II positive thymocytes were fully protected from insulitis and diabetes.[69] Therefore, *IDDM2*-mediated susceptibility is probably caused by the reduced thymic insulin expression and defective negative selection associated with the VNTR class I alleles.

4. *PTPN22*-ENCODED LYMPHOID-SPECIFIC PHOSPHATASE (LYP)

PTPN22 is located on chromosome 1p13, which encodes the lymphoid-specific phosphatase (LYP), a 110-kD protein tyrosine phosphatase (PTP) consisting of an N-terminal phosphatase domain and a long noncatalytic C-terminus with several praline-rich motifs.[70,71] LYP is important in negative control of T-cell activation and in T-cell development.[72,73] *PTPN22* is thus far the third confirmed susceptibility factor for T1D.

Bottini *et al.*[74] were the first to implicate *PTPN22* in the susceptibility to T1D. They found that *PTPN22* contains a single nucleotide polymorphism (SNP) at position 1858 in codon 620 that changes arginine (R) to tryptophan (W). R620 resides in the P1 proline-rich motif involved in binding to the SH3 domain of Csk; the R620W substitution disrupts the interaction between LYP and the protein tyrosine kinase Csk.[74] They used a case-control study design and demonstrated that the R620W substitution confers risk for T1D in both European-American and Sardinian cohorts.[74]

Subsequent association studies and linkage analyses from different research groups further confirmed the susceptibility of *PTPN22* to T1D in multiple populations.[75–82] The relative risk (RR) for the W620 variant is approximately 1.67–2.3, making *PTPN22* the most potent risk factor after *IDDM1* and *IDDM2*.[83] Of note, the susceptibility conferred by *PTPN22* is shared by several organ-specific and systemic autoimmune diseases, including rheumatoid arthritis,[81,82,84–94] juvenile idiopathic arthritis,[88,89] systemic lupus erythematosus (SLE),[81,84,86,95] Graves' disease,[75,84,96,97] vitiligo,[98] and Hashimoto's thyroiditis[84]; but not multiple sclerosis,[84,89,99,100] Crohn's disease,[101,102] Sjogren's syndrome,[103] inflammatory bowel disease (IBD),[104] psoriasis,[105] and celiac disease.[106] Interestingly, the disease-associated allele (W620) is absent in Asian populations.[107,108]

Although the accumulating genetic data implicate the *PTPN22* W620 variant as a determinant of T1D susceptibility, the functional basis of the association remains elusive. Previous studies have shown that the interaction between LYP and protein tyrosine kinase Csk enables these effectors to synergistically inhibit T-cell activation; it is therefore speculated that the R620W substitution may biologically translate to the potential for hyperreactive T-cell responses.[109] This assumption is based on the known roles of PTPs in attenuating signaling pathways integral to immune cellular reactivity. However, aberrant LYP function has also been found to be associated with impaired antigen receptor–evoked apoptosis of murine B cells *in vitro* as well as augmented lymphoid germinal center development and hypergammaglobulinemia *in vivo*.[109] These defects are potentially associated with the B-cell autoreactivity and autoantibody production typical of autoimmune diseases.[72,110] Furthermore, LYP may exert at least some of its biological effects through activities in myeloid or monocytic cells, as it has been shown to dephosphorylate several myeloid cell signaling effectors and inhibit myeloid cell growth.[111]

More recently, a follow-up study indicated that the disease-predisposing allele (LYP-Trp620) is a gain-of-function mutant which is a more efficient inhibitor of T-cell activation by increasing the threshold for T-cell receptor (TCR) signaling.[112] Based on these observations, two possible mechanisms by which the R620W substitution is implicated in T1D pathogenesis are proposed: (1) the increase of threshold for TCR signaling could result in the positive selection of thymocytes that would otherwise be deleted during negative selection, resulting in the appearance of potentially autoreactive T-cells in the periphery; and (2) the increased TCR signaling threshold could result in reduced signaling for Treg cells, and lead to deficiency in the regulation of autoreactive T-cells.[113]

5. *SUMO4*-ENCODED SUSCEPTIBILITY TO T1D (*IDDM5*)

The *IDDM5* locus located on chromosome 6q25 was initially reported by Davies and coworkers.[114] Significant linkage evidence was obtained through the fine mapping analysis of a larger Caucasian data set (MLS = 4.5) by our research group.[115,116] Subsequent studies from multiple groups further confirmed the existence of *IDDM5* in different populations with genetic heterogeneity in the UK population.[117-119] The locus was further narrowed to a region of 5 cM between the markers *D6S476* and *D6S473*.[115,116] Recently, we performed a fine mapping study using both case-control and LD approaches, and localized the *IDDM5* locus to a 197-kb region of genomic DNA.[120] A novel gene named *SUMO4* (small ubiquitin-like modifier 4) and a 163A > G mutation within the CUE domain of *SUMO4* were identified. The mutation results in the amino acid Met55-to-Val substitution (M55V). Met55 is evolutionarily conserved among diverse species, and we provided strong evidence indicating that the substituted allele, Val55, is associated with increased risk for T1D development.[120]

In our initial report, we also observed a significant association between *SUMO4* and T1D in two Asian populations (Chinese and Korean).[120,121] The association was independently confirmed in a Korean case-control cohort consisting of 386 individuals with T1D and 553 normal controls.[122] Consistent with our report, Park *et al.*[122] demonstrated that the GG and AG genotypes had a higher frequency in the affected individuals (62%) than in matched controls (52.1%), with a relative risk (RR) of 1.5 ($P < 0.003$).[121,122] More recently, Noso and coworkers[123] provided additional evidence supporting a significant association between T1D and *SUMO4* in Asian populations. They analyzed a large cohort of 1113 Japanese (472 cases and 641 controls) and 171 Korean (69 cases and 102 controls) subjects; once again, the frequency for the G allele (Val55) was significantly higher in Japanese diabetic patients than in controls (OR = 1.43; $P < 0.005$). A similar trend was also observed in Korean subjects (OR = 1.75). In combined data from Japanese and Korean subjects, the G allele was significantly associated with T1D (OR = 1.46; $P = 0.00083$).[123] We performed a meta-analysis using nonoverlapped case-control data sets from Asia, and demonstrated a highly significant association between *SUMO4* and T1D (OR $= 1.5$; $P = 1.5 \times 10^{-6}$); no heterogeneity was detected among the Asian populations.[121]

Given the results from the Asian populations, there is little doubt that *SUMO4* is a T1D susceptibility gene. However, the issue is less settled in Caucasian populations. We observed a significant genetic heterogeneity for the UK families and were unable to confirm our observation in these families.[120] A study including 478 families by Owerbach and coworkers,[124] of which 222 originated from the UK, suggested that the A allele (Met55) is associated with T1D susceptibility. This result is in contrast to our finding in the non-UK data set, but consistent with our finding in the UK data set.[120] Subsequently, Smyth *et al.*[125] and Qu *et al.*[126] were unable to confirm the association in their family-based studies among the UK and Canadian Caucasians, respectively. Likewise, in the large UK case-control data set studied by Smyth and coworkers,[125] there was a marginally significant association consistent with our initial finding.[121] Significant association for *SUMO4* was observed in several Caucasian populations studied in our initial report, including Florida, French, Spanish, and Mexican-Americans.[120] The strongest evidence was from the north-central Florida data set.[120,121] To confirm the association, we analyzed a new cohort consisting of 196 T1D patients and 1060 controls of European descent from north-central Florida. A significant association for the G allele (Val55) of *SUMO4* was obtained (OR $= 1.5$; $P = 0.01$). When we combined this new cohort with the initial Florida case-control data set (244 cases and 274 controls), the evidence was further enhanced (OR $= 1.6$; $P < 0.0001$).[120,121,127]

The above results provided strong evidence suggesting that *SUMO4* is also associated with T1D in Caucasians. The question, then, is why the association is not observed in some other populations. The discrepancies may be explained by two distinct possibilities. The first one is genetic heterogeneity, e.g. *SUMO4* is only implicated in T1D in some populations but not others. The human population is not homogeneous in terms of the risk of disease. Different patients or ethnic groups may have a different set of genes that in combination are responsible for their disease onset. Depending on the genetic background (gene–gene or gene–environment interactions), an etiological mutation may or may not exert its effect. Genetic heterogeneity has often been blamed for the inconsistent observations on disease association across different populations or ethnic groups. Few studies have truly distinguished genetic heterogeneity from false association. It remains to be determined whether genetic heterogeneity is truly responsible for the inconsistent association evidence for *SUMO4* and T1D. The second explanation for the inconsistent association is that the *IDDM5* interval may contain multiple T1D susceptibility genes, which may contain both susceptible and protective alleles that are in different LD patterns in different populations. This phenomenon has already been observed for the *DRB1* and *DQA1/DQB1* genes in the HLA region.[3] Further studies should be performed to determine whether a similar phenomenon also occurs in the *IDDM5* region.

Analysis of the functional properties of *SUMO4* provided additional evidence supporting its susceptibility in T1D pathogenesis. Unlike other SUMO members, SUMO4 expression is restricted to immune tissues and pancreatic islets,[1] and it is a potent regulator for immune response.[1,120] SUMO4 conjugates to IκBα and stabilizes IκBα from signal-induced degradation; as a result, SUMO4 negatively regulates NFκB, a pivotal transcription factor responsible for immune response. Further functional studies revealed that, upon signal-induced activation, NFκB on the one hand transcribes genes for immune response (e.g. inflammatory cytokines and costimulatory molecules); on the other hand, it activates the transcription for SUMO4 expression, which in turn switches off the immune responsive signals to prevent autoimmunity. Therefore, SUMO4 may function as a negative feedback regulator for the NFκB signaling pathway.[121]

We and others[120,124] reported strong evidence for a significantly reduced sumoylation function for the SUMO4*V55 variant, which has shown increased frequency in individuals with T1D.[1,120–123,127] HEK293 cells transfected with the SUMO4*V55 isoform showed 5.5-fold higher NFκB-dependent transcriptional activity than that of cells transfected with the SUMO4*M55 isoform upon IL-1β stimulation. Furthermore, the M55V substitution resulted in more than threefold higher *IL12-p40*, a NFκB-dependent gene expression *in vivo*.[120] Bohren *et al.*[128] have consistently demonstrated that M55V leads to a significantly reduced capability for SUMO4 to enhance HSF transcriptional activity.[128] All of these results indicate that the M55V substitution of SUMO4 results in a significantly reduced sumoylation function. These observations demonstrate a new pathway for T1D pathogenesis that involves SUMO4 sumoylation of IκBα and negative feedback regulation of NFκB activity. The M55V substitution of SUMO4 could result in a higher cellular immune response capacity to stimulation, thereby leading to higher levels of activated NFκB, which in turn activates transcription for genes implicated in the development of T1D. Our recent studies also suggested a role for SUMO4 in the regulation of oxidative stress during the destruction of pancreatic islets.[129] We further demonstrated that SUMO4 sumoylation represses the transcriptional activity for AP-1 and AP-2α, while it enhances DNA-binding activity for glucocorticoid receptor (GR). Therefore, the biological effect of SUMO4 on T1D pathogenesis may be far more complicated than what we had thought.

6. *CTLA4* SURROUNDING REGION (*IDDM12*) TO T1D SUSCEPTIBILITY

The chromosome 2q33 region surrounding *CTLA4* (cytotoxic T-lymphocyte antigen 4), designated as *IDDM12*, is another confirmed susceptibility locus for T1D. This region contains a cluster of genes, such as *CTLA4*, *CD28*, and *ICOS* (inducible costimulator). Both *CTLA4* and *CD28* function as members of the same regulatory pathway of T-lymphocytes: the CD28 signal activates T-cells, and CTLA4 acts as a negative regulator of T-cells.

Linkage evidence for a T1D locus in this chromosomal region was initially suggested with a microsatellite marker, $(AT)_n$, within the 3' UTR (untranslated region) of the *CTLA4* gene in 48 Italian families (LOD $= 3.6$).[130] Subsequent linkage analyses by our research group — using a multiethnic collection of families from Spain, France, China, Korea, and Mexican-Americans — provided confirmation of linkage with *CTLA4*.[131] The transmission/disequilibrium test (TDT) was used to evaluate the association for alleles at the $(AT)_n$ microsatellite and the A/G polymorphism at position 49 of the *CTLA4* gene (*CTLA4-A/G*). The overall evidence for transmission deviation of the *CTLA4-A/G* polymorphism remained significant after combining data from our study (669 multiplex and 357 simplex families) with data published earlier.[130] However, a significant heterogeneity was observed for the UK, Sardinian, and Chinese families; and only a weak deviation was observed for the US families.[131] Moreover, when those families with genetic heterogeneity were excluded, a significant deviation for transmission was obtained in the Italian, Spanish, French, Mexican-American, and Korean families ($P < 0.00005$),[131] providing strong support for a susceptibility gene in the region surrounding *CTLA4*. In addition to these family-based studies, significant evidence for association was also reported in several case-control studies including the Belgian population,[130] the German population,[132] and the Japanese population.[133]

A combined study of 3671 families from the US, UK, Finland, and other European populations excluded *CD28* and narrowed down the association to the 3' region of *CTLA4* and 5' end of *ICOS* (100-kb region of LD).[134]

Susceptibility was mapped to a polymorphism ($+6230G > A$ or CT60) in the noncoding 6.1-kb 3' region of *CTLA4* with a relative risk (RR) of 1.15. The CT60 polymorphism was indicated to control the ratio of soluble CTLA4 isoform (sCTLA4, lacking exon 3) over the full-length isoform.[134]

Ueda and coworkers[134] provided some evidence that the susceptible G allele of CT60 is associated with decreased levels of sCTLA4 expression, and that the protective A allele is associated with higher levels of sCTLA4 expression. However, contradictory results were obtained in subsequent studies. Using allele-specific single-nucleotide primer extension, Anjos and coworkers[135] were unable to detect the difference between mRNA transcripts derived from either CT60 allele in 11 heterozygous individuals in either of the two known *CTLA4* isoforms. In another study, Atabani and coworkers[136] studied the effect of CT60 on regulatory T-cells (Treg). They failed to observe significant genotype-associated differences in Treg-mediated suppression of $CD4^+$ T-cell proliferation or production of IFNγ. They were also unable to detect genotype-associated differences in CTLA4 mRNA or protein expression, or in the expression of Foxp3 (a transcription factor essential for the development and function of Treg). Furthermore, no significant differences in the expression of either sCTLA4 or the full-length isoform (flCTLA4) were observed in Treg associated with the CT60 genotypes.[136]

sCTLA4 has been found to be constitutively expressed in unstimulated human T-cells,[137] and is present in human serum.[138] Based on the findings by Ueda and coworkers[134] that the CT60-G allele is associated with a lower amount of sCTLA4 in peripheral blood lymphocytes than the disease-resistant CT60-A allele, one would predict lower sCTLA4 in the serum of T1D patients compared to controls. However, this prediction is in direct conflict with the observations in other autoimmune diseases, such as autoimmune thyroid disease,[139] systemic lupus erythematosus,[140] and myasthenia gravis,[141] in which the serum sCTLA4 levels are increased in patients compared to controls.

Indeed, our recent study[142] of serum samples from 218 diabetic subjects from the southeastern United States does not support the conclusion made by Ueda and coworkers. We found that sCTLA4 levels in diabetic patients (mean = 2.24 ng/mL; range = 0.0–10.1 ng/mL) and autoantibody-positive (AbP) subjects (mean = 2.17 ng/mL; range = 0.2–7.7 ng/mL) were slightly higher than those in autoantibody-negative (AbN) subjects (mean = 1.69 ng/mL; range = 0.0–11.5 ng/mL), although these differences were not statistically significant. After stratification of the data by phenotypic groups (T1D, AbP, and AbN) and CTLA4 CT60 genotypes (A/A, A/G, and G/G), no differences in sCTLA4 levels between CTLA4 genotypes ($P = 0.46$) or genotype/phenotype interactions ($P = 0.82$) were observed. Consistent with the higher serum sCTLA4 levels observed in other autoimmune diseases, our results suggested that sCTLA4 may be a risk factor for T1D. However, our results did not support the conclusion that the CT60 SNP controls the expression of sCTLA4.[142]

In summary, genetic evidence supporting the association of *CTLA4* with T1D is substantial. However, the functional basis for the CTLA4 association with autoimmunity is yet to be determined.

7. *IDDM4* AND *IDDM8*: GENOME SCAN AND CONFIRMATION OF LINKAGE

Evidence for two additional intervals, *IDDM4* on chromosome 11q13 and *IDDM8* on chromosome 6q27, has been suggested in multiple studies. *IDDM4* was initially suggested in two genome scans[114,143]: a logarithm of odds (LOD) score of 1.3 was obtained in a UK data set of 96 affected sib-pair families,[114] and an LOD score of 1.5 in a French data set of 231 families.[143] Using 265 Caucasian families, our research group analyzed several microsatellite markers in the 11q13 region and obtained an LOD score of 3.4.[115] Nakagawa and coworkers[144] further suggested that *IDDM4* may be a gene very close to the marker *D11S1917*. We performed fine mapping using 382 Caucasian families.[145] The markers with strong linkage evidence were located within an interval of approximately 6 cM between *D11S4205* and *GALN* (galanin). Analyses of the data by TDT and extended TDT (ETDT) did not provide any evidence of linkage with two candidate genes, *FADD* (Fas-associated death domain protein) and *GALN*. However, ETDT did reveal a significant linkage with the marker *D11S987* ($P = 0.0004$), suggesting that *IDDM4* may be in close proximity to *D11S987*.[145]

Linkage evidence for *IDDM8* was suggested by several independent studies,[114,115,117] and the LOD score of the combined data set was 3.6. By analyzing 523 multiplex families, Delepine and coworkers[118] suggested a significant linkage ($P = 0.00004$) for *IDDM8*. Based on these results, *IDDM8* was localized to a 5–7-cM region on chromosome 6q27, about 40 cM telomeric to *IDDM5*.[4] Owerbach[146] recently analyzed 266 families and suggested the location of *IDDM8* to the terminal 200-kb region of chromosome 6q27. This region harbors a cluster of three genes encoding proteasome subunit beta 1 (*PMSB1*), TATA-box binding protein (*TBP*), and a homolog of mouse programming cell death activator 2 (*PDCD2*). Chistiakov and coworkers analyzed 114 affected Russian simplex families, and suggested that *PDCD2* could be a good candidate gene for *IDDM8*.[147]

8. OTHER PUTATIVE T1D SUSCEPTIBILITY INTERVALS

To date, more than 20 genomic intervals have been suggested by complete or partial genomewide scans (Table 1). The first genomewide scan identified *IDDM1* as a major T1D risk locus. Subsequent studies provided evidence for non-HLA susceptibility loci on chromosomes 11p15.5 (*IDDM2*), 11q13 (*IDDM4*), 6q25 (*IDDM5*), 6q27 (*IDDM8*), 2q31–q33 (*IDDM12*), and 1p13 (*PTPN22*). Using both linkage and association approaches, more putative T1D susceptibility loci were identified on chromosomes 15q26 (*IDDM3*),[148,149] 18q12–q21 (*IDDM6*),[150,151] 2q33 (*IDDM7*),[152–154] 3q22–q25 (*IDDM9*),[155] 10p11–q11 (*IDDM10*),[156,157] 14q24–q31 (*IDDM11*),[158,159] 2q34–q35 (*IDDM13*),[160] 6q21 (*IDDM15*),[161] 14q32 (*IDDM16*),[162] 10q25 (*IDDM17*),[163] 5q33 (*IDDM18*),[164,165] 7p15–p13 (*GCK*),[166] 1q42,[167] 16q22–q24,[168] Xp11 (conditional on HLA-DR genotype),[169] 2p12,[170] 5p11–q32,[170] 16p11–p13,[170] and 8q11–q24.[171] Given the likely small contribution (low λ_s) of each non-HLA susceptibility locus, a large number of affected sibpair families is required for the identification of T1D susceptibility genes. Therefore, although statistical evidence supporting linkage for some of these regions was strong in the initial reports, most regions have not been clearly established in multiple populations.[172] Many of these intervals may represent false-positive linkage, which is expected by random chance in a genome scan if stringent linkage criteria are not applied.

It has been suggested that joint analyses of combined existing families is a powerful approach for clarification of the role of non-HLA–linked loci for T1D.[173] A joint analysis of data from three previous genomewide scans (US, UK, and Scandinavia) and 254 new families was recently performed by the Type 1 Diabetes Genetic Consortium (T1DGC).[172] The study included a total of 1435 multiplex families with an average map information content of 67% (from ~400 polymorphic microsatellite markers in each scan), and the family collection provided ~95% power to detect a locus with locus-specific $\lambda_s \geq 1.3$ and $P = 10^{-4}$. The strongest evidence for linkage to T1D was obtained on chromosome 6p21 (nominal $P = 2 \times 10^{-52}$) within the HLA region. The study also provided supporting evidence for the existence of 10 non-HLA–linked loci for T1D, including 2q31–2q33 (*IDDM7* and *IDDM12*), 3p13–p14, 6q21 (*IDDM15*), 9q33–q34, 10p14–q11 (*IDDM10*), 11p15 (*IDDM2*), 12q14–q12, 16p12–q11.1, 16q22–q24, and 19p13.3–13.2.

9. CHALLENGES FOR THE CHARACTERIZATION OF MORE MODEST T1D SUSCEPTIBILITY GENES

As the advancement in high-throughput genotyping technologies (microsatellite markers and SNPs) and computational tools was accompanied by the completion of the human genome sequence, many investigators in the field of T1D genetics and other complex diseases believed that systematically identifying all complex disease genes throughout the entire human genome would be relatively easy. However, the initial excitement gradually dissipated as they began to realize the difficulties of replicating the linkage findings and moving from linkage intervals of 5–20 cM (thousands of genes) to the identification of the specific disease-causing gene and the etiological mutation (often a single base-pair change). The literature is teeming with reports of linkages that either cannot be replicated or for which corroboration by linkage has been impossible to find.[174,175]

Explanations for this slow progress have been rehearsed. One major possibility is that a large number of genes probably influence T1D susceptibility, with each gene having a weak effect that is difficult to be detectable by linkage analysis. To further complicate matters, different patients, families, ethnic groups, or races may have a different subset of genes that in combination are responsible for their disease onset (known as genetic heterogeneity).[121] Gene/gene interaction and gene/environment interaction provide additional complications that contribute to the difficulty in identifying T1D susceptibility genes. Another possibility is the lack of sophisticated tools and standard for correlation between linkage/association results and the phenotypic difference caused by the real etiological mutation.

Identification of the defective phenotypes associated with the disease-causing alleles for complex diseases is not an easy task. As aforementioned, the etiological mutations are usually single nucleotide polymorphisms (SNPs), which have been found in one in every 566 nt within the genome[176]; several estimates suggest that, overall, 20% of them could have the ability to create a phenotypic difference.[177] Therefore, a sophisticated analytical tool is necessary to correlate the linkage/association results with the defective phenotype caused by the true disease-causing allele.

10. ASSOCIATION VERSUS LINKAGE: THE RIGHT CHOICE FOR FUTURE STUDY

For the past two decades, the dominant study design for the investigation of the genetic basis of inherited disease has been linkage analysis in families.

Since the early 1990s, a number of laboratories have carried out genome scans for linkage using affected sib pairs. Despite the success in identifying over 20 different genomic intervals (Table 1) showing suggestive or significant linkages, the susceptibility genes encoded in these regions remain largely elusive. This is because linkage analysis is preferential for the localization of a disease gene with high genotype relative risks ($g \geq 4$) and intermediate allele frequencies ($p = 0.05$–0.50),[178–182] such as *IDDM1* and *IDDM2*; however, the fact is that each of the non-HLA susceptibility genes for T1D only contributes a very small proportion of the total susceptibility (more modest relative risk, $g \leq 2$). Using realistic assumptions, several thousands to tens of thousands of affected sib pairs would be required to detect a significant linkage for those weaker genotype–phenotype correlations. Therefore, linkage analysis is not the best approach for the characterization of more modest susceptibility genes for T1D.

In contrast to linkage analysis, association (case-control) studies compare the frequencies of an allele or genotype at a locus in patient and matched control groups. The pattern of allele sharing among affected individuals within pedigrees is less striking than the pattern of allele sharing between unrelated affected individuals for modest-risk alleles contributing to a complex disease. Such differences make association analysis more powerful for the detection of common disease alleles that confer modest disease risks. It is suggested that association studies provide adequate power for genes with relative risks as low as 1.5, even when using a stringent significance level (5×10^{-8}).[180] A typical example demonstrating the power of association study design is the identification of *PTPN22* for T1D susceptibility. All of the previous genomewide scans failed to detect a linkage for the chromosomal 1p13 region (*PTPN22*). Furthermore, a recent study combining four genomewide scans in 1435 multiplex families still failed to obtain support evidence for linkage of this chromosomal region.[172] In contrast, Bottini and coworkers[74] applied a case-control design (293 cases and 395 controls of North American origin; 174 cases and 214 controls of Italian origin) and demonstrated strong association of *PTPN22* to T1D susceptibility.

Another major advantage of the case-control association design is the relative ease with which one can collect a large number of patients and matched controls from many different populations. Sampling different populations allows the effects of genetic heterogeneity to be considered. In addition, the disease gene and etiological mutation can be found with relative ease because association only occurs when the genetic marker is very close to an etiological mutation.

Although the case-control study design has many advantages, results from these studies must be interpreted with caution. The most serious concern is that the results could be explained by spurious associations due to mismatches in ethnicity or geographical regions between the control and patient groups (e.g. population stratification). Family-based tests of association, such as the transmission/disequilibrium test (TDT), can be used to find an association due to linkage with an etiological mutation rather than due to spurious associations. These tests work by examining the frequency with which a marker allele is transmitted from parent to affected offspring, and look for an allele to be transmitted more often than by chance. Such family-based approaches clearly complement the case-control design. Replication of association in multiple studies (especially in different populations, ethnic groups, and/or laboratories) is a very useful way to confirm a real association.

11. CANDIDATE GENES VERSUS GENOMEWIDE APPROACH: BACK TO THE FUTURE

Despite the above discussed advantages and the necessity for association mapping, this method suffers from one major limitation: association can only be detected by genetic markers in very close proximity to the etiological polymorphism. Association studies have been traditionally applied to specific candidate genes of presumed importance in the disease for this exact reason. The recent discovery of abundant SNPs throughout the entire human genome has provided opportunities for genomewide association studies. Although this opportunity is exciting, it suffers from the lack of affordable genotyping technologies necessary for the hundreds of thousands of SNPs (or SNP blocks) required for genomewide association studies.

Due to the limitations of both linkage and genomewide association studies, the traditional approach of testing association with functional SNP-based candidate genes is gaining favor.[183] The candidate gene approach focuses on testing specific genes that may play a role in the etiology of the disease. The approach takes advantage of functional SNPs characterized within candidate genes and the increased statistical power of association studies as well as the biological understanding of the disease etiology.

A brief review of the history of T1D susceptibility gene identification indicates that the candidate gene approach has been quite successful. Candidate gene testing has identified all T1D susceptibility genes (*HLA-DQB1*, *HLA-DQA1*, *HLA-DRB1*, *INS*, *PTPN22*, and *CTLA-4*), except the *SUMO4* gene (*IDDM5*). Therefore, association study design with appropriate selection of candidate genes combined with functional SNPs located within the candidate genes should significantly speed up the discovery of more modest T1D susceptibility genes.

12. SUMMARY AND CONCLUSION REMARKS

T1D is a polygenic and complex disorder. Genetic factors influence both the susceptibility and resistance to T1D. Among the inherited susceptibility loci that have been identified, *IDDM1* is the strongest risk factor. Other non-HLA susceptibility genes may contribute to diabetes risk in an epistatic way. Addressing the lack of consistent results across populations with regard to the genetic determinants of the disease is most critical at the present time. Functional SNP-based candidate gene association study is probably the best approach for the characterization of susceptibility genes with more modest relative risks to T1D. A large number of genes may contribute susceptibility to T1D, but the elucidation of their functions in the pathogenesis of T1D will be the most important factor towards understanding its molecular etiology.

REFERENCES

1. Li M, Guo D, Isales CM, *et al.* (2005) SUMO wrestling with type 1 diabetes. *J Mol Med* **83**:504–513.
2. She JX, Bui MM, Tian XH, *et al.* (1994) Additive susceptibility to insulin-dependent diabetes conferred by HLA-DQB1 and insulin genes. *Autoimmunity* **18**:195–203.
3. She JX. (1996) Susceptibility to type I diabetes: HLA-DQ and DR revisited. *Immunol Today* **17**:323–329.
4. She JX, Marron MP. (1998) Genetic susceptibility factors in type 1 diabetes: linkage, disequilibrium and functional analyses. *Curr Opin Immunol* **10**:682–689.
5. Pugliese A, Gianani R, Moromisato R, *et al.* (1995) HLA-DQB1*0602 is associated with dominant protection from diabetes even among islet cell antibody-positive first-degree relatives of patients with IDDM. *Diabetes* **44**:608–613.
6. Pugliese A, Eisenbarth GS. (2004) Type 1 diabetes mellitus of man: genetic susceptibility and resistance. *Adv Exp Med Biol* **552**:170–203.
7. Hirschhorn JN. (2003) Genetic epidemiology of type 1 diabetes. *Pediatr Diabetes* **4**:87–100.
8. Atkinson MA. (2005) Thirty years of investigating the autoimmune basis for type 1 diabetes: why can't we prevent or reverse this disease? *Diabetes* **54**:1253–1263.

9. Pugliese A. (2004) Genetics of type 1 diabetes. *Endocrinol Metab Clin North Am* **33**:1–16.

10. Nerup J, Platz P, Anderson OO, *et al.* (1974) HL-A antigens and diabetes mellitus. *Lancet* **2**:864–866.

11. Valdes AM, Thomson G, Erlich HA, Noble JA. (1999) Association between type 1 diabetes age of onset and HLA among sibling pairs. *Diabetes* **48**:1658–1661.

12. Demaine AG, Hibberd ML, Mangles D, Millward BA. (1995) A new marker in the HLA class I region is associated with the age at onset of IDDM. *Diabetologia* **38**:623–628.

13. Fennessy M, Metcalfe K, Hitman GA, *et al.* (1994) A gene in the HLA class I region contributes to susceptibility to IDDM in the Finnish population. Childhood Diabetes in Finland (DiMe) Study Group. *Diabetologia* **37**:937–944.

14. Langholz B, Tuomilehto-Wolf E, Thomas D, *et al.* (1995) Variation in HLA-associated risks of childhood insulin-dependent diabetes in the Finnish population: I. Allele effects at A, B, and DR loci. DiMe Study Group. Childhood Diabetes in Finland. *Genet Epidemiol* **12**:441–453.

15. Honeyman MC, Harrison LC, Drummond B, *et al.* (1995) Analysis of families at risk for insulin-dependent diabetes mellitus reveals that HLA antigens influence progression to clinical disease. *Mol Med* **1**:576–582.

16. Fujisawa T, Ikegami H, Kawaguchi Y, *et al.* (1995) Class I HLA is associated with age-at-onset of IDDM, while class II HLA confers susceptibility to IDDM. *Diabetologia* **38**:1493–1495.

17. Pugliese A, Bugawan T, Moromisato R, *et al.* (1994) Two subsets of HLA-DQA1 alleles mark phenotypic variation in levels of insulin autoantibodies in first degree relatives at risk for insulin-dependent diabetes. *J Clin Invest* **93**:2447–2452.

18. Bell GI, Horita S, Karam JH. (1984) A polymorphic locus near the human insulin gene is associated with insulin-dependent diabetes mellitus. *Diabetes* **33**:176–183.

19. Lie BA, Todd JA, Pociot F, *et al.* (1999) The predisposition to type 1 diabetes linked to the human leukocyte antigen complex includes at least one non–class II gene. *Am J Hum Genet* **64**:793–800.

20. Gambelunghe G, Ghaderi M, Cosentino A, *et al.* (2000) Association of MHC class I chain-related A (MIC-A) gene polymorphism with type I diabetes. *Diabetologia* **43**:507–514.

21. Cox NJ, Bell GI, Xiang KS. (1988) Linkage disequilibrium in the human insulin/insulin-like growth factor II region of human chromosome II. *Am J Hum Genet* **43**:495–501.

22. Deng GY, Muir A, Maclaren NK, She JX. (1995) Association of LMP2 and LMP7 genes within the major histocompatibility complex with insulin-dependent diabetes mellitus: population and family studies. *Am J Hum Genet* **56**:528–534.

23. Ronningen KS, Spurkland A, Iwe T, *et al.* (1991) Distribution of HLA-DRB1, -DQA1 and -DQB1 alleles and DQA1-DQB1 genotypes among

Norwegian patients with insulin-dependent diabetes mellitus. *Tissue Antigens* **37**:105–111.

24. Sanjeevi CB, Zeidler A, Shaw S, *et al.* (1993) Analysis of HLA-DQA1 and -DQB1 genes in Mexican Americans with insulin-dependent diabetes mellitus. *Tissue Antigens* **42**:72–77.

25. Lee KH, Wucherpfennig KW, Wiley DC. (2001) Structure of a human insulin peptide-HLA-DQ8 complex and susceptibility to type 1 diabetes. *Nat Immunol* **2**:501–507.

26. Nepom BS, Schwarz D, Palmer JP, Nepom GT. (1987) Transcomplementation of HLA genes in IDDM. HLA-DQ alpha- and beta-chains produce hybrid molecules in DR3/4 heterozygotes. *Diabetes* **36**:114–117.

27. Tait BD, Mraz G, Harrison LC. (1988) Association of HLA-DQw3 (TA10-) with type I diabetes occurs with DR3/4 but not DR1/4 patients. *Diabetes* **37**:926–929.

28. Nepom GT, Erlich H. (1991) MHC class-II molecules and autoimmunity. *Annu Rev Immunol* **9**:493–525.

29. Kockum I, Wassmuth R, Holmberg E, *et al.* (1993) HLA-DQ primarily confers protection and HLA-DR susceptibility in type I (insulin-dependent) diabetes studied in population-based affected families and controls. *Am J Hum Genet* **53**:150–167.

30. Awata T, Kuzuya T, Matsuda A, *et al.* (1992) Genetic analysis of HLA class II alleles and susceptibility to type 1 (insulin-dependent) diabetes mellitus in Japanese subjects. *Diabetologia* **35**:419–424.

31. Huang HS, Peng JT, She JY, *et al.* (1995) HLA-encoded susceptibility to insulin-dependent diabetes mellitus is determined by DR and DQ genes as well as their linkage disequilibria in a Chinese population. *Hum Immunol* **44**:210–219.

32. Huang HS, Huang MJ, Huang CC. (1988) A strong association of HLA-DR 3/4 heterozygotes with insulin-dependent diabetes among Chinese in Taiwan. *Taiwan Yi Xue Hui Za Zhi* **87**:1–6.

33. Lin LY, Hu C, Wang ZY, Hao ZP. (2004) Photocatalytic degradation of ethylene over titania-based photocatalysts. *Huan Jing Ke Xue* **25**:105–108.

34. Koeleman BP, Lie BA, Undlien DE, *et al.* (2004) Genotype effects and epistasis in type 1 diabetes and HLA-DQ trans dimer associations with disease. *Genes Immun* **5**:381–388.

35. Hu CY, Allen M, Chuang LM, *et al.* (1993) Association of insulin-dependent diabetes mellitus in Taiwan with HLA class II DQB1 and DRB1 alleles. *Hum Immunol* **38**:105–114.

36. Todd JA, Fukui Y, Kitagawa T, Sasazuki T. (1990) The A3 allele of the HLA-DQA1 locus is associated with susceptibility to type 1 diabetes in Japanese. *Proc Natl Acad Sci USA* **87**:1094–1098.

37. Penny MA, Jenkins D, Mijoric CH, *et al.* (1992) Susceptibility to IDDM in a Chinese population. Role of HLA class II alleles. *Diabetes* **41**:914–919.

38. Todd JA, Mijoric C, Fletcher J, *et al.* (1989) Identification of susceptibility loci for insulin-dependent diabetes mellitus by trans-racial gene mapping. *Nature* **338**:587–589.

39. Thorsby E, Ronningen KS. (1992) Role of HLA genes in predisposition to develop insulin-dependent diabetes mellitus. *Ann Med* **24**:523–531.

40. Huang W, She JX, Muir A, *et al.* (1994) High risk HLA-DR/DQ genotypes for IDD confer susceptibility to autoantibodies but DQB1*0602 does not prevent them. *J Autoimmun* **7**:889–897.

41. Erlich HA, Zeidler A, Chang J, *et al.* (1993) HLA class II alleles and susceptibility and resistance to insulin dependent diabetes mellitus in Mexican-American families. *Nat Genet* **3**:358–364.

42. Cucca F, Lampis R, Frau F, *et al.* (1995) The distribution of DR4 haplotypes in Sardinia suggests a primary association of type I diabetes with DRB1 and DQB1 loci. *Hum Immunol* **43**:301–308.

43. Cucca F, Muntoni F, Lampis R, *et al.* (1993) Combinations of specific DRB1, DQA1, DQB1 haplotypes are associated with insulin-dependent diabetes mellitus in Sardinia. *Hum Immunol* **37**:85–94.

44. Baisch JM, Weeks T, Giles R, *et al.* (1990) Analysis of HLA-DQ genotypes and susceptibility in insulin-dependent diabetes mellitus. *N Engl J Med* **322**:1836–1841.

45. Tait BD, Drummond BP, Varney MD, Harrison LC. (1995) HLA-DRB1*0401 is associated with susceptibility to insulin-dependent diabetes mellitus independently of the DQB1 locus. *Eur J Immunogenet* **22**:289–297.

46. Van der Auwera B, Van Waeyenbenge C, Schuit F, *et al.* (1995) DRB1*0403 protects against IDDM in Caucasians with the high-risk heterozygous DQA1*0301–DQB1*0302/DQA1*0501–DQB1*0201 genotype. Belgian Diabetes Registry. *Diabetes* **44**:527–530.

47. Erlich HA, Zeidler A, Chang J, *et al.* (1993) HLA class II alleles and susceptibility and resistance to insulin dependent diabetes mellitus in Mexican-American families. *Nat Genet* **3**:358–364.

48. Yasunaga S, Kimura A, Hamaguchi K, *et al.* (1996) Different contribution of HLA-DR and -DQ genes in susceptibility and resistance to insulin-dependent diabetes mellitus (IDDM). *Tissue Antigens* **47**:37–48.

49. Pociot F, McDermott MF. (2002) Genetics of type 1 diabetes mellitus. *Genes Immun* **3**:235–249.

50. Julier C, Hyer RN, Davies J, *et al.* (1991) Insulin-IGF2 region on chromosome 11p encodes a gene implicated in HLA-DR4–dependent diabetes susceptibility. *Nature* **354**:155–159.

51. Bain SC, Prins JB, Hearne CM, *et al.* (1992) Insulin gene region–encoded susceptibility to type 1 diabetes is not restricted to HLA-DR4–positive individuals. *Nat Genet* **2**:212–215.

52. Bell GI, Horita S, Karam JH. (1984) A polymorphic locus near the human insulin gene is associated with insulin-dependent diabetes mellitus. *Diabetes* **33**:176–183.

53. Lucassen AM, Julier C, Beressi JP, *et al.* (1993) Susceptibility to insulin dependent diabetes mellitus maps to a 4.1 kb segment of DNA spanning the insulin gene and associated VNTR. *Nat Genet* **4**:305–310.

54. Owerbach D, Gabbay KH. (1996) The search for IDDM susceptibility genes: the next generation. *Diabetes* **45**:544–551.

55. Heath VL, Moore NC, Parnell SM, Mason DW. (1998) Intrathymic expression of genes involved in organ specific autoimmune disease. *J Autoimmun* **11**:309–318.

56. Bennett ST, Lucassen AM, Gough SC, *et al.* (1995) Susceptibility to human type 1 diabetes at IDDM2 is determined by tandem repeat variation at the insulin gene minisatellite locus. *Nat Genet* **9**:284–292.

57. Owerbach D, Aagaard L. (1984) Analysis of a 1963-bp polymorphic region flanking the human insulin gene. *Gene* **32**:475–479.

58. Rotwein P, Yokoyama S, Didier DK, Chirgwin JM. (1986) Genetic analysis of the hypervariable region flanking the human insulin gene. *Am J Hum Genet* **39**:291–299.

59. Alleva DG, Crowe PD, Jin L, *et al.* (2001) A disease-associated cellular immune response in type 1 diabetics to an immunodominant epitope of insulin. *J Clin Invest* **107**:173–180.

60. Wegmann DR, Norbury-Glaser M, Daniel D. (1994) Insulin-specific T cells are a predominant component of islet infiltrates in pre-diabetic NOD mice. *Eur J Immunol* **24**:1853–1857.

61. Lucassen AM, Screaton GR, Julier C, *et al.* (1995) Regulation of insulin gene expression by the IDDM associated, insulin locus haplotype. *Hum Mol Genet* **4**:501–506.

62. Werdelin O, Cordes U, Jensen T. (1998) Aberrant expression of tissue-specific proteins in the thymus: a hypothesis for the development of central tolerance. *Scand J Immunol* **47**:95–100.

63. Pugliese A, Brown D, Garza D, *et al.* (2001) Self-antigen–presenting cells expressing diabetes-associated autoantigens exist in both thymus and peripheral lymphoid organs. *J Clin Invest* **107**:555–564.

64. Sospedra M, Ferrer-Francesch X, Dominguez O, *et al.* (1998) Transcription of a broad range of self-antigens in human thymus suggests a role for central mechanisms in tolerance toward peripheral antigens. *J Immunol* **161**:5918–5929.

65. Anjos S, Polychronakos C. (2004) Mechanisms of genetic susceptibility to type I diabetes: beyond HLA. *Mol Genet Metab* **81**:187–195.

66. Pugliese A, Zeller M, Fernandez A Jr, *et al.* (1997) The insulin gene is transcribed in the human thymus and transcription levels correlate with allelic variation at the INS VNTR-IDDM2 susceptibility locus for type 1 diabetes. *Nat Genet* **15**:293–297.

67. Vafiadis P, Bennett ST, Todd JA, *et al.* (1997) Insulin expression in human thymus is modulated by INS VNTR alleles at the IDDM2 locus. *Nat Genet* **15**:289–292.

68. Thebault-Baumont K, Dubois-Laforgue D, Krief P, *et al.* (2003) Acceleration of type 1 diabetes mellitus in proinsulin 2–deficient NOD mice. *J Clin Invest* **111**:851–857.

69. French MB, Allison J, Cram DS, *et al.* (1997) Transgenic expression of mouse proinsulin II prevents diabetes in nonobese diabetic mice. *Diabetes* **46**:34–39.

70. Gjorloff-Wingren A, Saxena M, Williams S, *et al.* (1999) Characterization of TCR-induced receptor-proximal signaling events negatively regulated by the protein tyrosine phosphatase PEP. *Eur J Immunol* **29**:3845–3854.

71. Cohen S, Dadi H, Shaoul E, *et al.* (1999) Cloning and characterization of a lymphoid-specific, inducible human protein tyrosine phosphatase, Lyp. *Blood* **93**:2013–2024.

72. Hasegawa K, Martin F, Huang G, *et al.* (2004) PEST domain–enriched tyrosine phosphatase (PEP) regulation of effector/memory T cells. *Science* **303**:685–689.

73. Hill RJ, Zozulya S, Lu YL, *et al.* (2002) The lymphoid protein tyrosine phosphatase Lyp interacts with the adaptor molecule Grb2 and functions as a negative regulator of T-cell activation. *Exp Hematol* **30**:237–244.

74. Bottini N, Musumeci L, Alonso A, *et al.* (2004) A functional variant of lymphoid tyrosine phosphatase is associated with type I diabetes. *Nat Genet* **36**:337–338.

75. Smyth D, Cooper JD, Collins JE, *et al.* (2004) Replication of an association between the lymphoid tyrosine phosphatase locus (LYP/PTPN22) with type 1 diabetes, and evidence for its role as a general autoimmunity locus. *Diabetes* **53**:3020–3023.

76. Onengut-Gumuscu S, Ewens KG, Spielman RS, Concannon P. (2004) A functional polymorphism (1858C/T) in the PTPN22 gene is linked and associated with type I diabetes in multiplex families. *Genes Immun* **5**:678–680.

77. Ladner MB, Bottini N, Valdes AM, Noble JA. (2005) Association of the single nucleotide polymorphism C1858T of the PTPN22 gene with type 1 diabetes. *Hum Immunol* **66**:60–64.

78. Criswell LA, Pfeiffer KA, Lum RF, *et al.* (2005) Analysis of families in the Multiple Autoimmune Disease Genetics Consortium (MADGC) collection: the PTPN22 620W allele associates with multiple autoimmune phenotypes. *Am J Hum Genet* **76**:561–571.

79. Zheng W, She JX. (2005) Genetic association between a lymphoid tyrosine phosphatase (PTPN22) and type 1 diabetes. *Diabetes* **54**:906–908.

80. Qu H, Tessier MC, Hudson TJ, Polychronakos C. (2005) Confirmation of the association of the R620W polymorphism in the protein tyrosine phosphatase PTPN22 with type 1 diabetes in a family based study. *J Med Genet* **42**:266–270.

81. Gomez LM, Anaya JM, Gonzalez CI, *et al.* (2005) PTPN22 C1858T polymorphism in Colombian patients with autoimmune diseases. *Genes Immun* **6**:628–631.

82. Zhernakova A, Eerligh P, Wijmenga C, *et al.* (2005) Differential association of the PTPN22 coding variant with autoimmune diseases in a Dutch population. *Genes Immun* **6**:459–461.

83. Maier LM, Wicker LS. (2005) Genetic susceptibility to type 1 diabetes. *Curr Opin Immunol* **17**:601–608.

84. Criswell LA, Pfeiffer KA, Lum RF, *et al.* (2005) Analysis of families in the Multiple Autoimmune Disease Genetics Consortium (MADGC) collection: the PTPN22 620W allele associates with multiple autoimmune phenotypes. *Am J Hum Genet* **76**:561–571.

85. Begovich AB, Carlton VE, Honigberg LA, *et al.* (2004) A missense single-nucleotide polymorphism in a gene encoding a protein tyrosine phosphatase (PTPN22) is associated with rheumatoid arthritis. *Am J Hum Genet* **75**:330–337.

86. Orozco G, Sanchez E, Gonzalez-Gay MA, *et al.* (2005) Association of a functional single-nucleotide polymorphism of PTPN22, encoding lymphoid protein phosphatase, with rheumatoid arthritis and systemic lupus erythematosus. *Arthritis Rheum* **52**:219–224.

87. Lee AT, Li W, Liew A, *et al.* (2005) The PTPN22 R620W polymorphism associates with RF positive rheumatoid arthritis in a dose-dependent manner but not with HLA-SE status. *Genes Immun* **6**:129–133.

88. Viken MK, Amundsen SS, Kvien TK, *et al.* (2005) Association analysis of the 1858C > T polymorphism in the PTPN22 gene in juvenile idiopathic arthritis and other autoimmune diseases. *Genes Immun* **6**:271–273.

89. Hinks A, Barton A, John S, *et al.* (2004) Association between the PTPN22 gene and rheumatoid arthritis and juvenile idiopathic arthritis in a UK population: further support that PTPN22 is an autoimmunity gene. *Arthritis Rheum* **52**:1694–1699.

90. Simkins HM, Merriman ME, Highton J, *et al.* (2005) Association of the PTPN22 locus with rheumatoid arthritis in a New Zealand Caucasian cohort. *Arthritis Rheum* **52**:2222–2225.

91. Carlton VE, Hu X, Chokkalingam AP, *et al.* (2005) PTPN22 genetic variation: evidence for multiple variants associated with rheumatoid arthritis. *Am J Hum Genet* **77**:567–581.

92. Wesoly J, van der Helm-van Mil AH, Toes RE, *et al.* (2005) Association of the PTPN22 C1858T single-nucleotide polymorphism with rheumatoid arthritis phenotypes in an inception cohort. *Arthritis Rheum* **52**:2948–2950.

93. Seldin MF, Shigeta R, Laiho K, *et al.* (2005) Finnish case-control and family studies support PTPN22 R620W polymorphism as a risk factor in rheumatoid arthritis, but suggest only minimal or no effect in juvenile idiopathic arthritis. *Genes Immun* **6**:720–722.

94. Reddy MV, Johansson M, Sturfelt G, *et al.* (2005) The R620W C/T polymorphism of the gene PTPN22 is associated with SLE independently of the association of PDCD1. *Genes Immun* **6**:658–662.

95. Wu H, Cantor RM, Graham DS, *et al.* (2005) Association analysis of the R620W polymorphism of protein tyrosine phosphatase PTPN22 in systemic lupus erythematosus families: increased T allele frequency in systemic lupus erythematosus patients with autoimmune thyroid disease. *Arthritis Rheum* **52**:2396–2402.

96. Velaga MR, Wilson V, Jennings CE, *et al.* (2004) The codon 620 tryptophan allele of the lymphoid tyrosine phosphatase (LYP) gene is a major determinant of Graves' disease. *J Clin Endocrinol Metab* **89**:5862–5865.

97. Skorka A, Bednarczuk T, Bar-Andziak E, *et al.* (2005) Lymphoid tyrosine phosphatase (PTPN22/LYP) variant and Graves' disease in a Polish population: association and gene dose-dependent correlation with age of onset. *Clin Endocrinol (Oxf)* **62**:679–682.

98. Canton I, Akhtar S, Gavalas NG, *et al.* (2005) A single-nucleotide polymorphism in the gene encoding lymphoid protein tyrosine phosphatase (PTPN22) confers susceptibility to generalised vitiligo. *Genes Immun* **6**:584–587.

99. Begovich AB, Caillier SJ, Alexander HC, *et al.* (2005) The R620W polymorphism of the protein tyrosine phosphatase PTPN22 is not associated with multiple sclerosis. *Am J Hum Genet* **76**:184–187.

100. Matesanz F, Rueda B, Orozco G, *et al.* (2005) Protein tyrosine phosphatase gene (PTPN22) polymorphism in multiple sclerosis. *J Neurol* **252**:994–995.

101. Wagenleiter SE, Klein W, Griga T, *et al.* (2005) A case-control study of tyrosine phosphatase (PTPN22) confirms the lack of association with Crohn's disease. *Int J Immunogenet* **32**:323–324.

102. van Oene M, Wintle RF, Liu X, *et al.* (2005) Association of the lymphoid tyrosine phosphatase R620W variant with rheumatoid arthritis, but not Crohn's disease, in Canadian populations. *Arthritis Rheum* **52**:1993–1998.

103. Ittah M, Gottenberg JE, Proust A, *et al.* (2005) No evidence for association between 1858 C/T single-nucleotide polymorphism of PTPN22 gene and primary Sjogren's syndrome. *Genes Immun* **6**:457–458.

104. Martin MC, Oliver J, Urcelay E, *et al.* (2005) The functional genetic variation in the PTPN22 gene has a negligible effect on the susceptibility to develop inflammatory bowel disease. *Tissue Antigens* **66**:314–317.

105. Nistor I, Nair RP, Stuart P, *et al.* (2005) Protein tyrosine phosphatase gene PTPN22 polymorphism in psoriasis: lack of evidence for association. *J Invest Dermatol* **125**:395–396.

106. Rueda B, Nunez C, Orozco G, *et al.* (2005) C1858T functional variant of PTPN22 gene is not associated with celiac disease genetic predisposition. *Hum Immunol* **66**:848–852.

107. Mori M, Yamada R, Kobayashi K, *et al.* (2005) Ethnic differences in allele frequency of autoimmune-disease-associated SNPs. *J Hum Genet* **50**:264–266.

108. Noso S. (2005) Genetic heterogeneity of SUMO4 M55V variant with susceptibility to type 1 diabetes. *Diabetes* **54**:3582–3586.

109. Siminovitch KA. (2004) PTPN22 and autoimmune disease. *Nat Genet* **36**:1248–1249.

110. Hasegawa K, Yajima H, Katagiri T, *et al.* (1999) Requirement of PEST domain tyrosine phosphatase PEP in B cell antigen receptor–induced growth arrest and apoptosis. *Eur J Immunol* **29**:887–896.

111. Chien W, Tidow N, Williamson EA, *et al.* (2003) Characterization of a myeloid tyrosine phosphatase, Lyp, and its role in the Bcr-Abl signal transduction pathway. *J Biol Chem* **278**:27413–27420.

112. Vang T, Congia M, Macis MD, *et al.* (2005) Autoimmune-associated lymphoid tyrosine phosphatase is a gain-of-function variant. *Nat Genet* **37**: 1317–1319.

113. Gregersen PK. (2005) Gaining insight into PTPN22 and autoimmunity. *Nat Genet* **37**:1300–1302.

114. Davies JL, Kawaguchi Y, Bennett ST, *et al.* (1994) A genome-wide search for human type 1 diabetes susceptibility genes. *Nature* **371**:130–136.

115. Luo DF, Buzzetti R, Rotter JI, *et al.* (1996) Confirmation of three susceptibility genes to insulin-dependent diabetes mellitus: IDDM4, IDDM5 and IDDM8. *Hum Mol Genet* **5**:693–698.

116. Luo DF, Bui MM, Muir A, *et al.* (1995) Affected-sib-pair mapping of a novel susceptibility gene to insulin-dependent diabetes mellitus (IDDM8) on chromosome 6q25–q27. *Am J Hum Genet* **57**:911–919.

117. Davies JL, Cucca F, Goy JV, *et al.* (1996) Saturation multipoint linkage mapping of chromosome 6q in type 1 diabetes. *Hum Mol Genet* **5**:1071–1074.

118. Delepine M, Pociot F, Habita C, *et al.* (1997) Evidence of a non-MHC susceptibility locus in type I diabetes linked to HLA on chromosome 6. *Am J Hum Genet* **60**:174–187.

119. Zhoucun A, Zhang S, Xiao C. (2001) Preliminary studies on associations of IDDM3, IDDM4, IDDM5 and IDDM8 with IDDM in Chengdu population. *Chin Med Sci J* **16**:120–122.

120. Guo D, Li M, Zhang Y, *et al.* (2004) A functional variant of SUMO4, a new I kappa B alpha modifier, is associated with type 1 diabetes. *Nat Genet* **36**:837–841.

121. Wang C, Podolsky R, She J. (2006) Genetic and functional evidence supporting SUMO4 as a type 1 diabetes susceptibility gene. *Ann N Y Acad Sci* **1079**:257–267.

122. Park Y, Park S, Kang J, *et al.* (2005) Assessing the validity of the association between the SUMO4 M55V variant and risk of type 1 diabetes. *Nat Genet* **37**:112–113.

123. Noso S, Ikegami H, Fujisawa T, *et al.* (2005) Genetic heterogeneity in association of the SUMO4 M55V variant with susceptibility to type 1 diabetes. *Diabetes* **54**:3582–3586.

124. Owerbach D, Pina L, Gabbay KH. (2004) A 212-kb region on chromosome 6q25 containing the TAB2 gene is associated with susceptibility to type 1 diabetes. *Diabetes* **53**:1890–1893.

125. Smyth DJ, Howson JM, Lowe CE, *et al.* (2005) Assessing the validity of the association between the SUMO4 M55V variant and risk of type 1 diabetes. *Nat Genet* **37**:110–111.

126. Qu H, Bharaj B, Liu XQ, *et al.* (2005) Assessing the validity of the association between the SUMO4 M55V variant and risk of type 1 diabetes. *Nat Genet* **37**:111–112.

127. Wang CY, Yang P, She JX. (2005) Assessing the validity of the association between the SUMO4 M55V variant and risk of type 1 diabetes. *Nat Genet* **37**:112–113.

128. Bohren KM, Nadkarni V, Song JH, *et al.* (2004) A M55V polymorphism in a novel SUMO gene (SUMO-4) differentially activates heat shock transcription factors and is associated with susceptibility to type I diabetes mellitus. *J Biol Chem* **279**:27233–27238.

129. Guo D, Han J, Adam BL, *et al.* (2005) Proteomic analysis of SUMO4 substrates in HEK293 cells under serum starvation-induced stress. *Biochem Biophys Res Commun* **337**:1308–1318.

130. Nistico L, Buzzetti R, Pritchard LE, *et al.* (1996) The CTLA-4 gene region of chromosome 2q33 is linked to, and associated with, type 1 diabetes. Belgian Diabetes Registry. *Hum Mol Genet* **5**:1075–1080.

131. Marron MP, Raffel LJ, Garchon HJ, *et al.* (1997) Insulin-dependent diabetes mellitus (IDDM) is associated with CTLA4 polymorphisms in multiple ethnic groups. *Hum Mol Genet* **6**:1275–1282.

132. Donner H, Rau H, Walfish PG, *et al.* (1997) CTLA4 alanine-17 confers genetic susceptibility to Graves' disease and to type 1 diabetes mellitus. *J Clin Endocrinol Metab* **82**:143–146.

133. Awata T, Kurihara S, Iitaka M, *et al.* (1998) Association of CTLA-4 gene A-G polymorphism (IDDM12 locus) with acute-onset and insulin-depleted IDDM as well as autoimmune thyroid disease (Graves' disease and Hashimoto's thyroiditis) in the Japanese population. *Diabetes* **47**:128–129.

134. Ueda H, Howson JM, Esposito L, *et al.* (2003) Association of the T-cell regulatory gene CTLA4 with susceptibility to autoimmune disease. *Nature* **423**:506–511.

135. Anjos SM, Shao W, Marchand L, Polychronakos C. (2005) Allelic effects on gene regulation at the autoimmunity-predisposing CTLA4 locus: a re-evaluation of the $3' + 6230G > A$ polymorphism. *Genes Immun* **6**:305–311.

136. Atabani SF, Thio CL, Divanovic S, *et al.* (2005) Association of CTLA4 polymorphism with regulatory T cell frequency. *Eur J Immunol* **35**:2157–2162.

137. Magistrelli G, Jeannin P, Herbault N, *et al.* (1999) A soluble form of CTLA-4 generated by alternative splicing is expressed by nonstimulated human T cells. *Eur J Immunol* **29**:3596–3602.

138. Oaks MK, Hallett KM, Penwell RT, *et al.* (2000) A native soluble form of CTLA-4. *Cell Immunol* **201**:144–153.

139. Oaks MK, Hallett KM. (2000) Cutting edge: a soluble form of CTLA-4 in patients with autoimmune thyroid disease. *J Immunol* **164**:5015–5018.

140. Liu MF, Wang CR, Chen PC, Fung LL. (2003) Increased expression of soluble cytotoxic T-lymphocyte-associated antigen-4 molecule in patients with systemic lupus erythematosus. *Scand J Immunol* **57**:568–572.

141. Wang XB, Kakoulidou M, Giscombe R, *et al.* (2002) Abnormal expression of CTLA-4 by T cells from patients with myasthenia gravis: effect of an AT-rich gene sequence. *J Neuroimmunol* **130**:224–232.

142. Purohit S, Podolsky R, Collins C, *et al.* (2005) Lack of correlation between the levels of soluble cytotoxic T-lymphocyte associated antigen-4 (CTLA-4) and the CT-60 genotypes. *J Autoimmune Dis* **2**:8.

143. Hashimoto L, Habita C, Beressi JP, *et al.* (1994) Genetic mapping of a susceptibility locus for insulin-dependent diabetes mellitus on chromosome 11q. *Nature* **371**:161–164.

144. Nakagawa Y, Kawaguchi Y, Twells RC, *et al.* (1998) Fine mapping of the diabetes-susceptibility locus, IDDM4, on chromosome 11q13. *Am J Hum Genet* **63**:547–556.

145. Eckenrode S, Marron MP, Nicholls R, *et al.* (2000) Fine-mapping of the type 1 diabetes locus (IDDM4) on chromosome 11q and evaluation of two candidate genes (FADD and GALN) by affected sibpair and linkage-disequilibrium analyses. *Hum Genet* **106**:14–18.

146. Owerbach D. (2000) Physical and genetic mapping of IDDM8 on chromosome 6q27. *Diabetes* **49**:508–512.

147. Chistiakov DA, Seryogin YA, Turakulov RI, *et al.* (2005) Evaluation of IDDM8 susceptibility locus in a Russian simplex family data set. *J Autoimmun* **24**:243–250.

148. Field LL, Tobias R, Magnus T. (1994) A locus on chromosome 15q26 (IDDM3) produces susceptibility to insulin-dependent diabetes mellitus. *Nat Genet* **8**:189–194.

149. Zamani M, Pociot F, Raeymaekers P, *et al.* (1996) Linkage of type I diabetes to 15q26 (IDDM3) in the Danish population. *Hum Genet* **98**:491–496.

150. Merriman T, Twells R, Merriman M, *et al.* (1997) Evidence by allelic association-dependent methods for a type 1 diabetes polygene (IDDM6) on chromosome 18q21. *Hum Mol Genet* **6**:1003–1010.

151. Merriman TR, Eaves IA, Twells RC, *et al.* (1998) Transmission of haplotypes of microsatellite markers rather than single marker alleles in the mapping of a putative type 1 diabetes susceptibility gene (IDDM6). *Hum Mol Genet* **7**: 517–524.

152. Esposito L, Hill NJ, Pritchard LE, *et al.* (1998) Genetic analysis of chromosome 2 in type 1 diabetes: analysis of putative loci IDDM7, IDDM12, and IDDM13 and candidate genes NRAMP1 and IA-2 and the interleukin-1 gene cluster. IMDIAB Group. *Diabetes* **47**:1797–1799.

153. Kristiansen OP, Pociot F, Bennett EP, *et al.* (2000) IDDM7 links to insulin-dependent diabetes mellitus in Danish multiplex families but linkage is not explained by novel polymorphisms in the candidate gene GALNT3. The Danish Study Group of Diabetes in Childhood and The Danish IDDM Epidemiology and Genetics Group. *Hum Mutat* **15**:295–296.

154. Cox NJ, Wapelhorst B, Morrison VA, *et al.* (2001) Seven regions of the genome show evidence of linkage to type 1 diabetes in a consensus analysis of 767 multiplex families. *Am J Hum Genet* **69**:820–830.

155. Paterson AD, Rahman P, Petronis A. (1999) IDDM9 and a locus for rheumatoid arthritis on chromosome 3q appear to be distinct. *Hum Immunol* **60**:883–885.

156. Reed P, Cucca F, Jenkins S, *et al.* (1997) Evidence for a type 1 diabetes susceptibility locus (IDDM10) on human chromosome 10p11–q11. *Hum Mol Genet* **6**:1011–1016.

157. Chistiakov DA, Seryogin Y, Savost'anov KV, *et al.* (2004) Evidence for a type 1 diabetes susceptibility locus (IDDM10) on chromosome 10p11–q11 in a Russian population. *Scand J Immunol* **60**:316–323.

158. Field LL, Tobias R, Thomson G, Plon S. (1996) Susceptibility to insulin-dependent diabetes mellitus maps to a locus (IDDM11) on human chromosome 14q24.3–q31. *Genomics* **33**:1–8.

159. Corder EH, Woodbury MA, Manton KG, Field LL. (2001) Grade-of-membership sibpair linkage analysis maps IDDM11 to chromosome 14q24.3–q31. *Ann Hum Genet* **65**:387–394.

160. Fu J, Ikegami H, Kawaguchi Y, *et al.* (1998) Association of distal chromosome 2q with IDDM in Japanese subjects. *Diabetologia* **41**:228–232.

161. Concannon P, Gogolin-Ewens KJ, Hinds DA, *et al.* (1998) A second-generation screen of the human genome for susceptibility to insulin-dependent diabetes mellitus. *Nat Genet* **19**:292–296.

162. Field LL, Larsen Z, Pociot F, *et al.* (2002) Evidence for a locus (IDDM16) in the immunoglobulin heavy chain region on chromosome 14q32.3 producing susceptibility to type 1 diabetes. *Genes Immun* **3**:338–344.

163. Eller E, Vardi P, Daly MJ, *et al.* (2004) IDDM17: polymorphisms in the AMACO gene are associated with dominant protection against type 1A diabetes in a Bedouin Arab family. *Ann NY Acad Sci* **1037**:145–149.

164. Morahan G, Huang D, Ymer SI, *et al.* (2001) Linkage disequilibrium of a type 1 diabetes susceptibility locus with a regulatory IL12B allele. *Nat Genet* **27**:218–221.

165. Bergholdt R, Ghandil P, Johannesen J, *et al.* (2004) Genetic and functional evaluation of an interleukin-12 polymorphism (IDDM18) in families with type 1 diabetes. *J Med Genet* **41**:e39.

166. Rowe RE, Wapelhorst B, Bell GI, *et al.* (1995) Linkage and association between insulin-dependent diabetes mellitus (IDDM) susceptibility and markers near the glucokinase gene on chromosome 7. *Nat Genet* **10**: 240–242.

167. Ewens KG, Johnson LN, Wapelhorst B, *et al.* (2002) Linkage and association with type 1 diabetes on chromosome 1q42. *Diabetes* **51**:3318–3325.

168. Kristiansen OP, Larsen ZM, Johannesen J, *et al.* (1999) No linkage of P187S polymorphism in NAD(P)H: quinone oxidoreductase (NQO1/DIA4) and type 1 diabetes in the Danish population. DIEGG and DSGD. Danish IDDM Epidemiology and Genetics Group and The Danish Study Group of Diabetes in Childhood. *Hum Mutat* **14**:67–70.

169. Zavattari P, Esposito L, Nutland S, *et al.* (2000) Transmission-ratio distortion at Xp11.4–p21.1 in type 1 diabetes. *Am J Hum Genet* **66**:330–332.

170. Nerup J, Pociot F. (2001) A genomewide scan for type 1-diabetes susceptibility in Scandinavian families: identification of new loci with evidence of interactions. *Am J Hum Genet* **69**:1301–1313.

171. Sale MM, FitzGerald LM, Charlesworth JC, *et al.* (2002) Evidence for a novel type 1 diabetes susceptibility locus on chromosome 8. *Diabetes* **51** (Suppl 3):S316–S319.

172. Concannon P, Erlich HA, Julier C, *et al.* (2005) Type 1 diabetes: evidence for susceptibility loci from four genome-wide linkage scans in 1,435 multiplex families. *Diabetes* **54**:2995–3001.

173. Cox NJ, Wapelhorst B, Morrison VA, *et al.* (2001) Seven regions of the genome show evidence of linkage to type 1 diabetes in a consensus analysis of 767 multiplex families. *Am J Hum Genet* **69**:820–830.

174. Terwilliger JD, Weiss KM. (1998) Linkage disequilibrium mapping of complex disease: fantasy or reality? *Curr Opin Biotechnol* **9**:578–594.

175. Weiss KM, Terwilliger JD. (2000) How many diseases does it take to map a gene with SNPs? *Nat Genet* **26**:151–157.

176. Conde L, Vaquerizas JM, Santoyo J, *et al.* (2004) PupaSNP Finder: a web tool for finding SNPs with putative effect at transcriptional level. *Nucleic Acids Res* **32**:W242–W248.

177. Sunyaev S, Ramensky V, Koch I, *et al.* (2001) Prediction of deleterious human alleles. *Hum Mol Genet* **10**:591–597.

178. Risch N, Merikangas K. (1996) The future of genetic studies of complex human diseases. *Science* **273**:1516–1517.

179. Carlson CS, Eberle MA, Kruglyak L, Nickerson DA. (2004) Mapping complex disease loci in whole-genome association studies. *Nature* **429**:446–452.

180. Risch NJ. (2000) Searching for genetic determinants in the new millennium. *Nature* **405**:847–856.

181. Cardon LR, Bell JI. (2001) Association study designs for complex diseases. *Nat Rev Genet* **2**:91–99.

182. Marchini J, Donnelly P, Cardon LR. (2005) Genome-wide strategies for detecting multiple loci that influence complex diseases. *Nat Genet* **37**: 413–417.

183. Tabor HK, Risch NJ, Myers RM. (2002) Opinion: candidate-gene approaches for studying complex genetic traits: practical considerations. *Nat Rev Genet* **3**:391–397.

Chapter 25

Genetic Susceptibility to Human Obesity

Yong-Jun Liu[*,§] and Hong-Wen Deng[*,†,‡]

*Departments of Orthopedic Surgery & Basic Medical Sciences
School of Medicine, University of Missouri–Kansas City
Kansas City, MO 64108, USA*

†*Key Laboratory of Biomedical Information Engineering of the
Ministry of Education & Institute of Molecular Genetics
School of Life Science and Technology, Xi'an Jiaotong University
Xi'an 710049, P. R. China*

‡*Laboratory of Molecular and Statistical Genetics
College of Life Sciences, Hunan Normal University
Changsha, Hunan 410081, P. R. China*

Obesity is a condition of excess body fat that causes or exacerbates several major public health problems. There has been considerable success in elucidating the molecular basis of monogenic forms of obesity in both rodents and humans. The most common form of obesity, however, is considered to be a polygenic disorder arising from the interaction of multiple genetic and environmental factors. The identification and characterization of susceptibility genes to obesity will contribute to a greater understanding of the pathogenesis of obesity, and will ultimately assist to develop better strategies for prevention and therapeutic intervention. In this review article, we provide an overview of the recent findings in the genetic dissection of obesity. We also consider emerging strategies for gene identification. Some major problems are addressed, concluding with a brief discussion on the future perspective of research in the genetics of obesity.

---

§ Correspondence author.

Current Topics in Human Genetics

1. INTRODUCTION

Obesity is a disease condition with excess body fat that adversely affects health.[1] Obesity increases the risk of multiple conditions, including cardiovascular disease, type 2 diabetes, cancer, and premature death.[1] The prevalence of excess weight is rapidly increasing across the United States, and today close to 65% of the adult population are overweight [body mass index (BMI, the ratio of weight to the square of height) ≥ 25 kg/m^2] or obese (BMI ≥ 30 kg/m^2).[2] Close to 300 000 deaths each year in the United States may be attributable to obesity,[3] making obesity the second leading cause of preventable death in the country.[4] The direct and indirect annual costs associated with obesity in the United States alone are $117 billion.[5]

Obesity has long been known to run in families. Although the current environment characterized by ample food supply and sedentary lifestyle promotes the obesity rate,[6] genetic risk factors also play an important role in the development of obesity; this is reflected by the high heritability of many components of obesity.[7] The past years have witnessed rapid progress in understanding the genetics of obesity. Most of the previously existing mutations in mouse obesity genes that segregate as Mendelian traits have been cloned, and a number of homologous mutations have been discovered as rare causes of human obesity.[8] These genes (e.g. leptin and leptin receptor) and their products have provided molecular focal points for new biochemical and physiological pathways that are conserved between mice and humans. However, in contrast to monogenic obesity, the most common form of obesity is a polygenic disorder resulting from the interaction among multiple genetic and environmental factors.[9,10] Several approaches, e.g. genomewide linkage scan and candidate gene association studies, have been developed and widely applied in the exploration of genetic variants contributing to common variations in body weight and adiposity. To date, however, convincing results have not been obtained, due largely to the complicated etiology of obesity and the potential limitations of current study designs.

Lately, with the complete human genome sequence unraveled, functional genomics becomes realistic and will help disentangle the organization and control of genetic pathways that come together to define obesity. Advanced technologies are being developed that allow simultaneous determination of the expression of thousands of genes at the mRNA (transcriptomics) and protein (proteomics) levels. Functional genomics essentially complements genetic mapping approaches (including linkage scan and association mapping), and is believed to help clarify the complicated biochemical mechanisms underlying obesity.[11,12] An improved understanding of molecular genetic mechanisms underlying body weight regulation will eventually provide promising opportunities for the prevention and therapeutic intervention of obesity.[13]

2. LESSONS FROM ANIMAL MODELS

Because of the difficulties inherent in the genetic analysis of human obesity, animal models (particularly rodents) have been intensively studied for clues to the analogous genes in humans. This task has been accomplished through the production of several mutant lines of mice and genetically modified mice.[14,15] These animal models (monogenic or polygenic) have provided powerful tools to permit the deciphering and understanding of energy and lipid metabolism as well as the development of potential treatments for human obesity.

2.1. *Monogenic Models*

Over the past few years, the genes responsible for most of the known cases of monogenic forms of obesity in mice have been cloned, and have generated a number of interesting candidate genes for human obesity.[16] Among these, six single-gene obesity mutations — *agouti, obese (ob), diabetic (db), fat, tubby,* and *mahogany* mutant mouse strains — have been frequently reviewed.[8,16] These six genes encode molecules that appear to interact in physiological pathways influencing body fat storage.[16]

So far, the best described monogenic animal obesity models are the *ob* mutants[17] and *db* mutants,[18] which are deficient in circulating the adipostatic hormone leptin and its cognate receptor (leptin receptor), respectively. The discovery of leptin as a satiety factor in the *ob* mouse line stimulated research into the genetics of obesity, promoting the dissection of the leptin pathway controlling food intake and energy homeostasis.[16] In humans, it has been shown that circulating serum leptin levels are highly correlated to body fat content, and mutations in this gene seem to cause extreme obesity at an early age.[16]

2.2. *Knockout and Transgenic Models*

With the advent of recombinant DNA technology, transgene expression (i.e. the introduction and overexpression of specific genes in mice) has become one of the most frequently used approaches in obesity research. Transgenic technology has provided several models of obese mice that have helped to clarify the importance and function of specific genes and proteins involved in the physiological pathway of adiposity.[8,19] However, the close interaction between polygenes underlying obesity and environmental factors was, for some time, a limiting factor to this conventional transgenic methodology.

Gene knockout has offered new opportunities in obesity research.[20] This is a more direct and refined method, which basically consists of the disruption (knockout) of specific endogenous genes. This permits us to assess the relative importance of the individual components of the genetic framework of body weight regulation through the association between the resulting abnormalities and the disrupted gene.[20] A number of mice obesity genes have been knocked out,[8] allowing researchers to reveal the actual role of putative obesity candidate genes.

2.3. *Polygenic Mouse Models*

As with monogenic mice models, quantitative genetic approaches have provided important insights into the fundamental aspects of genetic architecture.[15,21] The vast spectrum of existing genetic variation, associated with a short generation time and low husbandry cost, make mice well suited to this kind of approach. To date, a total of 183 animal quantitative trait loci (QTLs) related to body fat content, body weight, and adiposity have been mapped.[8] Several of these QTLs are near the mapped position for single-gene mutations that cause obesity. For example, the chromosome 6 locus identified in BSB mice is very close to the *ob* gene.[22] Similarly, the chromosome 4 locus identified in the SWR/J × AKR/J cross contains the *db* gene.[18] An extensive equivalence between animal QTLs and specific regions in the human genome has been postulated,[8] some of them containing human obesity candidate genes. Examples of these are the QTLs mouse obesity 1 (Mob1) and obesity qt 1 (Obq1), which are supposed to be collocated with the uncoupling proteins 2 and 3 (UCP2 and UCP3) genes.[23]

More recently, the combination of conventional QTL mapping and transgene methodologies has provided more power to QTL analysis.[15] The traditional QTL mapping in F2 crosses and backcross populations is now corroborated by the utilization of genetically identical lines of individuals. This permits the production of as many genetically identical animals as necessary for a reliable identification of QTLs in polygenic models.

3. THE SEARCH FOR HUMAN OBESITY GENES

3.1. *Monogenic Human Obesity*

Obesity or abnormal fat distribution characterizes some Mendelian syndromes, such as Prader–Willi syndrome, Bardet–Biedl syndrome, Cohen syndrome, and Wilson–Turner syndrome.[8] In these cases, obesity could be a related, but not a dominant, clinical feature. Most of these syndromic forms of obesity have been genetically mapped to different chromosomal regions, but causative genes have not yet been isolated due to the extreme rarity of these mutations.

By screening human subjects for mutations in candidate genes derived from mouse studies, a few cases of single-gene obesity syndrome have been reported.[8] To date, rare mutations on the genes of leptin receptor (*LEPR*), leptin (*LEP*), proopiomelanocortin (*POMC*), melanocortin 4 receptor (*MC4R*), proprotein convertase subtilisin/kexin type 1 (*PCSK1*), peroxisome proliferative activated receptor γ (*PPAR-γ*), *Drosophila* single-minded (*SIm1*), corticotropin-releasing factor receptor 1 (*CRHR1*), corticotropin-releasing factor receptor 2 (*CRHR2*), and G protein–coupled receptor 24 (*GPR24*) have been found to result in obesity as a dominant clinical feature.[8] These mutations are typically associated with early-onset and severe obesity. For example, in the *LEP* gene, a homozygous frameshift mutation involving the deletion of a single guanine nucleotide in codon 133 was identified in two severely obese children,[24] and another mutation of C-T substitution at codon 105 was found in three morbidly obese individuals.[25] Since these mutations are rare in the general human population and usually exist in extremely obese subjects, they may not be responsible for common human obesity.

3.2. *Common Human Obesity*

3.2.1. *Heritability*

Heritability is a measure of the relative contribution of genetic factors to trait variability.[26] During the past decades, a strong genetic control for obesity has been consistently reported. Twin studies have suggested a heritability of fat mass of 40%–70% with a concordance of 0.7–0.9 between monozygotic twins compared with 0.35–0.45 between dizygotic twins.[7,27,28] Family studies and adoption studies looking at the heritability of obesity have also drawn overall positive conclusions regarding the familiarity of obesity.[7] In general, all of the studies identified a considerable proportion (20%–50%) of the phenotypic variation being attributable to genetic factors. In a study in American Caucasian families, heritability estimates of BMI, fat mass, percentage fat mass, and lean mass were in the range of 52%–57%.[29] Moreover, segregation analyses in several populations suggested that major genes with recessive effects might account for 35%–45% of the variation in obesity-related phenotypes, after adjustment for sex and age.[30–32]

3.2.2. *Approaches*

Currently, two complementary approaches are commonly used to search for human obesity genes. The first one is the linkage approach, which seeks to identify loci that cosegregate with a trait within families. To improve the likelihood that a gene influencing obesity might be identified, investigators search the genome, testing polymorphic markers evenly spaced on all chromosomes. The genomewide linkage screen attempts to map genes purely by genomic position, thus requiring no presumptions on the function of genes at the susceptibility loci. Linkage analyses for complex diseases are commonly performed using affected sibling pairs or other types of affected relative pairs. In the case of obesity, these might be relatives diagnosed with obesity (e.g. BMI > 30 kg/m^2). However, it is likely that the use of a quantitative phenotype (e.g. BMI) rather than a dichotomous phenotype is more powerful for linkage mapping. The linkage approach has been successfully used to locate susceptibility genes underlying Mendelian disorders, such as Duchenne muscular dystrophy, cystic fibrosis, and Huntington's disease; however, its utility in locating complex disease genes has proved complicated.

The second approach is candidate gene association analysis, which tests the association between a specific genetic variant and a phenotype variation. In the case of obesity, the candidate genes are selected based on their involvement in the regulation of energy homeostasis, adipose tissue metabolism, or other aspects that may cause obesity. One of the major limitations with this approach is our limited understanding of the physiopathology of obesity; hence, different investigators may have different opinions regarding which gene(s) should be selected as appropriate candidate(s). Moreover, the association approach tests only association, which may not interpret causation; thus, association alone may not be sufficient to identify susceptibility genes underlying complex diseases.

3.2.3. *Genomewide Linkage Screens*

The past few years have witnessed enthusiasm for genomewide scans for human obesity genes, resulting in the publication of more than 50 genome scans since the first one in 1997. For example, an earlier genomewide linkage study in 630 subjects from 53 Caucasian pedigrees contained $>10\,000$ relative pairs (including 1249 sib pairs) and identified several potential QTLs for obesity-related phenotypes.[33] In the subsequent study, denser markers were genotyped at these regions in an expanded sample of 1816 subjects from 79 pedigrees and found partial support for the earlier findings.[34]

Genomewide linkage screens for obesity genes have been conducted in populations of diverse ethnic backgrounds, including Mexican-Americans,[35,36] Pima Indians,[37,38] French families,[39] American Caucasians,[33,40–42] Canadian Caucasians,[43,44] African-Americans,[45] Japanese and Chinese,[46] as well as isolated populations such as Old Order Amish[47] and Finnish.[48,49] Up to now, more than 200 QTLs underlying normal variation of human obesity phenotypes have been reported from around 50 screens.[8] Some of the screens involved reanalysis of the same cohort with different phenotypes. The number of loci reported is large, and at least some may well be false-positive. A few regions, such as 1p36, 2p21, 3q21–27, 6p21, 10p12–p11, 11q23–q24, 17p12, 18q21, 20q11–q13, and Xq23–24, have been replicated by multiple studies (summarized in Table 1). However, these regions are generally large ($\sim$10–30 cM) and poorly defined; thus, fine mapping these regions to a small interval ($\sim$1 cM) is an essential step to eventually identify the causal gene(s).

3.2.4. *Candidate Gene Association Studies*

An inherent limitation with the linkage approach is the lack of power to detect modest genetic effects. For a QTL of relatively high genetic effect (e.g. $h^2 > 15\%$), it is realistic to expect linkage analysis to provide statistical evidence for detection; however, for more modest genetic effects (e.g. $h^2 < 5\%$), linkage analysis is not likely to offer such evidence except in unrealistically large samples. In contrast, the candidate gene association approach is more powerful for the identification of such modest effects.[50] Association studies test the correlation between a specific genetic variant and the variation underlying a trait of interest. Unlike linkage, association depends on the presence of linkage disequilibrium (LD) among genes or markers in adjacent genomic locations; thus, confounding such as variation in LD patterns and population stratification could pose potential problems.

A long list of candidate genes has arisen from the molecules involved in energy homeostasis and adipose metabolism. To date, a number of studies have reported positive associations with 113 candidate genes for obesity-related phenotypes.[8] Here, we review several prominent genes that have been most frequently examined, including *LEPR*, *PPAR-γ*, and beta-3 adrenergic receptor (*ADRB3*). In addition, we briefly address major issues that may cause inconsistent and/or controversial results.

Table 1 Putative Genomic Regions Replicated in Multiple Linkage Studies

Genomic Region	Marker	Candidate Gene	Phenotype	LOD Score or P value	Population	Sample Size*	Reference
1p36	D1S508, 468	*TNFR2*	BMI	LOD = 2.2–2.5	Caucasian	994; 37 pedigrees	133
	D1S468		BMI	LOD = 2.75	Caucasian	630; 53 pedigrees	33
2p22–24	D2S1788	*POMC*	Fat mass/Leptin	LOD = 4.9/2.8	Mexican-American	458; 10 pedigrees	35
			Leptin, BMI	$0.008 < P < 0.03$	African-American	720; 230 families	134
	D2S165, 367		Leptin	LOD = 2.4/2.7	French families	514; 18 nuclear families	39
3q27	D3S2427	*Adiponectin*	BMI/WC	LOD = 3.3/2.4	Caucasian	2209; 507 families	40
			BMI	LOD = 3.4	Combined	6849; 4 ethnic groups	46
			BMI	LOD = 1.8	African-American	618; 202 families	45
6p21	D6S1959	*TNF*	PFM	LOD = 2.7	African-American	618; 202 families	45
	D6S276		Eating behavior	LOD = 2.1	Old Order Amish	624; 28 families	135
	D6S271		Leptin	LOD = 2.1	Pima Indians	770; 239 families	136
10p12–p11	D10S197	*GAD2*	Obesity	LOD = 4.9	French families	514; 18 nuclear families	39
	D10S220		Leptin	LOD = 2.7	Old Order Amish	672; 28 families	47
	D10S204, 193		Obesity	$1.1 < LOD < 2.5$	Caucasian	286; 93 families	137
	D10S582, 107		Obesity	$0.0005 < P < 0.03$	Mixed	862; 170 families	138

(Continued)

Table 1 *(Continued)*

Genomic Region	Marker	Candidate Gene	Phenotype	LOD Score or P value	Population	Sample Size*	Reference
11q23–q24	D11S976	APOA4	24-h EE	LOD = 2.0	Pima Indians	236 sib pairs; 82 nuclear families	38
	D11S4464	DRD2	BMI	2.6 < LOD < 2.8	Caucasian	994; 37 pedigrees	133
	D11S912		BMI	LOD = 3.6	Pima Indians	1766 sib pairs; 264 families	37
	D11S1998		BMI	LOD = 2.7	Pima Indians	1526 sib pairs	139
17p12	D17S947	SREBF1	Leptin	LOD = 5.0	Caucasian	2209; 507 families	40
			Adiponectin	LOD = 1.7	Caucasian	1100; 170 families	140
			BMI	LOD = 2.5	Combined	6849; 4 ethnic groups	46
18q21	D18S877	MC4R	Fat-free mass	LOD = 3.6	Caucasian	336 sib pairs	141
			PFM	LOD = 2.3	Pima Indians	451 sib pairs	38
	D18S115		Obesity	LOD = 2.4	Finnish	193 sib pairs	48
20q11–q13	D20S107, 211, 149	ASIP	BMI, PFM	3.0 < LOD < 3.2	Caucasian	423 sib pairs	41
	D20S601	GHRH	24-h RQ	LOD = 3.0	Pima Indians	236 sib pairs; 82 nuclear families	38
	D20S478, 481	ADA	BMI	2.0 < LOD < 2.2	Caucasian	994; 37 pedigrees	133
Xq23–24	DXS1057	HTR2C	BMI	LOD = 2.0	Caucasian	994; 37 pedigrees	133
	DXS6804	PWLSX	Obesity	LOD = 3.1	Finnish	193 sib pairs	48

*In terms of number of subjects, unless otherwise stated.

Note: LOD, logarithm of odds; PFM, percentage fat mass; WC, waist circumference; EE, energy expenditure; RQ, respiratory quotient.

a. *LEPR*

Leptin is a fat tissue–derived hormone that reports nutritional information and regulates energy expenditure by binding and activating its specific
receptor, LEPR, in the hypothalamus.[51] LEPR is a single transmembrane
protein belonging to the superfamily of cytokine receptors, and signals
through the Janus kinase/signal transducer and activator of transcription
(JAK-STAT) pathway.[52] Three common amino acid variants — Lys109Arg,
Gln223Arg, and Lys656Asn — have been described and most frequently
examined. These variants are located in the extracellular binding domain
of the receptor, and may have potential effects on the signaling capacity of
LEPR.

   In young Dutch adults,[53] the Arg109 allele had higher leptin levels in
weight gainers than noncarriers of the allele. In a group of overweight and
obese postmenopausal women, carriers of the Asn656 allele had increased
hip circumference, total abdominal fat, and subcutaneous fat measured by
CT scan.[54] A significant interaction effect between the menopausal state
and the Lys109Arg genotype was seen for BMI, and with the Lys656Asn
genotype for subcutaneous and total abdominal fat.[54] For the Gln223Arg
variant, association was observed in the Quebec Family Study[55] and the
HERITAGE family study.[56] Further support comes from the studies in
a Mediterranean population,[57] postmenopausal Caucasian women,[58] and
young Dutch adults.[53] Our group reported significant association between
the Lys656Asn variant and obesity phenotypes in 405 Caucasian families,
wherein subjects carrying allele G at the Lys656Asn site had, on average,
3.16% higher lean mass and 2.71% higher fat mass than those without it.[59]

b. *PPAR-γ*

PPAR-γ is a nuclear hormone receptor that is involved in adipogenesis and
insulin signaling. Of the two main isoforms, γ1 is ubiquitously expressed,
whereas γ2 is found mainly in adipose tissue. Current studies have focused
on a common variant, Pro12Ala, in which a proline is substituted for alanine
at codon 12 of the γ2 isoform.[60] This variant was shown to cause not only
decreased ability of the receptor to bind to PPAR-γ responsive elements,
a hexanucleotide sequence AGGTCA found in the promoters of PPAR-γ
responsive genes, but also promotion of transcription.[61]

   The Pro12Ala variant has been consistently associated with increased
insulin sensitivity and decreased risk of type 2 diabetes.[61,62] However, its
effects on body weight and adiposity remain less clear: the Pro12Ala variant
has been associated with increased BMI,[63,64] decreased BMI,[61] or neither.[65]

A possible explanation for this discrepancy is the interaction between the Pro12Ala variant and other genes, diet, or even obesity itself. It was shown that this variant might, in its homozygous form, interact with various combinations of genetic and environmental factors in lean and obese subjects to cause divergent modulating effects on BMI and long-term body weight control.[66] In another study, weight regain during a 12-month follow-up was greater in women with the Ala allele than in women homozygous for the Pro allele.[67] Moreover, the synergic effects of PPAR-γ2 Pro12Ala and ADRB3 Trp60Arg on obesity were reported,[68] reinforcing the idea that gene effects will be underestimated if only their individual effects are considered.

c. *ADRB3*

ADRB3 is expressed primarily in visceral adipose tissue, and is involved in the regulation of lipolysis and thermogenesis. It is mapped on chromosome 8p12–p11, which has shown linkage to obesity phenotypes. A common missense mutation, Trp64Arg, is most frequently examined. Trp64Arg is located at the beginning of the first intracellular loop of the protein. The Arg64 allele tends to lower lipolysis activity, likely due to its inability to link to G proteins.[69]

In Pima Indians, individuals homozygous for the Arg64 allele had a lower resting metabolic rate and earlier onset of type 2 diabetes.[70] In Finns, this allele contributed to abdominal obesity, resistance to insulin, and earlier onset of type 2 diabetes.[71] The variant was also associated with greater obesity and weight gain in obese French whites,[72] and increased BMI and fat mass in Mexican-Americans.[73] Further evidence supporting the Trp64Arg variant as a contributor to increased BMI comes from several meta-analyses.[74] Although the effect of this variant by itself is modest, it may interact with other genes, such as PPAR-γ[68] and uncoupling protein 1,[75] to intensify the observed phenotypes.

3.2.5. *Potential Reasons for Inconsistent/Conflicting Results in Linkage and Association Studies*

Despite the extensive utilization of both linkage and association approaches in searching for obesity genes, a frustrating fact is that initial findings are often difficult to be replicated, leading to concerns about the reliability of the results and the value of these mapping strategies. In general, there are two major sources that lead to this situation. The first involves poor study design, which may cause putative results. The second concerns the complicated nature of the etiology of obesity (such as genetic heterogeneity, epistasis, low penetrance, variable expressivity and pleiotropy), which poses special challenges for gene discovery.

In different populations, all of these factors could be different for the same disease/trait. Even when the causal variant is under investigation, it may be more or less important in different populations, especially if the variant has low relative risk, variable penetrance, and variable allele frequencies. From this perspective, it is sometimes hard or even impossible to replicate significant findings across populations from different ethnic backgrounds. However, it is reasonable to expect that significant results obtained from the same population or from populations of the same or similar ethnicity are largely reproducible. Indeed, the linkage and association approaches are powerful tools if they are used properly.[76,77]

a. *Heterogeneity and interaction*

The effects of genetic heterogeneity and interactions (including gene–gene and gene–environment) on the development of obesity have been explicitly revealed. For instance, interactions between the *PPAR-γ*, *ADRB3*, and *UCP1* genes were observed to affect body weight in a synergic way.[68,75] Epistatic interactions between loci on chromosomes 20 and 10 play a role in extreme human obesity.[78] The effects of nutrient–gene interaction on the susceptibility, onset, and/or severity of obesity have also been demonstrated.[79] Heterogeneity and interaction, if existing but not accounted for, may interfere with the detection of significant effects and specific genes, because the effects of a gene depend on the specific background of other genetic and environmental factors. However, this issue is usually overlooked, and only a small fraction of the human genetics literature specifically reports on investigations of such complexity.

Heterogeneity and interaction are among the most complex and perplexing aspects of genetic studies, and are known to be difficult to study without first identifying the potentially significant factors that may interact or that may render heterogeneity. There are two conventional strategies in dealing with heterogeneity and/or interactions. The first one is to directly model interaction terms as product (or other appropriate) forms in multivariate analyses as covariates; their significance can be examined by testing for the significance of the interaction terms. The second one is to stratify the sample into subgroups by potential sources of heterogeneity (e.g. subgroups of pedigrees, genders, or age groups) and interaction, and then apply the analyses to each subgroup. Other methods are also applicable, such as the

M-test, the β-test, the admixture test, and most recently ordered subset analysis (OSA).[80] However, none of the methodologies alone is superior in all respects for the range of complicating factors that might be present in any given data set. Thus, there is an urgent need for extensive re-evaluation of the existing methodologies for heterogeneity and interaction, as well as for massive efforts in new method development.

b. *Insufficient power*

Currently, the first and foremost factor is lack of sufficient power to generate reliable and reproducible results. Like other complex diseases, obesity is under the control of multiple genetic factors, each having small or moderate effects. However, most studies were based on small sample sizes, which may not provide sufficient power to detect individual effects.

As an illustration, we present the sample sizes required for mapping QTLs of various effect sizes in a classical sib-pair linkage study (Table 2). It can be seen that to reliably detect a QTL of large heritability with decent power, at least 1000 sib pairs — the most informative relative pair for linkage analysis — are needed. With this in consideration, we reviewed the genomewide screens for obesity published by the year 2005 and found that, of the reports, 72% were based on samples of <1000 sib pairs. One would expect that logarithm of odds (LOD) scores based on small sample sizes would be unstable and contribute to lack of replication. It should be noted that this is a simplified demonstration, as a few studies employed

Table 2   Sample Sizes Required to Achieve 80% Power under an LOD Score of 3.0 for a Sib-Pair Linkage Study

h_q^2	Additive		Dominant		Recessive	
	$\theta = 0.00$	$\theta = 0.05$	$\theta = 0.00$	$\theta = 0.05$	$\theta = 0.00$	$\theta = 0.05$
0.05	22 380	34 130	23 290	36 820	24 610	38 090
0.10	6179	9437	6442	10 260	6800	10 580
0.15	3007	4602	3141	5036	3311	5178
0.20	1838	2819	1922	3103	2024	3182
0.25	1269	1952	1329	2159	1398	2209
0.30	945	1458	991	1619	1042	1654

Power was calculated using Genetic Power Calculator (http://statgen.iop.kcl.ac.uk/gpc/).
h_q^2 is the heritability due to the QTL; θ is the recombination fraction between the QTL and a marker.
Power computation was performed under an ideal condition without considering genetic interaction, heterogeneity, etc.

ascertainment strategy by recruiting subjects with extreme phenotypic values (e.g. BMI > 30 kg/m^2), which may enhance the statistical power.[81,82] Without sufficient statistical power, it will be a challenge to resolve the continuously appearing significant, yet largely inconsistent, results from genomewide screen studies.

It is worth mentioning that in linkage studies, the important feature of sample size is not merely the total number of subjects, but also the number of informative relative pairs. Although neither large multiplex pedigrees nor small families are fully ideal for all situations, preferring not just small but minimal family structures (such as sib pairs or nuclear family triads) is almost certainly detrimental.[83] Generally, extended pedigrees can be more powerful due to the increasingly large numbers of relative pairs informative for linkage analysis.

In association studies, apart from sample size, the power of a study can be affected by many other factors, including the extent and degree of LD between the markers tested and the causal variants as well as the allele frequency differences between them.[84,85] Any of these conditions may individually or interactively affect the power of a specific study, as illustrated in Fig. 1. Thus, before embarking on a study, statistical power should be estimated and, for an association study, it is necessary to characterize the LD pattern and haplotype structure around the gene(s) of interest.

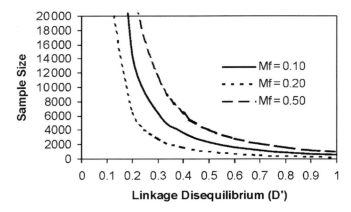

Fig. 1. Influence of mismatch between marker and causal variant allele frequencies and extent of LD on the power of a family-based association study. The figure presents the sample sizes required for achieving 80% power ($\alpha = 0.001$) to detect a QTL (effect size = 5%; allele frequency = 0.20) (y-axis). Patterns for causal variant allele frequencies (Mf) of 0.10, 0.20, and 0.50 are shown. We assume that the QTL is under additive inheritance; and that the LD between the QTL and the marker, in terms of D$'$, is from 0.1 to 1.0 (x-axis).

c. *Population stratification*

For association studies, population stratification may cause spurious outcomes — not only false-positive results, but also false-negative results.[86,87] Although the actual impact of population stratification on association studies has been a matter of some debate,[87,88] we believe that it (to a certain degree) contributes to nonreplication of association and, unless thoroughly addressed, remains a potential source of bias.[89,90] It is notable that, of the hundreds of association studies on BMI reported by 2005, most did not address this critical issue.

The family-based approaches, such as transmission/disequilibrium test (TDT), completely obviate concerns about population stratification and are gaining popularity.[91,92] Our group employed TDT to test the association and/or linkage between several candidate genes (e.g. *APOE* and *LEPR*) and obesity phenotypes, and obtained interesting findings.[59,93] A limitation of TDT is its reliance on heterozygous parents, which may decrease statistical power. With the recent extensions, this limitation is somewhat alleviated.[94,95] Other methods that may be unbiased to population stratification — such as genomic control (GC)[96] and structured association (SA)[97] — can be considered for population-based association studies, and merit further application and development.

d. *Others*

The potential confounding factors that lead to the lack of replication of genetic studies are not confined to the abovementioned situations. Other issues such as poor data quality and multiple testing are also believed to be the most likely reasons.

4. PROSPECTS FOR GENE DISCOVERY IN OBESITY

With the completion of the Human Genome Project and the construction of a dense single nucleotide polymorphism (SNP) map, new opportunities are being presented for unraveling the complex genetic basis of obesity. Functional genomics — the quantitative determination of the spatial and temporal accumulation patterns of specific mRNA, proteins, and important metabolites — is becoming the focus of researches. These methods are complementary to the classical linkage and association approaches, and will facilitate the identification of genes and their interactions underlying obesity in the postgenome era (Fig. 2).

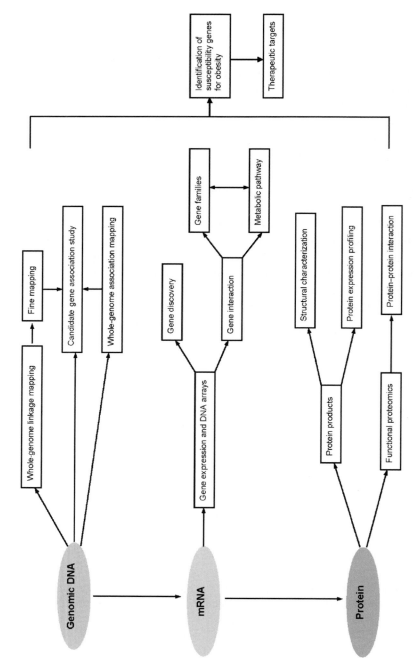

Fig. 2. A schematic display of an integrated approach for the identification of susceptibility genes for obesity.

4.1. *Fine Mapping*

Genomewide linkage screens have suggested several genomic regions that may harbor genes underlying obesity. Once these data are firm, the next step is to fine map these regions. Generally, the regions are very large, encompassing ~30 cM of genomic DNA with hundreds of genes, which is not feasible for physical mapping. Therefore, fine mapping these regions to a small interval (~1 cM) is an essential step to identify the causal gene(s).

It has been recognized that saturation of a candidate interval with ever-denser markers contributes little to its narrowing by linkage mapping.[98,99] An alternative approach is regional LD mapping. There are two commonly used regional LD mapping strategies: the positional candidate approach, in which specific genes or variants are examined on the basis of proposed relationships with the phenotype; and the positional cloning approach, in which markers are selected for evaluation purely on the basis of their proximity to one another in a chromosome region.[76] By examining allelic variations across a region of known linkage, it will be possible to see variations in the strength of associations between these markers and the phenotype, enabling the trait-associated locus and causative mutation to be mapped.[76]

Due to their high abundance, their low mutation rate, and the accessibility of high-throughput genotyping, SNP markers are preferred for LD mapping.[100] Currently, a major problem is the limited understanding of LD patterns throughout the human genome. Consistent with the complex evolutionary history of any set of haplotypes, there is marked genomic variability in LD across the genome. The genome has been portrayed as a series of high-LD regions separated by short discrete segments of very low LD.[101] Those high-LD regions also exhibit limited haplotype diversity, so that a small number of distinct haplotypes account for most of the chromosomes in the population, and they are now termed as "haplotype blocks".[102–104] Within haplotype blocks, allelic dependence yields redundancy among markers and improves the chances of detecting association when only a fraction of the markers, called haplotype-tagging SNPs, is tested.[101] By contrast, in low-LD regions, low correlation between markers means that these regions can only be adequately characterized by typing many or even all markers.

So far, no practical fine mapping study has been reported in terms of obesity. However, successful application of LD mapping has been seen in the refinement of genomic regions underlying some other complex diseases,[105–107] shedding light on the gene hunting of obesity. It is worthwhile to remember that, whatever genomic evidence has been shown, definitive biological evidence of a role for a mutation or polymorphism will require functional studies.

4.2. *Whole-Genome Association Mapping*

The completion of the human genome sequence as well as recent progress in SNP identification and high-throughput genotyping technology have made whole-genome association mapping increasingly practical. Whole-genome association mapping in case-control or family-based studies using haplotypes generated from SNPs can be much more powerful than linkage analysis and also yields a finer map resolution.[50,108] A major problem with association mapping is the appropriate choice of SNP and marker density. This largely depends on LD patterns across the human genome, which may vary across different genomic regions and different ethnic groups.[103,109] The International HapMap Project — a genomewide catalog of common haplotype blocks in multiple human populations — is underway, and is anticipated to provide an important shortcut to carry out whole-genome association mapping.[110]

There are some caveats in the practical application of whole-genome association mapping. First, recent studies suggest that, because haplotype blocks can arise from several causes, simply identifying them does not ensure either their conservation within/between populations or their utility for mapping genes associated with a disease.[111] Another problem is multiple testing, as massive statistical tests will be performed for very large numbers of markers. Risch and Merikangas[50] suggested that a significance threshold of $p = 5 \times 10^{-8}$ would produce a genomewide false-positive rate of 5%; however, this guideline was based on a Bonferroni correction and seems to be too conservative, and may decrease statistical power. Finally, the use of LD for mapping relies on the assumption that common genetic variants are responsible for the susceptibility to common diseases, which is still under debate.[112,113]

4.3. *Gene Expression and DNA Arrays*

DNA microarray technology is a miniaturized hybridization technology that allows simultaneous analyses of the expression of a large number of genes.[114] One of the benefits of analyzing the gene expression patterns of tens of thousands of messages at the same time is the ability to carry out a very detailed analysis of a disease or biological state without having to rely on preconceived ideas about which genes to study. Thus, microarrays provide the opportunity to discover abnormalities in the expression or sequence of genes that were never thought to be involved in a disease. In addition, it is possible to analyze entire groups of genes that perform similar functions in a cell, to determine the most affected gene groups in a disease, or to analyze entire groups of genes present in close proximity (at the same cytogenetic locus) in order to rapidly determine the most affected functional or genetic regions.

Because of the recognized role of adipose tissue in the etiology of obesity, DNA array technology has largely been applied to the studies of adipose gene expression. Many *in vitro* studies analyzed gene expression during adipogenesis by using 3T3-L1 cells, a white preadipocyte cell line derived from segregated mouse embryos. However, it has been shown that the gene expression profiles of 3T3-L1 cells display distinct differences from the gene expression of stromal cells (including preadipocytes) and adipocytes isolated from adipose tissue, suggesting that this preadipocyte cell line might not truly reflect adipogenesis *in vivo*.[115]

There are several published studies on adipose tissue gene expression *in vivo* in humans[116,117] and rodents.[118] All of these studies have led to the identification of a vast number of genes expressed in adipose tissue whose function within this tissue is not yet established. Although there are many adipose tissue gene expression profile studies, interpretation of the data is difficult because adipose tissue is not a clearly defined, homogeneous tissue; rather, it consists of different depots with considerable functional and morphologic differences. In addition, it is not clear whether depot- or sex-dependent variability in adipose tissue gene expression is more important than individual variability.

Because the brain plays a major role in the control of eating behavior, the differential gene expression in the human hypothalami of obese and lean donors has been assessed.[119] Gene expression studies for insulin resistance have also been conducted using the skeletal muscle, where the majority of insulin-stimulated glucose uptake occurs.[120]

Despite the promising potential of microarray technology, currently there are major limitations to this approach. There are as yet no available microarrays that can interrogate all genes of the genome and all alternatively spliced transcript isoforms. The resulting gene expression profiles do not necessarily reflect similar levels of translated proteins. This methodology is tissue-specific, thus limiting its applicability to organs accessible to biopsy and marginally accessible to postmortem sampling. Furthermore, sample and subject variability, experimental design, and false-positive and negative results can be statistically challenging.[121]

4.4. *Proteomics*

Although microarray studies can monitor gene expression at the mRNA level, they do not illuminate molecular and genetic mechanisms of the final functional products of gene translation — proteins.[122] This is largely due to alternative mRNA splicing and complicated posttranslational modifications. Proteomics is an approach to profile protein expression for thousands of proteins in parallel, including protein abundance, protein–protein interactions, and posttranslational modifications.[123] It has been enhanced by the development of two critical technologies, electrospray ionization (ESI) and matrix-assisted laser desorption ionization (MALDI), for the evaporation of peptides and proteins and their analysis by mass spectrometry (MS).

There are three major types of proteomics: expression proteomics, functional proteomics, and structural proteomics.[124] Expression proteomics concerns quantitative analysis and the identification of differentially expressed proteins from expression protein profiles between the case and the control groups. Functional proteomics involves systemwide understanding of protein–protein interactions. Since drug targets are usually the key proteins involved in the pathways that lead to a disease, functional proteomics is a feasible approach to hunting drug targets. The prerequisite of drug design is unraveling the space structure of drug targets, and structural proteomics — by mapping out the structure of protein complexes or proteins in a specific cellular organelle[125] — contributes to fulfilling this requirement.

Two-dimensional gel electrophoresis (2DE) followed by mass spectrometry (MS) is currently the most prevailing approach in expression proteomics. 2DE is applied to separate the total proteins extracted from the interested subject. The application of 2DE proteomics on adipocytes can be dated back to 1979.[126] Due to limitations in technology and knowledge at that time, the characteristics and identity of the proteins could not be acquired. With the recent development in technologies, many of these roadblocks have been removed. In a recent *in vitro* pharmacoproteomics study, Welsh *et al.*[127] followed the changes in protein expression profile that occur during the differentiation of 3T3-L1 fibroblasts into adipocytes in response to dexamethasone, isobutyl methyl xanthine, and insulin; several important signaling molecules were identified. In another study, the proteomics approach was successfully used to characterize differences at the hepatic proteome level between lean and obese diabetic mice, to map metabolic pathways affected by treatment, and to discriminate between effects caused by treatment with agonists of the closely related PPAR-α and PPAR-γ receptors.[128] So far, data on protein–protein interactions concerning adiposity are quite scarce.

Unlike DNA microarray study, proteomics currently does not have the equivalent of the polymerase chain reaction to enhance the signal. Thus, careful consideration must be made of the expected copy number of a protein in the cell type being studied. Another problem with proteomics is that there is currently no working technology that can readily display complete proteomes qualitatively or quantitatively.

5. OBESITY MANAGEMENT FROM A GENETIC PERSPECTIVE

Recent research advances have highlighted the importance of genetic factors in determining individual susceptibility to obesity. Our current genetic knowledge corresponds to only some parts of the whole puzzle. Nevertheless, there is a lot of expectation in terms of the therapeutic potential of these discoveries. The identification of the hormone leptin and its role in body weight regulation has been one of the most promising events in the recent history of obesity research. While the leptin pathway is still not fully characterized, its role as a key element of the body weight regulation system is widely recognized.[51] Leptin acts as part of a feedback loop to maintain constant stores of fat.[51] Compounding this loop are different molecules (hormones, neurotransmitters, and neuropeptides) whose function and pharmacological potential have been analyzed.[13]

The utilization of transgenic animal models has been particularly helpful in assessing the efficacy and determination of the action mode of potential new therapeutic agents.[19] Transgenic animals expressing human target proteins are valuable for assessing the efficacy of newly developed compounds and for ensuring that these compounds will act as expected in humans.[19] The combination of gene-manipulated methodologies with conventional pharmacological manipulation has opened unprecedented insight into the underlying mechanism of obesity and, consequently, resulted in new treatments.[129] Exemplifying this situation is the *POMC* gene knockout mouse model. Researchers have observed that treatment of obese *POMC*-deficient mice with α-MSH injections may induce a rapid loss of weight.[130] Since equivalent genetic defects related to obesity have also been described in humans, the potential of α-MSH agonists in obesity treatment is now under investigation.

The growing recognition that the expression of appetite is chemically coded in the hypothalamus has opened a broad path for the development of therapeutic strategies.[13] Numerous genetic studies have been conducted to elucidate genomic regions governing cell specificity and physiological regulation of neurosecretory gene expression. Peptides like NPY, galanin, leptin, and their corresponding receptors are examples of molecules under study (or already available) for therapeutic intervention against obesity.[13] Some other targets are involved in pathways through an increase in energy expenditure and/or fat oxidation via central mechanisms or directly on peripheral tissues. These candidates include β_3-adrenoreceptor agonists, which target the human receptor and are receiving much attention[131]; among other potential obesity targets are PPAR-γ, peroxisome proliferation-activated receptor γ coactivator 1 (PGC1), ghrelin, and UCPs.[13] Of course, these molecules represent only a small fraction of the complicated matter of appetite and energy homeostasis regulation. The incessant search for novel molecules involved in the biochemical process of metabolism as well as the regulation of energy homeostasis and adipogenesis will undoubtedly lead, in the next few years, to new tools and new classes of drugs that should help in the treatment and prevention of obesity.

In the future, it is likely that the treatment of obesity will be characterized by greater individuality and sophistication. This may involve the application of pharmacogenetics, which studies how genetic differences influence the variability of patients' responses to drugs.[132] Through the use of pharmacogenetics, we will be able to profile variations between individuals' DNA in order to predict responses to a particular medicine. Actually, this area has not been extensively practiced for obesity, largely due to the lack of efficient drugs that are currently available. However, with the tremendous efforts to be devoted, it is anticipated that individualized medicine will be designed, thereby improving the efficacy and safety of treatment.

ACKNOWLEDGMENTS

The investigators were partially supported by grants from the National Institutes of Health (NIH). The study also benefited from the support (to H.-W. Deng) of the 211 State Key Research Fund by Xi'an Jiaotong University, Hunan Province, the Chinese National Science Foundation, the Ministry of Education of P. R. China, and the Huo Ying-Dong Education Foundation.

REFERENCES

1. Kopelman PG. (2000) Obesity as a medical problem. *Nature* **404**:635–643.
2. Flegal KM, Carroll MD, Ogden CL, Johnson CL. (2002) Prevalence and trends in obesity among US adults, 1999–2000. *JAMA* **288**:1723–1727.
3. Allison DB, Fontaine KR, Manson JE, *et al.* (1999) Annual deaths attributable to obesity in the United States. *JAMA* **282**:1530–1538.
4. National Heart, Lung, and Blood Institute. (1998) Clinical guidelines on the identification, evaluation, and treatment of overweight and obesity in adults: the evidence report. National Institutes of Health. *Obes Res* **6**(Suppl 2):51S–209S.
5. US Department of Health and Human Services. (2001) *The Surgeon General's Call to Action to Prevent and Decrease Overweight and Obesity*. US Department of Health and Human Services, Rockville, MD.
6. Hill JO, Peters JC. (1998) Environmental contributions to the obesity epidemic. *Science* **280**:1371–1374.
7. Maes HH, Neale MC, Eaves LJ. (1997) Genetic and environmental factors in relative body weight and human adiposity. *Behav Genet* **27**:325–351.
8. Perusse L, Rankinen T, Zuberi A, *et al.* (2005) The human obesity gene map: the 2004 update. *Obes Res* **13**:381–490.
9. Barsh GS, Farooqi IS, O'Rahilly S. (2000) Genetics of body-weight regulation. *Nature* **404**:644–651.
10. Comuzzie AG, Allison DB. (1998) The search for human obesity genes. *Science* **280**:1374–1377.
11. Lockhart DJ, Winzeler EA. (2000) Genomics, gene expression and DNA arrays. *Nature* **405**:827–836.
12. Pandey A, Mann M. (2000) Proteomics to study genes and genomes. *Nature* **405**:837–846.
13. Chiesi M, Huppertz C, Hofbauer KG. (2001) Pharmacotherapy of obesity: targets and perspectives. *Trends Pharmacol Sci* **22**:247–254.
14. Robinson SW, Dinulescu DM, Cone RD. (2000) Genetic models of obesity and energy balance in the mouse. *Annu Rev Genet* **34**:687–745.
15. Brockmann GA, Bevova MR. (2002) Using mouse models to dissect the genetics of obesity. *Trends Genet* **18**:367–376.
16. Echwald SM. (1999) Genetics of human obesity: lessons from mouse models and candidate genes. *J Intern Med* **245**:653–666.
17. Zhang Y, Proenca R, Maffei M, *et al.* (1994) Positional cloning of the mouse obese gene and its human homologue. *Nature* **372**:425–432.
18. Tartaglia LA, Dembski M, Weng X, *et al.* (1995) Identification and expression cloning of a leptin receptor, OB-R. *Cell* **83**:1263–1271.
19. Livingston JN. (1999) Genetically engineered mice in drug development. *J Intern Med* **245**:627–635.
20. Butler AA, Cone RD. (2001) Knockout models resulting in the development of obesity. *Trends Genet* **17**:S50–S54.
21. Pomp D. (1997) Genetic dissection of obesity in polygenic animal models. *Behav Genet* **27**:285–306.

22. Warden CH, Fisler J, Shoemaker SM, *et al.* (1995) Identification of four chromosomal loci determining obesity in a multifactorial mouse model. *J Clin Invest* **95**:1545–1552.

23. York B, Truett AA, Monteiro MP, *et al.* (1999) Gene–environment interaction: a significant diet-dependent obesity locus demonstrated in a congenic segment on mouse chromosome 7. *Mamm Genome* **10**:457–462.

24. Montague CT, Farooqi IS, Whitehead JP, *et al.* (1997) Congenital leptin deficiency is associated with severe early-onset obesity in humans. *Nature* **387**:903–908.

25. Strobel A, Issad T, Camoin L, *et al.* (1998) A leptin missense mutation associated with hypogonadism and morbid obesity. *Nat Genet* **18**:213–215.

26. Lynch M, Walsh B. (1998) *Genetics and Data Analysis of Quantitative Traits.* Sinauer, Sunderland, MA.

27. Stunkard AJ, Harris JR, Pedersen NL, McClearn GE. (1990) The body-mass index of twins who have been reared apart. *N Engl J Med* **322**:1483–1487.

28. Stunkard AJ, Foch TT, Hrubec Z. (1986) A twin study of human obesity. *JAMA* **256**:51–54.

29. Deng HW, Lai DB, Conway T, *et al.* (2001) Characterization of genetic and lifestyle factors for determining variation in body mass index, fat mass, percentage of fat mass, and lean mass. *J Clin Densitom* **4**:353–361.

30. Moll PP, Burns TL, Lauer RM. (1991) The genetic and environmental sources of body mass index variability: the Muscatine Ponderosity Family Study. *Am J Hum Genet* **49**:1243–1255.

31. Borecki IB, Bonney GE, Rice T, *et al.* (1993) Influence of genotype-dependent effects of covariates on the outcome of segregation analysis of the body mass index. *Am J Hum Genet* **53**:676–687.

32. Comuzzie AG, Blangero J, Mahaney MC, *et al.* (1995) Major gene with sex-specific effects influences fat mass in Mexican Americans. *Genet Epidemiol* **12**:475–488.

33. Deng HW, Deng H, Liu YJ, *et al.* (2002) A genomewide linkage scan for quantitative-trait loci for obesity phenotypes. *Am J Hum Genet* **70**: 1138–1151.

34. Liu YJ, Xu FH, Shen H, *et al.* (2004) A follow-up linkage study for quantitative trait loci contributing to obesity-related phenotypes. *J Clin Endocrinol Metab* **89**:875–882.

35. Comuzzie AG, Hixson JE, Almasy L, *et al.* (1997) A major quantitative trait locus determining serum leptin levels and fat mass is located on human chromosome 2. *Nat Genet* **15**:273–276.

36. Mitchell BD, Cole SA, Comuzzie AG, *et al.* (1999) A quantitative trait locus influencing BMI maps to the region of the beta-3 adrenergic receptor. *Diabetes* **48**:1863–1867.

37. Hanson RL, Ehm MG, Pettitt DJ, *et al.* (1998) An autosomal genomic scan for loci linked to type II diabetes mellitus and body-mass index in Pima Indians. *Am J Hum Genet* **63**:1130–1138.

38. Norman RA, Tataranni PA, Pratley R, *et al.* (1998) Autosomal genomic scan for loci linked to obesity and energy metabolism in Pima Indians. *Am J Hum Genet* **62**:659–668.

39. Hager J, Dina C, Francke S, *et al.* (1998) A genome-wide scan for human obesity genes reveals a major susceptibility locus on chromosome 10. *Nat Genet* **20**:304–308.

40. Kissebah AH, Sonnenberg GE, Myklebust J, *et al.* (2000) Quantitative trait loci on chromosomes 3 and 17 influence phenotypes of the metabolic syndrome. *Proc Natl Acad Sci USA* **97**:14478–14483.

41. Lee JH, Reed DR, Li WD, *et al.* (1999) Genome scan for human obesity and linkage to markers in 20q13. *Am J Hum Genet* **64**:196–209.

42. Rice T, Chagnon YC, Perusse L, *et al.* (2002) A genomewide linkage scan for abdominal subcutaneous and visceral fat in black and white families: the HERITAGE Family Study. *Diabetes* **51**:848–855.

43. Chagnon YC, Rice T, Perusse L, *et al.* (2001) Genomic scan for genes affecting body composition before and after training in Caucasians from HERITAGE. *J Appl Physiol* **90**:1777–1787.

44. Perusse L, Rice T, Chagnon YC, *et al.* (2001) A genome-wide scan for abdominal fat assessed by computed tomography in the Quebec Family Study. *Diabetes* **50**:614–621.

45. Zhu X, Cooper RS, Luke A, *et al.* (2002) A genome-wide scan for obesity in African-Americans. *Diabetes* **51**:541–544.

46. Wu X, Cooper RS, Borecki I, *et al.* (2002) A combined analysis of genomewide linkage scans for body mass index from the National Heart, Lung, and Blood Institute Family Blood Pressure Program. *Am J Hum Genet* **70**:1247–1256.

47. Hsueh WC, Mitchell BD, Schneider JL, *et al.* (2001) Genome-wide scan of obesity in the Old Order Amish. *J Clin Endocrinol Metab* **86**:1199–1205.

48. Ohman M, Oksanen L, Kaprio J, *et al.* (2000) Genome-wide scan of obesity in Finnish sibpairs reveals linkage to chromosome Xq24. *J Clin Endocrinol Metab* **85**:3183–3190.

49. Perola M, Ohman M, Hiekkalinna T, *et al.* (2001) Quantitative-trait-locus analysis of body-mass index and of stature, by combined analysis of genome scans of five Finnish study groups. *Am J Hum Genet* **69**:117–123.

50. Risch N, Merikangas K. (1996) The future of genetic studies of complex human diseases. *Science* **273**:1516–1517.

51. Friedman JM, Halaas JL. (1998) Leptin and the regulation of body weight in mammals. *Nature* **395**:763–770.

52. Bates SH, Myers MG Jr. (2003) The role of leptin receptor signaling in feeding and neuroendocrine function. *Trends Endocrinol Metab* **14**:447–452.

53. van Rossum CT, Hoebee B, van Baak MA, *et al.* (2003) Genetic variation in the leptin receptor gene, leptin, and weight gain in young Dutch adults. *Obes Res* **11**:377–386.

54. Wauters M, Mertens I, Chagnon M, *et al.* (2001) Polymorphisms in the leptin receptor gene, body composition and fat distribution in overweight and obese women. *Int J Obes Relat Metab Disord* **25**:714–720.

55. Chagnon YC, Chung WK, Perusse L, *et al.* (1999) Linkages and associations between the leptin receptor (LEPR) gene and human body composition in the Quebec Family Study. *Int J Obes Relat Metab Disord* **23**:278–286.

56. Chagnon YC, Wilmore JH, Borecki IB, *et al.* (2000) Associations between the leptin receptor gene and adiposity in middle-aged Caucasian males from the HERITAGE family study. *J Clin Endocrinol Metab* **85**:29–34.

57. Yiannakouris N, Yannakoulia M, Melistas L, *et al.* (2001) The Q223R polymorphism of the leptin receptor gene is significantly associated with obesity and predicts a small percentage of body weight and body composition variability. *J Clin Endocrinol Metab* **86**:4434–4439.

58. Mattevi VS, Zembrzuski VM, Hutz MH. (2002) Association analysis of genes involved in the leptin-signaling pathway with obesity in Brazil. *Int J Obes Relat Metab Disord* **26**:1179–1185.

59. Liu YJ, Rocha-Sanchez SM, Liu PY, *et al.* (2004) Tests of linkage and/or association of the LEPR gene polymorphisms with obesity phenotypes in Caucasian nuclear families. *Physiol Genomics* **17**:101–106.

60. Yen CJ, Beamer BA, Negri C, *et al.* (1997) Molecular scanning of the human peroxisome proliferator activated receptor gamma (hPPAR gamma) gene in diabetic Caucasians: identification of a Pro12Ala PPAR gamma 2 missense mutation. *Biochem Biophys Res Commun* **241**:270–274.

61. Deeb SS, Fajas L, Nemoto M, *et al.* (1998) A Pro12Ala substitution in PPARgamma2 associated with decreased receptor activity, lower body mass index and improved insulin sensitivity. *Nat Genet* **20**:284–287.

62. Altshuler D, Hirschhorn JN, Klannemark M, *et al.* (2000) The common PPARgamma Pro12Ala polymorphism is associated with decreased risk of type 2 diabetes. *Nat Genet* **26**:76–80.

63. Li WD, Lee JH, Price RA. (2000) The peroxisome proliferator-activated receptor gamma 2 Pro12Ala mutation is associated with early onset extreme obesity and reduced fasting glucose. *Mol Genet Metab* **70**:159–161.

64. Hasstedt SJ, Ren QF, Teng K, Elbein SC. (2001) Effect of the peroxisome proliferator-activated receptor-gamma 2 Pro^{12}Ala variant on obesity, glucose homeostasis, and blood pressure in members of familial type 2 diabetic kindreds. *J Clin Endocrinol Metab* **86**:536–541.

65. Schaffler A, Barth N, Schmitz G, *et al.* (2001) Frequency and significance of Pro12Ala and Pro115Gln polymorphism in gene for peroxisome proliferation-activated receptor-gamma regarding metabolic parameters in a Caucasian cohort. *Endocrine* **14**:369–373.

66. Ek J, Urhammer SA, Sorensen TI, *et al.* (1999) Homozygosity of the Pro12Ala variant of the peroxisome proliferation-activated receptor-gamma2 (PPAR-gamma2): divergent modulating effects on body mass index in obese and lean Caucasian men. *Diabetologia* **42**:892–895.

67. Nicklas BJ, van Rossum EF, Berman DM, *et al.* (2001) Genetic variation in the peroxisome proliferator-activated receptor-gamma2 gene (Pro12Ala) affects metabolic responses to weight loss and subsequent weight regain. *Diabetes* **50**:2172–2176.

68. Hsueh WC, Cole SA, Shuldiner AR, *et al.* (2001) Interactions between variants in the beta3-adrenergic receptor and peroxisome proliferator-activated receptor-gamma2 genes and obesity. *Diabetes Care* **24**:672–677.

69. Umekawa T, Yoshida T, Sakane N, *et al.* (1999) Trp64Arg mutation of beta3-adrenoceptor gene deteriorates lipolysis induced by beta3-adrenoceptor agonist in human omental adipocytes. *Diabetes* **48**:117–120.

70. Walston J, Silver K, Bogardus C, *et al.* (1995) Time of onset of non-insulin-dependent diabetes mellitus and genetic variation in the beta 3-adrenergic-receptor gene. *N Engl J Med* **333**:343–347.

71. Widen E, Lehto M, Kanninen T, *et al.* (1995) Association of a polymorphism in the beta 3-adrenergic-receptor gene with features of the insulin resistance syndrome in Finns. *N Engl J Med* **333**:348–351.

72. Clement K, Vaisse C, Manning BS, *et al.* (1995) Genetic variation in the beta 3-adrenergic receptor and an increased capacity to gain weight in patients with morbid obesity. *N Engl J Med* **333**:352–354.

73. Mitchell BD, Blangero J, Comuzzie AG, *et al.* (1998) A paired sibling analysis of the beta-3 adrenergic receptor and obesity in Mexican Americans. *J Clin Invest* **101**:584–587.

74. Shuldiner AR, Sabra M. (2001) Trp64Arg beta3-adrenoceptor: when does a candidate gene become a disease-susceptibility gene? *Obes Res* **9**:806–809.

75. Clement K, Ruiz J, Cassard-Doulcier AM, *et al.* (1996) Additive effect of A → G (−3826) variant of the uncoupling protein gene and the Trp64Arg mutation of the beta 3-adrenergic receptor gene on weight gain in morbid obesity. *Int J Obes Relat Metab Disord* **20**:1062–1066.

76. Cardon LR, Bell JI. (2001) Association study designs for complex diseases. *Nat Rev Genet* **2**:91–99.

77. Colhoun HM, McKeigue PM, Davey SG. (2003) Problems of reporting genetic associations with complex outcomes. *Lancet* **361**:865–872.

78. Dong C, Wang S, Li WD, *et al.* (2003) Interacting genetic loci on chromosomes 20 and 10 influence extreme human obesity. *Am J Hum Genet* **72**:115–124.

79. Perusse L, Bouchard C. (2000) Gene–diet interactions in obesity. *Am J Clin Nutr* **72**:1285S–1290S.

80. Hauser ER, Watanabe RM, Duren WL, *et al.* (2004) Ordered subset analysis in genetic linkage mapping of complex traits. *Genet Epidemiol* **27**:53–63.

81. Risch N, Zhang H. (1995) Extreme discordant sib pairs for mapping quantitative trait loci in humans. *Science* **268**:1584–1589.

82. Deng HW, Li J. (2002) The effects of selected sampling on the transmission disequilibrium test of a quantitative trait locus. *Genet Res* **79**:161–174.

83. Spence MA, Greenberg DA, Hodge SE, Vieland VJ. (2003) The emperor's new methods. *Am J Hum Genet* **72**:1084–1087.

84. Hirschhorn JN, Lohmueller K, Byrne E, Hirschhorn KA. (2002) A comprehensive review of genetic association studies. *Genet Med* **4**:45–61.

85. Zondervan KT, Cardon LR. (2004) The complex interplay among factors that influence allelic association. *Nat Rev Genet* **5**:89–100.

86. Deng HW. (2001) Population admixture may appear to mask, change or reverse genetic effects of genes underlying complex traits. *Genetics* **159**:1319–1323.

87. Thomas DC, Witte JS. (2002) Point: population stratification: a problem for case-control studies of candidate-gene associations? *Cancer Epidemiol Biomarkers Prev* **11**:505–512.

88. Wacholder S, Rothman N, Caporaso N. (2002) Counterpoint: bias from population stratification is not a major threat to the validity of conclusions from epidemiological studies of common polymorphisms and cancer. *Cancer Epidemiol Biomarkers Prev* **11**:513–520.

89. Freedman ML, Reich D, Penney KL, *et al.* (2004) Assessing the impact of population stratification on genetic association studies. *Nat Genet* **36**:388–393.

90. Marchini J, Cardon LR, Phillips MS, Donnelly P. (2004) The effects of human population structure on large genetic association studies. *Nat Genet* **36**:512–517.

91. Spielman RS, Ewens WJ. (1996) The TDT and other family-based tests for linkage disequilibrium and association. *Am J Hum Genet* **59**:983–989.

92. Spielman RS, McGinnis RE, Ewens WJ. (1993) Transmission test for linkage disequilibrium: the insulin gene region and insulin-dependent diabetes mellitus (IDDM). *Am J Hum Genet* **52**:506–516.

93. Long JR, Liu PY, Liu YJ, *et al.* (2003) APOE and TGF-beta1 genes are associated with obesity phenotypes. *J Med Genet* **40**:918–924.

94. Horvath S, Laird NM. (1998) A discordant-sibship test for disequilibrium and linkage: no need for parental data. *Am J Hum Genet* **63**:1886–1897.

95. Spielman RS, Ewens WJ. (1998) A sibship test for linkage in the presence of association: the sib transmission/disequilibrium test. *Am J Hum Genet* **62**:450–458.

96. Devlin B, Roeder K. (1999) Genomic control for association studies. *Biometrics* **55**:997–1004.

97. Pritchard JK, Rosenberg NA. (1999) Use of unlinked genetic markers to detect population stratification in association studies. *Am J Hum Genet* **65**:220–228.

98. Atwood LD, Heard-Costa NL. (2003) Limits of fine-mapping a quantitative trait. *Genet Epidemiol* **24**:99–106.

99. Feakes R, Sawcer S, Chataway J, *et al.* (1999) Exploring the dense mapping of a region of potential linkage in complex disease: an example in multiple sclerosis. *Genet Epidemiol* **17**:51–63.

100. Gray IC, Campbell DA, Spurr NK. (2000) Single nucleotide polymorphisms as tools in human genetics. *Hum Mol Genet* **9**:2403–2408.

101. Cardon LR, Abecasis GR. (2003) Using haplotype blocks to map human complex trait loci. *Trends Genet* **19**:135–140.

102. Daly MJ, Rioux JD, Schaffner SF, *et al.* (2001) High-resolution haplotype structure in the human genome. *Nat Genet* **29**:229–232.

103. Gabriel SB, Schaffner SF, Nguyen H, *et al.* (2002) The structure of haplotype blocks in the human genome. *Science* **296**:2225–2229.

104. Patil N, Berno AJ, Hinds DA, *et al.* (2001) Blocks of limited haplotype diversity revealed by high-resolution scanning of human chromosome 21. *Science* **294**:1719–1723.

105. Horikawa Y, Oda N, Cox NJ, *et al.* (2000) Genetic variation in the gene encoding calpain-10 is associated with type 2 diabetes mellitus. *Nat Genet* **26**:163–175.

106. Rioux JD, Daly MJ, Silverberg MS, *et al.* (2001) Genetic variation in the 5q31 cytokine gene cluster confers susceptibility to Crohn disease. *Nat Genet* **29**:223–228.

107. Zhang Y, Leaves NI, Anderson GG, *et al.* (2003) Positional cloning of a quantitative trait locus on chromosome 13q14 that influences immunoglobulin E levels and asthma. *Nat Genet* **34**:181–186.

108. Weiss KM, Clark AG. (2002) Linkage disequilibrium and the mapping of complex human traits. *Trends Genet* **18**:19–24.

109. Reich DE, Cargill M, Bolk S, *et al.* (2003) Linkage disequilibrium in the human genome. *Nature* **411**:199–204.

110. The International HapMap Consortium. (2003) The International HapMap Project. *Nature* **426**:789–796.

111. Phillips MS, Lawrence R, Sachidanandam R, *et al.* (2003) Chromosome-wide distribution of haplotype blocks and the role of recombination hot spots. *Nat Genet* **33**:382–387.

112. Pritchard JK, Cox NJ. (2002) The allelic architecture of human disease genes: common disease-common variant ... or not? *Hum Mol Genet* **11**:2417–2423.

113. Weiss KM, Terwilliger JD. (2000) How many diseases does it take to map a gene with SNPs? *Nat Genet* **26**:151–157.

114. Lipshutz RJ, Fodor SP, Gingeras TR, Lockhart DJ. (1999) High density synthetic oligonucleotide arrays. *Nat Genet* **21**:20–24.

115. Soukas A, Socci ND, Saatkamp BD, *et al.* (2001) Distinct transcriptional profiles of adipogenesis *in vivo* and *in vitro*. *J Biol Chem* **276**: 34167–34174.

116. Yang YS, Song HD, Li RY, *et al.* (2003) The gene expression profiling of human visceral adipose tissue and its secretory functions. *Biochem Biophys Res Commun* **300**:839–846.

117. Gabrielsson BL, Carlsson B, Carlsson LM. (2000) Partial genome scale analysis of gene expression in human adipose tissue using DNA array. *Obes Res* **8**:374–384.

118. Nadler ST, Stoehr JP, Schueler KL, *et al.* (2000) The expression of adipogenic genes is decreased in obesity and diabetes mellitus. *Proc Natl Acad Sci USA* **97**:11371–11376.

119. Tataranni PA, DelParigi A. (2003) Functional neuroimaging: a new generation of human brain studies in obesity research. *Obes Rev* **4**:229–238.

120. Yang X, Pratley RE, Tokraks S, *et al.* (2002) Microarray profiling of skeletal muscle tissues from equally obese, non-diabetic insulin-sensitive and insulin-resistant Pima Indians. *Diabetologia* **45**:1584–1593.

121. Churchill GA. (2002) Fundamentals of experimental design for cDNA microarrays. *Nat Genet* **32**(Suppl):490–495.

122. Abbott A. (1999) A post-genomic challenge: learning to read patterns of protein synthesis. *Nature* **402**:715–720.

123. Shilling PD, Kelsoe JR. (2002) Functional genomics approaches to understanding brain disorders. *Pharmacogenomics* **3**:31–45.

124. Graves PR, Haystead TA. (2002) Molecular biologist's guide to proteomics. *Microbiol Mol Biol Rev* **66**:39–63.

125. Blackstock WP, Weir MP. (1999) Proteomics: quantitative and physical mapping of cellular proteins. *Trends Biotechnol* **17**:121–127.

126. Sidhu RS. (1979) Two-dimensional electrophoretic analyses of proteins synthesized during differentiation of 3T3–L1 preadipocytes. *J Biol Chem* **254**:11111–11118.

127. Welsh GI, Griffiths MR, Webster KJ, *et al.* (2004) Proteome analysis of adipogenesis. *Proteomics* **4**:1042–1051.

128. Edvardsson U, von Lowenhielm HB, Panfilov O, *et al.* (2003) Hepatic protein expression of lean mice and obese diabetic mice treated with peroxisome proliferator-activated receptor activators. *Proteomics* **3**:468–478.

129. Inui A. (2000) Transgenic approach to the study of body weight regulation. *Pharmacol Rev* **52**:35–61.

130. Yaswen L, Diehl N, Brennan MB, Hochgeschwender U. (1999) Obesity in the mouse model of pro-opiomelanocortin deficiency responds to peripheral melanocortin. *Nat Med* **5**:1066–1070.

131. Weyer C, Gautier JF, Danforth E Jr. (1999) Development of beta 3-adrenoceptor agonists for the treatment of obesity and diabetes — an update. *Diabetes Metab* **25**:11–21.

132. Roses AD. (2000) Pharmacogenetics and the practice of medicine. *Nature* **405**:857–865.

Chapter 26

Genes in Estrogen Metabolism Pathway and Breast Cancer

Ji-Rong Long

Department of Medicine and Vanderbilt–Ingram Cancer Center
Vanderbilt University, Nashville, TN 37232, USA

1. INTRODUCTION

Breast cancer is the leading cause of death for women and the second leading cancer cause of death (after lung cancer) in the USA. Caucasian women have a higher risk of developing breast cancer than African-American, Asian, or Hispanic women. One in 7 white women will receive a diagnosis of breast cancer during her lifetime, and 1 in 30 will die of this disease. In 2004, an estimated 215 990 new cases of breast cancer were diagnosed in American women, and 40 922 women died of the disease. At present, there are slightly over two million women living in the USA who have been diagnosed with and treated for breast cancer.

Abundant evidence indicates that estrogens, including estrone (E1) and ß-estradiol (E2), play a key role in the pathogenesis and progression of breast cancer. Epidemiologic studies have indicated that breast cancer risk is higher in women with early menarche and late menopause, i.e. those who have longer exposure to sex hormones.[1] Long-term use of the antiestrogen tamoxifen reduces the incidence of breast cancer; and adjuvant treatment with the aromatase inhibitor anastrozole, which reduces estrogen synthesis, reduces the incidence of contralateral breast cancer by more than 80%.[2] An overview analysis of nine prospective studies[3] found that circulating levels of several steroid hormones, including estrogens, androgens, and their precursors, are directly related to the risk of breast cancer in postmenopausal women. Specifically, women with circulating estradiol levels in the highest quintile were estimated to have twice the risk of breast cancer compared to

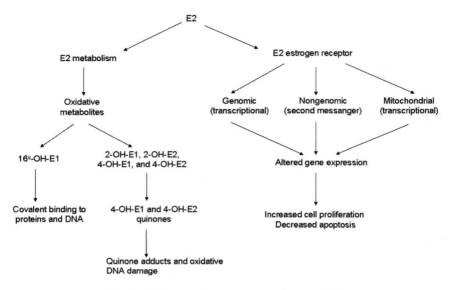

Fig. 1.　Pathways for estrogen carcinogenesis.[57]

women with levels in the lowest quintile. An increased risk of breast cancer was also associated with increased circulating levels of the precursors and metabolites of estradiol. All of these evidences support the hypothesis that cumulative, excessive exposure to endogenous estrogen across a woman's lifespan contributes to and may be a causal factor in breast cancer.

The mechanism of estrogen carcinogenesis in breast cancer has been mainly explained by enhancing receptor-mediated cell proliferation.[4] Now, there is increasing evidence that estrogen metabolites have indirect and direct genotoxiciy (Fig. 1).

Hydroxylation, an important elimination step for estrogens, occurs via two main competing pathways — at C2 and C4 — to generate catechol estrogens (reviewed in Ref. 5). The 4-hydroxyestrogen may generate free radicals from reductive–oxidative cycling with the corresponding semiquinone/quinone forms, thus causing DNA damage. In contrast to 4-hydroxyestrogen, 2-hydroxyestrogen is not carcinogenic, although it might also undergo metabolic redox cycling to generate free radicals and semiquinone/quinone intermediates. 2-Hydroxyestrogen is methylated to generate nongenotoxic methylethers at a faster rate than 4-hydroxyestrogen, and 2-methoxyestrogen has a potent inhibitory effect on the growth of tumor cells and on angiogenesis.[6–8] Therefore, factors favoring

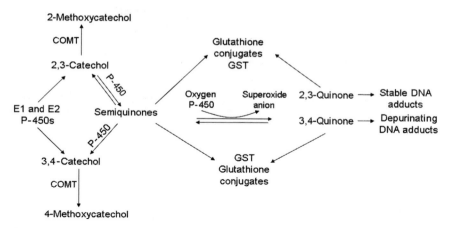

Fig. 2. Major genes and associated enzymes involved in estrogen biosynthesis and metabolism.[58]

2-hydroxylation at the expense of 4-hydroxylation as well as those favoring the inactivation of catechol estrogens may be related to a decreased risk and better progression of breast cancer. Over a dozen genes and enzymes are involved in estrogen synthesis, metabolism, binding, and signal transduction (Fig. 2). In the following, we will focus on the genetic studies of breast cancer genes.

2. ESTROGEN BIOSYNTHESIS

CYP17A1 encodes the P450C 17α enzyme (17-hydroxylase/17,20-lyase), which catalyzes two sequential reactions that yield dehydroepiandrosterone and androstenedione, precursors of estrogens and testosterone, respectively.[9] A common polymorphism in the 5′ promoter region (−34T>C) has been suggested to create an additional binding site for the transcription factor Sp1.[10] This could theoretically lead to increased levels of the enzyme, but use of the extra binding site has not been experimentally confirmed.[11] Several studies have shown that the A2 allele (C) is associated with elevated levels of estrogen, progesterone, and testosterone as well as earlier menarche, though such associations were not detected in others (reviewed in Ref. 12). Surprisingly, with a few exceptions, most studies found no association with breast cancer risk. A recent meta-analysis of 15 case-control studies did not find any overall association,[13] but this analysis was criticized by Feigelson and colleagues,[14] who found a borderline significant association between the *CYP17A1* polymorphism and advanced breast cancer. Interestingly, it was reported that women with the A2/A2 genotype were about half as likely as women with the A1/A1 genotype to use hormone replacement therapy (HRT), an important risk factor for breast cancer; but results have been inconsistent (reviewed in Ref. 12).

CYP19A1 encodes aromatase, the main enzyme that catalyzes the final and rate-limiting step of estrogen biosynthesis, i.e. aromatization of androstenedione and testosterone to estrone and estradiol, respectively. A direct effect of aromatase on *in situ* estrogen synthesis in the breast has been reported.[15] Elevated levels of aromatase expression have been observed in breast tumors, relative to normal breast tissue.[16] This evidence indicates a potential role for the *CYP19A1* gene in the development and progression of breast cancer. The importance of aromatase in the pathogenesis of breast cancer has also been clearly demonstrated in a clinical setting, as inhibitors of this enzyme have been regularly used in the treatment of postmenopausal breast cancer.[17] A recent study suggested that aromatase inhibitors might be more effective than modulators of the estrogen receptor in slowing tumor progression.[18] The $(TTTA)_{10}$ allele of the tetranucleotide repeat polymorphism in intron 4 has been shown to have a correlation with increased breast cancer risk in a pool analysis of several studies.[19] Recently, a comprehensive haplotype analysis of CYP19 and breast cancer risk was performed in the Multiethnic Cohort and the researcher observed significant haplotype effects in block 2 [haplotypes 2b (OR = 1.23; 95% CI, 1.07–1.40), 2d (OR = 1.28; 95% CI, 1.01–1.62)].[20] Our recent study showed that SNPs in haplotype block 2 and the nonsynonymous SNP rs700519 are associated with breast cancer survival.[21]

HSD17B1 encodes 17ß-hydroxysteroid dehydrogenase 1, whose primary function is to catalyze the final step of estradiol biosynthesis, converting estrone to the more biologically active estradiol.[22] Its complement, 17HSD2 (encoded by *HSD17B2*), is the enzyme that predominates in the reverse reaction, the oxidation of estradiol to estrone. The balance of these two enzymes, in part, regulates estrogen concentrations in breast tissue. 17HSD2 activity predominates in normal breast tissue, whereas 17HSD1 activity predominates in malignant breast tissue.[23] *HSD17B1* amplification in breast tumors correlates with poorer prognosis, especially among women with estrogen receptor (ER)-positive tumors.[23] Recently, a comprehensive haplotype analysis was performed to test the association of this gene with breast cancer risk in a sample from five large cohorts in the Breast and Prostate Cancer Cohort Consortium, and no significant results were observed;

however, two common haplotypes were associated with ER-negative tumors.[24]

The *CYP11A1* gene encodes the cholesterol side-chain cleavage enzyme (P450scc), which catalyzes the conversion of cholesterol to pregnenolone, the first and rate-limiting step for the biosynthesis of all steroid hormones, including estrogens, progesterones, and androgens (Fig. 2). Studies on rabbits with a spontaneous *CYP11A1* gene deletion showed that P450scc is the only enzyme that can convert cholesterol to pregnenolone.[25] A pentanucleotide repeat $[(TAAAA)_n]$ polymorphism in the promoter region, located at 529 bp upstream from the translation start site (487 bp upstream from exon 1), was identified in this gene. No association was found between this polymorphism and blood sex hormone levels in two studies[2,26,27]; however, a significant association with breast cancer risk was observed.[28] Women who carried the eight-repeat allele had a greater than 50% elevated risk of breast cancer, while the risk was increased nearly threefold in those who were homozygous for this allele.[28]

3. ESTROGEN HYDROXYLATION

Hydroxylation, an important elimination step for estrogens, is catalyzed by a number of enzymes (e.g. CYP1A1, CYP1A2, CYP1B1, CYP3A4, and CYP3A5). However, CYP1A2, CYP3A4, and CYP3A5 are expressed primarily in the liver[29]; whereas CYP1A1 and CYP1B1 are the principal estrogen-hydroxylation enzymes in the breast. The catalytic efficiency of CYP1B1 for estrogen 4-hydroxylation has been reported to be 18- to 20-fold higher than that of CYP1A2 and CYP3A4 for estrogen 2-hydroxylation.[30] Furthermore, the potentially carcinogenic 4-hydroxy catechol estrogen (4-OHE) is lipophilic and hydrophobic, confining it to the cell of origin. Therefore, it is unlikely that 4-OHE produced in the liver could be released into the circulation and reach the breast. For these reasons, we believe that CYP1A1 and CYP1B1 are among the most important in estrogen hydroxylation in the breast.

For the *CYP1A1* gene, four polymorphisms (3801T → C, Ile462Val, 3205T → C, and Thr461Asp) have been studied in relation to breast cancer. The functional significance of these polymorphisms is unclear. Meta-analysis found no significant risk for these polymorphisms.[31] In our recent study, we found that the CC genotype of SNP 3801T → C is significantly associated with both disease-free survival and overall survival among breast cancer patients.[32] CYP1A1 is also a key enzyme in phase I bioactivation of xenobiotics, catalyzing the first step of polycyclic aromatic hydrocarbons (PAHs) to diol epoxides capable of binding covalently to DNA, thus potentially initiating the carcinogenic process (reviewed in Ref. 31). Future studies should explore possible interactions between *CYP1A1* and sources of PAHs, markers of estrogen exposure, other lifestyle factors influencing hormonal levels, and other genes involved in PAH metabolism or hormonal biosynthesis.

Six common polymorphisms have been described for the *CYP1B1* gene, of which four result in amino acid substitutions: Arg → Gly in codon 48, Ala → Ser in codon 119, Leu → Val in codon 432, and Asn → Ser in codon 453.[33] These polymorphisms have been shown to have 2.4- to 3.4-fold higher catalytic activity than the wild-type enzyme.[34] Although it is biologically reasonable to hypothesize that the minor alleles of these SNPs are associated with higher breast cancer risk and worse survival, meta-analyses did not find overall associations of breast cancer risk with any of these polymorphisms.[33] No association was observed of these polymorphisms with breast cancer survival, either overall survival or disease-free survival, in our recent study.[32]

4. ESTROGEN CONJUGATION

Estrogen and estrogen metabolites can be deactivated through conjugative metabolism, catalyzed by phase II enzymes. Of these, sulfotransferases (SULTs), UPD-glucuronosyltransferases (UGTs), and catechol *O*-methyltransferases (COMTs) are the key players.

SULTs can be classified into two families: phenol SULTs and hydroxysteroid SULTs. Estrogen sulfates have no biological activity and are more water-soluble than their unconjugated parent compounds, and thus are easier to be excreted from urine. SULT1A1, the most abundant enzyme in the SULT family, has been shown to be highly expressed in breast cancer cell lines.[35] The nonsynonymous SNP Arg213His in this gene has been shown to significantly influence the enzymatic activity; individuals with two His alleles had only 15% of SULT1A1 activity compared with carriers of the Arg allele.[36] However, studies involving different ethnic populations that have explored the potential association between this SNP and breast cancer risk have shown inconsistent results.[37]

SULT1E1 exhibits the highest affinity for estrogens among SULTs,[38] indicating that it is active at physiologically significant concentrations of estrogens. Moreover, SULT1E1 is highly expressed in normal human mammary epithelial cells[35] and might play an important role in estrogen-driven breast cancer development. Although SULT1E1 seems rarely expressed in breast cancer cell lines,[35] its expression has been detected in human breast carcinomas, which in turn is associated with a decreased risk of recurrence or improved prognosis of breast cancer.[39] Recently, an association was observed between the SNP 959G>A in this gene with breast cancer risk and disease-free survival of breast cancer in Korean women.[37]

Uridine diphosphoglucuronosyltransferases (UGTs) catalyze the glucuronidation of estrogen and many other endogenous and exogenous compounds, in most cases giving rise to water-soluble and less reactive metabolites.[40] UGT1A1 is a major member of the UGT1 family. A common polymorphism [A(TA)6TAA (allele *1) to A(TA)7TAA (allele *28) change] in the TATA-box of the promoter region of the *UGT1A1* gene has been reported to have possible influences on the transcription. Subjects who carry the variant *28 allele have a significant decrease in the expression of enzymatic activity of hepatic UGT, resulting in higher serum bilirubin levels compared to those homozygous for the common *1 allele.[41] However, inconsistent results were reported in the association of the *28 allele with breast cancer risk among different ethnic groups.[42]

The *COMT* gene encodes the enzyme catalyzing *O*-methylation, the most active inactivation pathway for catechol estrogens. If this inactivation process is incomplete, catechol estrogens may undergo metabolic redox cycling to generate free radicals and reactive semiquinone/quinone intermediates, which cause DNA damage.[43] Studies have shown that the Met/Met genotype of the nonsynonymous SNP Val158Met in the *COMT* gene is linked to a twofold-to-fourfold decreased enzyme activity.[44–46] Therefore, it is conceivable that patients carrying the genotype AA may have increased risk of breast cancer and worse progression. However, meta-analyses showed no overall association of this SNP with breast cancer risk,[33] but a significant association with breast cancer survival was observed in our recent study.[32] Women carrying the Met/Met genotype had significantly decreased disease-free survival when compared with the other patients.

5. ESTROGEN BINDING AND RECEPTOR

Sex hormone–binding globulin (SHBG) binds to estrogens and reduces their biological availability to target tissues such as breast. In addition, SHBG has recently been found to function as an active regulator of the steroid-signaling system in target tissues.[47] In breast cancer cells, SHBG — through

its specific membrane receptor (SHBG-R) and second messenger system
(cyclic AMP and protein kinase A)— not only effectively inhibits estradiol-
induced cell proliferation, but also controls progesterone receptor expression
at both the mRNA and protein levels and influences its function (receptor-
binding capacity).[48] Epidemiologic studies have shown that blood SHBG
levels are inversely associated with breast cancer risk.[49] Thus, it is conceivable that functional polymorphisms in the *SHBG* gene may be related to
the risk of breast cancer. Recently, a large case-control study showed that
the variant *Asn* allele of the nonsynonymous SNP Asp327Asn is associated with increased SHBG levels and a reduced estradiol-to-SHBG ratio in
postmenopausal women,[33,50] and this allele has been found to be related to
a reduced risk of breast cancer among postmenopausal women in another
large-scale, population-based case-control study.[51]

Estrogen influences the growth, differentiation, and function of many
target tissues, including the breast, uterus, vagina, ovary, testis, epididymis,
and prostate. The biological effect of estrogens, such as stimulating the
growth and differentiation of normal mammary tissue, is mediated primarily
through high-affinity binding to estrogen receptors (ERs). ERs are nuclear
receptor proteins that have an estrogen-binding domain and a DNA-binding
domain. Two ERs have been identified: estrogen receptor alpha (ESR1) and
estrogen receptor beta (ESR2). High expression of both ER genes has been
found in uterus, ovarian, and other hormone-sensitive tissues.

ERs have been demonstrated to be significant prognostic factors for
breast cancer because their protein levels are elevated in premalignant and
malignant breast cells. Consequently, inhibition of ESR1 has become one
of the major strategies for the prevention and treatment of breast cancer.[52]
The *ESR1* gene is located on chromosome 6q25.1 and has been extensively
studied. A number of common polymorphisms in this gene have been
reported, and the association of genetic polymorphisms in this gene and
risk of diseases (including breast cancer) has been the subject of increasing
interest. Several studies have been conducted to test the associations of
breast cancer with polymorphisms in the *ESR1* gene, such as *XbaI*, *PvuII*,
and GT repeat; however, the results have been inconsistent.[53,54]

The *ESR2* gene was recently identified, and it shares about 95% homology in the DNA-binding domain and 55% homology in the ligand-binding
domain. *ESR2* and *ESR1* have distinct cellular distributions, regulate separate sets of genes, and oppose each other's actions when regulating some
genes. *ESR2* is widely expressed in both normal and malignant breast,
and there are proliferating cells in the breast that express *ESR2*.[55] It is

hypothesized that certain sequence variants of the *ESR2* gene are associated with an increased risk for breast cancer, particularly among women who have a high level of and long-term estrogen exposure. Associations with breast cancer risk have been observed at two SNPs, C14206T and C33390G, in combination with a high level of steroid sex hormone or a low level of SHBG in a population-based case-control study.[56] Recently, haplotype analyses also defined several risk haplotypes in this gene for breast cancer.[3,55]

In summary, estrogens play a key role in the pathogenesis and progression of breast cancer. The genes involved in the estrogen metabolism pathway may be associated with breast cancer risk and progression, and may have significant clinical implications for breast cancer prevention and treatment.

REFERENCES

1. Henderson BE, Bernstein L. (1991) The international variation in breast cancer rates: an epidemiological assessment. *Breast Cancer Res Treat* **18**(Suppl 1):S11–S17.
2. Baum M, Budzar AU, Cuzick J, *et al.* (2002) Anastrozole alone or in combination with tamoxifen versus tamoxifen alone for adjuvant treatment of postmenopausal women with early breast cancer: first results of the ATAC randomised trial. *Lancet* **359**:2131–2139.
3. Key T, Appleby P, Barnes I, Reeves G. (2002) Endogenous sex hormones and breast cancer in postmenopausal women: reanalysis of nine prospective studies. *J Natl Cancer Inst* **94**:606–616.
4. Henderson BE, Feigelson HS. (2000) Hormonal carcinogenesis. *Carcinogenesis* **21**:427–433.
5. Tsuchiya Y, Nakajima M, Yokoi T. (2005) Cytochrome P450-mediated metabolism of estrogens and its regulation in human. *Cancer Lett* **227**: 115–124.
6. Dawling S, Hachey DL, Roodi N, Parl FF. (2004) *In vitro* model of mammary estrogen metabolism: structural and kinetic differences between catechol estrogens 2- and 4-hydroxyestradiol. *Chem Res Toxicol* **17**:1258–1264.
7. Lakhani NJ, Sarkar MA, Venitz J, Figg WD. (2003) 2-Methoxyestradiol, a promising anticancer agent. *Pharmacotherapy* **23**:165–172.
8. Brueggemeier RW, Bhat AS, Lovely CJ, *et al.* (2001) 2-Methoxymethylestradiol: a new 2-methoxy estrogen analog that exhibits antiproliferative activity and alters tubulin dynamics. *J Steroid Biochem Mol Biol* **78**:145–156.
9. Einarsdottir K, Rylander-Rudqvist T, Humphreys K, *et al.* (2005) CYP17 gene polymorphism in relation to breast cancer risk: a case-control study. *Breast Cancer Res* **7**:R890–R896.

10. Carey AH, Waterworth D, Patel K, *et al.* (1994) Polycystic ovaries and premature male pattern baldness are associated with one allele of the steroid metabolism gene CYP17. *Hum Mol Genet* **3**:1873–1876.

11. Nedelcheva Kristensen V, Haraldsen EK, Anderson, KB, *et al.* (1991) CYP17 and breast cancer risk: the polymorphism in the $5'$ flanking area of the gene does not influence binding to Sp-1. *Cancer Res* **59**:2825–2828.

12. Sharp L, Cardy AH, Cotton SC, Little J. (2004) P17 gene polymorphisms: prevalence and associations with hormone levels and related factors. A HuGE review. *Am J Epidemiol* **160**:729–740.

13. Ye Z, Parry JM. The CYP17 MspA1 polymorphism and breast cancer risk: a meta-analysis. *Mutagenesis* **17**:119–126.

14. Feigelson HS, McKean-Cowdin R, Henderson BE. (2002) Concerning the CYP17 MspA1 polymorphism and breast cancer risk: a meta-analysis. *Mutagenesis* **17**:445–446.

15. Maggiolini M, Bonofiglio D, Pezzi V, *et al.* (2002) Aromatase overexpression enhances the stimulatory effects of adrenal androgens on MCF7 breast cancer cells. *Mol Cell Endocrinol* **193**:13–18.

16. Irahara N, Miyoshi Y, Taguchi T, *et al.* (2006) Quantitative analysis of aromatase mRNA expression derived from various promoters (I.4, I.3, PII and I.7) and its association with expression of TNF-alpha, IL-6 and COX-2 mRNAs in human breast cancer. *Int J Cancer* **118**:1915–1921.

17. Mouridsen HT, Robert NJ. (2005) Benefit with aromatase inhibitors in the adjuvant setting for postmenopausal women with breast cancer. *MedGenMed* **7**:20.

18. Dowsett M, Cuzick J, Wale C, *et al.* (2005) Retrospective analysis of time to recurrence in the ATAC trial according to hormone receptor status: an hypothesis-generating study. *J Clin Oncol* **23**:7512–7517.

19. Dumitrescu RG, Cotarla I. (2005) Understanding breast cancer risk — where do we stand in 2005? *J Cell Mol Med* **9**:208–221.

20. Haiman CA, Stram DO, Pike MC, *et al.* (2003) A comprehensive haplotype analysis of CYP19 and breast cancer risk: the Multiethnic Cohort. *Hum Mol Genet* **12**:2679–2692.

21. Long JR, Kataoka N, Shu XO, *et al.* (2006) Genetic polymorphisms of the CYP19A1 gene and breast cancer survival. *Cancer Epidemiol Biomarkers Prev* **15**:2115–2122.

22. Oduwole OO, Li Y, Isomaa VV, *et al.* (2004) 17beta-hydroxysteroid dehydrogenase type 1 is an independent prognostic marker in breast cancer. *Cancer Res* **64**:7604–7609.

23. Gunnarsson C, Ahnstrom M, Kirschner K, *et al.* (2003) Amplification of HSD17B1 and ERBB2 in primary breast cancer. *Oncogene* **22**:34–40.

24. Feigelson HS, Cox DG, Cann HM, *et al.* (2006) Haplotype analysis of the HSD17B1 gene and risk of breast cancer: a comprehensive approach to multicenter analyses of prospective cohort studies. *Cancer Res* **66**:2468–2475.

25. Yang X, Iwamoto K, Wang M, *et al.* (1993) Inherited congenital adrenal hyperplasia in the rabbit is caused by a deletion in the gene encoding

cytochrome P450 cholesterol side-chain cleavage enzyme. *Endocrinology* **132**:1977–1982.

26. Garcia-Closas M, Herbstman J, Schiffman M, *et al.* (2002) Relationship between serum hormone concentrations, reproductive history, alcohol consumption and genetic polymorphisms in pre-menopausal women. *Int J Cancer* **102**:172–178.

27. San Millan JL, Sancho J, Calvo RM, Escobar-Morreale HF. (2001) Role of the pentanucleotide $(TTTTA)_n$ polymorphism in the promoter of the CYP11a gene in the pathogenesis of hirsutism. *Fertil Steril* **75**:797–802.

28. Zheng W, Gao YT, Shu XO, *et al.* (2004) Population-based case-control study of CYP11A gene polymorphism and breast cancer risk. *Cancer Epidemiol Biomarkers Prev* **13**:709–714.

29. Williams JA, Phillips DH. (2000) Mammary expression of xenobiotic metabolizing enzymes and their potential role in breast cancer. *Cancer Res* **60**:4667–4677.

30. Waxman DJ, Lapenson DP, Aoyama T, *et al.* (1991) Steroid hormone hydroxylase specificities of eleven cdNA-expressed human cytochrome P450s. *Arch Biochem Biophys* **290**:160–166.

31. Masson LF, Sharp L, Cotton SC, Little J. (2005) Cytochrome P-450 1A1 gene polymorphisms and risk of breast cancer: a HuGE review. *Am J Epidemiol* **161**:901–915.

32. Long JR, Cai Q, Shu XO, *et al.* (2007) Genetic polymorphisms in estrogen-metabolizing genes and breast cancer survival. *Pharmacogenet Genomics* **17**:311–338.

33. Wen W, Cai Q, Shu XO, *et al.* (2005) Cytochrome P450 1B1 and catechol-*O*-methyltransferase genetic polymorphisms and breast cancer risk in Chinese women: results from the Shanghai Breast Cancer Study and a meta-analysis. *Cancer Epidemiol Biomarkers Prev* **14**:329–335.

34. Hanna IH, Dawling S, Roodi N, *et al.* (2000) Cytochrome P450 1B1 (CYP1B1) pharmacogenetics: association of polymorphisms with functional differences in estrogen hydroxylation activity. *Cancer Res* **60**:3440–3444.

35. Falany JL, Pilloff DE, Leyh TS, Falany CN. (2006) Sulfation of raloxifene and 4-hydroxytamoxifen by human cytosolic sulfotransferases. *Drug Metab Dispos* **34**:361–368.

36. Ozawa S, Tang YM, Yamazoe Y, *et al.* (1998) Genetic polymorphisms in human liver phenol sulfotransferases involved in the bioactivation of N-hydroxy derivatives of carcinogenic arylamines and heterocyclic amines. *Chem Biol Interact* **109**:237–248.

37. Choi JY, Lee KM, Park SK, *et al.* (2005) Genetic polymorphisms of SULT1A1 and SULT1E1 and the risk and survival of breast cancer. *Cancer Epidemiol Biomarkers Prev* **14**:1090–1095.

38. Adjei AA, Weinshilboum RM. (2002) Catecholestrogen sulfation: possible role in carcinogenesis. *Biochem Biophys Res Commun* **292**:402–408.

39. Suzuki T, Nakata T, Miki Y, *et al.* (2003) Estrogen sulfotransferase and steroid sulfatase in human breast carcinoma. *Cancer Res* **63**:2762–2770.

40. Belanger A, Hum DW, Beaulieu M, *et al.* (1998) Characterization and regulation of UDP-glucuronosyltransferases in steroid target tissues. *J Steroid Biochem Mol Biol* **65**:301–310.

41. Lin JP, Cupples LA, Wilson PW, *et al.* (2003) Evidence for a gene influencing serum bilirubin on chromosome 2q telomere: a genomewide scan in the Framingham Study. *Am J Hum Genet* **72**:1029–1034.

42. Adegoke OJ, Shu XO, Gao YT, *et al.* (2004) Genetic polymorphisms in uridine diphospho-glucuronosyltransferase 1A1 (UGT1A1) and risk of breast cancer. *Breast Cancer Res Treat* **85**:239–245.

43. Zhu BT, Conney AH. (1998) Functional role of estrogen metabolism in target cells: review and perspectives. *Carcinogenesis* **19**:1–27.

44. Lachman HM, Papolos DF, Saito T, *et al.* (1996) Human catechol-*O*-methyltransferase pharmacogenetics: description of a functional polymorphism and its potential application to neuropsychiatric disorders. *Pharmacogenetics* **6**:243–250.

45. Dawling S, Roodi N, Mernaugh RL, *et al.* (2001) Catechol-*O*-methyltransferase (COMT)-mediated metabolism of catechol estrogens: comparison of wild-type and variant COMT isoforms. *Cancer Res* **61**:6716–6722.

46. Goodman JE, Jensen LT, He P, Yager JD. (2002) Characterization of human soluble high and low activity catechol-*O*-methyltransferase catalyzed catechol estrogen methylation. *Pharmacogenetics* **12**:517–528.

47. Kahn SM, Hryb DJ, Nakhla AM, *et al.* (2002) Sex hormone–binding globulin is synthesized in target cells. *J Endocrinol* **175**:113–120.

48. Fazzari A, Catalano MG, Comba A, *et al.* (2001) The control of progesterone receptor expression in MCF-7 breast cancer cells: effects of estradiol and sex hormone–binding globulin (SHBG). *Mol Cell Endocrinol* **172**:31–36.

49. Key T, Appleby P, Barnes I, Reeves G. (2002) Endogenous sex hormones and breast cancer in postmenopausal women: reanalysis of nine prospective studies. *J Natl Cancer Inst* **94**:606–616.

50. Dunning AM, Dowsett M, Healey CS, *et al.* (2004) Polymorphisms associated with circulating sex hormone levels in postmenopausal women. *J Natl Cancer Inst* **96**:936–945.

51. Cui Y, Shu XO, Cai Q, *et al.* (2005) Association of breast cancer risk with a common functional polymorphism (Asp327Asn) in the sex hormone–binding globulin gene. *Cancer Epidemiol Biomarkers Prev* **14**:1096–1101.

52. Sommer S, Fuqua SA. (2001) Estrogen receptor and breast cancer. *Semin Cancer Biol* **11**:339–352.

53. Cai Q, Gao YT, Wen W, *et al.* (2003) Association of breast cancer risk with a GT dinucleotide repeat polymorphism upstream of the estrogen receptor-alpha gene. *Cancer Res* **63**:5727–5730.

54. Cai Q, Shu XO, Jin F, *et al.* (2003) Genetic polymorphisms in the estrogen receptor alpha gene and risk of breast cancer: results from the Shanghai Breast Cancer Study. *Cancer Epidemiol Biomarkers Prev* **12**:853–859.

55. Poola I, Fuqua SA, De Witty RL, *et al.* (2005) Estrogen receptor alpha–negative breast cancer tissues express significant levels of estrogen-independent transcription factors, ERbeta1 and ERbeta5: potential molecular targets for chemoprevention. *Clin Cancer Res* **11**:7579–7585.

56. Zheng SL, Zheng W, Chang BL, *et al.* (2003) Joint effect of estrogen receptor beta sequence variants and endogenous estrogen exposure on breast cancer risk in Chinese women. *Cancer Res* **63**:7624–7629.

57. Yager JD, Davidson NE. (2006) Estrogen carcinogenesis in breast cancer. *N Engl J Med* **354**:270–282.

58. Tsuchiya Y, Nakajima M, Yokoi T. (2005) Cytochrome P450-mediated metabolism of estrogens and its regulation in human. *Cancer Lett* **227**: 115–124.

Chapter 27

Hereditary Breast and Ovarian Cancers

Wentao Yang[*,‡], Daren Shi[*] and Jinsong Liu[†,‡]

[*]Department of Pathology, Cancer Hospital, Fudan University
Shanghai 200031, P. R. China

[†]Department of Pathology
The University of Texas M. D. Anderson Cancer Center
1515 Holcombe Boulevard, Houston
TX 77030-4095, USA

Breast cancer is a major disease affecting women in industrialized countries, for whom the lifetime risk exceeds 10%. In the US, breast cancer affects one in eight women.[1] Ovarian cancer is substantially less common than breast cancer, with a lifetime risk of 1.6% for women in industrialized countries[2]; however, it is one of the most lethal gynecological malignancies. It is estimated that 5%–10% of breast and invasive ovarian cancer cases are hereditary and attributable to mutations in several highly penetrant susceptibility genes.[3,4] Inherited susceptibility has already been recognized as a significant risk factor for cancers of the breast and female genital organs.

1. HEREDITARY SYNDROMES ASSOCIATED WITH BREAST AND OVARIAN CANCERS

Fewer than 1% of all breast cancers are associated with hereditary syndromes. Individuals who have the disorders described here may have a higher-than-normal breast cancer risk.

[‡] Correspondence authors.

1.1. *Hereditary Breast and Ovarian Cancer (HBOC) Syndrome*

HBOC refers to cases of hereditary breast cancer in families that may also have a history of ovarian cancer. Features of the HBOC syndrome include premenopausal breast cancer, ovarian cancer (at any age), bilateral breast cancer, both breast and ovarian cancer in the same person, and male breast cancer. The cancer history is usually reported in several generations of a family related through the same bloodline, either maternal or paternal.

Because breast cancer is unfortunately quite common, many people have at least one relative who has had breast cancer. In larger families, it is not unusual to have several relatives with cancer. This pattern of incidence may result from a shared genetic predisposition, but it may also be a consequence of shared environmental exposures or sociocultural risk factors.

Family history is the strongest risk factor for ovarian cancer. In the general population, the lifetime risk for developing ovarian cancer is 1.6%. However, women who have one first-degree relative with ovarian cancer have an approximately 5% risk, and women who have two first-degree relatives with ovarian cancer have a 7% risk.[5,6] Three clinical manifestations of hereditary ovarian cancer have been recognized: (1) site-specific ovarian cancer, (2) breast and ovarian cancer syndrome, and (3) hereditary nonpolyposis colorectal cancer (HNPCC) syndrome.[7,8] The first two types of hereditary ovarian cancer are associated with germline mutations in the *BRCA1* and *BRCA2* genes; whereas HNPCC is associated with germline mutations in the DNA mismatch repair (*MMR*) genes, primarily *hMLH1* and *hMSH2*. Up to now, no gene that confers increased susceptibility to ovarian cancer alone has been found, so site-specific ovarian cancer and HBOC syndrome are considered part of the same spectrum of disease. It is currently accepted that at least 10% of all epithelial ovarian cancers are hereditary, with mutations in the *BRCA* genes accounting for about 90% of cases and most of the remaining 10% related to HNPCC.[9,10]

1.2. *Cowden Syndrome (CS)*

CS, also called multiple hamartoma syndrome, is caused by mutations of the *PTEN* gene[11] and accounts for a small proportion (<1%) of hereditary breast cancers. CS is characterized by an excess of breast cancer, gastrointestinal malignancies, and thyroid disease, both benign and malignant.[12] The majority (75%) of women with CS have benign breast disease (fibroadenoma or fibrocystic breasts); their risk for breast cancer is greater than normal and ranges from 25% to 50%.[13,14] As in other forms of hereditary breast cancer, onset often occurs at a young age and may be bilateral.[15]

CS is associated with physical features that are pathognomonic, including facial trichilemmoma, acral keratoses, and oral papillomatous papules.[16] Benign thyroid tumors (goiter and adenoma) are also common; and the risk for nonmedullary thyroid cancer, especially the follicular type, may be as high as 10%.[17,18] Cerebral dysplastic gangliocytoma is considered to be one of the major criteria of CS, but its frequency is uncertain. More recently, endometrial cancer has been included in the CS cancer spectrum, with risks approaching 5%–10%. Germline mutations in *PTEN*, a protein tyrosine phosphatase gene located on chromosome 10q23, are responsible for this syndrome. Loss of heterozygosity at the *PTEN* locus, which has been observed in a high proportion of related cancers, suggests that *PTEN* functions as a tumor suppressor gene.

1.3. *Li–Fraumeni Syndrome (LFS)*

Breast cancer is also a component of the LFS, in which germline mutations of the *TP53* gene on chromosome 17p have been documented.[19] LFS is a rare cancer predisposition syndrome associated with early onset (<40 years of age) of soft tissue sarcoma; leukemia; osteosarcoma; melanoma; and cancers of the colon, breast, pancreas, adrenal cortex, and brain. Individuals with LFS are at increased risk for developing multiple primary cancers.

Age-specific cancer risks have been calculated. Frequently, the first diagnosis of cancer is made in childhood or early adulthood, with an estimated cancer risk of 50% by the age of 30 years. The criteria used to identify an affected individual in a Li–Fraumeni family are (1) occurrence of sarcoma before the age of 45, (2) at least one first-degree relative with any cancer before age 45, and/or (3) a second- or first-degree relative with cancer before age 45 or a sarcoma at any age.[19,20] Within LFS families, breast cancer accounts for up to one third of all cancers and occurs at an average age of 36, but LFS likely accounts for less than 1% of all cases of breast cancer.[21]

1.4. *Peutz–Jeghers Syndrome (PJS)*

PJS is characterized by the association of gastrointestinal polyposis and mucocutaneous pigmentation. Peutz–Jeghers-type hamartomatous polyps are most prevalent in the small intestine (jejunum, ileum, and duodenum), but can also occur in the stomach and large bowel. Gastrointestinal polyps can result in chronic bleeding and anemia, or cause recurrent obstruction and intussusception requiring repeated laparotomies and bowel resections. Mucocutaneous hyperpigmentation presents in childhood as dark blue to dark brown mucocutaneous macules around the mouth, eyes, and nostrils; in the perianal area; and on the buccal mucosa. Hyperpigmented macules on the fingers are common; the macules may fade in puberty and adulthood. Females are at risk for sex cord tumors with annular tubules (SCTAT) of the ovary. Females can also present with adenoma malignum of the cervix, a rare aggressive cancer. Males occasionally develop calcifying Sertoli cell tumors of the testis, which secrete estrogen and can lead to gynecomastia.

The most commonly reported malignancies in PJS kindreds are those of the colon and breast, with a mean patient age at diagnosis of about 45 years. The risk estimates for breast and ovarian cancer are similar to those observed in women carrying *BRCA1* or *BRCA2* mutations. Only mutations in the gene *STK11* (alternatively denoted *LKB1*) have been identified as a cause of PJS.[22] Mutations in *STK11/LKB1* are lacking in many individuals with clinical PJS, suggesting the presence of other susceptibility genes, genetic mosaicism, or limitations in current gene mutation analysis.

1.5. *Ataxia-Telangiectasia (AT)*

AT is an autosomal recessive condition that results in cerebellar ataxia, immune defects, telangiectasias, radiosensitivity, and a predisposition to cancer, especially leukemia and lymphoma.[23] AT is caused by mutations in the *ATM* gene, which is involved in maintaining genome stability.[24] Studies of obligate heterozygote females have revealed that, relative to the population as a whole, they carry a greater relative risk (between five and seven times) for breast cancer. However, studies searching for an excess of *ATM* mutations among breast cancer patients have provided conflicting results. If heterozygous carrier status is found to be associated with breast cancer, it could account for a significant proportion of hereditary breast cancer.

1.6. *HNPCC*

HNPCC, or Lynch II syndrome, was first described in 1966.[25] In patients with HNPCC, the risk of colorectal cancer is 80%. The syndrome is characterized by autosomal dominant inheritance of predominantly right-sided colonic cancer in the absence of colonic polyposis. However, women in HNPCC families also have a lifetime risk of endometrial cancer of 20%–40% and a risk of ovarian cancer of 12%.[26]

2. GENETICS OF HEREDITARY BREAST AND OVARIAN CANCERS

The genetic basis of hereditary breast and ovarian cancer (HBOC) has become clear with the cloning of two major disease susceptibility genes, *BRCA1* and *BRCA2*. The majority (80%–90%) of HBOCs are caused by mutations of the *BRCA1* and *BRCA2* genes. Other genes, such as *p53*, *CHEK2*, and *ATM*, have also been demonstrated to be associated with increased breast cancer risk.[27,28]

2.1. *BRCA1*

Although an autosomal dominant predisposition to breast cancer was suspected for many years, the formal demonstration of its existence was not easy. Hall and colleagues[29] demonstrated that one associated locus was linked to chromosomal region 17q21. After an intense multinational effort, the breast cancer 1 or *BRCA1* gene was isolated in 1994.[30] Mutations of this gene are transmitted in an autosomal dominant pattern, and account for about half of hereditary breast cancers and approximately 80% of cases in families with both hereditary breast and ovarian cancers.[31] The 24 exons of the *BRCA1* gene (22 coding exons and alternative 5′ UTR [untranslated region] exons) span an 81-kb chromosomal region, which has an unusually high density of Alu-repetitive DNA.[30,32]

The BRCA1 protein consists of 1863 amino acids and is expressed in most proliferating cells. The C-terminus of BRCA1 contains an amino-acid sequence motif, now known as the BRCT domain, recognized in many DNA repair proteins. This appears to be a site of protein–protein reactions. In the N-terminus of BRCA1, there is a RING-finger domain, also allowing for protein–protein interactions, but thought to be involved in protein ubiquitination.[33] Recent studies suggest that the BRCA1 protein is required for maintenance of chromosomal stability, thereby protecting the genome from damage. New data also show that the BRCA1 transcriptionally regulates some genes involved in DNA repair, the cell cycle, and apoptosis.[34]

BRCA1 colocalizes with the BRCA2 and RAD51 proteins in discrete foci during S phase. Within minutes of DNA damage, the histone H2A family member H2AX becomes extensively phosphorylated and forms foci at break sites. BRCA1 is recruited to these foci several hours before other factors such as RAD50 and RAD51. This suggests that H2AX and BRCA1 initiate repair by modifying local chromatin structure, thereby allowing DNA repair proteins access to the damaged site. DNA damage leads to hyperphosphorylation of BRCA1, dispersal of the BRCA1/BRCA2/RAD51 nuclear foci, and their relocalization to proliferating cell nuclear antigen (PCNA)-containing DNA replication structures. In meiotic cells, BRCA1, BRCA2, and RAD51 colocalize on the axial elements of developing synaptonemal complexes.[35] A large protein complex consisting of other tumor suppressor and DNA repair proteins, known as BRCA1-associated genome surveillance complex (BASC), has been identified. Among these, it has been demonstrated that BRCA1 partially colocalizes with RAD50, MRE11, and BLM in nuclear foci analogous to those of BRCA2 and RAD51.[36] In addition to its interactions with BRCA2, RAD51, and BASC, the BRCA1 protein has been shown to form complexes with several other proteins involved in diverse cellular functions including DNA repair, DNA transcription, chromatin remodeling, and protein ubiquination.[37]

In adult mice, *BRCA1* and *BRCA2* expressions are induced during mammary gland ductal proliferation, morphogenesis and differentiation occurring at puberty and again during proliferation of the mammary epithelium during pregnancy.[38] Consistent with its role as a tumor suppressor gene, the wild-type allele of *BRCA1* is lost in the majority of tumors from patients with inherited mutations, presumably leading to an absence of normal protein.[39] In sporadic cancer, BRCA1 protein expression is absent or reduced in the majority of high-grade breast carcinomas and sporadic ovarian tumors.[40,41]

Over 300 unique mutations have been documented in *BRCA1*, most of which result in a shortened, presumably nonfunctioning protein product.[42] Several founder mutations have been identified in the *BRCA1* gene in various populations. Among them, two founder mutations are relatively common in Ashkenazi Jewish individuals (185delAG and 5382insC in *BRCA1*). Combined with 6174delT in *BRCA2*, the prevalence of these three mutations reached 2.3% in this population.[43]

2.2. *BRCA2*

The discovery of *BRCA2*, the second gene responsible for hereditary predisposition to breast and ovarian cancer, was reported in 1995. The *BRCA2* gene encodes a 10.4-kb transcript composed of 27 exons. It codes for a large protein of 3418 amino acids, making a 380-kDa protein.[44] *BRCA2* is a tumor suppressor gene involved in the repair of chromosomal damage. The BRCA2 protein is normally located in the nucleus and contains phosphorylated residues.

BRCA1 and BRCA2 proteins appear to share a number of functional similarities, which may suggest why mutations in these genes lead to a specific hereditary predisposition to breast and ovarian cancer. Like *BRCA1*, *BRCA2* is expressed in most tissues and cell types analyzed, indicating that gene expression does not account for the tissue-restricted phenotype of breast and ovarian cancer. *BRCA2* transcription is induced late in the G_1 phase of the cell cycle and remains elevated during the S phase, indicating it has some role in DNA synthesis. *BRCA2* also appears to be involved in the DNA repair process. The BRCA2 protein binds to and regulates the protein produced by the *RAD51* gene to fix breaks in DNA. These breaks can be caused by natural and medical radiation or radiation from other environmental sources, but they also occur when chromosomes exchange genetic material during a special type of cell division that creates sperm and eggs (meiosis). Perhaps through their mutual association with RAD51, BRCA1 and BRCA2 come to associate with each other at sites of DNA synthesis after DNA damage is induced.

To study the function of *BRCA2*, homozygous knockout mice have been created. In most cases, complete loss of function of *BRCA2* results in embryonic fatality characterized by a lack of cell proliferation. Cells derived from mouse embryos lacking *BRCA2* are defective in their repair of DNA damage and are hypersensitive to radiation and radiomimetics, which may have implications for both mammographic screening and treatment modalities. Finally, *BRCA2* knockout mice can be partially rescued by crossing with a *p53* knockout strain, suggesting that these genes interact with the *p53*-mediated DNA damage checkpoint. Therefore, the available evidence indicates that *BRCA2* is a "caretaker," like *p53*, which serves to maintain genomic integrity. When this function is lost, it probably allows for the accumulation of other genetic defects that are themselves directly responsible for cancer formation.

Additional studies have attempted to attribute specific biochemical functions to the *BRCA2* gene product. The BRCA2 protein contains regions that are capable of inducing transcription and has histone acetyltransferase activity potentially supporting its role in DNA repair, RNA transcription, or both. It is likely that the BRCA2 protein will eventually be implicated in a variety of cellular processes, only some of which will be related to the etiology of breast and ovarian cancers.[45]

Most *BRCA2* mutations reported to date are frameshift deletions, insertions, or nonsense mutations leading to premature truncation of protein transcription, consistent with the loss of function that is expected with clinically significant mutations in tumor suppressor genes. As with *BRCA1*, hundreds of *BRCA2* mutations have been identified, although the number is somewhat lower (~450 for *BRCA2* vs. >600 for *BRCA1*). Because *BRCA2* was cloned later than *BRCA1* and is more difficult to screen, it is likely that the range of mutations in *BRCA2* will ultimately be found to be comparable to that in *BRCA1*.[33,46]

2.3. *p53*

The *p53* gene is one of the most commonly mutated genes described in human neoplasia, with mutations estimated to occur in up to 50% of all cancers.[47] However, the occurrence of mutations is less common in sporadic breast carcinomas than in other cancers, with an overall frequency of about 20%. There is evidence that *p53* is mutated at a significantly higher frequency in breast carcinomas arising in carriers of germline *BRCA1* and *BRCA2* mutations.[48] Germline mutation in the *p53* gene has also been identified in more than 50% of families exhibiting LFS; inheritance is autosomal dominant, with a penetrance of at least 50% by age 50 years.[49]

The *p53* gene is located on chromosome 17p, and encodes a 53-kDa nuclear phosphoprotein that binds DNA sequences and functions as a negative regulator of cell growth and proliferation in the setting of DNA damage. In response to DNA damage, the p53 protein arrests cells in the G_1 phase of the cell cycle, allowing DNA repair mechanisms to proceed before DNA synthesis. The p53 protein is also an active component of programmed cell death. Inactivation of the *p53* gene or disruption of the protein product is thought to allow the persistence of damaged DNA and the possible development of malignant cells. Evidence also exists that patients undergoing chemotherapy or radiation therapy for a *p53*-related tumor may be at risk of a treatment-related second malignancy.

Evidence from conditional knockout mice suggests that loss of *BRCA1* in mammary cells leads to incomplete proliferation as well as apoptosis and tumors at a low frequency. However, additional heterozygous mutation in *p53* in these mice leads to many more mammary tumors, most of which have lost the remaining *p53* allele.[50] Some of the *p53* mutants identified in *BRCA1* and *BRCA2* mutation carriers have not been described or have only been infrequently reported in sporadic human cancers. In the proposed model, mutant *p53* would inactivate a cell cycle checkpoint and lead to uncontrolled proliferation and invasive growth. Functional characterization of such mutants in various systems has revealed that they frequently possess properties not commonly associated with mutations occurring in sporadic cases: they retain apoptosis-inducing, transactivating, and growth-inhibitory activities similar to the wild-type protein; yet are compromised for transformation suppression and also possess an independent transforming phenotype. The occurrence of such mutants in familial breast cancer implies the operation of distinct selective pressures during tumorigenesis in *BRCA*-associated breast cancers.[47]

2.4. *ATM*

The *ATM* gene is located on human chromosome 11q22–23. The gene is large, spanning 150 kb of genomic DNA. It encodes a ubiquitously expressed transcript of approximately 13 kb, consisting of 66 exons, giving a 350-kDa protein of 3056 amino acids. The initiation codon falls within exon 4. The last exon is 3.8 kb long, and contains the stop codon and a 3' UTR of about 3600 nucleotides.[51]

The classic form of AT results from the presence of two truncating *ATM* mutations leading to loss of function of the ATM protein,[52] whereas milder forms are associated with a leaky splice-site *ATM* mutation[53–55] or the presence of missense mutations.[56] The first four exons of the gene, which fall within the 5' UTR, undergo extensive alternative splicing. Differential polyadenylation results in 3' UTRs of various lengths. These structural features suggest that ATM expression might be subject to complex post-transcriptional regulation.[57]

The ATM protein plays a key role in the detection and repair of DNA double-strand breaks, which can arise endogenously during cellular processes such as VDJ recombination or meiosis, or after exposure to DNA-damaging agents such as ionizing radiation. ATM is held inactive in undamaged cells as a dimer or higher-order multimer, with the kinase domain bound to a region surrounding serine 1981, which is contained within the protein's FAT domain. The kinase domain contains the signature motifs of phosphatidylinositol 3-kinases. ATM's kinase activity is itself enhanced in response to DNA double-strand breaks, resulting in a phosphorylation cascade activating many proteins, each of which in turn affects a specific signaling pathway. These substrates include the protein products of several well-characterized tumor suppressor genes including *p53, BRCA1*, and *CHEK2*, which play important roles in triggering cell-cycle arrest, DNA repair, or apoptosis.[58]

Mutations in the *ATM* gene have been reported in cancers outside the setting of AT families. Frequent *ATM* inactivation is found in sporadic lymphoid tumors of a mature phenotype.[59] Epidemiological studies have consistently shown that the female relatives of A-T patients are at increased risk of breast cancer,[60,61] However, germline *ATM* mutations were found to be present in only 2 of 401 women with early-onset breast cancer compared with 2 of 202 controls. It was concluded that heterozygous mutations do not confer genetic predisposition to early-onset breast cancer.[62] However, the results of that study are consistent with a moderately elevated risk of breast cancer because the confidence interval was large.

2.5. *CHEK2*

Newly reported breast cancer susceptibility genes confer smaller breast cancer risks than those associated with mutations in *BRCA1* and *BRCA2*. *CHEK2*, which is also known as *CHK2*, is a gene involved in the DNA damage-repair response pathway. The gene was initially evaluated as a potential cause of LFS.[63] While it does not appear to be a common cause of LFS, results from numerous studies suggest that a single mutation, 1100delC, may be a rare, low-penetrance susceptibility allele.[64–67] Vahteristo *et al.*[68] found that 5.5% of 507 patients with familial breast cancer harbored the variant *CHEK2**1100delC allele compared with 1.4% of healthy controls, resulting in a fourfold higher breast cancer risk associated with 1100delC. In both Europe and the United States (where the mutation appears to be slightly less common), additional studies have detected the mutation in 4%–11% of familial cases of breast cancer, and overall have found that the mutation confers an approximately 1.5-fold to 2-fold increased risk for female breast cancer.[69–71]

Because of its low frequency, no single study has had sufficient power to detect a statistically significant risk among unselected breast cancer cases. A multicenter combined analysis–reanalysis of nearly 20 000 subjects from 10 case-control studies, however, has verified a significant, 2.3-fold excess of breast cancer among mutation carriers.[72] Although the initial report suggested that male mutation carriers were at a significantly increased risk for breast cancer, several follow-up studies have failed to confirm the association.[73,74] Additional, larger studies will be required to more precisely define the absolute risk of female breast and other cancers in individuals who carry germline *CHEK2* variants.

3. HISTOPATHOLOGY AND PROGNOSIS OF FAMILIAL BREAST AND OVARIAN CANCERS

As detailed in this section, *BRCA1*-related tumors show an excess of medullary histopathology, are of higher histological grade, and are more likely to be estrogen receptor–negative and progesterone receptor–negative than other breast cancers. At the molecular level, these cancers also have a higher frequency of *p53* mutations and less HER2/c-erb-B2/neu over-expression. Some of these features are prognostically favorable, while others are adverse. Information regarding *BRCA2*-related tumors is more limited, but they do not seem to have a characteristic histopathology and are at least as likely to be hormone receptor–positive as control tumors.

3.1. *Pathology of Breast Cancer with BRCA1 Mutation*

Certain types of breast cancer — including medullary carcinoma, tubular carcinoma, and lobular carcinoma *in situ* — have been reported to be more common in hereditary breast cancer. Patients with *BRCA1* mutations have more medullary carcinomas than do controls.[75] Tumors in *BRCA1* mutation carriers are generally of a higher grade than those in patients with sporadic cancers. However, in a multifactorial analysis conducted by the Breast Cancer Linkage Consortium (BCLC), the only features significantly associated with *BRCA1* were total mitotic count, continuous pushing margins, and lymphocytic infiltrate. The BCLC also found fewer cases of ductal carcinoma *in situ* (DCIS) among *BRCA1* mutation carriers than in a control group.[76] *BRCA1*-associated tumors are more likely to be estrogen and progesterone receptor– and erb-B2–negative.[77] *BRCA1*-linked tumors show a higher frequency of *p53* mutations and *p53* expression than sporadic breast cancers.[48]

3.2. *Pathology of Breast Cancer with BRCA2 Mutation*

Up to now, no specific histological type is thought to be associated with *BRCA2*-related breast cancer. A report from Iceland, where a *BRCA2* founder mutation (999del5) accounts for nearly all hereditary breast cancers, found less tubule formation, more nuclear pleomorphism, and higher mitotic rates in *BRCA2*-related tumors than in sporadic controls.[78] In a multifactorial analysis of data from North America and Europe, the only factors found to be significantly associated with *BRCA2* mutation were fewer tubules, fewer mitoses, and continuous pushing margins.[79] *BRCA2*-related breast cancers are of an overall higher grade than sporadic cancers. The estrogen and progesterone receptor status of *BRCA2* tumors is similar to that of sporadic cancers. *BRCA2*-related tumors do not show a higher frequency of *p53* mutation and/or higher *p53* expression than sporadic breast cancers.

3.3. *Pathology of Non-BRCA1/BRCA2–associated Breast Cancers*

Lakhani and colleagues[80] reviewed 82 breast cancers from non-*BRCA1/BRCA2* families and compared these with 149 cancers from *BRCA1* mutation carriers, 88 from *BRCA2* mutation carriers, and 715 control breast cancers. They noted that the familial non-*BRCA1/BRCA2* breast cancers showed a significantly lower grade, fewer pleiomorphic features, and a lower mitotic count than the *BRCA*-associated familial cancers. Palacios *et al.*[81] examined the morphological characteristics of 37 breast tumors from women with familial breast cancer but no *BRCA1/BRCA2* mutation, and compared these with breast tumors from 20 *BRCA1* carriers and 18 *BRCA2* carriers. Familial non-*BRCA1/BRCA2* tumors again seemed, on average, to be of lower grade, more frequently estrogen and progesterone receptor–positive, and more often *p53*-negative than *BRCA1*-associated breast cancers. In both studies, the main difference between *BRCA2* tumors and the non-*BRCA1/BRCA2* familial tumors was the former group's higher average grade.

The long-term outlook for non-*BRCA1/BRCA2* familial tumors would be expected to be more favorable than average, particularly compared with that of *BRCA1*-associated tumors. These observations can also provide a guide as to whether genetic analysis might be worth pursuing — a family with predominantly low-grade, estrogen receptor–positive breast cancers would not be a strong candidate for harboring a *BRCA1* mutation. Furthermore, selection of familial cases with similar types of breast tumors not likely to be *BRCA1*- or *BRCA2*-associated could provide a more focused approach to linkage studies looking for new candidate moderate-risk and high-risk susceptibility genes.

3.4. *Pathology of Hereditary Ovarian Cancer*

Ovarian cancer arising in women with *BRCA1* mutations is more likely to be invasive serous adenocarcinoma.[82] More than 90% of tumors in women with *BRCA1* cancer-predisposing mutations are serous, compared with approximately 50% of tumors in women without a *BRCA1* cancer-predisposing mutation.[83,84] Both primary ovarian carcinomas and primary peritoneal carcinomas have a higher-than-average incidence of somatic *p53* mutations and exhibit relatively aggressive features, including higher grade, *p53* overexpression, and more frequent bilaterality. Although serous carcinomas predominate, other histological types such as endometrioid carcinomas are more frequent than first thought.[82,85] Even if occasionally described, borderline and mucinous tumors are uncommon,[83,86] suggesting that *BRCA* mutations do not play a role in their development. In no study was a particular mutation found to be associated with a specific histological type. Occasional reports exist of siblings and first-degree relatives with malignant germ cell ovarian tumors,[87,88] but a hereditary predisposition for such neoplasms has not been proven.

In patients undergoing prophylactic oophorectomy for a known family history of ovarian cancer, the ovaries are found to contain more surface epithelial inclusion cysts and surface micropapillae than ovaries removed from women lacking such a history.[89,90] However, nuclear abnormalities that might represent early preneoplastic changes, such as larger and more irregular nuclei, are difficult to identify reliably with light microscopy and are generally detected by image analysis. In fallopian tubes removed prophylactically, precursor lesions such as dysplasia and atypical hyperplasia have been found. The histopathologic profile of *BRCA2*-related ovarian cancer has not been well defined.

A retrospective study of 80 ovarian cancer patients who were members of HNPCC families revealed that their tumors occurred at an earlier age than those in the general population.[91] More than 90% of the tumors were carcinomas, with borderline tumors comprising only 4.1% of the epithelial cancers. Most carcinomas were well or moderately differentiated, and 85% were FIGO (*Fédération Internationale de Gynécologie Obstétrique* [International Federation of Gynecology and Obstetrics]) stage I or II at diagnosis.

3.5. *Breast Cancer Histopathology in Other Cancer Syndromes*

No typical pathological features have been identified for breast cancer in other cancer syndromes.

3.6. *Prognosis of Hereditary Breast Cancer*

In order to explore the influence of familial and hereditary factors on the prognosis of breast cancer, Chappuis *et al.*[92] reviewed the English literature through Medline between 1976 and February 1999. Publications were divided into three categories: family history–based studies, linkage studies, and mutation-based studies. Eighteen articles of family history–based studies were reviewed. Among them, four studies showed a statistically significant better survival in patients with a family history of breast cancer, and two studies demonstrated a significantly worse prognosis; the remaining articles showed no significant difference. In linkage studies, two studies based on linkage to *BRCA1* found that overall survival was better in linked families; a third one found a worse outcome in *BRCA2*-linked tumors. In 10 mutation-based studies, the association between germline mutations in *BRCA1/BRCA2* and clinical outcomes was reviewed; 8 articles reported no significant difference in outcome, whereas 2 studies showed a worse outcome in patients with mutations. Despite the conflicting data, no studies revealed a survival advantage for *BRCA1* mutation carriers. The authors indicated that *BRCA1*-related breast cancer is not associated with a survival advantage, and certain *BRCA1* germline mutations may confer a worse prognosis.[92]

After that review, more papers about the prognosis of hereditary breast cancer were published. Two European studies reported survival rates that were similar to or worse than sporadic cases, with a significantly increased risk of contralateral breast cancer. A case series report found higher rates of both ipsilateral and contralateral breast cancers among *BRCA1/2* mutation carriers than among mutation-negative cases.[93] Higher rates of ipsilateral and contralateral breast cancers were also seen in a cohort of Ashkenazi women with founder mutations[94]; however, that study failed to show a significantly different disease-free or overall survival at 5 years compared to nonmutation carriers. A retrospective cohort study of 496 Ashkenazi breast cancer patients from two centers compared the relative survival among the 56 *BRCA1/2* mutation carriers followed for a median of 116 months. *BRCA1* mutations were independently associated with worse disease-specific survival; there was no effect for *BRCA2*. In summary, there is a growing consensus that *BRCA1*-related breast cancer presents with a more aggressive phenotype. Most clinical outcome studies are consistent with there being a somewhat worse prognosis, especially among women who do not receive chemotherapy.[95]

However, there are some conflicting reports. The results of a comparative study showed that the prognosis of the family history–positive (FHP) group was significantly better than that of the family history–negative (FHN) group. But, when those patients were classified according to their menopausal status at onset, there were no significant differences in survival rates between the FHP and FHN groups with onset before menopause, whereas the survival rate of the FHP group was significantly higher than that of the FHN group with onset after menopause.[96]

Most available data derived from retrospective or indirect data are based on small numbers (<50 cases), and are probably confounded by different biases and by lack of appropriate controls. For example, in most studies of breast cancer prognosis, molecular genetic testing was not performed in the control group and controls were not matched to cases for stage at diagnosis. Some investigators have suggested that matching for stage at the time of diagnosis may mask real biological differences between *BRCA1/BRCA2*-related tumors and sporadic tumors.[97] To adequately answer this question, more efficient molecular tools to identify all of the genetic changes responsible for breast cancer predisposition as well as large cohort studies to evaluate their clinical consequences are needed.

3.7. *Prognosis of Hereditary Ovarian Cancer*

Although the *BRCA1/BRCA2*-associated cancers are more likely to be of a higher stage and grade than the sporadic cancers, several studies have reported that the survival of patients with *BRCA*-associated ovarian cancer is better compared with that of sporadic ovarian cancer. This is most likely because of the increased sensitivity of the *BRCA*-deficient tumor cells to DNA-damaging agents such as cisplatin.[98] One study found that the survival of 43 *BRCA1* carriers with advanced ovarian cancer was significantly better than that of matched sporadic cases. Median survival was 77 months in the *BRCA1* carriers vs. 29 months in noncarriers.[83] A nationwide, population-based case-control study in Israel found 3-year survival rates to be significantly better for ovarian cancer patients with *BRCA* founder mutations compared with controls.[99] In a retrospective, hospital-based study ($n = 71$), Ashkenazi *BRCA* heterozygotes had a better response to platinum-based chemotherapy, as measured by response to primary therapy, disease-free survival, and overall survival, compared with sporadic cases.[100] A similar study in Japanese patients also found a survival advantage in stage III *BRCA1*-associated ovarian cancers treated with cisplatin regimens compared with nonhereditary cancers treated in a similar manner.[101] In a study of 103 ovarian cancers unselected for family history, the cancers in the 11 *BRCA1* mutation carriers had prognostic features consistent with better survival, including a high rate of negative second-look surgeries after the primary chemotherapy and more well-differentiated pathologic features.[102] The most recent study of consecutive cases of ovarian cancers, which compared *BRCA*-associated cancer with sporadic ovarian cancer from the same institution, found that *BRCA* mutation status was a favorable and independent predictor of survival for women with advanced disease.[103]

However, there have been negative studies as well. A population-based study from Sweden noted an initial survival advantage in *BRCA1* mutant groups, but this advantage did not persist after 3 or 4 years.[104] Similarly, a case-control study from the Netherlands found an improvement in short-term (up to 5 years) survival among women with familial ovarian cancer compared to sporadic controls, but no difference in longer-term survival.[105] A case-control study at the University of Iowa failed to find any survival advantage for women with *BRCA1* inactivation, whether by germline mutation, somatic mutation, or *BRCA1* promoter silencing.[106] The discrepancies between these studies may be due to the insensitivity of *BRCA* mutation testing, difficulties in the selection of control groups, and differences between the stage of disease compared.[107] Further large studies with appropriate control populations will be required to determine whether there is a survival advantage in either *BRCA1*- or *BRCA2*-related ovarian cancers.

4. IDENTIFICATION OF HBOC FAMILIES

4.1. *Risk Assessment*

Risk assessment clinics provide women with an estimation of their risk of developing breast cancer as well as the likelihood that this risk can be explained by one of the known breast cancer susceptibility genes. For unaffected women, breast cancer risks can be estimated using quantitative models developed by Gail and Claus.

The most widely applicable model for general risk assessment is the Gail model.[108] The Gail model was used to determine eligibility for the Breast Cancer Prevention Trial, and has since been modified (in part to adjust for race) and made available on the National Cancer Institute website (http://bcra.nci.nih.gov/brc/q1.htm). It incorporates personal reproductive history, breast biopsy history, history of atypical hyperplasia, and number of first-degree relatives diagnosed with breast cancer. The Gail model has been found to be reasonably accurate at predicting breast cancer risk in large groups of white women who undergo annual screening mammography.

Although the Gail model is widely used, it has several important deficits. It uses only a limited amount of family history information; does not consider breast cancer in second-degree relatives, family history of contralateral cancer, or the age at which relatives developed breast cancer; and does not consider the findings of ovarian cancer or lobular carcinoma *in situ*. The Claus model was developed to address some of these deficits. The model is based on family history, and places emphasis on the age at breast cancer diagnosis. It incorporates either a maternal or paternal family

Table 1   ASCO Principles for Genetic Risk Assessment (2003)[115,116]

- Clinical oncologists should document a family history of cancer in their patients, provide counseling regarding familial cancer risk and options for prevention and early detection, and recognize those families for whom genetic testing may serve as an aid in counseling.
- Genetic testing for cancer susceptibility should be perfomed in the setting of long-term outcome studies.
- Physicians should learn methods of quantitative cancer risk assessment, genetic testing, and pretest and posttest genetic counseling so that they may more responsibly integrate genetic counseling and testing into the practice of clinical and preventive oncology.
- Informed consent must be given by the patient as an integral part of the process of genetic predisposition testing, whether such testing is offered on a clinical or a research basis.

history. Unlike the Gail model, the Claus model cannot be used for a woman without a family history of breast cancer.[109]

The American Society of Clinical Oncology (ASCO) suggests that genetic testing for cancer predisposition be considered when the family history is suggestive of a hereditary predisposition, when the test results can be adequately interpreted, and when the results from the testing will affect the patient's medical care (see also Table 1).[110] Other models of breast cancer risk assessment are also aimed at high-risk subsets of patients, and thus are less widely applicable than the Gail model. These include the Couch,[111] Shattuck–Eidens,[112] Frank,[113] and Berry[114] models.

4.2. *Genetic Counseling*

Genetic counseling is the process of providing individuals and families with information about the nature, inheritance, and implications of genetic disorders to help them be informed of medical and personal decisions. Appropriate genetic counseling can improve their knowledge and perception of absolute risk for breast and ovarian cancer, and can often reduce anxiety. Counseling for breast cancer risk typically involves individuals with family histories of cancers that may be attributable to *BRCA1* or *BRCA2*; it may also include individuals with family histories of LFS, AT, CS, or PJS. However, the inclusion criteria vary. The inclusion criteria used for genetic counseling recommended by the German Consortium[117] is as the following: (1) at least two relatives (mother, daughter, sister, or patient herself) with breast cancer and/or ovarian cancer, one of which was diagnosed before the age of 50; (2) one female relative (mother, daughter, sister, or patient herself) with breast cancer on one side prior to age 30; (3) one female relative (mother, daughter, sister, or patient herself) with breast cancer on both sides prior to age 40; (4) one female relative (mother, daughter, sister, or patient herself) with ovarian cancer prior to age 30; (5) one female relative (mother, daughter, sister, or patient herself) with breast cancer and ovarian cancer prior to age 40; and (6) one male relative with breast cancer.

Women considering genetic testing for *BRCA1* and *BRCA2* mutations can be assessed using various published risk models, which generate a pretest probability that an individual or family carries a deleterious mutation in either gene. The complex issues surrounding genetic counseling for HBOCs have been the subject of extensive recent research. The estimated risks for people with *BRCA1/2* mutations to develop breast or ovarian cancer have been confirmed. Management strategies for gene mutation carriers may involve decisions about the nature, frequency, and timing of screening and surveillance procedures; chemoprevention; risk-reducing surgery; and use of hormone replacement therapy. Counseling also includes consideration of related psychosocial concerns as well as discussions about why and how to warn other family members of the possibility that they may be at increased risk for breast, ovarian, and other cancers. Published descriptions of counseling programs for *BRCA1* (and subsequently for *BRCA2*) testing include strategies for gathering a family history, assessing eligibility for testing, and communicating the considerable volume of relevant information about breast/ovarian cancer genetics as well as associated medical and psychosocial risks and benefits; these reports also discuss the special ethical considerations about confidentiality and family communication that arise in these instances.[118–122]

4.3. *Psychosocial Issues Related to Predictive Testing for Breast and Ovarian Cancer Genes*

The impact of breast and ovarian cancers on patients and their families can be sizable. Investigations into the emotional impact of breast cancer testing have been undertaken. Attitudes toward testing varied by ethnicity, previous exposure to genetic information, age, optimism, and information style. Audrain *et al.*[123] characterized the psychological status of women with a family history of breast or ovarian cancer who self-referred for genetic counseling and *BRCA1* testing. Participants were 256 women aged 18 years and older who had at least one first-degree relative with breast and/or ovarian cancer. The study suggested that self-referred genetic counseling participants may be psychologically vulnerable, and may benefit from interventions designed to decrease distress and the perceived absence of control over developing breast cancer.[123]

In general, genetic counseling and subsequent genetic testing for *BRCA* mutations have not been shown to lead to undue psychological stress.[124] Many professional groups have made recommendations regarding informed consent,[125] but there is some evidence that not all practitioners are aware of or follow these guidelines. Research shows that many *BRCA1/2* genetic testing consent forms do not fulfill recommendations by professional groups about the 11 areas that should be addressed,[125] and that the forms omit highly relevant points of information. In a study of women with a history of breast or ovarian cancer, interviews revealed that the women reported feeling inadequately prepared for the ethical dilemmas they encountered when imparting genetic information to family members.[126] These data suggest that the emotional burden of disseminating genetic information to family members can be alleviated by providing the patient with additional information and coaching on how to deliver the message.

In addition, psychosocial effects exist in women when they are recommended for prophylactic surgery after genetic testing. Lerman *et al.*[127] reported that, among women who underwent appropriate genetic counseling and were members of families with *BRCA1* or *BRCA2* mutation, 48% considered surgery 1 month after counseling; however, only 2% actually underwent this surgery 6 months later. Clearly, the removal of ovaries resulting in premature menopause could have a profound effect on a woman's body image and lifestyle.[127]

In a recent study, Lynch *et al.*[128] reported that, compared with noncarriers without cancer, a significantly higher percentage of carriers felt guilty about passing a mutation on to their children, were worried about developing additional cancer or their children developing cancer, and were concerned about health insurance discrimination. Despite these psychological consequences, carriers and noncarriers reported a positive attitude toward genetic testing.[128]

5. MANAGEMENT OF HBOCs

5.1. *Management of Breast Cancer Risk*

5.1.1. *Close Observation*

It is uncertain whether a program of close observation alone in *BRCA* mutation carriers contributes to a decrease in death from breast cancer. In patients choosing this option, it is recommended that education and teaching regarding proper monthly breast self-examination be provided. According to the Cancer Genetics Studies Consortium Consensus

Statement in 1999,[126,129] surveillance for breast cancer in female *BRCA* mutation carriers should include (1) beginning by the age of 18, monthly breast self-examinations; and (2) beginning at the age of 25 or 10 years earlier than the age at which a family member was first diagnosed with breast cancer, annual or semiannual clinical breast examinations and annual mammograms. Other methods of surveillance — including magnetic resonance imaging (MRI) of the breast — are being evaluated, and could be superior to mammography and clinical breast examination. Several small studies have suggested that MRI is significantly more sensitive for detecting breast cancers in high-risk women than mammography or sonography (Table 2).[130–132]

5.1.2. *Prophylactic Chemoprevention*

Chemoprevention is an area of uncertainty in general, and even more so in the population of *BRCA*-mutated women. Despite mixed results in its initial large prevention trials, tamoxifen has been shown to be effective at reducing risk in high-risk women,[133–136] with a risk reduction of approximately 38% when all of the tamoxifen prevention trials to date were taken into account.[137] The National Surgical Adjuvant Breast and Bowel Project (NSABP) showed that chemoprevention with tamoxifen decreased the risk of breast cancer by 49% in "high-risk women", that is, those with a 5-year breast cancer risk of 1.7% or more (including those with a family history of breast cancer).[132] The reduction of breast cancer in the NSABP trial was seen with estrogen receptor–positive tumors, but not with estrogen receptor–negative tumors.

Narod *et al.*[138] studied tamoxifen use and the incidence of contralateral breast cancer in a case-control study of *BRCA1* and *BRCA2* mutation carriers. Tamoxifen protected against contralateral breast cancer for carriers of *BRCA1* and *BRCA2* mutations; in women who used tamoxifen for 2–4 years, the risk of contralateral breast cancer was reduced by 75%.[138] However, important questions remain regarding the use of tamoxifen, including at what risk level and patient age such an intervention should be initiated. Moreover, it is reported that tamoxifen was found to reduce breast cancer incidence among healthy *BRCA2* carriers, but not in healthy *BRCA1* carriers.[139]

5.1.3. *Prophylactic Surgery*

The patient's decision about prophylactic surgery is highly personal and should be respected. The incidence of breast cancer in *BRCA1* carriers is maximal at the age of 44–55 years, and then declines slightly thereafter. This observation suggests that ovarian hormones may have a promoting role in breast carcinogenesis. In support of this hypothesis, oophorectomy has been protective against breast cancer in *BRCA1/2* mutation carriers in several studies. Rebbeck *et al.*[140] compared the breast cancer risk in a historical cohort of *BRCA1* mutation carriers, some of whom had undergone an oophorectomy and some of whom had both ovaries intact. The estimated relative risk of breast cancer in women who had an oophorectomy was approximately half that of the other women[140]; this result was confirmed by prospective follow-up studies later on.[141–143] These findings suggest that oophorectomy might be used as a strategy to decrease the risk of breast cancer among *BRCA1* mutation carriers. The combination of oophorectomy and tamoxifen has been shown to reduce the risk of contralateral breast cancer.[138] This result implies that the combination of tamoxifen and oophorectomy may be more effective than either treatment alone, and that the two prevention strategies may be complementary.

It is not surprising that mastectomy is an effective way to prevent breast cancer. Hartmann *et al.*[144] evaluated the efficacy of bilateral prophylactic mastectomy in a retrospective cohort analysis of 639 women with moderate-to-high risk for breast cancer (based on family history) at the Mayo Clinic. In that study, bilateral prophylactic mastectomy had yielded a 90% reduction in breast cancer risk after 14 years of follow-up. More recently, Meijers-Heijboer *et al.*[145] reported the results of a prospective study of 139 *BRCA1/2* mutation carriers. Seventy-six women underwent prophylactic mastectomy, and the remainder chose close surveillance. After a mean follow-up of 3 years, no breast cancers were detected in the prophylactic mastectomy group and eight were detected in the surveillance group.[145] Schrag *et al.*[146] estimated that women 30 years old who have *BRCA* mutations might gain 2.9–5.3 years of life expectancy from such a procedure, with gains decreasing with the age at time of surgery and becoming minimal by 60 years of age. Rebbeck *et al.*[147] studied 483 women with disease-associated germline *BRCA1/2* mutations. Breast cancer was diagnosed in 2 (1.9%) of 105 women who had bilateral prophylactic mastectomy and in 184 (48.7%) of 378 matched controls who did not have the procedure, with a mean follow-up of 6.4 years. Bilateral prophylactic mastectomy reduced the risk of breast cancer by approximately 95% in women with prior or concurrent bilateral prophylactic oophorectomy.[147] Currently, total mastectomy is recommended over subcutaneous or nipple-sparing mastectomy; however, technical advances in skin-sparing techniques such as muscle-containing flaps or implantable prostheses have broadened the surgical options available to women considering these procedures.[148]

Breast-conserving therapy (BCT), consisting of lumpectomy and radiation therapy, has been demonstrated to be safe and effective for nonhereditary or sporadic forms of early breast cancer. However, for hereditary breast cancers, the use of BCT is controversial because of conflicting data about increased risk of recurrence in the treated breast and development of new tumors in the untreated breast. For example, Eccles *et al.*[149] reported that ipsilateral recurrence occurred in 22.2% of family history (FH)+ patients compared with 24.1% of FH− patients ($p = 0.774$) who underwent BCT; however, there was a striking excess of contralateral breast cancers in the FH+ group (35.9% vs. 16%, $p = 0.0007$), with a cumulative risk of contralateral cancer of 36% at 10 years. This leaves women with *BRCA* mutations at a disadvantage when choosing between BCT and bilateral mastectomy.

Brekelmans *et al.*[150] studied the impact of a family history of breast cancer on the local recurrence (LR) risk after BCT within the framework of a large, multicenter matched case-control study. Family history was assessed for 218 breast cancer patients with LR and 480 patients without LR as control. The risk of LR for patients with a positive family history was similar to or less than that of nonfamilial patients. They concluded that a positive family history did not appear to be a contraindication for BCT.[150]

Robson and colleagues[151] evaluated long-term cancer risk in 87 women diagnosed with breast cancer and *BRCA* mutations who underwent BCT. They found no increase in the risk of cancer recurrence in the treated breast compared with the risk for young women without these mutations. Ten years after their initial diagnosis, 13.6% of the women with a genetic mutation had experienced a recurrence; this rate is similar to previously published recurrence rates for women with nonhereditary breast cancers receiving BCT. However, the researchers noted that more than half of the women suffered a cancer-related event (a recurrence or a second primary cancer) within 10 years of their initial diagnosis, including 37.6% who experienced a new cancer in the untreated breast. No clinical risk factors were linked to an increased risk of cancer. The authors concluded that BCT is a reasonable option for women with *BRCA* mutations, and that the indications for unilateral mastectomy should be the same for both hereditary and nonhereditary breast cancers. They cautioned, however, that "discussion of bilateral mastectomy is warranted by the significant contralateral breast cancer risk."[151]

5.2. *Management of Ovarian Cancer Risk*

5.2.1. *Close Observation*

Although serial serum CA125 and transvaginal ultrasound examinations have not proven to be sensitive means of detecting stage I and stage II ovarian cancers in the general or low-risk population, it is unclear whether they are equally ineffective in high-risk individuals, specifically in patients with *BRCA* mutations.[126,152] In one study, seven incident ovarian or peritoneal cancers were identified in a historical cohort of 33 *BRCA* mutation carriers who underwent regular screening examinations. Six of the seven cases were at stage III at the time of diagnosis. In the majority of cases, the ultrasound findings were normal prior to diagnosis and the women presented with pain or abdominal distension.[153] Despite the lack of proven efficacy of these modalities, little else is available in the way of practical screening methods for high-risk women. Until the results of ongoing prospective trials give more guidance, annual or semiannual serum CA125 analysis and transvaginal color Doppler sonography remain the best options for early detection of cancer in *BRCA* mutation carriers from high-risk families. It has been suggested that surveillance begin at age 25–35 years (Table 2).

5.2.2. *Prophylactic Chemoprevention*

Oral contraceptives, independent of their specific formulation, have been shown to reduce the incidence of epithelial ovarian cancer by 40% in the general population. This protective effect was seen with as little as 3 consecutive months of contraceptive use, and increased as the duration of use increased.[154] In patients with *BRCA* mutations, oral contraceptive use was

Table 2 Suggested Surveillance Program for Individuals at Risk for HBOC[126,128]

Site	Modality	Frequency	Age to Begin (years)
Breast	Breast self-examination	Monthly	18
	Clinical breast examination	q6 months	25
	Mammography	Yearly	25
Male Breast	Breast self-examination	Monthly	Not defined
	Clinical examination	Not defined	Not defined
	Mammography	Consider annual	Not defined
Ovary	Transvaginal sonography	q6–12 months	30–35
	CA125 testing	q6–12 months	30–35

associated with a decreased risk of ovarian cancer (up to 60% after 6 years of use) in two studies,[155,156] but not in another.[157]

The question remains whether the risk of breast cancer is increased in oral contraceptive users who are also *BRCA* mutation carriers.[158,159] Some studies suggested that the use of oral contraceptives may enhance the risk of breast cancer in *BRCA1* or *BRCA2* carriers,[160] but most of these studies were based on contraceptive use before the mid-1970s. A more recent population-based study has found no evidence that mutation carriers have an increased risk of breast cancer associated with the use of current formulations of oral contraceptives.[161]

5.2.3. *Prophylactic Surgery: Salpingo-Oophorectomy, Hysterectomy, and Tubal Ligation*

In 1995, a National Institutes of Health consensus conference concluded that the risk of ovarian cancer in women from families with hereditary ovarian cancer syndromes is sufficiently high to "recommend prophylactic oophorectomy in these women at 35 years of age or when childbearing is completed".[152] This approach is believed to reduce the risk of ovarian cancer by 85%–96%, and also reduces the risk of breast cancer by up to 70% compared with close observation.[141,162] Despite surgery, these women still have a small risk of developing peritoneal carcinomatosis.

The incidence of occult malignancy at the time of prophylactic oophorectomy appears to be between 2% and 4%.[163] Studies combining prophylactic mastectomy and oophorectomy for patients with *BRCA* mutations have shown that life expectancy can be extended up to 5 years.[163] Prophylactic oophorectomy performed in women undergoing abdominal surgery for other indications is also associated with a significant reduction in the risk of ovarian cancer.[152]

Although there is no convincing documentation that women with *BRCA* mutations have an increased incidence of endometrial cancer, some authors have recommended concomitant hysterectomy and salpingo-oophorectomy[164] because of concerns over the risk of fallopian tube cancer arising from the small amount of intramyometrial tubal tissue left after salpingo-oophorectomy. To date, no cases of fallopian tube cancer after bilateral salpingo-oophorectomy have been reported.

Some recent data have suggested a possible association between uterine papillary serous carcinomas and *BRCA* mutations.[165–167] These studies have been small and have had conflicting results. Therefore, further studies will be needed before any recommendation can be made regarding prophylactic hysterectomy.[168] Conversely, because of the increased risk of uterine cancer (mainly endometrial cancer, but also uterine sarcoma) in patients receiving tamoxifen, hysterectomy at the time of prophylactic oophorectomy could be considered for selected individuals.

Tubal ligation has been found to be protective against ovarian cancer in the general population and among *BRCA1* mutation carriers, but the reported efficacy is lower than that seen with prophylactic oophorectomy.[156] Narod and colleagues[156] also reported that the combination of tubal ligation and oral contraceptives was estimated to reduce the risk of ovarian cancer by 72%; however, no effect of tubal ligation was seen in *BRCA2* mutation carriers.

REFERENCES

1. Feuer EJ, Wun LM, Boring CC, *et al.* (1993) The lifetime risk of developing breast cancer. *J Natl Cancer Inst* **85**:892–897.
2. Prat J, Ribe A, Gallardo A. (2005) Hereditary ovarian cancer. *Hum Pathol* **36**:861–870.
3. Collaborative Group on Hormonal Factors in Breast Cancer. (2001) Familial breast cancer: collaborative reanalysis of individual data from 52 epidemiological studies including 58,209 women with breast cancer and 101,986 women without the disease. *Lancet* **358**:1389–1399.
4. Claus EB, Schildkraut JM, Thompson WD, Risch NJ. (1996) The genetic attributable risk of breast and ovarian cancer. *Cancer* **77**:2318–2324.
5. Werness BA, Eltabbakh GH. (2001) Familial ovarian cancer and early ovarian cancer: biologic, pathologic, and clinical features. *Int J Gynecol Pathol* **20**:48–63.
6. Pharoah PD, Ponder BAJ. (2002) The genetics of ovarian cancer. *Best Pract Res Clin Obstet Gynaecol* **16**:449–468.
7. Boyd J. (2001) Molecular genetics of hereditary ovarian cancer. In: *Ovarian Cancer,* 2nd ed., Rubin SC, Sutton GP (eds.), Lippincott Williams & Wilkins, Philadelphia, PA, pp. 3–17.
8. Lynch HT, Smyrk T, Lynch J. (1997) An update of HNPCC (Lynch syndrome). *Cancer Genet Cytogenet* **93**:84–99.
9. Boyd J. (2003) Specific keynote: hereditary ovarian cancer: what we know. *Gynecol Oncol* **88**:S8–S10.
10. Risch HA, McLaughlin J, Cole DEC, *et al.* (2001) Prevalence and penetrance of germline BRCA1 and BRCA2 mutations in a population series of 649 women with ovarian cancer. *Am J Hum Genet* **68**:700–710.
11. Nelen MR, Padberg GW, Peeters EA, *et al.* (1996) Localization of the gene for Cowden disease to chromosome 10q22–23. *Nat Genet* **13**:114–116.

12. Tsou HC, Teng DH, Ping XL, *et al.* (1997) The role of MMAC1 mutations in early-onset breast cancer: causative in association with Cowden syndrome and excluded in BRCA1–negative cases. *Am J Hum Genet* **61**:1036–1043.

13. Eng C. (1997) Cowden syndrome. *J Genet Couns* **6**:181–191.

14. Starink T, van der Veen JP, Arwert F, *et al.* (1986) The Cowden syndrome: a clinical and genetic study in 21 patients. *Clin Genet* **29**:222–233.

15. Olopade OI, Weber BL. (1998) Breast cancer genetics: toward molecular characterization of individuals at increased risk for breast cancer: part I. In: *Principles and Practice of Oncology Updates*, Vol. 12, DeVita V Jr, Hellman S, Rosenbeng S (eds.), JB Lipincott, Philadelphia, PA, pp. 1–12.

16. Eng C. (2000) Will the real Cowden syndrome please stand up: revised diagnostic criteria. *J Med Genet* **37**:828–830.

17. Longy M, Lacombe D. (1996) Cowden disease. Report of a family and review. *Ann Genet* **39**:35–42.

18. Dahia PL, March DJ, Zheng Z, *et al.* (1997) Somatic deletions and mutations in the Cowden disease gene, PTEN, in sporadic thyroid tumors. *Cancer Res* **57**(21):4710–4713.

19. Garber JE, Goldstein AM, Kantor AF, *et al.* (1991) Follow-up study of twenty-four families with Li–Fraumeni syndrome. *Cancer Res* **51**:6094–6097.

20. Li FP, Fraumeni JF Jr, Mulvihill JJ, *et al.* (1988) A cancer family syndrome in twenty-four kindreds. *Cancer Res* **48**:5358–5362.

21. Sidransky D, Tokino T, Helzlsouer K, *et al.* (1992) Inherited p53 gene mutations in breast cancer. *Cancer Res* **52**:2984–2986.

22. Jenne DE, Reimann H, Nezu J, *et al.* (1998) Peutz–Jeghers syndrome is caused by mutations in a novel serine threonine kinase. *Nat Genet* **18**: 38–43.

23. Izatt L, Greenman J, Hodgson S, *et al.* (1999) Identification of germline missense mutations and rare allelic variants in the ATM gene in early-onset breast cancer. *Genes Chromosomes Cancer* **26**:286–294.

24. Savitsky K, Sfez S, Tagle DA, *et al.* (1995) The complete sequence of the coding region of the ATM gene reveals similarity to cell cycle regulators in different species. *Hum Mol Genet* **4**:2025–2032.

25. Lynch HT, Shaw MW, Magnuson CW, *et al.* (1966) Hereditary factors in cancer. Study of two large Midwestern kindreds. *Arch Intern Med* **117**: 206–212.

26. Aarnio M, Sankila R, Pukkala E, *et al.* (1999) Cancer risk in mutation carriers of DNA-mismatch repair genes. *Int J Cancer* **81**:214–218.

27. Ford D, Easton DF, Stratton M, *et al.* (1998) Genetic heterogeneity and penetrance analysis of the BRCA1 and BRCA2 genes in breast cancer families. The Breast Cancer Linkage Consortium. *Am J Hum Genet* **62**:676–689.

28. Tavassoli FA, Devilee P. (2003) *World Health Organization Classification of Tumours: Pathology and Genetics of Tumours of the Breast and Female Genital Organs*. IARC Press, Lyon, France.

29. Hall JM, Lee MK, Newman B, *et al.* (1990) Linkage of early-onset familial cancer to chromosme 17q21. *Science* **250**:1684–1689.

30. Miki Y, Swensen J, Shattuck-Eidens D, *et al.* (1994) A strong candidate for the breast and ovarian cancer susceptibility gene BRCA1. *Science* **266**:66–71.

31. Easton DF, Bishop DT, Ford D, Crockford GP. (1993) Genetic linkage analysis in familial breast and ovarian cancer: results from 214 families. The Breast Cancer Linkage Consortium. *Am J Hum Genet* **52**:678–701.

32. Smith TM, Lee MK, Szabo Cl, *et al.* (1996) Complete genomic sequence and analysis of 117 kb of human DNA containing the gene BRCA1. *Genome Res* **6**:1029–1049.

33. Powell SN, Kachnic LA. (2003) Roles of BRCA1 and BRCA2 in homologous recombination, DNA replication fidelity and the cellular response to ionizing radiation. *Oncogene* **22**:5784–5791.

34. Yoshida K, Miki Y. (2004) Role of BRCA1 and BRCA2 as regulators of DNA repair, transcription, and cell cycle in response to DNA damage. *Cancer Sci* **95**:866–871.

35. Scully R, Chen J, Ochs RL, *et al.* (1997) Dynamic changes of BRCA1 subnuclear location and phosphorylation state are initiated by DNA damage. *Cell* **90**:425–435.

36. Wang Y, Cortez D, Yazdi P, *et al.* (2000) BASC, a supercomplex of BRCA1-associated proteins involved in the recognition and repair of aberrant DNA structures. *Genes Dev* **14**:927–939.

37. Venkitaraman AR. (2002) Cancer susceptibility and the functions of BRCA1 and BRCA2. **108**:171–182.

38. Rajan JV, Marquis ST, Gardner HP, Chodosh LA. (1997) Developmental expression of Brca2 colocalizes with Brca1 and is associated with proliferation and differentiation in multiple tissues. *Dev Biol* **184**:385–401.

39. Cornelis RS, Neuhausen SL, Johansson O, *et al.* (1995) High allele loss rates at 17q12–q21 in breast and ovarian tumors from BRCA1-linked families. The Breast Cancer Linkage Consortium. *Genes Chromosomes Cancer* **13**:203–210.

40. Russell PA, Pharoah PD, De Foy K, *et al.* (2000) Frequent loss of BRCA1 mRNA and protein expression in sporadic ovarian cancers. *Int J Cancer* **87**:317–321.

41. Wilson CA, Ramos L, Villasenor MR, *et al.* (1999) Localization of human BRCA1 and its loss in high-grade, non-inherited breast carcinomas. *Nat Genet* **21**:236–240.

42. Taylor MR. (2001) Genetic testing for inherited breast and ovarian cancer syndromes: important concepts for the primary care physician. *Postgrad Med J* **77**:11–15.

43. Hartge P, Struewing JP, Wacholder S, *et al.* (1999) The prevalence of common BRCA1 and BRCA2 mutations among Ashkenazi Jews. *Am J Hum Genet* **64**:963–970.

44. Tonin P, Weber B, Offit K, *et al.* (1996) Frequency of recurrent BRCA1 and BRCA2 mutations in Ashkenazi Jewish breast cancer families. *Nat Med* **2**:1179–1183.

45. Deng CX, Brodie SG. (2001) Knockout mouse models and mammary tumorigenesis. *Semin Cancer Biol* **11**:387–394.

46. Rudkin TM, Foulkes WD. (2005) BRCA2: breaks, mistakes and failed separations. *Trends Mol Med* **11**:145–148.

47. Gasco M, Yulug IG, Crook T. (2003) Mutations in familial breast cancer: functional aspects. *Hum Mutat* **21**:301–306.

48. Phillips KA, Nichol K, Ozcelik H, *et al.* (1999) Frequency of p53 mutations in breast carcinomas from Ashkenazi Jewish carriers of BRCA1 mutations. *J Natl Cancer Inst* **91**:469–473.

49. Varley JM, Evans DGR, Birch JM. (1997) Li–Fraumeni syndrome: a molecular and clinical review. *Br J Cancer* **76**:1–14.

50. Welcsh PL, King MC. (2001) BRCA1 and BRCA2 and the genetics of breast and ovarian cancer. *Hum Mol Genet* **10**:705–713.

51. Uziel T, Savitsky K, Platzer M, *et al.* (1996) Genomic organization of the ATM gene. *Genomics* **33**:317–320.

52. Gilad S, Khosravi R, Shkedy D, *et al.* (1996) Predominance of null mutations in ataxia-telangiectasia. *Hum Mol Genet* **5**:433–439.

53. McConville CM, Stankovic T, Byrd PJ, *et al.* (1996) Mutations associated with variant phenotypes in ataxia-telangiectasia. *Am J Hum Genet* **59**: 320–330.

54. Stewart GS, Last JL, Stankovic T, *et al.* (2001) Residual ataxia telangiectasia mutated protein function in cells from ataxia telangiectasia patients, with 5762ins137 and 727IT->G mutations, showing a less severe phenotype. *J Biol Chem* **276**:30133–30141.

55. Sutton IJ, Last JL, Ritchie SJ, *et al.* (2004) Adult-onset ataxia telangiectasia due to ATM 5762ins137 mutation homozygosity. *Ann Neurol* **55**:891–895.

56. Stankovic T, Kidd AM, Sutcliffe A, *et al.* (1998) ATM mutations and phenotypes in ataxia-telangiectasia families in the British Isles: expression of mutant ATM and the risk of leukemia, lymphoma and breast cancer. *Am J Hum Genet* **62**:334–345.

57. Savitsky K, Platzer M, Uziel T, *et al.* (1997) Ataxia-telangiectasia: structural diversity of untranslated sequences suggests complex posttranscriptional regulation of ATM gene expression. *Nucleic Acids Res* **25**: 1678–1684.

58. Shiloh Y, Kastan MB. (2001) ATM: genome stability, neuronal development, and cancer cross paths. *Adv Cancer Res* **83**:209–254.

59. Stankovic T, Stewart GS, Byrd P, *et al.* (2002) ATM mutations in sporadic lymphoid tumours. *Leuk Lymphoma* **43**:1563–1571.

60. Swift M, Morrell D, Massey RB, Chase CL. (1991) Incidence of cancer in 161 families affected by ataxia-telangiectasia. *N Engl J Med* **325**:1831–1836.

61. Athma P, Rappaport R, Swift M. (1996) Molecular genotyping shows that ataxia-telangiectasia heterozygotes are predisposed to breast cancer. *Cancer Genet Cytogenet* **92**:130–134.

62. FitzGerald MG, Bean JM, Hegde SR, *et al.* (1997) Heterozygous ATM mutations do not contribute to early onset of breast cancer. *Nat Genet* **15**: 307–310.

63. Bell DW, Varley JM, Szydlo TE, *et al.* (1999) Heterozygous germ line hCHK2 mutations in Li–Fraumeni syndrome. *Science* **286**:2528–2531.

64. Meijers-Heijboer H, van den Ouweland A, Klijn J, *et al.* (2002) Low-penetrance susceptibility to breast cancer due to CHEK2(*)1100delC in noncarriers of BRCA1 or BRCA2 mutations. *Nat Genet* **31**:55–59.

65. Kuschel B, Auranen A, Gregory CS, *et al.* (2003) Common polymorphisms in checkpoint kinase 2 are not associated with breast cancer risk. *Cancer Epidemiol Biomarkers Prev* **12**:809–812.

66. Sodha N, Bullock S, Taylor R, *et al.* (2002) CHEK2 variants in susceptibility to breast cancer and evidence of retention of the wild type allele in tumours. *Br J Cancer* **87**:1445–1448.

67. Meijers-Heijboer H, Wijnen J, Vasen H, *et al.* (2003) The CHEK2 1100delC mutation identifies families with a hereditary breast and colorectal cancer phenotype. *Am J Hum Genet* **72**:1308–1314.

68. Vahteristo P, Bartkova J, Eerola H, *et al.* (2002) A CHEK2 genetic variant contributing to a substantial fraction of familial breast cancer. *Am J Hum Genet* **71**:432–438.

69. Offit K, Pierce H, Kirchhoff T, *et al.* (2003) Frequency of CHEK2*1100delC in New York breast cancer cases and controls. *BMC Med Genet* **4**:1.

70. Neuhausen S, Dunning A, Steele L, *et al.* (2004) Role of CHEK2*1100delC in unselected series of non-BRCA1/2 male breast cancers. *Int J Cancer* **108**:477–478.

71. Ohayon T, Gal I, Baruch RG, *et al.* (2004) CHEK2*1100delC and male breast cancer risk in Israel. *Int J Cancer* **108**:479–480.

72. CHEK2 Breast Cancer Case-Control Consortium. (2004) CHEK2*1100delC and susceptibility to breast cancer: a collaborative analysis involving 10,860 breast cancer cases and 9,065 controls from 10 studies. *Am J Hum Genet* **74**:1175–1182.

73. Osorio A, Rodriguez-López R, Diez O, *et al.* (2004) The breast cancer low-penetrance allele 1100delC in the CHEK2 gene is not present in Spanish familial breast cancer population. *Int J Cancer* **108**:54–56.

74. Syrjäkoski K, Kuukasjärvi T, Auvinen A, Kallioniemi OP. (2004) CHEK2 1100delC is not a risk factor for male breast cancer population. *Int J Cancer* **108**:475–476.

75. Armes JE, Egan AJM, Southey MC, *et al.* (1998) The histologic pheno-types of breast carcinoma occurring before age 40 years in women with and without BRCA1 and BRCA2 germline mutations. *Cancer* **83**:2335–2345.

76. Breast Cancer Linkage Consortium. (1997) Pathology of familial breast cancer: differences between breast cancers in carriers of BRCA1 or BRCA2 mutations and sporadic cases. *Lancet* **349**:1505–1510.

77. Moller P, Borg A, Heimdal K, *et al.* (2001) The BRCA1 syndrome and other inherited breast or breast-ovarian cancers in a Norwegian prospective series. *Eur J Cancer* **37**:1027–1032.

78. Agnarsson BA, Jonasson JG, Björnsdottir IB, *et al.* (1998) Inherited BRCA2 mutation associated with high grade breast cancer. *Breast Cancer Res Treat* **47**:121–127.

79. Lakhani SR, Jacquemier J, Sloane JP, *et al.* (1998) Multifactorial analysis of differences between sporadic breast cancers and cancers involving BRCA1 and BRCA2 mutations. *J Natl Cancer Inst* **90**:1138–1145.

80. Lakhani SR, Gusterson BA, Jacquemier J, *et al.* (2000) The pathology of familial breast cancer: histological features of cancers in families not attributable to mutations in BRCA1 or BRCA2. *Clin Cancer Res* **6**:782–789.

81. Palacios J, Honrado E, Osorio A, *et al.* (2003) Immunohistochemical characteristics defined by tissue microarray of hereditary breast cancer not attributable to BRCA1 or BRCA2 mutation carriers. *Clin Cancer Res* **9**:3606–3614.

82. Lakhani SR, Manek S, Penault-Llorca F, *et al.* (2004) Pathology of ovarian cancers in BRCA1 and BRCA2 carriers. *Clin Cancer Res* **10**:2473–2481.

83. Rubin SC, Benjamin I, Behbakht K, *et al.* (1996) Clinical and pathological features of ovarian cancer in women with germ-line mutations of BRCA1. *N Engl J Med* **335**:1413–1416.

84. Lakhani SR. (1999) The pathology of familial breast cancer: morphological aspects. *Breast Cancer Res* **1**:31–35.

85. Werness BA, Ramus S, Whittemore AS, *et al.* (2000) Histopathology of familial ovarian tumors in women from families with and without germline BRCA1 mutations. *Hum Pathol* **31**:1420–1424.

86. Zweemer RP, Verheijen RH, Gille JJ, *et al.* (1998) Clinical and genetic evaluation of thirty ovarian cancer families. *Am J Obstet Gynecol* **178**: 85–90.

87. Mandel M, Toren A, Kende G, *et al.* (1994) Familial clustering of malignant germ cell tumors and Langerhans' histiocytosis. *Cancer* **73**:1980–1983.

88. Stettner A, Hartenbach E, Schink J. (1999) Familial ovarian germ cell cancer: report and review. *Am J Med Genet* **84**:43–46.

89. Werness BA, Afify AM, Bielat KL, *et al.* (1999) Altered surface and cyst epithelium of ovaries removed prophylactically from women with a family history of ovarian cancer. *Hum Pathol* **30**:151–157.

90. Casey ML, Bewtra C, Hoehne L, *et al.* (2000) Histology of prophylactically removed ovaries from BRCA1 and BRCA2 mutation carriers compared with noncarriers in hereditary breast ovarian cancer syndrome kindreds. *Gynecol Oncol* **78**:278–287.

91. Watson P, Butzow R, Lynch HT, *et al.* (2001) The clinical features of ovarian cancer in hereditary nonpolyposis colorectal cancer. *Gynecol Oncol* **82**: 223–228.

92. Chappuis PO, Rosenblatt J, Foulkes WD. (1999) The influence of familial and hereditary factors on the prognosis of breast cancer. *Ann Oncol* **10**:1163–1170.

93. Haffty BG, Harrold E, Khan AJ, *et al.* (2002) Outcome of conservatively managed early-onset breast cancer by BRCA1/2 status. *Lancet* **359**: 1471–1477.

94. Robson M, Levin D, Federici M, *et al.* (1999) Breast conservation therapy for invasive breast cancer in Ashkenazi women with BRCA gene founder mutations. *J Natl Cancer Inst* **91**:2112–2117.

95. Robson ME, Chappuis PO, Satagopan J, *et al.* (2004) A combined analysis of outcome following breast cancer: differences in survival based on BRCA1/BRCA2 mutation status and administration of adjuvant treatment. *Breast Cancer Res* **6**:R8–R17.

96. Kinoshita T, Fukutomi T, Iwamoto E, Akashi-Tanaka S. (2004) Prognosis of breast cancer patients with familial history classified according to their menopausal status. *Breast J* **10**:218–222.

97. National Cancer Institute. Genetics of Breast and Ovarian Cancer. Health Professional Version. Available at http://www.cancer.gov/cancertopics/pdq/genetics/breast-and-ovarian/HealthProfessional/

98. Husain A, He G, Venkatraman ES, Spriggs DR. (1998) BRCA1 up-regulation is associated with repair-mediated resistance to *cis*-diamminedichloroplatinum(II). *Cancer Res* **58**:1120–1123.

99. Ben David Y, Chetrit A, Hirsh-Yechezkel G, *et al.* (2002) Effect of BRCA mutations on the length of survival in epithelial ovarian tumors. *J Clin Oncol* **20**:463–466.

100. Cass I, Baldwin RL, Varkey T, *et al.* (2003) Improved survival in women with BRCA-associated ovarian carcinoma. *Cancer* **97**:2187–2195.

101. Aida H, Takakuwa K, Nagata H, *et al.* (1998) Clinical features of ovarian cancer in Japanese women with germ-line mutations of BRCA1. *Clin Cancer Res* **4**:235–240.

102. Berchuck A, Heron KA, Carney ME, *et al.* (1998) Frequency of germline and somatic BRCA1 mutations in ovarian cancer. *Clin Cancer Res* **4**:2433–2437.

103. Boyd J, Sonoda Y, Federici MG, *et al.* (2000) Clinicopathologic features of BRCA-linked and sporadic ovarian cancer. *J Am Med Assoc* **283**:2260–2265.

104. Johannsson OT, Ranstam J, Borg A, Olsson H. (1998) Survival of BRCA1 breast and ovarian cancer patients: a population-based study from southern Sweden. *J Clin Oncol* **16**:397–404.

105. Zweemer RP, Verheijen RH, Coebergh JW, *et al.* (2001) Survival analysis in familial ovarian cancer, a case control study. *Eur J Obstet Gynecol Reprod Biol* **98**:219–223.

106. Buller RE, Shahin MS, Geisler JP, *et al.* (2002) Failure of BRCA1 dysfunction to alter ovarian cancer survival. *Clin Cancer Res* **8**:1196–1202.

107. Sowter HM, Ashworth A. (2005) BRCA1 and BRCA2 as ovarian cancer susceptibility genes. *Carcinogenesis* **10**:1651–1656.

108. Gail MH, Brinton LA, Byar DP, *et al.* (1989) Projecting individualized probabilities of developing breast cancer for white females who are being examined annually. *J Natl Cancer Inst* **81**:1879–1886.

109. Claus EB, Risch N, Thompson WD. (1994) Autosomal dominant inheritance of early-onset breast cancer. Implications for risk prediction. *Cancer* **73**:643–651.

110. Thull DL, Vogel VG. (2004) Recognition and management of hereditary breast cancer syndromes. *Oncologist* **9**:13–24.

111. Couch FJ, DeShano ML, Blackwood MA, *et al.* (1997) BRCA1 mutations in women attending clinics that evaluate the risk of breast cancer. *N Engl J Med* **336**:1409–1415.

112. Shattuck-Eidens D, Oliphant A, McClure M, *et al.* (1997) BRCA1 sequence analysis in women at high risk for susceptibility mutations: risk factor analysis and implications for genetic testing. *J Am Med Assoc* **278**:1242–1250.

113. Frank TS, Manley SA, Olopade OI, *et al.* (1998) Sequence analysis of BRCA1 and BRCA2: correlation of mutations with family history and ovarian cancer risk. *J Clin Oncol* **16**:2417–2425.

114. Berry DA, Parmigiani G, Sanchez J, *et al.* (1997) Probability of carrying a mutation of breast-ovarian cancer gene BRCA1 based on family history. *J Natl Cancer Inst* **89**:227–238.

115. American Society of Clinical Oncology. (1996) Statement of the American Society of Clinical Oncology: genetic testing for cancer susceptibility, adopted on February 20, 1996. *J Clin Oncol* **14**:1730–1736; discussion 1737–1740.

116. American Society of Clinical Oncology. (2003) American Society of Clinical Oncology policy statement update: genetic testing for cancer susceptibility. *J Clin Oncol* **21**:2397–2406.

117. Schmutzler RK, Beckmann MW, Kiechle M. (2002) Familiäres mamma- und ovarialkarzinom. *Dtsch Ärztebl* **99**:A-1372–1378.

118. Richards MP, Hallowell N, Green JM, *et al.* (1995) Counseling families with hereditary breast and ovarian cancer: a psychosocial perspective. *J Genet Couns* **4**:219–233.

119. Hoskins KF, Stopfer JE, Calzone KA, *et al.* (1995) Assessment and counseling for women with a family history of breast cancer. A guide for clinicians. *JAMA* **273**:577–585.

120. Cummings S, Olopade O. (1998) Predisposition testing for inherited breast cancer. *Oncology (Williston Park)* **12**:1227–1241; discussion 1241–1242.

121. Lipkus IM, Klein WM, Rimer BK. (2001) Communicating breast cancer risks to women using different formats. *Cancer Epidemiol Biomarkers Prev* **10**:895–898.

122. Butow PN, Lobb EA. (2004) Analyzing the process and content of genetic counseling in familial breast cancer consultations. *J Genet Couns* **13**: 403–424.

123. Audrain J, Schwartz MD, Lerman C, *et al.* (1997) Psychological distress in women seeking genetic counseling for breast-ovarian cancer risk: the contributions of personality and appraisal. *Ann Behav Med* **19**:370–377.

124. Marteau T, Croyle R. (1998) Psychological response to genetic testing. *BMJ* **316**:693–695.

125. Burke W, Daly M, Garber J, *et al.* (1997) Recommendations for follow-up care of individuals with an inherited predisposition to cancer. II. BRCA1 and BRCA2. Cancer Genetics Studies Consortium. *JAMA* **277**:997–1003.

126. Hallowell N, Foster C, Eeles R, *et al.* (2003) Balancing autonomy and responsibility: the ethics of generating and disclosing genetic information. *J Med Ethics* **29**:74–79; discussion 80–83.

127. Lerman C, Hughes C, Lemon S, *et al.* (1997) Outcomes study of BRCA1/2 testing in members of hereditary breast ovarian cancer families. *San Antonio Breast Cancer Symposium,* San Antonio, Texas.

128. Lynch HT, Snyder C, Lynch JF, *et al.* (2006) Patient responses to the disclosure of BRCA mutation tests in hereditary breast-ovarian cancer families. *Cancer Genet Cytogenet* **165**:91–97.

129. Robson ME. (2002) Clinical considerations in the management of individuals at risk for hereditary breast and ovarian cancer. *Cancer Control* **9**: 457–465.

130. Kriege M, Brekelmans CT, Boetes C, *et al.* (2004) Efficacy of MRI and mammography for breast-cancer screening in women with a familial or genetic predisposition. *N Engl J Med* **51**:427–437.

131. Lehman CD, Blume JD, Weatherall P, *et al.* (2005) Screening women at high risk for breast cancer with mammography and magnetic resonance imaging. *Cancer* **103**:1898–1905.

132. Warner E, Plewes DB, Hill KA, *et al.* (2004) Surveillance of BRCA1 and BRCA2 mutation carriers with magnetic resonance imaging, ultrasound, mammography, and clinical breast examination. *JAMA* **292**:1317–1325.

133. Fisher B, Costantino JP, Wickerham DL, *et al.* (1998) Tamoxifen for prevention of breast cancer: report of the National Surgical Adjuvant Breast and Bowel Project P-1 Study. *J Natl Cancer Inst* **90**:1371–1388.

134. Powles TJ, Eeles R, Ashley S, *et al.* (1998) Interim analysis of the incidence of breast cancer in the Royal Marsden Hospital tamoxifen randomised chemoprevention trial. *Lancet* **352**:98–101.

135. Veronesi U, Maisonneuve P, Costa A, *et al.* (1998) Prevention of breast cancer with tamoxifen: preliminary findings from the Italian randomised trial among hysterectomised women. Italian Prevention Study. *Lancet* **352**: 93–97.

136. Cuzick J, Forbes J, Edwards R, *et al.* (2002) First results from the International Breast Cancer Intervention Study (IBIS-1): a randomised prevention trial. *Lancet* **360**:817–824.

137. Cuzick J, Powles T, Veronesi U, *et al.* (2003) Overview of the main outcomes in breast-cancer prevention trials. *Lancet* **361**:296–300.

138. Narod SA, Brunet JS, Ghadirian P, *et al.* (2000) Tamoxifen and risk of contralateral breast cancer in BRCA1 and BRCA2 mutation carriers: a case-control study. Hereditary Breast Cancer Clinical Study Group. *Lancet* **356**:1876–1881.

139. King MC, Wieand S, Hale K, *et al.* (2001) Tamoxifen and breast cancer incidence among women with inherited mutations in BRCA1 and BRCA2. *JAMA* **286**:2251–2256.

140. Rebbeck TR, Lynch HT, Neuhausen SL, *et al.* (2002) Prophylactic oophorectomy in carriers of BRCA1 or BRCA2 mutations. *N Engl J Med* **346**: 1616–1622.

141. Kramer JL, Velazquez IA, Chen BE, *et al.* (2005) Prophylactic oophorectomy reduces breast cancer penetrance during prospective, long-term follow-up of BRCA1 mutation carriers. *J Clin Oncol* **23**:8629–8635.

142. Kauff ND, Satagopan JM, Robson ME, *et al.* (2002) Risk-reducing salpingo-oophorectomy in women with a BRCA1 or BRCA2 mutation. *N Engl J Med* **346**:1609–1615.

143. Wirk B. (2005) The role of ovarian ablation in the management of breast cancer. *Breast J* **11**:416–424.
144. Hartmann LC, Schaid DJ, Woods JE, *et al.* (1999) Efficacy of bilateral prophylactic mastectomy in women with a family history of breast cancer. *N Engl J Med* **340**:77–84.
145. Meijers-Heijboer H, van Geel B, van Putten WLJ, *et al.* (2001) Breast cancer after prophylactic bilateral mastectomy in women with a BRCA1 or BRCA2 mutation. *N Engl J Med* **345**:159–164.
146. Schrag D, Kuntz KM, Garber JE, Weeks JC. (1997) Decision analysis — effects of prophylactic mastectomy and oophorectomy on life expectancy among women with BRCA1 or BRCA2 mutations. *N Engl J Med* **336**:1465–1471. Erratum in: *N Engl J Med* **337**:434.
147. Rebbeck TR, Friebel T, Lynch HT, *et al.* (2004) Bilateral prophylactic mastectomy reduces breast cancer risk in BRCA1 and BRCA2 mutation carriers: the PROSE study group. *J Clin Oncol* **22**:1055–1062.
148. Levine DA, Gemignani ML. (2003) Prophylactic surgery in hereditary breast/ovarian cancer syndrome. *Oncology (Williston Park)* **17**:932–941.
149. Eccles D, Simmonds P, Goddard J, *et al.* (2001) Familial breast cancer: an investigation into the outcome of treatment for early stage disease. *Fam Cancer* **1**:65–72.
150. Brekelmans CT, Voogd AC, Botke G, *et al.* (1999) Family history of breast cancer and local recurrence after breast-conserving therapy. The Dutch Study Group on Local Recurrence after Breast Conservation (BORST). *Eur J Cancer* **354**:620–626.
151. Robson M, Svahn T, McCormick B, *et al.* (2005) Appropriateness of breast-conserving treatment of breast carcinoma in women with germline mutations in BRCA1 or BRCA2: a clinic-based series. *Cancer* **103**:44–51.
152. NIH Consensus Conference. (1995) Ovarian cancer. Screening, treatment, and follow-up. *JAMA* **273**:491–497.
153. Liede A, Karlan BY, Baldwin RL, *et al.* (2002) Cancer incidence in a population of Jewish women at risk of ovarian cancer. *J Clin Oncol* **20**:1570–1577.
154. The Cancer and Steroid Hormone Study of the Centers for Disease Control and the National Institute of Child Health and Human Development. (1987) The reduction in risk of ovarian cancer associated with oral-contraceptive use. *N Engl J Med* **316**:650–655.
155. Narod SA, Risch H, Moslehi R, *et al.* (1998) Oral contraceptives and the risk of hereditary ovarian cancer. *N Engl J Med* **339**:424–428.
156. Narod SA, Sun P, Ghadirian P, *et al.* (2001) Tubal ligation and risk of ovarian cancer in carriers of BRCA1 or BRCA2 mutations: a case-control study. *Lancet* **357**:1467–1470.
157. Modan B, Hartge P, Hirsh-Yechezkel G, *et al.* (2001) Parity, oral contraceptives, and the risk of ovarian cancer among carriers and noncarriers of a BRCA1 or BRCA2 mutation. *N Engl J Med* **345**:235–240.
158. Ursin G, Henderson BE, Haile RW, *et al.* (1997) Does oral contraceptive use increase the risk of breast cancer in women with BRCA1/BRCA2 mutations more than in other women? *Cancer Res* **57**:3678–3681.

159. Narod SA, Dube MP, Klijin J, *et al.* (2002) Oral contraceptives and the risk of breast cancer in BRCA1 and BRCA2 mutation carriers. *J Natl Cancer Inst* **94**:1773–1779.

160. Collaborative Group on Hormonal Factors in Breast Cancer. (1996) Breast cancer and hormonal contraceptives: collaborative reanalysis of individual data on 53,297 women with breast cancer and 100,239 women without breast cancer from 54 epidemiological studies. Collaborative Group on Hormonal Factors in Breast Cancer. *Lancet* **347**:1713–1727.

161. Milne RL, Knight JA, John EM, *et al.* (2005) Oral contraceptive use and risk of early-onset breast cancer in carriers and noncarriers of BRCA1 and BRCA2 mutations. *Cancer Epidemiol Biomarkers Prev* **14**:350–356.

162. Rebbeck TR, Levin AM, Eisen A, *et al.* (1999) Breast cancer risk after prophylactic oophorectomy in BRCA1 mutation carriers. *J Natl Cancer Inst* **91**:1475–1479.

163. Levine DA, Gemignani ML. (2003) Prophylactic surgery in hereditary breast/ovarian cancer syndrome. *Oncology* **17**:932–941.

164. Paley PJ, Swisher EM, Garcia RL, *et al.* (2001) Occult cancer of the fallopian tube in BRCA-1 germline mutation carriers at prophylactic oophorectomy: a case for recommending hysterectomy at surgical prophylaxis. *Gynecol Oncol* **80**:176–180.

165. Lavie O, Hornreich G, Ben-Arie A, *et al.* (2004) BRCA germline mutations in Jewish women with uterine serous papillary carcinoma. *Gynecol Oncol* **92**:521–524.

166. Goshen R, Chu W, Elit L, *et al.* (2000) Is uterine papillary serous adenocarcinoma a manifestation of the hereditary breast-ovarian cancer syndrome? *Gynecol Oncol* **79**:477–481.

167. Goldman NA, Goldberg GL, Runowicz CD, *et al.* (2002) BRCA mutations in women with concurrent breast carcinoma and uterine papillary serous carcinoma. *Proc Am Soc Clin Oncol* **21**:221a.

168. Karlan BY. (2004) Defining cancer risks for BRCA germline mutation carriers: implications for surgical prophylaxis. *Gynecol Oncol* **92**:519–520.

Chapter 28

Activation of Wnt Signaling in Tumor Development

Zhimin Lu

Departments of Neuro-Oncology & Molecular Genetics
The University of Texas M. D. Anderson Cancer Center
The University of Texas Graduate School of
Biomedical Sciences at Houston, Houston, TX 77030, USA

The Wnt signaling pathway regulates cell growth and differentiation, and plays an important role in embryonic development. Aberrant activation of Wnt signaling results in the transactivation of β-catenin/ T-cell factor/lymphoid enhancer factor 1, which leads to downstream gene transcription and causes tumor formation and development. As a key element in the Wnt pathway, β-catenin can be upregulated in its activity by stabilizing its protein expression via the inhibition of glycogen synthase kinase 3β or via the mutation of Wnt pathway components such as β-catenin, adenomatous polyposis coli, and Axin. β-catenin activity can also be enhanced by increasing its nuclear translocation without altering its protein turnover.

1. INTRODUCTION TO THE Wnt SIGNALING PATHWAY

Wnts constitute a large family of cysteine-rich secreted glycoproteins that are involved in the development of organisms ranging from nematode worms to mammals. The intracellular signaling pathway of Wnt is evolutionarily conserved and regulates cellular proliferation, cell morphology, cell motility, axis formation, and organ development.[1,2] Wnts act in a paracrine fashion by activating diverse signaling cascades inside the target cells. Wnt signaling can be categorized into three major pathways[3]:

1. The canonical or classic pathway activates target genes by stabilizing cytosolic β-catenin. β-Catenin functions in embryonic development, and abnormalities of this pathway lead to tumor formation.

2. The planar cell polarity pathway regulates the polarity of cells through effects on their cytoskeletal organization. This pathway involves RhoA and c-Jun NH_2-terminal kinase (JNK). The main role of this pathway is in the temporal and spatial control of embryonic development, as exemplified by the polar arrangement of cuticular hairs in *Drosophila* or the convergent extension movements in *Xenopus* embryos.

3. The Wnt/Ca^{2+} pathway is initiated by Wnt5a and Wnt11, and involves an increase in intracellular Ca^{2+} and activation of Ca^{2+}-sensitive signaling components. The Wnt/Ca^{2+} pathway can counteract the canonical Wnt pathway and is involved in regulating cell adhesion. It is unclear whether this pathway is conserved in mammals.

Because the involvement of the planar cell polarity and Wnt/Ca^{2+} pathways in tumor development is not yet clarified, this chapter focuses on the canonical pathway.

2. REGULATION OF THE CANONICAL Wnt SIGNALING PATHWAY

The discovery of the common origin of the *Drosophila* segment polarity gene *Wingless* and the murine proto-oncogene *Int-1* laid the groundwork for elucidation of the canonical Wnt cascade (Fig. 1). Human and mouse genomes encode 19 *WNT* and 18 *Wnt* genes, respectively.[4] Wnts are divided into functional classes, based on their ability to induce a secondary body axis in *Xenopus* embryos and to activate certain signaling cascades. Members of the Wnt1 class are inducers of a secondary body axis in *Xenopus* and include Wnt1, Wnt2, Wnt3, Wnt3a, Wnt8, and Wnt8b; this class activates the canonical Wnt/β-catenin pathway. The Wnt5a class cannot induce secondary body axis formation in *Xenopus* and includes Wnt4, Wnt5a, Wnt5b, Wnt6, Wnt7a, and Wnt11; these Wnts activate the planar cell polarity and Wnt/Ca^{2+} pathways.[5] The secretion of Wnt proteins can be promoted by the protein Wntless (Wls), also named Evenness interrupted (Evi), which is a conserved multipass transmembrane protein. That loss of Wls/Evi leads to an accumulation of the Wnt protein in Wnt-producing cells indicates the essential role of Wls/Evi in the secretion of Wnts.[6,7]

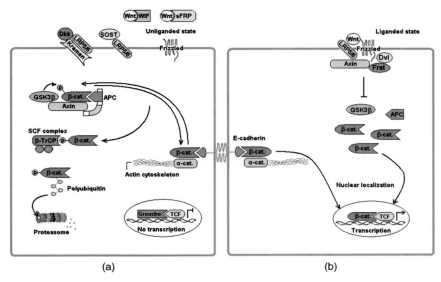

Fig. 1. The canonical Wnt signaling pathway. (a) In the absence of a Wnt signal, free cytoplasmic β-catenin, which is in equilibrium with β-catenin at the cell–cell adherens junctions, is recruited to a "destruction complex" containing APC, Axin, and GSK-3β. GSK-3β phosphorylates β-catenin, allowing the latter to be recognized and polyubiquitinated by an SCF complex containing the F-box protein β-TrCP, leading β-catenin to be degraded by the proteasome. The corepressor Groucho prevents the activation of Tcf-responsive genes in the absence of β-catenin. Wnt signaling can also be blocked by interactions between Wnt and its inhibitory factors, such as WIF and sFRP, or by the interaction between LRP5/6 and the Dkk/Kremen complex or sclerostin. (b) In the presence of Wnt, Frizzled in complex with LRP6 is activated, thus inhibiting GSK-3β activity through mechanisms involving Axin, Frat, and Dvl. β-Catenin accumulates and translocates to the nucleus, where it acts as a coactivator for Tcf/Lef1-responsive genes.

Signaling is initiated when Wnt ligands engage their cognate receptor complex. This complex consists of a serpentine receptor of the Frizzled family, which has seven transmembrane domain receptors, and a member of the Wnt coreceptor family, the low-density lipoprotein (LDL) receptor–related proteins 5 and 6 (LRP5/6).[8] Frizzled receptors have an extracellular cysteine-rich domain (CRD) and an intracellular carboxyl tail.[9] Ten Frizzled receptors have been identified in mammalian cells.[10] Nineteen Wnts and 10 Frizzled receptors can potentially give rise to a large number of different combinations, and the specificity of binding of Wnts to their receptors is not yet completely understood. Other secreted factors, such as WIF, Cerberus, and FrzB, antagonize Wnt binding to Frizzled receptors; whereas Dickkopf (Dkk) and sclerostin (the *SOST* gene product) inhibit

LRP activation. Another membrane-associated protein, Kremen, binds to Dkk and induces endocytosis of LRP, thereby antagonizing Wnt signaling.[3,11] In addition, there are a number of soluble Frizzled-related proteins (secreted Frizzled-related proteins [sFRP] 1–4, Crescent, and Sizzled) that have the CRD of the Frizzled receptors, but not the transmembrane domains. They compete for binding with native Frizzled receptors for Wnts, thus reducing the response to Wnt signaling.[10]

In the absence of Wnts, the tumor suppressors adenomatous polyposis coli (APC) and Axin bind to newly synthesized β-catenin (Fig. 1). These tumor suppressors are scaffolding proteins in the destruction complex that also contains Axin2/Conductin, glycogen synthase kinase 3β (GSK-3β), and Diversin, which links casein kinase 1α (CK1α) to the complex. CK1α and GSK-3β then sequentially phosphorylate a set of conserved serine and threonine residues in the N-terminus of β-catenin. β-Catenin is phosphorylated by CK1α at Ser45, which enables GSK-3β to phosphorylate Thr41, Ser37, and Ser33. Thr41 functions to bridge sequential phosphorylation from Ser45 to Ser37.[12] Phosphorylation of Ser37 and Ser33 recruits the F-box protein β-transducing repeat-containing protein (β-TrCP)-containing E3 ubiquitin ligase, which targets β-catenin for proteasomal degradation.[3,8]

Besides CK1α/GSK-3β–dependent regulation of β-catenin, GSK-3β phosphorylation-independent degradation of β-catenin, which is not regulated by Wnt engagement, has also been reported.[13,14] Siah, a mammalian homolog of *Drosophila* Sina, binds to the ubiquitin-conjugating enzyme through its N-terminal RING domain and forms a complex with Ebi through SIP and Skp1. The complex of SIP, Skp1, and Ebi functions as an ubiquitin ligase. Ebi, an F-box protein, binds to β-catenin independently of the phosphorylation sites recognized by β-TrCP, and recruits β-catenin to the Siah-1/SIP/Skp1 complex for polyubiquitination and subsequent proteasome-mediated degradation.[13] DNA damage and p53 activation induce the expression of Siah, which binds to the C-terminal region of APC and downregulates β-catenin in an APC-dependent manner.[14] Thus, β-catenin degradation can be dependent on or independent of phosphorylation by GSK-3β and mediated by binding to F-box protein β-TrCP or Ebi, respectively.

In the presence of Wnts, the kinase activity of the destruction complex is inhibited by an incompletely understood mechanism involving the direct interaction of Axin with LRP5/6 and the actions of the Axin-binding molecule Dishevelled (Dvl). Wnt treatment induces the rapid CK1γ-mediated phosphorylation of LRP6, which in turn promotes the recruitment of the scaffold protein Axin.[15] As a consequence, the GSK-3β–dependent phosphorylation and the ubiquitination-required degradation of β-catenin are inhibited. The GSK-3β–regulated degradation of β-catenin can be counteracted by phosphorylation at Thr393 by CK2[16] and at Ser675 by cyclic AMP-dependent protein kinase (protein kinase A).[17] However, it is contradictory to another report that the phosphorylation of Ser675 by protein kinase A increases β-catenin transactivation in a manner unrelated to GSK-3β.[18]

Stabilized β-catenin protein accumulates and travels into the nucleus (Fig. 1), where it engages the N-termini of DNA-binding proteins of the T-cell factor (Tcf)/lymphoid enhancer factor (Lef) family and complexes with various activators, such as the histone acetyl transferase CBP/p300 and the chromatin-remodeling SWI/SNF complex.[3,19] The vertebrate genome encodes four highly similar Tcf/Lef proteins: Tcf1, Lef1, Tcf3, and Tcf4. In the absence of a Wnt signal, Tcf/Lef1 proteins repress target genes by directly associating with corepressors such as Groucho. The interaction with β-catenin transiently converts Tcf/Lef1 into transcriptional activators.[3,20] *Drosophila* genetics has recently identified two additional nuclear components, Pygopus and Bcl9/Legless, which are conserved in vertebrates. Pygopus is essential for transcriptional activation of Tcf/Lef1 target genes, whereas Bcl9 seems to bridge Pygopus to Tcf-bound β-catenin.[8,21] Tcf/Lef1 proteins contain DNA-binding high-mobility group (HMG) domains, which have a high affinity for the DNA sequence (A/T)(A/T)CAA(A/T)GG. Tcf/Lef1 activated by β-catenin stimulates the expression of various genes, including *c-myc, cyclin D1, fra1, c-jun, peroxisome proliferator-activated receptor δ, matrilysin, CD44*, and *urokinase-type plasminogen activator receptor*, all of which contain Tcf/Lef1-binding sites in their promoters.[2,20]

As one of the target genes of Wnt signaling, *c-Jun*, a member of the AP1 transcription factor family, appears to play an important role in *APC* mutation-induced formation of intestinal tumors in mice. c-Jun phosphorylated at the N-terminus by JNK interacts with Tcf4 to form a ternary complex containing c-Jun, Tcf4, and β-catenin. c-Jun and TCF4 cooperatively bind to and activate the *c-jun* promoter. Genetic abrogation of c-Jun N-terminal phosphorylation or gut-specific conditional *c-jun* inactivation reduces the number and size of tumors, and prolongs the life span of mice with the *APC* mutation.[15]

In the canonical pathway, the mutation of a component responsible for the degradation of β-catenin leads to the stabilization of β-catenin, its abnormal accumulation in the nucleus, and the activation of Tcf/Lef1-mediated gene expression. Activating mutations of Wnt pathway components — such as loss-of-function mutations of *APC* or gain-of-function mutations of *CTNNB1* (which encodes β-catenin); and less frequently mutations in *AXIN*, *AXIN2/conductin*, and *hTRCP1* (which encodes βTrCP1) — have been found in numerous human cancers (Table 1). The most frequently mutated of these tumors are of gastrointestinal origin.[21,22] The fact that only one of these genes is usually mutated in a given tumor sample indicates that defects in this pathway are obligatory, and that mutation of a single component is sufficient for tumorigenesis. For instance, colon tumors with mutations in *APC* generally have wild-type *CTNNB1* and vice versa (i.e. tumors with mutations in *CTNNB1* have wild-type *APC*).[21] In addition to the enhanced stability and trans-activation of β-catenin induced by genetic defect, degradation-independent transactivation of β-catenin is also involved in tumor development.[22] In the following section, some components of the Wnt pathway that are involved in tumor development are discussed.

3. β-CATENIN

3.1. *β-Catenin Structure*

Originally identified as a component of cell–cell adhesion, β-catenin interacts with the cytoplasmic domain of cadherin and links cadherin to α-catenin, which in turn mediates the anchorage of the cadherin complex to the cortical actin cytoskeleton (Fig. 1).[23–25] Genetic and embryologic studies have revealed that β-catenin is also a component of the Wnt signaling pathway and exhibits signaling functions.[26] The primary structure of β-catenin is characterized by a central region that contains 12 repeats of 42 amino acids, known as armadillo repeats (R1–R12) (Fig. 2). The armadillo repeats of β-catenin interact with cadherin and α-catenin on plasma membranes, and associate with Axin and APC in the cytoplasm. α-Catenin binds to β-catenin at the junction of the N-terminus and R1; whereas the central armadillo repeats bind to Axin and APC, and are required

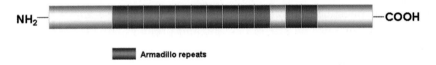

Fig. 2.   Schematic structure of β-catenin.

for both cadherin binding and localization to the adherens junction.[27,28] In the nucleus, β-catenin binds Tcf/Lef1 through R3–R10.

In addition to the Tcf/Lef1 binding domain, two other domains are important for Wnt signaling, namely, R1 and R11C (C-terminus).[27,28] It was shown that R1–R4 are necessary and sufficient for Bcl9/Legless binding,[28,29] and that R11C is important for the binding of other cofactors which contribute to transcription activation. These cofactors include TATA-binding protein (TBP), Brahma/Brg1 (a component of the SWI/SNF and Rsc chromatin-remodeling complexes), CBP/p300, MED12 (a component of the Mediator complex that links transcriptional regulators to RNA polymerase II), and Hyrax/Parafibromin (the *Drosophila* and human homologs of yeast Cdc73p, respectively).[30–35] Cdc73p is a component of the polymerase-associated factor 1 complex. Mutations in or deletions of the R1 and R11C domains result in reduced signaling activity of β-catenin.[28] Thus, β-catenin has different binding partners in distinct subcellular fractions, which regulate the cellular functions of β-catenin.

3.2. β-Catenin Mouse Models

The fact that β-catenin knockout is embryonically lethal and β-catenin null-mutant embryos have defects in the development of the embryonic ectoderm at day 7 postcoitus demonstrates that β-catenin is specifically required in the ectodermal cell layer, although it is expressed rather ubiquitously.[36] When β-catenin is conditionally mutated in the epidermis and hair follicles during embryogenesis, the formation of placodes that generate hair follicles is blocked. If β-catenin is deleted after hair follicles have formed, skin stem cells fail to differentiate into follicular keratinocytes and instead adopt an epidermal fate.[37]

The contribution of activated β-catenin to tumorigenesis has also been demonstrated in a number of transgenic mouse models. A deletion mutant of exon 3 encoding the phosphorylation sites of β-catenin by GSK-3β and a truncated mutant β-catenin that lacks N-terminal 131 amino acids lead to intestinal tumorigenesis,[38,39] demonstrating that activation of the Wnt signaling pathway can cause intestinal and colonic tumors. Mice expressing stable β-catenin controlled by an epidermal promoter develop human-like epithelioid cysts, trichofolliculomas, and eventually pilomatricomas.[40] Stabilized β-catenin is able to induce high-grade prostate intraepithelial neoplasias reminiscent of early human prostate cancer.[41] Compared with control mice, transgenic mice in which stabilized β-catenin is expressed in mesenchymal cells develop aggressive fibromatoses and hyperplastic gastrointestinal polyps after 3 months of transgene induction and heal with hyperplastic cutaneous wounds.[42] The expression of a stabilized and transcriptionally active form of β-catenin, lacking the N-terminal 89 amino acids in the mammary gland, induces multiple aggressive adenocarcinomas early in the life of mice.[43] Taken together, these findings show that blocking the degradation of β-catenin is sufficient for tumorigenesis in multiple organs and tissues in mice.

3.3. β-Catenin Mutation in Human Cancer

Mutations in the *CTNNB1* gene that affect specific Ser and Thr residues and their adjacent amino acids, which are essential for the targeted degradation of β-catenin, are found in a wide variety of human cancers, including colon cancer, desmoid tumor, gastric cancer, hepatocarcinoma, medulloblastoma, melanoma, ovarian cancer, pancreatic cancer, and prostate cancer (Table 1).[2] For instance, mutations in the β-catenin gene (*CTNNB1*) are present in approximately 10% of colorectal cancers and 50% of colorectal

Table 1 Mutations of Wnt Signaling Components in Selected Human Cancers (more detailed information and more compressive references are available in Ref. 21)

Cancer Type	Gene	Mutation Frequency
Colorectal cancer	*APC*	80%
	CTNNB1	10%
Small intestinal adenocarcinoma	*N/A*	48%
Fundic gland polyps (gastric)	*CTNNB1*	64%–91%
Gastric carcinoma	*CTNNB1*	8%–26%
Gastric cancer (intestinal-like)	*CTNNB1*	27%
Gastric adenoma (without associated adenocarcinoma)	*APC*	76%
Gastrointestinal carcinoid tumor	*CTNNB1*	38%
Esophageal adenocarcinoma	*APC* and *CTNNB1*	2%
Juvenile nasopharyngeal angiofibroma	*CTNNB1*	75%
Melanoma	*CTNNB1*	2%–22%
	APC	3%–7%
Pilomatricoma	*CTNNB1*	2%–100%
Lung adenocarcinoma	*CTNNB1*	3% (3 of 93)
Ovarian carcinoma (endometrioid type)	*CTNNB1*	16%–54%
	APC	2%
	AXIN1	4%
	AXIN2	2%

(Continued)

Table 1   *(Continued)*

Cancer Type	Gene	Mutation Frequency
Ovarian carcinoma (mucinous type)	*CTNNB1*	14%
Uterine endometrial cancer	*CTNNB1*	10%–45%
Breast fibromatosis	*CTNNB1*	45%
	APC	33%
Prostate cancer	*CTNNB1*	5%–9%
	APC	14%
	HTRCP1	9%
Thyroid carcinoma	*CTNNB1*	46%
Hepatoblastoma	*CTNNB1*	13%–89%
	APC	0%–69%
	AXIN1	7%
Hepatocellular carcinoma	*CTNNB1*	12%–34%
	AXIN1	8%–10%
	AXIN2	3%
Hepatocellular carcinoma associated with hepatitis C	*CTNNB1*	41%
Medulloblastoma	*CTNNB1*	4%–18%
	APC	2%–4%
	AXIN1	4%–12%
Desmoid tumor	*CTNNB1*	25%–58%
	APC	21%–50%
Wilms' tumor (kidney)	*CTNNB1*	14%–15%
Pancreatic cancer (nonductal solid pseudopapillary)	*CTNNB1*	83%–90%
Pancreatic cancer (nonductal acinar cell)	*CTNNB1*	6%
	APC	18%
Pancreatoblastoma	*CTNNB1*	55%
Synovial sarcoma	*CTNNB1*	8% (4 of 49)
	APC	8% (4 of 49)

cancers without *APC* mutations; it has been suggested that these mutations occur mainly in microsatellite-unstable tumors.[44]

Most mutations in cancers occur in or around exon 3 of *CTNNB1,* abrogating the phosphorylation-dependent interaction of β-catenin with β-TrCP and thereby stabilizing β-catenin. The β-catenin mutations frequently occur at Asp32, Ser33, Gly34, Ser37, Thr41, and Ser45.[2] As has already been described, Ser45 is a phosphorylation site of CK1α; and Ser33, Ser37, and Thr41 are phosphorylation sites of GSK-3β. Asp32 and Gly34 are necessary for the interaction of β-catenin with β-TrCP. In contrast to those mutations in colorectal cancers, *CTNNB1* mutations in preneoplastic lesions in the colon are spread over the entire exon 3. It has been speculated that activating mutations of *CTNNB1* are selected during colon carcinogenesis.[45]

3.4. *Mutation-Independent β-Catenin Activation in Human Cancer*

The mutation of Wnt pathway proteins that alters the stability of β-catenin is not the only factor contributing to β-catenin activation. For instance, in 12 (60%) of 20 endometrial cancers, β-catenin was found to accumulate in the nucleus, which is a hallmark of β-catenin activation; whereas there were only 2 instances of mutations in the *CTNNB1* gene.[46] Similarly, only 1 of 65 primary melanomas had detectable *CTNNB1* mutations, whereas a third of the cases displayed nuclear accumulation of β-catenin.[47] Moreover, nearly 50% of hepatocellular carcinomas (in which the *APC* gene is rarely mutated) revealed nuclear accumulation of the β-catenin protein, and genetic alterations in *CTNNB1* were detected only in 16%–26% of the tumors.[48–51] Clearly, more than one mechanism regulates the activity of β-catenin.[22]

Wnt overexpression in human cancers is rare. Nevertheless, β-catenin/Tcf/Lef1 transcriptional activity, which can function independently of Wnt ligand engagement, can be enhanced by growth factors such as epidermal growth factor (EGF), hepatocyte growth factor/scatter factor, insulin-like growth factors I and II, and insulin.[52–55] During insulin signaling, phosphatidylinositol (PI) 3-kinase–activated PKB/AKT phosphorylates GSK-3β at Ser9, thus inactivating GSK-3β and augmenting β-catenin/Tcf/Lef1 transcriptional activity.[56,57]

In addition to its role in Wnt/Wingless signal transduction, β-catenin is an important structural component of E-cadherin–mediated and Ca^{2+}-dependent cell–cell adherens junctions. β-Catenin links E-cadherin to α-catenin, which in turn binds to vinculin, α-actinin, ZO-1, and actin; and thus connects E-cadherin to the actin cytoskeleton.[24,25,58] In response to EGF stimulation, β-catenin dissociates from the E-cadherin adherens complex and accumulates in the nucleus, where it forms an active complex with Tcf/Lef1.[53] Strikingly, this process does not depend on inhibition of GSK-3β because there is no change in the phosphorylation of β-catenin at the GSK-3β sites, the phosphorylation of GSK-3β itself at the site that inhibits its kinase activity, or the half-life of β-catenin in response to EGF treatment.[53]

Intriguingly, caveolin-1 — which negatively regulates EGF receptor, platelet-derived growth factor receptor (PDGFR), Neu/ErbB2, Ha-Ras, c-Src, endothelial nitric oxide synthase (eNOS), and PI 3′-kinase[59,60] — plays an essential role in regulating EGF-mediated transactivation of β-catenin. Caveolin-1 is a principal component of caveolae, surface structures that can endocytose. Short-term EGF treatment induces

E-cadherin endocytosis in a manner that depends on the integrity of caveolae, and this internalization contributes to the dissociation of the E-cadherin/β-catenin complex and the release of β-catenin from cell–cell adherens junctions. During prolonged EGF treatment, the expression of caveolin-1 is downregulated — thereby reducing the expression of the E-cadherin protein — probably through upregulation of the E-cadherin transcriptional repressor Snail, which causes E-cadherin downregulation and epithelial-to-mesenchymal transition.[53,61]

In the absence of E-cadherin, β-catenin is not sequestered by the adherens protein complex that assembles at cell–cell junctions. These factors may help maintain an increased abundance of β-catenin, which is localized in the cytoplasm and nucleus in EGF-treated cells. Downregulation of E-cadherin, higher activity of signaling molecules due to depletion of caveolin-1, and increased β-catenin transactivation contribute to EGF-induced tumor cell invasion.[53] In line with the requirement for downregulation of caveolin-1 in β-catenin transactivation, overexpression of caveolin-1 blocks EGF-induced β-catenin/Tcf/Lef1 transcriptional activity. The fact that β-catenin is recruited to caveolar membranes when caveolin-1 is overexpressed[62] supports the role of caveolin-1 in this process.

In addition to the increased abundance of β-catenin in the nucleus after EGF stimulation, the release of signaling proteins such as Ras — which is sequestered in an inactive state in the caveolae — might also contribute to β-catenin activation as well as the reduced transcription of caveolin-1 and E-cadherin. For instance, activation of the ERK/MAP kinase pathway, which is commonly constitutively activated in tumor cells, leads to downregulation of caveolin-1 transcription.[63] In addition, the coexpression of catalytically inactive GSK-3β (R85GSK-3β) and activated V12Ras synergistically enhances Tcf/Lef1 transcriptional activity.[52] Therefore, β-catenin/Tcf/Lef1 transcriptional activity is regulated through at least two distinct mechanisms: the regulation of β-catenin stability by β-TrCP in response to Wnt/Wingless signaling or PKB/AKT activation, or by Ebi in response to DNA damage and p53 activation; and the regulation of β-catenin cellular distribution mediated through EGF signaling by downregulation of caveolin-1.

EGF receptor overexpression has been reported in many human tumors (including lung, colon, breast, prostate, brain, head and neck, thyroid, ovarian, and bladder tumors), gliomas, and renal carcinomas.[64–69] Detailed mechanistic studies to decipher how caveolin-1 regulates β-catenin/Tcf/Lef1 transactivation, including the intracellular translocation and possible posttranslational modification of themselves or of a β-catenin–binding regulator, should further illuminate the phenomenon of EGF receptor/ β-catenin–related tumor development.

4. APC

4.1. *APC Structure and Function in β-Catenin Regulation*

APC has a broad spectrum of functions, ranging from control of the Wnt signal transduction pathway to cell adhesion, cell migration, apoptosis, and chromosomal segregation at mitosis.[70] The *APC* gene encodes a 300-kDa multifunctional protein containing an N-terminal oligomerization domain, followed by seven repeats of an armadillo motif, three successive 15-amino-acid repeats, seven related but distinct 20-amino-acid repeats integrated with three SAMP repeats, a basic amino acid cluster region, and the C-terminal S/TXV motif (Fig. 3). APC interacts with a protein phosphatase 2A regulatory subunit, B56, the expression of which reduces the abundance of β-catenin as well as inhibits the transcription of β-catenin target genes in mammalian cells and *Xenopus* embryo explants.[71]

The armadillo motif of APC binds to Asef, a GDP/GTP exchange protein for the small G-protein Rac,[72] and KIF3, which binds to kinesin and regulates vesicle transport on microtubules.[73] Both the 15-amino-acid and 20-amino-acid repeats are able to bind independently to β-catenin.[70] The regulator of the G-protein signaling (RGS) domain of Axin binds to the three SAMP repeats of APC between the third and fourth 20-amino-acid repeats, the fourth and fifth 20-amino-acid repeats, and after repeat 7; and it promotes GSK-3β–dependent phosphorylation and degradation of β-catenin.[74] In the complex, GSK-3β bound to Axin efficiently phosphorylates APC, which enhances the binding of β-catenin to APC.[75,76] Expression of wild-type APC or the fragment of APC containing the 20-amino-acid repeats, but not an APC fragment with mutated β-catenin- or Axin-binding sites, reduces β-catenin levels.[70,77] Thus, APC requires interaction with

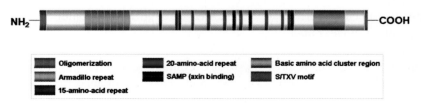

Fig. 3.   Schematic structure of APC.

Axin and β-catenin to downregulate β-catenin. The basic amino acid cluster region of APC binds to microtubules[78] and the microtubule-binding protein EB1.[79] The S/TXV motif at the C-terminus of APC interacts with the PDZ domain-containing DLG, the human homolog of the *Drosophila* discs large tumor suppressor protein.[80]

4.2. *APC mutation in Human Cancer*

Mutations of *APC* are frequently identified in familial adenomatous polyposis (FAP) and colorectal cancer, and less frequently in other cancers (Table 1).[70] Inactivation of the APC function seems to underlie both tumor initiation and tumor promotion in the large intestine. Germline mutations in the *APC* gene result in FAP,[70] one of the principal hereditary predispositions to colorectal cancer. Somatic *APC* mutations are also found in about 80% of sporadic colorectal tumors.[81] Most *APC* mutations result in truncated proteins that lack all Axin-binding motifs and a variable number of the 20-amino-acid repeats, and remove most of the β-catenin regulatory domain, thereby stabilizing β-catenin.[26]

In FAP, germline mutations are inherited in one allele of *APC* and, after loss of heterozygosity, result in the development of hundreds of polyps in the colon. Most of the mutations are nonsense or truncating frameshift mutations occurring at two hot spots (around codons 1061 and 1309) at the N-terminal part; by contrast, most somatic mutations are in a mutation cluster region between codons 1286 and 1513 that corresponds to the β-catenin/Axin-binding domain.[3,26] Selective pressure is directed against the presence of Axin-binding sites because the presence of these sites is critical to the regulation of β-catenin level by APC. Furthermore, the remaining N-terminal truncated form of APC may affect cell migration by activating Asef, thereby leading to metastasis.[26]

Methylation of the promoter region of the *APC* gene constitutes an alternative mechanism for gene inactivation in colon cancer and other tumors of the gastrointestinal tract. The hypermethylated *APC* promoter in 18% of primary sporadic colorectal carcinomas and adenoma is associated with loss of transcription of this allele. This hypermethylation occurs rarely in FAP samples and more frequently in sporadic colorectal cancers at the wild-type allele. As with *APC* mutation, aberrant *APC* methylation occurs early in colorectal carcinogenesis. Thus, hypermethylation of *APC* provides an important mechanism for impairing APC function and underscores the importance of the APC pathway in gastrointestinal tumorigenesis.[82,83]

5. AXIN

5.1. *Axin Structure and Function in β-Catenin Regulation*

Axin and its homolog (Axin2 in humans, conductin in mice, and Axil in rats) consist of an N-terminal RGS domain, a central region, and a C-terminal DIX domain, which is related to a segment of Dvl and mediates the dimerization of Axin (Fig. 4).[3] In addition to the DIX domain, Axin contains D (amino acids 601–633) and I (amino acids 667–751) domains for homodimerization; defects in homodimer formation have no effect on β-catenin signaling, but fail to activate JNK.[84] Axin and conductin share an overall 45% identity in amino acids. Axin appears to be ubiquitously expressed during embryonic development and in adult tissues, whereas conductin shows a more restricted expression pattern.[85,86]

Axin binds to various components of the Wnt signaling pathway, and functions as a scaffold protein for phosphorylation of β-catenin by approximating all needed components including GSK-3β, CK1α, APC, and β-catenin itself. The RGS domain of Axin binds to APC. GSK-3β, β-catenin, and CK1α interact with different sites of the central region of Axin. Dvl binds to the C-terminal region of Axin, including the DIX domain,[26] and this interaction is required for Axin-promoted β-catenin degradation.[87] Dvl also binds to CK1ε, which mediates Wnt3a-dependent phosphorylation of Dvl. Phosphorylated Dvl has a high affinity for Frat, which binds to and inhibits GSK-3β. Therefore, when Wnt binds its receptor, Frat bound to phosphorylated Dvl may inhibit GSK-3β — which has interacted with Axin — to phosphorylate β-catenin.[88,89] Axin significantly enhances β-catenin phosphorylation by GSK-3β *in vitro*, and GSK-3β inefficiently phosphorylates β-catenin in the absence of Axin.[90] When overexpressed, Axin promotes the degradation of cytosolic β-catenin,[74,86,90,91] prevents the Wnt-induced accumulation of β-catenin, and blocks the transcription of a Tcf-dependent reporter gene.[92] Thus, Axin, functioning as a negative regulator, plays an important role in the Wnt signaling pathway.

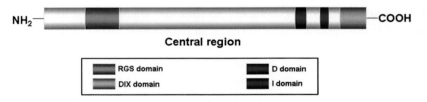

Fig. 4.   Schematic structure of Axin.

5.2. *Axin Mutations in Cancers*

The function of Axin as a tumor suppressor that downregulates β-catenin signaling was revealed by its biallelic inactivation in human hepatocellular cancers and cell lines.[49] Like the mutually exclusive relationship between mutations of *APC* and the β-catenin gene, those hepatocellular cancers containing Axin gene (*AXIN1*) mutations lack activating mutations in *CTNNB1* and *APC*. All of these mutations have been predicted to truncate the Axin protein in a manner that eliminates the β-catenin–binding sites.[49] The *AXIN1* mutation corresponding to the GSK-3β–binding site[49] and large deletions of *AXIN1*[93] have been detected in medulloblastomas. Infrequent *AXIN1* mutations have also been found in oral squamous cell carcinomas[94] and pituitary adenomas.[95]

Like overexpression of Axin, overexpression of Axin2/Conductin promotes β-catenin degradation. Nevertheless, Axin2/Conductin is upregulated in colorectal carcinomas, liver tumors, and ovarian endometrioid adenocarcinomas, which exhibit activated Wnt signaling.[96,97] Moreover, the promoter of the *Axin2/Conductin* gene contains functional Tcf-binding sites, and Axin2/Conductin expression is upregulated by activation of the Wnt pathway.[96–99] These results suggest that Axin2/Conductin provides negative feedback in response to Wnt signaling, whereas Axin is a constitutive negative regulator of β-catenin activity.

6. β-TrCP

The β-TrCP family proteins, which are highly conserved, include *Drosophila melanogaster* Slimb72, *Xenopus laevis* β-TrCP, and mammalian β-TrCP1 (also termed FBW1a or FWD1) and β-TrCP2 (also known as FBW1b or HOS).[100] β-TrCPs are members of about 70 F-box family proteins identified in humans, and function together with other components of the SCF complex as a ubiquitin E3 ligase.[100] The SCF complex consists of the invariable components Skp1, Cul1, and Rbx1; and a variable component, which is an F-box protein that binds to Skp1 through its F-box motif and is responsible for substrate recognition.[101] In combination with the core complex of Skp1, Cul1, and Rbx1 as well as associated E2 proteins, F-box proteins provide the basis for multiple substrate-specific ubiquitylation pathways.

Many proteins, such as β-catenin, IκBα, IκBβ, IκBϵ, Emi1, Vpu, Cdc25A, Wee1A, ATF4, NF-κB p105, NF-κB p100, and hDlg are targets of β-TrCP–mediated degradation.[101] Nuclear β-catenin, IκBα, IκBβ, and

Emi1 are accumulated in β-TrCP1$^{-/-}$ cells.[102,103] Full-length β-TrCP1 or a dominant-negative F-box deletion mutant of β-TrCP1, which inhibits β-catenin ubiquitination and proteasomal degradation, was transgenetically expressed in the intestine, liver, and kidney; 46% of the transgenic mice developed intestinal adenomas (10 of 35), hepatic tumors (4 of 35), or urothelial tumors (2 of 35).[104] The stabilization of β-catenin in tumors generated by transgenic expression of wild-type β-TrCP1 is likely due to the sequestration of a limited amount of Skp1 or other SCF components by overexpressed β-TrCP1; thus, overexpressed wild-type β-TrCP1 can also dominant-negatively function to inhibit the SCF complex and cause β-catenin stabilization instead of degradation.[104,105] Consistent with this observation, 25 (56%) of 45 primary human colorectal carcinomas and 23 (100%) of 23 primary hepatoblastomas had increased mRNA and protein levels of β-TrCP1, and increased β-TrCP1 levels were significantly associated with β-catenin activation and decreased apoptosis.[106,107]

In addition to aberrant expression, β-*TrCP* gene mutations were also detected in human tumors. A nucleotide substitution in β-*TrCP2* that led to F462S amino acid substitution in the seventh WD repeat domain was identified in a gastric cancer cell line (1 of 17 cell lines and primary cancers). F462 was conserved among β-TrCP derived from human, mouse, *Xenopus laevis*, and *Drosophila melanogaster*; WD repeats are the substrate-recognition motif for binding β-catenin.[108] A study of human prostate cancer cell lines, xenografts, and primary tumors revealed deletions in β-*TrCP1* transcripts in 2 of 22 samples (one cell line and one xenograft); in that study, alterations in the *CTNNB1*, *APC*, and *hTRCP1* genes were mutually exclusive.[109] These findings are consistent with the equivalent effects of the genetic alterations of these Wnt components on β-catenin stability, and imply an instrumental role of β-catenin signaling in the development of a significant fraction of prostate cancers.

7. CONCLUSION

Active Wnt signaling causes cancer. Tumor promotion by this pathway can proceed through a number of genetic defects. Wnt signaling activation, which is not caused by genetic defects of components of this pathway, can result from the integration of activation of other signaling molecules such as growth factor receptors. The manifestation of cancer by aberrant Wnt signaling most likely results from inappropriate gene activation mediated by stabilized β-catenin. The elucidation of the molecular and cellular mechanisms underlying β-catenin–driven tumorigenesis will identify the critical cellular functions affected as well as the specifically associated gene expression profiles that contribute to the altered cellular functions leading to tumorigenesis and tumor promotion. This knowledge will open the way for targeted preventive and therapeutic interventions for human cancer.

ACKNOWLEDGMENTS

I thank Theresa Willis for her critical reading of the manuscript, and Dexing Fang and Yanhua Zheng for preparation of the figures. This work was supported by the National Cancer Institute (1R01CA109035-01A1), the Pediatric Brain Tumor Foundation, The Charlotte Geyer Foundation, and an Institutional Research Grant from the M. D. Anderson Cancer Center.

REFERENCES

1. Wodarz A, Nusse R. (1998) Mechanisms of Wnt signaling in development. *Annu Rev Cell Dev Biol* **14**:59–88.
2. Polakis P. (2000) Wnt signaling and cancer. *Genes Dev* **14**:1837–1851.
3. Lustig B, Behrens J. (2003) The Wnt signaling pathway and its role in tumor development. *J Cancer Res Clin Oncol* **129**:199–221.
4. Miller JR. (2002) The Wnts. *Genome Biol* **3**:REVIEWS3001.
5. Westendorf JJ, Kahler RA, Schroeder TM. (2004) Wnt signaling in osteoblasts and bone diseases. *Gene* **341**:19–39.
6. Banziger C, Soldini D, Schutt C, *et al.* (2006) Wntless, a conserved membrane protein dedicated to the secretion of Wnt proteins from signaling cells. *Cell* **125**:509–522.
7. Bartscherer K, Pelte N, Ingelfinger D, Boutros M. (2006) Secretion of Wnt ligands requires Evi, a conserved transmembrane protein. *Cell* **125**: 523–533.
8. Reya T, Clevers H. (2005) Wnt signalling in stem cells and cancer. *Nature* **434**:843–850.
9. Hsieh JC, Rattner A, Smallwood PM, Nathans J. (1999) Biochemical characterization of Wnt–Frizzled interactions using a soluble, biologically active vertebrate Wnt protein. *Proc Natl Acad Sci USA* **96**:3546–3551.
10. Widelitz R. (2005) Wnt signaling through canonical and non-canonical pathways: recent progress. *Growth Factors* **23**:111–116.
11. Krishnan V, Bryant HU, Macdougald OA. (2006) Regulation of bone mass by Wnt signaling. *J Clin Invest* **116**:1202–1209.
12. Wu G, He X. (2006) Threonine 41 in beta-catenin serves as a key phosphorylation relay residue in beta-catenin degradation. *Biochemistry* **45**: 5319–5323.

13. Matsuzawa SI, Reed JC. (2001) Siah-1, SIP, and Ebi collaborate in a novel pathway for beta-catenin degradation linked to p53 responses. *Mol Cell* **7**:915–926.

14. Liu J, Stevens J, Rote CA, *et al.* (2001) Siah-1 mediates a novel beta-catenin degradation pathway linking p53 to the adenomatous polyposis coli protein. *Mol Cell* **7**:927–936.

15. Davidson G, Wu W, Shen J, *et al.* (2005) Casein kinase 1 gamma couples Wnt receptor activation to cytoplasmic signal transduction. *Nature* **438**:867–872.

16. Song DH, Dominguez I, Mizuno J, *et al.* (2003) CK2 phosphorylation of the armadillo repeat region of beta-catenin potentiates Wnt signaling. *J Biol Chem* **278**:24018–24025.

17. Hino S, Tanji C, Nakayama KI, Kikuchi A. (2005) Phosphorylation of beta-catenin by cyclic AMP-dependent protein kinase stabilizes beta-catenin through inhibition of its ubiquitination. *Mol Cell Biol* **25**:9063–9072.

18. Taurin S, Sandbo N, Qin Y, *et al.* (2006) Phosphorylation of beta-catenin by cyclic AMP-dependent protein kinase. *J Biol Chem* **281**:9971–9976.

19. Eastman Q, Grosschedl R. (1999) Regulation of LEF-1/TCF transcription factors by Wnt and other signals. *Curr Opin Cell Biol* **11**:233–240.

20. Hurlstone A, Clevers H. (2002) T-cell factors: turn-ons and turn-offs. *EMBO J* **21**:2303–2311.

21. Giles RH, van Es JH, Clevers H. (2003) Caught up in a Wnt storm: Wnt signaling in cancer. *Biochim Biophys Acta* **1653**:1–24.

22. Lu Z, Hunter T. (2004) Wnt-independent beta-catenin transactivation in tumor development. *Cell Cycle* **3**:571–573.

23. Rimm DL, Koslov ER, Kebriaei P, *et al.* (1995) Alpha 1(E)-catenin is an actin-binding and -bundling protein mediating the attachment of F-actin to the membrane adhesion complex. *Proc Natl Acad Sci USA* **92**:8813–8817.

24. Vasioukhin V, Fuchs E. (2001) Actin dynamics and cell–cell adhesion in epithelia. *Curr Opin Cell Biol* **13**:76–84.

25. Nagafuchi A. (2001) Molecular architecture of adherens junctions. *Curr Opin Cell Biol* **13**:600–603.

26. Kikuchi A. (2003) Tumor formation by genetic mutations in the components of the Wnt signaling pathway. *Cancer Sci* **94**:225–229.

27. Orsulic S, Peifer M. (1996) An *in vivo* structure–function study of armadillo, the beta-catenin homologue, reveals both separate and overlapping regions of the protein required for cell adhesion and for wingless signaling. *J Cell Biol* **134**:1283–1300.

28. Stadeli R, Hoffmans R, Basler K. (2006) Transcription under the control of nuclear Arm/beta-catenin. *Curr Biol* **16**:R378–R385.

29. Kramps T, Peter O, Brunner E, *et al.* (2002) Wnt/Wingless signaling requires BCL9/Legless-mediated recruitment of Pygopus to the nuclear beta-catenin–TCF complex. *Cell* **109**:47–60.

30. Barker N, Hurlstone A, Musisi H, *et al.* (2001) The chromatin remodelling factor Brg-1 interacts with beta-catenin to promote target gene activation. *EMBO J* **20**:4935–4943.

31. Hecht A, Litterst CM, Huber O, Kemler R. (1999) Functional characterization of multiple transactivating elements in beta-catenin, some of which interact with the TATA-binding protein *in vitro*. *J Biol Chem* **274**:18017–18025.

32. Hecht A, Vleminckx K, Stemmler MP, *et al.* (2000) The p300/CBP acetyltransferases function as transcriptional coactivators of beta-catenin in vertebrates. *EMBO J* **19**:1839–1850.

33. Kim S, Xu X, Hecht A, Boyer TG. (2006) Mediator is a transducer of Wnt/beta-catenin signaling. *J Biol Chem* **281**:14066–14075.

34. Mosimann C, Hausmann G, Basler K. (2006) Parafibromin/Hyrax activates Wnt/Wg target gene transcription by direct association with beta-catenin/Armadillo. *Cell* **125**:327–341.

35. Takemaru KI, Moon RT. (2000) The transcriptional coactivator CBP interacts with beta-catenin to activate gene expression. *J Cell Biol* **149**:249–254.

36. Haegel H, Larue L, Ohsugi M, *et al.* (1995) Lack of beta-catenin affects mouse development at gastrulation. *Development* **121**:3529–3537.

37. Huelsken J, Vogel R, Erdmann B, *et al.* (2001) beta-Catenin controls hair follicle morphogenesis and stem cell differentiation in the skin. *Cell* **105**:533–545.

38. Harada N, Tamai Y, Ishikawa T, *et al.* (1999) Intestinal polyposis in mice with a dominant stable mutation of the beta-catenin gene. *EMBO J* **18**:5931–5942.

39. Romagnolo B, Berrebi D, Saadi-Keddoucci S, *et al.* (1999) Intestinal dysplasia and adenoma in transgenic mice after overexpression of an activated beta-catenin. *Cancer Res* **59**:3875–3879.

40. Gat U, DasGupta R, Degenstein L, Fuchs E. (1998) *De Novo* hair follicle morphogenesis and hair tumors in mice expressing a truncated beta-catenin in skin. *Cell* **95**:605–614.

41. Gounari F, Signoretti S, Bronson R, *et al.* (2002) Stabilization of beta-catenin induces lesions reminiscent of prostatic intraepithelial neoplasia, but terminal squamous transdifferentiation of other secretory epithelia. *Oncogene* **21**:4099–4107.

42. Cheon SS, Cheah AY, Turley S, *et al.* (2002) beta-Catenin stabilization dysregulates mesenchymal cell proliferation, motility, and invasiveness and causes aggressive fibromatosis and hyperplastic cutaneous wounds. *Proc Natl Acad Sci USA* **99**:6973–6978.

43. Imbert A, Eelkema R, Jordan S, *et al.* (2001) Delta N89 beta-catenin induces precocious development, differentiation, and neoplasia in mammary gland. *J Cell Biol* **153**:555–568.

44. Kitaeva MN, Grogan L, Williams JP, *et al.* (1997) Mutations in beta-catenin are uncommon in colorectal cancer occurring in occasional replication error-positive tumors. *Cancer Res* **57**:4478–4481.

45. Yamada Y, Oyama T, Hirose Y, *et al.* (2003) beta-Catenin mutation is selected during malignant transformation in colon carcinogenesis. *Carcinogenesis* **24**:91–97.

46. Ashihara K, Saito T, Mizumoto H, *et al.* (2002) Mutation of beta-catenin gene in endometrial cancer but not in associated hyperplasia. *Med Electron Microsc* **35**:9–15.

47. Rimm DL, Caca K, Hu G, *et al.* (1999) Frequent nuclear/cytoplasmic localization of beta-catenin without exon 3 mutations in malignant melanoma. *Am J Pathol* **154**:325–329.

48. de La Coste A, Romagnolo B, Billuart P, *et al.* (1998) Somatic mutations of the beta-catenin gene are frequent in mouse and human hepatocellular carcinomas. *Proc Natl Acad Sci USA* **95**:8847–8851.

49. Satoh S, Daigo Y, Furukawa Y, *et al.* (2000) AXIN1 mutations in hepatocellular carcinomas, and growth suppression in cancer cells by virus-mediated transfer of AXIN1. *Nat Genet* **24**:245–250.

50. Miyoshi Y, Iwao K, Nagasawa Y, *et al.* (1998) Activation of the beta-catenin gene in primary hepatocellular carcinomas by somatic alterations involving exon 3. *Cancer Res* **58**:2524–2527.

51. Ihara A, Koizumi H, Hashizume R, Uchikoshi T. (1996) Expression of epithelial cadherin and alpha- and beta-catenins in nontumoral livers and hepatocellular carcinomas. *Hepatology* **23**:1441–1447.

52. Desbois-Mouthon C, Cadoret A, Blivet-Van Eggelpoel MJ, *et al.* (2001) Insulin and IGF-1 stimulate the beta-catenin pathway through two signalling cascades involving GSK-3beta inhibition and Ras activation. *Oncogene* **20**:252–259.

53. Lu Z, Ghosh S, Wang Z, Hunter T. (2003) Downregulation of caveolin-1 function by EGF leads to the loss of E-cadherin, increased transcriptional activity of beta-catenin, and enhanced tumor cell invasion. *Cancer Cell* **4**:499–515.

54. Morali OG, Delmas V, Moore R, *et al.* (2001) IGF-II induces rapid beta-catenin relocation to the nucleus during epithelium to mesenchyme transition. *Oncogene* **20**:4942–4950.

55. Muller T, Bain G, Wang X, Papkoff J. (2002) Regulation of epithelial cell migration and tumor formation by beta-catenin signaling. *Exp Cell Res* **280**:119–133.

56. Cross DA, Alessi DR, Cohen P, *et al.* (1995) Inhibition of glycogen synthase kinase-3 by insulin mediated by protein kinase B. *Nature* **378**:785–789.

57. Weston CR, Davis RJ. (2001) Signal transduction: signaling specificity — a complex affair. *Science* **292**:2439–2440.

58. Rimm DL, Koslov ER, Kebriaei P, *et al.* (1995) Alpha 1(E)-catenin is an actin-binding and -bundling protein mediating the attachment of F-actin to the membrane adhesion complex. *Proc Natl Acad Sci USA* **92**:8813–8817.

59. Okamoto T, Schlegel A, Scherer PE, Lisanti MP. (1998) Caveolins, a family of scaffolding proteins for organizing "preassembled signaling complexes" at the plasma membrane. *J Biol Chem* **273**:5419–5422.

60. Garcia-Cardena G, Martasek P, Masters BS, *et al.* (1997) Dissecting the interaction between nitric oxide synthase (NOS) and caveolin. Functional significance of the NOS caveolin binding domain *in vivo*. *J Biol Chem* **272**:25437–25440.

61. Yokoyama K, Kamata N, Fujimoto R, *et al.* (2003) Increased invasion and matrix metalloproteinase-2 expression by Snail-induced mesenchymal transition in squamous cell carcinomas. *Int J Oncol* **22**:891–898.

62. Galbiati F, Volonte D, Brown AM, *et al.* (2000) Caveolin-1 expression inhibits Wnt/beta-catenin/Lef-1 signaling by recruiting beta-catenin to caveolae membrane domains. *J Biol Chem* **275**:23368–23377.

63. Engelman JA, Zhang XL, Razani B, *et al.* (1999) p42/44 MAP kinase-dependent and -independent signaling pathways regulate caveolin-1 gene expression. Activation of Ras-MAP kinase and protein kinase A signaling cascades transcriptionally down-regulates caveolin-1 promoter activity. *J Biol Chem* **274**:32333–32341.

64. Berger MS, Greenfield C, Gullick WJ, *et al.* (1987) Evaluation of epidermal growth factor receptors in bladder tumours. *Br J Cancer* **56**:533–537.

65. Gullick WJ. (1991) Prevalence of aberrant expression of the epidermal growth factor receptor in human cancers. *Br Med Bull* **47**:87–98.

66. Lemoine NR, Hughes CM, Gullick WJ, *et al.* (1991) Abnormalities of the EGF receptor system in human thyroid neoplasia. *Int J Cancer* **49**:558–561.

67. Libermann TA, Nusbaum HR, Razon N, *et al.* (1985) Amplification, enhanced expression and possible rearrangement of EGF receptor gene in primary human brain tumours of glial origin. *Nature* **313**:144–147.

68. Salomon DS, Brandt R, Ciardiello F, Normanno N. (1995) Epidermal growth factor–related peptides and their receptors in human malignancies. *Crit Rev Oncol Hematol* **19**:183–232.

69. Tillotson JK, Rose DP. (1991) Endogenous secretion of epidermal growth factor peptides stimulates growth of DU145 prostate cancer cells. *Cancer Lett* **60**:109–112.

70. Fodde R, Smits R, Clevers H. (2001) APC, signal transduction and genetic instability in colorectal cancer. *Nat Rev Cancer* **1**:55–67.

71. Seeling JM, Miller JR, Gil R, *et al.* (1999) Regulation of beta-catenin signaling by the B56 subunit of protein phosphatase 2A. *Science* **283**:2089–2091.

72. Kawasaki Y, Senda T, Ishidate T, *et al.* (2000) Asef, a link between the tumor suppressor APC and G-protein signaling. *Science* **289**:1194–1197.

73. Jimbo T, Kawasaki Y, Koyama R, *et al.* (2002) Identification of a link between the tumour suppressor APC and the kinesin superfamily. *Nat Cell Biol* **4**:323–327.

74. Kishida S, Yamamoto H, Ikeda S, *et al.* (1998) Axin, a negative regulator of the Wnt signaling pathway, directly interacts with adenomatous polyposis coli and regulates the stabilization of beta-catenin. *J Biol Chem* **273**:10823–10826.

75. Ikeda S, Kishida M, Matsuura Y, *et al.* (2000) GSK-3beta–dependent phosphorylation of adenomatous polyposis coli gene product can be modulated by beta-catenin and protein phosphatase 2A complexed with Axin. *Oncogene* **19**:537–545.

76. Rubinfeld B, Albert I, Porfiri E, *et al.* (1996) Binding of GSK3beta to the APC-beta-catenin complex and regulation of complex assembly. *Science* **272**:1023–1026.

77. Kawahara K, Morishita T, Nakamura T, *et al.* (2000) Down-regulation of beta-catenin by the colorectal tumor suppressor APC requires association with Axin and beta-catenin. *J Biol Chem* **275**:8369–8374.

78. Smith KJ, Levy DB, Maupin P, *et al.* (1994) Wild-type but not mutant APC associates with the microtubule cytoskeleton. *Cancer Res* **54**:3672–3675.

79. Su LK, Burrell M, Hill DE, *et al.* (1995) APC binds to the novel protein EB1. *Cancer Res* **55**:2972–2977.

80. Matsumine A, Ogai A, Senda T, *et al.* (1996) Binding of APC to the human homolog of the *Drosophila* discs large tumor suppressor protein. *Science* **272**:1020–1023.

81. Powell SM, Zilz N, Beazer-Barclay Y, *et al.* (1992) APC mutations occur early during colorectal tumorigenesis. *Nature* **359**:235–237.

82. Hiltunen MO, Alhonen L, Koistinaho J, *et al.* (1997) Hypermethylation of the APC (adenomatous polyposis coli) gene promoter region in human colorectal carcinoma. *Int J Cancer* **70**:644–648.

83. Esteller M, Sparks A, Toyota M, *et al.* (2000) Analysis of adenomatous polyposis coli promoter hypermethylation in human cancer. *Cancer Res* **60**:4366–4371.

84. Luo W, Zou H, Jin L, *et al.* (2005) Axin contains three separable domains that confer intramolecular, homodimeric, and heterodimeric interactions involved in distinct functions. *J Biol Chem* **280**:5054–5060.

85. Zeng L, Fagotto F, Zhang T, *et al.* (1997) The mouse Fused locus encodes Axin, an inhibitor of the Wnt signaling pathway that regulates embryonic axis formation. *Cell* **90**:181–192.

86. Behrens J, Jerchow BA, Wurtele M, *et al.* (1998) Functional interaction of an Axin homolog, Conductin, with beta-catenin, APC, and GSK3beta. *Science* **280**:596–599.

87. Kishida S, Yamamoto H, Hino S, *et al.* (1999) DIX domains of Dvl and Axin are necessary for protein interactions and their ability to regulate beta-catenin stability. *Mol Cell Biol* **19**:4414–4422.

88. Kishida M, Hino S, Michiue T, *et al.* (2001) Synergistic activation of the Wnt signaling pathway by Dvl and casein kinase Iepsilon. *J Biol Chem* **276**:33147–33155.

89. Lee E, Salic A, Kirschner MW. (2001) Physiological regulation of [beta]-catenin stability by Tcf3 and CK1epsilon. *J Cell Biol* **154**:983–993.

90. Ikeda S, Kishida S, Yamamoto H, *et al.* (1998) Axin, a negative regulator of the Wnt signaling pathway, forms a complex with GSK-3beta and beta-catenin and promotes GSK-3beta–dependent phosphorylation of beta-catenin. *EMBO J* **17**:1371–1384.

91. Hart MJ, de los Santos R, Albert IN, *et al.* (1998) Downregulation of beta-catenin by human Axin and its association with the APC tumor suppressor, beta-catenin and GSK3 beta. *Curr Biol* **8**:573–581.

92. Sakanaka C, Weiss JB, Williams LT. (1998) Bridging of beta-catenin and glycogen synthase kinase-3beta by Axin and inhibition of beta-catenin–mediated transcription. *Proc Natl Acad Sci USA* **95**:3020–3023.

93. Dahmen RP, Koch A, Denkhaus D, *et al.* (2001) Deletions of AXIN1, a component of the WNT/Wingless pathway, in sporadic medulloblastomas. *Cancer Res* **61**:7039–7043.

94. Iwai S, Katagiri W, Kong C, *et al.* (2005) Mutations of the APC, beta-catenin, and Axin 1 genes and cytoplasmic accumulation of beta-catenin in oral squamous cell carcinoma. *J Cancer Res Clin Oncol* **131**:773–782.

95. Sun C, Yamato T, Kondo E, *et al.* (2005) Infrequent mutation of APC, AXIN1, and GSK3B in human pituitary adenomas with abnormal accumulation of CTNNB1. *J Neurooncol* **73**:131–134.

96. Leung JY, Kolligs FT, Wu R, *et al.* (2002) Activation of AXIN2 expression by beta-catenin–T cell factor. A feedback repressor pathway regulating Wnt signaling. *J Biol Chem* **277**:21657–21665.

97. Lustig B, Jerchow B, Sachs M, *et al.* (2002) Negative feedback loop of Wnt signaling through upregulation of Conductin/Axin2 in colorectal and liver tumors. *Mol Cell Biol* **22**:1184–1193.

98. Jho EH, Zhang T, Domon C, *et al.* (2002) Wnt/beta-catenin/Tcf signaling induces the transcription of Axin2, a negative regulator of the signaling pathway. *Mol Cell Biol* **22**:1172–1183.

99. Yan D, Wiesmann M, Rohan M, *et al.* (2001) Elevated expression of axin2 and hnkd mRNA provides evidence that Wnt/beta-catenin signaling is activated in human colon tumors. *Proc Natl Acad Sci USA* **98**:14973–14978.

100. Nakayama KI, Nakayama K. (2006) Ubiquitin ligases: cell-cycle control and cancer. *Nat Rev Cancer* **6**:369–381.

101. Nakayama KI, Nakayama K. (2005) Regulation of the cell cycle by SCF-type ubiquitin ligases. *Semin Cell Dev Biol* **16**:323–333.

102. Guardavaccaro D, Kudo Y, Boulaire J, *et al.* (2003) Control of meiotic and mitotic progression by the F box protein beta-Trcp1 *in vivo*. *Dev Cell* **4**:799–812.

103. Nakayama K, Hatakeyama S, Maruyama S, *et al.* (2003) Impaired degradation of inhibitory subunit of NF-kappa B (I kappa B) and beta-catenin as a result of targeted disruption of the beta-TrCP1 gene. *Proc Natl Acad Sci USA* **100**:8752–8757.

104. Belaidouni N, Peuchmaur M, Perret C, *et al.* (2005) Overexpression of human beta TrCP1 deleted of its F box induces tumorigenesis in transgenic mice. *Oncogene* **24**:2271–2276.

105. Sadot E, Simcha I, Iwai K, *et al.* (2000) Differential interaction of plakoglobin and beta-catenin with the ubiquitin–proteasome system. *Oncogene* **19**:1992–2001.

106. Ougolkov A, Zhang B, Yamashita K, *et al.* (2004) Associations among beta-TrCP, an E3 ubiquitin ligase receptor, beta-catenin, and NF-kappaB in colorectal cancer. *J Natl Cancer Inst* **96**:1161–1170.

107. Koch A, Waha A, Hartmann W, *et al.* (2005) Elevated expression of Wnt antagonists is a common event in hepatoblastomas. *Clin Cancer Res* **11**:4295–4304.

108. Saitoh T, Katoh M. (2001) Expression profiles of betaTRCP1 and beta-TRCP2, and mutation analysis of betaTRCP2 in gastric cancer. *Int J Oncol* **18**:959–964.

109. Gerstein AV, Almeida TA, Zhao G, *et al.* (2002) APC/CTNNB1 (beta-catenin) pathway alterations in human prostate cancers. *Genes Chromosomes Cancer* **34**:9–16.

Chapter 29

Genetics of Autoimmune Diseases: MHC and Beyond

Yahuan Lou

Department of Diagnostic Sciences Dental Branch
The University of Texas Health Science Center at Houston
6516 M. D. Anderson Blvd, Houston, TX 77030, USA

The contribution of major histocompatibility complex (MHC) genes to autoimmune diseases has been well demonstrated. However, less is known regarding the roles of non-MHC susceptibility genes in these diseases. Nonetheless, recent studies have not only suggested their critical roles in pathogenesis, but have also identified several characteristics of their influence on the diseases. First, because multiple genes participate in the pathogenesis of an autoimmune disease, only a statistically weak linkage to any particular gene in a disease could be detected. Second, because these genes exhibit complicated epistatic interactions, expression of a disease phenotype may only be correlated with susceptibility genes using the "liability threshold" model. Third, it remains unclear which factors, genetic or environmental, play a more dominant role in pathogenesis. These characteristics continue to make mapping susceptibility genes in autoimmune diseases very challenging. Although only a few disease-associated genes have been identified thus far, thanks to the Human Genome Project and other new technologies, the pace of gene discovery is accelerating.

1. INTRODUCTION

Autoimmune disease is a category of diseases in which the host's immune system inappropriately attacks its own tissues or cells, resulting in tissue damage and/or dysfunction. It is believed that susceptibility to autoimmune disease is determined not only by genetic aspects, but also by environmental factors or even stochastic events.

Although all types of autoimmune diseases have been linked to major histocompatibility complex (MHC) genes, MHC genes alone are not sufficient to cause the disease. On the other hand, the identification of non-MHC susceptibility genes has proven to be a challenging task. The most problematic issue is that genomic linkage studies failed to reveal a strong statistical association between DNA segments and diseases, especially in the human. Nevertheless, several characteristics of genetic influence on autoimmune disease have been revealed. First, multiple genes participate in the autoimmune process, and different combinations of genes are capable of causing a similar disease. Second, these genes exhibit complicated epistatic interactions; thus, expression of a disease phenotype may be explained by the so-called "liability threshold" model. Third, it remains controversial which factor, environmental or genetic, plays a more dominant role in autoimmune disease.

Identification of susceptibility genes and elucidation of their function must be the ultimate goal of genetic studies. Fortunately, a few genes associated with autoimmune diseases have already been identified. The pace of gene discovery is accelerating, thanks to the Human Genome Project and other techniques.

2. WHAT IS AUTOIMMUNE DISEASE?

The immune system prepares for the elimination of any foreign invader such as a virus, bacterium, parasite, or even an organ graft. During the process, the first step — which is one of the most striking features of the immune system — is to discriminate "nonself" from "self". As a result, it mounts a vigorous assault on nonself–foreign invaders, but shows tolerance toward self-antigens.[1,2] Autoimmune diseases arise if the immune system loses self-tolerance and wrongly responds to the self-antigens normally present in the body.[1] Consequently, the body's immune system attacks its own cells or tissues, resulting in their damage or dysfunction. How immune self-tolerance is broken still remains the most puzzling question in immunology. It is clear that T-cells play a pivotal role in both self-tolerance and autoimmune pathogenesis.

Currently, there are more than 40 human diseases classified as either definite or probable autoimmune diseases, and these diseases affect 5%–7% of the population (Table 1).[1] Interestingly, the number of human autoimmune diseases is steadily increasing. Almost all autoimmune diseases appear

Table 1   Common Autoimmune Diseases and Their Associated MHC Alleles and Related Animal Models

Disease	Target	MHC Allele	Animal Model
Systemic lupus erythematosus (SLE)	Connective and vascular tissues	DR3	MRL/*lpr*, NZW/NZBF1 mice (spontaneous)
Rheumatoid arthritis (RA)	Joint connective tissues	DR4	CIA in mice
Sjögren's syndrome (SS)	Salivary/Lacrimal glands	DR3	MRL/*lpr* (spontaneous); inducible by Ro/La
Insulin-dependent diabetes mellitus (IDDM)	Pancreas β-islet	DR3, DR4	NOD mouse and BB rat (spontaneous)
Multiple sclerosis (MS)	CNS	DR2	EAE in mice or rats
Myasthenia gravis (MG)	Nerve/Muscle	DR3	EAMG inducible by AChr
Inflammatory bowel disease; Crohn's disease	Intestine	B27	C3H/HeJBir mice (spontaneous)
Graves' disease (hyperthyroidism)	Thyroid	DR3	TcR transgenic mice (spontaneous)
Hashimoto's disease (hypothyroidism)	Thyroid	DR5	Inducible by thyroglobin, EAT
Acute anterior uveitis	Eye	B27	EAU, inducible by S-Ag
Addison's disease	Adrenal gland	DR3 DR4-DQ8 DR3-DQ2	Inducible
Goodpasture's syndrome	Glomerulus (kidney) and lung	DR2	Inducible by GBM T-cell epitope or antibody

Abbreviations: BB, biobreeding; EAE, experimental allergic encephalomyelitis; EAMG, experimental allergic myasthenia gravis; EAT, experimental allergic thyroiditis; EAU, experimental allergic uveitis; GBM, glomerular basement membrane; S-Ag, S-arrestin antigen in retina; TcR, T-cell receptor; Ro/La, two nuclear antigens for SS.

without warning or apparent cause, and most patients first suffer from fatigue. Human autoimmune diseases can be classified into two categories: systemic and organ-specific. Systemic autoimmune diseases are exemplified by systemic lupus erythematosus (SLE), rheumatoid arthritis (RA), and Sjögren's syndrome (SS); while common organ-specific diseases are insulin-dependent diabetes mellitus (IDDM), multiple sclerosis (MS) of central nerve system, several autoimmune diseases in the thyroid including Hashimoto's thyroiditis (HT) and Graves' disease, myasthenia gravis (MG) of skeletal muscles, and Goodpasture's syndrome or antiglomerular basement membrane (anti-GBM) disease of the kidney.

Women tend to be affected more often by autoimmune disorders than men, with females representing nearly 79% of all autoimmune disease patients in the USA.[3,4] They tend to be afflicted during or shortly after puberty. It is not known why this is the case, although sex hormone levels have been shown to affect the severity of some autoimmune diseases such as MS.[5] Other causes may be associated with the presence of fetal cells in the maternal bloodstream.[6]

Most autoimmune diseases are chronic and incurable at the present time. The causes of autoimmune diseases are still obscure, as we still do not know how self-tolerance, especially of T-cells, is lost. It has been demonstrated that T-cells reactive to autoantigens play a central role in many autoimmune diseases. Numerous studies have suggested that the expression of human autoimmune diseases is multifactoral — genetic background, environmental factors, and dysfunction of immune regulation (Fig. 1).[1,7–10] For

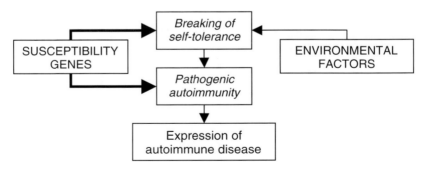

Fig. 1.   Genetic factors in the development of human autoimmune diseases. Combined effects of disease susceptibility genes and environmental factors such as infection lead to the breaking of immune self-tolerance, and autoimmunity occurs. Susceptibility genes also affect the pathogenicity of autoimmunity. Only pathogenic autoimmunity will result in autoimmune diseases.

example, a genetic predisposition in the presence of an infection may result in an autoimmune disease.

3. GENETIC PREDISPOSITION TO HUMAN AUTOIMMUNE DISEASES

Although the pathogenic mechanisms responsible for breakdown of self-tolerance and initiation/development of human autoimmune diseases remain poorly understood, mounting evidence from various types of studies has demonstrated that genetic predisposition may be a major factor in autoimmune disease susceptibility. First, a common feature of human autoimmune diseases is their propensity to occur in families, suggesting an underlying genetic susceptibility. For example, psoriasis, a chronic skin disease of scaling and inflammation, can occur among several members of the same family, suggesting that a specific gene or set of genes predisposes a family member to psoriasis.[11] The significant impact of genetic predisposition on susceptibility to human autoimmune diseases has been demonstrated for a number of autoimmune diseases through epidemiological studies using sophisticated statistical analyses. These analyses of genetic contributions in human autoimmunity have often been based on family studies, particularly twin studies comparing disease incidence between monozygotic twins and dizygotic twins or nontwin siblings.[12–26] Earlier twin studies have reported significantly higher concordance rates of 24%–50% in autoimmune diseases such as RA and SLE among monozygotic twins.[12–16,20–22]

However, genetic contribution to a predisposition to human autoimmune disease remains controversial. Despite their significant data, those earlier twin studies have been criticized for including a higher proportion of monozygotic, female, and disease-concordant volunteers.[20,27,28] More recent population-based twin studies in RA detect, at most, a 12%–15% concordance rate among monozygotic twins, and continue to be criticized for their lack of validated classification criteria for disease confirmation.[24–26] A recent study among twins in Denmark concluded that there is only a minor contribution of genetic factors to the risk of developing RA[29]; however, this study only included a small number of twin pairs, and thus is not sufficiently powered to conclude with confidence that genetics contributes little to the risk of developing RA.[30] As a result of the inconsistent observations, some researchers now suggest that there may be a more important contribution from environmental factors than genetic factors in the development of RA. This may also apply to thyroid autoimmune diseases, including

Graves' disease, in which the relative contribution of genetic and environmental factors to the pathogenesis of these diseases remains unclear.[31] At this time, it is generally accepted that genetic factors must play a certain role in human autoimmune pathogenesis, although it is unknown how critical the genetic factors are for disease development.

Animal models, on the other hand, have provided a much clearer picture of genetic influences on autoimmune diseases. Although there are significant differences in the pathogenesis and manifestations of a disease between human and other mammals, there is no doubt that animal models for autoimmune diseases are critical and, sometimes, irreplacable tools. The clear impact of genetic predisposition on susceptibility to autoimmune diseases has been well demonstrated in numerous animal models using inbred strains with either spontaneous or inducible autoimmune disorders or their related congenic strains.[32–39]

Rather surprisingly, the actual identification of specific disease-susceptibility genes has proven to be an extremely difficult task even in those animal models which have a uniform genetic background. As will be discussed later, the genetics of many autoimmune diseases, such as SLE, are complex and involve interaction among multiple genes.[10,30,33] Furthermore, the genes involved in various diseases may be very different. In spite of decades of intense study, only a few genes that are associated with autoimmune diseases and that may underlie the pathogenic mechanism are actually known. More commonly, allelic variants of chromosomal regions have been linked to an autoimmune disease risk.

The most potent genetic influences on susceptibility to autoimmune diseases are the major histocompatibility complex (MHC) genes.[40] However, recent studies have revealed that non-MHC genes may play more important roles than MHC genes in the pathogenesis of different autoimmune diseases. Although only a few have been identified or studied at this time, this chapter will focus primarily on these non-MHC genes.

4. MHC GENES AND AUTOIMMUNE DISEASES

T-cells, especially of the CD4$^+$ type, play a central role in an immune response or immune tolerance toward a specific antigen. The antigens for T-cells are proteins. Protein antigens must be fragmented into peptides within antigen-presenting cells (APCs), and the fragmented peptides further bind to the cleft of MHC molecules, which are then expressed on the surface of APCs. Unlike antibody or B-cells, T-cells only recognize the MHC–peptide complex on APCs through their surface T-cell receptor (TcR).

There are two groups of MHC molecules: class I and class II. T-cells are grouped functionally according to which class of the MHC–peptide complex they recognize. Helper T-cells (CD4$^+$ T-cells) recognize those peptides associated with class II MHC molecules, while cytotoxic T-cells (CD8$^+$ T-cells) react with those associated with class I MHC molecules. Class II MHC molecules are composed of α and β chains, while class I molecules consist of a single-chain molecule stabilized with β_2-microglobulin.[42] MHC molecules are highly polymorphic in their peptide-binding cleft (Fig. 2).[42]

Due to the highly polymorphic shape of its peptide-binding cleft, a given type of MHC molecule accommodates only peptides which can fit into the given motif. Therefore, whether T-cells can recognize an antigen (in the form of peptide) depends fully on the particular type of MHC molecule. This phenomenon is called "MHC restriction".[41] In a broader sense, MHC restriction determines whether the host is able to mount an immune response to a given antigen. This same mechanism is also applied to the T-cell's response to a self-antigen (i.e. autoantigen). The type of MHC in an individual will govern whether peptides from a given self-antigen bind to it and, thus, are presented to T-cells. Without being presented, a self-antigen would never have a chance to be recognized by T-cells, and there can be no possibility for the occurrence of autoimmunity.

In the human, MHC genes are clustered on the short arm of chromosome 6.[44] The human MHC occupies a large segment of DNA (4.75 Mbp or 4 cM), which is designated as human leukocyte antigen or HLA (Fig. 3).[44,45] The HLA region has been recently sequenced.[45] Other genes beyond MHC molecules are also located in this region, and some of them may also be associated with susceptibility to autoimmune diseases. However, those genes which have been described in several review papers will be excluded from this discussion.[46]

Human MHC class II genes are located closest to the centromere in the order of DP, DQ, and DR. There may be multiple functional β chains for a class II locus, but typically only one functional α chain exists or is expressed. The use of more than one β chain gene allows each class II locus to be expressed in more than two allelic forms. In addition, an $\alpha\beta$ heterodimer can be formed by the combination of an α and a β chain from different alleles, which explains why it is common for one individual to express 10 to 20 different class II gene products. Class I genes include A, B, and C loci, which means that a heterozygous individual can express six different polymorphic alleles. The polymorphism of human MHC has been recently shown to be rich in single nucleotide polymorphisms (SNPs) and deletion/insertion polymorphisms (DIPs).[43] The corresponding genomic regions

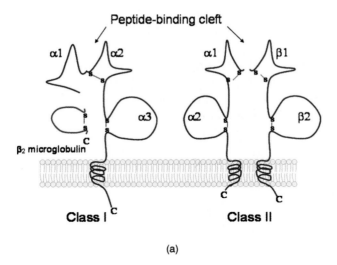

(a)

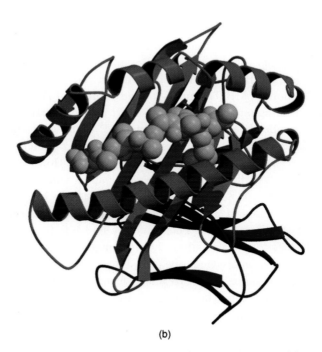

(b)

Fig. 2. Molecular structures of MHC and MHC–peptide complex. (a) Schematic diagram of MHC class I and class II molecules. Both molecules belong to the immunoglobulin gene superfamily. (b) *En face* view of an antigenic peptide (green) that binds to the cleft of an MHC class II molecule (red) (courtesy of Dr Roger Miesfeld, University of Arizona).

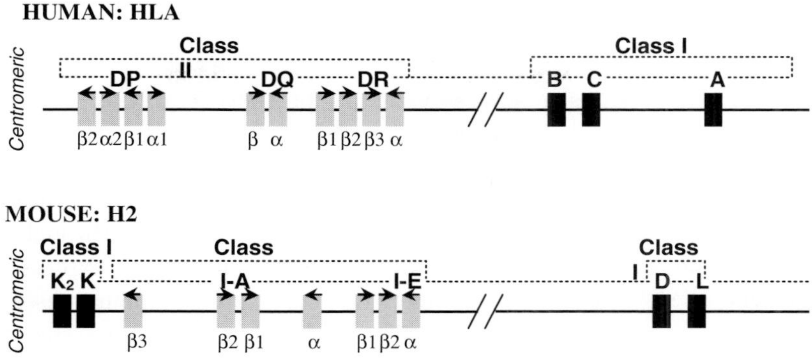

Fig. 3. Genomic organization of human and mouse MHC loci. The MHC locus is called HLA for human (on chromosome 6) and H2 for mouse (on chromosome 17). The organization of the MHC genes is similar in the two species. The number of α and β genes in each class II locus varies among alleles. Each gene consists of multiple exons that are not shown here. Non-MHC related genes between class I and class II are omitted.

encoding MHC in mice and rats are called H2 and RT1, respectively. The arrangement of class I and class II genes in these rodents is similar to those for the human (Fig. 3).

The linkage between MHC and human autoimmune diseases has been of interest for autoimmune disease research since the 1970s, shortly after the discovery of the MHC. Numerous studies have demonstrated that each type of autoimmune disease can be linked to one or multiple HLA types, either class I or class II (Table 1).[46] Based on these findings, it is generally accepted by basic researchers and the medical community that a strong linkage with certain type(s) of MHC is one of the most important criteria that determines whether an unknown human disease can be classified as an autoimmune disease. So far, MHC is the most potent genetic factor influencing susceptibility to autoimmune diseases.[47–53] Table 1 lists the linkages between MHC alleles and some representative human autoimmune diseases.

Although the MHC gene complex is associated with most, if not all, of the common autoimmune diseases, it has been difficult to identify the precise MHC gene(s) associated with autoimmunity, mainly because of the strong linkage disequilibrium as a result of recombination over the region.[54,55] However, several novel methods have been developed in an attempt to overcome these problems. For example, multiple fully sequenced MHC haplotypes in combination with large well-documented patient/ control groups and sophisticated statistical methods have led to the creation of high-density HapMaps. These advances hold promise to provide a comprehensive view of the genetics of the HLA region and autoimmune disease susceptibility.

Obviously, MHC is a necessary factor for the occurrence of human autoimmune diseases because only a certain type of MHC can provide an opportunity to present a special self-antigen to T-cells. Yet, an MHC gene alone is not sufficient for the occurrence of autoimmune diseases because self-reactive T-cells are either eliminated or controlled through tolerance mechanisms. For instance, among mice strains which express the same type of MHC, only a few (if any) will develop autoimmune diseases; similarly, the overwhelming majority of people with an MHC-type susceptibility remain free of disease. Thus, non-MHC genes must play an important role as the sufficient factors. For this reason, the identification of non-MHC susceptibility genes has become one of the prime foci for current autoimmune disease research.

5. NON-MHC GENES LINKED TO AUTOIMMUNE DISEASES

The search for non-MHC susceptibility genes involved in autoimmune disease has been one of the major foci in autoimmune disease research. Yet, after a decade of frustrating and extensive classic linkage studies, the identification of non-MHC susceptibility alleles has proved much more difficult than originally anticipated. It has become very clear that susceptibility to complex autoimmune diseases is a multigenic phenotype affected by a variety of genetic and environmental or stochastic factors.[1,10] These variables — which include extensive genetic heterogeneity and possible epistatic interactions among multiple genes, and the strong influence of environmental factors — will be described in this section.

Nevertheless, progress has been made in elucidating the genetic mechanisms that influence the inheritance of susceptibility, and several key techniques have been developed. First, the completion of the human genome sequence, coupled with mouse genome data, is expected to provide key information that will dramatically accelerate the identification of genes associated with human autoimmune diseases. Second, genomewide DNA microarray detection of differentially expressed genes in autoimmune diseases, especially in combination with SNPs, will rapidly narrow the number of candidate genes. In addition, many animal (mainly mouse) models of either spontaneous or inducible autoimmune diseases, which closely mimic human diseases, have been established. The most studied animal models for genetic studies include those for SLE, experimental allergic encephalomyelitis (EAE) model for MS, and nonobese diabetic (NOD) mouse model for IDDM (Table 1).[56–59] These models have been established for more than two decades, and numerous congenic strains have been constructed. Studies using these models have provided precious results, which will be discussed throughout this section.

5.1. *Predisposition and Problems*

As described earlier, the results from statistical analyses on autoimmune twins and other human populations are controversial. Earlier genetic analyses indicated a powerful influence of genetic background on autoimmune susceptibility.[60–62] Those results led to an optimistic belief by many researchers that genomewide linkage analyses would soon allow the identification of many potent non-MHC autoimmune disease susceptibility genes, similar to what had been seen in many other human diseases. However, more detailed and better controlled analyses have exposed new difficulties in determining which factor, genetic (excluding MHC) or environmental, is more critical for the occurrence of autoimmune diseases. In spite of more than a decade's studies with expanded knowledge of the genome, the inheritance mechanism of autoimmune disease susceptibility continues to be elusive.

It appears that current techniques and methodologies may not be readily amenable to genetic analyses to identify a disease-causing gene or genes in such a heterogeneous population, which may also be influenced by strong environmental factors. First, genomic scans in several important autoimmune diseases have consistently led to the identification of multiple genomic segments, each of which exhibits only a weak statistical association with disease susceptibility. In terms of logarithmic odds (LOD) score, such weak association with disease susceptibility for an individual genomic segment ranges from 2.0 to occasionally $5.0^{10,63-66}$; yet, it is well known that the LOD score for a fully Mendelian disease locus should be easily close to 30. Second, genetic studies carried out by independent groups of investigators on the same autoimmune disease at times did not detect the same loci or segments attributed with susceptibility of the same disease. There may be several explanations for this: each independent study was carried out using different populations and controls, making genetic comparison between those populations more difficult or impossible; subtle linkages of these genomic regions with the disease could be easily affected by the environment; and/or earlier studies lacked more sophisticated methodology, which has become available today. Thus, the above factors raise the issue of reproducibility for many previous linkage studies of human autoimmune diseases, and the data may have to be revisited and reviewed.

Many animal models have been developed during the past two decades in an attempt to overcome problems associated with the highly diverse genetic background in human populations.[56-59] However, despite the extremely uniform genetic backgrounds of animal models, similar tendencies as seen in human autoimmune disease have also been observed. Although studies on animal models have provided a much clearer pattern of genetic influence on autoimmune disease, it has been challenging to pinpoint the disease-causing genes and to elucidate just how they are involved in autoimmune pathogenesis. Nevertheless, the previous studies — either from human studies or animal models — have not only raised problems in the genetic study of autoimmune disease, but have also revealed several unique features of the genetics of autoimmune disease, which will be described in this section.

5.2. *Association of Multiple Genes with Autoimmune Diseases*

Genetic heterogeneity is one of the most important features for autoimmune diseases both in humans and animal models. First, multiple genes may be involved in the pathogenesis of an autoimmune disease. Alternatively, multiple combinations of genes within the genome are capable of causing a similar or identical disease phenotype. It simply reflects the fact that multiple genes participate in the development of the complex phenotype of autoimmune diseases, and that different combinations of genetic abnormalities can lead to similar outcomes. This situation mimics a complicated electronic circuit with multiple parallel and serial switches, in which each switch would correspond to a susceptibility gene (Fig. 4); to switch on or off depends on environmental or other factors.

In several animal models used for the analysis of genetic heterogeneity, investigators have generated multiple independent congenic or combination strains of animals (usually mouse) that develop the same or a similar disease phenotype. Comparisons of the genomic location of susceptibility genes in independent animal models for SLE, NOD, EAE, and RA models clearly show that the chromosomal location of some susceptibility genes varies between models of the same disease.[67-70] For example, a linkage analysis of the SLE-prone BXSB mouse strain revealed that only two of five intervals

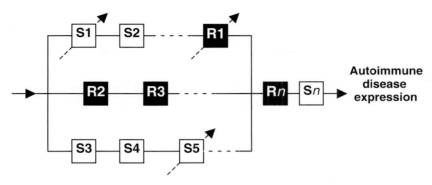

Fig. 4. Complex electric circuit mimicking interactions among susceptibility genes in autoimmune diseases. An open block with S represents a susceptibility gene which allows "current" to pass, while a closed block with R is a resistant gene which blocks the "current". Arrows that cross the block indicate the variable influence of environmental factors and/or other genes, which may "switch" the susceptibility genes "on" or "off". Occurrence of autoimmunity is the overall outcome of the circuit.

overlapped with those in other strain combinations.[70] This result indicates that lupus susceptibility loci in BXSB are not involved in disease susceptibility in other SLE-prone mouse strains such as NZB/NZW or MRL/*lpr*. Investigators had hypothesized that the same susceptibility genes might be found in different autoimmune diseases; however, this more recent finding challenges the original hypothesis, since most genomic segments previously shown to be associated with disease susceptibility are not shared between different disease models.

Although it may be more difficult to prove, investigators suggest that genetic heterogeneity could be the mechanism of genetic influence on human autoimmune diseases as well. Classical association studies for the identification of candidate susceptibility loci or genes in many autoimmune diseases, such as SLE and IDDM, frequently observed significant variations among different ethnic groups in the disease association of specific alleles and disease phenotypes.[71,72] Linkage analyses have revealed an increase in statistical association with specific genomic intervals only within specific ethnic group(s).[64,73] Although these results suggest distinct variations in the genetic basis for predisposition to autoimmune diseases between ethnic groups, the final conclusion must await the identification and analysis of specific susceptibility alleles. In summary, autoimmune disease susceptibility is mediated predominantly by a heterogeneous array of genes, and a certain combination of multiple genes with a different genetic background may also be sufficient to lead to the same or a similar disease phenotype.

5.3. *Epistatic Interactions of Susceptibility Genes*

As described above, an autoimmune disease phenotype is believed to be mediated by a heterogeneous array of genes. Epistasis, i.e. genetic interaction in which the genotype at one locus affects the phenotypic expression of the genotype at another locus, is a frequently seen phenomenon. In fact, it may be the most critical genetic mechanism leading to an autoimmune disease. Investigation of epistasis in autoimmune disease has been the focus of many research groups. As a result, evidence for epistatic interactions among disease susceptibility genes or loci has been provided by both animal models and human genetic linkage studies of autoimmune diseases. However, the most convincing evidence has been obtained from the analysis of animal models.[74-76] Two types of epistases have been characterized: synergistic and suppressive interactions.

Synergism between susceptibility alleles (genes) is clearly demonstrated in animal models using different congenic strains. This is best exemplified by the analysis of several congenic mouse strains for SLE. Serial B6 congenic strains such as B6.*sle1* and B6.*yaa* carry either a *Sle1* or a *Yaa* susceptibility gene. Although each strain spontaneously produces antinuclear autoantibodies (ANAs) which are characteristic of SLE, they fail to develop severe autoimmune pathologies such as glomerulonephritis. A severe systemic autoimmune disease develops only if the two susceptibility genes are combined in a bicongenic strain.[76]

Unlike the first type, the second type of epistasis manifests itself as an autoimmune phenotype of susceptibility alleles (genes) that are suppressed by an epistatic modifier. This has been clearly demonstrated in animal models, especially for SLE. The presence of suppressive alleles has been demonstrated through analysis of the disease phenotype known to be mediated by several susceptibility genes when they were introduced into different genetic backgrounds. For example, the genes *Sle1*, *Sle2*, and *Sle3* have been identified in SLE-prone NZM2410 as the primary genes (or cluster of genes) responsible for lupus glomerulonephritis, which is the most severe form of this disease. Thus, a triple congenic strain of B6 background carrying all three genes develops fatal glomerulonephritis. Interestingly, although all three genes are derived from genomes of the strain NZM, NZM itself exhibits only a very benign phenotype of the disease in females older than 12 months.[77] Although it remains unknown which allele(s) suppresses the disease expression, it is clear that the expression of the three genes is significantly suppressed in NZM mice. Thus, it would be especially useful to identify such genes because of their potential value in designing a novel method for the prevention of autoimmune diseases.

5.4. *Influence of Environmental Factors on Genetic Factors*

The extraordinarily critical role of environmental factors in the penetrance of autoimmune diseases is well known. As repeatedly emphasized, unlike many other hereditary diseases, it remains to be determined which factor — environmental or genetic — plays a more dominant role in triggering autoimmune diseases. In other words, the fundamental question is how environmental and/or stochastic factors are involved in the pathogenesis of autoimmune diseases, and how they affect the expression of the diseases based on genetic factors.

The first evidence for the role of nongenetic factors in the initiation of human autoimmune diseases was provided by the incomplete penetrance of the disease expression among monozygotic twin pairs.[12–16,24,60,61] Although concordance can be relatively high in twins, it was far lower than 100%, which is typically seen in many other hereditary diseases. The critical role of environmental factors in autoimmune disease was also observed in animal models, which have a uniform genetic background. The incomplete penetrance of spontaneous disease was observed in several animal models prone to the development of diabetes or lupus.[56–59] Thus, genetically predisposed individuals may or may not develop autoimmune disease, contingent upon other factors affecting their health.

The following investigations have provided strong evidence supporting the contingency of the occurrence of autoimmune diseases as modulated by environmental factors. Several studies revealed that the incidence of MS is distributed geographically in a nonrandom fashion, with individuals at higher latitudes being at greater risk.[78–80] Although it is still unknown why the local environment significantly affects the incidence of the disease, infection by certain microbes is one of the most likely triggers for initiation of autoimmune diseases. Microbial infections have been implicated in a number of autoimmune diseases presumably through molecular mimicry or nonspecific activation of lymphocytes.[81] In a retrospective study of a large cohort of patients and matched controls for human SLE, susceptibility was associated with positive seroconversion for Epstein–Barr virus (EBV) antibodies.[82] It is well known that EBV infection has a strong impact on B-cell activation; thus, EBV infection may result in nonspecific activation of B-cells, including those which may produce pathogenic ANAs. However, more direct evidence will be required to establish a linkage between EBV and SLE. Infection as a trigger for autoimmune disease has also been demonstrated in well-controlled animal studies. For example, a transgenic mouse strain bore a uniform TcR specific for a pathogenic T-cell epitope derived from the known disease-causing autoantigen myelin basic protein (MBP); when housed in one research facility, the mice did not develop EAE unless they were immunized with the T-cell epitope, but the same mice spontaneously developed severe EAE when they were transferred to another facility.[83] It was subsequently demonstrated that the immune response to environmental microbes triggered the disease.

Studies using several disease-prone inbred animals, however, argue against the importance of environmental factors. These animal models provide an exceptional tool for testing the potential role of environmental factors in initiation of autoimmune disease, and thus the results from those studies should be much more accurate and easier to interpret. Yet, as we have mentioned previously, inbred strains of rodents that are susceptible to spontaneous autoimmune diseases in general exhibit an incomplete penetrance of diseases despite highly controlled biological (sometimes pathogen-free) and physical environments in animal facilities. Those models include NOD mice and BB rats for IDDM, and MRL/*lpr* and NZM2410 strains for lupus. At least in those models, the results strongly suggested that environmental variables do not account for the incomplete penetrance of the disease; rather, stochastic events are more likely the trigger for initiation of autoimmune diseases. More interestingly and unexpectedly is the fact that the incidence of diabetes dramatically increased in colonies of NOD mice housed in a specific pathogen-free facility[84]; this implied that infection in this model seems to impart protection from rather than trigger the disease process.

Despite those confounding results, it is generally accepted that the impact of environmental factors on the occurrence of autoimmune disease may be different from disease to disease, or even among individuals with a similar disease phenotype. The incomplete penetrance of disease in animal models as described above suggests that, in some cases, disease can be initiated in disease-susceptible individuals simply by a random (or stochastic) event that occurs during the normal functioning of immune systems or specific tissues. This phenomenon is referred to as the "bystander" effect. In fact, the phenomenon, in which the disease is triggered by a stochastic event during normal physiological activity or development, has also been observed in other genetic diseases. The questions are, what is the stochastic event and how does it occur? One example for the first question is "unfortunate" bystander activation of a self-reactive T-cell during an immune response or an infection.[85] From the viewpoint of disease prevention, it is most useful for researchers to discover the specific stochastic events. It is believed that multiple susceptibility genes will create a tendency in the immune system for the activation of autoreactive T- or B-cells; the cells are activated only if they encounter a specific autoantigen which has been appropriately presented within a stimulatory environment such as bystander activation. However, it is unclear whether the presentation of a self-antigen in a stimulatory environment is a stochastic event or is controlled by other specific genetic factors.

In summary, the expression of disease phenotype in an individual with disease susceptibility genes appears to be well controlled or inhibited until a triggering event occurs; an environmental factor or a stochastic event during normal physiological functioning may act as triggers for disease expression. Therefore, the probability of occurrence of the stochastic event will help to predict the relative risk of disease in those individuals with disease susceptibility genes.

6. HYPOTHETICAL MODEL: RISK OF AUTOIMMUNE DISEASE VERSUS SUSCEPTIBILITY GENES

There are three possible conclusions regarding the genetic control of autoimmune diseases based on the above discussions. First, autoimmune disease susceptibility is controlled by multiple genes beyond the MHC. Second, complicated interactions, either synergistic or suppressive, among the genes lead to final expression or inhibition of various disease phenotypes. Third, although the disease susceptibility genes create the potential for occurrence of autoimmunity (i.e. activation of autoreactive T- or B-cells), it must be triggered by environmental factors or a stochastic event, and this combination of circumstances determines the ultimate risk for the disease. Thus, autoimmune disease susceptibility is determined by inheritance of multifactorial traits as well as environmental/ stochastic triggers, making autoimmune pathogenesis an extremely complex process.

In many hereditary diseases (or patterns), multifactorial traits are considered continuous because there is a bell-shaped distribution of these traits in the population. However, in autoimmune diseases, the traits are more discontinuous because there is a cutoff or "threshold" of both genetic and

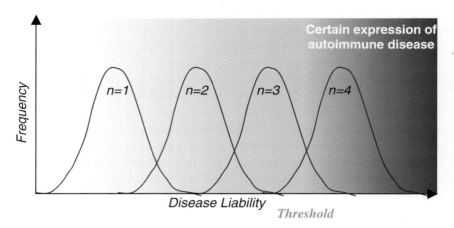

Fig. 5.   Liability threshold model for multigenetic control of autoimmune diseases (modified from Wanstrat and Wakeland[10]). In this model, the x-axis is disease liability or tendency of disease expression, while the y-axis is frequency of liability. Each distribution curve represents the frequency of disease liability in individuals with a certain number of susceptibility genes as indicated. The shape of the distribution curve is determined by environmental factors. Overall, the greater the number of susceptibility genes, the more the curve will shift to the right. Eventually, the curve will locate beyond the threshold, whereby expression of the disease is certain.

environmental risks that must be crossed before the trait can occur. In both cases, the genetic and environmental factors that are involved in the occurrence of the condition are referred to as "liability". The concept of "threshold liability" in multifactorial inheritance was first described in a guinea pig model for a hereditary phenomenon called polydactyly.[86] It was hypothesized that the penetrance of polygenic, qualitative phenotypes would increase in relation to the number of susceptibility genes present in the genome of an individual.

Based on the threshold liability concept, a similar hypothetical mechanism has recently been proposed for the inheritance of autoimmune diseases by Wanstrat and Wakeland, as shown in Fig. 5.[10] In this figure, the x-axis represents progressive liability of the disease, which is expressed as increasing numbers; the disease threshold (i.e. the point beyond which the disease is certain to occur) is indicated by a vertical line. The y-axis shows the frequency of liability. The disease liability for each individual with a particular genetic construction forms a normal distribution curve because of variations in influences of environmental factors or stochastic events. The more disease susceptibility genes an individual has, the more his or her disease liability curve will shift to the right (toward higher disease liability);

as a result, a certain number of susceptibility genes within an individual will predict a move of the liability curve beyond the threshold, whereby occurrence of autoimmunity becomes certain.

It must be pointed out, however, that the graph describing the inheritance of autoimmune disease is presented in an extremely simplistic additive fashion. Each additional susceptibility gene appears to incrementally move an individual the same distance toward the disease threshold; in reality, the interactions among genes, as described previously, are much more complicated. It is predictable that different epistatic interactions, whether synergistic or suppressive, would significantly modify the incremental shifting of an individual's disease liability as seen in this simplified model. In addition, the liability distribution curve would display a totally different shape in the presence of strong influences from environmental factors.

In the presence of complicated epistatic interactions among the susceptible genes in human autoimmune diseases, it seems difficult to distinguish additive inheritance from multiplicative models. For example, linkage analyses in test crosses of disease-prone animals have found a proportional correlation between the relative risk and the number of susceptibility genes present in their genomes.[75,87–89] However, the goodness of fit for additive vs. multiplicative models has not been statistically established.

In summary, the liability threshold model is possibly applicable to most (if not all) autoimmune diseases, although the modulation toward the disease threshold would be highly variable based on circumstances or individuals. Finally, it must be remembered that, although the liability model gives a general relationship between susceptibility genes and disease risk, it does not address the mechanism which leads to pathogenic autoimmunity. Thus, the identification of susceptibility genes with special emphasis on their functional interactions will continue to be the most promising focus of future studies on the genetics of autoimmunity.

7. IDENTIFICATION OF DISEASE-CAUSING GENES: TODAY AND TOMORROW

7.1. *Identified Susceptibility Genes*

All genetic studies on autoimmune diseases aim to identify a gene or genes and their potential mechanisms which provoke autoimmunity. Despite many decades of classical linkage studies, to date, only a handful of susceptibility genes have been identified due primarily to a lack of sophisticated technology. In most cases, the identification of genes has been hampered by complex interactions among genes associated with autoimmune diseases and the highly diverse genetic background in the human populations. It is predictable, however, that the process for identification of susceptibility genes will now be greatly accelerated due to the completion of human and rodent genome projects and the advancing sophistication of biotechnology.

So far, pinpointing a specific gene from an identified locus is one of the most difficult tasks confronting researchers. Locating the disease suscepti-bility alleles with enough precision to limit the number of potential candi-date genes is the first step of this process. Unfortunately, most of the current linkage studies of autoimmune diseases, with few exceptions, have failed to yield a robust statistical correlation with specific loci. Therefore, the pre-cision with which susceptibility alleles are positioned is extremely poor, and the resulting large size of genomic DNA may contain an unmanagable number of genes (from hundreds to thousands). For human autoimmune diseases, since 1999, transmission/disequilibrium testing (TDT) has been the best strategy for narrowing the interval size to identify potential disease-causing genes, and it has been used to successfully identify or exclude several strong candidates.[90–92]

A handful of non-MHC susceptibility genes in human diseases such as IDDM and SLE have been identified. The first identified non-MHC allele was a variable-number tandem repeat polymorphism in the promoter of the insulin gene that affects susceptibility to IDDM.[93,94] Although it remains unknown how this polymorphism is associated with susceptibility to dia-betes, some studies have described multiple functional consequences of this polymorphism on the expression of both insulin and insulin-like growth factors.[95] Some studies suggested that the polymorphism affects the expres-sion level of those molecules in the thymus, which in turn interferes with the induction of self-tolerance to the molecules.[96] Another unusual example is the identification of deficiency in the complement component *C1q* in SLE patients from a unique subset of families.[97] Unlike susceptibility alleles or genes in other autoimmune diseases, the degree of penetrance of SLE in the absence of *C1q* was higher than 90%, suggesting that a deficiency of *C1q* is sufficient to mediate disease in a monogenic fashion, a condition which is rarely seen in human autoimmune diseases. Although the mecha-nism remains unknown, some studies have suggested critical roles of *C1q* in early B-cell development and tolerance and in the clearance of apoptotic cells.[98–100] However, it must be emphasized that the *Cq1* deficiency is not responsible for all cases of SLE.

Novel techniques and strategies have accelerated the identification of susceptibility genes. First, through the development of dense genetic maps of highly informative microsatellite loci obtained by using polymerase chain reaction (PCR) analysis, a genomewide linkage study in IDDM has identified the *FGF3* gene on chromosome 11q to be associated with the disease.[101] Second, by applying positional cloning strategy and SNP typing, a susceptibility locus for Crohn's disease (an autoimmune inflammatory disease in the gut mucosa) has been narrowed down to gene *NOD2*; a frameshift variant and two missense mutations in this gene are associated with disease susceptibility.[102,103] Crohn's disease has long been linked to abnormal immune response to intestinal bacteria. Considering its known function in the activation of nuclear factor-κB and conferring responsiveness to bacterial lipopolysaccharide (LPS), the inactivated *NOD2* gene may lead to an abnormal immune response. Third, using an approach to focus on disease inheritance in an isolated human subpopulation with limited ethnic heterogeneity, mutations in the novel gene *AIRE*, which encodes for a putative nuclear protein featuring two PHD zinc-finger motifs, has been identified to be responsible for autoimmune polyendocrinopathy-candidiasis-ectodermal dystrophy (APECED).[104,105] Thus, this disorder may be initiated at the transcription level. Fourth, very recently, using a case-control study by SNP-based linkage disequilibrium (LD) mapping together with *in vitro* testing, a Japanese group determined that a particular variant of the gene *FCRL3* may be responsible for multiple autoimmune disorders such as RA, SLE, and thyroid diseases.[106] *FCRL3* encodes a receptor which may be involved in the regulation of the immune system, and its variant may affect the binding of an important transcription factor called NF-κB.

Studies on the identification of disease susceptibility genes in animal models have been much more fruitful. In general, the most successful strategy for gene identification in animal models has been the genetic dissection of various congenic strains. As with heterogeneity that occurs in nature, the construction of a set of congenic strains, in which each carries one susceptibility gene, allows the investigation to separately analyze each individual gene or locus. Using this approach, susceptibility genes have been identified or narrowed down to a few candidate genes in several important models, including NOD mouse, those for SLE, EAE, RA, and autoimmune oophoritis (an autoimmune disease in the ovary). Identified genes can be sorted into two categories: molecules directly associated with or not associated with the immune system. Several example genes for each category will be described below.

Most of the identified genes are directly associated with the immune system. With its well-known important function in the immune response, it is not surprising that the *IL2* gene has been identified as a candidate for *idd3*, which is one of the susceptibility loci for diabetes in the NOD model.[107,108] A single mutation affecting glycosylation in NOD mouse is believed to greatly affect the function of *IL2*. Another example is the identification of the *FAS* gene as the disease susceptibility gene in *lpr* mouse for SLE.[109] In fact, one of the most striking characteristics of the *lpr* mouse is uncontrolled lymphocyte proliferation. Deficiency in its *FAS* gene leads to expansion of autoreactive B- or T-cells, which otherwise would be eliminated through apoptosis. There are many other molecules such as β2-microglobulin in NOD models and IL18 in inflammatory bowel disease.[110,111] Recently, the E3 ubiquitin ligase *Cb1b* gene was identified as a major susceptibility gene for Komeda diabetes-prone rats, a model for IDDM.[112] Interestingly, it has been demonstrated that *Cb1b* plays a critical role in the regulation of CD28 dependence on T-cell activation.[113] In autoimmune oophoritis, the disease susceptibility locus *Bphs* has been narrowed down to the gene encoding histamine receptor H1 (*Hrh1*).[114] Although the mechanism remains unclear, alleles of *Hrh1* are believed to control both the autoimmune T-cell and vascular responses.

A few identified susceptibility genes belong to the second category (i.e. not directly associated with the immune system). Genetic defects in the regulation of autoantigens such as GAD have been blamed for diabetes in NOD mouse, although some studies showed opposite results.[115,116] Autoimmunity may be affected by abnormal regulation of autoantigen presentation. *MHC2TA* (MHC class II transactivator), which regulates MHC expression, has been shown to be associated with susceptibility to RA, MS, and myocardial infarction in the rat model.[117]

Although animal models provide a unique tool for rapid gene discoveries, there is continuing concern as to whether the susceptibility genes identified in animal models are relevant to human autoimmune diseases. Nevertheless, at this time, the genes identified from animal models will undoubtedly contribute to our understanding of the nature and mechanism of not only human autoimmune diseases, but also normal immune regulation as a whole.

7.2. Are There Common Susceptibility Genes Among Autoimmune Diseases?

Since all autoimmune diseases are caused by dysregulation of the immune system, it has always been interesting to ask, do autoimmune diseases share the same susceptibility genes? The same question has also been asked for many other human diseases such as cancers. Although many previous studies suggested that it is an unlikely scenario for autoimmune diseases, some investigators believe that a defect in the common genetic pathways in the immune system may contribute to immune dysregulation, which in turn may lead to occurrence of autoimmunity.[10,118,119] Some preliminary results seem to support this hypothesis. The most favorable result was observed in the colocalization of susceptibility loci in genomewide scans in both mouse and human studies. Several distinct non-MHC autoimmune susceptibility clusters have been identified in independent human and mouse genome scans.[10] For example, one cluster (extended from 16q13 to 16q24) in human chromosome 16 is shared among four types of diseases including SLE, RA, Crohn's disease, and type 1 diabetes.[118-120] More interestingly, clusters for susceptibility loci in human and mouse have been mapped to the corresponding genomic regions of the two species.[10]

Is this a good indication for the existence of common susceptibility genes among multiple diseases or between different species? It seems premature to answer this question at the present time. The frequent detection of the genomic clustering of susceptibility loci or genes may be explained in two other ways beyond the "common gene" hypothesis. First, as discussed previously, most autoimmune diseases are multifactorial traits with weak statistically significant linkage for each locus; clustering in genomewide scans may have included several closely associated loci and, thus, represent an ascertainment bias due to weak linkages for each locus. Second, the detection of clusters may indicate an organizational feature of mammalian genomes, in which genes involved in multiple fundamental pathways in the immune system tend to cluster in certain regions; thus, each type of disease does not necessarily share the same pathways or genes. Clustering of functionally related genes of the immune system has been revealed by the Human Genome Project[121]; thus, it is possible that susceptibility of several autoimmune diseases may be associated with the defect of the same regulatory pathway, but determined by different genes involving the same pathways. In summary, it seems more likely that several autoimmune diseases could be caused by a defect in the same pathways in the immune system. However, a different disease or the same disease in each individual could involve a deficiency of a different gene (or genes) in the same pathways.

7.3. Future Studies: How to Pinpoint Genes?

The ultimate goal of studies on autoimmune genetics is the identification of disease-causing genes. The major barrier to achieving this goal is the transition from current linkage analysis and modeling to gene identification; this is due to the complex nature of autoimmune genetics and a lack of powerful analytical methods. There are two issues in overcoming these barriers. First, more powerful techniques in combination with classical methods need to be developed to accelerate the gene discovery process. In fact, some techniques have become available and will be described in the next paragraph. The second critical issue is how to seek or identify special populations more suitable for genetic analysis, especially for human studies. It is critical for future studies to define a larger patient population with a relatively homogeneous ethnic background. Although the potential for success in autoimmune disease gene identification in animal models has been excellent, those models, especially congenic animals, will continue evolving. For example, the dissection of serial congenic strains of mice has been able to narrow the susceptibility interval to as little as 800–1000 kb in NOD mouse for type 1 diabetes and in the models for SLE.

As mentioned in the previous paragraph, several new technologies have been established or are under development. First, as a rapidly increasing number of SNP markers in the human genome have been identified, SNP technology has already distinguished itself as an extremely powerful tool for quickly identifying abnormal disease-causing genes in general. However, SNPs alone may not be sufficient to overcome the complexities introduced into the analysis by epistasis and genetic heterogeneity in autoimmune diseases. Second, the use of a large-scale gene expression microarray on disease-prone and disease-resistant populations will identify differentially expressed genes. In many cases, dysregulation of its expression, rather than polymorphism in the gene itself, may be directly related to disease susceptibility; thus, it will be necessary to investigate not only the gene structure, but also its regulation mechanism. The combination of classical linkage studies with this technique has the potential to revolutionize the gene mapping strategy. Third, the completion of the Human and Mouse Genome Projects will provide quality molecular and physical maps of the genome for these species. Furthermore, the complementation of bacterial artificial chromosomes (BACs) or radiation hybrid (RH) maps will be able to directly link the chromosomal location of a locus of interest to an actual genomic DNA region.

REFERENCES

1. Ermann J, Fathman CG. (2001) Autoimmune diseases: genes, bugs, and failed regulation. *Nat Immunol* **2**(9):759–761.

2. Rose NR, Mackay IR (eds.). (1999) *The Autoimmune Diseases*, 3rd ed. Academic Press, San Diego, CA.

3. Whitacre CC. (2001) Sex differences in autoimmune disease. *Nat Immunol* **2**(9):777–780.

4. Fairweather D, Rose NR. (2004) Women and autoimmune diseases. *Emerg Infect Dis* **10**(11):2005–2011.

5. Whitacre CC, Reingold SC, O'Looney PA. (1999) A gender gap in autoimmunity. *Science* **283**:1277–1278.

6. Adams KM, Nelson JL. (2004) Microchimerism: an investigative frontier in autoimmunity and transplantation. *JAMA* **291**:1127–1131.

7. Pollard KM, Hultman P, Kono DH. (2005) Immunology and genetics of induced systemic autoimmunity. *Autoimmun Rev* **4**(5):282–288.

8. Akerblom HK, Vaarala O, Hyoty H, *et al.* (2002) Environmental factors in the etiology of type 1 diabetes. *Am J Med Genet* **115**:18–29.

9. Luppi P, Rossiello MR, Faas S, Trucco M. (1995) Genetic background and environment contribute synergistically to the onset of autoimmune diseases. *J Mol Med* **73**:381–393.

10. Wanstrat A, Wakeland E. (2001) The genetics of complex autoimmune diseases: non-MHC susceptibility genes. *Nat Immunol* **2**(9):802–829.

11. Espinoza LB, Bombardier C, Gaylord SW, *et al.* (1980) Histocompatibility studies in psoriasis vulgaris: family studies. *J Rheumatol* **7**(4):445–452.

12. Redondo MJ, Rewers M, Yu L, *et al.* (1999) Genetic determination of islet cell autoimmunity in monozygotic twin, dizygotic twin, and non-twin siblings of patients with type 1 diabetes: prospective twin study. *Br Med J* **318**:698–702.

13. Chan OT, Madaio MP, Shlomchik MJ. (1999) B cells are required for lupus nephritis in the polygenic, Fas-intact MRL model of systemic autoimmunity. *J Immunol* **163**:3592–3596.

14. Gulko P, Winchester R. (1999) Genetics of systemic lupus erythematosus. In: *Lupus: Molecular and Cellular Pathogenesis*, Kammer G, Tsokos GC (eds.), Humana Press, Totowa, NJ, pp. 101–123.

15. Silman AJ, MacGregor AJ, Thomson W, *et al.* (1993) Twin concordance rates for rheumatoid arthritis: results from a nationwide study. *Br J Rheumatol* **32**:903–907.

16. Kalman B, Lublin FD. (1999) The genetics of multiple sclerosis. A review. *Biomed Pharmacother* **53**:358–370.

17. Hasegawa Y, Ohi M, Fujisaku A. (2000) Autoimmune cholangitis (AIC) associated with myositis in monozygotic twins. *Ryumachi* **40**(6): 898–903.

18. Bolstad AI, Haga HJ, Wassmuth R, Jonsson R. (2000) Monozygotic twins with primary Sjogren's syndrome. *J Rheumatol* **27**(9):2264–2266.

19. Russell GA, Coulter JB, Isherwood DM, *et al.* (1991) Autoimmune Addison's disease and thyrotoxic thyroiditis presenting as encephalopathy in twins. *Arch Dis Child* **66**(3):350–352.

20. Harvald B, Hauge M. (1965) Hereditary factors elucidated by twin studies. In: *Genetics and the Epidemiology of Chronic Disease*, Neel JV, Shaw

MV, Schull WJ (eds.), Department of Health, Education and Welfare, Washington, DC, pp. 64–76.

21. Jawaheer D, Thomson W, MacGregor AJ, *et al.* (1994) "Homozygosity" for the HLA-DR shared epitope contributes the highest risk for rheumatoid arthritis concordance in identical twins. *Arthritis Rheum* **37**(5):681–686.

22. Bellamy N, Duffy D, Martin N, Mathews J. (1992) Rheumatoid arthritis in twins: a study of aetiopathogenesis based on the Australian Twin Registry. *Ann Rheum Dis* **51**(5):588–593.

23. Deapen D, Escalante A, Weinrib L. (1992) A revised estimate of twin concordance in systemic lupus erythematosus. *Arthritis Rheum* **35**:311–318.

24. Silman AJ, MacGregor AJ, Thomson W, *et al.* (1993) Twin concordance rates for rheumatoid arthritis: results from a nationwide study. *Br J Rheumatol* **32**(10):903–907.

25. Aho K, Koskeuvuo M, Tuominen J, Kaprio J. (1986) Occurrence of rheumatoid arthritis in a nationwide series of twins. *J Rheumatol* **13**(5): 899–902.

26. Jones MA, Silman AJ, Whiting S, *et al.* (1996) Occurrence of rheumatoid arthritis is not increased in the first degree relatives of a population based inception cohort of inflammatory polyarthritis. *Ann Rheum Dis* **55**(2): 89–93.

27. Lykken DT, Tellegen A, DeRubeis R. (1978) Volunteer bias in twin research: the rule of two-thirds. *Soc Biol* **25**(1):1–9.

28. Phillips DI. (1993) Twin studies in medical research: can they tell us whether diseases are genetically determined? *Lancet* **341**(8851):1008–1009.

29. Svendsen AJ, Holm NV, Kyvik K, *et al.* (2002) Relative importance of genetic effects in rheumatoid arthritis: historical cohort study of Danish nationwide twin population. *Br Med J* **324**(7332):264–266.

30. Dooley MA, Hogan SL. (2003) Environmental epidemiology and risk factors for autoimmune disease. *Curr Opin Rheumatol* **15**(2):99–103.

31. Weetman AP. (2002) Determinants of autoimmune thyroid disease. *Nat Immunol* **2**(9):769–770.

32. Wandstrat AE, Nguyen C, Limaye N, *et al.* (2004) Association of extensive polymorphisms in the SLAM/CD2 gene cluster with murine lupus. *Immunity* **21**(6):769–780.

33. Morahan G, Morel L. (2002) Genetics of autoimmune diseases in humans and in animal models. *Curr Opin Immunol* **14**(6):803–811.

34. Meeker ND, Hickey WF, Korngold R, *et al.* (1995) Multiple loci govern the bone marrow–derived immunoregulatory mechanism controlling dominant resistance to autoimmune orchitis. *Proc Natl Acad Sci USA* **92**(12): 5684–5688.

35. Todd JA, Aitman TJ, Cornall RJ, *et al.* (1991) Genetic analysis of autoimmune type 1 diabetes mellitus in mice. *Nature* **351**(6327):542–547.

36. Parfrey NA, Prud'homme GJ, Colle E, *et al.* (1989) Immunologic and genetic studies of diabetes in the BB rat. *Crit Rev Immunol* **9**(1):45–65.

37. Santiago-Raber ML, Laporte C, Reininger L, Izui S. (2004) Genetic basis of murine lupus. *Autoimmun Rev* **3**(1):33–39.

38. Mountz JD, Yang P, Wu Q, *et al.* (2005) Genetic segregation of spontaneous erosive arthritis and generalized autoimmune disease in the BXD2 recombinant inbred strain of mice. *Scand J Immunol* **61**(2):128–138.

39. Wang J, Yoshida T, Nakaki F, *et al.* (2005) Establishment of NOD-Pdcd1$^{-/-}$ mice as an efficient animal model of type I diabetes. *Proc Natl Acad Sci USA* **102**(33):11823–11828.

40. Wu J, Longmate JA, Adamus G, *et al.* (1986) Interval mapping of quantitative trait loci controlling humoral immunity to exogenous antigens: evidence that non-MHC immune response genes may also influence susceptibility to autoimmunity. *J Immunol* **157**(6):2498–2505.

41. Babbitt BP, Allen PM, Matsueda G, *et al.* (1985) Binding of immunogenic peptides to Ia histocompatibility molecules. *Nature* **317**(6035):359–361.

42. Abbas AK, Lichtman AH, Pober JS. (1991) The major histocompatibility complex. In: *Cellular and Molecular Immunology*, W. B. Saunders Co., Philadelphia, PA, pp. 99–114.

43. Miretti MM, Walsh EC, Ke X, *et al.* (2005) A high-resolution linkage-disequilibrium map of the human major histocompatibility complex and first generation of tag single-nucleotide polymorphisms. *Am J Hum Genet* **76**:634–646.

44. Bjorkman PJ, Parham P. (1990) Structure, function, and diversity of class I major histocompatibility complex molecules. *Ann Rev Biochem* **59**:253–288.

45. Stewart CA, Horton R, Allcock RJ, *et al.* (2000) Complete MHC haplotype sequencing for common disease gene mapping. *Genome Res* **14**:1176–1187.

46. Lechler R, Warrens A (eds.). (2000) *HLA in Health and Diseases*, 2nd ed. Academic Press, San Diego, CA.

47. Todd JA, Mijoric C, Fletcher J, *et al.* (1989) Identification susceptibility loci for insulin-dependent diabetes mellitus by trans-racial gene mapping. *Nature* **338**(6216):587–589.

48. Bugawan TL, Angelini G, Larrick J, *et al.* (1989) A combination of a particular HLA-DP beta allele and an HLA-DQ heterodimer confers susceptibility to coeliac disease. *Nature* **339**(6224):470–473.

49. Nath SK, Kilpatrick J, Harley JB. (2004) Genetics of human systemic lupus erythematosus: the emerging picture. *Curr Opin Immunol* **16**(6):794–800.

50. Bougacha-Elleuch N, Rebai A, Mnif M, *et al.* (2004) Analysis of MHC genes in a Tunisian isolate with autoimmune thyroid diseases: implication of TNF-308 gene polymorphism. *J Autoimmun* **23**(1):75–80.

51. Tait KF, Gough SC. (2003) The genetics of autoimmune endocrine disease. *Clin Endocrinol* **59**(1):1–11.

52. Redondo MJ, Eisenbarth GS. (2002) Genetic control of autoimmunity in type I diabetes and associated disorders. *Diabetologia* **45**(5):605–622.

53. Nakken B, Jousson R, Brokstad KA, *et al.* (2001) Associations of MHC class II alleles in Norwegian primary Sjogren's syndrome patients: implications for development of autoantibodies to the Ro52 autoantigen. *Scand J Immunol* **54**(4):428–433.

54. Trowsdale J. (2005) HLA genomics in the third millennium. *Curr Opin Immunol* **17**(5):489–504.

55. Pauppi K, Sajantilla A, Jeffreys AJ. (2003) Recombination hotspots rather than population history dominate linkage disequilibrium in MHC class II region. *Hum Mol Genet* **12**:33–40.

56. Beighlie-poole D, Teplitz RL. (1978) NZB/WF1 hybrid autoimmune disease: a genetic analysis. *J Rheumatol* **5**(2):129–135.

57. Pisetsky DS, McCarty GA, Peters DV. (1980) Mechanisms of autoantibody production in autoimmune MRL mice. *J Exp Med* **152**(5):1302–1310.

58. Bernard CC. (1976) Experimental autoimmune encephalomyelitis in mice: genetic control of susceptibility. *J Immunogenet* **3**(4):263–274.

59. Makino S, Kunimoto K, Muraoka Y, *et al.* (1980) Breeding of a non-obese, diabetic strain of mice. *Jikken Dobutsu* **29**(1):1–13.

60. Kofler R, Noonan DJ, Levy DE, *et al.* (1985) Genetic elements used for a murine lupus anti-DNA autoantibody are closely related to those for antibodies to exogenous antigens. *J Exp Med* **161**(4):805–815.

61. Silman AJ, MacGregor AJ, Thomson W, *et al.* (1999) Twin concordance rates for rheumatoid arthritis: results from a nationwide study. *Br J Rheumatol* **32**:903–907.

62. Gulko P, Winchester R. (1999) Genetics of systemic lupus erythematosus. In: *Lupus: Molecular and Cellular Pathogenesis*, Kammer G, Tsokos GC (eds.), Humana Press, Totowa, NJ, pp. 101–123.

63. Kalman B, Lublin FD. (1999) The genetics of multiple sclerosis: a review. *Biomed Pharmacother* **53**:358–370.

64. Davies JL, Kawaguchi Y, Bennett ST, *et al.* (1994) A genome-wide search for human type 1 diabetes susceptibility genes. *Nature* **371**:130–136.

65. Moser KL, Neas BR, Salmon JE, *et al.* (1998) Genome scan of human systemic lupus erythematosus: evidence for linkage on chromosome 1q in African-American pedigrees. *Proc Natl Acad Sci USA* **95**:14869–14874.

66. Mein CA, Esposito L, Dunn MG, *et al.* (1998) A search for type 1 diabetes susceptibility genes in families from the United Kingdom. *Nat Genet* **19**:297–300.

67. Lindqvist AK, Steinsson K, Johanneson B, *et al.* (2000) A susceptibility locus for human systemic lupus erythematosus (hSLE1) on chromosome 2q. *J Autoimmun* **14**:169–178.

68. Prochazka M, Serreze DV, Frankel WN, Leiter EH. (1992) NOR/Lt mice: MHC-matched diabetes-resistant control strain for NOD mice. *Diabetes* **41**:98–106.

69. Rozzo SJ, Vyse TJ, Drake CG, Kotzin BL. (1996) Effect of genetic background on the contribution of New Zealand black loci to autoimmune lupus nephritis. *Proc Natl Acad Sci USA* **93**:15164–15168.

70. McDuffie M. (1998) Genetics of autoimmune diabetes in animal models. *Curr Opin Immunol* **10**:704–709.

71. Jawaheer D, Seldin MF, Amos CI, *et al.* (2001) A genomewide screen in multiplex rheumatoid arthritis families suggests genetic overlap with other autoimmune diseases. *Am J Hum Genet* **68**:927–936.

72. Salmon JE, Millard S, Schachter LA, *et al.* (1996) Fc γRIIA alleles are heritable risk factors for lupus nephritis in African Americans. *J Clin Invest* **97**:1348–1354.

73. Leech NJ, Kitabchi AE, Gaur LK, *et al.* (1995) Genetic and immunological markers of insulin dependent diabetes in Black Americans. *Autoimmunity* **22**:27–32.

74. Hogarth MB, Slingsby JH, Allen PJ, *et al.* (1998) Multiple lupus susceptibility loci map to chromosome 1 in BXSB mice. *J Immunol* **161**:2753–2761.

75. Prins JB, Todd JA, Rodrigues NR, *et al.* (1993) Linkage on chromosome 3 of autoimmune diabetes and defective Fc receptor for IgG in NOD mice. *Science* **260**:695–698.

76. Sundvall M, Jirholt J, Yang HT, *et al.* (1995) Identification of murine loci associated with susceptibility to chronic experimental autoimmune encephalomyelitis. *Nat Genet* **10**:313–317.

77. Morel L, Croker BP, Blenman KR, *et al.* (2000) Genetic reconstitution of systemic lupus erythematosus immunopathology with polycongenic murine strains. *Proc Natl Acad Sci USA* **97**:6670–6675.

78. Willer CJ, Ebers GC. (2000) Susceptibility to multiple sclerosis: interplay between genes and environment. *Curr Opin Neurol* **13**:241–247.

79. Kurtzke JF, Delasnerie-Laupretre N, Wallin MT. (1998) Reflection on the geographic distribution of multiple sclerosis in France. *Acta Neurol Scand* **93**:110–117.

80. Kalman B, Lublin FD. (1999) The genetics of multiple sclerosis. A review. *Biomed Pharmacother* **53**:358–370.

81. Barnaba V, Sinigaglia F. (1997) Molecular mimicry and T cell–mediated autoimmune disease. *J Exp Med* **185**(9):1529–1531.

82. Harley JB, James JA. (1999) Epstein–Barr virus infection may be an environmental risk factor for systemic lupus erythematosus in children and teenagers. *Arthritis Rheum* **42**:1782–1783.

83. Lafaille JJ, Nagashima K, Katsuki M, Tonegawa S. (1994) High incidence of spontaneous autoimmune encephalomyelitis in immunodeficient anti-myelin basic protein T cell receptor transgenic mice. *Cell* **78**(3):399–408.

84. Ehl S, Hombach J, Aichele P, *et al.* (1997) Bystander activation of cytotoxic T cells: studies on the mechanism and evaluation of *in vivo* significance in a transgenic mouse model. *J Exp Med* **185**(7):1241–1251.

85. Murali-Krishna K, Altman JD, Suresh M, *et al.* (1998) *In vivo* dynamics of anti-viral CD8 T cell responses to different epitopes. An evaluation of bystander activation in primary and secondary responses to viral infection. *Adv Exp Med Biol* **452**:123–142.

86. Wright S. (1933) An analysis of variability in number of digits in an inbred strain of guinea pigs. *Genetics* **19**:506–536.

87. McAleer MA, Reifsnyder P, Palmer SM, *et al.* (1995) Crosses of NOD mice with the related NON strain: a polygenic model for type 1 diabetes. *Diabetes* **44**:1186–1195.

88. Drake CG, Rozzo SJ, Hirschfeld HF, *et al.* (1995) Analysis of the New Zealand Black contribution to lupus-like renal disease. Multiple genes that operate in a threshold manner. *J Immunol* **154**:2441–2447.

89. Morel L, Rudofsky UH, Longmate JA, *et al.* (1994) Polygenic control of susceptibility to murine systemic lupus erythematosus. *Immunity* **1**:219–229.

90. Tsao BP, Cantor RM, Grossman JM, *et al.* (1999) PARP alleles within the linked chromosomal region are associated with systemic lupus erythematosus. *J Clin Invest* **103**(8):1135–1140.

91. Wu H, Cantor RM, Graham DS, *et al.* (2005) Association analysis of the R620W polymorphism of protein tyrosine phosphatase PTPN22 in systemic lupus erythematosus families: increased T allele frequency in systemic lupus erythematosus patients with autoimmune thyroid disease. *Arthritis Rheum* **52**(8):2396–2402.

92. Payne F, Smyth DJ, Pask R, *et al.* (2004) Haplotype tag single nucleotide polymorphism analysis of the human orthologues of the rat type 1 diabetes genes Ian4 (Lyp/Iddm1) and Cblb. *Diabetes* **53**(2):505–509.

93. Bell GI, Horita S, Karam JH. (1984) A polymorphic locus near the human insulin gene is associated with insulin-dependent diabetes mellitus. *Diabetes* **33**:176–183.

94. Bell GI, Karam JH, Raffel LJ, *et al.* (1985) Recessive inheritance for the insulin linked IDDM predisposing gene. *Am J Hum Genet* **37**:188A.

95. Paquette J, Giannoukakis N, Polychronakos C, *et al.* (1998) The INS 5′ variable number of tandem repeats is associated with IGF2 expression in humans. *J Biol Chem* **273**:14158–14164.

96. Pugliese A, Zeller M, Fernandez A Jr, *et al.* (1997) The insulin gene is transcribed in the human thymus and transcription levels correlated with allelic variation at the INS VNTR-IDDM2 susceptibility locus for type 1 diabetes. *Nat Genet* **15**:293–297.

97. Slingsby JH, Norsworthy P, Pearce G, *et al.* (1996) Homozygous hereditary C1q deficiency and systemic lupus erythematosus. A new family and the molecular basis of C1q deficiency in three families. *Arthritis Rheum* **39**(4):663–670.

98. Taylor PR, Carugati A, Fadok VA, *et al.* (2000) A hierarchical role for classical pathway complement proteins in the clearance of apoptotic cells *in vivo*. *J Exp Med* **192**:359–366.

99. Rossbacher J, Shlomchik MJ. (2003) The B cell receptor itself can activate complement to provide the complement receptor 1/2 ligand required to enhance B cell immune responses *in vivo*. *J Exp Med* **198**(4):591–602.

100. Gommerman JL, Carroll MC. (2000) Negative selection of B lymphocytes: a novel role for innate immunity. *Immunol Rev* **173**:120–130.

101. Hashimoto L, Habita C, Beressi JP, *et al.* (1994) Genetic mapping of a susceptibility locus for insulin-dependent diabetes mellitus on chromosome 11q. *Nature* **371**(6493):161–164.

102. Hugot JP, Chamaillard M, Zouali H, *et al.* (2001) Association of NOD2 leucine-rich repeat variants with susceptibility to Crohn's disease. *Nature* **411**:599–603.

103. Ogura Y, Bonen DK, Inohara N, *et al.* (2001) A frameshift mutation in NOD2 associated with susceptibility to Crohn's disease. *Nature* **411**:603–606.

104. [Anonymous]. (1997) An autoimmune disease, APECED, caused by mutations in a novel gene featuring two PHD-type zinc-finger domains. The

Finnish–German APECED Consortium. Autoimmune Polyendocrinopathy-Candidiasis-Ectodermal Dystrophy. *Nat Genet* **17**:399–403.

105. Nagamine K, Peterson P, Scott HS, *et al.* (1997) Positional cloning of the APECED gene. *Nat Genet* **17**:393–398.

106. Kochi Y, Yamada R, Suzuki A, *et al.* (2005) A functional variant in FCRL3, encoding Fc receptor-like 3, is associated with rheumatoid arthritis and several autoimmunities. *Nat Genet* **37**(5):478–485.

107. Lord CJ, Bohlander SK, Hopes EA, *et al.* (1995) Mapping the diabetes polygene *Idd3* on mouse chromosome 3 by use of novel congenic strains. *Mamm Genome* **6**:563–570.

108. Podolin PL, Wilusz MB, Cubbon RM, *et al.* (2000) Differential glycosylation of interleukin 2, the molecular basis for the NOD *Idd3* type 1 diabetes gene? *Cytokine* **12**:477–482.

109. Watanabe-Fukunaga R, Brannan CI, Copeland NG, *et al.* (1992) Lympho-proliferation disorder in mice explained by defects in Fas antigen that mediates apoptosis. *Nature* **356**(6367):314–317.

110. Serreze DV, Bridgett M, Chapman HD, *et al.* (1998) Subcongenic analysis of the *Idd13* locus in NOD/Lt mice: evidence for several susceptibility genes including a possible diabetogenic role for β2–microglobulin. *J Immunol* **160**:1472–1478.

111. Kozaiwa K, Sugawara K, Smith MF Jr, *et al.* (2003) Identification of a quantitative trait locus for ileitis in a spontaneous mouse model of Crohn's disease: SAMP1/YitFc. *Gastroenterology* **125**(2):477–490.

112. Yokoi N, Komeda K, Wang HY, *et al.* (2002) Cblb is a major susceptibility gene for rat type 1 diabetes mellitus. *Nat Genet* **31**(4):391–394.

113. Li D, Gal I, Vermes C, *et al.* (2004) Cutting edge: Cbl-b: one of the key molecules tuning CD28- and CTLA-4-mediated T cell costimulation. *J Immunol* **173**(12):7135–7139.

114. Ma RZ, Gao J, Meeker ND, *et al.* (2002) Identification of Bphs, an autoimmune disease locus, as histamine receptor H1. *Science* **297**(5581): 620–623.

115. Martignat L, Elmansour A, Andrain M, *et al.* (1995) Pancreatic expression of antigens for islet cell antibodies in non-obese diabetic mice. *J Autoimmun* **8**(4):465–482.

116. Yamamoto T, Yamato E, Tashiro F, *et al.* (2004) Development of autoimmune diabetes in glutamic acid decarboxylase 65 (GAD65) knockout NOD mice. *Diabetologia* **47**(2):221–224.

117. Swanberg M, Lidman O, Padyukov L, *et al.* (2005) MHC2TA is associated with differential MHC molecule expression and susceptibility to rheumatoid arthritis, multiple sclerosis and myocardial infarction. *Nat Genet* **37**(5): 486–494.

118. Becker KG. (1999) Comparative genetics of type 1 diabetes and autoimmune disease: common loci, common pathways? *Diabetes* **48**:1353–1358.

119. Griffiths MM, Encinas JA, Remmers EF, *et al.* (1999) Mapping autoimmunity genes. *Curr Opin Immunol* **11**:689–700.

120. She JX, Marron MP. (1998) Genetic susceptibility factors in type 1 diabetes: linkage, disequilibrium and functional analyses. *Curr Opin Immunol* **10**:682–689.

121. Caron H, van Schaik B, van der Mec M, *et al.* (2001) The human transcriptome map: clustering of highly expressed genes in chromosomal domains. *Science* **291**:1289–1292.

Chapter 30

Genetics of Substance Dependence

Joel Gelernter[*,‡] and Henry R. Kranzler[†]

[*]*Department of Psychiatry, Genetics, and Neurobiology*
Division of Human Genetics, Yale University School of Medicine
New Haven, CT 06520, USA; & VA Connecticut Healthcare System
West Haven, CT 06516, USA

[†]*Department of Psychiatry, University of Connecticut School of Medicine*
Farmington, CT 06030, USA

1. INTRODUCTION

Like other important psychiatric traits, substance dependence (SD) is genetically complex. (The "substances" considered for the purposes of this chapter will mostly be alcohol, tobacco, and illegal drugs, primarily cocaine and opioids.) As is the case for other complex traits, SD risk is influenced by both genetic and environmental factors, but with a twist: there is a necessary component of gene-by-environment interaction. One cannot become substance-dependent without exposure to the substance, regardless of one's genetic constitution. This places a trait like opioid dependence (you cannot be opioid-dependent if you do not have access to opioids) in contrast to a trait like schizophrenia, which, so far as we know, does not require any special environmental exposure to develop.

In this chapter, we will discuss the progress in mapping genes that influence risk for SD. But, how do we know that genes are important? SD is known to be genetically influenced, first, on the basis of genetic epidemiology. We will briefly discuss the evidence supporting the genetic contribution to each of the major forms of SD. Then, we will discuss the results from

---

[‡] Correspondence author.

genetic linkage analyses, which provide chromosomal localization of susceptibility genes but not the actual genes. Finally, we will discuss some of the evidence supporting the relationship of specific candidate genes to phenotype.

Overall, the last few years have seen remarkable developments in our understanding of the genetic basis of SD. A series of consistent, replicated findings have emerged that are providing new insight into the biological mechanisms of these disorders.

2. EVIDENCE THAT GENES ARE IMPORTANT FOR DRUG-DEPENDENCE RISK

We will discuss here four substance use disorders that are among the greatest public health problems in the US: alcohol dependence (AD); cocaine dependence (CD); opioid dependence (OD); and tobacco, or nicotine, dependence (ND). We will refer to the last three (CD, OD, and ND) as drug dependence (DD). All of these disorders influence many facets of American society, cutting across geography, race, ethnicity, and socioeconomic status. These disorders are of varying importance internationally, with alcohol and tobacco being the most consistently available substances and, consequently, alcohol and nicotine dependence the most consistent problems worldwide.

Support for the importance of genetic factors in risk for a disorder usually comes from twin and adoption studies. Family studies can demonstrate that disorders are familial and show the range of related phenotypes, but cannot demonstrate heritability *per se.*

2.1. *Heritability of Alcohol Dependence*

Genetic factors are important for the development of AD, as established by twin, family, and adoption studies. The largest twin studies have settled on heritability estimates in the range of 50%–60%, i.e. half or more of the risk for AD is genetic.[1–3] A particularly noteworthy study[2] considered the intersection between the Swedish Twin Registry, which logged almost all twins born from 1902 to 1949 (about 9000 male pairs), and Swedish temperance board registrations from 1929 to 1974 (about 2500 twins). Subjects came to the attention of temperance boards mostly from physicians and law enforcement agencies because of alcoholism or crimes related to alcohol. Temperance board registration was a proxy measure for the diagnosis of AD. Because multiple cohorts of twins were assessed, in this study, the researchers were able to consider not only the heritability of AD *per se*, but also the stability of that heritability over time, in an epidemiological sample.

These authors found that, although the prevalence was similar overall in monozygotic (MZ) and dizygotic (DZ) twins, the concordance rate was significantly higher in MZ than in DZ twins. Moreover, heritability was in fact quite stable over time, suggesting that the environmental contributions to the risk for AD were somewhat consistent as well (in magnitude if not in their exact nature) and that major social and historical changes did not affect environmentally determined AD risk, at least in this sample. Thus, the diagnostic constructs used for AD for which genetic liability can be estimated are valid and meaningful in terms of the consequences and outcomes for individuals.

2.2. Heritability of Drug Dependence

Tsuang *et al.*[4] reported on data from the Vietnam Era Twin Registry, which includes more than 3000 male twin pairs, among whom drug abuse was defined as the at least weekly use of any of a variety of drugs. This study has yielded detailed and comprehensive information regarding the heritability of a range of substance use disorders. Significant pairwise concordance rates showed a familial basis for every one of the drugs considered. The difference in pairwise concordance rates for MZ and DZ twins was significant for abuse of or dependence on marijuana, stimulants, cocaine, and all drugs combined. For OD, the estimated (additive genetic) heritability was 0.43. Despite the large sample, there were relatively few OD twins; the MZ concordance for this disorder was 13.3% and DZ concordance, 2.9% (with sample sizes of around 30 pairs in both cases). For stimulant abuse, the estimated heritability was 0.44. The MZ concordance was 14.1% (21/149) and DZ concordance, 5.3% (6/113); the difference between proportions was significant at the $p < 0.05$ level. Using the Vietnam Era Twin Registry, Tsuang *et al.*[5] also examined the genetic risk for the co-occurrence of different forms of substance use disorders. They determined that there exist genetic factors both specific to certain individual drugs of abuse, including stimulants, and general to multiple forms of abuse or dependence.

Kendler has published twin study data demonstrating remarkably high heritability for cocaine dependence. In a sample of female twins, heritabilities for cocaine abuse and dependence were estimated at 0.79 and 0.65, respectively.[6] Data from a sample of male twins showed the heritability of CD to be 0.79.[7] For this study, however, the best-fit model for cocaine abuse did not include an additive genetic (a^2) term; Kendler *et al.*[7] pointed out that the power to distinguish between models from this analysis was limited. The twin samples used in these studies were large, but since they were epidemiological samples, in which the prevalence of the disorders is low, the actual numbers of twin pairs from which heritabilities were estimated were considerably smaller. Nonetheless, these data are the best and most specific to date addressing genetic liability for cocaine abuse and dependence.

Many behaviors related to nicotine dependence have been shown to be heritable. For example, the Vietnam Era Twin Registry has also shed light on ND heritability, which was estimated at >60% in this sample.[8] Heritability of AD was estimated as 0.55 in the same sample, with genetic correlation between the disorders and the genetic correlation between ND and AD to be 0.68. Madden *et al.*[9] also found the genetic risk for these disorders to be correlated. Heritability of regular tobacco use was estimated at >60%, based on analysis of a large sample of Swedish twin pairs reared together or apart.[10] (For this investigation, the phenotype was defined by the subject's response that he smoked or used snuff regularly.) In a large Finnish twin study, the heritability for age of smoking initiation was 0.59 for males and 0.36 for females; for amount smoked, 0.54 in males and 0.61 in females; and for smoking cessation, 0.58 in males and 0.50 in females.[11]

Li *et al.*[12] conducted a meta-analysis of twin studies for phenotypes related to ND and reported that, based on reported twin study information, smoking initiation heritability could be estimated at 0.37 for males and 0.55 for females, and smoking persistence heritability at 0.59 for males and 0.46 for females. Lessov *et al.*[13] specifically addressed the issue of ND-related trait definition for the purpose of gene mapping. They considered DSM-IV[14] trait definition and alternate definitions based on other phenotypic measures.

2.3. *Summary: Genetic Effects and Substance Dependence*

Both AD and DD are familial and genetically influenced. The data on these disorders are consistent with a model in which there are both general and specific risk factors — i.e. some loci act to influence the risk to specific kinds of substance dependence, and others act more generally to influence the risk for more than one kind. Establishing genetic bases for these disorders provides a clear rationale for efforts to identify the genes that underlie the risk for their development.

3. GENOMEWIDE STUDIES

Genomewide linkage studies, the traditional approach to identifying risk loci, do not require an *a priori* physiological hypothesis. Linkage studies are family-based studies and require that markers mapping throughout the entire genome be genotyped to allow identification of a genomic region where markers are coinherited with the phenotype of interest. Genomewide linkage scans have been completed for alcohol, cocaine, opioid, and nicotine dependence. Genomewide association scans involve a newer methodology in which closely spaced markers are studied throughout the genome in an effort to discover those that vary in frequency in cases compared to controls. In this kind of study, the goal is to genotype enough markers so that they are within linkage disequilibrium distance of any point in the genome. At the time of this writing, no such studies have been completed for SD, but some such studies are on the horizon; we may expect such data to be available in the years ahead.

Successful genomewide lineage studies give chromosomal locations of risk loci, but generally do not point to specific genes. Successful genomewide association studies can implicate specific genes.

3.1. *Genomewide Linkage Studies of Alcohol Dependence*

Genome scan linkage mapping projects[15,16] have identified promising map positions for AD susceptibility loci, some of which have already led to the discovery of disease-influencing loci. Linkage studies of AD published by the Collaborative Study on the Genetics of Alcoholism (COGA)[16,17] and investigators in the intramural program of the US National Institute on Alcohol Abuse and Alcoholism (NIAAA)[15] provided several chromosomal locations with promising logarithm of odds (LOD) scores that may harbor loci influencing the risk of AD. Interestingly, both groups reported data consistent with loci influencing the risk for AD mapping close to an alcohol dehydrogenase (*ADH*) gene cluster on the long arm of chromosome 4 (see below). Several other linkage peaks, even fairly unimpressive ones that did not meet standard criteria for genomewide significance, have led to the identification of likely disease-influencing loci. Part of the Genetic Analysis Workshop (GAW) 11 was devoted to analysis of the COGA data set; this resulted in several analyses of the AD data using the transmission/disequilibrium test (TDT) and related methods,[18–22] revealing several additional areas of interest.

Wilhelmsen *et al.*[23] reported a genomewide linkage study for an alcohol-related trait, low-level response to alcohol. Although this study considered a relatively small sample (139 sibling pairs), several suggestive linkages were identified. Ehlers *et al.*[24] conducted linkage analysis using 791 microsatellite markers in 243 Mission Indians in the Southwest US. Although the diagnosis of alcohol dependence yielded no LOD scores suggestive of linkage, several LOD scores in excess of 2.0 were identified for the phenotypes of alcohol severity (chromosomes 4 and 12) and alcohol withdrawal (chromosomes 6, 15, and 16).

3.2. Genomewide Linkage Studies of Drug Dependence

There have been numerous genomewide linkage scans for ND and related traits, and one each for CD and OD.

3.2.1. Genomewide Linkage Study of Cocaine Dependence

We studied a sample of small nuclear families with at least one subject affected with CD that included 528 full and 155 half sib pairs, 45.5% European-American (EA) and 54.5% African-American (AA). We completed a genomewide linkage scan for CD diagnosis, cocaine-induced paranoia, and six cocaine-related subtypes that were derived using cluster analytic methods. For the diagnosis of CD, we found suggestive linkage signals on chromosome 10 in the full sample and at different locations on chromosome 3 in the EA part of the sample. The strongest results were obtained for the cluster-derived subtypes, including an LOD score of 4.66 for membership in the "heavy-use, cocaine-predominant" cluster on chromosome 12 (in EAs only) and a LOD score of 3.35 for membership in the "moderate cocaine and opioid abuse" cluster on chromosome 18. Finally, among AA families, we observed a genomewide-significant LOD score of 3.65 on chromosome 9 for the trait of cocaine-induced paranoia.

3.2.2. Genomewide Linkage Study of Opioid Dependence

We studied a sample of small nuclear families (393 families, including 250 full and 46 half sib pairs, overlapping with the sample studied in the CD linkage study described above), each with at least one individual affected with OD. We completed a genomewide linkage scan (409 markers) for the DSM-IV diagnosis of OD and, analogous to the CD study, for two cluster-defined phenotypes represented by >250 families, a heavy-opioid-use cluster and a non–opioid-using cluster. Further exploratory analyses were completed for the other cluster-defined phenotypes. The strongest results statistically were, again, seen with the cluster-defined traits: for the "heavy opioid users" cluster, we observed an LOD score of 3.06, for EA and AA subjects combined, on chromosome 17 (empirical pointwise $p = 0.0002$); and for the "non-opioid users" cluster, we observed an LOD score of 3.46 (empirical pointwise $p = 0.00002$; uncorrected for multiple traits studied) elsewhere on chromosome 17, for EA subjects only.

3.2.3. *Genomewide Linkage Studies of Nicotine Dependence and Related Traits*

There have been several previous genomewide linkage scans for smoking and related phenotypes, as reviewed by Li *et al.*[12] Many studies have used patient samples recruited for an index trait other than ND. A wide range of possible linkages have been identified. We recently reported genomewide-significant linkage of the Fagerstrom Test for Nicotine Dependence (FTND) score to chromosome 5 markers in AA subjects.

4. CANDIDATE GENE STUDIES

The following will summarize some of the most notable candidate gene studies in substance dependence. Many of the genes discussed below are associated with both drug and alcohol dependence.

4.1. *Alcohol Dependence Candidate Gene Studies*

4.1.1. *Alcohol-Metabolizing Enzymes*

The influence of genetic polymorphism at some loci encoding acetaldehyde dehydrogenases and alcohol dehydrogenases on the risk of AD in some populations is well established, and the mechanism is very clear. Ethanol is metabolized to acetaldehyde by alcohol dehydrogenases, for which the most historically important loci for AD research are *ADH1B* (previously known as *ADH2*) and *ADH1C* (previously known as *ADH3*). Acetaldehyde is metabolized primarily by acetaldehyde dehydrogenases, for which the relevant locus to be considered is *ALDH2*. Acetaldehyde is toxic and produces a "flushing reaction" characterized by a set of uncomfortable symptoms including flushing, lightheadedness, palpitations, and nausea. Thus, deviation from the normal metabolism of ethanol that results in increased exposure to acetaldehyde would be expected to create an aversive effect of ethanol use, which might decrease alcohol consumption and, thus, the risk of AD.[25] Indeed, a variant that greatly reduces or eliminates ALDH function (thereby decreasing the elimination of acetaldehyde), which occurs mostly in Asian populations, has long been known to be protective against AD; and *ADH* variants that increase function may also be protective by increasing the production of acetaldehyde.[26–28]

As noted above, several genomewide linkage scans have implicated a region of chromosome 4q that contains an *ADH* gene cluster, and this has prompted more intensive investigation of the ADHs. *ADH4*[29,30] has now been identified as one of several disease-influencing loci in this cluster. In adults, the π subunit encoded by *ADH4* — which mainly contributes to liver ADH activity[31–33] — demonstrated that −75A, at a promoter polymorphic site, has promoter activity that is more than twice that of the −75C allele; this variant could affect π subunit regulation.[33] Despite the important role of the *ADH4* gene in alcohol metabolism, it was initially largely overlooked with respect to this disorder. We have reported strong associations of *ADH4* markers to AD using a range of methods, including Hardy–Weinberg disequilibrium (HWD) analysis, structured association, and family-based association.[29,34]

We reported that *ADH* and *ALDH2* variants are associated with AD.[35] We genotyped 16 markers within the *ADH* gene cluster, 4 markers within the *ALDH2* gene, and 38 unlinked ancestry-informative markers (AIMs) in a case-control sample. Associations between markers and diseases were analyzed by a Hardy–Weinberg equilibrium (HWE) test, a conventional case-control comparison, structured association (SA) analysis, and a novel diplotype trend regression (DTR) analysis. All markers were found to be in HWE in controls, but some markers showed HWD in cases. Genotypes of many markers were associated with AD. DTR showed that *ADH5* genotypes, as well as diplotypes of *ADH1A*, *ADH1B*, *ADH7*, and *ALDH2*, were associated with AD in EAs, AAs, or both. In this study, the risk-influencing alleles were fine-mapped and were found to coincide with some well-known functional variants.

Edenberg *et al.*[30] genotyped SNPs across the *ADH* genes on chromosome 4q in a set of families from COGA. The most consistent evidence of association to AD was observed for *ADH4*, extending into the 3′ untranslated region of that gene bordering on the *ADH5* gene. Haplotype analysis using tag SNPs that covered the entire *ADH4* gene along with both upstream and downstream regions of the gene also showed significant evidence for association. In addition, suggestive evidence of association with AD was obtained by these investigators for the *ADH1A* and *ADH1B* genes.

4.1.2. *NPY*

Using a case-control study, we found that a functional variant at the neuropeptide Y (*NPY*) locus was associated with the risk for AD.[36]

Quantitative trait locus studies and observations in animals manipulated for the *NPY* gene suggested that variation within this gene may contribute to alcoholism. A population study[37] had suggested that the Pro7 allele of a functional *NPY* polymorphism (Leu7Pro) was associated with increased alcohol consumption. Our objective was to test whether the Pro7 allele was associated with AD in European-Americans (EAs). We studied population stratification potential and diagnostic specificity by genotyping individuals from additional populations and psychiatric diagnostic classes. Two independently collected samples of EA alcohol-dependent subjects and a sample of psychiatrically screened EA controls were studied. In addition, we studied eight population samples, including AAs, and four samples of individuals with other psychiatric phenotypes to evaluate the specificity of gene effects.

The frequency of the Pro7 allele was higher in the alcohol-dependent subjects compared to the controls in both samples. Furthermore, we found no significant evidence that the association of the Pro7 allele to AD was attributable to an association with one of the other psychiatric disorders studied. These results suggested that the *NPY* Pro7 allele is a risk factor for AD. Subsequently, Zhu *et al.*[38] compared Pro7 allele frequencies in Finnish ($n = 135$) and Swedish ($n = 472$) alcohol-dependent individuals and ethnically matched controls (Finns: $N = 213$; Swedes: $N = 177$); they found no evidence for the association of the Pro7 allele with AD in this population.

4.1.3. *GABRA2*

Some of the most compelling recent work supporting associations between specific loci (although not yet specific alleles) with the risk of AD or related traits comes from the COGA group. Work reported by these investigators[39] demonstrated, first, linkage of EEG beta frequencies (a quantitative trait) to a region of chromosome 4p; then, linkage disequilibrium to a GABA$_A$ receptor cluster (which maps to chromosome 4p), in a sample ascertained through pedigrees with multiple alcohol-dependent individuals. Fine mapping showed allelic and haplotypic association to *GABRA2*, one of four GABA$_A$ genes in this region.[40] Using case-control samples, we[41,42] and Fehr *et al.*[43] have replicated this major finding, with evidence of association of AD to a haplotype at *GABRA2* in three different populations. Although we know that the association is consistent and replicable, no specific causative variant has been reported, nor is a mechanism of action known. Interestingly, however, Pierucci-Lagha *et al.*[44] found an allelic association of *GABRA2* with the subjective response to alcohol in a sample of

healthy subjects. Subjects homozygous for the more common A-allele of
SNP rs279858 (a variant that was informative in association studies[40,41])
showed greater subjective effects of alcohol than did individuals with one
or two copies of the AD-associated G-allele.

These data provide preliminary evidence that the risk of alcoholism
associated with *GABRA2* alleles may be related to differences in the subjective response to alcohol. Two other genes encoding GABA$_A$ receptor
subunits, *GABRA1* and *GABRG3*, have also been shown to be associated
to AD, although the evidence supporting these loci is presently substantially weaker than that for *GABRA2*.[45,46]

4.1.4. *CHRM2*

Another recent report provides moderately strong support for muscarinic
acetylcholine receptor M2 (genetic locus *CHRM2*) as an AD risk locus.[47]
Like the *ADH* cluster loci and *GABRA2*, *CHRM2* maps to a chromosomal
region that had been identified as being of interest based on an AD linkage
study.[16] Several *CHRM2* SNPs were reported to be significantly associated to AD; two intron 4 SNPs at this locus were also associated to major
depression. In addition, we found evidence of association of markers at this
locus with the same two phenotypes.[48] This raises the interesting question
of exactly which phenotype or phenotypes are influenced by polymorphic
variation at the locus of interest, and whether the intermediate phenotypes
or endophenotypes are common to AD and depression in this case.

4.2. *Drug Dependence Candidate Gene Studies*

4.2.1. *OPRM1*

The mu-opioid receptor has been implicated in the pathogenesis of dependence on opioids, alcohol, nicotine, and cocaine. Studies examining the association of the mu-opioid receptor gene (genetic locus *OPRM1*) with SD have
focused on the A118G polymorphism, which encodes an Asn40Asp amino
acid substitution. Arias *et al.*,[49] using meta-analysis, examined the association of Asn40Asp with SD in 22 published articles describing 28 distinct
samples and over 8000 subjects. They included a variety of factors (i.e.
ethnicity, type of SD, rigor with which controls were screened, severity of
SD among cases) as potential moderators of the association. Four studies showed a significantly higher frequency of the Asp40 allele among SD
cases, while three studies showed a significantly higher frequency of the

Asp40 allele among controls. There was no significant association between Asn40Asp and SD (OR = 1.01; 95% CI = 0.86–1.19), nor was there substantial evidence of a moderator effect.

We,[50] in the most comprehensive study of the locus to date, examined 13 SNPs spanning the *OPRM1* coding region among 382 European-Americans affected with AD and/or DD and 338 healthy controls. These SNPs delineated two haplotype blocks: genotype distributions for all SNPs were in Hardy–Weinberg equilibrium (HWE) in controls; but in some cases, four SNPs in block I and three SNPs in block II showed deviation from HWE. Significant differences were found between cases and controls in allele and/or genotype frequencies for six SNPs in block I and two SNPs in block II. Frequency distributions of haplotypes (constructed by five tag SNPs) differed significantly for cases and controls ($P < 0.001$ for both AD and DD). Logistic regression analyses confirmed the association between *OPRM1* variants and SD when the sex and age of subjects as well as the alleles, genotypes, and haplotypes or diplotypes of the five tag SNPs were considered, and when population structure analyses excluded population stratification artifact. Additional supporting evidence for the association between *OPRM1* and AD was obtained in a smaller sample of Russian subjects. These findings suggest that *OPRM1* intronic variants play a role in susceptibility to AD and DD in populations of European ancestry. Although the opioid-dependent subsample was comparatively small, some of the most robust associations were observed in that sample.

4.2.2. *DDC*

DDC is an enzyme of major importance for dopamine synthesis, and also plays a role in the serotonin biosynthetic pathway.[51] Ma *et al.*[52] studied *DDC*, the gene encoding this protein, in ND based on its physiology and map location in an ND linkage peak that they identified. They showed an association between long-range *DDC* haplotypes and three correlated quantitative traits of smoking behavior (smoking quantity, heaviness of smoking index, and the Fagerstrom Test for Nicotine Dependence [FTND] score) using family-based association tests (FBATs) in families ascertained through an affected sib pair (ASP) for ND.

We[53] reported an association of alleles and haplotypes at *DDC* with the DSM-IV diagnosis of ND or FTND score. We genotyped 18 SNPs spanning a region of approximately 210 kb, including *DDC* and the genes immediately flanking it, in a large sample of affected sibling pairs recruited for OD or

CD. Evidence of association (using family-based tests) was observed with several SNPs for both traits; the most significant result was obtained for the relationship of FTND score to an SNP that maps to the same intron as the splice site for a neuronal isoform of human *DDC* lacking exons 10–15. These findings were consistent with the findings of Ma *et al.*,[52] and served to localize the causative variants to the 3′ end of the coding region.

It is noteworthy that several of the genes discussed above (*GABRA2, DDC*) were identified as risk loci based on their map positions in proximity to linkage peaks, identified in genomewide linkage analysis.

5. GENE–ENVIRONMENT INTERACTION

There is a growing appreciation for the fact that gene-by-environment interaction is not only an important factor for modulating the risk for psychiatric phenotypes, but that the magnitude of such effects can be large enough to be detected reliably. For example, Caspi *et al.*,[54] in a prospective longitudinal study of a representative birth cohort, found that individuals with one or two copies of the short allele of a common functional polymorphism, 5-HTTLPR, in the serotonin transporter gene (*SLC6A4*) reported more depressive symptoms, depressive disorder, and suicidality in relation to stressful life events than individuals homozygous for the long allele. Kaufman *et al.*[55] reported consistent results, but in an adolescent population.

Covault *et al.*[56] recently reported finding a similar gene-by-environment interaction in which the 5-HTTLPR polymorphism, together with negative life events, moderated drinking and drug use in college students. Specifically, these investigators found that individuals homozygous for the short 5-HTTLPR allele who experienced multiple negative life events reported more frequent drinking and heavy drinking and greater nonprescribed drug use; in contrast, drinking and drug use among individuals homozygous for the long allele were unaffected by negative life events. Heterozygous subjects showed drinking outcomes that were intermediate to the two homozygous groups.

Kaufman *et al.*[57] examined genetic and environmental predictors of early alcohol use, but in adolescent subjects, and reported results consistent with those of Covault *et al.*[56] In this study, predictors of early alcohol use included maltreatment, SD family loading, and 5-HTTLPR genotype. Maltreated children and matched community controls participated. The rate of alcohol use in the maltreated children was more than seven times the rate observed in controls. Maltreated children also initiated drinking, on average, more than 2 years earlier than controls. Early alcohol use was predicted by maltreatment, 5-HTTLPR, and a gene-by-environment interaction, with increased risk for early alcohol use associated with the short allele.

6. DISCUSSION: GENES THAT INFLUENCE RISK AND THE NATURE OF THEIR EFFECTS

Studies in genetic epidemiology have shown that alcohol and drug dependence are genetically influenced. Since SD traits are genetically complex, identifying risk genes has been a difficult task. Nevertheless, chromosomal locations for risk genes have been identified through the use of genetic linkage studies, and risk genes have been found based on the linkage peaks. Further, candidate genes based on physiology have been identified reliably as risk genes for AD, DD, or both. There are now several cases where detailed replications have been published. In fact, SD traits present some of the more encouraging examples of in-progress genetic dissection of sets of complex traits.

In addition to the identification of other loci that contribute to the risk of these disorders, considerable work is needed to elucidate the mechanisms by which the genes that are identified exert their effects. Such insights, when available, will provide a firm basis for efforts at prevention, early identification, and treatment of these common and often destructive disorders.

ACKNOWLEDGMENTS

This work was supported in part by NIDA grants R01-DA12849, R01-DA12690, and K24-DA15105; NIAAA grants R01-AA11330 and K24-AA13736; and funds from the US Department of Veterans Affairs (the VA Medical Research Program and the VA Connecticut–Massachusetts Mental Illness Research, Education and Clinical Center [MIRECC]).

REFERENCES

1. Kendler KS, Heath AC, Neale MC, *et al.* (1992) A population-based twin study of alcoholism in women. *JAMA* **268**(14):1877–1882.
2. Kendler KS, Prescott CA, Neale MC, Pedersen NL. (1997) Temperance board registration for alcohol abuse in a national sample of Swedish male twins, born 1902 to 1949. *Arch Gen Psychiatry* **54**(2):178–184.

3. Prescott CA, Kendler KS. (1999) Genetic and environmental contributions to alcohol abuse and dependence in a population-based sample of male twins. *Am J Psychiatry* **156**(1):34–40.

4. Tsuang MT, Lyons MJ, Eisen SA, *et al.* (1996) Genetic influences on DSM-III-R drug abuse and dependence: a study of 3,372 twin pairs. *Am J Med Genet* **67**(5):473–477.

5. Tsuang MT, Lyons MJ, Meyer JM, *et al.* (1998) Co-occurrence of abuse of different drugs in men: the role of drug-specific and shared vulnerabilities. *Arch Gen Psychiatry* **55**(11):967–972.

6. Kendler KS, Prescott CA. (1998) Cocaine use, abuse and dependence in a population-based sample of female twins. *Br J Psychiatry* **173**:345–350.

7. Kendler KS, Karkowski LM, Neale MC, Prescott CA. (2000) Illicit psychoactive substance use, heavy use, abuse, and dependence in a US population-based sample of male twins. *Arch Gen Psychiatry* **57**(3):261–269.

8. True WR, Xian H, Scherrer JF, *et al.* (1999) Common genetic vulnerability for nicotine and alcohol dependence in men. *Arch Gen Psychiatry* **56**(7): 655–661.

9. Madden PA, Bucholz KK, Martin NG, Heath AC. (2000) Smoking and the genetic contribution to alcohol-dependence risk. *Alcohol Res Health* **24**(4): 209–214.

10. Kendler KS, Thornton LM, Pedersen NL. (2000) Tobacco consumption in Swedish twins reared apart and reared together. *Arch Gen Psychiatry* **57**(9):886–892.

11. Broms U, Silventoinen K, Madden PA, *et al.* (2006) Genetic architecture of smoking behavior: a study of Finnish adult twins. *Twin Res Hum Genet* **9**(1):64–72.

12. Li MD, Ma JZ, Beuten J. (2004) Progress in searching for susceptibility loci and genes for smoking-related behaviour. *Clin Genet* **66**(5):382–392.

13. Lessov CN, Martin NG, Statham DJ, *et al.* (2004) Defining nicotine dependence for genetic research: evidence from Australian twins. *Psychol Med* **34**(5):865–879.

14. American Psychiatric Association. (1994) *Diagnostic and Statistical Manual of Mental Disorders*, 4th ed. American Psychiatric Association, Washington, DC.

15. Long JC, Knowler WC, Hanson RL, *et al.* (1998) Evidence for genetic linkage to alcohol dependence on chromosomes 4 and 11 from an autosome-wide scan in an American Indian population. *Am J Med Genet* **81**(3):216–221.

16. Reich T, Edenberg HJ, Goate A, *et al.* (1998) Genome-wide search for genes affecting the risk for alcohol dependence. *Am J Med Genet* **81**(3):207–215.

17. Foroud T, Edenberg HJ, Goate A, *et al.* (2000) Alcoholism susceptibility loci: confirmation studies in a replicate sample and further mapping. *Alcohol Clin Exp Res* **24**(7):933–945.

18. Curtis D, Miller MB, Sham PC. (1999) Combining the sibling disequilibrium test and transmission/disequilibrium test for multiallelic markers. *Am J Hum Genet* **64**(6):1785–1786.

19. Camp NJ, Bansal A. (1999) A low density genome-wide search for loci involved in alcohol dependence using the transmission/disequilibrium test, sib-TDT, and two combined tests. *Genet Epidemiol* **17**:S85–S90.

20. Page G, King T, Barnholtz J, *et al.* (1999) Genome scans for genetic predisposition to alcoholism by use of TDT analyses. *Genet Epidemiol* **17**(Suppl 1):s277–s281.

21. Nielsen D, Zaykin D. (1999) Novel tests for marker-disease association using the Collaborative Study on the Genetics of Alcoholism data. *Genet Epidemiol* **17**(Suppl 1):S265–S270.

22. Sun F, Flanders WD, Yang Q, Khoury MJ. (1999) Transmission disequilibrium test (TDT) when only one parent is available: the 1–TDT. *Am J Epidemiol* **150**(1):97–104.

23. Wilhelmsen KC, Schuckit M, Smith TL, *et al.* (2003) The search for genes related to a low-level response to alcohol determined by alcohol challenges. *Alcohol Clin Exp Res* **27**(7):1041–1047.

24. Ehlers CL, Gilder DA, Wall TL, *et al.* (2004) Genomic screen for loci associated with alcohol dependence in Mission Indians. *Am J Med Genet B Neuropsychiatr Genet* **129**(1):110–115.

25. Goedde HW, Harada S, Agarwal DP. (1979) Racial differences in alcohol sensitivity: a new hypothesis. *Hum Genet* **51**(3):331–334.

26. Thomasson HR, Edenberg HJ, Crabb DW, *et al.* (1991) Alcohol and aldehyde dehydrogenase genotypes and alcoholism in Chinese men. *Am J Hum Genet* **48**(4):677–681.

27. Hasin D, Aharonovich E, Liu X, *et al.* (2002) Alcohol and ADH2 in Israel: Ashkenazis, Sephardics, and recent Russian immigrants. *Am J Psychiatry* **159**(8):1432–1434.

28. Konishi T, Smith JL, Lin KM, Wan YJ. (2003) Influence of genetic admixture on polymorphisms of alcohol-metabolizing enzymes: analyses of mutations on the CYP2E1, ADH2, ADH3 and ALDH2 genes in a Mexican-American population living in the Los Angeles area. *Alcohol Alcohol* **38**(1):93–94.

29. Luo X, Kranzler HR, Zuo L, *et al.* (2005) ADH4 gene variation is associated with alcohol and drug dependence: results from family controlled and population-structured association studies. *Pharmacogenet Genomics* **15**(11): 755–768.

30. Edenberg HJ, Xuei X, Chen HJ, *et al.* (2006) Association of alcohol dehydrogenase genes with alcohol dependence: a comprehensive analysis. *Hum Mol Genet* **15**(9):1539–1549.

31. Li TK, Bosron WF, Dafeldecker WP, *et al.* (1977) Isolation of pi-alcohol dehydrogenase of human liver: is it a determinant of alcoholism? *Proc Natl Acad Sci USA* **74**(10):4378–4381.

32. Li TK, Bosron WF. (1987) Distribution and properties of human alcohol dehydrogenase isoenzymes. *Ann NY Acad Sci* **492**:1–10.

33. Edenberg HJ, Jerome RE, Li M. (1999) Polymorphism of the human alcohol dehydrogenase 4 (ADH4) promoter affects gene expression. *Pharmacogenetics* **9**(1):25–30.

34. Luo X, Kranzler HR, Zuo L, *et al.* (2006) ADH4 gene variation is associated with alcohol dependence and drug dependence in European Americans: results from HWD tests and case-control association studies. *Neuropsychopharmacology* **31**(5):1085–1095.

35. Luo X, Kranzler HR, Zuo L, *et al.* (2006) Diplotype trend regression analysis of the ADH gene cluster and the ALDH2 gene: multiple significant associations with alcohol dependence. *Am J Hum Genet* **78**(6):973–987.

36. Lappalainen J, Kranzler HR, Malison R, *et al.* (2002) A functional neuropeptide Y Leu7Pro polymorphism associated with alcohol dependence in a large population sample from the United States. *Arch Gen Psychiatry* **59**(9): 825–831.

37. Kauhanen J, Karvonen MK, Pesonen U, *et al.* (2000) Neuropeptide Y polymorphism and alcohol consumption in middle-aged men. *Am J Med Genet* **93**(2):117–121.

38. Zhu G, Pollak L, Mottagui-Tabar S, *et al.* (2003) NPY Leu7Pro and alcohol dependence in Finnish and Swedish populations. *Alcohol Clin Exp Res* **27**(1): 19–24.

39. Porjesz B, Almasy L, Edenberg HJ, *et al.* (2002) Linkage disequilibrium between the beta frequency of the human EEG and a GABAA receptor gene locus. *Proc Natl Acad Sci USA* **99**(6):3729–3733.

40. Edenberg HJ, Dick DM, Xuei X, *et al.* (2004) Variations in GABRA2, encoding the alpha 2 subunit of the GABA$_A$ receptor, are associated with alcohol dependence and with brain oscillations. *Am J Hum Genet* **74**(4):705–714.

41. Covault J, Gelernter J, Hesselbrock V, *et al.* (2004) Allelic and haplotypic association of GABRA2 with alcohol dependence. *Am J Med Genet B Neuropsychiatr Genet* **129**(1):104–109.

42. Lappalainen J, Krupitsky E, Remizov M, *et al.* (2005) Association between alcoholism and gamma-amino butyric acid alpha2 receptor subtype in a Russian population. *Alcohol Clin Exp Res* **29**(4):493–498.

43. Fehr C, Sander T, Tadic A, *et al.* (2006) Confirmation of association of the GABRA2 gene with alcohol dependence by subtype-specific analysis. *Psychiatr Genet* **16**(1):9–17.

44. Pierucci-Lagha A, Covault J, Feinn R, *et al.* (2005) GABRA2 alleles moderate the subjective effects of alcohol, which are attenuated by finasteride. *Neuropsychopharmacology* **30**(6):1193–1203.

45. Dick DM, Edenberg HJ, Xuei X, *et al.* (2004) Association of GABRG3 with alcohol dependence. *Alcohol Clin Exp Res* **28**(1):4–9.

46. Dick DM, Plunkett J, Wetherill LF, *et al.* (2006) Association between GABRA1 and drinking behaviors in the Collaborative Study on the Genetics of Alcoholism sample. *Alcohol Clin Exp Res* **30**(7):1101–1110.

47. Wang JC, Hinrichs AL, Stock H, *et al.* (2004) Evidence of common and specific genetic effects: association of the muscarinic acetylcholine receptor M2 (CHRM2) gene with alcohol dependence and major depressive syndrome. *Hum Mol Genet* **13**(17):1903–1911.

48. Luo X, Kranzler HR, Zuo L, *et al.* (2005) CHRM2 gene predisposes to alcohol dependence, drug dependence and affective disorders: results from an

extended case-control structured association study. *Hum Mol Genet* **14**(16): 2421–2434.

49. Arias A, Feinn R, Kranzler HR. (2006) Association of an Asn40Asp (A118G) polymorphism in the mu-opioid receptor gene with substance dependence: a meta-analysis. *Drug Alcohol Depend* **83**(3):262–268.

50. Zhang H, Luo X, Kranzler HR, *et al.* (2006) Association between two micro-opioid receptor gene (OPRM1) haplotype blocks and drug or alcohol dependence. *Hum Mol Genet* **15**(6):807–819.

51. Cooper JR, Bloom FE, Roth RH. (2003) Dopamine. In: *The Biochemical Basis of Neuropharmacology*, 8 ed., Oxford University Press, Oxford, England, pp. 225–270.

52. Ma JZ, Beuten J, Payne TJ, *et al.* (2005) Haplotype analysis indicates an association between the DOPA decarboxylase (DDC) gene and nicotine dependence. *Hum Mol Genet* **14**(12):1691–1698.

53. Yu Y, Panhuysen C, Kranzler H, *et al.* (2006) Intronic variants in the dopa decarboxylase (DDC) gene are associated with smoking behavior in European-Americans and African-Americans. *Hum Mol Genet* **15**:2192–2199.

54. Caspi A, Sugden K, Moffitt TE, *et al.* (2003) Influence of life stress on depression: moderation by a polymorphism in the 5-HTT gene. *Science* **301**(5631): 386–389.

55. Kaufman J, Yang BZ, Douglas-Palumberi H, *et al.* (2004) Social supports and serotonin transporter gene moderate depression in maltreated children. *Proc Natl Acad Sci USA* **101**(49):17316–17321.

56. Covault J, Tennen H, Herman A, *et al.* (2007) Interactive effects of the serotonin transporter 5-HTTLPR polymorphism and stressful life events on college student drinking and drug use. *Biol Psychiatry* **61**:609–616.

57. Kaufman J, Yang B-Z, Douglas-Palumberi H, *et al.* (2006) Genetic and environmental predictors of early alcohol use. *Biol Psychiatry*, November 20 [Epub ahead of print].

Chapter 31

Insulin Resistance and Metabolic Syndrome

Wei-Dong Li* and R. Arlen Price

*Department of Psychiatry, University of Pennsylvania
TRL2214, 125 S 31st Street, Philadelphia, PA 19104, USA*

Metabolic syndrome (MS) is a common disorder that affects 25% of the American population. MS is characterized by obesity, hyperglycemia, hypertriglyceridemia, low HDL cholesterol levels, and hypertension. In general, MS is the result of insulin resistance and is linked to the pathogenesis of type 2 diabetes and cardiovascular disease (CVD). Genetic factors play important roles in the pathogenesis of MS and its endophenotypes. Ample studies have been carried out to dissect the genetics of MS, including hundreds of genome scans, animal quantitative trait loci (QTLs), and association studies. Like in other complex traits, gene–gene and gene–environment interactions, low penetrance, and genetic heterogeneity have made the MS gene hunting very challenging; on the other hand, factor analyses, replicated linkage results, and high-throughput association studies have yielded promising results. In this chapter, we integrate pieces of the genetics of MS; technical issues such as linkage disequilibrium (LD) mapping and the parent-of-origin effect are also discussed.

1. INTRODUCTION

1.1. *Epidemiology of Insulin Resistance and Metabolic Syndrome*

Insulin resistance (IR) is defined as an abnormal response to insulin. IR can be found in type 2 diabetes and prediabetes patients, obese patients,

---

*Correspondence author.

and many IR-related syndromes like polycystic ovary syndrome (PCOS); in some cases, it can even be observed in presumably "healthy" individuals. Insulin resistance (or sensitivity) is also associated with physical activities, pregnancy, food intake, and other physiological statuses and environmental factors. IR is a predictor of life span (in animals) and longevity (in human).[1-3] Compared with type 2 diabetes, IR is a status rather than a disease.

Prediabetes is a condition that occurs when a person's blood glucose levels are higher than normal, but not high enough for a diagnosis of type 2 diabetes. It is estimated that at least 54 million Americans have prediabetes, in addition to the 20.8 million with diabetes (American Diabetes Association; http://www.diabetes.org/about-diabetes.jsp). Most prediabetes individuals have IR. Understanding IR in prediabetes is important in both the etiology and prevention of type 2 diabetes.

Metabolic syndrome (MS) is a cluster of symptoms associated with IR, including central obesity, dyslipidemia, hyperglycemia, and hypertension.[4,5] According to the Adult Treatment Panel III (ATP III) definition,[6] MS can be diagnosed if a patient meets any three or more of the following five criteria: (1) waist circumference > 102 cm in men and > 88 cm in women; (2) serum triglyceride ≥ 1.7 μmol/L; (3) blood pressure $\geq 130/85$ mm Hg; (4) HDL < 1.0 μmol/L in men and < 1.3 μmol/L in women; and (5) fasting glucose ≥ 6.1 μmol/L. MS is a major risk factor of type 2 diabetes and cardiovascular disease (CVD).

The prevalence of MS varies in different populations.[7] In general, the prevalence of MS in the American population is around 25%. The incidence is higher in Native Americans than in non-Hispanic European-Americans. The prevalence of MS also increases with age: more than 40% of people in the USA aged 60 years and above have MS, compared with only 7% in those aged 20–29 years.[8]

Recently, more references use the term "insulin resistance syndrome" rather than "metabolic syndrome". The clustered MS symptoms are more or less correlated with IR; however, many other phenotypes (especially newly identified adipokines) are also associated with IR. IR is associated with many diseases and statuses, such as diabetes, prediabetes, aging, pregnancy, and infection. Moreover, similar statuses also exist in model organisms (*C. elegans* and yeasts). Indeed, an understanding of IR involves far more than just predicting the risk of CVD and type 2 diabetes from several related phenotypes.

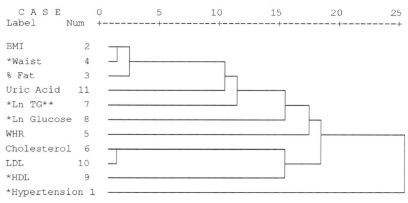

```
C A S E       0       5       10      15      20      25
Label    Num  +---------+---------+---------+---------+---------+

BMI          2
*Waist       4
% Fat        3
Uric Acid   11
*Ln TG**     7
*Ln Glucose  8
WHR          5
Cholesterol  6
LDL         10
*HDL         9
*Hypertension 1
```

* Phenotypes of metabolic syndrome.
** ln TG: log-transformed triglyceride.

Fig. 1.   Hierarchical cluster analysis for IR-related phenotypes in obesity nuclear families (2800 subjects).

1.2. *Phenotypes Associated with Insulin Resistance*

1.2.1. *Components of the Insulin Resistance Syndrome*

A factor analysis[9] examined phenotypes of MS and found that IR factors (insulin and glucose), obesity factors, and lipid factors represent MS better than hypertension. Recently, controversies were raised on the components of MS[10]; however, it seems that certain phenotypes are still clustered as risk factors of CVD and type 2 diabetes.[11]

We performed a cluster analysis in a cohort of obesity nuclear families that segregated extremely obese and normal weight.[12] A total of 400 nuclear families (with extremely obese proband, body mass index [BMI] > 40 kg/m,[2] and obese- and normal-weight siblings) and 300 parent-proband trios were selected. In this cohort (2800 subjects), we collected body-weight-related phenotypes (BMI, body fat percentage, waist circumference, and waist/hip ratio), lipid-related phenotypes (total cholesterol, triglyceride, HDL, and LDL), fasting plasma glucose, plasma uric acid, and affection status of hypertension. The dendrogram in Fig. 1 shows the clustering of MS phenotypes.

Clearly, hypertension is in a separate cluster; this tendency has been found by a previous factor analysis[9] and other studies.[10] Body-weight-related phenotypes — including BMI, waist circumference, and % fat — are highly correlated. Interestingly, triglyceride is closer to the body weight cluster than the lipid cluster (total cholesterol, LDL, and HDL). Further evidence showed that triglyceride could be an independent risk factor of CVD. All five major components of MS (waist circumference, triglyceride, fasting glucose, HDL, and hypertension) are highly representative in each cluster. The waist/hip ratio reflects the adipose tissue distribution between visceral and subcutaneous fat.

Compared with other body weight measurements, the waist/hip ratio is closer to insulin-sensitivity-related phenotypes (fasting glucose). The relationship between abdominal obesity and IR has been well documented.[13,14] Visceral adipose tissue mass poses a greater risk of insulin sensitivity; in fact, visceral adipose tissue is more metabolically active. Gene expression profiles are quite different between peripheral and visceral adipotissues. A visceral fat-specific adipokine, visfatin, binds to insulin and IGF1 receptor[14] and regulates NAD^+-dependent protein deacetylase.

The waist/hip ratio is a better index of visceral fat tissue than the waist circumference, while computed tomography (CT) and magnetic resonance imaging (MRI) can provide more accurate measurements. Other factors, such as plasma uric acid, are also closer correlated with MS components than the hypertension factor. This suggests that the definition of "metabolic syndrome" needs to be modified; however, major components of MS still stand.

Clustering phenotypes together as a single syndrome will help us understand the population susceptibility to IR. Those overlapped phenotypes could be regulated by similar gene networks. Identifying IR individuals will be an important step for the early prevention and treatment of type 2 diabetes and CVD, although the definition of MS may need some modification. In particular, after summarizing the results of factor analyses, fat distribution, fasting glucose, and dyslipidemia are found to be closer to IR. Common genetic factors may lie beneath those correlated phenotypes.

1.2.2. *Insulin Resistance Is the Key Connector Between Obesity and Type 2 Diabetes*

Obesity is the most common disorder in human beings and has a strong connection to type 2 diabetes,[15] hypertension, CVD, and cancer. The average BMI of the US population is in the overweight range (BMI $= 28.1$ kg/m^2), and more than 30% of people in the US are obese (BMI > 30 kg/m^2).[16] Similarly, more than 300 million people in the world and 18.2 million people in the US (6.3% of the population, including 17 million type 2 diabetes patients) have diabetes.

Obesity is a major risk factor of type 2 diabetes. More than 70% of type 2 diabetes patients are overweight. The incidence of type 2 diabetes rises dramatically with an increase in BMI. Obesity, especially central obesity (waist/hip ratio > 0.90 for female and > 0.85 for male), is a strong predictor of type 2 diabetes. Obesity-induced IR plays an important role in the pathogenesis of type 2 diabetes.

There are strong associations among high hyperglycemia, hyperinsulinemia, type 2 diabetes, and obesity. Edwards *et al.*[17] showed that insulin/glucose could be an independent factor that contributes to IR. A factor analysis by Shen *et al.*[9] showed that fasting plasma glucose and insulin are strongly correlated with body weight; thus, fasting glucose and insulin could be classified as related phenotypes and are indexes for insulin sensitivity.

There are also some functional connections among obesity, IR, and type 2 diabetes. Adipose tissue produces certain regulatory factors such as free fatty acid, adiponectin, TNFα, and resistin, which regulate insulin sensitivity. Proinflammatory cytokines (including TNFα and IL1β) are now widely recognized as central components in the pathogenesis of obesity-related IR.

Although the connections among obesity, IR, and type 2 diabetes are very complicated, a simplified model could explain the sequential changes in obese individuals: Once obesity develops, adipokines (secreted by adipocytes) act on skeleton muscle cells and reduce glucose uptake; as a result, pancreatic islets have to produce more insulin to stimulate glucose transportation into muscle cells. In the long term, IR is established when the elevated insulin production and the decreased glucose uptake reach a balance; however, when the exhausted pancreatic islets can no longer secrete enough insulin, type 2 diabetes develops.

1.2.3. *Proinflammatory Factors and Other Components of IR Syndrome*

Proinflammatory factors and adipokines are associated with IR and the risk of CVD. C-reactive protein (CRP), a pentraxin-like inflammatory factor, has high affinity to chromatins. Plasma CRP levels have been found to be associated with obesity,[18] type 2 diabetes,[19,20] insulin resistance,[19,21] and the risk of coronary heart disease.[22] Other adipokines are produced by adipose tissue, including IL6, IL1β, TNFα, resistin,[23] and adiponectin,[24] which also have physiological and epidemiological ties to IR.[25] These proinflammatory factors are important endophenotypes of MS and IR. The correlation between elevated uric acid and MS, CVD, and type 2 diabetes has been well documented.[26] Significant correlations between body weight and uric acid have also been found in many samples, including our obesity cohort.[27]

1.2.4. *Lipid Factors and IR*

Plasma triglyceride (TG) has been approved to be an independent risk factor of CVD. It is important to know the relationship between IR and other lipid-related phenotypes, especially LDL. Although several familial hyperlipidemia are combined, more studies suggested that plasma LDL, HDL, and triglyceride levels and even subtypes of lipoproteins are determined by different genetic factors. Recently, a case-control test from the Dallas Heart Study showed that individuals with lifetime moderate LDL reduction (less than 30%, due to *PSCK9* gene mutations) can significantly reduce the risk of CVD (up to 88% reduction).[28] Until now, it is hard to tell whether LDL has less connection with IR than HDL and/or TG. Needless to say, more large-scale epidemiology studies are needed to reappraise the major components of MS.

1.2.5. *Measuring Insulin Resistance*

There are several different methods for insulin sensitivity measurement, including euglycemic clamp, oral glucose tolerance test (OGTT), fasting glucose, and insulin. Euglycemic clamp is the most accurate method, but it needs special equipment and is not suitable for population-based studies. On the other hand, it is necessary to distinguish IR in different organs/tissues, such as skeletal muscle, liver, and adipocytes.

It would be difficult to perform OGTT as a follow-up study. Studies showed that fasting insulin levels could be a better index of insulin sensitivity than OGTT.[29] So far, several indexes based on fasting plasma glucose and insulin have been used as simple and accurate measurements of insulin sensitivity in large epidemiological studies, including homeostasis model assessment (HOMA)[30,31] and quantitative insulin sensitivity check index (QUICKI).[32]

QUICKI can be calculated by $1/[\log(I_0) + \log(G_0)]$, where I_0 and G_0 are fasting plasma insulin and glucose levels, respectively. Katz *et al.*[32]

showed that QUICKI has a higher correlation with the glucose clamp study ($r = 0.78$) than does minimal model analysis of a frequently sampled intravenous glucose tolerance test (FESIVGTT) and HOMA [calculated as $(I_0 \times G_0)/22.5$]. QUICKI is a simple, accurate method based on fasting glucose and insulin levels for assessing insulin sensitivity.

Although noninvasive methods (HOMA and QUICKI) are suitable for mass screen, glucose-induced IR (such as minimal model–based insulin sensitivity) is a more direct and accurate measurement. A study in the Insulin Resistance Atherosclerosis Study (IRAS) cohort showed that the minimal model–based insulin sensitivity index (S_I) has almost twofold higher heritability than HOMA.[33]

2. GENETIC FACTORS AND IR

2.1. *Genetics of MS Components*

Over 100 years of research have suggested that genetic factors play very important roles in the etiology of major components of MS. More recent studies showed that "gold standards" (glucose tolerance and other IR indices) are also controlled by genetic factors. Using a bivariate segregation analysis, Mitchell *et al.*[34] showed that the same gene (or the same set of genes) may account for the fasting and 2-hour insulin as well as other MS phenotypes (except blood pressure).

Although environmental factors and environment–gene interactions (food intake, lack of physical activities) contribute to the ever-increasing incidence of type 2 diabetes, the ability to develop IR is mainly controlled by genetic factors. Bouchard *et al.*[35] performed an elegant twin study for weight gain; 12 pairs of male monozygotic (MZ) twins were overfed for 8 weeks (additional 1000 kcal/day), and the difference in weight gain within twin pairs was much smaller than among twin pairs. Similarly, Mayer *et al.*[36] compared fasting insulin among MZ and dizygotic (DZ) twins, and the correlation of fasting insulin in MZ twins was much larger than in DZ twins ($r = 0.64$ in MZ twins vs. 0.40 in DZ twins). In family studies, siblings of type 2 diabetes patients showed a narrower range of IR index within families than among families.

2.1.1. *Genetics of Obesity*

Based on over 80 family and twin studies, the heritability of obesity is 40%–60%. Segregation analyses have also suggested the genetic etiology, with or without major gene effects.

Family studies showed that obesity and thinness follow family lines; moreover, as the severity of obesity increases, so does the relative risk. Studies examining several obesity thresholds found that the relative risk increases from 1.5 to 3.0 for moderate obesity (BMI $\geq$ 30 kg/m^2), and from 3.0 to 9.0 for extreme obesity (BMI $\geq$ 40 kg/m^2).[37,38] Twin and adoption studies found that identical twins are more highly correlated in body composition than fraternal twins, whose correlations are similar to those of other first-degree relatives.[39,40] Adoptees closely resembled their biological relatives in body composition, but not members of their adoptive families[41–43]; and studies of twins reared apart confirmed this pattern.[44,45] From this research, it is estimated that genes account for as much as two thirds of individual differences in obesity, with the remainder due to idiosyncratic influences from outside the family.

A number of linkage studies have focused on obesity and related phenotypes. Perusse and colleagues,[46] in their unselective annual obesity gene map update, reported more than 600 genetic loci linked or associated with obesity in humans and animal models. These include 10 major genes in humans, 49 genetic syndromes, 221 animal quantitative trait loci (QTLs), 113 candidate genes with reported associations in humans, and more than 204 loci with at least one report of possible linkage. All studies considered, every human chromosome except the Y chromosome has been implicated.

The linkage results above came from more than 200 separate publications, including 50 genome scans, some of which involved reanalysis of the same cohort with different phenotypes. Thirty-eight of the linkages achieved at least minimal replication through an independent report of linkage to the same region. However, it is difficult to know how much confidence to place in many of the results. It is encouraging that the replications are concentrated in a few regions, including 2p22, 3q27, 6p21.3–p21.1, 7q22–7qtel, 10p12–p11, 11q23–q24, 17p12, and 18q21.[46]

2.1.2. *Genetic Factors of Fasting Glucose and Risks of Type 2 Diabetes*

By definition, any individual can be diagnosed as diabetic when fasting plasma glucose $\geq$ 126 mg/dL or 2-hour plasma glucose $\geq$ 200 mg/dL by oral glucose tolerance test (OGTT). In theory, the affection status of diabetes acts as a discrete trait of fasting glucose; however, the biological meaning is much more than a threshold-selected phenotype. Type 2 diabetes is a status with irreversible IR. Hyperglycemia is only a diagnostic criterion of type 2 diabetes rather than the sole reason or result of that disease. Genetic factors for fasting glucose and type 2 diabetes may overlap, but the latter are more complicated.

Like obesity, the connection between genetic factors and type 2 diabetes is well recognized, including through twin studies,[47] segregation analyses,[48] and population-based studies.[15] More than 25 genome scans for type 2 diabetes and at least 10 for metabolic syndrome have been performed. Several chromosome regions, including 1q21–25,[49–54] 12q,[55–61] and 20q13,[56,61–66] were among the best replicated regions for type 2 diabetes and related phenotyes. More than a dozen other chromosome regions were found with significant linkages to type 2 diabetes, and even more were found with suggestive linkages.[67,68]

Fasting glucose is a major component of MS and the key index of type 2 diabetes. In general, we can consider type 2 diabetes as a dichotomous trait of plasma glucose. Fasting glucose levels are moderately heritable, with heritabilities averaging around 30%.[69,70] Previous genome scans have found linkage signals for fasting glucose in several chromosome regions (1p, 1q, 3p, 10q, and 17p).[53,71,72]

In our second-generation genome scan of 320 obese nuclear families, we detected a significant linkage (close to D18S1371 at 116 cM; logarithm of odds [LOD] = 6.59) for fasting plasma glucose on chromosome 18q22–23 (1583 subjects, 320 families).[73] Linkages and associations of diabetic nephropathy[74,75] were found in the 18q22–23 region (between markers D18S469 and D18S58, about 1 Mb from our linkage peak D18S1371). More recently, QTLs influencing plasma TNFα and IL1β (proinflammatory cytokines related to IR) were colocalized in 18q22–23[76]; both linkages were between markers ATA82B02 and D18S1371 (LOD scores for TNFα and IL1β were 3.1 and 4.0, respectively).

2.1.3. *Serum TG as an Independent Risk Factor for Type 2 Diabetes and CVD*

CVD and type 2 diabetes are among the leading causes of death in the United States. About one million (950 000) Americans die of cardiovascular diseases each year, while about 61 million Americans have certain forms of CVD (Centers for Disease Control and Prevention, CDC; http://www.cdc.gov/heartdisease).

Although etiologies of CVD, obesity, MS, and type 2 diabetes are complicated, elevated serum triglyceride (TG) level is common to both obese and type 2 diabetes individuals.[77–80] As an endophenotype of MS, hypertriglyceridemia acts as a key index of IR. Many epidemiologic studies indicated that TG is an independent risk factor of CVD[81–87] and stroke.[88] In a case-control study from the National Heart, Lung, and Blood Institute (NHLBI) Family Heart Study, Hopkins *et al.*[89] found that elevated TG levels in familial combined hyperlipidemia (FCHL) and familial hypertriglyceridemia (FHTG) are strongly related to MS and are associated with the risk of coronary artery disease (CAD).[89]

Recently, functional evidence was presented that increased TG might also induce leptin resistance at the blood–brain barrier,[90] indicating that TG is an important factor connecting food intake, leptin resistance, and obesity. Needless to say, understanding the mechanism of hypertriglyceridemia will provide useful information for etiologies of CVD and type 2 diabetes.

2.1.4. *Genetic Bases of Hypertriglyceridemia and Related Diseases*

Serum TG levels are largely controlled by genetic factors. More than 60 studies during the past 30 years have shown that the heritability of plasma TG ranges from 20%–75%, with most studies indicating heritabilities of 30%–40%.[91–96]

FHTG[97] is also classified as hyperlipoproteinemia IV (elevated serum triglyceride, but normal cholesterol). Actually, five out of six types of hyperlipoproteinemia (all except type IIa) have moderate to severely increased triglyceride levels, but most of them also have elevated cholesterol. FHTG, FCHL, and chylomicronemia are all diseases exhibiting inherited abnormal TG.

In a 20-year follow-up study, Austin *et al.*[83] found that baseline TG is associated with CVD mortality risk independent of total cholesterol among relatives in FHTG families[83]; interestingly, this tendency was not found in FCHL families. A more recent study has shown an association between elevated TG and risk of CAD in both FCHL and FHTG families related to MS.[89]

Several genetic bases for hypertriglyceridemia have been identified. For example, mutations of the *ABC1* gene cause Tangier disease, an autosomal recessive disease characterized by hypertriglyceridemia, hypocholesterolemia, and absence of normal HDL in plasma (MIM 205400).[98] Lipoprotein lipase (LPL) deficiency can cause hypertriglyceridemia. Association studies found plasma TG levels to be associated with apolipoprotein A5 gene

(*APOA5*) variability.[99] *APOA4* gene polymorphisms were also found to be related to quantitative plasma lipid risk factors of coronary heart disease.[100] In addition, mutations of the lecithin-cholesterol acyltransferase (*LCAT*) gene cause fish eye syndrome,[101] including corneal opacities, HDL cholesterol < 10 mg/dL, normal plasma cholesteryl esters, and elevated TG.

The genetic factors of common forms of hypertriglyceridemia, however, remain unclear. Certain diseases such as severe diabetes, hypothyroidism, and Gaucher's disease can cause secondary hypertriglyceridemia; while environmental factors such as a high-fat diet also increase serum TG levels.

a. *Genome scans and linkage analyses for TG*

Genome scans (whole-genome linkage analyses) for common forms of TG found linkages on several regions across the human genome, including 1q,[102] 2q,[103–105] 3p,[106] 5q,[104,105] 6q,[107] 7q,[108–111] 8p,[107] 8q,[102] 9p,[103] 10p,[112] 11p,[107] 11q,[104,113] 15p,[114] 15q,[108] 16q,[102,109] 19q,[115] 20q,[109,111] and 22p.[105] Our group also performed a genome scan 4 years ago for lipid-related phenotypes in 75 nuclear families. We found suggestive linkages for TG on chromosomes 1, 3, 7, 10, 12, 17, and 20.[110]

b. *Evidence of linkages for TG to chromosome region 7q35–7qtel*

So far, at least five genome scans for TG found linkages on the 7q36–7qtel region (Table 1).[102,105,109,111,116] Shearman *et al.*[109] found linkage for log-transformed TG/HDL cholesterol on D7S3070 (LOD = 2.5) in the Framingham Heart Study (1702 subjects), and Sonnenberg *et al.*[116] obtained LOD = 3.7 on D7S3058 in 507 Caucasian families (2209 subjects). Linkages for plasma HDL cholesterol and TG/HDL cholesterol ratio were also found in this region.[108–111] Cholesterol, however, has not been linked to 7q36–7qtel. Rice *et al.*[117] found LOD = 2.53 for abdominal total fat (response to training) in African-Americans. TG-related phenotypes are strongly related to 7q, having been replicated more than any linkage to any other chromosome region.

2.1.5. *Total Cholesterol, LDL, and CVD*

Cholesterol is a component of cell membranes, and functions as a precursor to bile acids and steroid hormones. The synthesis, metabolism, and transportation of cholesterol are well documented. Cholesterol travels in the blood with lipoproteins (LDL, HDL, and VLDL). The LDL, HDL, and

Table 1   Summary of Linkages Found on 7q36–7qtel Region for TG and Related Phenotypes

Reference	Subjects	Location/Marker	LOD	Phenotype
108	Mexican-Americans, 418 subjects	D7S1824/(139.4 Mb)	1.86	Log (TG)
109	1702 subjects from Framingham Heart Study	D7S3070 (151.0 Mb, 163 cM)	1.8 2.5	Log (TG) Log (TG/HDL)
111	1702 subjects from Framingham Heart Study	D7S3058 (154.0 Mb, 174 cM)	1.7	Log (TG)
110	75 obesity nuclear families	D7S530	2.44 (*Z* score)	TG
118	Old Order Amish, 672 subjects	About 20 cM upstream of leptin (164 cM)	1.77	BMI-adjusted leptin
116	507 Caucasian families (2209 subjects)	D7S3058 (154.0 Mb, 174 cM)	3.7	Log (TG)
73	1514 subjects from 320 families (260 European-American families, 58 African-American families)	D7S3070 (151.0 Mb, 163 cM)	1.78 (variance components); 1.02 (family regression)	Log (glucose)
117	105 black HERITAGE families	D7S3070 (151.0 Mb, 163 cM)	2.52	ATF (abdominal total fat), response to training
119	1514 subjects from 320 families (260 European-American families, 58 African-American families)	D7S3058 (154.0 Mb, 174 cM)	3.52	Log (TG)

VLDL cholesterols make up 60%–70%, 20%–30%, and 10%–15% of the total serum cholesterol, respectively.

Decades of research have shown that total cholesterol (TC) and LDL are the major risk factors of coronary heart disease (CHD),[120,121] while HDL is the major protecting factor.[122] In a follow-up study of 7733 Framingham Heart Study subjects,[123] TC levels were associated with a lifetime risk of CHD: for those individuals with TG $<$ 200 mg/dL, 200–239 mg/dL, and $\geq$ 240 mg/dL, the risks in men of developing CHD over the next 40 years were 31%, 43%, and 57%, respectively; the risks in women were 15%, 26%, and 33%, respectively.

As a result, cholesterol, especially LDL, has become the key target of CHD treatment and prevention. TC- and LDL-lowering drugs could reduce the incidence of CHD.[124] Finding genes associated with or regulating hypercholesterolemia/hyperlipoproteinemia would help us identify high-risk individuals, and should therefore lead to more individualized therapies and more effective prevention strategies.

2.1.6. *Genetic Factors and Hypercholesterolemia*

Several types of hypercholesterolemia and hyperlipoproteinemia are monogenic disorders. LDL receptor,[125] *APOB*,[126] *PCSK9*,[127] and *ARH*[128] mutations have been found in familial hypercholesterolemia patients. Although hypercholesterolemia, decreased HDL cholesterol, and even hypertriglyceridemia are related, the genetic backgrounds of these disorders are different.

FCHL occurs in 1% of the general population and is characterized by increased LDL, VLDL, and TG, and reduced HDL. Linkages of FCHL were found in several chromosome regions, including 1q21,[129,130] 11p,[131] and 16q.[132]

In general populations, serum TC and LDL levels are largely controlled by genetic factors. More than 100 twin and population studies conducted during the past 30 years have shown that the heritability of TC and LDL ranges from 15%–88%, with most studies indicating heritabilities of 30%–40%.[91,92,96,133–138]

a. *Genome scans for cholesterol, LDL, and HDL*

More than 40 genome scans have been performed for hyperlipidemia and related phenotypes. Linkages of TC and/or LDL were found in several chromosome regions, including two regions on 1q,[104,110] 2p23, 3p25,[113] 10q11,[112] chromosome 11,[139] 15q26, 17q21, 19q13,[104] and 19p.[113,140]

Based on our genome scan data (382 microsatellite markers and 320 families; 260 European-American and 58 African-American), we carried out quantitative analyses for lipid-related phenotypes, including TC, TG, LDL, HDL, and cholesterol/HDL ratio. Using family regression (Merlin_regress),[141,142] we found significant linkage (LOD = 4.36, $p < 0.00001$) on the 2q34 region for TC and suggestive linkage of LDL (LOD = 1.65, $p = 0.003$) on the marker D2S2944 (210.4 cM) (Li *et al.*, unpublished data).

Wang *et al.*[143] found a linkage on chromosome region 2q34 at D2S2944 (LOD = 3.08) for carotid artery intima-media thickness (CIMT) in a genome scan of Mexican-American CAD families. CIMT is a subclinical measure of atherosclerosis and is associated with CHD. In a meta-analysis of CHD genome scans in four populations (Finnish, Mauritian, European, and Australian families),[144–147] Chiodini and Lewis[148] found the strongest evidence of linkages on 3q26–27 ($p = 0.0001$) and 2q34–37 ($p = 0.009$). In our genome scan, we found linkages for age-adjusted TC on the same chromosome region (D2S2944, LOD = 4.36). A suggestive linkage for LDL was also found in this 2q34 region. The elevated TC and/or LDL could be associated with CIMT[149]; gene(s) harbors in 2q34 could contribute to both hypercholesterolemia and CIMT, and therefore act as a risk factor(s) of CHD.

The 2q34 QTL is among the most significant and well-localized linkages in our genome scans. The identification of genes for the 2q34 QTL should lead to more individualized therapies and more effective prevention strategies.

2.2. MS as a Combined Phenotype

2.2.1. Monogenic Disorders Associated with IR

Both insulin[150] and insulin receptor[151] mutations cause IR/diabetes. Mutations of the insulin cascade genes, such as *IRS1*,[152,153] PI3 kinase,[154] and *PPP1R3A*,[155] can also develop severe IR. Compared with the total number of IR individuals, single-gene mutation-induced IR is rare. However, these monogenic mutations provided both biological and genetic information to understand the complexity of IR in human populations.

2.2.2. Single-Gene Mutations and MODY (Maturity-Onset Diabetes of the Young)

Maturity-onset diabetes of the young (MODY) is a group of early-onset (before 25 years of age), autosomal-dominant inheritance of diabetes.

Table 2   Single-Gene Mutations and Subtypes of MODY

MODY Type	Gene	Chromosome Region	Function
MODY1	*HNF4A*	20q13.12	Transcription factor
MODY2	*GCK* (glucokinase)	7p13	Glycolysis
MODY3	*HNF1A (TCF1)*	12q24.31	Transcription factor
MODY4	*IPF1 (PDX1)*	13q12.2	Transcription factor
MODY5	*HNF1B (TCF2)*	17q12	Transcription factor
MODY6	*NeuroD1*	2q31.3	Transcription factor

Table 2 shows single-gene mutations and subtypes of MODY. Most of these MODY genes are transcriptional factors, so that downstream effects of MODY genes could be very profound.

Mutations and polymorphisms of MODY genes were also found in common diabetes.[156–158] Usually, these common forms of MODY gene variants are mild and are not totally loss-of-function mutations. Like other association studies, there is no lack of negative association among MODY genes and type 2 diabetes.[159,160]

Besides germline mutations, several genes were found associated with type 2 diabetes, including calpain 10,[161] *PPARγ*,[162,163] and *FABP2* (fatty acid–binding protein 2).[164]

2.2.3. *Familial Partial Lipodystrophy (FPLD)*

Familial partial lipodystrophy (FPLD) is a monogenic form of IR syndrome characterized by abnormal fat distribution, hyperinsulinemia, hypertriglyceridemia, and low levels of HDL. Two subtypes of FPLD, FPLD2 and FPLD3, are caused by mutations of *LMNA* (lamin A/C, R482Q or R482W)[165] and *PPARγ* (F388L),[166] respectively. Although the clinical characterizations of FPLD2 and FPLD3 are quite similar, there are still distinguishable differences between these two subtypes, including the severity of IR, adipose tissue redistribution, hypertension, and early onset of type 2 diabetes. Interestingly, compared with the much more common *Pro12Ala* mutation, the *PPARγ* F388L mutation is a loss-of-function mutation.

2.2.4. *Genetic Factors and Risks of MS*

Although hundreds of linkages and associations were found for MS/IR components, consistent results were found when MS was treated as a single phenotype. Several chromosome regions, including 1q21–25,[167,168] 2q,[169–171]

3q27,[172] 7q21–31,[173,174] 11q,[175–177] 12q,[175,178] and 17p,[179] were among the best replicated regions for MS.

McCarthy *et al.*[180] screened 110 candidate genes for MS in CAD patients. They found eight genes — *LDLR*, *GBE1*, *IL1R1*, *TGFB1*, *IL6*, *COL5A2*, *SELE*, and *LIPC* — that were associated with MS. Ample references suggested strong connections between *PPARγ* polymorphisms and MS.[181,182] *PPARα*[183] and *PGC1*[181] genes are also associated with MS. In addition, knockout mice of *HSD11B1*[184] and *NEIL1*[185] genes showed MS-like phenotypes.

Although monogenic mutations only account for a very small proportion of IR, we learned that single-gene mutation can cause a series of phenotype changes. Moreover, these changes are correlated and act as a result of IR. Hegele and Pollex[186] summarized the germline *LMNA* and *PPARγ* mutations and the incidence of phenomic changes in FPLD patients. Fat redistribution seemed to be the earliest event, followed by dyslipidemia, hypertension, type 2 diabetes, and atherosclerosis. FPLD can be used as a model of IR in general populations. Indeed, single-gene mutations are human knockouts and provide valuable insights into the complex MS.

3. DIFFERENT APPROACHES TO MS GENE HUNTING

3.1. *Combined Phenotypes vs. Endophenotypes*

Phenotyping is always the biggest issue for any linkage or linkage disequilibrium (LD) study. Since obesity, hyperglycemia, dyslipidemia, and other IR-related phenotypes (CRP, adiponectin, etc.) are highly correlated, and single-gene mutations can induce a series of phenotype changes (such as *LMNA* and *PPARγ* mutations in FPLD), it is possible to treat IR as a single phenotype.

McQueen *et al.*[106] defined "metabolic syndrome score" as a combined phenotype. The composite MS trait yielded a high heritability (0.61) in samples of the Framingham Heart Study, although the combined factor did not show significant linkage in their genome scan. Stein *et al.*[187] integrated MS-related phenotypes with genetic marker information into their genome scan using the multivariate structural equation model (SEM). In the San Antonio Family Heart Study, Cai *et al.*[188] identified 4 major factors — body size–adiposity, insulin–glucose, blood pressure, and lipid levels — out of 14 IR-related phenotypes using factor analyses. Factor-specific linkages were found in four different chromosome regions in their genome scan. Indeed, multiple related metabolic abnormalities may be localized in the same chromosome region; the 1q21–25 linkage[167] is among the most replicated linkages for multiple IR phenotypes.

Other covariates, including age, sex, race, and environmental factors, could have a direct impact on IR. In many genome scans, MS-related phenotypes are adjusted by the linear effects of age. Atwood *et al.*[189] carried out sex- and age-specific genome scans of BMI in the Framingham Study; all subjects had six examinations during a 28-year period. They found significant LOD differences among young and old intersections. Since the incidences of central obesity, dyslipidemia, and hyperglycemia may not happen at a single time-point, multiple-time[170] or even lifetime approaches are needed for linkage and/or fine mapping.

3.2. *Moderate vs. Extreme Phenotypes*

There are several key issues that need to be considered in human QTL mapping, including sample size, family structure, heterogeneity, and phenotyping. On the other hand, the methodology of linkage mapping has a direct impact on sample selection. Highly selected (extreme) or moderate phenotype is the first study design issue we need to face even before the phenotype collection starts.

Selected sampling has been widely used in type 2 diabetes, hypertension, and obesity genome scans. In theory, individuals with extreme phenotypes are likely to have single-gene mutations. O'Rahilly[190] summarized single-gene mutations in extreme human phenotypes, including leptin, leptin receptor, *PC1*, *MC4R*, and *POMC* mutations in extreme obesity; and insulin, insulin receptor, and *PPARγ* mutations in syndromes of severe IR. So far, most of the rare achievements in identifying IR genes have been based on extreme phenotypes.

Selecting family samples with extreme phenotypes should have a larger genetic relative risk (GRR). In a study of the National Association to Advance Fat Acceptance (NAAFA) and National Health and Nutrition Examination Survey III samples, Lee *et al.*[38] found that the risk of extreme obesity (BMI ≥ 40 kg/m^2) in relatives of extremely obese women (BMI ≥ 40 kg/m^2) was more than five times greater than in the population.

Because of the lack of powerful quantitative linkage tools, discrete analyses were common in the earlier stage of genome scans. Interestingly, as the most straightforward design, extreme sampling has been proved to be powerful in large-scale case-control association studies. Transmission/disequilibrium test (TDT)[191] and other TDT-related methods in LD mapping are also based on threshold-cut (extreme) sampling.

One of the major drawbacks of extreme sampling is the skewness of normal distribution. For quantitative traits like fasting glucose, HDL, and TG levels, dichotomous design somehow has less power. Almost all quantitative linkage analyses, including variance components,[192,193] Haseman–Elston regression, and family regression[142] methods, need normal distribution of traits. The variance components approach implemented in the SOLAR program[193] allows ascertainment correction of probands (with extreme phenotypes). When trait mean and variance are given, the family regression method[142] is not sensitive to skewed distribution. It is necessary to check the departure of normal distribution before using quantitative-trait linkage analyses. For instance, log transformation is usually needed for plasma TG levels.

We should keep in mind that one should not "cut off toes to fit into shoes" in the study design. Pursuing a higher GRR should outweigh the methodology difficulties.

3.3. *Density of Linkage Mapping*

There are long-lasting controversies on the marker density of linkage analyses and genome scans for complex traits. In theory, saturation mapping and haplotype analyses are necessary for monogenic disorders. However, researchers have failed to narrow down confidence intervals of linkages of IR (common form)-related phenotypes by genotyping more microsatellite markers.

After we found the original linkage of percent fat (% fat) and BMI on 20q13,[194] we added up to 20 microsatellite markers to a 25-cM region for saturated linkage mapping. However, dual peaks were found in the 20q13 region, and the overall nonparametric linkage (NPL) score did not improve as we had expected. A similar tendency was found in the FUSION study for type 2 diabetes[63,64] and in many other studies for complex traits.

According to recent studies, increasing the marker density from 10 cM to 1 cM, on average, decreases confidence intervals (CIs) by about 20%–40%.[195,196] According to the empirical genome data reported by Matise et al.,[197] increasing the marker density should improve the linkage confidence interval: 10-cM, 5-cM, and 1-cM marker intervals gave 95% CIs of 19 cM, 12 cM, and 8 cM, respectively. Our pilot data using the GeneFinder program[198] for the 7q22 QTL found CI reductions of 34.5%–40% after adding fine mapping markers[199] (Li et al., unpublished data).

However, fine linkage mapping is not always successful for complex traits, especially when multiple candidate genes harbor in the same QTL.

Another reason is the limitation of linkage analysis. Atwood and Heard-Costa[196] simulated fine mapping for quantitative traits with 2-, 1-, and 0.5-cM intervals; they found that the fine mapping at 0.5 cM did not differ from the 1-cM results.

For complex traits like IR, ranking linkage signals based on family-specific LOD scores could provide useful information for further fine mapping. We found significant linkage for TC (LOD = 4.4) and suggestive linkage for fasting glucose (LOD = 3.1) on the marker D2S2944 (Li *et al.*, unpublished data). After we ranked family-specific LOD scores by MER-LIN (variance components method),[141] only 35% of families showed family-specific LOD > 0.05 for both phenotypes. It is suggested that different genes or mutations may account for the changes in TC and glucose. Identifying high-ranked families may save a lot of effort in the next step of fine mapping, linkage disequilibrium (LD).

We should be aware of the existence of heterogeneity among different studies. Replication is the key and is still the gold standard for evaluating linkages. However, those hard-to-repeat linkages are not equal to artifacts. Even if the overall designs of sample collection for different linkage analyses are similar, the population heterogeneity could be quite significant. Linkage signals may be enriched in different sample collections; however, the linkage signal might be diluted in a combined sample analysis. There are ample examples of lower linkage signal in combined studies, although the pooled analysis should have much larger power.

Meta-analyses provided important information for MS-related phenotypes, including dyslipidemia,[200] type 2 diabetes,[201] and obesity.[202] Recently, more linkage mapping and genome scans were performed by commercial or National Institutes of Health (NIH)-supported genotyping services (e.g. Marshfield Center for Medical Genetics, Center for Inherited Disease Research, allowing us to compare results among different groups since similar microsatellite markers were used. However, linkages which were not on these "backbone markers" were much less replicated. For example, a linkage for BMI on 10p12 was found by three independent groups[203–205]; since the Marshfield genome scan marker set (set 11) has a 14-cM gap in that region, we failed to replicate that linkage in our genome scan unless we had genotyped the marker D10S197.[178] Interestingly, the 10p12 linkage increased almost threefold after the marker D10S197 was added.[178]

Compared with 10 years ago, we now have much more information about linkage mapping. More replicated linkages have been found for IR phenotypes. Although some of the linkages could be false-positive, we still need to be aware of false-negatives for those less replicated signals.

3.4. *LD Mapping and Methodology*

3.4.1. *Transmission/Disequilibrium Tests for Discrete Traits*

The transmission/disequilibrium test (TDT) was first developed by Spielman *et al.*[191] This method tabulates transmitted and untransmitted alleles from heterozygous parents, and tests the values for deviation from random segregation. The program also tests combinations of adjacent markers making up haplotypes. TDT was implemented in computer programs like Genehunter 2.1. Up to four marker haplotypes will be constructed and examined by Genehunter 2.1.

3.4.2. *LD Analyses of Quantitative Phenotypes*

The family-based association test (FBAT) is a recently developed method for quantitative trait association studies. This method extends the approach for testing described by Rabinowitz and Laird[206] to handle multiple tightly linked markers. This computer program package is one of the most flexible ones in that it is applicable to dichotomous, quantitative, and censored data. It handles multiple siblings, missing parents, and mixed family types. More recently, the program has been modified to handle multilocus haplotypes (haplotype FBAT).[207] A related program, PBAT, can be used to estimate power under mixed designs.

The computer program QTDT can also be used for quantitative TDT.[208] In QTDT, variance components are used to construct a test that utilizes information from all available offspring, but that is not biased in the presence of linkage or familial resemblance. QTDT requires normal distributions of quantitative traits.

3.4.3. *Detection of Linkage and LD Together*

Since certain markers genotyped for linkage analyses are very close to candidate genes, certain alleles may account for both linkage and association. It is necessary to detect both linkage and LD together. A recently developed method, the genotype identity-by-descent (IBD) sharing test (GIST), can detect whether or not a certain allele accounts for both linkage and association (LD). Among affected sib pairs, GIST tests weights of genotypes under recessive, dominant, and additive models.[209] In affected sibs, weighted genotypes of recessive, dominant, and additive models are calculated by the SAS program.

Li *et al.*[210] developed a model-based method using a maximum likelihood model to extract information on genetic linkage and association from samples of unrelated individuals, sib pairs, sib trios, and larger pedigrees. This method has been implemented in the LAMP program.

3.4.4. *Correction of Multiple Tests*

Multiple genes with real associations may harbor in certain QTLs. For a fine mapping study with hundreds of single nucleotide polymorphisms (SNPs), determining significant levels for multiple tests could be problematic. There are several tiers of correction that could be employed, but none of them can be considered as a gold standard:

(1) Bonferroni correction — P values are adjusted by the number of SNPs and phenotypes. Since phenotypes could be highly correlated and SNPs may be in strong LD, Bonferroni correction is too conservative; however, all positive associations after Bonferroni corrections ($P < 0.05$) could be given the highest priority for candidate gene screening.
(2) Adjusted by the number of haplotype blocks — Since SNPs in certain haplotype blocks are in strong LD, P values could be adjusted with the number of haplotype blocks. Genes could be annotated if multiple haplotypes are associated with the phenotype.
(3) Adjusted by the number of phenotypes — A large number of phenotypes are intercorrelated (like BMI, % fat, and waist circumference). It is important to verify the significant associations among highly related phenotypes.

The false discovery rate (FDR) could be estimated to determine the threshold for statistical significance.[211–212] The ideal threshold that limits false-positives with a minimum loss of power has been suggested to be $p < 0.10$ for two-stage studies.

No matter how significant the association obtained from fine mapping, biological evidence is the key to validate the plausible associations. It is almost the rule rather than the exception that negative reports will be found once the association is identified, no matter how strong the original result was. There are some long-lasting controversies on calpain 10 with type 2 diabetes,[161,213] *GAD2* with obesity,[214,215] or even the most replicated association: *PPARγ* Pro12Ala polymorphism.

3.4.5. *A Common Mutation, a Very Complicated Story: The Example of PPARγ Pro12Ala Polymorphism*

The *PPARγ* codon 12 Pro/Ala polymorphism is one of the most reported sequence variants associated with obesity and IR. Interestingly, although the majority of publications showed that the Pro12Ala polymorphism (the 12Ala allele) was associated with higher BMI,[162,216] it was associated with lower BMI in certain studies.[217,218] A meta-analysis[219] showed that the 12Ala allele was correlated with higher BMI in samples with BMI ≥ 27 kg/m,[2] but no such tendency could be detected in carriers with BMI < 27 kg/m.[2] Moreover, the 12Ala allele was associated with reduced risk of type 2 diabetes.[162,220,221] The *PPARγ* codon 12 Pro/Ala polymorphism was also associated with the ratio of dietary polyunsaturated fat to saturate fat intake[222] and of glucose response to exercise.[223]

It seems that the 12Ala allele could increase insulin sensitivity, but the *in vitro* test showed ambiguous results. Using site-directed mutagenesis, we transfected the 12Ala into 3T3L1 preadipocytes. However, no significant differences were found in adipocyte differentiation and glucose uptake between the 12Ala and 12Pro alleles (Li *et al.*, unpublished data). The association among *PPARγ* gene polymorphisms, body weight, and type 2 diabetes is well established; however, the story is very complicated even for the most replicated associations. Both gain-of-function and loss-of-function polymorphisms were found in the *PPARγ* gene, in addition to gene–gene,[224] gene–nutrient,[222,225] and gene–environment interactions.[223] It seems that the Pro12Ala mutation is too enigmatic to decipher.

3.5. *Whole-Genome LD Mapping or Candidate Gene Approach*

3.5.1. *Haplotype Structure of the Human Genome and Measurements*

Several groups of researchers examined the distribution of haplotype "blocks".[226] It was found that, in European-Americans and Asians, more than half of the genomic regions examined are in haplotype blocks of 50 kb or greater.[227–229] In many regions of the human genome, even larger haplotype blocks (>100 kb) have been found.[230] However, we caution here that different definitions and different parameter settings within the same definition may lead to different identified blocks.

Although selecting SNPs based on physical distance could be sufficient, LD levels vary across even short distances in the genome. It is necessary to measure the LD levels using a metric system other than physical distance. Maniatis *et al.*[231] defined the linkage disequilibrium unit (LDU) system. One LDU represents the decay of LD between two SNPs by about 37% of its maximum value when fitted to the Malecot model.[231] Based on empirical results, 0.30 LDU is roughly equal to a pairwise D' value > 0.85.

Haplotype R^2 was introduced by Stram *et al.*[232] for accessing LD among multiple SNPs, which are based on haplotypes. Unlike the pairwise comparison r^2 (statistical coefficient of determination), R^2 is based on multiple comparisons. According to Chapman *et al.*,[233] haplotype R^2 is an appropriate numerical index for the haplotype diversity.

Several SNP browsing softwares, including the Applied Biosystems SNP Browser (Foster City, CA) and the Perlegen Genotype Browser (Mountain View, CA), have made it possible to select validated SNPs when the chromosome region and marker density are given. The ABI SNP Browser was based on the genotyping data of 160 000 validated SNPs in 45 unrelated Caucasians, 45 African-Americans (Coriell Institute Repository), 45 Chinese, and 45 Japanese.

With the development of human haplotype mapping and the ever-reducing cost of SNP genotyping, whole-genome associations (LD mapping) have become possible.

3.5.2. *Power and Sample Size*

The power of detecting IR-related genes is determined by two major factors: the genetic relative risk (GRR) and the number of affected genes (heterogeneity). Compared with inherited diseases with single-gene mutations, the GRR of IR is low and the number of candidate genes could easily exceed 300. Low penetrance, gene–gene interactions, and gene–environment interactions make the genetic background of IR even more complex.

Risch and Merikangas[234] calculated the sample size needed for affected sib-pair (ASP) linkage analysis and association study by TDT. Given a reasonable sample size (less than 1000 trios or ASPs), GRR > 2 is required for linkage analysis and GRR > 1.5 for TDT. For example, plenty of positive associations were identified among the *PPARγ* Pro12Ala mutation (GRR $<$ 2 in general samples) and IR-related phenotypes; however, linkages were seldom found in the 3p25 region (where the *PPARγ* gene locates).

For a complex trait with GRR $= 1.2$, we need more than 4600 trios to have 80% power to detect association in a genomewide search.[234] In the case of MS, different genetic factors may account for single-phenotype or single-gene variants, many of them associated with multiple pathological factors (body weight/fat distribution, dyslipidemia, and blood pressure).

Although positional cloning of IR-related genes is theoretically difficult, more and more replicated linkages linkages and QTLs have been found across the human genome. Recently, relatively common alleles/haplotypes of the *TCF7L2* gene were found in type 2 diabetes patients with a population-attributable risk of 21%.[235] The number of replicated candidate genes (replicated in at least three independent studies) is approaching 100. Most importantly, the incidence of IR is more than 25% in the general population, and the attributive risk is quite significant.

4. THE CHALLENGE OF COMPLEX TRAITS

4.1. *Gene–Gene Interactions*

Given the number of obesity-related genes, gene–gene interactions may be common. Family studies estimate the genetic heritability of obesity at ~40%; and twin studies place the figure higher, at ~65%. Thus, about one third of the genetic variance of obesity is nonadditive (gene–gene interaction may account for the difference).[236] Our group identified interactions (correlations between IBD sharing in different loci) among several chromosome regions in obesity nuclear families, including chromosome regions 20q and 10q[237] for BMI as well as loci on 2p25–24 and 13q13–21[238] for BMI, % fat, and waist circumference.

Considering the number of genes involved in the pathogenesis of obesity, the interaction between genes could be enormous. The pairwise combination of 400 loci is approaching 8×10^4; the total numbers of combinations for 3-way, 4-way, 5-way, and 6-way interactions are 1×10^7, 1×10^9, 8×10^{10}, and 6×10^{12}, respectively (only biallelic markers are considered; we assume that the marker and the disease-causing allele are in complete LD, pairwise correlation $r^2 = 1$).

4.1.1. *Statistical Methods for Gene–Gene Interactions*

Multiple testing is the major challenge of gene–gene interaction study. Obviously, the Bonferroni correction is too conservative for the genomewide interaction. Recently, several dimensional reduction methods were developed, including multifactor dimensionality reduction (MDR),[239] patterning and recursive partitioning (PRP),[240] random forest,[241] and focused interaction testing framework (FITF).[242]

MDR[239] is a method that reduces the dimensionality of multilocus information to identify polymorphisms associated with an increased risk of disease. This approach takes multilocus genotypes, and develops a model for defining disease risk by pooling high-risk genotype combinations into one group and low-risk combinations into another (therefore, degrees of freedom are reduced). Cross-validation and permutation testing are used to identify optimal models. MDR is suitable for case-control samples or matched discordant sib pairs. Martin *et al.*[243] merged the MDR method with the pedigree disequilibrium test (PDT). MDR-PDT allows identification of single-locus effects or joint effects of multiple loci in families of diverse structure (e.g. discordant sib pairs or two affected and one unaffected sibling). One limitation of MDR is the zero tolerance of missing data; however, a modified method — extended MDR (EMDR)[244] — has solved the missing data problem.

The random forest method[241] is suitable for detecting interactions among clusters of SNPs in candidate genes. Dichotomous environmental factors, such as smoking, depression, and dieting histories, can also be tested for gene–environment interactions.

4.1.2. *Pathway-driven Gene–Gene Interactions*

The statistical genotype interaction is quite different from physical and biological interactions. However, several candidate genes for MS showed significant interactions in both aspects. Although the combination of pairwise genotype interactions is an astronomy number, several genotype interactions were found in obesity and type 2 diabetes.

Recently, we selected 350 obesity trios for TDT[191] analyses in a separate study of fine mapping on chromosomes 10, 12, and 20 (unpublished data). More than 100 SNPs in 30 candidate genes were genotyped. We obtained significant associations ($P < 0.01$) in about 40% of the candidate genes. All positive associations were consistent between single-point and haplotype TDTs.

We performed a pathway-driven gene–gene interaction prediction using INGENUITY Pathways Analysis. INGENUITY is a bioinformatics platform that extracts millions of pathway interactions from the literature. More than 10 900 human genes have been covered in the database. It provides a nice tool to explore reported biology interactions. We found several genes clustered in common pathways. We further compared the observed genotype frequencies with the joint frequencies under random combination

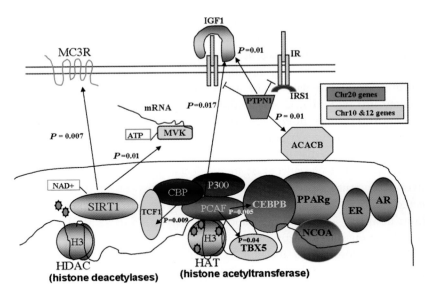

Fig. 2. Genotype interactions among candidate genes on chromosomes 10, 12, and 20. Arrows indicate P values of genotype interactions (observed vs. predicted genotype combinations, chi-square tests).

at two SNPs in genes in the same reported pathway; 10 genes showed significant interactions ($P < 0.05$). Furthermore, phenotype differences (BMI) were found among individuals with different genotypes. Interestingly, several key genes in the histone acetyltransferase (HAT)/nuclear receptor complex showed significant genotype interactions (Fig. 2). A similar tendency was found in a much larger study for weight change and obesity (4300 SNPs, 75 genes): genotype interactions matched very well with reported biological interactions, including genes in the insulin and nuclear receptor cascades.

It is hard to predict if genotype interactions reflect a "gene network" in the pathogenesis of IR. The statistical genotype interactions are different from biology interactions (physical interactions of proteins in the same pathway). However, understanding the population genetic background of IR will pave the road for mass prevention and individualized treatment.

4.2. Parent-of-Origin Effect and Gene Imprinting

Gene imprinting is a phenomenon in which the maternal and paternal alleles (genes) are not equally expressed. So far, about 1%–3% of genes in the human genome were found to have gene imprinting. Interestingly, it seems that a high percentage of IR and body-weight-related genes are imprinted. In animal models, the *IGF2* gene controlling body weight and fat thickness in pigs is imprinted.[245] The Prader–Willi syndrome (PWS) is characterized by diminished fetal activity, obesity, muscular hypotonia, mental retardation, short stature, hypogonadotropic hypogonadism, and small hands and feet; maternal genes in the 15q imprinting region are inactivated in PWS patients. In studies of Pima Indians, paternal and maternal imprinting were found for birth weight (chromosome 11),[246] type 2 diabetes (chromosomes 5 and 14), and BMI (chromosomes 5, 6, and 10).[247] Also, parent-of-origin effects were found in six chromosome regions for BMI in a separate study of children and young adults.[248] In our previous studies, we found significant parent-of-origin effects on chromosome regions 10p12, 12q24, and 13q34, based on our genome scan data.[249]

Epigenetic heterogeneity in the imprinting region is quite common in normal populations[250]; more imprinting regions were found in the human genome.[251] Greally[252] investigated the distribution of imprinted genes in the human genome, and found that the content of short interspersed transposable element (SINE) repeat sequences (including Alu sequences) could be a predictor of imprinted genes. According to his prediction, up to 15% of all genes have been annotated. Although the real number of imprinted genes could be much less than 15%, predicting imprinted genes based on SINE content provides a useful tool for whole-genome parent-of-origin study. We should be able to test 3000–5000 genes in families with IR patients. Recently, we found an allelic parent-of-origin effect in trios on two 10p12 genes, *GPR158* and *TCF8* (Dong *et al.*, unpublished data). Usually, well-recognized imprinted regions are quite large and harbor gene clusters (like the PWS region on 15q); however, a whole-genome parent-of origin search is needed for smaller and isolated imprinted genes. In an unpublished study (4000 SNPs), we found that almost 6% of SNPs showed parent-of-origin effects.

The regulation of genomic imprinting is complicated, including DNA methylation. Interestingly, common mechanisms may account for epigenetic phenomena such as imprinting and interchromosomal interactions. Ling *et al.*[253] found that two imprinted loci in different chromosomes, *Igf2/H19* and *Wsb1/Nf1*, were colocalized together with the existence of the CCCTC-binding factor (CTCF). Obviously, we need much larger whole-genome studies to evaluate the importance of the parent-of-origin effect in human obesity and IR.

5. FUTURE TOPICS

5.1. *Integrated Data Analyses and Meta-Analyses*

More than 100 genome scans have been performed for IR-related pheno-types. Numerous phenotyping and genotyping data have been collected. More than a dozen studies were performed by Marshfield Genotyping Service, which allows researchers to compare and replicate linkage results. When we compare studies with similar designs, methodologies, and marker selections, it is not difficult to find replications. DNA samples as well as the phenotyping and genotyping data of several well-studied cohorts, such as the Framingham Heart Study and the Genetics of Non–Insulin-Dependent Diabetes Mellitus (NIDDM) (GENNID) Study, have become available to the public. Finding linkages and associations for complex traits is no longer like "finding a needle in a haystack." Data sharing is an important step, although there are still many technical difficulties to overcome.

5.2. *Whole-Genome Association*

Whole-genome association studies were discussed 10 years ago,[234] but were not feasible until only recently.[254] Given a GRR $= 1.5$, the genotyping cost is as low as 1 cent/SNP; a whole-genome case-control study will cost more than US\$3 million (based on the sample size needed as calculated by Risch and Merikangas[234]). Although this cost is still beyond the budget of an ordinary National Institutes of Health Research Project Grant (NIH R01), it has nevertheless become much more accessible than it was several years ago.

5.3. *Expansion of Phenotypes*

Compared with genotyping millions of SNPs, exploring new genotypes could be cost-effective. For those cohorts which already have intensive genotyp-ing data, expanding phenotypes will open the door for new findings. It is impossible to have all phenotypes collected at the inauguration stage of the study, but achieved plasma samples can be used to detect new biochemical factors such as adipokines. On the other hand, analyzing multiple related phenotypes together will reduce rather than increase the type I error of linkage and LD mapping.

5.4. *Validation of Statistical Associations and Interactions*

No matter how strong the association is, biological evidence is needed to make connections between the DNA sequence variant and the status of

IR. Unfortunately, there is no universal way (like whole-genome genotyping) to explore the function connection. However, high-throughput screens are available for certain targeted functions such as transcription factors, chromatin remodeling, and adipocyte differentiation. Somehow, the biological validation should be considered to solve the puzzle of multiple test correction.

6. CONCLUSIONS

MS is a cluster of IR-related phenotypes, including central obesity, hyperglycemia, dyslipidemia, and hypertension. Cluster analyses showed that most of the major components of MS reflect the status of IR; however, some of the factors (especially hypertension) may need reappraisal. Family and twin studies showed that genetic factors play important roles in IR and MS. Years of linkage and association studies have yielded hundreds of linkages, QTLs, and positive associations; at least some of them have been replicated by separate studies. The QTLs for MS on 1q and linkages for TG on 7q36 are among the most replicated results of MS genome scans.

Because of the relatively low GRR of each gene and the total number of IR susceptibility genes, the traditional positional cloning strategy could be challenging for identifying IR- or MS-related genes. Also, epigenetic factors, including gene–gene interactions and parent-of-origin effects, need to be considered. Recently, new statistical methods have been developed to decipher gene–gene and gene–environment interactions. Also, genome scans of parent-of-origin effects (imprinting) are necessary for IR-related phenotypes. Exploring the connection between genotype and biological interactions may lead to a better understanding of the genetic background of IR in general populations.

ACKNOWLEDGMENTS

This work was supported in part by research grants from the National Institutes of Health (DK44073, DK48095, and DK56210 to R. A. Price) and the American Heart Association Scientist Development Grant (AHA 0630188N to W.-D. Li).

REFERENCES

1. Clancy DJ, Gems D, Harshman LG, *et al.* (2001) Extension of life-span by loss of CHICO, a *Drosophila* insulin receptor substrate protein. *Science* **292**: 104–106.

2. Bonafe M, Barbieri M, Marchegiani F, *et al.* (2003) Polymorphic variants of insulin-like growth factor I (IGF-I) receptor and phosphoinositide 3-kinase genes affect IGF-I plasma levels and human longevity: cues for an evolutionarily conserved mechanism of life span control. *J Clin Endocrinol Metab* **88**:3299–3304.

3. Kloting N, Bluher M. (2005) Extended longevity and insulin signaling in adipose tissue. *Exp Gerontol* **40**:878–883.

4. Avogaro P, Crepaldi G, Enzi G, Tiengo A. (1965) [Metabolic aspects of essential obesity]. *Epatologia* **11**:226–238.

5. Reaven GM. (1988) Banting lecture 1988. Role of insulin resistance in human disease. *Diabetes* **37**:1595–1607.

6. Grundy SM, Brewer HB Jr, Cleeman JI, *et al.* (2004) Definition of metabolic syndrome: report of the National Heart, Lung, and Blood Institute/ American Heart Association conference on scientific issues related to definition. *Circulation* **109**:433–438.

7. Cameron AJ, Shaw JE, Zimmet PZ. (2004) The metabolic syndrome: prevalence in worldwide populations. *Endocrinol Metab Clin North Am* **33**:vi, 351–375.

8. Ford ES, Giles WH, Dietz WH. (2002) Prevalence of the metabolic syndrome among US adults: findings from the third National Health and Nutrition Examination Survey. *JAMA* **287**:356–359.

9. Shen BJ, Todaro JF, Niaura R, *et al.* (2003) Are metabolic risk factors one unified syndrome? Modeling the structure of the metabolic syndrome X. *Am J Epidemiol* **157**:701–711.

10. Kahn R, Buse J, Ferrannini E, Stern M. (2005) The metabolic syndrome: time for a critical appraisal: joint statement from the American Diabetes Association and the European Association for the Study of Diabetes. *Diabetes Care* **28**:2289–2304.

11. Gotto AM Jr, Blackburn GL, Dailey GE 3rd, *et al.* (2006) The metabolic syndrome: a call to action. *Coron Artery Dis* **17**:77–80.

12. Price RA, Reed DR, Lee JH. (1998) Obesity related phenotypes in families selected for extreme obesity and leanness. *Int J Obes Relat Metab Disord* **22**:406–413.

13. Marin P, Andersson B, Ottosson M, *et al.* (1992) The morphology and metabolism of intraabdominal adipose tissue in men. *Metabolism* **41**: 1242–1248.

14. Fukuhara A, Matsuda M, Nishizawa M, *et al.* (2005) Visfatin: a protein secreted by visceral fat that mimics the effects of insulin. *Science* **307**: 426–430.

15. Knowler WC, Pettitt DJ, Savage PJ, Bennett PH. (1981) Diabetes incidence in Pima Indians: contributions of obesity and parental diabetes. *Am J Epidemiol* **113**:144–156.

16. Friedman JM. (2003) A war on obesity, not the obese. *Science* **299**:856–858.

17. Edwards KL, Newman B, Mayer E, *et al.* (1997) Heritability of factors of the insulin resistance syndrome in women twins. *Genet Epidemiol* **14**: 241–253.

18. Park HS, Park JY, Yu R. (2005) Relationship of obesity and visceral adiposity with serum concentrations of CRP, TNF-alpha and IL-6. *Diabetes Res Clin Pract* **69**:29–35.

19. Bahceci M, Tuzcu A, Ogun C, *et al.* (2005) Is serum C-reactive protein concentration correlated with HbA1c and insulin resistance in type 2 diabetic men with or without coronary heart disease? *J Endocrinol Invest* **28**: 145–150.

20. King DE, Mainous AG 3rd, Buchanan TA, Pearson WS. (2003) C-reactive protein and glycemic control in adults with diabetes. *Diabetes Care* **26**: 1535–1539.

21. Behre CJ, Fagerberg B, Hulten LM, Hulthe J. (2005) The reciprocal association of adipocytokines with insulin resistance and C-reactive protein in clinically healthy men. *Metabolism* **54**:439–444.

22. Albert MA, Glynn RJ, Ridker PM. (2003) Plasma concentration of C-reactive protein and the calculated Framingham Coronary Heart Disease Risk Score. *Circulation* **108**:161–165.

23. Steppan CM, Bailey ST, Bhat S, *et al.* (2001) The hormone resistin links obesity to diabetes. *Nature* **409**:307–312.

24. Combs TP, Berg AH, Obici S, *et al.* (2001) Endogenous glucose production is inhibited by the adipose-derived protein Acrp30. *J Clin Invest* **108**: 1875–1881.

25. Eckel RH, Grundy SM, Zimmet PZ. (2005) The metabolic syndrome. *Lancet* **365**:1415–1428.

26. Hayden MR, Tyagi SC. (2004) Uric acid: a new look at an old risk marker for cardiovascular disease, metabolic syndrome, and type 2 diabetes mellitus: the urate redox shuttle. *Nutr Metab (Lond)* **1**:10.

27. Reed DR, Price RA. (2000) X-linkage does not account for the absence of father–son similarity in plasma uric acid concentrations. *Am J Med Genet* **92**:142–146.

28. Cohen JC, Boerwinkle E, Mosley TH Jr, Hobbs HH. (2006) Sequence variations in PCSK9, low LDL, and protection against coronary heart disease. *N Engl J Med* **354**:1264–1272.

29. Laakso M. (1993) How good a marker is insulin level for insulin resistance? *Am J Epidemiol* **137**:959–965.

30. Matthews DR, Hosker JP, Rudenski AS, *et al.* (1985) Homeostasis model assessment: insulin resistance and beta-cell function from fasting plasma glucose and insulin concentrations in man. *Diabetologia* **28**: 412–419.

31. Haffner SM, Kennedy E, Gonzalez C, *et al.* (1996) A prospective analysis of the HOMA model. The Mexico City Diabetes Study. *Diabetes Care* **19**: 1138–1141.

32. Katz A, Nambi SS, Mather K, *et al.* (2000) Quantitative insulin sensitivity check index: a simple, accurate method for assessing insulin sensitivity in humans. *J Clin Endocrinol Metab* **85**:2402–2410.

33. Bergman RN, Zaccaro DJ, Watanabe RM, *et al.* (2003) Minimal model-based insulin sensitivity has greater heritability and a different genetic

basis than homeostasis model assessment or fasting insulin. *Diabetes* **52**: 2168–2174.

34. Mitchell BD, Kammerer CM, Mahaney MC, *et al.* (1996) Genetic analysis of the IRS. Pleiotropic effects of genes influencing insulin levels on lipoprotein and obesity measures. *Arterioscler Thromb Vasc Biol* **16**:281–288.

35. Bouchard C, Tremblay A, Despres JP, *et al.* (1990) The response to long-term overfeeding in identical twins. *N Engl J Med* **322**:1477–1482.

36. Mayer EJ, Newman B, Austin MA, *et al.* (1996) Genetic and environmental influences on insulin levels and the insulin resistance syndrome: an analysis of women twins. *Am J Epidemiol* **143**:323–332.

37. Price RA, Lee JH. (2001) Risk ratios for obesity in families of obese African-American and Caucasian women. *Hum Hered* **51**:35–40.

38. Lee JH, Reed DR, Price RA. (1997) Familial risk ratios for extreme obesity: implications for mapping human obesity genes. *Int J Obes Relat Metab Disord* **21**:935–940.

39. Maes HH, Neale MC, Eaves LJ. (1997) Genetic and environmental factors in relative body weight and human adiposity. *Behav Genet* **27**:325–351.

40. Grilo CM, Pogue-Geile MF. (1991) The nature of environmental influences on weight and obesity: a behavior genetic analysis. *Psychol Bull* **110**: 520–537.

41. Price RA, Cadoret RJ, Stunkard AJ, Troughton E. (1987) Genetic contributions to human fatness: an adoption study. *Am J Psychiatry* **144**:1003–1008.

42. Stunkard AJ, Sorensen TI, Hanis C, *et al.* (1986) An adoption study of human obesity. *N Engl J Med* **314**:193–198.

43. Sorensen TI, Price RA, Stunkard AJ, Schulsinger F. (1989) Genetics of obesity in adult adoptees and their biological siblings. *BMJ* **298**:87–90.

44. Stunkard AJ, Harris JR, Pedersen NL, McClearn GE. (1990) The body-mass index of twins who have been reared apart. *N Engl J Med* **322**:1483–1487.

45. Price RA, Gottesman II. (1991) Body fat in identical twins reared apart: roles for genes and environment. *Behav Genet* **21**:1–7.

46. Perusse L, Rankinen T, Zuberi A, *et al.* (2005) The human obesity gene map: the 2004 update. *Obes Res* **13**:381–490.

47. Barnett AH, Eff C, Leslie RD, Pyke DA. (1981) Diabetes in identical twins. A study of 200 pairs. *Diabetologia* **20**:87–93.

48. Baekkeskov S, Aanstoot HJ, Christgau S, *et al.* (1990) Identification of the 64K autoantigen in insulin-dependent diabetes as the GABA-synthesizing enzyme glutamic acid decarboxylase. *Nature* **347**:151–156.

49. Hanson RL, Ehm MG, Pettitt DJ, *et al.* (1998) An autosomal genomic scan for loci linked to type II diabetes mellitus and body-mass index in Pima Indians. *Am J Hum Genet* **63**:1130–1138.

50. Elbein SC, Hoffman MD, Teng K, *et al.* (1999) A genome-wide search for type 2 diabetes susceptibility genes in Utah Caucasians. *Diabetes* **48**: 1175–1182.

51. Vionnet N, Hani EH, Dupont S, *et al.* (2000) Genomewide search for type 2 diabetes-susceptibility genes in French whites: evidence for a novel susceptibility locus for early-onset diabetes on chromosome 3q27-qter and

independent replication of a type 2-diabetes locus on chromosome 1q21–q24. *Am J Hum Genet* **67**:1470–1480.

52. Wiltshire S, Hattersley AT, Hitman GA, *et al.* (2001) A genomewide scan for loci predisposing to type 2 diabetes in a U.K. population (the Diabetes UK Warren 2 Repository): analysis of 573 pedigrees provides independent replication of a susceptibility locus on chromosome 1q. *Am J Hum Genet* **69**:553–569.

53. Meigs JB, Panhuysen CI, Myers RH, *et al.* (2002) A genome-wide scan for loci linked to plasma levels of glucose and HbA_{1c} in a community-based sample of Caucasian pedigrees: the Framingham Offspring Study. *Diabetes* **51**:833–840.

54. Hsueh WC, St Jean PL, Mitchell BD, *et al.* (2003) Genome-wide and fine-mapping linkage studies of type 2 diabetes and glucose traits in the Old Order Amish: evidence for a new diabetes locus on chromosome 14q11 and confirmation of a locus on chromosome 1q21–q24. *Diabetes* **52**:550–557.

55. Mahtani MM, Widen E, Lehto M, *et al.* (1996) Mapping of a gene for type 2 diabetes associated with an insulin secretion defect by a genome scan in Finnish families. *Nat Genet* **14**:90–94.

56. Bowden DW, Sale M, Howard TD, *et al.* (1997) Linkage of genetic markers on human chromosomes 20 and 12 to NIDDM in Caucasian sib pairs with a history of diabetic nephropathy. *Diabetes* **46**:882–886.

57. Shaw JT, Lovelock PK, Kesting JB, *et al.* (1998) Novel susceptibility gene for late-onset NIDDM is localized to human chromosome 12q. *Diabetes* **47**:1793–1796.

58. Ehm MG, Karnoub MC, Sakul H, *et al.* (2000) Genomewide search for type 2 diabetes susceptibility genes in four American populations. *Am J Hum Genet* **66**:1871–1881.

59. Lindgren CM, Mahtani MM, Widen E, *et al.* (2002) Genomewide search for type 2 diabetes mellitus susceptibility loci in Finnish families: the Botnia Study. *Am J Hum Genet* **70**:509–516.

60. Wiltshire S, Frayling TM, Groves CJ, *et al.* (2004) Evidence from a large U.K. family collection that genes influencing age of onset of type 2 diabetes map to chromosome 12p and to the MODY3/NIDDM2 locus on 12q24. *Diabetes* **53**:855–860.

61. Rotimi CN, Chen G, Adeyemo AA, *et al.* (2004) A genome-wide search for type 2 diabetes susceptibility genes in West Africans: the Africa America Diabetes Mellitus (AADM) Study. *Diabetes* **53**:838–841.

62. Zouali H, Hani EH, Philippi A, *et al.* (1997) A susceptibility locus for early-onset non-insulin dependent (type 2) diabetes mellitus maps to chromosome 20q, proximal to the phosphoenolpyruvate carboxykinase gene. *Hum Mol Genet* **6**:1401–1408.

63. Ghosh S, Watanabe RM, Hauser ER, *et al.* (1999) Type 2 diabetes: evidence for linkage on chromosome 20 in 716 Finnish affected sib pairs. *Proc Natl Acad Sci USA* **96**:2198–2203.

64. Ghosh S, Watanabe RM, Valle TT, *et al.* (2000) The Finland–United States Investigation of Non–Insulin-Dependent Diabetes Mellitus Genetics

(FUSION) Study. I. An autosomal genome scan for genes that predispose to type 2 diabetes. *Am J Hum Genet* **67**:1174–1185.

65. Klupa T, Malecki MT, Pezzolesi M, *et al.* (2000) Further evidence for a susceptibility locus for type 2 diabetes on chromosome 20q13.1–q13.2. *Diabetes* **49**:2212–2216.

66. Mori Y, Otabe S, Dina C, *et al.* (2002) Genome-wide search for type 2 diabetes in Japanese affected sib-pairs confirms susceptibility genes on 3q, 15q, and 20q and identifies two new candidate loci on 7p and 11p. *Diabetes* **51**:1247–1255.

67. Florez JC, Hirschhorn J, Altshuler D. (2003) The inherited basis of diabetes mellitus: implications for the genetic analysis of complex traits. *Annu Rev Genomics Hum Genet* **4**:257–291.

68. Barroso I. (2005) Genetics of type 2 diabetes. *Diabet Med* **22**:517–535.

69. Henkin L, Bergman RN, Bowden DW, *et al.* (2003) Genetic epidemiology of insulin resistance and visceral adiposity. The IRAS Family Study design and methods. *Ann Epidemiol* **13**:211–217.

70. Freeman MS, Mansfield MW, Barrett JH, Grant PJ. (2002) Heritability of features of the insulin resistance syndrome in a community-based study of healthy families. *Diabet Med* **19**:994–999.

71. Watanabe RM, Ghosh S, Langefeld CD, *et al.* (2000) The Finland–United States Investigation of Non–Insulin-Dependent Diabetes Mellitus Genetics (FUSION) Study. II. An autosomal genome scan for diabetes-related quantitative-trait loci. *Am J Hum Genet* **67**:1186–1200.

72. Bray MS, Boerwinkle E, Hanis CL. (1999) Linkage analysis of candidate obesity genes among the Mexican-American population of Starr County, Texas. *Genet Epidemiol* **16**:397–411.

73. Li WD, Dong C, Li D, *et al.* (2004) A quantitative trait locus influencing fasting plasma glucose in chromosome region 18q22–23. *Diabetes* **53**: 2487–2491.

74. Vardarli I, Baier LJ, Hanson RL, *et al.* (2002) Gene for susceptibility to diabetic nephropathy in type 2 diabetes maps to 18q22.3–23. *Kidney Int* **62**:2176–2183.

75. Halama N, Yard-Breedijk A, Vardarli I, *et al.* (2003) The Kruppel-like zinc-finger gene ZNF236 is alternatively spliced and excluded as susceptibility gene for diabetic nephropathy. *Genomics* **82**:406–411.

76. Proffitt JM, Cai G, Azim D, *et al.* (2007) Quantitative trait loci influencing plasma TNF- and IL-1 in humans co-localize to chromosome 18. *Obes Res* (in press).

77. Krentz AJ. (2003) Lipoprotein abnormalities and their consequences for patients with type 2 diabetes. *Diabetes Obes Metab* **5**(Suppl 1):S19–S27.

78. Howard BV, Knowler WC, Vasquez B, *et al.* (1984) Plasma and lipoprotein cholesterol and triglyceride in the Pima Indian population. Comparison of diabetics and nondiabetics. *Arteriosclerosis* **4**:462–471.

79. Ford S Jr, Bozian RC, Knowles HC Jr. (1968) Interactions of obesity, and glucose and insulin levels in hypertriglyceridemia. *Am J Clin Nutr* **21**: 904–910.

80. Thelle DS, Shaper AG, Whitehead TP, *et al.* (1983) Blood lipids in middle-aged British men. *Br Heart J* **49**:205–213.
81. Austin MA, Hokanson JE, Edwards KL. (1998) Hypertriglyceridemia as a cardiovascular risk factor. *Am J Cardiol* **81**:7B–12B.
82. Castelli WP. (1986) The triglyceride issue: a view from Framingham. *Am Heart J* **112**:432–437.
83. Austin MA, McKnight B, Edwards KL, *et al.* (2000) Cardiovascular disease mortality in familial forms of hypertriglyceridemia: a 20-year prospective study. *Circulation* **101**:2777–2782.
84. Hokanson JE, Austin MA. (1996) Plasma triglyceride level is a risk factor for cardiovascular disease independent of high-density lipoprotein cholesterol level: a meta-analysis of population-based prospective studies. *J Cardiovasc Risk* **3**:213–219.
85. Assmann G, Schulte H, von Eckardstein A. (1996) Hypertriglyceridemia and elevated lipoprotein(a) are risk factors for major coronary events in middle-aged men. *Am J Cardiol* **77**:1179–1184.
86. Jeppesen J, Hein HO, Suadicani P, Gyntelberg F. (1998) Triglyceride concentration and ischemic heart disease: an eight-year follow-up in the Copenhagen Male Study. *Circulation* **97**:1029–1036.
87. Fontbonne A, Eschwege E, Cambien F, *et al.* (1989) Hypertriglyceridaemia as a risk factor of coronary heart disease mortality in subjects with impaired glucose tolerance or diabetes. Results from the 11-year follow-up of the Paris Prospective Study. *Diabetologia* **32**:300–304.
88. Tanne D, Koren-Morag N, Graff E, Goldbourt U. (2001) Blood lipids and first-ever ischemic stroke/transient ischemic attack in the Bezafibrate Infarction Prevention (BIP) Registry: high triglycerides constitute an independent risk factor. *Circulation* **104**:2892–2897.
89. Hopkins PN, Heiss G, Ellison RC, *et al.* (2003) Coronary artery disease risk in familial combined hyperlipidemia and familial hypertriglyceridemia: a case-control comparison from the National Heart, Lung, and Blood Institute Family Heart Study. *Circulation* **108**:519–523.
90. Banks WA, Coon AB, Robinson SM, *et al.* (2004) Triglycerides induce leptin resistance at the blood–brain barrier. *Diabetes* **53**:1253–1260.
91. Austin MA, King MC, Bawol RD, *et al.* (1987) Risk factors for coronary heart disease in adult female twins. Genetic heritability and shared environmental influences. *Am J Epidemiol* **125**:308–318.
92. Rice T, Vogler GP, Laskarzewski PM, *et al.* (1991) Familial aggregation of lipids and lipoproteins in families ascertained through random and nonrandom probands in the Stanford Lipid Research Clinics Family Study. *Am J Med Genet* **39**:270–277.
93. Abney M, McPeek MS, Ober C. (2001) Broad and narrow heritabilities of quantitative traits in a founder population. *Am J Hum Genet* **68**: 1302–1307.
94. Edwards KL, Mahaney MC, Motulsky AG, Austin MA. (1999) Pleiotropic genetic effects on LDL size, plasma triglyceride, and HDL cholesterol in families. *Arterioscler Thromb Vasc Biol* **19**:2456–2464.

95. Brenn T. (1994) Genetic and environmental effects on coronary heart disease risk factors in northern Norway. The cardiovascular disease study in Finnmark. *Ann Hum Genet* **58**(Pt 4):369–379.

96. Christian JC, Feinleib M, Hulley SB, *et al.* (1976) Genetics of plasma cholesterol and triglycerides: a study of adult male twins. *Acta Genet Med Gemellol (Roma)* **25**:145–149.

97. Goldstein JL, Schrott HG, Hazzard WR, *et al.* (1973) Hyperlipidemia in coronary heart disease. II. Genetic analysis of lipid levels in 176 families and delineation of a new inherited disorder, combined hyperlipidemia. *J Clin Invest* **52**:1544–1568.

98. Brooks-Wilson A, Marcil M, Clee SM, *et al.* (1999) Mutations in ABC1 in Tangier disease and familial high-density lipoprotein deficiency. *Nat Genet* **22**:336–345.

99. Kao JT, Wen HC, Chien KL, *et al.* (2003) A novel genetic variant in the apolipoprotein A5 gene is associated with hypertriglyceridemia. *Hum Mol Genet* **12**:2533–2539.

100. Wang GQ, DiPietro M, Roeder K, *et al.* (2003) Cladistic analysis of human apolipoprotein a4 polymorphisms in relation to quantitative plasma lipid risk factors of coronary heart disease. *Ann Hum Genet* **67**:107–124.

101. Klein HG, Santamarina-Fojo S, Duverger N, *et al.* (1993) Fish eye syndrome: a molecular defect in the lecithin-cholesterol acyltransferase (LCAT) gene associated with normal alpha-LCAT–specific activity. Implications for classification and prognosis. *J Clin Invest* **92**:479–485.

102. Zhang X, Wang K. (2003) Bivariate linkage analysis of cholesterol and triglyceride levels in the Framingham Heart Study. *BMC Genet* **4** (Suppl 1):S62.

103. Newman DL, Abney M, Dytch H, *et al.* (2003) Major loci influencing serum triglyceride levels on 2q14 and 9p21 localized by homozygosity-by-descent mapping in a large Hutterite pedigree. *Hum Mol Genet* **12**:137–144.

104. Bosse Y, Chagnon YC, Despres JP, *et al.* (2004) Genome-wide linkage scan reveals multiple susceptibility loci influencing lipid and lipoprotein levels in the Quebec Family Study. *J Lipid Res* **45**:419–426.

105. Horne BD, Malhotra A, Camp NJ. (2003) Comparison of linkage analysis methods for genome-wide scanning of extended pedigrees, with application to the TG/HDL-C ratio in the Framingham Heart Study. *BMC Genet* **4**(Suppl 1):S93.

106. McQueen MB, Bertram L, Rimm EB, *et al.* (2003) A QTL genome scan of the metabolic syndrome and its component traits. *BMC Genet* **4**(Suppl 1):S96.

107. Naoumova RP, Bonney SA, Eichenbaum-Voline S, *et al.* (2003) Confirmed locus on chromosome 11p and candidate loci on 6q and 8p for the triglyceride and cholesterol traits of combined hyperlipidemia. *Arterioscler Thromb Vasc Biol* **23**:2070–2077.

108. Duggirala R, Blangero J, Almasy L, *et al.* (2000) A major susceptibility locus influencing plasma triglyceride concentrations is located on chromosome 15q in Mexican Americans. *Am J Hum Genet* **66**:1237–1245.

109. Shearman AM, Ordovas JM, Cupples LA, *et al.* (2000) Evidence for a gene influencing the TG/HDL-C ratio on chromosome 7q32.3-qter: a genome-wide scan in the Framingham Study. *Hum Mol Genet* **9**:1315–1320.

110. Reed DR, Nanthakumar E, North M, *et al.* (2001) A genome-wide scan suggests a locus on chromosome 1q21–q23 contributes to normal variation in plasma cholesterol concentration. *J Mol Med* **79**:262–269.

111. Lin JP. (2003) Genome-wide scan on plasma triglyceride and high density lipoprotein cholesterol levels, accounting for the effects of correlated quantitative phenotypes. *BMC Genet* **4**(Suppl 1):S47.

112. Pajukanta P, Terwilliger JD, Perola M, *et al.* (1999) Genomewide scan for familial combined hyperlipidemia genes in Finnish families, suggesting multiple susceptibility loci influencing triglyceride, cholesterol, and apolipoprotein B levels. *Am J Hum Genet* **64**:1453–1463.

113. Pollin TI, Hsueh WC, Steinle NI, *et al.* (2004) A genome-wide scan of serum lipid levels in the Old Order Amish. *Atherosclerosis* **173**:89–96.

114. Austin MA, Edwards KL, Monks SA, *et al.* (2003) Genome-wide scan for quantitative trait loci influencing LDL size and plasma triglyceride in familial hypertriglyceridemia. *J Lipid Res* **44**:2161–2168.

115. Elbein SC, Hasstedt SJ. (2002) Quantitative trait linkage analysis of lipid-related traits in familial type 2 diabetes: evidence for linkage of triglyceride levels to chromosome 19q. *Diabetes* **51**:528–535.

116. Sonnenberg GE, Krakower GR, Martin LJ, *et al.* (2004) Genetic determinants of obesity-related lipid traits. *J Lipid Res* **45**:610–615.

117. Rice T, Chagnon YC, Perusse L, *et al.* (2002) A genomewide linkage scan for abdominal subcutaneous and visceral fat in black and white families: the HERITAGE Family Study. *Diabetes* **51**:848–855.

118. Hsueh WC, Mitchell BD, Schneider JL, *et al.* (2001) Genome-wide scan of obesity in the Old Order Amish. *J Clin Endocrinol Metab* **86**: 1199–1205.

119. Li WD, Dong C, Li D, *et al.* (2005) A genome scan for serum triglyceride in obese nuclear families. *J Lipid Res* **46**:432–438.

120. Stamler J, Wentworth D, Neaton JD. (1986) Is relationship between serum cholesterol and risk of premature death from coronary heart disease continuous and graded? Findings in 356,222 primary screenees of the Multiple Risk Factor Intervention Trial (MRFIT). *JAMA* **256**:2823–2828.

121. Pekkanen J, Linn S, Heiss G, *et al.* (1990) Ten-year mortality from cardiovascular disease in relation to cholesterol level among men with and without preexisting cardiovascular disease. *N Engl J Med* **322**:1700–1707.

122. Gordon DJ, Probstfield JL, Garrison RJ, *et al.* (1989) High-density lipoprotein cholesterol and cardiovascular disease. Four prospective American studies. *Circulation* **79**:8–15.

123. Lloyd-Jones DM, Larson MG, Beiser A, Levy D. (1999) Lifetime risk of developing coronary heart disease. *Lancet* **353**:89–92.

124. Law MR, Wald NJ, Thompson SG. (1994) By how much and how quickly does reduction in serum cholesterol concentration lower risk of ischaemic heart disease? *BMJ* **308**:367–372.

125. Goldstein JL, Brown MS. (1974) Binding and degradation of low density lipoproteins by cultured human fibroblasts. Comparison of cells from a normal subject and from a patient with homozygous familial hypercholesterolemia. *J Biol Chem* **249**:5153–5162.

126. Higgins MJ, Lecamwasam DS, Galton DJ. (1975) A new type of familial hypercholesterolaemia. *Lancet* **2**:737–740.

127. Abifadel M, Varret M, Rabes JP, *et al.* (2003) Mutations in PCSK9 cause autosomal dominant hypercholesterolemia. *Nat Genet* **34**:154–156.

128. Garcia CK, Wilund K, Arca M, *et al.* (2001) Autosomal recessive hypercholesterolemia caused by mutations in a putative LDL receptor adaptor protein. *Science* **292**:1394–1398.

129. Pajukanta P, Nuotio I, Terwilliger JD, *et al.* (1998) Linkage of familial combined hyperlipidaemia to chromosome 1q21–q23. *Nat Genet* **18**:369–373.

130. Eurlings PM, van der Kallen CJ, Geurts JM, *et al.* (2001) Genetic dissection of familial combined hyperlipidemia. *Mol Genet Metab* **74**:98–104.

131. Aouizerat BE, Allayee H, Cantor RM, *et al.* (1999) A genome scan for familial combined hyperlipidemia reveals evidence of linkage with a locus on chromosome 11. *Am J Hum Genet* **65**:397–412.

132. Pajukanta P, Allayee H, Krass KL, *et al.* (2003) Combined analysis of genome scans of Dutch and Finnish families reveals a susceptibility locus for high-density lipoprotein cholesterol on chromosome 16q. *Am J Hum Genet* **72**:903–917.

133. Heiberg A. (1974) The heritability of serum lipoprotein and lipid concentrations. A twin study. *Clin Genet* **6**:307–316.

134. Hunt SC, Hasstedt SJ, Kuida H, *et al.* (1989) Genetic heritability and common environmental components of resting and stressed blood pressures, lipids, and body mass index in Utah pedigrees and twins. *Am J Epidemiol* **129**:625–638.

135. Heller DA, de Faire U, Pedersen NL, *et al.* (1993) Genetic and environmental influences on serum lipid levels in twins. *N Engl J Med* **328**:1150–1156.

136. Mitchell BD, Kammerer CM, Blangero J, *et al.* (1996) Genetic and environmental contributions to cardiovascular risk factors in Mexican Americans. The San Antonio Family Heart Study. *Circulation* **94**:2159–2170.

137. Perusse L, Rice T, Despres JP, *et al.* (1997) Familial resemblance of plasma lipids, lipoproteins and postheparin lipoprotein and hepatic lipases in the HERITAGE Family Study. *Arterioscler Thromb Vasc Biol* **17**:3263–3269.

138. Mathias RA, Roy-Gagnon MH, Justice CM, *et al.* (2003) Comparison of year-of-exam- and age-matched estimates of heritability in the Framingham Heart Study data. *BMC Genet* **4**(Suppl 1):S36.

139. Coon H, Eckfeldt JH, Leppert MF, *et al.* (2002) A genome-wide screen reveals evidence for a locus on chromosome 11 influencing variation in LDL cholesterol in the NHLBI Family Heart Study. *Hum Genet* **111**:263–269.

140. Imperatore G, Knowler WC, Pettitt DJ, *et al.* (2000) A locus influencing total serum cholesterol on chromosome 19p: results from an autosomal genomic scan of serum lipid concentrations in Pima Indians. *Arterioscler Thromb Vasc Biol* **20**:2651–2656.

141. Abecasis GR, Cherny SS, Cookson WO, Cardon LR. (2002) Merlin — rapid analysis of dense genetic maps using sparse gene flow trees. *Nat Genet* **30**:97–101.

142. Sham PC, Purcell S, Cherny SS, Abecasis GR. (2002) Powerful regression-based quantitative-trait linkage analysis of general pedigrees. *Am J Hum Genet* **71**:238–253.

143. Wang D, Yang H, Quinones MJ, *et al.* (2005) A genome-wide scan for carotid artery intima-media thickness: the Mexican-American Coronary Artery Disease Family Study. *Stroke* **36**:540–545.

144. Pajukanta P, Cargill M, Viitanen L, *et al.* (2000) Two loci on chromosomes 2 and X for premature coronary heart disease identified in early- and late-settlement populations of Finland. *Am J Hum Genet* **67**:1481–1493.

145. Francke S, Manraj M, Lacquemant C, *et al.* (2001) A genome-wide scan for coronary heart disease suggests in Indo-Mauritians a susceptibility locus on chromosome 16p13 and replicates linkage with the metabolic syndrome on 3q27. *Hum Mol Genet* **10**:2751–2765.

146. Broeckel U, Hengstenberg C, Mayer B, *et al.* (2002) A comprehensive linkage analysis for myocardial infarction and its related risk factors. *Nat Genet* **30**:210–214.

147. Harrap SB, Zammit KS, Wong ZY, *et al.* (2002) Genome-wide linkage analysis of the acute coronary syndrome suggests a locus on chromosome 2. *Arterioscler Thromb Vasc Biol* **22**:874–878.

148. Chiodini BD, Lewis CM. (2003) Meta-analysis of 4 coronary heart disease genome-wide linkage studies confirms a susceptibility locus on chromosome 3q. *Arterioscler Thromb Vasc Biol* **23**:1863–1868.

149. Kent SM, Coyle LC, Flaherty PJ, *et al.* (2004) Marked low-density lipoprotein cholesterol reduction below current national cholesterol education program targets provides the greatest reduction in carotid atherosclerosis. *Clin Cardiol* **27**:17–21.

150. Haneda M, Polonsky KS, Bergenstal RM, *et al.* (1984) Familial hyperinsulinemia due to a structurally abnormal insulin. Definition of an emerging new clinical syndrome. *N Engl J Med* **310**:1288–1294.

151. Odawara M, Kadowaki T, Yamamoto R, *et al.* (1989) Human diabetes associated with a mutation in the tyrosine kinase domain of the insulin receptor. *Science* **245**:66–68.

152. Clausen JO, Hansen T, Bjorbaek C, *et al.* (1995) Insulin resistance: interactions between obesity and a common variant of insulin receptor substrate-1. *Lancet* **346**:397–402.

153. Esposito DL, Mammarella S, Ranieri A, *et al.* (1996) Deletion of Gly723 in the insulin receptor substrate-1 of a patient with noninsulin-dependent diabetes mellitus. *Hum Mutat* **7**:364–366.

154. Dib K, Whitehead JP, Humphreys PJ, *et al.* (1998) Impaired activation of phosphoinositide 3-kinase by insulin in fibroblasts from patients with severe insulin resistance and pseudoacromegaly. A disorder characterized by selective postreceptor insulin resistance. *J Clin Invest* **101**:1111–1120.

155. Savage DB, Agostini M, Barroso I, *et al.* (2002) Digenic inheritance of severe insulin resistance in a human pedigree. *Nat Genet* **31**:379–384.
156. Weedon MN, Owen KR, Shields B, *et al.* (2004) Common variants of the hepatocyte nuclear factor-4alpha P2 promoter are associated with type 2 diabetes in the U.K. population. *Diabetes* **53**:3002–3006.
157. Love-Gregory LD, Wasson J, Ma J, *et al.* (2004) A common polymorphism in the upstream promoter region of the hepatocyte nuclear factor-4 alpha gene on chromosome 20q is associated with type 2 diabetes and appears to contribute to the evidence for linkage in an Ashkenazi Jewish population. *Diabetes* **53**:1134–1140.
158. Silander K, Mohlke KL, Scott LJ, *et al.* (2004) Genetic variation near the hepatocyte nuclear factor-4 alpha gene predicts susceptibility to type 2 diabetes. *Diabetes* **53**:1141–1149.
159. Winckler W, Burtt NP, Holmkvist J, *et al.* (2005) Association of common variation in the HNF1alpha gene region with risk of type 2 diabetes. *Diabetes* **54**:2336–2342.
160. Winckler W, Graham RR, de Bakker PI, *et al.* (2005) Association testing of variants in the hepatocyte nuclear factor 4alpha gene with risk of type 2 diabetes in 7,883 people. *Diabetes* **54**:886–892.
161. Horikawa Y, Oda N, Cox NJ, *et al.* (2000) Genetic variation in the gene encoding calpain-10 is associated with type 2 diabetes mellitus. *Nat Genet* **26**:163–175.
162. Li WD, Lee JH, Price RA. (2000) The peroxisome proliferator–activated receptor gamma 2 Pro12Ala mutation is associated with early onset extreme obesity and reduced fasting glucose. *Mol Genet Metab* **70**:159–161.
163. Hara K, Okada T, Tobe K, *et al.* (2000) The Pro12Ala polymorphism in PPAR gamma2 may confer resistance to type 2 diabetes. *Biochem Biophys Res Commun* **271**:212–216.
164. Georgopoulos A, Aras O, Tsai MY. (2000) Codon-54 polymorphism of the fatty acid–binding protein 2 gene is associated with elevation of fasting and postprandial triglyceride in type 2 diabetes. *J Clin Endocrinol Metab* **85**:3155–3160.
165. Cao H, Hegele RA. (2000) Nuclear lamin A/C R482Q mutation in Canadian kindreds with Dunnigan-type familial partial lipodystrophy. *Hum Mol Genet* **9**:109–112.
166. Hegele RA, Cao H, Frankowski C, *et al.* (2002) PPARG F388L, a transactivation-deficient mutant, in familial partial lipodystrophy. *Diabetes* **51**:3586–3590.
167. Ng MC, So WY, Lam VK, *et al.* (2004) Genome-wide scan for metabolic syndrome and related quantitative traits in Hong Kong Chinese and confirmation of a susceptibility locus on chromosome 1q21–q25. *Diabetes* **53**:2676–2683.
168. Langefeld CD, Wagenknecht LE, Rotter JI, *et al.* (2004) Linkage of the metabolic syndrome to 1q23–q31 in Hispanic families: the Insulin Resistance Atherosclerosis Study Family Study. *Diabetes* **53**:1170–1174.

169. Deng HW, Deng H, Liu YJ, *et al.* (2002) A genomewide linkage scan for quantitative-trait loci for obesity phenotypes. *Am J Hum Genet* **70**:1138–1151.

170. North KE, Martin LJ, Dyer T, *et al.* (2003) HDL cholesterol in females in the Framingham Heart Study is linked to a region of chromosome 2q. *BMC Genet* **4**(Suppl 1):S98.

171. Martin LJ, North KE, Dyer T, *et al.* (2003) Phenotypic, genetic, and genome-wide structure in the metabolic syndrome. *BMC Genet* **4** (Suppl 1): S95.

172. Kissebah AH, Sonnenberg GE, Myklebust J, *et al.* (2000) Quantitative trait loci on chromosomes 3 and 17 influence phenotypes of the metabolic syndrome. *Proc Natl Acad Sci USA* **97**:14478–14483.

173. Loos RJ, Katzmarzyk PT, Rao DC, *et al.* (2003) Genome-wide linkage scan for the metabolic syndrome in the HERITAGE Family Study. *J Clin Endocrinol Metab* **88**:5935–5943.

174. Arya R, Blangero J, Williams K, *et al.* (2002) Factors of insulin resistance syndrome–related phenotypes are linked to genetic locations on chromosomes 6 and 7 in nondiabetic Mexican-Americans. *Diabetes* **51**:841–847.

175. Rich SS, Bowden DW, Haffner SM, *et al.* (2004) Identification of quantitative trait loci for glucose homeostasis: the Insulin Resistance Atherosclerosis Study (IRAS) Family Study. *Diabetes* **53**:1866–1875.

176. Panhuysen CI, Cupples LA, Wilson PW, *et al.* (2003) A genome scan for loci linked to quantitative insulin traits in persons without diabetes: the Framingham Offspring Study. *Diabetologia* **46**:579–587.

177. Silander K, Scott LJ, Valle TT, *et al.* (2004) A large set of Finnish affected sibling pair families with type 2 diabetes suggests susceptibility loci on chromosomes 6, 11, and 14. *Diabetes* **53**:821–829.

178. Li WD, Dong C, Li D, *et al.* (2004) An obesity-related locus in chromosome region 12q23–24. *Diabetes* **53**:812–820.

179. Rich SS, Bowden DW, Haffner SM, *et al.* (2005) A genome scan for fasting insulin and fasting glucose identifies a quantitative trait locus on chromosome 17p: the Insulin Resistance Atherosclerosis Study (IRAS) Family Study. *Diabetes* **54**:290–295.

180. McCarthy JJ, Meyer J, Moliterno DJ, *et al.* (2003) Evidence for substantial effect modification by gender in a large-scale genetic association study of the metabolic syndrome among coronary heart disease patients. *Hum Genet* **114**:87–98.

181. Sookoian S, Garcia SI, Porto PI, *et al.* (2005) Peroxisome proliferator–activated receptor gamma and its coactivator-1 alpha may be associated with features of the metabolic syndrome in adolescents. *J Mol Endocrinol* **35**:373–380.

182. Meirhaeghe A, Cottel D, Amouyel P, Dallongeville J. (2005) Association between peroxisome proliferator–activated receptor gamma haplotypes and the metabolic syndrome in French men and women. *Diabetes* **54**:3043–3048.

183. Robitaille J, Brouillette C, Houde A, *et al.* (2004) Association between the PPARalpha-L162V polymorphism and components of the metabolic syndrome. *J Hum Genet* **49**:482–489.

184. Masuzaki H, Paterson J, Shinyama H, *et al.* (2001) A transgenic model of visceral obesity and the metabolic syndrome. *Science* **294**:2166–2170.

185. Vartanian V, Lowell B, Minko IG, *et al.* (2006) The metabolic syndrome resulting from a knockout of the NEIL1 DNA glycosylase. *Proc Natl Acad Sci USA* **103**:1864–1869.

186. Hegele RA, Pollex RL. (2005) Genetic and physiological insights into the metabolic syndrome. *Am J Physiol Regul Integr Comp Physiol* **289**:R663–R669.

187. Stein CM, Song Y, Elston RC, *et al.* (2003) Structural equation model–based genome scan for the metabolic syndrome. *BMC Genet* **4**(Suppl 1):S99.

188. Cai G, Cole SA, Freeland-Graves JH, *et al.* (2004) Principal component for metabolic syndrome risk maps to chromosome 4p in Mexican Americans: the San Antonio Family Heart Study. *Hum Biol* **76**:651–665.

189. Atwood LD, Heard-Costa NL, Fox CS, *et al.* (2006) Sex and age specific effects of chromosomal regions linked to body mass index in the Framingham Study. *BMC Genet* **7**:7.

190. O'Rahilly S. (2002) Insights into obesity and insulin resistance from the study of extreme human phenotypes. *Eur J Endocrinol* **147**:435–441.

191. Spielman RS, McGinnis RE, Ewens WJ. (1993) Transmission test for linkage disequilibrium: the insulin gene region and insulin-dependent diabetes mellitus (IDDM). *Am J Hum Genet* **52**:506–516.

192. Amos CI, Elston RC, Bonney GE, *et al.* (1990) A multivariate method for detecting genetic linkage, with application to a pedigree with an adverse lipoprotein phenotype. *Am J Hum Genet* **47**:247–254.

193. Almasy L, Blangero J. (1998) Multipoint quantitative-trait linkage analysis in general pedigrees. *Am J Hum Genet* **62**:1198–1211.

194. Lee JH, Reed DR, Li WD, *et al.* (1999) Genome scan for human obesity and linkage to markers in 20q13. *Am J Hum Genet* **64**:196–209 [published erratum appears in *Am J Hum Genet* 2000;**66**(4):1472].

195. Atwood LD, Heard-Costa NL. (2003) Limits of fine-mapping a quantitative trait. *Genet Epidemiol* **24**:99–106.

196. Schulze TG, Chen YS, Badner JA, *et al.* (2003) Additional, physically ordered markers increase linkage signal for bipolar disorder on chromosome 18q22. *Biol Psychiatry* **53**:239–243.

197. Matise TC, Sachidanandam R, Clark AG, *et al.* (2003) A 3.9-centimorgan-resolution human single-nucleotide polymorphism linkage map and screening set. *Am J Hum Genet* **73**:271–284.

198. Liang KY, Chiu YF, Beaty TH. (2001) A robust identity-by-descent procedure using affected sib pairs: multipoint mapping for complex diseases. *Hum Hered* **51**:64–78.

199. Li WD, Li D, Wang S, *et al.* (2003) Linkage and linkage disequilibrium mapping of genes influencing human obesity in chromosome region 7q22.1–7q35. *Diabetes* **52**:1557–1561.

200. Malhotra A, Coon H, Feitosa MF, *et al.* (2005) Meta-analysis of genome-wide linkage studies for quantitative lipid traits in African Americans. *Hum Mol Genet* **14**:3955–3962.

201. Demenais F, Kanninen T, Lindgren CM, *et al.* (2003) A meta-analysis of four European genome screens (GIFT Consortium) shows evidence for a novel region on chromosome 17p11.2–q22 linked to type 2 diabetes. *Hum Mol Genet* **12**:1865–1873.

202. Johnson L, Luke A, Adeyemo A, *et al.* (2005) Meta-analysis of five genome-wide linkage studies for body mass index reveals significant evidence for linkage to chromosome 8p. *Int J Obes (Lond)* **29**:413–419.

203. Hager J, Dina C, Francke S, *et al.* (1998) A genome-wide scan for human obesity genes reveals a major susceptibility locus on chromosome 10. *Nat Genet* **20**:304–308.

204. Hinney A, Ziegler A, Oeffner F, *et al.* (2000) Independent confirmation of a major locus for obesity on chromosome 10. *J Clin Endocrinol Metab* **85**:2962–2965.

205. Price RA, Li WD, Bernstein A, *et al.* (2001) A locus affecting obesity in human chromosome region 10p12. *Diabetologia* **44**:363–366.

206. Rabinowitz D, Laird N. (2000) A unified approach to adjusting association tests for population admixture with arbitrary pedigree structure and arbitrary missing marker information. *Hum Hered* **50**:211–223.

207. Horvath S, Xu X, Lake SL, *et al.* (2004) Family-based tests for associating haplotypes with general phenotype data: application to asthma genetics. *Genet Epidemiol* **26**:61–69.

208. Abecasis GR, Cookson WO, Cardon LR. (2000) Pedigree tests of transmission disequilibrium. *Eur J Hum Genet* **8**:545–551.

209. Li C, Scott LJ, Boehnke M. (2004) Assessing whether an allele can account in part for a linkage signal: the genotype–IBD sharing test (GIST). *Am J Hum Genet* **74**:418–431.

210. Li M, Boehnke M, Abecasis GR. (2005) Joint modeling of linkage and association: identifying SNPs responsible for a linkage signal. *Am J Hum Genet* **76**:934–949.

211. Benjamini Y, Drai D, Elmer G, *et al.* (2001) Controlling the false discovery rate in behavior genetics research. *Behav Brain Res* **125**:279–284.

212. Benjamini Y, Yekutieli D. (2005) Quantitative trait loci analysis using the false discovery rate. *Genetics* **171**:783–790.

213. Song Y, Niu T, Manson JE, *et al.* (2004) Are variants in the CAPN10 gene related to risk of type 2 diabetes? A quantitative assessment of population and family-based association studies. *Am J Hum Genet* **74**:208–222.

214. Boutin P, Dina C, Vasseur F, *et al.* (2003) GAD2 on chromosome 10p12 is a candidate gene for human obesity. *PLoS Biol* **1**:E68.

215. Swarbrick MM, Waldenmaier B, Pennacchio LA, *et al.* (2005) Lack of support for the association between GAD2 polymorphisms and severe human obesity. *PLoS Biol* **3**:e315.

216. Beamer BA, Yen CJ, Andersen RE, *et al.* (1998) Association of the Pro12Ala variant in the peroxisome proliferator–activated receptor-gamma2 gene with obesity in two Caucasian populations. *Diabetes* **47**:1806–1808.

217. Deeb SS, Fajas L, Nemoto M, *et al.* (1998) A Pro12Ala substitution in PPARgamma2 associated with decreased receptor activity, lower body mass index and improved insulin sensitivity. *Nat Genet* **20**:284–287.

218. Doney A, Fischer B, Frew D, *et al.* (2002) Haplotype analysis of the PPARgamma Pro12Ala and C1431T variants reveals opposing associations with body weight. *BMC Genet* **3**:21.

219. Masud S, Ye S. (2003) Effect of the peroxisome proliferator activated receptor-gamma gene Pro12Ala variant on body mass index: a meta-analysis. *J Med Genet* **40**:773–780.

220. Altshuler D, Hirschhorn JN, Klannemark M, *et al.* (2000) The common PPARgamma Pro12Ala polymorphism is associated with decreased risk of type 2 diabetes. *Nat Genet* **26**:76–80.

221. Memisoglu A, Hu FB, Hankinson SE, *et al.* (2003) Prospective study of the association between the proline to alanine codon 12 polymorphism in the PPARgamma gene and type 2 diabetes. *Diabetes Care* **26**:2915–2917.

222. Luan J, Browne PO, Harding AH, *et al.* (2001) Evidence for gene–nutrient interaction at the PPARgamma locus. *Diabetes* **50**:686–689.

223. Adamo KB, Sigal RJ, Williams K, *et al.* (2005) Influence of Pro12Ala per-oxisome proliferator–activated receptor gamma2 polymorphism on glucose response to exercise training in type 2 diabetes. *Diabetologia* **48**:1503–1509.

224. Hsueh WC, Cole SA, Shuldiner AR, *et al.* (2001) Interactions between vari-ants in the beta3-adrenergic receptor and peroxisome proliferator–activated receptor-gamma2 genes and obesity. *Diabetes Care* **24**:672–677.

225. Nicklas BJ, van Rossum EF, Berman DM, *et al.* (2001) Genetic variation in the peroxisome proliferator–activated receptor-gamma2 gene (Pro12Ala) affects metabolic responses to weight loss and subsequent weight regain. *Diabetes* **50**:2172–2176.

226. Daly MJ, Rioux JD, Schaffner SF, *et al.* (2001) High-resolution haplotype structure in the human genome. *Nat Genet* **29**:229–232.

227. Abecasis GR, Noguchi E, Heinzmann A, *et al.* (2001) Extent and distribu-tion of linkage disequilibrium in three genomic regions. *Am J Hum Genet* **68**:191–197.

228. Gabriel SB, Schaffner SF, Nguyen H, *et al.* (2002) The structure of haplo-type blocks in the human genome. *Science* **296**:2225–2229.

229. Phillips MS, Lawrence R, Sachidanandam R, *et al.* (2003) Chromosome-wide distribution of haplotype blocks and the role of recombination hot spots. *Nat Genet* **33**:382–387.

230. Crawford DC, Carlson CS, Rieder MJ, *et al.* (2004) Haplotype diversity across 100 candidate genes for inflammation, lipid metabolism, and blood pressure regulation in two populations. *Am J Hum Genet* **74**:610–622.

231. Maniatis N, Collins A, Xu CF, *et al.* (2002) The first linkage disequilibrium (LD) maps: delineation of hot and cold blocks by diplotype analysis. *Proc Natl Acad Sci USA* **99**:2228–2233.

232. Stram DO, Haiman CA, Hirschhorn JN, *et al.* (2003) Choosing haplotype-tagging SNPs based on unphased genotype data using a preliminary sample

of unrelated subjects with an example from the Multiethnic Cohort Study. *Hum Hered* **55**:27–36.

233. Chapman JM, Cooper JD, Todd JA, Clayton DG. (2003) Detecting disease associations due to linkage disequilibrium using haplotype tags: a class of tests and the determinants of statistical power. *Hum Hered* **56**:18–31.

234. Risch N, Merikangas K. (1996) The future of genetic studies of complex human diseases. *Science* **273**:1516–1517.

235. Grant SF, Thorleifsson G, Reynisdottir I, *et al.* (2006) Variant of transcription factor 7-like 2 (TCF7L2) gene confers risk of type 2 diabetes. *Nat Genet* **38**:320–323.

236. Price RA. (2002) Genetics and common obesities: background, current status, strategies and future prospects. In: *Obesity, Theory and Therapy*, Wadden T (ed.), Guilford Publications, New York, NY, pp. 73–94.

237. Dong C, Wang S, Li WD, *et al.* (2003) Interacting genetic loci on chromosomes 20 and 10 influence extreme human obesity. *Am J Hum Genet* **72**:115–124.

238. Dong C, Li WD, Li D, Price RA. (2005) Interaction between obesity-susceptibility loci in chromosome regions 2p25–p24 and 13q13–q21. *Eur J Hum Genet* **13**:102–108.

239. Ritchie MD, Hahn LW, Roodi N, *et al.* (2001) Multifactor-dimensionality reduction reveals high-order interactions among estrogen-metabolism genes in sporadic breast cancer. *Am J Hum Genet* **69**:138–147.

240. Bastone L, Reilly M, Rader DJ, Foulkes AS. (2004) MDR and PRP: a comparison of methods for high-order genotype–phenotype associations. *Hum Hered* **58**:82–92.

241. Lunetta KL, Hayward LB, Segal J, Van Eerdewegh P. (2004) Screening large-scale association study data: exploiting interactions using random forests. *BMC Genet* **5**:32.

242. Millstein J, Conti DV, Gilliland FD, Gauderman WJ. (2006) A testing framework for identifying susceptibility genes in the presence of epistasis. *Am J Hum Genet* **78**:15–27.

243. Martin ER, Bass MP, Gilbert JR, *et al.* (2003) Genotype-based association test for general pedigrees: the genotype-PDT. *Genet Epidemiol* **25**:203–213.

244. Mei H, Ma D, Ashley-Koch A, *et al.* (2005) Extension of multifactor dimensionality reduction for identifying multilocus effects in the GAW14 simulated data. *BMC Genet* **6**(Suppl 1):S145.

245. Jeon JT, Carlborg O, Tornsten A, *et al.* (1999) A paternally expressed QTL affecting skeletal and cardiac muscle mass in pigs maps to the IGF2 locus. *Nat Genet* **21**:157–158.

246. Lindsay RS, Kobes S, Knowler WC, Hanson RL. (2002) Genome-wide linkage analysis assessing parent-of-origin effects in the inheritance of birth weight. *Hum Genet* **110**:503–509.

247. Lindsay RS, Kobes S, Knowler WC, *et al.* (2001) Genome-wide linkage analysis assessing parent-of-origin effects in the inheritance of type 2 diabetes and BMI in Pima Indians. *Diabetes* **50**:2850–2857.

248. Gorlova OY, Amos CI, Wang NW, *et al.* (2003) Genetic linkage and imprinting effects on body mass index in children and young adults. *Eur J Hum Genet* **11**:425–432.

249. Dong C, Li WD, Geller F, *et al.* (2005) Possible genomic imprinting of three human obesity-related genetic loci. *Am J Hum Genet* **76**:427–437.

250. Sakatani T, Wei M, Katoh M, *et al.* (2001) Epigenetic heterogeneity at imprinted loci in normal populations. *Biochem Biophys Res Commun* **283**: 1124–1130.

251. Okita C, Meguro M, Hoshiya H, *et al.* (2003) A new imprinted cluster on the human chromosome 7q21–q31, identified by human–mouse monochromosomal hybrids. *Genomics* **81**:556–559.

252. Greally JM. (2002) Short interspersed transposable elements (SINEs) are excluded from imprinted regions in the human genome. *Proc Natl Acad Sci USA* **99**:327–332.

253. Ling JQ, Li T, Hu JF, *et al.* (2006) CTCF mediates interchromosomal colocalization between Igf2/H19 and Wsb1/Nf1. *Science* **312**:269–272.

254. Hinds DA, Stuve LL, Nilsen GB, *et al.* (2005) Whole-genome patterns of common DNA variation in three human populations. *Science* **307**: 1072–1079.

Chapter 32

Genetics of Intracellular Traffic Jams: Multi-organellar Defects

Wei Li

Human Genetics, Institute of Genetics and Developmental Biology
Chinese Academy of Sciences, Beijing 100101, P. R. China

1. TRAFFIC JAMS IN EUKARYOTES

The secretory pathway was described in 1970 by George E. Palade, who published his classic work showing that proteins traffic between intracellular compartments[1] and thereafter won the Nobel Prize in 1974. Although this general phenomenon was clear by the mid-1970s, the details remained obscure. Two other laboratories led by James E. Rothman and Randy W. Schekman, respectively, dedicated to translate this descriptive field into one of clarity with molecular details by using a biochemical and genetic approach during the past three decades. The cellular trafficking machinery they elucidated underlies various vital processes, such as insulin releasing from pancreatic cells, neuronal communication, kiss of death at the immunological synapse, skin color development, clotting after platelet releasing, organ development in embryos stimulated by growth factors, and virus infection.

The overview of intracellular trafficking pathways is outlined in Fig. 1.[2,3] Alterations in these pathways explain a plethora of pathological processes due to traffic jams, including various inherited syndromes (Table 1), hyperlipidemia, diabetes, cancer, neurodegeneration, and infections such as AIDS and tetanus. Scientists are attempting to develop drugs that target the trafficking system in order to treat an increasing number of illnesses, particularly those immune and psychiatric disorders.

In mammalian cells, proteins are known to be transported between compartments by vesicles. These vesicles pinch off (or bud) from one membrane,

Table 1 Syndromes with Traffic Jams in Humans

Human Locus	Location	Syndrome	OMIM	Encoded Protein	Function	Mouse Mutant	Reference
BIG2/ ARFGEF2	20q13.13	ARPHM	608097	ARF GEF2	Guanine nucleotide exchange	N/A	10
CHS1/LYST	1q42.3	CHS	214500	CHS1p	Vesicle fusion	Beige (bg)	11
GS1/ MYO5A	15q21.2	GS1 ES	214450 256710	Myosin Va	Vesicle transport	Dilute (d)	12
GS2/ RAB27A	15q21.3	GS2 HS	607624	Rab27a	Vesicle transport	Ashen (ash)	13, 14
GS3/MLPH	2q37.3	GS3	609227	Melanophilin	Vesicle transport	Leaden (ln)	15
HPS1	10q24.2	HPS1	604982	HPS1p	BLOC-3 subunit	Pale ear (ep)	16
HPS2/ AP3B1	5q14.1	HPS2	608233	AP3 beta 3A subunit	AP-3 subunit	Pearl (pe)	17, 18
HPS3	3q24	HPS3	606118	HPS3p	BLOC-2 subunit	Cocoa (coa)	19, 20
HPS4	22q12.1	HPS4	606682	HPS4p	BLOC-3 subunit	Light ear (le)	21
HPS5	11p15.1	HPS5	607521	HPS5p	BLOC-2 subunit	Ruby eye-2 (ru2)	22
HPS6	10q24.32	HPS6	607522	HPS6p	BLOC-2 subunit	Ruby eye (ru)	22
HPS7/ DTNBP1	6p22.3	HPS7 SCZD3?	607145 600511	Dysbindin	BLOC-1 subunit	Sandy (sdy)	23

(Continued)

Table 1   *(Continued)*

Human Locus	Location	Syndrome	OMIM	Encoded Protein	Function	Mouse Mutant	Reference
HPS8/ BLOC1S3	19q13.32	HPS8	609762	Blos3	BLOC-1 subunit	Reduced pigmentation (*rp*)	24, 25
MCFD1/ LMAN1/ ERGIC53	18q21.32	F5F8D	601567	MCFD2-LMAN1 complex subunit	ER to Golgi transport	N/A	26
MCFD2	2p21	F5F8D	607788	MCFD2-LMAN1 complex subunit	ER to Golgi transport	N/A	27
SARA2	5q31.1	CMRD	246700	ARF, small	COPII vesicle trafficking	N/A	28
		ANDD CMRD-MSS	607689 607692	GTPase			
SNAP29	22q11.21	CEDNIK	609528	t-SNARE protein	Vesicle fusion	N/A	29
VPS33B	15q26.1	ARC	208085	Class C VPS33p	Sec1/Munc18 protein	N/A	30

ARC: arthrogryposis–renal dysfunction–cholestasis syndrome; ANDD: Anderson disease; ARPHM: autosomal recessive periventricular heterotopia with microcephaly; CEDNIK: cerebral dysgenesis, neuropathy, ichthyosis, and palmoplantar keratoderma syndrome; CHS: Chediak–Higashi syndrome; CMRD: chylomicron retention disease; CMRD-MSS: CMRD associated with Marinesco–Sjogren syndrome; ES: Elejalde syndrome; GS: Griscelli syndrome; F5F8D: combined deficiency of factor V and factor VIII; HPS: Hermansky–Pudlak syndrome; HS: hemophagocytic syndrome; SCZD: schizophrenia.

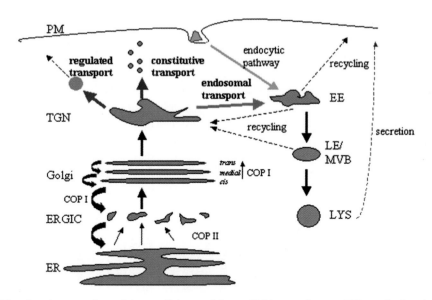

Fig. 1.   An overview of intracellular vesicle trafficking pathways. ER: endoplasmic reticulum; ERGIC: ER–Golgi intermediate compartment; Golgi: Golgi apparatus; TGN: trans-Golgi network; EE: early endosome; LE: late endosome; LYS: lysosome; MVB: multivesicular body.

and merge (or fuse) with the next. Multiple types of vesicles selectively transport cargo to different sites. Secreted proteins are captured into endoplasmic reticulum (ER) and travel to the cell surface by way of a structure called the Golgi, where posttranslational modifications occur. The antegrade transport from ER to Golgi is mediated by COPII-coated vesicles, while the retrograde transport from the Golgi to ER is thought to be mediated by COPI-coated vesicles.[2]

Through the trans-Golgi network (TGN), the fate of the cargo proteins is determined by three different pathways: (1) regulated transport for the secreted proteins under specific stimuli, (2) constitutive transport for the plasma membrane protein, and (3) endosomal transport for the organellar proteins (Fig. 1).[3] Various types of vesicles are responsible for these transport routes, such as the clathrin-coated AP-1 or AP-3 vesicles that mediate cargo transport via the endosomal transport. The well-known endocytotic pathway is mediated by clathrin-coated AP-2 vesicles; the coat proteins of these vesicles selectively recognize their cargo proteins. A variety of molecules regulate or direct the transport events such as vesicle budding, docking, and fusion; these molecules include GTP, Ca^{2+}, G proteins, molecular motors, phosphoinositides, Rab proteins, and SNARE complexes.

A GTP-binding protein is required for budding. The GTP–GDP switch provides the underpinnings for coat formation and budding. After vesicle formation, the GTP-binding protein converts its GTP to GDP. Calcium is required for most of the regulated secretion of proteins. Calcium release from the lumen of mammalian endocytic compartments and lysosomes mediates membrane fusion and efficient membrane trafficking in the endocytic pathway. G proteins are involved in this calcium-dependent signaling transduction. Molecular motors are a large collection of proteins (e.g. dynenin, myosin, actin, microtubule) that supply power for vesicle motion. Phosphoinositides serve as direct local modulators or recruiters of the protein machineries that control vesicle trafficking. Distinct phosphoinositide effectors are localized in different compartments; these effectors include regulators of small GTPases in coat assembly, dynamin in clathrin-coated vesicle formation, and FYVE finger proteins in endocytic membrane traffic.[4]

Rab proteins are Ras-like GTPases that regulate membrane trafficking events in eukaryotic cells. Human cells contain more than 60 Rab proteins, which are localized to distinct compartments and are specific for different transport pathways. Rabs switch between GDP-bound (inactive) and GTP-bound (active) forms by the actions of GDP/GTP exchange factor (GEF) and GTPase-activating protein (GAP). Membrane-associated Rabs are prenylated at the C-termini via Rab GGTase and Rab escort protein (REP). GDP-bound Rabs associate with donor or target membrane by the GDP dissociation inhibitor (GDI) cycle.[5]

During vesicle fusion, soluble N-ethylmaleimide–sensitive factor (NSF) attachment protein (SNAP) binds to membranes, and then NSF attaches to SNAP. SNAP receptors, or SNAREs, are required for vesicle fusion, and all vesicle and target membranes carry proteins (called v- and t-SNAREs, respectively) that recognize each other. The t-SNARE and v-SNARE components wrap around each other, pulling together the membranes in which they are embedded, thus prompting fusion. SNARE pairings contain enough information to direct specificity for membrane fusion events. The location of SNARE partners among cellular membranes therefore governs fusion events within the cell.[6,7]

The unveiling of the key machinery and mechanisms of vesicle trafficking is critical for numerous physiological events; and when it malfunctions, illness occurs. In studies of the vesicle trafficking pathway, mutants of the key players of trafficking are invaluable resources to dissect the building blocks of this system. In yeast, 23 genes involved in secretion were identified using standard genetic techniques, and the order in which the *SEC* genes acted was determined.[8] In *Drosophila*, some mutants with eye color defects (e.g. *rb, g, car*) were identified as traffic jams in melanosome biogenesis. In mouse, at least 16 mouse models of the Hermansky–Pudlak syndrome were described as multi-organellar defects due to the dysfunction of lysosome-related organelles (LROs) such as melanosomes and platelet dense granules.[9]

In human, a variety of syndromes with traffic jams have been identified both genetically and molecularly, as shown in Table 1. The clinical symptoms could be very different from each other in the listed syndromes. The traffic jams happen in various stages through the secretory pathways, such as from ER to Golgi (F5F8D, CMRD), from Golgi to lysosome (HPS, CHS, ARC), from microtubule-organizing center (MTOC) to cell surface (GS), in the process of vesicle budding or fusion events (ARPHM, CEDNIK), etc. The traffic jams of endocytosis or proteosomal degradation are not included in this category. The effects of the traffic jams on individuals may be either essential to development or tissue-specific symptoms. The most prominent trafficking disorder is the Hermansky–Pudlak syndrome (HPS), which will be discussed in more detail below.

Medical applications based on new information about trafficking extend much further. For example, depletion of AP-3 by siRNA treatment resulted in reduced release of HIV-1 virus particles, which implicates a novel therapeutic target for AIDS.[31] In type 2 diabetes, molecules that normally reside on the cell surface and import the sugar are stranded in vesicles inside; some patients are already using insulin-sensitizing drugs which restore fusion so that the transport molecules can reach the surface for proper glucose ingestion. Drugs that shift the balance between neurotransmitter uptake and export might someday provide novel therapies for cognitive and behavioral disorders. A vast number of interdisciplinary disciplines — including neurobiology, immunology, endocrinology, virology, and embryology — are expected to further expand their impact for a better understanding of vesicle trafficking pathways and traffic jams.

2. INTRODUCTION TO HERMANSKY–PUDLAK SYNDROME (HPS)

Albinism is caused by defects in pigment production or melanosomal biogenesis due to recessive single-gene mutation; and affects skin, eye, and hair. It is a genetically heterogeneous disease. In mice, more than 100 coat color loci have been described; while in humans, 17 genes have been identified as albinism genes, which can be categorized into the following two major groups:

(1) Nonsyndromic

    (a) Oculocutaneous (OCA): *OCA1, OCA2, OCA3, OCA4*

    (b) Ocular albinism (OA): *OA1*

(2) Syndromic

    (a) Hermansky–Pudlak syndrome (HPS): *HPS1~HPS8*

    (b) Chediak–Higashi syndrome (CHS): *CHS1*

    (c) Griscelli syndrome (GS): *GS1, GS2, GS3*

Except for OA as an X-linked Mendelian trait, all of the other types of albinism present autosomal recessive traits. I herein describe a recently well-recognized syndromic form of albinism: Hermansky–Pudlak syndrome (OMIM 203300).[32]

HPS is an autosomal recessive, genetically heterogeneous disorder characterized by a triad — oculocutaneous albinism, bleeding tendency, and ceroid deposition — which may cause lung fibrosis, colitis, and cardiomyopathy. Patients with HPS often die during the third-to-fifth decade.[33] The key pathological aspect of both human and mouse HPS is the disrupted biogenesis and/or function of both specialized lysosome-related organelles (LROs) and conventional lysosomes.[34,35]

2.1. *Genetics*

HPS is now known as a genetically heterogeneous (*HPS1, HPS2,...,
HPS8*), autosomal recessive inherited disorder.

2.2. *Incidence*

HPS occurs in many countries, with more than 800 patients reported worldwide. This figure is almost certainly underestimated because of misdiagnosis or undiagnosis. About 490 cases were reported in northwest Puerto Rico with founder effect (incidence rate, 1:1800; carrier rate, 1:21). HPS1 and HPS3 are the two common types of HPS in this region.

2.3. *Pathogenesis*

The pathogenesis underlying HPS results from defects in the synthesis of LROs.

2.4. *Clinical Manifestation*

The most common symptoms of HPS are hypopigmentation, loss of visual acuity, prolonged bleeding, colitis, and in many cases fatal lung disease.

2.5. *Prognosis*

Prolonged bleeding often requires multiple platelet transfusions, and fibrotic lung disease may lead to death in midlife.

2.6. *Therapy*

There is presently no cure. Only symptomatic (e.g. sunscreen, avoid aspirin) treatment of the disease exists.

2.7. *Diagnosis*

The "gold" indication of HPS is the absence of platelet dense granules upon electron microscopy. Symptoms of hypopigmentation and bleeding support the diagnosis. Molecular diagnoses are now available with the identification of eight HPS genes in humans.

For more details of the molecular features and phenotypes of each HPS gene or HPS-related gene, we have developed a HPS database (HPSD, http://liweilab.genetics.ac.cn/HPSD/) with a unique gene-oriented file (GOF) format.[36]

3. POSITIONAL CANDIDATE CLONING OF HPS GENES

Over the past 30 years, Dr Richard T. Swank's laboratory has identified at least 16 mouse HPS mutants.[35] Through positional candidate cloning, the first human HPS gene, *HPS1*, was identified in 1996.[16] This prompted the identification of the first murine HPS gene, *Hps1/ep*,[37] and the cloning of 15 other HPS genes in mouse or human thereafter.[9] The cloning of HPS genes has accelerated since the completion of the human and mouse genome projects (Table 2). The mouse mutants provide a powerful resource to dissect such a heterogeneous disorder as HPS.

With the identification of 16 HPS genes, their functional studies prelude a new era of the intracellular vesicle trafficking pathways at systems biology level. Currently, the 16 cloned mouse HPS genes fall into two classes. The first class, presented both in lower and higher eukaryotic species, includes

Table 2   The Cloned 16 Mouse and 8 Human HPS Genes

Mouse Loci	Human Loci	Year	Major Contributors (Laboratory)
Pale ear (*ep*)	*HPS1*	1996	Oh J (Dr Spritz)[16]; Gardner JM (Dr Brilliant)[37]
Mocha (*mh*)	*AP3D*	1998	Kantheti P (Dr Burmeister)[38]
Pearl (*pe*)	*HPS2/AP3B1*	1999	Feng L (Dr Swank) and Seymour AB (Dr Gorin)[18]; Dell'Angelica EC (Dr Bonifacino)[17]
Pallid (*pa*)	*PLDN*	1999	Huang L (Dr Gitschier)[39]
Gunmetal (*gm*)	*RABGGTA*	2000	Detter JC (Dr Kingsmore) and Zhang Q (Dr Swank)[40]
Ashen (*ash*)	*RAB27A*	2000	Wilson SM (Dr Jenkins)[14]
Cocoa (*coa*)	*HPS3*	2001	Anikster Y (Dr Gahl)[19]; Suzuki T (Dr Spritz) and Li W (Dr Swank)[20]
Muted (*mu*)	*MU*	2002	Zhang Q and Li W (Dr Swank)[41]
Light ear (*le*)	*HPS4*	2002	Suzuki T (Dr Spritz) and Li W (Dr Swank)[21]
Ruby eye 2 (*ru2*)	*HPS5*	2003	Zhang Q, Zhao B, Li W (Dr Swank), and Oiso N (Dr Spritz)[22]
Ruby eye (*ru*)	*HPS6*	2003	Zhang Q, Zhao B, Li W (Dr Swank), and Oiso N (Dr Spritz)[22]
Buff (*bf*)	*VPS33A*	2003	Suzuki T, Oiso N (Dr Spritz), and Gautam R (Dr Swank)[42]
Cappuccino (*cno*)	*CNO*	2003	Ciciotte SL (Dr Peters)[43]
Sandy (*sdy*)	*HPS7*	2003	Li W, Zhang Q (Dr Swank), and Oiso N (Dr Spritz)[23]
Reduced pigmentation (*rp*)	*HPS8/BLOC1S3*	2004 2006	Starcevic M (Dr Dell'Angelica)[25] Morgan NV (Dr Maher)[24]
Subtle gray (*sut*)	*SLC7A11*	2005	Chintala S and Li W (Dr Swank)[44]

genes well known to encode components or regulators of intracellular vesicle trafficking,[45] such as subunits of AP-3 adaptor complex, *Rab27a*, *VPS33a*, and Rab geranylgeranyl transferase alpha subunit.[40,46,47] This makes perfect sense of the organellar abnormalities in HPS patients, and enters into knowledge of the molecular and cellular mechanisms of HPS. The second class (first proposed by Zhang *et al.*[41]), presented only in higher eukaryotic organisms, is composed of at least 10 novel HPS genes. This category is supposed to be involved in regulating the synthesis of very specialized mammalian subcellular LROs such as platelet dense granules, melanosomes, MHC II compartments, lytic granules, and synaptic vesicles (Fig. 2).[9] Another known gene, *Slc7a11*, which is responsible for the subtle gray (*sut*) mutant, has been recently defined as a gene regulating both the production of pheomelanin pigment and the proliferation of cultured cells.[44]

Current Topics in Human Genetics

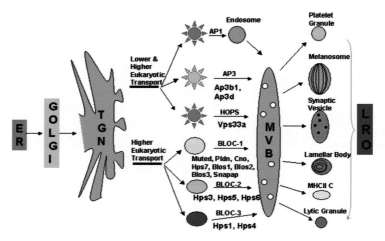

Fig. 2.   Biogenesis of LROs. A novel trafficking pathway has been defined in higher eukaryotes only that is mediated by the BLOC complexes. The known HPS genes involved in these pathways are listed under the complex names. ER: endoplasmic reticulum; TGN: trans-Golgi network; MVB: multivesicular body.

Regarding the functional aspects of the novel HPS genes, several novel biogeneses of LRO complexes (BLOC-1, BLOC-2, and BLOC-3) have been recently identified.[9,48] These include *dysbindin, muted, pallidin, blos1, blos2, blos3, snapin,* and *cno* in BLOC-1[23,25,43,49]; *coa/HPS3, ru2/HPS5,* and *ru/HPS6* in BLOC-2[22,50]; and *ep/HPS1* and *le/HPS4* in BLOC-3.[21,51] The biochemical features and assembling machineries of these complexes remain to be defined, although some pioneer studies have revealed the binding domains in building these BLOC(k)s[52] and the interactome of the BLOC network.[48]

4. GENES REGULATING THE BIOGENESIS OF LROs

Lysosomes are membrane-bound cytoplasmic organelles that are found in all mammalian cells, and contain hydrolases and lipases required for protein and membrane degradation. They are characterized by soluble acid–dependent hydrolases and a set of highly glycosylated integral membrane proteins such as LAMPs. Most of the properties of lysosomes are shared with a group of cell type–specific compartments referred to as lysosome-related organelles (LROs), which include melanosomes, lytic granules, MHC class II compartments, platelet dense granules, basophil granules, and azurophil granules (Table 3). In addition to lysosomal proteins, these

Table 3   Lysosome-related Organelles

Organelle	Tissue Distribution
Lysosomes	Ubiquitous
Melanosomes	Melanocytes, retinal pigment epithelia
Platelet dense granules	Platelets, megakaryocytes
Lamellar bodies	Alveolar type II epithelial cells
Lytic granules	NK cells, cytotoxic T-lymphocytes
MHC II compartments	Antigen-presenting cells
Basophilic granules	Mast cells
Azurophilic granules	Neutrophils, eosinophils
Osteoclast granules	Osteoclasts
Weibel–Palade bodies	Endothelial cells
Insulin granules	Pancreatic islets
Renin granules	Juxtaglomerular cells
Synaptic vesicles	Neurons

organelles contain cell type–specific components that are responsible for their specialized functions.[34] LROs present cell type–specific morphological features and functions, undergoing regulated secretion (also called secretory lysosomes). Most LROs are of the hematopoietic lineage, except for melanosomes and lamellar bodies. They can coexist with lysosomes in some cells like melanocytes and platelets.

The recent identification of genetic disorders which cause combined defects in several of these organelles indicates that they share common biogenetic pathways (Table 4). Studies of one of these disorders, the Hermansky–Pudlak syndrome (HPS), have provided helpful insights into the molecular machinery involved in LRO biogenesis.[53]

Table 4   Syndromic Dysgenesis of LROs

Specialized LRO	Dysfunction	Syndromes
Melanosomes	Oculocutaneous albinism (hypopigmentation)	HPS, CHS, ES, GS
Platelet granules	Bleeding diathesis	HPS, CHS, SPD, GPS
Synaptic vesicles	Abnormal behaviors; neurological symptoms	HPS, CHS, ES, GS
Lytic granules	Immunodeficiency	HPS, CHS, GS
MHC II compartments	Immunodeficiency	HPS, CHS
Azurophil granules	Neutropenia	HPS, CHS
Lamellar bodies	Lung fibrosis	HPS

CHS: Chediak–Higashi syndrome; ES: Elejalde syndrome; GPS: gray platelet syndrome; GS: Griscelli syndrome; HPS: Hermansky–Pudlak syndrome; SPD: storage pool deficiency.

HPS gene products play fundamental roles in the biogenesis of LROs. Several complexes formed by these HPS gene products have been demonstrated to mediate the synthesis of several LROs (Fig. 2), including the AP-3 complex; the Class C Vps (HOPS) complex; and BLOC-1, BLOC-2, and BLOC-3.[9,48] However, with the exception of the well-defined AP-3 complex, the HOPS complex, and the Rab27a tripartite protein complex, relatively little is known of the mechanisms by which other complexes like BLOCs regulate vesicle trafficking and organellar biogenesis. Whether these complexes interact with each other in governing the biogenesis of LROs remains to be defined. As shown in the tissue-specific LROs, the roles of HPS genes are involved in a variety of physiological functions.[9] Better understanding of the biochemical and functional properties of these complexes requires cutting-edge technology and interdisciplinary studies of HPS gene functions.

ACKNOWLEDGMENTS

This work was supported in part by the National Basic Research Program of China (No. 2006CB504100) and the National Natural Science Foundation of China (30525007).

REFERENCES

1. Palade G. (1975) Intracellular aspects of the process of protein synthesis. *Science* **169**:347–358.
2. Bonifacino JS, Glick BS. (2004) The mechanisms of vesicle budding and fusion. *Cell* **116**:153–166.
3. Roche PA. (1999) Intracellular protein traffic in lymphocytes: "how do I get THERE from HERE"? *Immunity* **11**:391–398.
4. Corvera S, D'Arrigo A, Stenmark H. (1999) Phosphoinositides in membrane traffic. *Curr Opin Cell Biol* **11**:460–465.
5. Seabra MC, Mules EH, Hume AN. (2002) Rab GTPases, intracellular traffic and disease. *Trends Mol Med* **8**:23–30.
6. Jahn R, Lang T, Sudhof TC. (2003) Membrane fusion. *Cell* **112**:519–533.
7. Rothman JE. (1994) Mechanisms of intracellular protein transport. *Nature* **372**:55–63.
8. Novick P, Field C, Shekman R. (1980) Identification of 23 complementation groups required for post-translational events in secretory pathway. *Cell* **21**: 205–215.
9. Li W, Rusiniak M, Chintala S, *et al.* (2004) Murine Hermansky–Pudlak syndrome: genes which control lysosome-related organelles. *Bioessays* **26**: 616–628.

10. Sheen VL, Ganesh VS, Topcu M, *et al.* (2004) Mutations in ARFGEF2 impli-
    cate vesicle trafficking in neural progenitor proliferation and migration in the
    human cerebral cortex. *Nat Genet* **36**:69–76.
11. Barbosa MD, Nguyen QA, Tchernev VT, *et al.* (1996) Identification of
    the homologous beige and Chediak–Higashi syndrome genes. *Nature* **382**:
    262–265.
12. Pastural E, Barrat FJ, Dufourcq-Lagelouse R, *et al.* (1997) Griscelli disease
    maps to chromosome 15q21 and is associated with mutations in the myosin-
    Va gene. *Nat Genet* **16**:289–292.
13. Menasche G, Pastural E, Feldmann J, *et al.* (2000) Mutations in RAB27A
    cause Griscelli syndrome associated with haemophagocytic syndrome. *Nat
    Genet* **25**:173–176.
14. Wilson SM, Yip R, Swing DA, *et al.* (2000) A mutation in Rab27a causes
    the vesicle transport defects observed in ashen mice. *Proc Natl Acad Sci USA*
    97:7933–7938.
15. Menasche G, Ho CH, Sanal O, *et al.* (2003) Griscelli syndrome restricted
    to hypopigmentation results from a melanophilin defect (GS3) or a MYO5A
    F-exon deletion (GS1). *J Clin Invest* **112**:450–456.
16. Oh J, Bailin T, Fukai K, *et al.* (1996) Positional cloning of a gene for
    Hermansky–Pudlak syndrome, a disorder of cytoplasmic organelles. *Nat
    Genet* **14**:300–306.
17. Dell'Angelica EC, Shotelersuk V, Aguilar RC, *et al.* (1999) Altered trafficking
    of lysosomal proteins in Hermansky–Pudlak syndrome due to mutations in
    the beta 3A subunit of the AP-3 adaptor. *Mol Cell* **3**:11–21.
18. Feng L, Seymour AB, Jiang S, *et al.* (1999) The beta3A subunit gene (Ap3b1)
    of the AP-3 adaptor complex is altered in the mouse hypopigmentation
    mutant pearl, a model for Hermansky–Pudlak syndrome and night blind-
    ness. *Hum Mol Genet* **8**:323–330.
19. Anikster Y, Huizing M, White J, *et al.* (2001) Mutation of a new gene causes
    a unique form of Hermansky–Pudlak syndrome in a genetic isolate of central
    Puerto Rico. *Nat Genet* **28**:376–380.
20. Suzuki T, Li W, Zhang Q, *et al.* (2001) The gene mutated in cocoa mice,
    carrying a defect of organelle biogenesis, is a homologue of the human
    Hermansky–Pudlak syndrome-3 gene. *Genomics* **78**:30–37.
21. Suzuki T, Li W, Zhang Q, *et al.* (2002) Hermansky–Pudlak syndrome is
    caused by mutations in HPS4, the human homolog of the mouse light-ear
    gene. *Nat Genet* **30**:321–324.
22. Zhang Q, Zhao B, Li W, *et al.* (2003) Ru2 and Ru encode mouse orthologs
    of the genes mutated in human Hermansky–Pudlak syndrome types 5 and 6.
    Nat Genet **33**:145–154.
23. Li W, Zhang Q, Oiso N, *et al.* (2003) Hermansky–Pudlak syndrome type 7
    (HPS-7) results from mutant dysbindin, a member of the biogenesis of
    lysosome-related organelles complex 1 (BLOC-1). *Nat Genet* **35**:84–89.
24. Morgan NV, Pasha S, Johnson CA, *et al.* (2006) A germline mutation
    in BLOC1S3/reduced pigmentation causes a novel variant of Hermansky–
    Pudlak syndrome (HPS8). *Am J Hum Genet* **78**:160–166.

25. Starcevic M, Dell'Angelica EC. (2004) Identification of snapin and three novel proteins (BLOS1, BLOS2, and BLOS3/reduced pigmentation) as subunits of biogenesis of lysosome-related organelles complex-1 (BLOC-1). *J Biol Chem* **279**:28393–28401.

26. Nichols WC, Seligsohn U, Zivelin A, *et al.* (1998) Mutations in the ER–Golgi intermediate compartment protein ERGIC-53 cause combined deficiency of coagulation factors V and VIII. *Cell* **93**:61–70.

27. Zhang B, Cunningham MA, Nichols WC, *et al.* (2003) Bleeding due to disruption of a cargo-specific ER-to-Golgi transport complex. *Nat Genet* **34**: 220–225.

28. Jones B, Jones EL, Bonney SA, *et al.* (2003) Mutations in a Sar1 GTPase of COPII vesicles are associated with lipid absorption disorders. *Nat Genet* **34**:29–31.

29. Sprecher E, Ishida-Yamamoto A, Mizrahi-Koren M, *et al.* (2005) A mutation in SNAP29, coding for a SNARE protein involved in intracellular trafficking, causes a novel neurocutaneous syndrome characterized by cerebral dysgenesis, neuropathy, ichthyosis, and palmoplantar keratodermia. *Am J Hum Genet* **77**:242–251.

30. Gissen P, Johnson CA, Morgan NV, *et al.* (2004) Mutations in VPS33B, encoding a regulator of SNARE-dependent membrane fusion, cause arthrogryposis–renal dysfunction–cholestasis (ARC) syndrome. *Nat Genet* **36**:400–404.

31. Dong X, Li H, Derdowski A, *et al.* (2005) AP-3 directs the intracellular trafficking of HIV-1 Gag and plays a key role in particle assembly. *Cell* **120**: 663–674.

32. Hermansky F, Pudlak P. (1959) Albinism associated with hemorrhagic diathesis and unusual pigmented reticular cells in the bone marrow: report of two cases with histochemical studies. *Blood* **14**:162–169.

33. Huizing M, Gahl WA. (2002) Disorders of vesicles of lysosome lineage: the Hermansky–Pudlak syndromes. *Curr Mol Med* **2**:451–467.

34. Dell'Angelica EC, Mullins C, Caplan S, Bonifacino JS. (2000) Lysosome-related organelles. *FASEB J* **14**:1265–1278.

35. Swank RT, Novak EK, McGarry MP, *et al.* (1998) Mouse models of Hermansky–Pudlak syndrome: a review. *Pigment Cell Res* **11**:60–80.

36. Li W, He M, Zhou HL, *et al.* (2006) Mutational data integration in gene-oriented files of Hermansky–Pudlak syndrome. *Hum Mutat* **27**:402–407.

37. Gardner JM, Wildenberg SC, Keiper NM, *et al.* (1997) The mouse pale ear (ep) mutation is the homologue of human Hermansky–Pudlak syndrome.*Proc Natl Acad Sci USA* **94**:9238–9243.

38. Kantheti P, Qiao X, Diaz ME, *et al.* (1998) Mutation in AP-3 delta in the mocha mouse links endosomal transport to storage deficiency in platelets, melanosomes, and synaptic vesicles. *Neuron* **21**:111–122.

39. Huang L, Kuo YM, Gitschier J. (1999) The pallid gene encodes a novel syntaxin 13–interacting protein involved in platelet storage pool deficiency. *Nat Genet* **14**:300–306.

40. Detter JC, Zhang Q, Mules EH, *et al.* (2000) Rab geranygeranyl transferase alpha mutation in the gunmetal mouse reduces Rab prenylation and platelet synthesis. *Proc Natl Acad Sci USA* **97**:4144–4149.

41. Zhang Q, Li W, Novak EK, *et al.* (2002) The gene for the muted (mu) mouse, a model for Hermansky–Pudlak syndrome, defines a novel protein which regulates vesicle trafficking. *Hum Mol Genet* **11**:697–706.

42. Suzuki T, Oiso N, Gautam R, *et al.* (2003) The mouse organellar biogenesis mutant buff results from a mutation in Vps33a, a homologue of yeast vps33 and *Drosophila* carnation. *Proc Natl Acad Sci USA* **100**:1146–1150.

43. Ciciotte SL, Gwynn B, Moriyama K, *et al.* (2003) Cappuccino, a mouse model of Hermansky–Pudlak syndrome, encodes a novel protein that is part of the pallidin-muted complex (BLOC-1). *Blood* **101**:4402–4407.

44. Chintala S, Li W, Lamoreux ML, *et al.* (2005) Slc7a11 controls the production of pheomelanin pigment and the proliferation of cultured cells. *Proc Natl Acad Sci USA* **102**:10964–10969.

45. Swank RT, Novak EK, McGarry MP, *et al.* (2000) *Pigment Cell Res* **13**: 59–67.

46. Li W, Detter JC, Weiss HJ, *et al.* (2000) 5′-UTR structural organization, transcript expression and single nucleotide polymorphisms of human Rab geranylgeranyl transferase alpha subunit (RABGGTA) gene. *Mol Genet Metab* **71**:599–608.

47. Zhang Q, Zhen L, Li W, *et al.* (2002) Cell specific abnormal prenylation of Rab proteins in platelets and melanocytes of the gunmetal mouse. *Br J Haematol* **117**:414–423.

48. Di Pietro SM, Dell'Angelica EC. (2005) The cell biology of Hermansky–Pudlak syndrome: recent advances. *Traffic* **6**:525–533.

49. Falcon-Perez JM, Starcevic M, Gautam R, Dell'Angelica EC. (2002) BLOC-1, a novel complex containing the pallidin and muted proteins involved in the biogenesis of melanosomes and platelet-dense granules. *J Biol Chem* **277**:28191–28199.

50. Gautum R, Chintala S, Li W, *et al.* (2004) The Hermansky–Pudlak syndrome 3 (cocoa) protein is a component of the biogenesis of lysosome-related organelle complex 2 (BLOC-2). *J Biol Chem* **279**:12935–12942.

51. Chiang PW, Oiso N, Gautam R, *et al.* (2003) The Hermansky–Pudlak syndrome 1 (HPS1) and HPS4 proteins are components of two complexes, BLOC-3 and BLOC-4, involved in the biogenesis of lysosome-related organelles. *J Biol Chem* **278**:20332–20337.

52. Dell'Angelica EC. (2004) The building BLOC(k)s of lysosomes and related organelles. *Curr Opin Cell Biol* **16**:458–464.

53. Bonifacino JS. (2004) Insights into the biogenesis of lysosome-related organelles from the study of the Hermansky–Pudlak syndrome. *Ann NY Acad Sci* **1038**:103–114.

AUTHOR INDEX

Subject Index